Lecture Notes in Computer Science 5171

Commenced Publication in 1973
Founding and Former Series Editors:
Gerhard Goos, Juris Hartmanis, and Jan van Leeuwen

Ashish Goel   Klaus Jansen
José D.P. Rolim   Ronitt Rubinfeld (Eds.)

Approximation, Randomization and Combinatorial Optimization

Algorithms and Techniques

11th International Workshop, APPROX 2008
and 12th International Workshop, RANDOM 2008
Boston, MA, USA, August 25-27, 2008
Proceedings

 Springer

Volume Editors

Ashish Goel
Stanford University, Department of Management Science and Engineering
and (by courtesy) Computer Science
Terman 311, Stanford, CA, 94305, USA
E-mail: ashishg@stanford.edu

Klaus Jansen
University of Kiel, Institute for Computer Science
Olshausenstrasse 40, 24118 Kiel, Germany
E-mail: kj@informatik.uni-kiel.de

José D.P. Rolim
Centre Universitaire d'Informatique, Battelle Bâtiment A
Route de Drize 7, 1227 Carouge, Geneva, Switzerland
E-mail: rolim@cui.unige.ch

Ronitt Rubinfeld
Massachusetts Insitute of Technology
Computer Science and Artificial Intelligence Laboratory
32 Vassar Street, Building 32-G698, Cambridge, MA 02139, USA
E-mail: ronitt@csail.mit.edu

Library of Congress Control Number: Applied for

CR Subject Classification (1998): F.2, G.2, G.1

LNCS Sublibrary: SL 1 – Theoretical Computer Science and General Issues

ISSN   0302-9743
ISBN-10   3-540-85362-6 Springer Berlin Heidelberg New York
ISBN-13   978-3-540-85362-6 Springer Berlin Heidelberg New York

Springer is a part of Springer Science+Business Media

springer.com

© Springer-Verlag Berlin Heidelberg 2008
Printed in Germany

Typesetting: Camera-ready by author, data conversion by Scientific Publishing Services, Chennai, India
Printed on acid-free paper   SPIN: 12456543   06/3180   5 4 3 2 1 0

Preface

This volume contains the papers presented at the 11th International Workshop on Approximation Algorithms for Combinatorial Optimization Problems (APPROX 2008) and the 12th International Workshop on Randomization and Computation (RANDOM 2008), which took place concurrently at the MIT (Massachusetts Institute of Technology) in Boston, USA, during August 25–27, 2008. APPROX focuses on algorithmic and complexity issues surrounding the development of efficient approximate solutions to computationally difficult problems, and was the 11th in the series after Aalborg (1998), Berkeley (1999), Saarbrücken (2000), Berkeley (2001), Rome (2002), Princeton (2003), Cambridge (2004), Berkeley (2005), Barcelona (2006), and Princeton (2007). RANDOM is concerned with applications of randomness to computational and combinatorial problems, and was the 12th workshop in the series following Bologna (1997), Barcelona (1998), Berkeley (1999), Geneva (2000), Berkeley (2001), Harvard (2002), Princeton (2003), Cambridge (2004), Berkeley (2005), Barcelona (2006), and Princeton (2007).

Topics of interest for APPROX and RANDOM are: design and analysis of approximation algorithms, hardness of approximation, small space, sub-linear time, streaming, algorithms, embeddings and metric space methods, mathematical programming methods, combinatorial problems in graphs and networks, game theory, markets, economic applications, geometric problems, packing, covering, scheduling, approximate learning, design and analysis of randomized algorithms, randomized complexity theory, pseudorandomness and derandomization, random combinatorial structures, random walks/Markov chains, expander graphs and randomness extractors, probabilistic proof systems, random projections and embeddings, error-correcting codes, average-case analysis, property testing, computational learning theory, and other applications of approximation and randomness.

The volume contains 20 contributed papers, selected by the APPROX Program Committee out of 42 submissions, and 27 contributed papers, selected by the RANDOM Program Committee out of 50 submissions.

We would like to thank all of the authors who submitted papers and the members of the Program Committees:

APPROX 2008

Matthew Andrews, Bell Labs
Timothy Chan, University of Waterloo
Julia Chuzhoy, Toyota Technological Institute at Chicago
Uriel Feige, Weizmann Institute
Ashish Goel, Stanford University (Chair)

RANDOM 2008

We would also like to thank the external subreferees:

Dana Ron, Mark Rudelson, Atri Rudra, Matthias Ruhl, Tamas Sarlos, Thomas Sauerwald, Gabriel Scalosub, Mathias Schacht, Christian Schaffner, Christian Scheideler, Florian Schoppmann, C. Seshadhri, Ronen Shaltiel, Asaf Shapira, Alexander Sherstov, Anastosios Sidiroupoulos, Shakhar Smorodinsky, Sagi Snir, Christian Sohler, Xiaoming Sun, Maxim Sviridenko, Chaitanya Swamy, Mario Szegedy, Kunal Talwar, Anush Taraz, Iannis Tourlakis, Chris Umans, Falk Unger, Paul Valiant, Santosh Vempala, Danny Vilenchik, Emanuele Viola, Tandy Warnow, Enav Weinreb, Udi Wieder, Sergey Yekhanin, Yitong Yin, Shengyu Zhang, David Zuckerman, and Hamid Zarrabi-Zadeh.

We gratefully acknowledge the support from Microsoft Research, the Department of Management Science and Engineering of Stanford University, the Department of Electrical Engineering and Computer Science and the Laboratory for Computer Science and Artificial Intelligence at the Massachusetts Institute of Technology, the Institute of Computer Science of the Christian-Albrechts-Universität zu Kiel and the Department of Computer Science of the University of Geneva.

August 2008

Ashish Goel
Ronitt Rubinfeld
Klaus Jansen
José D.P. Rolim

Table of Contents

Contributed Talks of APPROX

Approximating Optimal Binary Decision Trees.................... 1
  Micah Adler and Brent Heeringa

Santa Claus Meets Hypergraph Matchings 10
  Arash Asadpour, Uriel Feige, and Amin Saberi

Ordinal Embedding: Approximation Algorithms and Dimensionality
Reduction.. 21
  *Mihai Bǎdoiu, Erik D. Demaine, MohammadTaghi Hajiaghayi,
  Anastasios Sidiropoulos, and Morteza Zadimoghaddam*

Connected Vertex Covers in Dense Graphs 35
  Jean Cardinal and Eythan Levy

Improved Approximation Guarantees through Higher Levels of SDP
Hierarchies... 49
  Eden Chlamtac and Gyanit Singh

Sweeping Points 63
  Adrian Dumitrescu and Minghui Jiang

Constraint Satisfaction over a Non-Boolean Domain: Approximation
Algorithms and Unique-Games Hardness 77
  Venkatesan Guruswami and Prasad Raghavendra

Fully Polynomial Time Approximation Schemes for Time-Cost Tradeoff
Problems in Series-Parallel Project Networks 91
  Nir Halman, Chung-Lun Li, and David Simchi-Levi

Efficient Algorithms for Fixed-Precision Instances of Bin Packing and
Euclidean TSP 104
  David R. Karger and Jacob Scott

Approximating Maximum Subgraphs without Short Cycles 118
  Guy Kortsarz, Michael Langberg, and Zeev Nutov

Deterministic 7/8-Approximation for the Metric Maximum TSP 132
  Lukasz Kowalik and Marcin Mucha

Inapproximability of Survivable Networks 146
  Yuval Lando and Zeev Nutov

Approximating Single Machine Scheduling with Scenarios   153
    Monaldo Mastrolilli, Nikolaus Mutsanas, and Ola Svensson

Streaming Algorithms for k-Center Clustering with Outliers and with
Anonymity   165
    Richard Matthew McCutchen and Samir Khuller

A General Framework for Designing Approximation Schemes for
Combinatorial Optimization Problems with Many Objectives Combined
into One   179
    Shashi Mittal and Andreas S. Schulz

The Directed Minimum Latency Problem   193
    Viswanath Nagarajan and R. Ravi

A Simple LP Relaxation for the Asymmetric Traveling Salesman
Problem   207
    Thành Nguyen

Approximating Directed Weighted-Degree Constrained Networks   219
    Zeev Nutov

A Constant Factor Approximation for Minimum λ-Edge-Connected
k-Subgraph with Metric Costs   233
    MohammadAli Safari and Mohammad R. Salavatipour

Budgeted Allocations in the Full-Information Setting   247
    Aravind Srinivasan

Contributed Talks of RANDOM

Optimal Random Matchings on Trees and Applications   254
    Jeff Abrahamson, Béla Csaba, and Ali Shokoufandeh

Small Sample Spaces Cannot Fool Low Degree Polynomials   266
    Noga Alon, Ido Ben-Eliezer, and Michael Krivelevich

Derandomizing the Isolation Lemma and Lower Bounds for Circuit
Size   276
    V. Arvind and Partha Mukhopadhyay

Tensor Products of Weakly Smooth Codes Are Robust   290
    Eli Ben-Sasson and Michael Viderman

On the Degree Sequences of Random Outerplanar and Series-Parallel
Graphs   303
    Nicla Bernasconi, Konstantinos Panagiotou, and Angelika Steger

Improved Bounds for Testing Juntas   317
    Eric Blais

The Complexity of Distinguishing Markov Random Fields.............   331
    Andrej Bogdanov, Elchanan Mossel, and Salil Vadhan

Reconstruction of Markov Random Fields from Samples: Some
Observations and Algorithms   343
    Guy Bresler, Elchanan Mossel, and Allan Sly

Tight Bounds for Hashing Block Sources   357
    Kai-Min Chung and Salil Vadhan

Improved Separations between Nondeterministic and Randomized
Multiparty Communication   371
    Matei David, Toniann Pitassi, and Emanuele Viola

Quantum and Randomized Lower Bounds for Local Search on
Vertex-Transitive Graphs   385
    Hang Dinh and Alexander Russell

On the Query Complexity of Testing Orientations for Being Eulerian ...   402
    *Eldar Fischer, Oded Lachish, Ilan Newman, Arie Matsliah, and
    Orly Yahalom*

Approximately Counting Embeddings into Random Graphs   416
    Martin Fürer and Shiva Prasad Kasiviswanathan

Increasing the Output Length of Zero-Error Dispersers   430
    Ariel Gabizon and Ronen Shaltiel

Euclidean Sections of ℓ_1^N with Sublinear Randomness and
Error-Correction over the Reals....................................   444
    Venkatesan Guruswami, James R. Lee, and Avi Wigderson

The Complexity of Local List Decoding   455
    Dan Gutfreund and Guy N. Rothblum

Limitations of Hardness vs. Randomness under Uniform Reductions   469
    Dan Gutfreund and Salil Vadhan

Learning Random Monotone DNF   483
    *Jeffrey C. Jackson, Homin K. Lee, Rocco A. Servedio, and
    Andrew Wan*

Breaking the ϵ-Soundness Bound of the Linearity Test over GF(2)......   498
    Tali Kaufman, Simon Litsyn, and Ning Xie

Dense Fast Random Projections and Lean Walsh Transforms   512
    Edo Liberty, Nir Ailon, and Amit Singer

Near Optimal Dimensionality Reductions That Preserve Volumes   523
    Avner Magen and Anastasios Zouzias

Sampling Hypersurfaces through Diffusion 535
    Hariharan Narayanan and Partha Niyogi

A 2-Source Almost-Extractor for Linear Entropy 549
    Anup Rao

Extractors for Three Uneven-Length Sources 557
    Anup Rao and David Zuckerman

The Power of Choice in a Generalized Pólya Urn Model 571
    Gregory B. Sorkin

Corruption and Recovery-Efficient Locally Decodable Codes 584
    David Woodruff

Quasi-randomness Is Determined by the Distribution of Copies of a
Fixed Graph in Equicardinal Large Sets 596
    Raphael Yuster

Author Index 603

Approximating Optimal Binary Decision Trees

Micah Adler[1] and Brent Heeringa[2]

[1] Department of Computer Science, University of Massachusetts, Amherst, 140 Governors Drive, Amherst, MA 01003
micah@cs.umass.edu
[2] Department of Computer Science, Williams College, 47 Lab Campus Drive, Williamstown, MA 01267
heeringa@cs.williams.edu

Abstract. We give a $(\ln n + 1)$-approximation for the decision tree (DT) problem. An instance of DT is a set of m binary tests $T = (T_1, \ldots, T_m)$ and a set of n items $X = (X_1, \ldots, X_n)$. The goal is to output a binary tree where each internal node is a test, each leaf is an item and the total external path length of the tree is minimized. Total external path length is the sum of the depths of all the leaves in the tree. DT has a long history in computer science with applications ranging from medical diagnosis to experiment design. It also generalizes the problem of finding optimal average-case search strategies in partially ordered sets which includes several alphabetic tree problems. Our work decreases the previous upper bound on the approximation ratio by a constant factor. We provide a new analysis of the greedy algorithm that uses a simple accounting scheme to spread the cost of a tree among pairs of items split at a particular node. We conclude by showing that our upper bound also holds for the DT problem with weighted tests.

1   Introduction

We consider the problem of approximating optimal binary decision trees. Garey and Johnson [8] define the decision tree (DT) problem as follows: given a set of m binary tests $T = (T_1, \ldots, T_m)$ and a set of n items $X = (X_1, \ldots, X_n)$, output a binary tree where each leaf is labeled with an item from X and each internal node is labeled with a test from T. If an item passes a test it follows the right branch; if it fails a test it follows the left branch. A path from the root to a leaf uniquely identifies the item labeled by that leaf. The depth of a leaf is the length of its path from the root. The total external path length of the tree is the sum of the depths of all the leaves in the tree. The goal of DT is to find a tree which minimizes the total external path length. An equivalent formulation of the problem views each item as an m-bit binary string where bit i is 1 if the item passes test T_i and 0 otherwise. We use instances of this type when discussing DT throughout this paper and denote them using the set of items X. If no two strings in X are identical, every feasible solution to DT has n leaves. In this paper we always assume the input is a set of unique strings since finding duplicate strings is easily computable in polynomial time. Decision trees have many natural applications (see [6,14,17] and references therein) including medical diagnosis (tests are symptoms) and experiment design (tests are experiments which determine some property). In fact, Hyafil and Rivest showed that DT was NP-complete precisely because "of the large

A. Goel et al. (Eds.): APPROX and RANDOM 2008, LNCS 5171, pp. 1–9, 2008.

amount of effort that [had] been put into finding efficient algorithms for constructing optimal binary decision trees" [11].

In this paper, we give a polynomial-time $(\ln n + 1)$-approximation for the decision tree problem. This improves the upper bound on the approximation ratio given by Kosaraju et al. [12] by a constant factor. More importantly, our work provides a substantially different analysis of the greedy algorithm for building decision trees. We employ an accounting scheme to spread the total cost of the tree among pairs of items split at internal nodes. The result is an elementary analysis that others may find of independent interest. In fact, our techniques have already been extended to the DT problem with weighted items [4]. We also consider the problem with weights associated with the tests (in contrast to the items) and show that the $(\ln n + 1)$-approximation remains intact.

1.1   Prior and Related Work

DT generalizes the problem of finding optimal search strategies in partially ordered sets when one wishes to minimize the *average* search time (assuming each item is desired with equal probability) as opposed to minimizing the *longest* search time [3]. The latter case corresponds to finding minimal height decision trees. This problem is known to have matching upper and lower bounds ($O(\log n)$ and $\Omega(\log n)$ respectively) on the approximation ratio [2,13,15]. However these results do not generally apply to DT because of the difference in the definition of cost. Additionally, DT generalizes several Huffman coding problems including numerous alphabetic tree problem [12].

The name decision tree also refers to a similar but subtly different problem which we call ConDT (for consistent decision tree) that is extremely hard to approximate. The input to ConDT is a set of n positive / negative labeled binary strings, each of length m, called examples[1]. The output is a binary tree where each internal node tests some bit i of the examples, and maps the example to its left child if i is a 0 and its right child if i is a 1. Each leaf is labeled either TRUE or FALSE. A consistent decision tree maps each positive example to a leaf labeled TRUE and each negative example to a leaf labeled FALSE. The size of a tree is the number of leaves. ConDT seeks the minimum size tree which is consistent with the examples.

Alekhnovich et. al. [1] show it is not possible to approximate size s decision trees by size s^k decision trees for any constant $k \geq 0$ unless NP is contained in DTIME$[2^{m^\epsilon}]$ for some $\epsilon < 1$. This improves a result from Hancock et. al. [9] which shows that no $2^{\log^\delta s}$-approximation exists for size s decision trees for any $\delta < 1$ unless NP is quasi-polynomial. These results hold for $s = \Omega(n)$.

Our results demonstrate that DT and ConDT – although closely related – are quite different in terms of approximability: ConDT has no $c \ln n$-approximation for any constant c (unless P = NP) whereas our results yield such an approximation for DT for $c > 1$. Also, we show that the lower bounds on learning decision trees of the ConDT type hold when minimizing total external path length instead of minimum size. Note that tree size is not an insightful measure for DT since all feasible solutions have n leaves. Thus, it is the difference in input and output, and not the difference in measure, that accounts for the difference in approximation complexity.

---

[1] Many papers take m to be the number of examples and take n to be the number of bits.

Moret [14] views DT and ConDT as unique instances of a general decision tree problem where each item is tagged with k possible labels. With DT there are always $k = n$ labels, but only one item per label. With ConDT, there are only two labels, but multiple items carry the same label. It appears then that labeling restrictions play a crucial role in the complexity of approximating decision trees.

DT shares some similarities with set cover. Since each pair of items is separated exactly once in any valid decision tree, one can view a path from the root to a leaf as a kind of covering of the items. In this case, each leaf defines a set cover problem where it must cover the remaining $n - 1$ items using an appropriate set of bits or tests. In fact, our analysis is inspired by this observation. However, in the decision tree problem, the n set cover problems defined by the leaves are not independent. For example, the bit at the root of an optimal decision tree appears in each of the n set cover solutions, but it is easy to construct instances of DT for which the optimal (independent) solutions to the n set cover instances have no common bits. More specifically, one can construct instances of DT where the n independent set cover problems have solutions of size 1, yielding a decision tree with cost $\Theta(n^2)$ but where the optimal decision tree has cost $O(n \log n)$. Hence, the interplay between the individual set cover problems appears to make the DT problem fundamentally different from set cover. Conversely, set cover instances naturally map to decision tree instances, however, the difference in cost between the two problems means that the optimal set cover is not necessarily the optimal decision tree.

The min-sum set cover (MSSC) problem is also similar to DT. The input to MSSC is the same as set cover (i.e., a universe of items X and a collection C of subsets of X), but the output is a linear ordering of the sets from 1 to $|C|$. If $f(x)$ gives the index of the first set in the ordering that covers x then the cost of the ordering is $\sum_{x \in X} f(x)$. This is similar, but not identical to the cost of the corresponding DT problem because the covered items must still be separated from one another, thus adding additional cost. Greedily selecting the set which covers the most remaining uncovered items yields a 4-approximation to MSSC [5,16]. This approximation is tight unless P=NP. As with set cover, we can think of DT as n instances of MSSC, but again, these instances are not independent so the problems inherent in viewing DT as n set cover problems remain when considering DT as n instances of MSSC.

In the following section we describe and analyze our approximation algorithm for DT. We then extend this analysis to the problem where weights are associated with the tests (but not the items). In Section 3 we show that the lower bounds on learning ConDTs hold for total external path length. Finally, we conclude with a discussion of some open problems including the gap between the upper and lower bounds on the approximation ratio.

2   Approximating DT

Given a set of binary m-bit strings S, choosing some bit i always partitions the items into two sets S^0 and S^1 where S^0 contains those items with bit $i = 0$ and S^1 contains those items with $i = 1$. A greedy strategy for splitting a set S chooses the bit i which minimizes the difference between the size of S^0 and S^1. In other words, it chooses the bit which most evenly partitions the set. Using this strategy, consider the following greedy algorithm for constructing decision trees of the DT type given a set of n items X:

GREEDY-DT(X)
1   **if** $X = \emptyset$
2   **then return** NIL
3   **else** Let i be the bit which most evenly partitions X into X^0 and X^1
4   Let T be a tree node with left child $left[T]$ and right child $right[T]$
5           $left[T] \leftarrow$ GREEDY-DT(X^0)
6           $right[T] \leftarrow$ GREEDY-DT(X^1)
7   **return** T

Fig. 1. A greedy algorithm for constructing decision trees

A straightforward implementation of this algorithm runs in time $O(mn^2)$. While the algorithm does not always give an optimal solution, it does approximate it within a factor of $\ln n + 1$.

Theorem 1. *If X is an instance of DT with n items and optimal cost C^* then* GREEDY-DT(X) *yields a tree with cost at most* $(\ln n + 1)C^*$

Proof. We begin with some notation. Let T be the tree constructed by GREEDY-DT on X with cost C. An unordered pair of items $\{x, y\}$ (hereafter just *pair of items*) is *separated* at an internal node S if x follows one branch and y follows the other. Note that each pair of items is separated exactly once in any valid decision tree. Conversely, each internal node S defines a set $\rho(S)$ of pairs of items separated at S. That is

$$\rho(S) = \{\{x, y\} \mid \{x, y\} \text{ is separated at } S\}$$

For convenience we also use S to denote the set of items in the subtree rooted at S. Let S^+ and S^- be the two children of S such that $|S^+| \geq |S^-|$. Note that $|S| = |S^+| + |S^-|$. The number of sets to which an item belongs equals the length of its path from the root, so the cost of T may be expressed as the sum of the sizes of each S:

$$C = \sum_{S \in T} |S|$$

Our analysis uses an accounting scheme to spread the total cost of the greedy tree among all unordered pairs of items. Since each set S contributes its size to the total cost of the tree, we spread its size uniformly among the $|S^+||S^-|$ pairs of items separated at S. Let c_{xy} be the pair cost assigned to each pair of items $\{x, y\}$ where

$$c_{xy} = \frac{1}{|S_{xy}^+|} + \frac{1}{|S_{xy}^-|}.$$

and S_{xy} separates x from y. Note that the greedy choice minimizes c_{xy}. We can now talk about the cost of a tree node S by the costs associated with the pairs of items separated at S. Summing the costs of these pairs is, by definition, exactly the size of S:

$$\sum_{\{x,y\} \in \rho(S)} c_{xy} = |S^+||S^-|\left(\frac{1}{|S^+|} + \frac{1}{|S^-|}\right) = |S|$$

Because two items are separated exactly once, $\mathcal{C}$ is exactly the sum of the all pair costs

$$\mathcal{C} = \sum_{\{x,y\}} c_{xy}.$$

Now consider the optimal tree $\mathcal{T}^*$ for X. If Z is an internal node of $\mathcal{T}^*$ then we also use Z to denote the set of items that are leaves of the subtree rooted at Z. Following our notational conventions, we let Z^+ and Z^- be the children of Z such that $|Z^+| \geq |Z^-|$ and $|Z| = |Z^+| + |Z^-|$. The cost of the optimal tree, $\mathcal{C}^*$, is

$$\mathcal{C}^* = \sum_{Z \in \mathcal{T}^*} |Z| \tag{1}$$

Since, every feasible tree separates each pair of items exactly once, we can rearrange the greedy pair costs according to the structure of the optimal tree:

$$\mathcal{C} = \sum_{Z \in \mathcal{T}^*} \sum_{\{x,y\} \in \rho(Z)} c_{xy} \tag{2}$$

If Z is a node in the optimal tree, then it defines $|Z^+||Z^-|$ pairs of items. Our goal is to show that the sum of the c_{xy} associated with the $|Z^+||Z^-|$ pairs of items split at Z (but which are defined with respect to the greedy tree) total at most a factor of $H(|Z|)$ more than $|Z|$ where $H(d) = \sum_{i=1}^{d} 1/i$ is the dth harmonic number. This is made precise in the following lemma:

Lemma 1. *For each internal node Z in the optimal tree:*

$$\sum_{\{x,y\} \in \rho(Z)} c_{xy} \leq |Z| H(|Z|)$$

where each c_{xy} is defined with respect to the greedy tree $\mathcal{T}$.

Proof. Consider any node Z in the optimal tree. For any unordered pair of items $\{x, y\}$ split at Z, imagine using the bit associated with the split at Z on the set S_{xy} separating x from y in the greedy tree. Call the resulting two sets $S_{xy}^{Z^+}$ and $S_{xy}^{Z^-}$ respectively. Since the greedy split at S_{xy} minimizes c_{xy}, we know

$$c_{xy} = \frac{1}{|S_{xy}^+|} + \frac{1}{|S_{xy}^-|} \leq \frac{1}{|S_{xy}^{Z^+}|} + \frac{1}{|S_{xy}^{Z^-}|} \leq \frac{1}{|S_{xy} \cap Z^+|} + \frac{1}{|S_{xy} \cap Z^-|}.$$

Hence

$$\sum_{\{x,y\} \in \rho(Z)} c_{xy} \leq \sum_{\{x,y\} \in \rho(Z)} \frac{1}{|S_{xy} \cap Z^+|} + \frac{1}{|S_{xy} \cap Z^-|}. \tag{3}$$

One interpretation of the sum in (3) views each item x in Z^+ as contributing

$$\sum_{y \in Z^-} \frac{1}{|S_{xy} \cap Z^-|}$$

to the sum and each node y in Z^- as contributing

$$\sum_{x \in Z^+} \frac{1}{|S_{xy} \cap Z^+|}$$

to the sum. For clarity, we can view Z as a complete bipartite graph where the items in Z^+ are one set of nodes and the items in Z^- is the other set. Letting $b_{xy} = 1/(|(S_{xy} \cap Z^-|)$ and $b_{yx} = 1/(|S_{xy} \cap Z^+|)$ we can think of every edge (x, y) where $x \in Z^+$ and $y \in Z^-$ as having two costs: one associated with x (b_{xy}) and the other associated with y (b_{yx}). Thus, the cost of Z is at most the sum of all the b_{xy} and b_{yx} costs. We can bound the total cost by first bounding all the costs associated with a particular node. In particular, we claim:

Claim. For any $x \in Z^+$ we have

$$\sum_{y \in Z^-} b_{xy} = \sum_{y \in Z^-} \frac{1}{|S_{xy} \cap Z^-|} \leq H(|Z^-|)$$

Proof. If Z^- has d items then let $(y_1, \ldots, y_d)$ be an ordering of Z^- in reverse order from when the items are split from x in the greedy tree (with ties broken arbitrarily). This means item y_1 is the last item split from x, y_d is the first item split from x, and in general y_{d-t+1} is the tth item split from x. When y_d is split from x there must be at least $|Z^-|$ items in S_{xy_d} — by our ordering the remaining items in Z^- must still be present — so $Z^- \subseteq S_{xy_d}$. Hence b_{xy_d}, the cost assigned to x on the edge (x, y_d), is at most $1/|(Z^-)|$ and in general, when y_t is separated from x there are at least t items remaining from Z^-, so the cost b_{xy_t} assigned to the edge (x, y_t) is at most $1/t$. This means, for any $x \in Z^+$

$$\sum_{y \in Z^-} b_{xy} \leq H(|Z^-|)$$

which proves the claim. $\square$

We can use the same argument to prove the analogous claim for all the items in Z^-. With these inequalities in hand we have

$$\sum_{\{x,y\} \in \rho(Z)} \frac{1}{|S_{xy} \cap Z^+|} + \frac{1}{|S_{xy} \cap Z^-|} \leq |Z^+| H(|Z^-|) + |Z^-| H(|Z^+|)$$

$$< |Z^+| H(|Z|) + |Z^-| H(|Z|)$$

$$= |Z| H(|Z|) \quad \text{(since } |Z^+| + |Z^-| = |Z|\text{)}$$

$\square$

Substituting this result into the initial inequality completes the proof of the theorem.

$$\sum_{Z \in T^*} \sum_{\{x,y\} \in \rho(Z)} c_{xy} \leq \sum_{Z \in T^*} |Z| H(|Z|) \leq \sum_{Z \in T^*} |Z| H(n) = H(n) \mathcal{C}^* \leq (\ln n + 1) \mathcal{C}^*$$

$\square$

2.1   Tests with Weights

In many applications, different tests may have different execution costs. For example, in experiment design, a single test might be a good separator of the items, but it may also be expensive. Running multiple, inexpensive tests may serve the same overall purpose, but at less cost. To model scenarios like these we associate a weight $w(k)$ with each bit k and without confusion take $w(S)$ to be the weight of the bit used at node S. We call this problem DT with weighted tests (in contrast to the DT problem with weighted items). In the original problem formulation, we can think of each test as having unit weight, so the cost of identifying an item is just the length of the path from the root to the item. When the tests have non-uniform weights, the cost of identifying an item is the sum of the weights of the tests along that path. We call this the path cost. The cost of the tree is the sum of the path costs of each item. When all the tests have equal weight, we choose the bit which most evenly splits the set of items into two groups. In other words, we minimize the pair cost c_{xy}. With equal weights, the cost of an internal node is just its size $|S|$. With unequal weights, the cost of an internal node is the weighted size $w(S)|S|$, so assuming S separates x from y the pair cost becomes

$$c_{xy} = \frac{w(S)}{|S^+|} + \frac{w(S)}{|S^-|} \tag{4}$$

and our new greedy algorithm recursively selects the bit which minimizes this quantity. This procedure yields a result equivalent to Theorem 1 for DT with weighted tests. A straightforward implementation on this algorithm still runs in time $O(mn^2)$.

Theorem 2. *The greedy algorithm which recursively selects the bit that minimizes Equation 4 yields a $(\ln n + 1)$-approximation to DT with weighted tests.*

Proof. Following the structure of the proof for Theorem 1 leads to the desired result. The key observation is that choosing the bit that minimizes Equation 4 yields the inequality

$$c_{xy} \leq w(Z)\left(\frac{1}{|S_{xy} \cap Z^+|} + \frac{1}{|S_{xy} \cap Z^-|}\right). \tag{5}$$

Since the weight term $w(Z)$ may be factored out of the summation

$$w(Z) \sum_{\{x,y\} \in \rho(Z)} \frac{1}{|S_{xy} \cap Z^+|} + \frac{1}{|S_{xy} \cap Z^-|}$$

we can apply the previous claim and the theorem follows:

$$\sum_{Z \in T^*} \sum_{\{x,y\} \in \rho(Z)} c_{xy} \leq \sum_{Z \in T^*} w(Z)|Z|H(n) \leq (\ln n + 1)C^*$$

Here $C^* = \sum_{Z \in T^*} w(Z)|Z|$ is the cost of the optimal tree.   □

Another natural extension to DT considers the problem with weighted items. Here, one weights each path length by the weight of the item which defines the path. Recently, Chakaravarthy et al. [4] extended our analysis to the DT problem with weighted items.

3   Hardness of Approximation for ConDT under Total External Path Length

Alekhnovich et. al. [1] showed it is not possible to approximate size s decision trees by size s^k decision trees for any constant $k \geq 0$ unless NP is contained in DTIME$[2^{m^\epsilon}]$ for some $\epsilon < 1$. Decision tree here refers to trees of the ConDT type and the measure is tree size. In this section we show that these hardness results also hold for ConDT under minimum total external path length. Our theorem relies on the observation that if I is an instance of ConDT with minimum total external path length s then I has minimum tree size at least $\Omega(\sqrt{s})$. If it didn't, a tree of smaller size would have smaller total external path length, a contradiction. The case where minimum total external path length s corresponds to minimum size $\Omega(\sqrt{s})$ is a cascading tree; that is, a tree with exactly one leaf at each depth save the deepest two.

Theorem 3. *If there exists an s^k approximation for some constant $k > 0$ to decision trees with minimum total external path length s then NP is contained in DTIME$[2^{m^\epsilon}]$ for some $\epsilon < 1$.*

Proof. Let I be an instance of ConDT with minimum total external path length $s = r^2$. It follows that I has minimum tree size at least $\Omega(r)$. Now, if an s^k approximation did exist for some k then there would exist an $\Omega(r^{2k}) = r^{k'}$ approximation for some constant k' for ConDT under minimum tree size; a contradiction.   □

4   Open Problems and Discussion

Our primary result in this paper is a $(\ln n + 1)$-approximation for the decision tree problem. The most prominent open problem is the gap between the upper and lower bounds on the approximation ratio of DT. The best lower bound on the approximation ratio in the unweighted items case is $2 - \epsilon$ for any $\epsilon > 0$ (modulo P$\neq$NP) [4]. This improves upon the no PTAS result from [10]. However, when the input has arbitrary weights on the items, then the lower bound on the approximation ratio becomes $\Omega(\log n)$.

Unfortunately, the $\Omega(\log n)$ lower bound of Laber and Nogueira [13] for decision trees of minimal height also does not apply to our problem. This is because *height* mirrors the notion of *size* in set cover problems.

Amplifying the $2 - \epsilon$ gap using techniques from [9] for ConDT does not work for DT. There, one squares an instance of ConDT, applies an α-approximation, and recovers a solution to the original instance which is a $\sqrt{\alpha}$-approximation. Repeating this procedure yields the stronger lower bound. This does not work for DT because the average path length only doubles when squaring the problem, so solving the squared problem with an α-approximation and recovering a solution to the original problem simply preserves (and unfortunately does not improve) the approximation ratio. The hardness results from [1] rely on the construction of a binary function which is difficult to approximate accurately when certain instances of a hitting-set problem have large solutions. These techniques do not appear to work for DT either.

The analysis of the greedy algorithm is also not known to be tight. We only know of instances where the approximation ratio of the greedy algorithm is not better than $\Omega(\frac{\log n}{\log \log n})$ of optimal [7,12].

Finally, we leave as an open question the problem of approximating DT with *both* arbitrary item weights and arbitrary test weights.

Acknowledgments. We thank the anonymous reviewers for their insightful and helpful comments.

References

1. Alekhnovich, M., Braverman, M., Feldman, V., Klivans, A.R., Pitassi, T.: Learnability and automatizability. In: Proceedings of the 45th Annual Symposium on Foundations of Computer Science, pp. 621–630. IEEE Computer Society Press, Los Alamitos (2004)
2. Arkin, E.M., Meijer, H., Mitchell, J.S.B., Rappaport, D., Skiena, S.: Decision trees for geometric models. International Journal of Computational Geometry and Applications 8(3), 343–364 (1998)
3. Carmo, R., Donadelli, J., Kohayakawa, Y., Laber, E.S.: Searching in random partially ordered sets. Theor. Comput. Sci. 321(1), 41–57 (2004)
4. Chakaravarthy, V.T., Pandit, V., Roy, S., Awasthi, P., Mohania, M.K.: Decision trees for entity identification: approximation algorithms and hardness results. In: Libkin, L. (ed.) Proceedings of the Twenty-Sixth ACM Symposium on Principles of Database Systems, pp. 53–62 (2007)
5. Feige, U., Lovász, L., Tetali, P.: Approximating min-sum set cover. Algorithmica 40(4), 219–234 (2004)
6. Garey, M.R.: Optimal binary identification procedures. SIAM Journal on Applied Mathematics 23(2), 173–186 (1972)
7. Garey, M.R., Graham, R.L.: Performance bounds on the splitting algorithm for binary testing. Acta Inf. 3, 347–355 (1974)
8. Garey, M.R., Johnson, D.S.: Computers and Intractability: A Guide to the Theory of NP-Completeness. W.H. Freeman, New York (1979)
9. Hancock, T., Jiang, T., Li, M., Tromp, J.: Lower bounds on learning decision lists and trees. Information and Computation 126(2), 114–122 (1996)
10. Heeringa, B.: Improving Access to Organized Information. PhD thesis. University of Massachusetts, Amherst (2006)
11. Hyafil, L., Rivest, R.: Constructing optimal binary decision trees is np-complete. Information Processing Letters 5(1), 15–17 (1976)
12. Rao Kosaraju, S., Przytycka, T.M., Borgstrom, R.S.: On an optimal split tree problem. In: Dehne, F.K.H.A., Gupta, A., Sack, J.-R., Tamassia, R. (eds.) WADS 1999. LNCS, vol. 1663, pp. 157–168. Springer, Heidelberg (1999)
13. Laber, E.S., Nogueira, L.T.: On the hardness of the minimum height decision tree problem. Discrete Applied Mathematics 144(1-2), 209–212 (2004)
14. Moret, B.M.E.: Decision trees and diagrams. ACM Comput. Surv. 14(4), 593–623 (1982)
15. Moshkov, M.J.: Greedy algorithm of decision tree construction for real data tables. In: Transactions on Rough Sets, pp. 161–168 (2004)
16. Munagala, K., Babu, S., Motwani, R., Widom, J.: The pipelined set cover problem. In: ICDT, pp. 83–98 (2005)
17. Murthy, K.V.S.: On growing better decision trees from data. PhD thesis, The Johns Hopkins University (1996)

Santa Claus Meets Hypergraph Matchings

Arash Asadpour[1,*], Uriel Feige[2], and Amin Saberi[3]

[1] Stanford University, Stanford CA 94305, USA
asadpour@stanford.edu
[2] Weizmann Institute, Rehovot 76100, Israel
uriel.feige@weizmann.ac.il
[3] Stanford University, Stanford CA 94305, USA
saberi@stanford.edu

Abstract. We consider the problem of max-min fair allocation of indivisible goods. Our focus will be on the restricted version of the problem in which there are m items, each of which associated with a non-negative value. There are also n players and each player is only interested in some of the items. The goal is to distribute the items between the players such that the least happy person is as happy as possible, i.e. one wants to maximize the minimum of the sum of the values of the items given to any player. This problem is also known as *the Santa Claus problem* [3]. Feige [9] proves that the integrality gap of a certain configuration LP, described by Bansal and Sviridenko [3], is bounded from below by some (unspecified) constant. This gives an efficient way to estimate the optimum value of the problem within a constant factor. However, the proof in [9] is nonconstructive: it uses the Lovasz local lemma and does not provide a polynomial time algorithm for finding an allocation. In this paper, we take a different approach to this problem, based upon local search techniques for finding perfect matchings in certain classes of hypergraphs. As a result, we prove that the integrality gap of the configuration LP is bounded by $\frac{1}{5}$. Our proof is nonconstructive in the following sense: it does provide a local search algorithm which finds the corresponding allocation, but this algorithm is not known to converge to a local optimum in a polynomial number of steps.

1 Introduction

Resource allocation problems, i.e. allocating limited resources to a number of players while satisfying some given constraints, have been studied extensively in computer science, operations research, economics, and the mathematics literature. Depending on whether the resource is divisible or not one can distinguish two main types of such problems. The divisible case has been considered mostly by combinatorists and measure theorists in the past century under the title of "Cake Cutting" problems [16,5]. On the other hand, the indivisible resource allocation problems have been mostly the focus of algorithmic lines of research. In

---

* The first and third authors were supported through NSF grants 0546889 and 0729586 and a gift from Google.

A. Goel et al. (Eds.): APPROX and RANDOM 2008, LNCS 5171, pp. 10–20, 2008.

such problems, often the set of resources R consists of m items. There is also a set P of n players. Each player i has a value function $f_i : 2^S \to \mathbb{R}$. For the sake of simplicity we define $v_{ij} = f_i(\{j\})$. The goal is to partition the set of items to subsets $S_1, S_2, \cdots, S_n$ and allocate each part to one of the players such that a certain objective function is optimized.

Depending on the objective functions, various indivisible resource allocation problems can be considered. For example, the problem of *maximizing social welfare* arises when we want to maximize $\sum_i f_i(S_i)$. See [6,8,10,17] for recent progress on this problem.

Minimizing the makespan is another example of indivisible resource allocation problems in which the goal is to minimize $\max_i f_i(S_i)$ and f_i's are linear functions, i.e. $f_i(S_i) = \sum_{j \in S_i} v_{ij}$. Lenstra, Shmoys and Tardos [15] provide a 2-approximation algorithm and also prove that the problem is hard to approximate within a factor of 1.5. Approximation ratios better than 2 are known for some very special cases of this problem [7].

Another interesting trend in indivisible resource allocation is *Max-min fair allocation problems*. Here, we aim to maximize $\min_i f_i(S_i)$ while f_i's are still linear functions. Although very similar at the first glance, this problem has turned out to be fundamentally different from minimizing the makespan and the techniques that are known to be useful there fail to give non-trivial results here. Most notably, the assignment LP used in [15] yields an additive approximation of $\max_{ij} v_{ij}$ [4]. It can be used to find a solution of value at least $\text{OPT} - \max_{ij} v_{ij}$, where OPT is the value of the optimal solution. Unfortunately, it offers no approximation guarantee in the most challenging cases of the problem when $\text{OPT} \leq \max_{ij} v_{ij}$.

Bansal and Sviridenko [3] studied this problem under the name of *the Santa Claus problem*, where Santa wants to distribute some presents among some kids and his goal is to do this in such a way that the least happy kid is as happy as possible. They considered a certain type of linear programming relaxation of the problem (known as *configuration* LP that we will explain shortly), and showed that it can be used to find a solution with value at least OPT/n. They also showed that the integrality gap of this LP is no better than $O(1/\sqrt{n})$. Asadpour and Saberi [2] showed how to round the solution of configuration LP to get a solution with value at least $\Omega(\text{OPT}/\sqrt{n}(\log n)^3)$.

Our focus here will be on a special case of the Max-min fair allocation problem, known as *restricted assignment problem*, in which each item j has an inherent value v_j and a set of players to which the item can be assigned. In other words, for each such player i, the value of v_{ij} is v_j and for all other players it is 0. Bezakova and Dani [4] showed that this problem is hard to approximate within a factor better than $\frac{1}{2}$. (In fact, this is also the best hardness result known for the general problem.) Bansal and Sviridenko [3] showed that it is possible to round the values of the configuration LP and get a feasible solution with value $\Omega(\text{OPT} \log \log \log n / \log \log n)$. Recently, Feige [9] proved that the optimal value of the configuration LP is in fact within a constant factor of OPT. Although [9] does not give a polynomial time algorithm to find a constant factor approximation solution, it does provide a constant factor estimation for the optimal value of the problem[1]. This is due to the fact that the configuration LP can be solved (up to arbitrary precision) in polynomial time, and its value is an upper bound on OPT. The main result of this paper can be summarized as the following:

Theorem 1. *In the restricted assignment problem, there is a polynomial time algorithm that estimates the optimal value of max-min allocation problem within a factor of $\frac{1}{5} - \epsilon$, where $\epsilon > 0$ is an arbitrarily small constant.*

The polynomial time algorithm referred to in the above theorem is simply the configuration LP. The proof of the $\frac{1}{5}$ estimation factor will follow from our proof that the optimal value of the configuration LP is at most 5OPT. There is a small loss of ϵ in the estimation factor because the known polynomial time algorithms [3] solve the configuration LP up to any desired degree of accuracy, but not necessarily exactly.

Our proof of Theorem 1 transforms the problem into a problem of finding a perfect matching in certain hypergraphs. We design a local search algorithm that finds such a perfect matching. It is inspired by the techniques of [11] which will be discussed in Sect.2. This method can be viewed as a generalization of *Hungarian method* [14] to the domain of hypergraphs.

Comparing our results to those in [9], our result has the advantage of providing an explicit bound (of $\frac{1}{5}$) on the integrality gap of the configuration LP. Also, our proof technique suggests an algorithmic approach to round the solution of the configuration LP. While in [9] multiple applications of the Lovasz local lemma are used, here we introduce a local search algorithm and prove that it ends up in a solution with value at least $\frac{\text{OPT}}{5}$. Although we cannot bound the running time within a polynomial, it puts the problem in the complexity class PLS[2] and proposes the open question of whether this local search (or a modified version of it) converges in polynomial time to an appropriate solution.

1.1   The Configuration LP

Fix a real number t and suppose that we want to see if it is possible to do the allocation in such a way that each player i receives a bundle of items S_i with $f_i(S_i) \geq t$. For any bundle S of items, let x_{iS} be the indicator 0/1 variable, representing if the whole bundle S is allocated to person i (in this case x_{iS} will be 1) or not ($x_{iS} = 0$). To provide a bundle with value at least t for every person, we need to solve the following integer program:

---

[1] We emphasize that all the results related to the hardness of approximation remains valid even for estimating the optimal value of the problem.

[2] The complexity class PLS consists of problems for which, given any input instance there exists a finite set of solutions and an efficient algorithm to compute a cost for each solution, and also a neighboring solution of lower cost provided that one exists. Then the problem is to find a solution, namely a *local optimum*, that has cost less than or equal to all its neighbors. For more information, see [12].

- Every player only accepts bundles with value at least t; $\forall i : x_{iS} = 0$ whenever $f_i(S) < t$.
- Every player receives one bundle; $\forall i : \sum_S x_{iS} = 1$.
- Every item is allocated to at most one player: $\forall j : \sum_{i,S | j \in S} x_{iS} \leq 1$.
- $x_{iS} \in \{0, 1\}$ for every player i and bundle S.

The configuration LP is the relaxation of the above integer program. The last constraint is replaced by $x_{iS} \geq 0$

If the LP is feasible for some t_0, then it is also feasible for all $t \leq t_0$. Let optLP be the maximum of all such values of t (it can be shown that such maximum exists). Every feasible allocation is a feasible solution of configuration LP. Hence optLP $\geq$ OPT. The value of optLP and a feasible solution to the configuration LP of value optLP can be approximated within any desired degree of accuracy in polynomial time, as shown in [3].

In this paper we show that any fractional solution of configuration LP corresponding to optLP can be rounded (though not necessarily in polynomial time) to an integral solution whose value is within a constant factor of optLP. We provide two versions of our proof. In Section 2 we show how this result can be deduced by combining (in a blackbox manner) a previous intermediate result of Bansel and Sviridenko [3] with a theorem of Haxell [11]. In Section 3 we provide our main result which is basically a local search that finds an integral solution with value at least $\frac{\text{optLP}}{5}$. The proof in Section 3 is inspired by the results of Section 2, but is presented in a self contained way. It circumvents the use of the intermediate result of [3], and extends the proof technique of [11] in certain ways. Any of the two sections 2 and 3 can be read and understood without needing to read the other section.

2   Matchings in Hypergraphs

Let $\mathcal{H} = (V, \mathcal{E})$ be a hypergraph. A *matching* in $\mathcal{H}$ is a set of pairwise disjoint edges. We denote by $\nu(\mathcal{H})$ the maximum size of a matching in $\mathcal{H}$. A matching is called perfect if any vertex appears in exactly one of its edges. Unlike the case for matchings in graphs, the problem of finding a perfect matching in hypergraphs is NP-complete. (A well known special case of this problem is the NP-hard problem of 3-dimensional matching. Note that 3-dimensional matching can also be cast as a special case of finding a perfect matching in a bipartite hypergraph, a problem that we shall describe below.) There are some sufficient conditions known for the existence of perfect matchings in hypergraphs. See for example [1] and [13]. Some of these sufficient conditions are not computable in polynomial time.

Here, we focus on the problem of finding a maximum matching in *bipartite* hypergraphs. A hypergraph $\mathcal{H} = (V, \mathcal{E})$ is called *bipartite* if the ground set V is the disjoint of sets U and V, and every $E \in \mathcal{E}$ satisfies $|E \cap U| = 1$. A perfect matching in a bipartite hypergraph is defined as a matching that saturates all the vertices in U. A *transversal* for hypergraph $\mathcal{H}$ is a subset $T \subseteq V$ with the property that $E \cap T \neq \emptyset$ for every $E \in \mathcal{E}$. Let $\tau(\mathcal{H})$ denote the minimum size of a transversal of $\mathcal{H}$. For a subset $C \subseteq U$, we write $\mathcal{E}_C = \{F \subseteq V : \{c\} \cup F \in \mathcal{E} \text{ for some } c \in C\}$, and let $\mathcal{H}_C$ be the hypergraph $(V, \mathcal{E}_C)$. The following theorem is proved by Haxell in [11].

Theorem 2. *(Haxell [11]) Let $\mathcal{H} = (U \cup V, \mathcal{E})$ be a bipartite hypergraph such that for every $E \in \mathcal{E}$ we have $|E \cap V| \leq r-1$, and also $\tau(\mathcal{H}_C) > (2r-3)(|C|-1)$ for every $C \subseteq U$. Then $\nu(\mathcal{H}) = |U|$.*

When $r = 2$, $\mathcal{H}$ becomes a graph, and Haxell's theorem reduces to Hall's theorem. The proof of Theorem 2 as described in [11] is not constructive.

2.1   A Constant Integrality Gap

In this section, we will consider a combinatorial conjecture (which is by now a theorem, by the results of [9]) which is equivalent up to constant factors to the restricted assignment problem, and prove it via Theorem 2. It reveals the intuition behind the relation between the restricted assignment problem and matchings in hypergraphs. Also, it is through this transformation that our local search appears.

Bansal and Sviridenko proved that if the following conjecture is true for some β, then it can be shown that the integrality gap of configuration LP relaxation for the restricted assignment problem is $\Omega(\beta)$.

> **Conjecture (by Bansal and Sviridenko [3]):** There is some universal constant $\beta > 0$ such that the following holds. Let $C_1, \cdots, C_p$ be collections of sets, $C_i = \{S_{i1}, \cdots, S_{il}\}$ for $i = 1, \cdots, p$, where each set S_{ij} is a k-element subset of some ground set and each element appears in at most l sets S_{ij}. Then there is a choice of some set $S_{i,f(i)} \in C_i$ for each $i = 1, \cdots, p$, and a choice $S'_i \subseteq S_{i,f(i)}$ with the property that $|S'_i| \geq \beta k$ and that each element occurs in at most one set in $\{S'_1, \cdots, S'_p\}$.

For every value k, it is not hard to see that the conjecture is true when $\beta = 1/k$. Feige [9] shows that the conjecture is true for some small enough universal constant β, for all values of k. Here, using Theorem 2 we prove that it is true even for $\beta = \frac{1}{5}$. (For every $k \geq 3$, our value of β is the largest number satisfying two constraints. Namely, that $(1 - \beta) \geq 2\beta$, which will be needed in the proof of Theorem 3, and that βk is an integer. Hence, $\beta = 1/3$ when k is divisible by 3, but might be as small as $\frac{1}{5}$ for $k = 5$.)

Theorem 3. *Conjecture 2.1 is true for any $\beta \leq \frac{\lfloor k/3 \rfloor}{k}$*

Proof. Consider the following bipartite hypergraph $\mathcal{H} = (U \cup V, \mathcal{E})$. Here, $V = \bigcup_{i,j} S_{i,j}$ and $U = \{a_1, a_2, \cdots, a_p\}$. Also $\mathcal{E} = \{S \cup \{a_i\} : S \subseteq S_{i,j} \text{ for some } j, |S| = \beta k\}$. Note that here $r = \beta k + 1$. By the construction of $\mathcal{H}$, it is enough to prove that $\mathcal{H}$ has a perfect matching (i.e. a matching with size $|U|$). We will do so by showing that $\mathcal{H}$ satisfies the conditions of Theorem 2.

Consider an arbitrary $C \subset U$ and a transversal set T in $\mathcal{H}_C$. Because T is a transversal set in H_C, it must have some intersection with all the edges in H_C.

But edges in H_C correspond to all subsets S of V with βk elements such that for some j and $a_i \in C$ it holds that $S \subseteq S_{i,j}$. It means that for any such i and j, at least $(1 - \beta)k$ elements of $S_{i,j}$ should be in T. (In fact, the number of elements of $S_{i,j}$ in T should be at least $(1 - \beta)k + 1$, but the extra $+1$ term does not appear to have a significant effect on the rest of the proof, so we omit it.)

Now, consider a bipartite graph $G = (V', E)$ such that $V' = U' \cup T$ where $U' = \bigcup_{i \in C} \{a_{i,1}, \cdots, a_{i,l}\}$ and $E = \{\{a_{i,j}, q\} : q \in S_{i,j}\}$. By the above discussion, $\deg(v) \geq (1 - \beta)k$, for all $v \in U'$. Hence, $|E| \geq (1 - \beta)k|C|l$. Also by the assumption of the conjecture, $\deg(v) \leq l$ for all $v \in V'$. Hence $|E| \leq l|T|$. Therefore,

$$l|T| \geq (1 - \beta)k|C|l.$$

Thus, $|T| \geq (1 - \beta)k|C| = \frac{1-\beta}{\beta}(r - 1)|C|$. Picking any $\beta \leq 1/3$, we have $|T| \geq 2(r-1)|C|$ which means that $\tau(\mathcal{H}_C) > (2r - 3)(|C| - 1)$ for every $C \subseteq U$. This completes the proof. □

3   A $\frac{1}{5}$-approximate Solution through a Local Search

In this section we prove that the integrality gap of the configuration LP is no worse than $\frac{1}{5}$.

Given a feasible solution $\{x_{iS}\}$ to the configuration LP, we modify it as follows. To simplify notation, scale values of all items so that we can assume that $t = 1$. Recall that $v_{ij} \in \{0, v_j\}$. Call an item j *fat* if $v_j > \frac{1}{5}$ and *thin* it $v_j \leq \frac{1}{5}$. (The value of $\frac{1}{5}$ is taken with hindsight, being the largest value p satisfying $2(p + p) \leq 1 - p$, needed later in the proof of Lemma 1.) For every fat item j, change v_j so that $v_j = 1$. Now modify the LP solution so as to make it *minimal*, by restricting players to choose bundles that are minimally satisfying for the player – dropping any item from the set reduces its value below 1. This can be achieved in polynomial time by dropping items from sets whenever possible. We are now left with an LP solution that uses only two types of sets:

- *Fat sets.* These are sets that contain only a single fat item and nothing else.
- *Thin sets.* These are sets that contain only thin items.

We call such a solution to the LP a *minimal solution*.

Construct a bipartite hypergraph based on the modified LP solution. The U side are the players. The V side are the items. For every player i put in hyperedges associated with those sets for which $x_{iS} > 0$ as follows. If $S = \{j\}$ is a fat set, include the hyperedge $\{i, j\}$. If S is a thin set, then for every *minimal* subset $S' \subset S$ of value at least $\frac{1}{5}$ (minimal in the sense that dropping any item from S' reduces its value below $\frac{1}{5}$), put in the hyperedge $\{i, S'\}$. Observe that by minimality, S' has weight at most $\frac{2}{5}$.

Theorem 4. *Given any minimal solution to the configuration LP, the bipartite hypergraph constructed above has a perfect matching (namely, a matching in which all vertices of U are covered).*

We note that Theorem 4 implies that there is an integer solution of value at least $\frac{1}{5}$, since every player can get either a fat set (that contains an item of value more than $\frac{1}{5}$), or a part of a thin set of value at least $\frac{1}{5}$.

Our proof of Theorem 4 is patterned a proof of [11], with some changes. The most significant of these changes is the use of Lemma 1.

For a set W of edges, we use the notation W_U to denote the vertices of U that are covered by W, and W_V to denote the vertices of V that are covered by W.

Proof. The proof is by induction on U. For $|U| = 1$, the theorem is obviously true (since the hypergraph has at least one edge). Hence assume that the theorem is true for $|U| = k$, and prove for $|U| = k + 1$.

Denote the vertices of U by $\{u_0, \ldots u_k\}$. By the inductive hypothesis, there is a matching of size k involving all U vertices except for u_0. (This is true because by removing u_0 from the hypergraph and all its edges, one obtains a hypergraph which corresponds to a minimal solution to an LP with one less player.) Pick an arbitrary such matching M. We present an algorithm that transforms this matching into a new matching of size $k + 1$. The algorithm is in some respects similar to the known algorithm for constructing matchings in bipartite graphs. It constructs an alternating tree in an attempt to find an alternating path. In the graph case, when such a path is found, the matching can be extended. In the hypergraph case, the situation is more complicated, and hence the proof will not provide a polynomial upper bound on the number of steps required until eventually the matching is extended.

In our alternating tree, there will be two types of edges. Edges of type A are edges that we try to add to the matching (A stands for *Add*). Edges of type B will be existing matching edges (hence $B \subset M$) that intersect edges of type A, and hence block us from adding edges of type A to the matching (B stands for *Block*). Every root to leaf path will be an alternating sequence of edges of type A and B.

The A edges will be numbered in the order in which they are added to the alternating tree. Hence their names will be $a_1, a_2, \ldots$, and these names are relative to a currently existing alternating tree (rather than being names that edges keep throughout the execution of the algorithm). For every $i \geq 1$, we associate with edge a_i an integer $m_i \geq 1$ that will correspond to the number of B edges that block a_i. The strict positivity of m_i implies that $|B| \geq |A|$.

Initially one needs to pick the first edge for the alternating tree. Pick an arbitrary edge e such that $e_U = u_0$. Let m_1 denote the number of edges from M that e_V intersects. If $m_1 = 0$, then *terminate*, because the edge e can be added to M, obtaining a perfect matching. If $m_1 > 0$, rename e as a_1, add a_1 to A, and add the m_1 matching edges that intersect a_1 to B.

Let $i \geq 2$ and assume that the alternating tree already contains $i - 1$ edges of type A (named as $a_1, \ldots, a_{i-1}$), and at least $i - 1$ edges of type B. We now pick an edge e such that $e_U \in (A \cup B)_U$ and e_V does not intersect $(A \cup B)_V$. The following lemma shows that such an edge must exist.

Lemma 1. *Let $H(U, V, E)$ be the hypergraph associated with a minimal solution to the configuration LP. Then given any alternating tree as described above, there always is an edge e such that $e_U \in (A \cup B)_U$ and e_V does not intersect $(A \cup B)_V$.*

Proof. Let ℓ denote the number of vertices of U in the alternating tree. Each hyperedge corresponds in a natural way either to a fat set or to (part of) a thin set. Let A_f (A_t, respectively) denote the collection of A edges in the alternating tree that correspond to fat sets (thin sets, respectively), and similarly for B_f and B_t with respect to B edges in the alternating tree. Observe that in an alternating tree necessarily $|A_f| + |A_t| = |A| < \ell$ and $|B_f| + |B_t| = |B| < \ell$. Moreover, $|A_f| = |B_f| = |(A_f \cup B_f)_V|$, because every fat edge of A contains exactly one vertex in V, this vertex is contained only in fat edges, and hence this fat edge is intersected by exactly one fat edge in B.

Consider now the restriction of the minimal solution to the LP to the set of players P represented by the ℓ vertices of $(A \cup B)_U$. Let S_f be the collection of fat sets and S_t be the collection of thin sets. Let $\alpha = \sum_{i \in P,\, S \in S_f} x_{iS}$ denote the total weight assigned by this restricted solution to fat sets, and let $\beta = \ell - \alpha = \sum_{i \in P,\, S \in S_t} x_{iS}$ denote the total weight assigned by this restricted solution to thin sets. If $\alpha > |(A_f \cup B_f)_V|$ then it must be the case that some fat set has positive weight in the restricted solution but is not part of the alternating tree. In this case, this fat set can contribute a hyperedge to the alternating tree. Hence it remains to deal with the case that $\alpha \leq |A_f|$. In this case, $2\beta \geq |A_t| + |B_t| + 2$. The hyperedges in the alternating tree that correspond to thin sets each take up value at most $\frac{2}{5}$. Hence even after removing all items appearing in the alternating tree, the sum of weights multiplied by respective remaining value of thin sets in the LP is

$$\sum_{i \in P,\, S \in S_t} x_{iS} \sum_{j \in S \setminus (A_t \cup B_t)} v_{ij} > \frac{\beta}{5}$$

This means than at least one thin set must have retained a value of at least $\frac{1}{5}$. Hence, this thin set can contribute a hyperedge to the alternating tree. $\square$

Pick an arbitrary hyperedge e satisfying Lemma 1 and let m_i denote the number of edges of M that e intersects. If $m_i > 0$, we call this an *extension* (the alternating tree grew larger), rename e as a_i, add a_i to A, and add the m_i matching edges that intersect a_i to B.

We now describe what to do when $m_i = 0$. If $e_U = u_0$, add edge e to the matching M, and *terminate*. If $e_U \neq u_0$, then let e' be the unique edge in B for which $e_U = e'_U$. Let a_j (here necessarily we will have $j < i$) be the unique edge in A that intersects e'. In the matching M, replace the matching edge e' by the matching edge e. Note that this still gives a valid matching of size k, because by construction, e does not intersect any edge of M except for sharing its U side vertex with e', which is removed from M. Update m_j by decreasing it by 1. If the new value of m_j is still positive, this step ends. However, if $m_j = 0$, then the above procedure is repeated with j replacing i (in particular, a_j will also become part of the matching M). Because $j < i$, the number of times the procedure can be repeated is finite, and hence eventually the step must end. We call such a step a *contraction* (the alternating tree becomes smaller).

This completes the description of the algorithm. Observe that the algorithm terminates only when we extend the matching M by one more edge. Hence it remains to show that the algorithm must terminate.

To see this, consider the evolution of vector $m_1, m_2, \ldots, m_j$. For simplicity of the argument, append at the end of each such vector a sufficiently large number ($|M| + 1$ would suffice). We call the resulting vector the *signature* of the alternating tree. We claim that the signatures of any two alternating trees are distinct. This is because ordering the signatures by the time in which they were generated sorts them in decreasing lexicographic order. For extension steps, this follows from the fact that we appended $|M| + 1$ at the end of the respective vector. For contraction steps, this follows from the fact that m_j decreases.

Since $\sum_i m_i \leq |M|$ and $m_i > 0$ (whenever m_i is defined), the number of possible signatures is $2^{|M|}$ (there is a one to one correspondence between these vectors and choices of after which items to place delimiters in a sequence of $|M|$ items), and hence the algorithm cannot have infinite executions. $\square$

The proof of Theorem 4 implicitly provides a local search algorithm to find an integral solution with value $\frac{1}{5}$. Its basic objects are the alternating trees. A basic step is that of adding an edge to the tree, resulting in either an *extension* step or a *contraction* step. The measure of progress of the local search is via the lexicographic value of the corresponding signature. Given a matching with $|M| < n$ edges (an allocation to M players), it will be extended after at most $2^{|M|}$ steps. Hence starting with the empty matching it takes at most $\sum_{|M|=0}^{n-1} 2^{|M|} < 2^n$ local search steps uptil a perfect matching is found. This corresponds to allocating disjoint bundles of value at least optLP/5 to all players. Noting that optLP is at least as large as the optimal solution, the following theorem is established.

Theorem 5. *After 2^n local moves, our algorithm finds a feasible integral $\frac{1}{5}$-approximate allocation.*

4   Open Directions

Characterizing the best possible approximation ratio for the max-min allocation problem is still open, both for the restricted assignment version and for the general version of the problem. We list here some research questions that are suggested by our work.

1. *Integrality gap.* We showed that the integrality gap of the configuration LP for the restricted assignment problem is no worse than 1/5. It was previously known to be no better than 1/2 (in particular, this follows from the NP-hardness result of [4]). Narrow the gap between these two bounds.
2. *Complexity of local search.* Our proof is based on a local search procedure. Can a locally optimal solution with respect to this local search be found in polynomial time? Is finding such a solution PLS-complete? These questions

apply also to similar local search procedures that find a perfect matching in hypergraphs satisfying the conditions of Theorem 2.

3. *Approximation algorithms.* Provide an approximation algorithm (that actually finds an allocation) with a constant approximation ratio for the restricted assignment problem.

4. *Hypergraph matchings.* Can the proof techniques used in our paper be used also for other problems? For example, can our approach be employed to prove that the integrality gap of configuration LP for general max-min fair allocation problem is $\Theta(\frac{1}{\sqrt{n}})$ (saving a $\log^3 n$ factor compared to [2])?

Acknowledgements

Part of this work was performed at Microsoft Research, Redmond, Washington.

References

1. Aharoni, R., Haxell, P.: Hall's theorem for hypergraphs. Journal of Graph Theory 35, 83–88 (2000)
2. Asadpour, A., Saberi, A.: An Approximation Algorithm for Max-Min Fair Allocation of Indivisible Goods. In: Proceedings of the ACM Symposium on Theory of Computing (STOC) (2007)
3. Bansal, N., Sviridenko, M.: The Santa Claus problem. In: Proceedings of the ACM Symposium on Theory of Computing (STOC) (2006)
4. Bezakova, I., Dani, V.: Allocating indivisible goods. SIGecom Exchanges (2005)
5. Brams, S.J., Taylor, A.D.: Fair division: from Cake Cutting to Dispute Resolution. Cambridge University Press, Cambridge (1996)
6. Dobzinski, S., Schapira, M.: An improved approximation algorithm for combinatorial auctions with submodular bidders. In: Proceedings of Symposium on Discrete Algorithms (SODA) (2006)
7. Ebenlendr, T., Krcal, M., Sgall, J.: Graph Balancing: A Special Case of Scheduling Unrelated Parallel Machines. In: Proceedings of Symposium on Discrete Algorithms (SODA) (2008)
8. Feige, U.: On maximizing welfare when utility functions are subadditive. In: Proceedings of the ACM Symposium on Theory of Computing (STOC) (2006)
9. Feige, U.: On allocations that maximize fairness. In: Proceedings of Symposium on Discrete Algorithms (SODA) (2008)
10. Feige, U., Vondrak, J.: Approximation algorithms for allocation problems: Improving the factor of 1 - 1/e. In: Proceedings of Foundations of Computer Science (FOCS) (2006)
11. Haxell, P.E.: A Condition for Matchability in Hypergraphs. Graphs and Combinatorics 11, 245–248 (1995)
12. Johnson, D.S., Papadimitriou, C.H., Yannakakis, M.: How easy is local search? Journal of Computer and System Sciences 37, 79–100 (1988)
13. Kessler, O.: Matchings in Hypergraphs. D.Sc. Thesis, Technion (1989)
14. Kuhn, H.W.: The Hungarian Method for the assignment problem. Naval Research Logistic Quarterly 2, 83–97 (1955)

15. Lenstra, J.K., Shmoys, D.B., Tardos, E.: Approximation algorithms for scheduling unrelated parallel machines. Mathematical Programming, Series A (1993)
16. Steinhaus, H.: The problem of fair division. Econometrica (1948)
17. Vondrak, J.: Optimal approximation for the Submodular Welfare Problem in the value oracle model. In: Proceedings of the ACM Symposium on Theory of Computing (STOC) (2008)

Ordinal Embedding: Approximation Algorithms and Dimensionality Reduction

Mihai Bădoiu[1], Erik D. Demaine[2,*], MohammadTaghi Hajiaghayi[3], Anastasios Sidiropoulos[2], and Morteza Zadimoghaddam[4]

[1] Google Inc.
mihai@theory.csail.mit.edu
[2] MIT Computer Science and Artificial Intelligence Laboratory
{edemaine,tasos}@mit.edu
[3] AT&T Labs — Research
hajiagha@research.att.com
[4] Department of Computer Engineering, Sharif University of Technology
zadimoghaddam@ce.sharif.edu

Abstract. This paper studies how to optimally embed a general metric, represented by a graph, into a target space while preserving the relative magnitudes of most distances. More precisely, in an ordinal embedding, we must preserve the relative order between pairs of distances (which pairs are larger or smaller), and not necessarily the values of the distances themselves. The relaxation of an ordinal embedding is the maximum ratio between two distances whose relative order is inverted by the embedding. We develop polynomial-time constant-factor approximation algorithms for minimizing the relaxation in an embedding of an unweighted graph into a line metric and into a tree metric. These two basic target metrics are particularly important for representing a graph by a structure that is easy to understand, with applications to visualization, compression, clustering, and nearest-neighbor searching. Along the way, we improve the best known approximation factor for ordinally embedding unweighted trees into the line down to 2. Our results illustrate an important contrast to optimal-distortion metric embeddings, where the best approximation factor for unweighted graphs into the line is $O(n^{1/2})$, and even for unweighted trees into the line the best is $\tilde{O}(n^{1/3})$. We also show that Johnson-Lindenstrauss-type dimensionality reduction is possible with ordinal relaxation and ℓ_1 metrics (and ℓ_p metrics with $1 \leq p \leq 2$), unlike metric embedding of ℓ_1 metrics.

1 Introduction

The maturing field of *metric embeddings* (see, e.g., [IM04]) originally grew out of the more classic field of *multidimensional scaling (MDS)*. In MDS, we are given a finite set of points and measured pairwise distances between them, and our goal is to embed the points into some target metric space while (approximately) preserving the distances. Originally, the MDS community considered embeddings

---

* Supported in part by NSF under grant number ITR ANI-0205445.

A. Goel et al. (Eds.): APPROX and RANDOM 2008, LNCS 5171, pp. 21–34, 2008.
© Springer-Verlag Berlin Heidelberg 2008

into an ℓ_p space, with the goal of aiding in visualization, compression, clustering, or nearest-neighbor searching; thus, the focus was on low-dimensional embeddings. An *isometric embedding* preserves all distances, while more generally, *metric embeddings* trade-off the dimension with the fidelity of the embeddings.

But the distances themselves are not essential in nearest-neighbor searching and many contexts of visualization, compression, and clustering. Rather, the *order* of the distances captures enough information; in order words, we only need an embedding of a monotone mapping of the distances into the target metric space. The early MDS literature considered such embeddings heavily under the terms *ordinal embeddings, nonmetric MDS*, or *monotone maps* [CS74, Kru64a, Kru64b, She62, Tor52].

While the early work on ordinal embeddings was largely heuristic, there has been some work with provable guarantees since then. Define a *distance matrix* to be any matrix of pairwise distances, not necessarily describing a metric. Shah and Farach-Colton [SFC04] have shown that it is NP-hard to decide whether a distance matrix can be ordinally embedded into an additive metric, i.e., the shortest-path metric in a tree. Define the *ordinal dimension* of a distance matrix to be the smallest dimension of a Euclidean space into which the matrix can be ordinally embedded. Bilu and Linial [BL04] have shown that every matrix has ordinal dimension at most $n - 1$. They also applied the methods of [AFR85] to show that (in a certain well-defined sense) almost every n-point metric space has ordinal dimension $\Omega(n)$. It is also known that ultrametrics have ordinal dimension exactly $n - 1$ [ABD+].

While ordinal embeddings and ordinal dimension provide analogs of exact isometric embedding with monotone mapping, Alon et al. [ABD+] introduced an ordinal analog of distortion to enable a broader range of embeddings. Specifically, a metric M' is an *ordinal embedding with relaxation* $\alpha \geq 1$ of a distance matrix M if $\alpha M[i, j] < M[k, l]$ implies $M'[i, j] < M'[k, l]$. In other words, the embedding must preserve the relative order of significantly different distances. Note that in an ordinary ordinal embedding, we must respect distance equality, while in an ordinal embedding with relaxation 1, we may break ties. The goal of the *ordinal relaxation problem* is to find an embedding of a given distance matrix into a target family of metric spaces while minimizing the relaxation. Here we optimize the confidence with which ordinal relations are preserved, rather than the number of ordinal constraints satisfied (as in [Opa79, CS98, SFC04]).

Our results. We develop polynomial-time constant-factor approximation algorithms for minimizing the relaxation in an embedding of an unweighted graph into a line metric and into a tree (additive) metric. These two basic target metrics are particularly important for representing a graph by a structure that is easy for humans to understand, with applications to visualization, compression, clustering, and nearest-neighbor searching.

Our 10/3-approximation for unweighted graphs into the line (Section 3) illustrates an important contrast to optimal-distortion metric embeddings, where the best approximation factor for unweighted graphs into the line is $O(n^{1/2})$, and even for unweighted trees into the line the best is $\tilde{O}(n^{1/3})$ [BDG+05]. This result significantly generalizes the previously known 3-approximation for minimum-relaxation ordinal embedding of unweighted trees into the line [ABD$^+$]. Along the way, we also improve this result to a 2-approximation. The main approach of our algorithm is to embed the given graph G into the line with additive distortion at most 4α (2α from expansion and 2α from contraction), where α is the minimum relaxation of an ordinal embedding of G into a tree. We show that this embedding has (multiplicative) ordinal relaxation at most 4α, a property not necessarily true of multiplicative distortion. When G is a tree, we show that the embedding is contractive, and thus we obtain a 2-approximation. For general graphs G, we modify the embedding by contracting certain distances to improve the (asymptotic) approximation factor to $10/3$.

Our 27-approximation for unweighted graphs into trees (Section 4) is in fact an approximation algorithm for both minimum-relaxation ordinal embedding and minimum-distortion metric embedding. We show that lower bounds on the ordinal relaxation (which are also lower bounds on metric distortion) provide new insight into the structure of both problems. Our result improves the best previous 100-approximation for metric distortion, and is also the first illustration that relaxation and distortion are within constant factors of each other in this context. The main approach of our algorithm is to construct a supergraph H of the given graph G such that (1) G can be embedded into H with distortion at most 9α, where α is the minimum relaxation of an ordinal embedding of G into a tree, and (2) H can be embedded into a spanning tree of H with distortion at most 3. The resulting embedding of distortion $27\,\alpha$ is a 27-approximation for both distortion and relaxation.

In each context where we obtain constant-factor approximations, e.g., ordinally embedding unweighted graphs into the line, it remains open to prove NP-hardness or inapproximability of minimizing relaxation.

Another topic of recent interest is dimensionality reduction. The famous Johnson-Lindenstrauss Theorem [JL84] guarantees low-distortion reduction to logarithmic dimension for arbitrary ℓ_2 metrics, but recently it was shown that the same is impossible without significant distortion for ℓ_1 metrics [BC05, LN04] (despite their usefulness and flexibility for representation). In contrast, we show in Section 5 that arbitrary ℓ_1 metrics can be ordinally embedded into logarithmic-dimension ℓ_1 space with relaxation $1+\varepsilon$ for any $\varepsilon > 0$. More generally, our analog of the Johnson-Lindenstrauss Theorem applies to ℓ_p metrics with $1 \le p \le 2$. We show that this result in fact follows easily from a combination of known results: the monotone property of ordinal embeddings, power transformations for making metrics Euclidean, the Johnson-Lindenstrauss Theorem, and Dvoretzky-type results to return to the desired ℓ_p space [FLM77, Ind07].

2 Definitions

In this section, we formally define ordinal embeddings and relaxation (as in [ABD$^+$]) as well as the contrasting notions of metric embeddings and distortion.

Consider a finite metric $D : P \times P \rightarrow [0, \infty)$ on a finite point set P—the *source metric*—and a class $\mathcal{T}$ of metric spaces $(T, d) \in \mathcal{T}$ where d is the distance function for space T—the *target metrics*. An *ordinal embedding (with no relaxation)* of D into $\mathcal{T}$ is a choice $(T, d) \in \mathcal{T}$ of a target metric and a mapping $\phi : P \rightarrow T$ of the points into the target metric such that every comparison between pairs of distances has the same outcome: for all $p, q, r, s \in P$, $D(p, q) \leq D(r, s)$ if and only if $d(\phi(p), \phi(q)) \leq d(\phi(r), \phi(s))$. Equivalently, ϕ induces a monotone function $D(p, q) \mapsto d(\phi(p), \phi(q))$. An *ordinal embedding with relaxation* α of D into $\mathcal{T}$ is a choice $(T, d) \in \mathcal{T}$ and a mapping $\phi : P \rightarrow T$ such that every comparison between pairs of distances not within a factor of α has the same outcome: for all $p, q, r, s \in P$ with $D(p, q) > \alpha\, D(r, s)$, $d(\phi(p), \phi(q)) > d(\phi(r), \phi(s))$. Equivalently, we can view a relaxation α as defining a partial order on distances $D(p, q)$, where two distances $D(p, q)$ and $D(r, s)$ are comparable if and only if they are not within a factor of α of each other, and the ordinal embedding must preserve this partial order on distances.

We pay particular attention to contrasts between relaxation in ordinal embedding relaxation and distortion in "standard" embedding, which we call "metric embedding" for distinction. A *contractive metric embedding with distortion* c of a source metric D into a class $\mathcal{T}$ of target metrics is a choice $(T, d) \in \mathcal{T}$ and a mapping $\phi : P \rightarrow T$ such that no distance increases and every distance is preserved up to a factor of c: for all $p, q \in P$, $1 \leq D(p, q)/d(\phi(p), \phi(q)) \leq c$. Similarly, we can define an *expansive metric embedding with distortion* c with the inequality $1 \leq d(\phi(p), \phi(q))/D(p, q) \leq c$. When $c = 1$, these two notions coincide to require exact preservation of all distances; such an embedding is called a *metric embedding with no distortion* or an *isometric embedding*. In general, $c^* = c^*(D, \mathcal{T})$ denotes the minimum possible distortion of a metric embedding of D into $\mathcal{T}$. (This definition is equivalent for both contractive and expansive metric embeddings, by scaling.)

3   Constant-Factor Approximations for Embedding Unweighted Graphs and Trees into the Line

In this section we give an asymptotically $10/3$-approximation algorithm for minimum-relaxation ordinal embedding of the shortest-path metric of an unweighted graph into the line. This result shows a sharp contrast from metric embedding, where the best known polynomial-time approximation algorithm for unweighted graphs into the line achieves an approximation ratio of just $O(n^{1/2})$, and even for unweighted trees into the line the best is $\tilde{O}(n^{1/3})$ [BDG+05]. Along the way, we give a 2-approximation algorithm for minimum-relaxation ordinal embedding of unweighted trees into the line, improving on the 3-approximation of [ABD+].

Let $G = (V, E)$ be the input unweighted graph. Suppose that there exists an ordinal embedding h of G into the line $\mathbb{R}$ with relaxation α. Let u and v be the vertices in G that h maps onto the leftmost and rightmost points, respectively, in the line. In other words, $h(u)$ and $h(v)$ are the minimum and maximum values taken by h. The algorithm guesses the vertices u and v, i.e., repeats the following procedure for all possible pairs of vertices u and v.

For a given guess of u and v, the algorithm computes an (arbitrary) shortest path P from u to v in G, say visiting vertices $u = v_0, v_1, v_2, \ldots, v_\delta = v$. Then it computes the Voronoi partition of the vertices $V = V_0 \cup V_1 \cup \cdots \cup V_\delta$ where the sites are the vertices $v_0, v_1, \ldots, v_\delta$ of the path P, i.e., for each $i \in \{0, 1, \ldots, \delta\}$ and for each $x \in V_i$, $D_G(x, v_i) = \min\{D_G(x, v_j) \mid v_j \in P\}$. In particular, $v_i \in V_i$. This partition defines a function $f : V \to \mathbb{R}$ by $f(x) = i$ for $x \in V_i$. This function will turn out to be a good embedding if G is a tree, but it will need further refinement for general graphs. We begin by deriving some properties of f.

Lemma 1. *For any $i \in \{0, 1, \ldots, \delta\}$ and any $x \in V_i$, we have $\alpha \geq D_G(x, v_i)$, and if G is a tree, we have $\alpha \geq D_G(x, v_i) + 1$.*

Proof. Suppose for contradiction that some vertex $x \in V_i$ has $\alpha < D_G(x, v_i)$. Consider the ordinal embedding h of G into $\mathbb{R}$ with relaxation α. By construction, $h(v_0) \leq h(x) \leq h(v_\delta)$, so some j with $0 \leq j < \delta$ has $h(v_j) \leq h(x) \leq h(v_{j+1})$. By assumption, $D_G(x, v_j) \geq D_G(x, v_i) > \alpha = \alpha\, D_G(v_j, v_{j+1}) = \alpha$. By definition of relaxation, $|h(x) - h(v_j)| > |h(v_j) - h(v_{j+1})|$, contradicting that $h(v_j) \leq h(x) \leq h(v_{j+1})$.

If G is a tree, we have the property that $|D_G(x, v_j) - D_G(x, v_{j+1})| = 1$. By construction, both $D_G(x, v_j)$ and $D_G(x, v_{j+1})$ are at least $D_G(x, v_i)$, and hence the larger is at least $D_G(x, v_i) + 1 > \alpha + 1$. The rest of the proof for trees is as above. $\qquad\square$

Lemma 2. *For any two vertices x_1 and x_2 in G, we have*

$$D_G(x_1, x_2) - 2\alpha \leq |f(x_1) - f(x_2)| \leq D_G(x_1, x_2) + 2\alpha,$$

and if G is a tree, we have

$$D_G(x_1, x_2) - 2(\alpha - 1) \leq |f(x_1) - f(x_2)| \leq D_G(x_1, x_2) + 2(\alpha - 1).$$

Proof. Suppose x_1 and x_2 are in V_{i_1} and V_{i_2}, respectively. By Lemma 1, $D_G(x_1, v_{i_1}) \leq \alpha$ and $D_G(x_2, v_{i_2}) \leq \alpha$. By the triangle inequality, $D_G(x_1, x_2) \leq D_G(x_1, v_{i_1}) + D_G(v_{i_1}, v_{i_2}) + D_G(v_{i_2}, x_2) \leq \alpha + |f(x_1) - f(x_2)| + \alpha$. We also have $|f(x_1) - f(x_2)| = D_G(v_{i_1}, v_{i_2}) \leq D_G(v_{i_1}, x_1) + D_G(x_1, x_2) + D_G(x_2, v_{i_2}) \leq \alpha + D_G(x_1, x_2) + \alpha$. If G is a tree, we can replace each α with $\alpha - 1$ by Lemma 1 and obtain the stronger inequalities. $\qquad\square$

Next we show the efficiency of f as an ordinal embedding for trees, improving on the 3-approximation of [ABD+]:

Theorem 1. *There is a polynomial-time algorithm which, given an unweighted tree T that ordinally embeds into the line with relaxation α, computes an ordinal embedding with relaxation $2\alpha - 1$.*

Proof. We prove that the function f defined above is an ordinal embedding with relaxation $2\alpha - 1$.

First we claim that, for any two vertices x and y, we have $|f(x) - f(y)| \leq D_T(x, y)$. Because T is a tree, there is a unique simple path Q from x to y. Suppose x and y belong to V_i and V_j, respectively. If $i = j$, then $f(x) = f(y)$, and the claim is trivial. Otherwise, Q must be the simple path from x to v_i to v_j (along P) to y. Therefore the length of Q is at least $|i - j| = |f(x) - f(y)|$. In other words, the embedding f does not increase the distance between x and y.

Next let x_1, x_2, x_3, x_4 be vertices of T with $D_T(x_1, x_2)/D_T(x_3, x_4) > 2\alpha - 1$. It remains to show that $|f(x_1) - f(x_2)| > |f(x_3) - f(x_4)|$. Because $\alpha \geq 1$ and $D_T(x_3, x_4) \geq 1$, we have $D_T(x_1, x_2) > (2\alpha - 1)D_T(x_3, x_4) \geq 2\alpha - 2 + D_T(x_3, x_4)$. By Lemma 2, we have $|f(x_1) - f(x_2)| \geq D_T(x_1, x_2) - 2\alpha + 2$, which is greater than $D_T(x_3, x_4)$. Above we proved that $D_T(x_3, x_4) \geq |f(x_3) - f(x_4)|$. Therefore $|f(x_1) - f(x_2)| > |f(x_3) - f(x_4)|$. □

Before we define our embedding for general unweighted graphs, we prove a final property of f:

Lemma 3. *For any $\varepsilon > 1/\alpha$, any vertex x, and any vertices y_1 and y_2 adjacent to x, we have either $\min\{f(y_1), f(y_2)\} > f(x) - \alpha(1+\varepsilon)$ or $\max\{f(y_1), f(y_2)\} < f(x) + \alpha(1+\varepsilon)$.*

Proof. Suppose for contradiction that there is a vertex x with neighbors y_1 and y_2 for which $f(y_1) \leq f(x) - \alpha(1+\varepsilon)$ and $f(x) \leq f(y_2) - \alpha(1+\varepsilon)$. Thus $|f(y_1) - f(y_2)| \geq 2\alpha(1+\varepsilon)$. But $D_G(y_1, y_2) \leq 2$, so by Lemma 2 we conclude $|f(y_1) - f(y_2)| \leq 2 + 2\alpha$, which is a contradiction for $\varepsilon > 1/\alpha$. □

Finally we can define our ordinal embedding $g : V \to \mathbb{R}$ for a general unweighted graph $G = (V, E)$, for any $\varepsilon > 0$:

$$g(x) = \begin{cases} f(x) - \alpha/3 & \text{if } x \text{ has a neighbor } y \text{ in } G \text{ with } f(y) \leq f(x) - \alpha(1+\varepsilon), \\ f(x) + \alpha/3 & \text{if } x \text{ has a neighbor } y \text{ in } G \text{ with } f(y) \geq f(x) + \alpha(1+\varepsilon), \\ f(x) & \text{otherwise.} \end{cases}$$

By Lemma 3, the embedding g is well-defined. It remains to bound its relaxation.

Lemma 4. *For any two vertices x_1 and x_2 in G, we have*

$$D_G(x_1, x_2) - 8\alpha/3 \leq |g(x_1) - g(x_2)| \leq D_G(x_1, x_2) + 8\alpha/3.$$

Proof. By construction, $|g(x) - f(x)| \leq \alpha/3$ for any vertex x. By Lemma 2,

$$D_G(x_1, x_2) - 2\alpha - 2\alpha/3 \leq |g(x_1) - g(x_2)| \leq D_G(x_1, x_2) + 2\alpha + 2\alpha/3. \quad □$$

Lemma 5. *For any $\varepsilon > 3/(2\alpha)$ and any edge $e = (x, y)$ in G, we have $|g(x) - g(y)| \leq (4/3 + \varepsilon)\alpha$.*

Proof. Without loss of generality, suppose that $f(x) \leq f(y)$. By Lemma 2, $|f(x) - f(y)| \leq 1 + 2\alpha$. If $f(x) < f(y) - \alpha(1+\varepsilon)$, then $g(x) = f(x) + \alpha/3$ and $g(y) = f(y) - \alpha/3$. In this case, we have

$$|g(x) - g(y)| = |f(x) - f(y)| - 2\alpha/3 \leq 2\alpha + 1 - 2\alpha/3 \leq (4/3 + \varepsilon)\alpha$$

for $\alpha \geq 1/\varepsilon$.

It remains to consider the case $f(x) \leq f(y) + (1 + \varepsilon)\alpha$. Observe that $g(x)$ is equal to one of the values $f(x) - \alpha/3$, $f(x)$, and $f(x) + \alpha/3$. There are also three cases for $g(y)$. So there are nine cases to consider. But the claim is clearly true for eight of them. The only case for which the claim is nontrivial is when $g(x) = f(x) - \alpha/3$ and $g(y) = f(y) + \alpha/3$.

In this case, we have $|g(x) - g(y)| = |f(x) - f(y)| + 2\alpha/3$. By definition of g, we conclude that there is a vertex x' adjacent to x in G such that $f(x') \leq f(x) - (1 + \varepsilon)\alpha$. Similarly, there is a vertex y' adjacent to y for which we have $f(y') \geq f(y) + (1 + \varepsilon)\alpha$. Thus $f(y') - f(x') \geq (2 + 2\varepsilon)\alpha$. But we know that $D_G(x', y') \leq 3$, and $|f(x') - f(y')|$ must be at most $3 + 2\alpha$, which is a contradiction for $\varepsilon > 3/(2\alpha)$. Therefore this case does not occur, and the claim is true for all nine cases. $\qquad\square$

Lemma 6. *The ordinal embedding g has relaxation $(10/3 + \varepsilon)\alpha + 1$ for $\varepsilon > 3/(2\alpha)$.*

Proof. Consider $x_1, x_2, x_3, x_4 \in V$ for which $D_G(x_1, x_2)/D_G(x_3, x_4) > (10/3 + \varepsilon)\alpha + 1$. It suffices to show that $|g(x_1) - g(x_2)| > |g(x_3) - g(x_4)|$. We consider two cases.

First suppose that $D_G(x_3, x_4) > 1$. Then

$$D_G(x_1, x_2) - D_G(x_3, x_4) > [(10/3 + \varepsilon)\alpha + 1 - 1]D_G(x_3, x_4) > 20\alpha/3.$$

By Lemma 4, $|g(x_1) - g(x_2)| \geq D_G(x_1, x_2) - 8\alpha/3$ and $|g(x_3) - g(x_4)| \leq D_G(x_3, x_4) + 8\alpha/3$. Thus

$$|g(x_1) - g(x_2)| - |g(x_3) - g(x_4)| \geq [D_G(x_1, x_2) - 8\alpha/3] - [D_G(x_3, x_4) + 8\alpha/3] \geq 1.$$

Therefore $|g(x_1) - g(x_2)| > |g(x_3) - g(x_4)|$.

In the second case, there is an edge between vertices x_3 and x_4. We also know that $D_G(x_1, x_2) > (10/3 + \varepsilon)\alpha + 1$. By Lemma 5, $|g(x_3) - g(x_4)| \leq (4/3 + \varepsilon)\alpha$. It suffices to prove that $|g(x_1) - g(x_2)| > (4/3 + \varepsilon)\alpha$. By Lemma 2, $|f(x_1) - f(x_2)| \geq D_G(x_1, x_2) - 2\alpha > (4/3 + \varepsilon)\alpha$. If $|g(x_1) - g(x_2)| \geq |f(x_1) - f(x_2)|$, the claim is true. On the other hand, if $|f(x_1) - f(x_2)| > (2 + \varepsilon)\alpha$, then because $|g(x_1) - g(x_2)| \geq |f(x_1) - f(x_2)| - 2\alpha/3$, we have $|g(x_1) - g(x_2)| > (4/3 + \varepsilon)\alpha$. So we can suppose that $|g(x_1) - g(x_2)| < |f(x_1) - f(x_2)|$ and that $|f(x_1) - f(x_2)| \in [(4/3 + \varepsilon)\alpha, (2 + \varepsilon)\alpha]$. Without loss of generality, we can suppose that $f(x_1) < f(x_2)$, and consequently $f(x_2) \in [f(x_1) + (4/3 + \varepsilon)\alpha, f(x_1) + (2 + \varepsilon)\alpha]$. Because $|g(x_1) - g(x_2)| < |f(x_1) - f(x_2)|$, and by the symmetry between x_1 and x_2, we can suppose that $g(x_1) = f(x_1) + \alpha/3$ and $g(x_2) \leq f(x_2)$.

We conclude that there exists a vertex x_5 for which $e = (x_1, x_5) \in E(G)$ and $f(x_1) + (1 + \varepsilon)\alpha \leq f(x_5) \leq f(x_1) + 2\alpha$. As a consequence, $D_G(x_5, x_2) \geq D_G(x_1, x_2) - 1 > (10/3 + \varepsilon)\alpha$ and $f(x_5) \in [f(x_1) + (1 + \varepsilon)\alpha, f(x_1) + 2\alpha]$. Therefore $|f(x_5) - f(x_2)| \leq \alpha$. But this inequality contradicts that $|f(x_5) - f(x_2)| \geq D_G(x_5, x_2) - 2\alpha \geq (4/3 + \varepsilon)\alpha$. We conclude that $|g(x_1) - g(x_2)| > (4/3 + \varepsilon)\alpha$, which completes the proof. $\qquad\square$

Substituting $\varepsilon = 3/(2\alpha) + \delta/\alpha$ in Lemma 6, we obtain the following result:

Theorem 2. *For any $\delta > 0$, there is a polynomial-time algorithm which, given an unweighted graph that ordinally embeds into the line with relaxation α, computes an ordinal embedding with relaxation $(10/3)\alpha + 5/2 + \delta$*

4   Constant-Factor Approximation for Embedding Unweighted Graphs into Trees

In this section, we develop a 27-approximation for the minimum-relaxation ordinal embedding of an arbitrary unweighted graph into a tree metric. Specifically, we give a polynomial-time algorithm that embeds a given unweighted graph G into a tree with (metric) distortion at most $27\,\alpha_G$, where α_G is the minimum relaxation needed to ordinally embed G into a tree. Because the relaxation of an embedding is always at most its distortion [ABD+, Proposition 1], we obtain the desired 27-approximation for minimum relaxation. Furthermore, because the optimal relaxation is also at most the optimal distortion, the same algorithm is a 27-approximation for minimum distortion. This result improves substantially on the 100-approximation for minimum-distortion metric embedding of an unweighted graph into a tree [BIS07]. Furthermore, we obtain that the minimum possible distortion c_G is $\Theta(\alpha_G)$ for any graph G, a property which is not true in many other cases [ABD+].

4.1   Lower Bound for Ordinal Embedding of Graphs into Trees

We start with a lower bound on the minimum relaxation needed to embed a graph with a special structure into any tree.

Theorem 3. *Any graph G has $\alpha_G \geq 2l/3$ if there are two vertices u and v and two paths P_1 and P_2 between them with the following properties:*

1. *P_1 is a shortest path between u and v; and*
2. *there is a vertex w on P_1 whose distance to any vertex on P_2 is at least l.*

Proof. Suppose that G can be ordinally embedded into a tree T with relaxation less than $2l/3$. Let $u = v_1, v_2, \ldots, v_m = v$ be the vertices of the path P_1 in G. By Property 2, we have $m \geq 2l$ because u and v are also two vertices on P_2. Note that in addition to u and v, P_1 and P_2 may have more vertices in common. Let v_i be mapped onto v'_i in this embedding, $v'_i \in V(T)$. Let P' be the unique path between v'_1 and v'_m in T. Also suppose that x_i is the first vertex on path P' that we meet when we are moving from v'_i to v'_m. Note that such a vertex necessarily exists because v'_m is a vertex on P' which we meet during our path in T, and there might be more vertices like v'_m. According to this definition, x_i is a vertex on P', and the vertices $v'_1 = x_1, x_2, \ldots, x_m = v'_m$ are not necessarily distinct. Let k be the maximum distance between two vertices x and y in T over all pairs (x, y) with the property that their representatives in G are adjacent.

Because there is exactly one path between any pair of vertices in T, we know that, if $x_i \neq x_{i+1}$, then the vertex x_i lies in the (shortest) path between v'_i and v'_{i+1} in T. Consequently, we have $d_T(v'_i, v'_{i+1}) = d_T(v'_i, x_i) + d_T(x_i, v'_{i+1})$ where $d_T(a, b)$ is the distance between a and b in T. Note that by definition of k, for any i where $x_i \neq x_{i+1}$, the sum of these two terms is at most k. This means that either $d_T(v'_i, x_i)$ or $d_T(x_i, v'_{i+1})$ is at most $k/2$. We use this fact frequently in the rest of proof.

Let w be the ith vertex on P_1. Equivalently, let w be v_i. In order to complete our proof, we consider two cases. At first, suppose that $x_{i-l/3} = x_{i-l/3+1} = \cdots = x_i = x_{i+1} = \cdots = x_{i+l/3}$. In this case, let i_1 and i_2 be respectively the minimum and maximum numbers for which we have $x_{i_1} = x_i = x_{i_2}$. We prove that either $d_T(v'_{i_1}, x_{i_1})$ or $d_T(x_{i_1}, v'_{i_1-1})$ is at most $k/2$. If $i_1 = 1$, then we have $x_{i_1} = v'_{i_1}$ and consequently $d_T(x_{i_1}, v'_{i_1}) = 0$. Otherwise, we have $x_{i_1} \neq x_{i_1-1}$ and therefore we deduce that either $d_T(v'_{i_1}, x_{i_1})$ or $d_T(x_{i_1}, v'_{i_1-1})$ is at most $k/2$. According to the symmetry of the case, we also have that either $d_T(v'_{i_2}, x_{i_2})$ or $d_T(x_{i_2}, v'_{i_2+1})$ is at most $k/2$. Note that $x_{i_1} = x_{i_2}$. Finally we conclude that there exist $j_1 \in \{i_1 - 1, i_1\}$ and $j_2 \in \{i_2, i_2 + 1\}$ such that $d_T(v'_{j_1}, v'_{j_2}) \leq k/2 + k/2 = k$. Note that the distance between v_{j_1} and v_{j_2} is at least $2l/3$ in G. Because there are two adjacent vertices in G such that their distance in T is k, we can say that the relaxation is at least $\frac{2l/3}{1} = 2l/3$.

Now we consider the second and final case. In this case, There exists a vertex $j_1 \in \{i+1-l/3, i+2-l/3, \ldots, i-1+l/3\}$ such that we have either $x_{j_1} \neq x_{j_1-1}$ or $x_{j_1} \neq x_{j_1+1}$. Using each of these inequalities, we reach the fact that there exists $j_2 \in \{j_1 - 1, j_1, j_1 + 1\}$ for which we have $d_T(v'_{j_2}, x_{j_1}) \leq k/2$. We define some similar terms for path P_2. Let $u = u_1, u_2, \ldots, u_{m'} = v$ be the vertices of the path P_2 in graph G. Let u_i is mapped onto u'_i in this embedding, $u'_i \in V(T)$. Suppose that y_i is the first vertex on path P' that we meet when we are moving from u'_i to u'_m. We know that either $x_{j_1} \neq v'_1$ or $x_{j_1} \neq v'_m$. Without loss of generality, suppose that $x_{j_1} \neq v'_1$. Now we know that $y_1 = v'_1$ lies before x_{j_1} on path P', and $y_{m'} = v'_m$ does not lie before x_{j_1} on this path. Therefore there exists a number j_3 for which y_{j_3} lies before x_{j_1} on P', and y_{j_3+1} does not lie before x_{j_1} on the path. Therefore x_{j_1} occurs in the (shortest) path between u'_{j_3} and u'_{j_3+1} in T. In the other words, we have $d_T(u'_{j_3}, u'_{j_3+1}) = d_T(u'_{j_3}, x_{j_1}) + d_T(x_{j_1}, u'_{j_3+1}) \leq k$. We can say that either $d_T(u'_{j_3}, x_{j_1})$ or $d_T(x_{j_1}, u'_{j_3+1})$ is at most $k/2$. Suppose that $d_T(u'_{j_3}, x_{j_1})$ is at most $k/2$. The proof in the other case is exactly the same.

Finally we reach the inequality $d_T(v'_{j_2}, u'_{j_3}) \leq d_T(v'_{j_2}, x_{j_1}) + d_T(x_{j_1}, u'_{j_3}) \leq k/2 + k/2 = k$. Note that the distance between v_{j_2} and $w = v_i$ is at most $l/3$ in G, and therefore the distance between v_{j_2} and u_{j_3} which is a vertex on path P_2 is at least $l - l/3 = 2l/3$ in G. Again we can say that there are two adjacent vertices in G such that their distance in T is k, and therefore the relaxation is at least $(2l/3)/1 = 2l/3$. $\qquad\square$

4.2   27-Approximation Algorithm

In this section we embed a given graph G into a tree with distortion (and hence relaxation) at most $27 \alpha_G$. We find the embedding in two phases. At first, we construct graph H from the given graph G only by adding some edges to G. Then we propose an algorithm which finds a spanning tree of H like T. Next, we prove that the distortion of embedding G into H is at most $O(\alpha_G)$. We also prove that the embedding H into T has distortion at most 3. Therefore the distortion of embedding G into T is $O(\alpha_G)$.

Let G be the given graph. We construct H as follows. Choose an arbitrary vertex v, and run a breadth-first search to find a tree T_v rooted at v in which the distance between each vertex and v is equal to their distance in G. The vertices of G occur in different levels of T_v. The ith level of this tree, L_i, consists of vertices whose distance to v is i. We have $L_0 = \{v\}$ and $V(G) = \bigcup_{i=0}^{n-1} L_i$. In constructing H from G, we add an edge between two vertices u_1 and u_2 if and only if u_1 and u_2 are in the same level such as L_i or in two consecutive levels such as L_i and L_{i+1}, and there is a path between u_1 and u_2 that does not use the vertices of levels $L_0, L_1, \ldots, L_{i-1}$. In the other words, there exists a path between u_1 and u_2 in graph $G[V - \bigcup_{j=0}^{i-1} L_j]$ where $G[X]$ is the subgraph of G induced by vertex set X. Using Lemma 3, we prove the following lemma.

Lemma 7. *The distortion of embedding G into H is at most $9\,\alpha_G$.*

Proof. Because we only add edges to G to form H, the distance between vertices does not increase. Therefore this metric embedding is contractive. The distortion of the embedding is thus $\max_{u,v \in V(G)=V(H)} d_G(u,v)/d_H(u,v)$. We also know that this maximum is equal to $\max_{(u,v) \in E(H)} d_G(u,v)/d_H(u,v)$ because, if we know that the distance between two vertices adjacent in H is at most k in G, then the distance between every pair of vertices in G is at most k times their distance in H. Therefore we just need to prove that, for each edge (u_1, u_2) that we add, the distance between u_1 and u_2 in G is at most $9\,\alpha_G$. In the rest of proof, when we talk about the distance between two vertices or a path between them, we consider all of them in graph G. Note that u_1 and u_2 are either in the same level such as L_i or in two consecutive levels L_i and L_{i+1}, and there is a path P_1 between them which uses only vertices in levels $L_i, L_{i+1}, \ldots$. Consider a shortest path P_2 between u_1 and u_2. There is also a unique path P_3 between u_1 and u_2 in the breadth-first-search tree rooted at v. Note that these paths are not necessarily disjoint. Let l be the length of P_2. We prove that $l \leq 9\,\alpha_G$. We consider two cases.

First suppose that there is a vertex in P_2 like w that is in $\bigcup_{j=0}^{i-l/6} L_j$. For $i < l/6$, $\bigcup_{j=0}^{i-l/6} L_j$ is empty. The distance between w and any vertex in P_1 is at least $l/6$ because the distance between v and w is at most $i - l/6$, and the distance between v and any vertex in P_1 is at least i. Applying Lemma 3 to P_2 as the shortest path, P_1 as the other path, and vertex w, G cannot be ordinally embedded into any tree with relaxation less than $\frac{2}{3} \cdot \frac{l}{6} = l/9$. Therefore $9\,\alpha_G \geq l$.

In the second case, all vertices of the path P_2 are in $\bigcup_{j=i+1-l/6}^{n-1} L_j$, including the vertex in the middle of P_2. Let w be the vertex in the middle of the P_2.

Because P_2 is a shortest path, the distance between w and u_1 and u_2 is at least $\frac{l-1}{2}$. We assert that the distance between w and any vertex in the path P_3 is at least $l/6$. Consider a vertex in P_3 like x. If x is in $\bigcup_{j=0}^{i+1-l/3} L_j$, the distance between w and x is at least $(i+1-l/6)-(i+1-l/3) = l/6$. Otherwise because of the special structure of path P_3, the distance between x and at least one of the vertices u_1 and u_2 is at most $i+1-(i+1-l/3+1) = l/3-1$. Because the distance between w and both u_1 and u_2 is at least $\frac{l-1}{2}$, we can say that the distance between w and x is at least $\frac{l-1}{2}-(l/3-1) \geq l/6$. Again applying Lemma 3 to P_2 as the shortest path, P_3 as the other path, and vertex w, G cannot be ordinally embedded into any tree with relaxation less than $\frac{2}{3}\cdot\frac{l}{6} = l/9$. Therefore $9\,\alpha_G \geq l$. □

Now we are going to find a spanning tree T of H with distortion at most 3. Before proposing the algorithm, we mention some important properties of H.

The subgraph $G[L_i]$ of H induced by vertices in level L_i is a union of some cliques. In fact, if there are two edges (a,b) and (b,c) in $G[L_i]$, then there must be a path between a and b in G that uses only vertices in $\bigcup_{j=i}^{n-1} L_j$, and also a path between b and c in G which uses only vertices in $\bigcup_{j=i}^{n-1} L_j$. Therefore there exists a path between a and c in G that uses only vertices in $\bigcup_{j=i}^{n-1} L_j$. Consequently we must have added an edge between a and c in constructing H from G. Because the connectivity relation in each level is transitive, each level is a union of some cliques. There is another important property of H. For any $a,b \in L_{i+1}$ and $c \in L_i$, if b is adjacent to both a and c in H, then there must be an edge between a and c in H. The claim is true because of the special definition of edges in H. Therefore, for each clique in level L_{i+1}, there exists a vertex in L_i that is adjacent to all vertices of that clique.

Now we find the tree T as follows. For any $i > 0$ and any clique C in level L_i, we just need to find a vertex v_C in L_{i-1} that is adjacent to all vertices in C, and then add all edges between vertex v_C and the vertices in C into the tree. Actually this tree is a breadth-first-search tree in graph H.

Lemma 8. *The distortion of embedding H into T is at most* 3.

Proof. It is clear that we obtain a spanning tree T. The embedding is expansive because T is a subgraph of H. In order to bound the distortion of this embedding, we must prove that, for each edge (x,y) in H, the distance between x and y is at most 3 in T. There are two kinds of edges in H: the edges between vertices in the same level and edges between vertices in two consecutive levels. If x and y are in the same level L_i, then they are connected to a vertex z in L_{i-1} in tree T. Therefore their distance in tree T is 2. Otherwise, suppose that x is in L_i and y is in L_{i-1}. Vertex x is connected to a vertex z in L_{i-1} in tree T. If $z = y$, then the claim is clear. If $y \neq z$, then by definition, there is an edge between y and z in H, and they are also in the same level L_{i-1}. Therefore the distance between y and z in T is 2, and consequently the distance between x and y is 3 in T. □

Combining Lemmas 7 and 8, we obtain the following result:

Theorem 4. *There is a polynomial-time algorithm that embeds a given graph* G *into a tree with distortion at most* $27\,\alpha_G$.

5   Dimensionality Reduction in ℓ_1

In this section, we prove that dimensionality reduction in ℓ_1, and indeed any ℓ_p space with $1 \le p \le 2$, is possible with ordinal embeddings of logarithmic dimension and relaxation $1 + \varepsilon$. This result sharply contrasts metric embedding distortion, where any embedding of an ℓ_1 metric of distortion c requires $n^{\Omega(1/c^2)}$ dimensions in ℓ_1 [BC05, LN04].

Theorem 5. *Any* ℓ_p *metric with* $1 \le p \le 2$ *can be embedded into* $O(\varepsilon^{-4} \lg n)$*-dimensional* ℓ_p *space with ordinal relaxation* $1 + \varepsilon$*, for any* $\varepsilon > 0$ *and positive integer* p.

Proof. First we take the $(p/2)$th power of the pairwise distances in the given ℓ_p metric D. The resulting metric D' is an ℓ_2 metric [Sch38, WW75]; see also [MN04]. Also, because $x \mapsto x^{p/2}$ is a monotone function, D' is an ordinal embedding of D (without relaxation). Next we apply Johnson-Lindenstrauss ℓ_2 dimensionality reduction [JL84] to obtain an $d = O((\log n)/\delta^2)$-dimensional ℓ_2 metric D'' with $1+\delta$ distortion relative to D'. Finally, we can embed this d-dimensional ℓ_2 metric into $O(d/\delta^2)$-dimensional ℓ_p space D''' with distortion $1 + \delta$ relative to D'' [FLM77]; see also [Ind07, JS03]. [**Is [FLM77] the right reference for** $O(1/\delta^2)$ **dimension blowup?**] Thus D''' is an $O((\log n)/\delta^4)$-dimensional ℓ_1 metric with distortion $(1 + \delta)^2$ relative to D'.

We claim that D''' is an ordinal embedding of D with relaxation at most $1 + \varepsilon$ for any desired $\varepsilon > 0$ and a suitable choice of δ. Suppose we have two distances $D[p,q]$ and $D[r,s]$ with $D[p,q]/D[r,s] \ge 1+\varepsilon$ for a desired $\varepsilon > 0$. Then $D'[p,q]/D'[r,s] = D'[p,q]^{2/p}/D'[r,s]^{2/p} = (D'[p,q]/D'[r,s])^{2/p} \ge (1 + \varepsilon)^{2/p} \ge 1 + (2/p)\varepsilon$. Thus, if we choose $\delta < \min\{\frac{2}{3}\varepsilon/p, 1\}$, then the distortion of D''' relative to D' is $(1 + \delta)^2 \le 1 + 3\delta < 1 + (2/p)\varepsilon \le D[p,q]/D'[r,s]$, so the embedding preserves the order of distances $D'''[p,q] > D'''[r,s]$. Therefore the relaxation of D''' relative to D is at most $1 + \varepsilon$ as desired. The dimension of the D''' embedding is $O((\log n)/\delta^4) = O((\log n)/\varepsilon^4)$. $\square$

This approach is pleasingly simple in its use of prior results as black boxes. By more involved arguments, it may be possible to improve the dependence on ε in the number of dimensions by directly analyzing with a modification of Johnson-Lindenstrauss [JL84] and avoiding the use of [FLM77].

Acknowledgments

We thank Bo Brinkman for suggesting the approach for the proof in Section 5 and Noga Alon, Martin Farach-Colton, Piotr Indyk, and Assaf Naor for helpful discussions.

References

[ABD+]   Alon, N., Bădoiu, M., Demaine, E.D., Farach-Colton, M., Hajiaghayi,
          M., Sidiropoulos, A.: Ordinal embeddings of minimum relaxation: Gen-
          eral properties, trees, and ultrametrics. ACM Transactions on Algorithms
          (to appear)
[AFR85]   Alon, N., Frankl, P., Rödl, V.: Geometrical realization of set systems and
          probabilistic communication complexity. In: Proceedings of the 26th An-
          nual Symposium on Foundations of Computer Science, pp. 277–280. Port-
          land, Oregon (1985)
[BC05]   Brinkman, B., Charikar, M.: On the impossibility of dimension reduction
          in l_1. Journal of the ACM (electronic) 52(5), 766–788 (2005)
[BDG+05] Bădoiu, M., Dhamdhere, K., Gupta, A., Rabinovich, Y., Raecke, H., Ravi,
          R., Sidiropoulos, A.: Approximation algorithms for low-distortion em-
          beddings into low-dimensional spaces. In: Proceedings of the 16th An-
          nual ACM-SIAM Symposium on Discrete Algorithms, Vancouver, January
          2005, British Columbia, Canada (2005)
[BIS07]   Bădoiu, M., Indyk, P., Sidiropoulos, A.: Approximation algorithms for em-
          bedding general metrics into trees. In: Proceedings of the 18th Symposium
          on Discrete Algorithms, January 2007, pp. 512–521 (2007)
[BL04]   Bilu, Y., Linial, N.: Monotone maps, sphericity and bounded second eigen-
          value. arXiv:math.CO/0401293 (January 2004)
[CS74]   Cunningham, J.P., Shepard, R.N.: Monotone mapping of similarities into
          a general metric space. Journal of Mathematical Psychology 11, 335–364
          (1974)
[CS98]   Chor, B., Sudan, M.: A geometric approach to betweennes. SIAM Journal
          on Discrete Mathematics 11(4), 511–523 (1998)
[FLM77]   Figiel, T., Lindenstrauss, J., Milman, V.D.: The dimension of almost spher-
          ical sections of convex bodies. Acta Mathematica 139(1-2), 53–94 (1977)
[IM04]   Indyk, P., Matoušek, J.: Low-distortion embeddings of finite metric spaces.
          In: Goodman, J.E., O'Rourke, J. (eds.) Handbook of Discrete and Compu-
          tational Geometry, 2nd edn., ch. 8, pp. 177–196. CRC Press, Boca Raton
          (2004)
[Ind07]   Indyk, P.: Uncertainty principles, extractors, and explicit embeddings of
          l2 into l1. In: Proceedings of the 39th Annual ACM Symposium on Theory
          of Computing, pp. 615–620 (2007)
[JL84]   Johnson, W.B., Lindenstrauss, J.: Extensions of lipshitz mapping into
          hilbert space. Contemporary Mathematics 26, 189–206 (1984)
[JS03]   Johnson, W.B., Schechtman, G.: Very tight embeddings of subspaces of
          L_p, $1 \leq p < 2$, into ℓ_p^n. Geometric and Functional Analysis 13(4), 845–851
          (2003)
[Kru64a] Kruskal, J.B.: Multidimensional scaling by optimizing goodness of fit to a
          nonmetric hypothesis. Psychometrika 29, 1–28 (1964)
[Kru64b] Kruskal, J.B.: Non-metric multidimensional scaling. Psychometrika 29,
          115–129 (1964)
[LN04]   Lee, J.R., Naor, A.: Embedding the diamond graph in L_p and dimension
          reduction in L_1. Geometric and Functional Analysis 14(4), 745–747 (2004)
[MN04]   Mendel, M., Naor, A.: Euclidean quotients of finite metric spaces. Advances
          in Mathematics 189(2), 451–494 (2004)

[Opa79]   Opatrny, J.: Total ordering problem. SIAM J. Computing 8, 111–114 (1979)

[Sch38]   Schoenberg, I.J.: Metric spaces and positive definite functions. Transactions of the American Mathematical Society 44(3), 522–536 (1938)

[SFC04]   Shah, R., Farach-Colton, M.: On the complexity of ordinal clustering. Journal of Classification (to appear, 2004)

[She62]   Shepard, R.N.: Multidimensional scaling with unknown distance function I. Psychometrika 27, 125–140 (1962)

[Tor52]   Torgerson, W.S.: Multidimensional scaling I: Theory and method. Psychometrika 17(4), 401–414 (1952)

[WW75]   Wells, J.H., Williams, L.R.: Embeddings and extensions in analysis. In: Ergebnisse der Mathematik und ihrer Grenzgebiete, Band 84, Springer-Verlag, New York (1975)

Connected Vertex Covers in Dense Graphs

Jean Cardinal and Eythan Levy[*]

Université Libre de Bruxelles, CP 212
B-1050 Brussels, Belgium
{jcardin,elevy}@ulb.ac.be

Abstract. We consider the variant of the minimum vertex cover problem in which we require that the cover induces a connected subgraph. We give new approximation results for this problem in dense graphs, in which either the minimum or the average degree is linear. In particular, we prove tight parameterized upper bounds on the approximation returned by Savage's algorithm, and extend a vertex cover algorithm from Karpinski and Zelikovsky to the connected case. The new algorithm approximates the minimum connected vertex cover problem within a factor strictly less than 2 on all dense graphs. All these results are shown to be tight. Finally, we introduce the *price of connectivity* for the vertex cover problem, defined as the worst-case ratio between the sizes of a minimum connected vertex cover and a minimum vertex cover. We prove that the price of connectivity is bounded by $2/(1 + \varepsilon)$ in graphs with average degree εn, and give a family of near-tight examples.

Keywords: approximation algorithm, vertex cover, connected vertex cover, dense graph.

1 Introduction

The Connected Vertex Cover Problem (CVC) is the variant of Vertex Cover (VC) in which we wish to cover all edges with a minimum-size set of vertices that induce a connected subgraph. The problem was first defined in 1977 by Garey and Johnson [1], who showed it to be NP-Hard even when restricted to planar graphs with maximum degree 4. Although CVC has been known since long, it has received far less attention than VC until the recent years. Most previous results are in the field of approximation and fixed-parameter tractability.

Regarding approximation algorithms, the first constant ratio is due to Carla Savage [2], who showed that the internal nodes of any depth-first search tree provide a 2-approximation for VC. Such a set of nodes always induces a connected subgraph, and, since the minimum connected vertex cover is always at least as large as the minimum vertex cover, the approximation ratio also applies to CVC. No better constant approximation ratio is known, and recent results [3] have shown that the problem is NP-hard to approximate within less

---

[*] This work was partially supported by the *Actions de Recherche Concertées (ARC)* fund of the *Communauté française de Belgique*.

A. Goel et al. (Eds.): APPROX and RANDOM 2008, LNCS 5171, pp. 35–48, 2008.

than $10\sqrt{5} - 21 \approx 1.36$. Another recent inapproximability result is the APX-hardness of CVC in bipartite graphs [4].

The constant ratio of 2 has recently been improved for several restricted classes of graphs. Escoffier et al. [4] have shown that CVC is polynomial in chordal graphs, admits a PTAS for planar graphs, and can be approximated within 5/3 for any class of graphs for which VC is polynomial.

Approximation results have also been proposed in the field of parallel computing. Fujito and Doi [5] have proposed two parallel 2-approximation algorithms. The first one is an NC algorithm running in time $\mathcal{O}\left(\log^2 n\right)$ using $\mathcal{O}\left(\Delta^2(m+n)/\log n\right)$ processors on an EREW-PRAM, and the second one an RNC algorithm running in $\mathcal{O}\left(\log n\right)$ expected time using $\mathcal{O}\left(m+n\right)$ processors on a CRCW-PRAM (with n, m and Δ standing for the number of vertices, the number of edges, and the maximum degree, respectively).

Several FPT algorithms have also been proposed [6,7,8], where the parameter is either the size of the optimum or the treewidth.

Density parameters such as the number of edges m and the minimum degree δ have been used as parameters for approximation ratios (see [9,10,11,12,13] for VC and [14,15] for DOMINATING SET and other problems). Most often, these ratios are expressed as functions of the normalized values of these parameters, namely $m^* = m/\binom{n}{2}$ and $\delta^* = \delta/n$. Currently, the best parameterized ratios for VC with parameters m^* and δ^* are $2/(2 - \sqrt{1 - m^*})$ and $2/(1 + \delta^*)$ [9], when only one parameter is allowed. Imamura and Iwama [12] later improved the $2/(2 - \sqrt{1 - m^*})$ result, by generalizing it to depend on both m^* and $\Delta^* = \Delta/n$.

Our Results. We present the first parameterized approximation ratios for CVC, with parameters m^* and δ^*. We first analyze Savage's algorithm, and prove ratios of $\min\{2, 1/(1 - \sqrt{1 - m^*})\}$ and $\min\{2, 1/\delta^*\}$. We then present a variant of Karpinski and Zelikovsy's algorithm which provides better ratios, namely $2/(2 - \sqrt{1 - m^*})$ and $2/(1 + \delta^*)$, and runs in time $\mathcal{O}\left(n^3\right)$ when m^* or δ^* are constant, against the $\mathcal{O}\left(n^2\right)$ complexity of Savage's algorithm. A summary of these results appears in Fig. 1.

Finally, we introduce a new graph invariant, the *price of connectivity* for VC, defined as the maximum ratio between the sizes of the optimal solutions of CVC and VC. We prove an upper bound of $2/(1 + m^*)$ for the price of connectivity, and present a family of nearly tight graphs.

Connected variants of classical covering problems such as VC and DOMINATING SET have recently found renewed interest, with many applications in wireless and ad-hoc networks among which fault location [16], wireless network design [17], broadcasting [18] and virtual backbone construction [19]. In such distributed settings, connectivity is a crucial issue, and the question of determining its price arises naturally.

Notations. We denote by σ and τ the sizes of the optimal solutions of CVC and VC respectively. We denote by m the number of edges in the graph, by δ its minimum degree, and by α the size of its maximum stable set. We use the * notation to denote normalized values of the graph parameters: $\tau^* = \tau/n$, $\sigma^* = \sigma/n$, $m^* = m/\binom{n}{2}$, $\delta^* = \delta/n$ and $\alpha^* = \alpha/n$.

We use the classical notations K_x, I_x and C_x for, respectively, a clique, a stable set and a cycle. We define the *join* $A \times B$ of two graphs A and B as the graph having the edges and vertices of A and B, as well as all possible edges joining both sets of vertices. Finally, we call *weakly dense* and *strongly dense* graphs those for which, respectively, m^* and δ^* is a constant. Throughout the sequel, OPT will denote an optimal solution and β the approximation ratio.

2   Savage's Algorithm

In 1982, Carla Savage [2] proposed a simple combinatorial algorithm that provides a 2-approximation to VC. It simply returns the internal nodes of an arbitrary depth-first search tree. As this algorithm returns always a connected solution, and $\sigma \geq \tau$, the 2-approximation is also valid for CVC.

Our first lemma provides lower bounds for σ.

Lemma 1. *The following lower bounds hold.*

$$\sigma^* \geq 1 - \sqrt{1 - m^*} + \mathcal{O}\left(\frac{1}{n}\right) \tag{1}$$

$$\sigma^* \geq \delta^* \tag{2}$$

Proof. We consider the first bound. In any graph, since at least $\binom{\alpha}{2}$ edges are missing, we have $m \leq \binom{n}{2} - \binom{\alpha}{2}$, hence $m^* \leq 1 - \alpha^{*2} + \mathcal{O}\left(\frac{1}{n}\right)$. Isolating α^* yields $\alpha^* \leq \sqrt{1 - m^*} + \mathcal{O}\left(\frac{1}{n}\right)$. Reverting to the normalized vertex cover $\tau^* = 1 - \alpha^*$ yields $\tau^* \geq 1 - \sqrt{1 - m^*} + \mathcal{O}\left(\frac{1}{n}\right)$. As $\sigma \geq \tau$, we obtain the desired result: $\sigma^* \geq 1 - \sqrt{1 - m^*} + \mathcal{O}\left(\frac{1}{n}\right)$.

The same kind of reasoning holds for bound (2). In any graph, since a vertex in a maximum stable set has at most $n - \alpha$ neighbors, we have $\delta \leq n - \alpha = \tau$, thus $\tau^* \geq \delta^*$. As $\sigma \geq \tau$, we obtain the desired result: $\sigma^* \geq \delta^*$.   □

The upper bounds on the ratio now follow immediately:

Theorem 1. *Savage's algorithm approximates* CVC *within a factor of*

$$\begin{cases} 2 & \text{if } m^* < \frac{3}{4} + o(1) \\ \frac{1}{1 - \sqrt{1 - m^*}} + o(1) & \text{otherwise.} \end{cases} \quad \text{(weak density)} \tag{3}$$

$$\begin{cases} 2 & \text{if } \delta^* < \frac{1}{2} + o(1) \\ \frac{1}{\delta^*} + o(1) & \text{otherwise.} \end{cases} \quad \text{(strong density)} \tag{4}$$

Proof. Since the ratio of 2 is known from Savage's result, and the value of $n - 1$ is the worst possible for any heuristic solution, we trivially have the following bound:

$$\min\left\{2, \frac{n-1}{\sigma}\right\} = \begin{cases} 2 & \text{if } \sigma < \frac{n}{2} \\ \frac{n-1}{\sigma} & \text{otherwise.} \end{cases} \tag{5}$$

Normalizing, we get a bound of $\min\left\{2, \frac{1}{\sigma^* + \mathcal{O}\left(\frac{1}{n}\right)}\right\}$. Plugging inequalities (1) and (2) immediately yields:

$$\frac{1}{\sigma^*} \le \frac{1}{1 - \sqrt{1 - m^*} + \mathcal{O}\left(\frac{1}{n}\right)} = \frac{1}{1 - \sqrt{1 - m^*}} + o(1)$$

$$\frac{1}{\sigma^*} \le \frac{1}{\delta^* + \mathcal{O}\left(\frac{1}{n}\right)} = \frac{1}{\delta^*} + o(1)$$

We can now easily compute when the minimum is 2:

$$\frac{1}{1 - \sqrt{1 - m^*}} + o(1) > 2 \iff m^* < \frac{3}{4} + o(1)$$

and

$$\frac{1}{\delta^*} + o(1) > 2 \iff \delta^* < \frac{1}{2} + o(1) \ . \qquad \Box$$

It should be noted that Theorem 1 applies to any 2-approximation algorithm for CVC, as its proof nowhere relies on the specific algorithm being used.

We define the *complete split graph* $\psi_{n,\alpha}$ as the join of a clique $K_{n-\alpha}$ and a stable set I_α. The following lemma expresses the result of Savage's algorithm on complete split graphs.

Lemma 2. *Let H be a worst-case solution returned by Savage's algorithm. Then*

$$|H\left(\psi_{n,\alpha}\right)| = \begin{cases} n - 1 & \text{if } \alpha < \frac{n}{2} \\ 2(n - \alpha) & \text{otherwise.} \end{cases}$$

Proof. Case 1: $\alpha < \frac{n}{2}$. One possible execution of the algorithm starts from a vertex in the clique, alternatively takes a vertex from the stable set and from the clique, then ends by taking all the remaining vertices in the clique. This execution yields a path of n vertices, hence a solution of size $n - 1$ (by removing the last vertex).

Case 2: $\alpha \ge \frac{n}{2}$. The worst possible execution of the algorithm starts from a vertex in the stable set, then alternatively takes a vertex from the clique and from the stable set. This induces a path of $2(n - \alpha) + 1$ vertices, hence a solution of size $2(n - \alpha)$. $\qquad \Box$

Theorem 2. *The bounds of theorem 1 are tight.*

Proof. We show that $\psi_{n,\alpha}$ are tight examples for both bounds (3) and (4).

Optimum. $\tau(\psi_{n,\alpha})$ is trivially $n - \alpha$, and the corresponding optimal solution is the clique $K_{n-\alpha}$. Since this optimal solution is connected, we have $\sigma\left(\psi_{n,\alpha}\right) = \tau\left(\psi_{n,\alpha}\right) = n - \alpha$. Combining this result with the result of Lemma 2 yields

$$\beta\left(\psi_{n,\alpha}\right) = \begin{cases} 2 & \text{if } \alpha > \frac{n}{2} \\ \frac{n-1}{n-\alpha} & \text{otherwise.} \end{cases} = \begin{cases} 2 & \text{if } \sigma^* < \frac{1}{2} + o(1) \\ \frac{1}{\sigma^*} + \mathcal{O}\left(\frac{1}{n}\right) & \text{otherwise.} \end{cases} \tag{6}$$

Bound 3. Since $\binom{\alpha}{2}$ edges are missing from $\psi_{n,\alpha}$, we have $m(\psi_{n,\alpha}) = \binom{n}{2} - \binom{\alpha}{2}$. Isolating α and normalizing yields $\alpha^*(\psi_{n,\alpha}) = \sqrt{1 - m^*} + \mathcal{O}\left(\frac{1}{n}\right)$. Finally, plugging $\sigma^* = 1 - \alpha^* = 1 - \sqrt{1 - m^*} + \mathcal{O}\left(\frac{1}{n}\right)$ into bound 6 immediately yields bound 3.

Bound 4. It is easy to see that $\delta(\psi_{n,\alpha}) = n - \alpha = \sigma$. Hence, plugging $\sigma^* = \delta^*$ into bound 6 immediately yields bound 4. □

This algorithm runs in time $\mathcal{O}(m)$, hence $\mathcal{O}(n^2)$ for fixed m^*. In the next section, we improve the approximation ratio at the expense of an increase in time complexity.

3   A Variant of Karpinski and Zelikovsky's Algorithm

Karpinski and Zelikovsky [9] proposed two approximation algorithms that ensure asymptotic approximation ratios of $\frac{2}{1+\delta^*}$ and $\frac{2}{2-\sqrt{1-m^*}}$ respectively. However, they do not always return a connected solution. We propose two variants of their algorithms for CVC, with the same asymptotic approximation factors.

The analysis relies on the following result.

Lemma 3. *Any solution H to* CVC *that consists of*
- a set $H_1 \subseteq OPT$ of size $\epsilon_1 n$,
- a 2-approximation H_2 of CVC *in $G[V - H_1]$ obtained with Savage's algorithm,*
- an additional set H_3 of $\epsilon_2 n$ vertices, with $|H_1| \geq |H_3|$,
approximates CVC *within a factor of*

$$\frac{2}{1 + \epsilon_1 - \epsilon_2} .$$

Proof. We compute the approximation ratio:

$$\beta = \frac{|H|}{|OPT|} = \frac{|H_1| + |H_2| + |H_3|}{|H_1| + |OPT'|} \quad \text{with } OPT' = OPT - H_1 .$$

Note that OPT' is a vertex cover of $G[V - H_1]$, and that H_2 is a 2-approximation of VC in $G[V - H_1]$ (as Savage's algorithm also 2-approximates VC). Hence $|H_2| \leq 2\tau(G[V - H_1]) \leq 2|OPT'|$ and therefore $|OPT'| \geq |H_2|/2$. This yields

$$\beta \leq \frac{|H_1| + |H_2| + |H_3|}{|H_1| + \frac{|H_2|}{2}} = \frac{\epsilon_1 n + |H_2| + \epsilon_2 n}{\epsilon_1 n + \frac{|H_2|}{2}} . \tag{7}$$

Differentiating (7) shows that it grows with $|H_2|$ when $\epsilon_1 \geq \epsilon_2$. Plugging the maximum possible value $|H_2| = n(1 - \epsilon_1 - \epsilon_2)$ into (7) yields

$$\beta \leq \frac{2}{1 + \epsilon_1 - \epsilon_2} . \qquad \square$$

Algorithm 1. A connected vertex cover algorithm for strongly-dense graphs

---

1: **procedure** P(W)   ▷ with $W \subseteq V$
2:    **for all** vertex $v \in W$ **do**
3:       $res(v) \leftarrow \{v\} \cup N(v)$
4:       **for all** connected components C of $G[V - \{\{v\} \cup N(v)\}]$ **do**
5:          Find a vertex $c \in C$ that has a neighbor in $N(v)$
6:          Let $Savage(c)$ be the result of Savage's algorithm in C, starting from c
7:          $res(v) \leftarrow res(v) \cup Savage(c)$
8:       **end for**
9:    **end for**
10:   $v_{min} \leftarrow \arg\min_{v \in W} |res(v)|$
11:   **return** $res(v_{min})$
12: **end procedure**
13: **return** $P(V)$

The algorithm of Karpinsky and Zelikovsky makes use of the trivial observation that if a vertex does not belong to a vertex cover, then all its neighbors do. Thus for each vertex v, it constructs the vertex cover made of the set $N(v)$ of neighbors of v, and of a 2-approximation on the remaining induced subgraph. Algorithm 1 implements this strategy. To ensure that the returned vertex cover induces a connected graph, we choose to start the execution of Savage's algorithm with a vertex that is connected to $N(v)$.

Theorem 3. *Algorithm 1 approximates* CVC *within a factor of* $\frac{2}{1+\delta^*}$.

Proof. It is easy to see that the algorithm returns a connected solution: $\{v \cup N(v)\}$ is connected, so are the 2-approximations computed in each component C, each of which are connected to $N(v)$ by their starting vertex c. Note that vertex c always exists since the graph is connected.

Furthermore, the returned solution has size at most $|res(v')|$, for some vertex $v' \notin OPT$. Since $v' \notin OPT$, we have $N(v') \subseteq OPT$. Thus Lemma 3 can be applied to $res(v')$, with $|H_1| = |N(v')| \geq \delta^* n$ and $|H_3| = |\{v'\}| = 1$, which immediately yields the desired result.   □

The second algorithm is based on the idea of choosing a set of vertices $W \subseteq V$ of size at least ρn all vertices of which have degree at least ρn for some well-chosen constant ρ. Then either $W \subseteq OPT$, or there exists a vertex w in W such that $N(w) \subseteq OPT$. In either case, a set of size at least ρn is included in OPT, and one can try them all. This original idea of Karpinski and Zelikovsky [9] does not always return a connected solution. In particular, if all vertices of W are in OPT, additional operations are needed, as W does not necessarily induce a connected subgraph. We show that connectivity can be achieved by adding a small set X of carefully chosen vertices (lines 10–14).

The analysis of Algorithm 2 relies heavily on the following lemma, which has been proved in [9].

Lemma 4 ([9]). *Let* $\rho = 1 - \sqrt{1 - m^*}$, *and let* W *be the set of* ρn *vertices with highest degree. Then every vertex of* W *has degree at least* $|W|$.

Algorithm 2. A connected vertex cover algorithm for weakly-dense graphs

---

1: $\rho \leftarrow 1 - \sqrt{1 - m^*}$
2: Let W be the set of ρn vertices with highest degree
3: $C_1 \leftarrow P(W)$ $\qquad\qquad\qquad\triangleright$ with $P(\cdot)$ the procedure defined in Algorithm 1
4: $C_2 \leftarrow W$
5: **for all** connected components C of $G[V - W]$ **do**
6: $\quad$ Find a vertex $c \in C$ that has a neighbor in W
7: $\quad$ Let $Savage(c)$ be the result of Savage's algorithm in C, starting from c
8: $\quad\quad C_2 \leftarrow C_2 \cup Savage(c)$
9: **end for**
10: $X \leftarrow \emptyset$
11: **while** $G[W \cup X]$ is not connected **do**
12: $\quad$ Find a vertex v in $V - W$ that is adjacent to the largest number of connected
$\quad$ components of $G[W \cup X]$
13: $\quad\quad X \leftarrow X \cup \{v\}$
14: **end while**
15: $C_2 \leftarrow C_2 \cup X$
16: **return** the set of minimum size among C_1 and C_2

Lemma 5. *There exists a set X of size $O(\log n)$ such that $G[W \cup X]$ is connected. Such a set is computed in lines 10–14.*

Proof. We construct X by iteratively choosing a vertex that connects the largest number of remaining connected components.

Let v_i be an arbitrary vertex of the ith connected component of $G[W]$. Let k_i be the size of this component, and k the total number of components. By Lemma 4, the degree of v_i is at least $|W|$, thus v_i has at least $|W| - k_i + 1$ neighbors in $V - W$. Summing, we get

$$\sum_{i=1}^{k} |W| - k_i + 1 = (k - 1)|W| + k .$$

Hence, by the pigeonhole principle there exists a vertex $v \in V - W$ that is connected to the following number of components:

$$\frac{(k-1)|W| + k}{n - |W|} = \frac{(k-1)\rho n + k}{n(1 - \rho)} = (k-1)\frac{\rho}{1 - \rho} + \frac{k}{n(1 - \rho)} > (k-1)\frac{\rho}{1 - \rho} .$$

We can thus add v to the set X and iterate this argument. Each time a new vertex is added to X, the number of connected components in $G[W \cup X]$ shrinks by a constant factor $1 - \frac{\rho}{1-\rho} = \frac{1-2\rho}{1-\rho}$. And since the initial number of connected components is at most $|W|$, we have

$$|X| \leq \log_{\frac{1-\rho}{1-2\rho}}(|W|) \leq \log_{\frac{1-\rho}{1-2\rho}}(\rho n) = O(\log n) .$$

$\square$

Note again that step 6 of the algorithm can always be done, otherwise the graph would not be connected.

Theorem 4. *Algorithm 2 approximates* CVC *within a factor of* $\frac{2}{2-\sqrt{1-m^*}}$.

Proof. Two cases can occur. If W contains a vertex $v \notin OPT$, in which case the proof is identical to that of Theorem 4, by plugging $|H_1| = |N(v)| \geq |W|$ and $|H_3| = 1$ into Lemma 3.

On the other hand, if $W \subseteq OPT$, we can again apply Lemma 3 with $|H_1| = |W|$ and, by Lemma 5, $|H_3| = O(\log n)$. The condition $|H_3| < |H_1|$ required by Lemma 3 holds asymptotically, and we have $\epsilon_2 = O(\log n)/n \rightarrow_{n \rightarrow \infty} 0$. Hence the approximation factor is $2/(1 + \epsilon_1 - \epsilon_2) \rightarrow_{n \rightarrow \infty} 2/(2 - \sqrt{1 - m^*})$. □

Theorems 3 and 4 now enable us to state the main corollary:

Corollary 1. CVC *is approximable within a factor strictly less than 2 in strongly and weakly dense graphs.*

Note that Algorithms 1 and 2 run in time $\mathcal{O}(nm)$, hence $\mathcal{O}(n^3)$ when m^* is fixed.

Theorem 5. *The bounds of Theorems 3 and 4 are tight.*

Proof. Tightness is witnessed by the following family of graphs: $\nu_{n,\alpha} = K_{n-2\alpha-1} \times (K_1 \times C_{2\alpha})$ (the join of a clique and a "wheel", see Fig. 1). We first show that $\sigma(\nu_{n,\alpha}) = n - \alpha$ and that both algorithms return $n - 1$ on $\nu_{n,\alpha}$. The ratio then follows naturally.

Optimum. One can easily check that taking the clique $K_{n-2\alpha-1}$, the center K_1 of the wheel, and every other vertex of the cycle $C_{2\alpha}$ yields a connected vertex cover of size $n - \alpha$, and that any smaller set would necessarily leave at least one edge uncovered.

Algorithm 1. If vertex v is in the clique or at the center of the wheel, then $\{\{v\} \cup N(v) = V\}$ and $|res(v)| = n - 1$. If on the other hand v is in the cycle $C_{2\alpha}$, Savage's algorithm is applied in line 6 to only one path $P_{2\alpha-3}$, yielding $|res(v)| = n - 1$.

Algorithm 2. We have $|C_1| = n - 1$ for the same reasons as above. Since $\sigma^* \geq 1 - \sqrt{1 - m^*} + o(1)$ (inequality 1), W contains at least the clique and the center of the wheel and hence already induces a connected subgraph. In the worst case, $V - W$ is therefore a path, which implies $|C_2| = n - 1$.

Ratio. Since only the vertices of the cycle $C_{2\alpha}$ have degree less than $n-1$, we have $\delta(\nu_{n,\alpha}) = n - 2\alpha$. Furthermore, $\nu_{n,\alpha}$ has all possible edges except the $\binom{2\alpha}{2} - 2\alpha$ edges missing from the cycle $C_{2\alpha}$, hence $m(\nu_{n,\alpha}) = \binom{n}{2} - 2\binom{\alpha}{2} + 2\alpha$. Solving the given expressions of $\delta(\nu_{n,\alpha})$ and $m(\nu_{n,\alpha})$ for α and inserting the results into our ratio of $(n - 1)/(n - \alpha)$ yields the bounds of Theorems 3 and 4. □

The family of graphs described in the above proof also provide tight examples for the original algorithms of Karpinski and Zelikovsky [9], provided they use

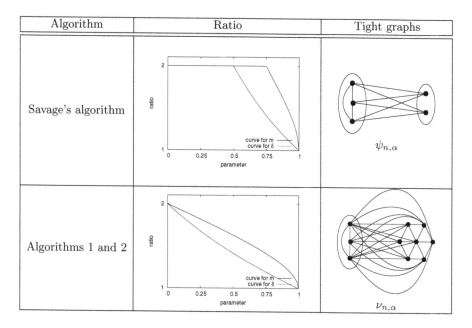

Algorithm	Ratio	Tight graphs
Savage's algorithm		$\psi_{n,\alpha}$
Algorithms 1 and 2		$\nu_{n,\alpha}$

Fig. 1. A comparison of the ratios and tight graphs of Savage's algorithm, Algorithm 1 and Algorithm 2. The second column compares the asymptotic approximation bounds as functions of parameters m^* and δ^* respectively, while the third column illustrates tight families of graphs for these bounds.

Savage's algorithm as a subroutine for the 2-approximation phase. This result is new, as the original article did not adress the issue of tightness. Figure 1 summarizes the results of Sects. 2 and 3.

A natural question to ask is wether we can use Theorem 1 to boost the approximation ratio of Algorithms 1 and 2. This is not immediately applicable, since we cannot guarantee that the subgraphs $G[V - \{\{v\} \cup N(v)\}]$ (in Algorithm 1) and $G[V - W]$ (in Algorithm 2) remain dense. Nevertheless, Imamura and Iwama [12] managed to apply the idea of Karpinski and Zelikovsky recursively, and obtained a randomized algorithm with a better ratio, depending on both parameters m^* and Δ (maximum degree). We believe this can be done for CVC as well and leave it as future work.

While we have shown that VC and CVC can both be approximated within the same ratio, as a function of m^* or δ^*, this does not settle the question of the price of connectivity, defined as the ratio between the optimal solutions of the 2 problems.

4   The Price of Connectivity

In the previous section, we showed that CVC is as well approximable as VC in dense graphs. The question of the maximum ratio between the connected vertex cover and the vertex cover then arises naturally, and is particularly relevant in networking applications for which connectivity is a crucial issue. This notion of *price of connectivity* is general and can similarly be defined for many graph covering problems.

4.1 Upper Bound

We denote by T an arbitrary optimal vertex cover, by $I = V - T$ the associated maximum stable set, by k the number of connected components in the subgraph induced by T, and by err the difference $\sigma - \tau$. Finally, we denote by S the additional vertices in a smallest connected vertex cover containing T, with size $s = |S|$.

Our first lemma expresses a simple relationship between err, s and k.

Lemma 6. *err $\leq s < k$.*

Proof. The first inequality, err $\leq s$, is straightforward as any s strictly smaller than err would imply the existence of a connected vertex cover of size smaller than σ. The second inequality, $s < k$, follows from the fact that, since S is a stable set, each one of its vertices necessarily decreases the number of connected components of T by at least one.   □

Our second lemma provides an upper bound on the degrees of the vertices in the maximum stable set I.

Lemma 7. *Every vertex of I is connected to at most $k-s+1$ different connected components of T.*

Proof. By contradiction, suppose that some vertex v in I were connected to at least $k-s+2$ connected components of T, then $T\cup\{v\}$ has at most $k-(k-s+1) = s-1$ connected components, hence the smallest subset X of I that contains v and such that $T \cup X$ is connected has size at most $s-1$, contradicting the minimality of S.   □

The last lemma bounds the number of edges by a function of (n, τ, k, s).

Lemma 8. *The following upper bound holds for m:*

$$m \leq \binom{\tau - k + 1}{2} + (n - \tau)(\tau - s + 1). \tag{8}$$

Proof. Let E_1 be the set of edges inside $G[T]$ and E_2 the set of edges between T and I. We bound the size of each set separately.

Clearly, E_1 is maximized when all the connected components in T are cliques. Furthermore, since the total number of edges in those cliques involves a sum of squares, it is maximized with one big clique of size $\tau - k + 1$ and $k - 1$ isolated vertices, by the concavity of the function x^2. Hence $|E_1| \leq \binom{\tau-k+1}{2}$.

We now consider E_2. As each vertex v in I is connected to at most $k - s + 1$ connected components of T (Lemma 7), there are at least $s - 1$ such connected components that v is not connected to, hence at least $s - 1$ vertices of T that v is not connected to. Hence v cannot have more than $\tau - (s - 1) = \tau - s + 1$ neighbors in T. Multiplying this upper bound of $\tau - s + 1$ by $n - \tau$, the size of I, yields $|E_2| \le (n - \tau)(\tau - s + 1)$.   □

Finally, Theorem 6 follows from first expressing bound 8 as a function of (n, β, τ), then bounding with respect to τ.

Theorem 6. *The ratio between the sizes of a minimum connected vertex cover and a minimum vertex cover in a graph with at least $m^*\binom{n}{2}$ edges is at most $\frac{2}{1+m^*} + o(1)$.*

Proof. Starting from the result of Lemma 8:

$$m \le \binom{\tau - k + 1}{2} + (n - \tau)(\tau - s + 1),$$

we can see that the bound is a decreasing function of both c and s. We therefore maximize it by taking the smallest possible values for k and s. These values are $s = \mathrm{err}$ and $k = \mathrm{err} + 1$, by Lemma 6. This yields:

$$m \le \binom{\tau - \mathrm{err}}{2} + (n - \tau)(\tau - \mathrm{err} + 1). \tag{9}$$

We define β as the ratio σ/τ. Since $\mathrm{err} = \sigma - \tau$, we have $\mathrm{err}/\tau = \beta - 1$ and $\mathrm{err} = \tau(\beta - 1)$. Plugging this into our last inequality yields:

$$m \le \binom{\tau(2 - \beta)}{2} + (n - \tau)(\tau(2 - \beta) + 1). \tag{10}$$

We now maximize the above expression with respect to τ. Differentiating bound (10) with respect to τ yields a unique maximum at

$$\tau_{opt} = \frac{n}{\beta} - \frac{4 - \beta}{2\beta(2 - \beta)} = \frac{n}{\beta} + \mathcal{O}(1) \text{ for each fixed } \beta.$$

Plugging our optimal τ_{opt} into 10 yields

$$m \le \binom{\tau_{opt}(2 - \beta)}{2} + (n - \tau_{opt})(\tau_{opt}(2 - \beta) + 1) = \frac{n^2(2 - \beta)}{2\beta} + \mathcal{O}(n).$$

Hence

$$m^* \le \frac{2 - \beta}{\beta} + o(1) \quad \text{and} \quad \beta \le \frac{2}{1 + m^*} + o(1).$$

□

4.2   Tightness

We now describe a family of graphs whose ratio almost matches the bound of Theorem 6. Let $G_{n,x}$, with $(n-x)$ a multiple of 3, be the graph composed of a clique of size x and $(n-x)/3$ paths P_3, all endpoints of which are totally joined to the clique. Figure 2(a) illustrates $G_{16,4}$.

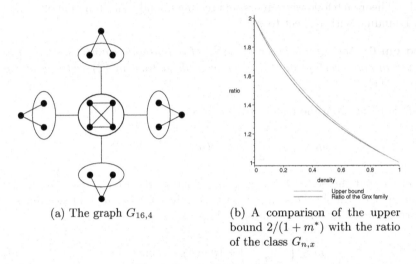

(a) The graph $G_{16,4}$

(b) A comparison of the upper bound $2/(1+m^*)$ with the ratio of the class $G_{n,x}$

Fig. 2. Nearly tight examples

The minimum vertex cover consists of the clique of size x and the center of each path, and hence has size $x + (n-x)/3 = (n+2x)/3$. On the other hand, the minimum connected vertex cover consists of the same vertices, augmented with one extra vertex per path, and hence has size $x + 2(n-x)/3 = (2n+x)/3$. We therefore have $\beta(G_{n,x}) = \frac{2n+x}{n+2x}$.

We express this bound as a function of the density m^*. The graph $G_{n,x}$ has $\binom{x}{2}$ edges in the clique, $x \cdot 2(n-x)/3$ edges between the paths and the clique, and $2(n-x)/3$ edges in the paths. Hence

$$m(G_{n,x}) = \binom{x}{2} + \frac{x \cdot 2(n-x)}{3} + \frac{2(n-x)}{3} = \frac{2nx}{3} - \frac{x^2}{6} - \frac{7x}{6} + \frac{n}{3}.$$

Normalizing yields

$$m^*(G_{n,x}) = \frac{m(G_{n,x})}{\binom{n}{2}} = \frac{4x^* - x^{*2}}{3} + \mathcal{O}\left(\frac{1}{n}\right), \quad \text{where } x^* = x/n .$$

Solving the above second-order equation for x^* yields $x^* = 2 \pm \sqrt{4 - 3m^*} + o(1)$, of which only the solution $x^* = 2 - \sqrt{4 - 3m^*} + o(1)$ must be kept in order to have x^* in $[0,1]$.

Plugging this value for x^* into our previous expression for β, it can be checked that

$$\beta(G_{n,x}) = \frac{2n+x}{n+2x} = \frac{4+2m^*+\sqrt{4-3m^*}}{3+4m^*} + o\,(1) \ .$$

This new bound is very close to the previous one, as shown by Fig. 2(b).In fact, the difference between the old and new ratios does not exceed 1.6% of the latter.

Acknowledgments. The initial conjectures behind several results of this paper were obtained with the help of the GraPHedron software [20] (http://www.graphedron.net), an online tool for obtaining tight inequalities among graph invariants.

References

1. Garey, M.R., Johnson, D.S.: The rectilinear steiner tree problem is NP complete. SIAM Journal of Applied Mathematics 32, 826–834 (1977)
2. Savage, C.D.: Depth-first search and the vertex cover problem. Inf. Process. Lett. 14(5), 233–237 (1982)
3. Fernau, H., Manlove, D.: Vertex and edge covers with clustering properties: Complexity and algorithms. In: Algorithms and Complexity in Durham 2006 - Proceedings of the Second ACiD Workshop, Durham, UK, 18-20 September 2006, pp. 69–84 (2006)
4. Escoffier, B., Gourvès, L., Monnot, J.: Complexity and approximation results for the connected vertex cover problem. In: Brandstädt, A., Kratsch, D., Müller, H. (eds.) WG 2007. LNCS, vol. 4769, pp. 202–213. Springer, Heidelberg (2007)
5. Fujito, T., Doi, T.: A 2-approximation NC algorithm for connected vertex cover and tree cover. Inf. Process. Lett. 90(2), 59–63 (2004)
6. Moser, H.: Exact algorithms for generalizations of vertex cover. Master's thesis, Institut für Informatik, Friedrich-Schiller Universität Jena (2005)
7. Guo, J., Niedermeier, R., Wernicke, S.: Parameterized complexity of vertex cover variants. Theory Comput. Syst. 41(3), 501–520 (2007)
8. Mölle, D., Richter, S., Rossmanith, P.: Enumerate and expand: Improved algorithms for connected vertex cover and tree cover. Theory Comput. Syst. 43(2), 234–253 (2008)
9. Karpinski, M., Zelikovsky, A.: Approximating dense cases of covering problems. In: Pardalos, P., Du, D. (eds.) Proc. of the DIMACS Workshop on Network Design: Connectivity and Facilites Location. DIMACS series in Disc. Math. and Theor. Comp. Sci, vol. 40, pp. 169–178 (1997)
10. Ibaraki, T., Nagamochi, H.: An approximation of the minimum vertex cover in a graph. Japan J. Indust. Appl. Math. 16, 369–375 (1999)
11. Cardinal, J., Labbé, M., Langerman, S., Levy, E., Mélot, H.: A tight analysis of the maximal matching heuristic. In: Wang, L. (ed.) COCOON 2005. LNCS, vol. 3595, pp. 701–709. Springer, Heidelberg (2005)
12. Imamura, T., Iwama, K.: Approximating vertex cover on dense graphs. In: Proc. of the 16th ACM-SIAM Symposium on Discrete Algorithms (SODA), pp. 582–589 (2005)
13. Cardinal, J., Langerman, S., Levy, E.: Improved approximation bounds for edge dominating set in dense graphs. In: Erlebach, T., Kaklamanis, C. (eds.) WAOA 2006. LNCS, vol. 4368, pp. 108–120. Springer, Heidelberg (2007)

14. Karpinski, M.: Polynomial time approximation schemes for some dense instances of NP-hard optimization problems. Algorithmica 30(3), 386–397 (2001)
15. Bar-Yehuda, R., Kehat, Z.: Approximating the dense set-cover problem. J. Comput. Syst. Sci. 69(4), 547–561 (2004)
16. Paul, S., Miller, R.E.: Locating faults in a systematic manner in a large heterogeneous network. In: Proc. IEEE INFOCOM 1995, The Conference on Computer Communications. Fourteenth Annual Joint Conference of the IEEE Computer and Communications Societies, pp. 522–529 (1995)
17. Grout, V.: Principles of cost minimisation in wireless networks. J. Heuristics 11(2), 115–133 (2005)
18. Wu, J., Lou, W., Dai, F.: Extended multipoint relays to determine connected dominating sets in MANETs. IEEE Trans. Computers 55(3), 334–347 (2006)
19. Thai, M.T., 0002, F.W., Liu, D., Zhu, S., Du, D.: Connected dominating sets in wireless networks with different transmission ranges. IEEE Trans. Mob. Comput 6(7), 721–730 (2007)
20. Mélot, H.: Facet Defining Inequalities among Graph Invariants: the system GraPHedron. Discrete Applied Mathematics (to appear, 2007)

Improved Approximation Guarantees through Higher Levels of SDP Hierarchies

Eden Chlamtac[1,*] and Gyanit Singh[2,**]

[1] Department of Computer Science, Princeton University, Princeton NJ 08544, USA
chlamtac@cs.princeton.edu
[2] Departmen of Computer Science & Engineering, University of Washington, Box 352350, Seattle, WA 98195-2350, USA
gyanit@cs.washington.edu

Abstract. For every fixed $\gamma \geq 0$, we give an algorithm that, given an n-vertex 3-uniform hypergraph containing an independent set of size γn, finds an independent set of size $n^{\Omega(\gamma^2)}$. This improves upon a recent result of Chlamtac, which, for a fixed $\varepsilon > 0$, finds an independent set of size n^ε in any 3-uniform hypergraph containing an independent set of size $(\frac{1}{2} - \varepsilon)n$. The main feature of this algorithm is that, for fixed γ, it uses the $\Theta(1/\gamma^2)$-level of a hierarchy of semidefinite programming (SDP) relaxations. On the other hand, we show that for at least one hierarchy which gives such a guarantee, $1/\gamma$ levels yield no non-trivial guarantee. Thus, this is a first SDP-based algorithm for which the approximation guarantee improves indefinitely as one uses progressively higher-level relaxations.

1 Introduction

Semidefinite Programming (SDP) has been one of the key tools in the development of approximation algorithms for combinatorial optimization problems since the seminal work of Goemans and Williamson [12] on MAXCUT. For a number of problems, including MAXCUT [12], MAX-3SAT [16,29], and Unique Games [6], SDPs lead to approximation algorithms which are essentially optimal under certain complexity-theoretic assumptions [13,18]. Howeve, for a host of other problems, large gaps between known hardness of approximation and approximation algorithmic guarantee persist.

One possibility for improvement on the approximation side is the use of so-called SDP hierarchies. In general, Linear Programming (LP) and SDP hierarchies give a sequence of nested (increasingly tight) relaxations for an integer $(0-1)$ program on n variables, where the nth level of the hierarchy is equivalent to the original integer program. Such hierarchies include LS and LS$_+$ (LP and SDP hierarchies, respectively), proposed by Lovász and Schrijver [22], a stronger

---

* Supported by Sanjeev Arora's NSF grants MSPA-MCS 0528414, CCF 0514993, ITR 0205594.
** Supported by NSF grant CCF 0514670.

A. Goel et al. (Eds.): APPROX and RANDOM 2008, LNCS 5171, pp. 49–62, 2008.

LP hierarchy proposed by Sherali and Adams [26], and the Lasserre [21] SDP hierarchy (see [20] for a comparison).

SDP hierarchies have been studied more generally in the context of optimization of polynomials over semi-algebraic sets [8,23]. In the combinatorial optimization setting, there has been quite a large number of negative results [2, 1, 25, 28, 11,5]. This body of work focuses on combinatorial problems for which the quality of approximation (integrality gap) of the hierarchies of relaxations (mostly LS, LS_+, and more recently Sherali-Adams) is poor (often showing no improvement over the simplest LP relaxation) even at very high levels.

On the other hand, there have been few positive results. For random graphs, Feige and Krauthgamer [9] have shown that $\Theta(\log n)$ rounds of LS_+ give a tight relaxation (almost surely) for Maximum Independent Set (a quasi-polynomial time improvement). De la Vega and Kenyon-Mathieu [28] showed that one obtains a polynomial time approximation scheme (PTAS) for MAXCUT in dense graphs using Sherali-Adams. SDP hierarchies at a constant level (where one can optimize in polynomial time) were used recently by Chlamtac [7], who examined the use of the Lasserre hierarchies for Graph Coloring and for Maximum Independent Set in 3-uniform hypergraphs. However, Chlamtac [7] used only the third level of the Lasserre hierarchy, whereas we exploit increasingly higher levels to get better approximation guarantees.

Our focus is on Maximum Independent Set in 3-uniform hypergraphs. *k-uniform hypergraphs* are collections of sets of size k ("hyperedges") over a vertex set. An independent set is a subset of the vertices which does not fully contain any hyperedge. The first SDP-based approximation algorithm for this problem was given by Krivelevich et al. [19], who showed that for any 3-uniform hypergraph on n vertices containing an independent set of size γn, one can find an independent set of size $\tilde{\Omega}(\min\{n, n^{6\gamma-3}\})$. This yielded no nontrivial guarantee for $\gamma \leq \frac{1}{2}$. Subsequently, it was shown by Chlamtac [7] that the SDP rounding of [19] finds an independent set of size $\Omega(n^\varepsilon)$ whenever $\gamma \geq \frac{1}{2} - \varepsilon$, for some fixed $\varepsilon > 0$, if one uses the third level of the Lasserre SDP hierarchy.

We improve upon [7] by giving two algorithms with a non-trivial approximation guarantee for every $\gamma > 0$. In 3-uniform hypergraphs containing an independent set of size γn, both algorithms find an independent set of size $\geq n^{\Omega(\gamma^2)}$. Our result is novel in that for every fixed $\gamma > 0$, the approximation guarantee relies on the $\Theta(1/\gamma^2)$-level of an SDP hierarchy (which can be solved in time $n^{O(1/\gamma^2)}$), and thus gives an infinite sequence of improvements at increasingly high (constant) levels.

For the first of the two hierachies we use, we also show that this guarantee cannot be achieved using a fixed constant level by giving a sequence of integrality gaps. The second hierarchy we consider, the Lasserre hierarchy, allows us to give a slightly better approximation guarantee, by use of an SDP rounding algorithm which uses vectors in the higher levels of the SDP relaxation (in contrast to the approach in [7], where the rounding algorithm was identical to that of [19], and the analysis only relied on the *existence* of vectors in the second and third level).

Note the discrepancy between our result, and the corresponding problem for *graphs*, where Halperin et al. [14] have shown how to find an independent set of size $n^{f(\gamma)}$ for some $f(\gamma) = 3\gamma - O(\gamma^2)$ when the graph contains an independent set of size γn.

The rest of the paper is organized as follows. In Section 2 we define the SDPs used in the various algorithms, and discuss some useful properties of these relaxations. In section 3 we describe a simple integrality gap, followed by a description of the various algorithms and their analyses. Finally, in Section 4, we discuss the possible implications of this result for SDP-based approximation algorithms.

2   SDP Relaxations and Preliminaries

2.1   Previous Relaxation for MAX-IS in 3-Uniform Hypergraphs

The relaxation proposed in [19] may be derived as follows. Given an independent set $I \subseteq V$ in a 3-uniform hypergraph $H = (V, E)$, for every vertex $i \in V$ let $x_i = 1$ if $i \in I$ and $x_i = 0$ otherwise. For any hyperedge $(i, j, l) \in E$ it follows that $x_i + x_j + x_l \in \{0, 1, 2\}$ (and hence $|x_i + x_j + x_l - 1| \leq 1$). Thus, we have the following relaxation (where vector v_i represents x_i, and $v_\emptyset$ represents 1:
MAX-KNS(H)

$$\text{Maximize } \sum_i \|v_i\|^2 \text{ s.t.   } v_\emptyset^2 = 1 \tag{1}$$

$$\forall i \in V \quad v_\emptyset \cdot v_i = v_i \cdot v_i \tag{2}$$

$$\forall (i, j, l) \in E \quad \|v_i + v_j + v_l - v_\emptyset\|^2 \leq 1 \tag{3}$$

2.2   Hypergraph Independent Set Relaxations Using LP and SDP Hierarchies

The Sherali-Adams Hierarchy. The Sherali-Adams hierarchy [26] is a sequence of nested linear programming relaxations for $0 - 1$ polynomial programs. These LPs may be expressed as a system of linear constraints on the variables $\{y_I \mid I \subseteq [n]\}$. To obtain a relaxed (non-integral) solution to the original problem, one takes $(y_{\{1\}}, y_{\{2\}}, \dots, y_{\{n\}})$.

Suppose $\{x_i^*\}$ is a sequence of n random variables over $\{0, 1\}$, and for all $I \subseteq [n]$ we have $y_I = \mathbb{E}[\prod_{i \in I} x_i^*] = \Pr[\forall i \in I : x_i^* = 1]$. Then by the inclusion-exclusion principle, for any disjoint sets $I, J \subseteq [n]$ we have

$$y_{I,-J} \stackrel{\text{def}}{=} \sum_{J' \subseteq J} (-1)^{|J'|} y_{I \cup J'} = \Pr[(\forall i \in I : x_i^* = 1) \wedge (\forall j \in J : x_j^* = 0)] \geq 0.$$

In fact, it is not hard to see that the constraints $y_{I,-J} \geq 0$ are a necessary and sufficient condition for the existence of a corresponding distribution on $\{0, 1\}$ variables $\{x_i^*\}$. Thinking of the intended solution $\{x_i^*\}$ as a set of indicator variables for a random independent set in a hypergraph $H = (V, E)$ motivates the following hierarchy of LP relaxations (assume $k \geq \max\{|e| \mid e \in E\}$): $\mathrm{IS}_k^{\mathrm{SA}}(H)$

$$y_\emptyset = 1 \qquad (4)$$

$$\forall I, J \subseteq V \text{ s.t. } I \cap J = \emptyset \text{ and } |I \cup J| \leq k \quad \sum_{J' \subseteq J} (-1)^{|J'|} y_{I \cup J'} \geq 0 \qquad (5)$$

$$\forall e \in E \quad y_e = 0 \qquad (6)$$

Note that if $\{y_I \mid |I| \leq k\}$ satisfy $\mathrm{IS}_k^{\mathrm{SA}}(H)$, then for any set of vertices $S \subseteq V$ of size k, there is a distribution over independent sets in H for which $\Pr[\forall i \in I : i \in \text{ind. set}] = y_I$ for all subsets $I \subseteq S$.

The Lasserre Hierarchy. The relaxations for maximum hypergraph independent set arising from the Lasserre hierarchy [21] are equivalent to those arising from the Sherali-Adams with one additional semidefiniteness constraint:

$$(y_{I \cup J})_{I,J} \succeq 0.$$

We will express these constraints in terms of the vectors $\{v_I \mid I \subseteq V\}$ arising from the Cholesky decomposition of the positive semidefinite matrix. In fact, we can express the constraints on $\{v_I\}$ in a more succinct form which implies the inclusion-exclusion constraints in Sherali-Adams but does not state them explicitly: $\mathrm{IS}_k^{\mathrm{Las}}(H)$

$$v_\emptyset^2 = 1 \qquad (7)$$

$$|I|, |J|, |I'|, |J'| \leq k \text{ and } I \cup J = I' \cup J' \Rightarrow v_I \cdot v_J = v_{I'} \cdot v_{J'} \qquad (8)$$

$$\forall e \in E \quad v_e^2 = 0 \qquad (9)$$

For convenience, we will henceforth write $v_{i_1 \ldots i_s}$ instead of $v_{\{i_1, \ldots, i_s\}}$. We will denote by $\mathrm{MAX\text{-}IS}_k^{\mathrm{Las}}(H)$ the SDP

$$\text{Maximize } \sum_i \|v_i\|^2 \text{ s.t. } \{v_I\}_I \text{ satisfy } \mathrm{IS}_k^{\mathrm{Las}}(H).$$

Since for any set S of size k all valid constraints on $\{v_I \mid I \subseteq S\}$ are implied by $\mathrm{IS}_k(H)$, this is, for all $k \geq 3$, a tighter relaxation than that of [19].

As in the Sherali-Adams hierarchy, for any set $S \subseteq V$ of size k, we may think of the vectors $\{v_I \mid I \subseteq S\}$ as representing a distribution on random $0-1$ variables $\{x_i^* \mid i \in S\}$, which can also be combined to represent arbitrary events (for example, we can write $v_{(x_i^*=0) \vee (x_j^*=0)} = v_\emptyset - v_{\{i,j\}}$). This distribution is made explicit by the inner-products. Formally, for any two events $\mathcal{E}_1, \mathcal{E}_2$ over the values of $\{x_i^* \mid i \in S\}$, we have $v_{\mathcal{E}_1} \cdot v_{\mathcal{E}_2} = \Pr[\mathcal{E}_1 \wedge \mathcal{E}_2]$.

Moreover, as in the Lovász-Schrijver hierarchy, lower-level relaxations may be derived by "conditioning on $x_i^* = \sigma_i$" (for $\sigma_i \in \{0,1\}$). In fact, we can condition on more complex events. Formally, for any event $\mathcal{E}_0$ involving $k_0 < k$ variables for which $\|v_{\mathcal{E}_0}\| > 0$, we can define

$$v_{\mathcal{E}}|_{\mathcal{E}_0} \stackrel{\text{def}}{=} v_{\mathcal{E} \wedge \mathcal{E}_0} / \|v_{\mathcal{E}_0}\|,$$

and the vectors $\{v_I|_{\mathcal{E}_0} \mid |I| \leq k - k_0\}$ satisfy $\mathrm{IS}_{k-k_0}(H)$.

An Intermediate Hierarchy. We will be primarily concerned with a hierarchy which combines the power of SDPs and Sherali-Adams local-integrality constraints in the simplest possible way: by imposing the constraint that the variables from the first two levels of a Sherali-Adams relaxation form a positive-semidefinite matrix. Formally, for all $k \geq 3$ and vectors $\{v_\emptyset\} \cup \{v_i \mid i \in V\}$ we have the following system of constraints:
$\mathrm{IS}_k^{\mathrm{mix}}(H)$

$$\exists \{y_I \mid |I| \leq k\} \text{ s.t.} \tag{10}$$

$$\forall I, J \subseteq V, |I|, |J| \leq 1 \ : \ v_I \cdot v_J = y_{I \cup J} \tag{11}$$

$$\{y_I\} \text{ satisfy } \mathrm{IS}_k^{\mathrm{SA}}(H) \tag{12}$$

As above, we will denote by MAX-$\mathrm{IS}_k^{\mathrm{mix}}(H)$ the SDP

$$\text{Maximize } \sum_i \|v_i\|^2 \text{ s.t. } \{v_\emptyset\} \cup \{v_i\} \text{ satisfy } \mathrm{IS}_k^{\mathrm{mix}}(H).$$

2.3 Gaussian Vectors and SDP Rounding

Recall that the *standard normal distribution* has density function $\frac{1}{\sqrt{2\pi}} e^{-x^2/2}$. A random vector $\zeta = (\zeta_1, \ldots, \zeta_n)$ is said to have the *n-dimensional standard normal distribution* if the components ζ_i are independent and each have the standard normal distribution. Note that this distribution is invariant under rotation, and its projections onto orthogonal subspaces are independent. In particular, for any unit vector $v \in \Re^n$, the projection $\zeta \cdot v$ has the standard normal distribution.

We use the following notation for the tail bound of the standard normal distribution: $N(x) \overset{\text{def}}{=} \int_x^\infty \frac{1}{\sqrt{2\pi}} e^{-\frac{t^2}{2}} dt$. The following property of the normal distribution ([10], Chapter VII) will be crucial.

Lemma 1. *For $s > 0$, we have $\frac{1}{\sqrt{2\pi}} \left(\frac{1}{s} - \frac{1}{s^3} \right) e^{-s^2/2} \leq N(s) \leq \frac{1}{\sqrt{2\pi}s} e^{-s^2/2}$.*

This implies the following corollary, which is at the core of the analysis of many SDP rounding schemes:

Corollary 1. *For any fixed constant $\kappa > 0$, we have $N(\kappa s) = \tilde{O}(N(s)^{\kappa^2})$.*

3 Integrality Gap and Algorithms

3.1 A Simple Integrality Gap

As observed in [26,7], MAX-KNS$(H) \geq \frac{n}{2}$ for any hypergraph H (even the complete hypergraph). In this section we will show the necessity of using increasingly many levels of the SDP hierarchy MAX-$\mathrm{IS}^{\mathrm{mix}}$ to yield improved approximations, by demonstrating a simple extention of the above integrality gap:

Theorem 1. *For every integer $k \geq 3$ and any 3-uniform hypergraph H, we have* MAX-$\mathrm{IS}_k^{\mathrm{mix}} \geq \frac{1}{k-1} n$.

Proof. Suppose $V(H) = [n]$ and let $v_\emptyset, u_1, \ldots, u_n$ be $n + 1$ mutually orthogonal unit vectors. For every $i \in V$ let $v_i = \frac{1}{k-1}v_\emptyset + \sqrt{\frac{1}{k-1} - \frac{1}{(k-1)^2}}u_i$, and $y_{\{i\}} = \frac{1}{k-1}$. Let $y_\emptyset = 1$ and for every pair of distinct vertices $i, j \in V$ let $y_{\{i,j\}} = \frac{1}{(k-1)^2}$. For all sets $I \subseteq V$ s.t. $3 \leq |I| \leq k$, let $y_I = 0$.

It is immediate that constraint (11) and the Sherali-Adams constraint (4) are satisfied. Since $y_I = 0$ for all sets I of size 3, Sherali-Adams constraint (6) is also satisfied. To verify Sherali-Adams constraints (5), it suffices to show, for any set $S \subseteq [n]$ of size k, a corresponding distribution on $0 - 1$ variables $\{x_i^* \mid i \in S\}$. Indeed, the following is such a distribution: Pick a pair of distinct vertices $i, j \in S$ uniformly at random. With probability $\frac{k}{2(k-1)}$, set $x_i^* = x_j^* = 1$ and for all other $l \in S$, set $x_l^* = 0$. Otherwise, set all $x_l^* = 0$. □

3.2 The Algorithm of Krivelevich, Nathaniel and Sudakov

We first review the algorithm and analysis given in [19]. Let us introduce the following notation: For all $l \in \{0, 1, \ldots, \lceil \log n \rceil\}$, let $T_l \stackrel{\text{def}}{=} \{i \in V \mid l/\log n \leq \|v_i\|^2 < (l+1)/\log n\}$. Also, since $\|v_i\|^2 = v_\emptyset \cdot v_i$, we can write $v_i = (v_\emptyset \cdot v_i)v_\emptyset + \sqrt{v_\emptyset \cdot v_i(1 - v_\emptyset \cdot v_i)}u_i$, where u_i is a unit vector orthogonal to $v_\emptyset$. They show the following two lemmas, slightly rephrased here:

Lemma 2. *If the optimum of* $\mathrm{KNS}(H)$ *is* $\geq \gamma n$, *there exists an index* $l \geq \gamma \log n - 1$ *s.t.* $|T_l| = \Omega(n/\log^2 n)$.

Lemma 3. *For index* $l = \beta \log n$ *and hyperedge* $(i, j, k) \in E$ *s.t.* $i, j, k \in T_l$, *constraint* (3) *implies*

$$\|u_i + u_j + u_k\|^2 \leq 3 + (3 - 6\beta)/(1 - \beta) + O(1/\log n). \tag{13}$$

Note that constraint (13) becomes unsatisfiable for constant $\beta > 2/3$. Thus, for such β, if $\mathrm{KNS}(H) \geq \beta n$, one can easily find an independent set of size $\tilde{\Omega}(n)$. Using the above notation, we can now describe the rounding algorithm in [19], which is applied to the subhypergraph induced on T_l, where l is as in Lemma 2.

KNS-Round$(H, \{u_i\}, t)$

- Choose $\zeta \in \mathbb{R}^n$ from the n-dimensional standard normal distribution.
- Let $V_\zeta(t) \stackrel{\text{def}}{=} \{i \mid \zeta \cdot u_i \geq t\}$. Remove all vertices in hyperedges fully contained in $V_\zeta(t)$, and return the remaining set.

The expected size of the remaining independent set can be bounded from below by $\mathbb{E}[|V_\zeta(t)|] - 3\mathbb{E}[|\{e \in E : e \subseteq V_\zeta(t)\}|]$, since each hyperedge contributes at most three vertices to $V_\zeta(t)$. If hyperedge (i, j, k) is fully contained in $V_\zeta(t)$, then we must have $\zeta \cdot (u_i + u_j + u_k) \geq 3t$, and so by Lemma 3, $\zeta \cdot \frac{u_i + u_j + u_k}{\|u_i + u_j + u_k\|} \geq (\sqrt{(3 - 3\gamma)/(2 - 3\gamma)} - O(1/\log n))t$. By Corollary 1, and linearity of expectation, this means the size of the remaining independent set is at least

$$\tilde{\Omega}(N(t)n) - \tilde{O}(N(t)^{(3-3\gamma)/(2-3\gamma)}|E|).$$

Choosing t appropriately then yields the guarantee given in [19]:

Theorem 2. *Given a 3-uniform hypergraph H on n vertices and m hyperedges containing an independent set of size $\geq \gamma n$, one can find, in polynomial time, an independent set of size $\tilde{\Omega}(\min\{n, n^{3-3\gamma}/m^{2-3\gamma}\})$.*

Note that m can be as large as $\Omega(n^3)$, giving no non-trivial guarantee for $\gamma \leq \frac{1}{2}$. Chlamtac [7] showed that when the vectors satisfy $\mathrm{IS}_3^{\mathrm{Las}}(H)$, the same rounding algorithm does give a non-trivial guarantee (n^ε) for $\gamma \geq \frac{1}{2} - \varepsilon$ (for some fixed $\varepsilon > 0$). However, it is unclear whether this approach can work for arbitrarily small $\gamma > 0$.

Let us note the following Lemma which was implicitly used in the above analysis, and which follows immediately from Corollary 1. First, we introduce the following notation for hyperedges e along with the corresponding vectors $\{u_i \mid i \in e\}$:

$$\alpha(e) \overset{\text{def}}{=} \tfrac{1}{|e|(|e|-1)} \sum_{i \in e} \sum_{j \in e \setminus \{i\}} u_i \cdot u_j$$

Lemma 4. *In algorithm KNS-Round, the probability that a hyperedge e is fully contained in $V_\zeta(t)$ is at most $\tilde{O}(N(t)^{|e|/(1+(|e|-1)\alpha(e))})$.*

3.3 Improved Approximation Via Sherali-Adams Constraints

Before we formally state our rounding algorithm, let us motivate it with an informal overview.

Suppose $\|v_i\|^2 = \gamma$ for all $i \in V$. A closer examination of the above analysis reveals the reason the KNS rounding works for $\gamma > \frac{1}{2}$: For every hyperedge $e \in E$ we have $\alpha(e) < 0$. Thus, the main obstacle to obtaining a large independent set using KNS-Round is the presence of many pairs i, j with large inner-product $u_i \cdot u_j$. As we shall see in section 3.4, we can use higher-moment vectors in the Lasserre hierarchy to turn this into an advantage. However, just using local integrality constraints, we can efficiently isolate a large set of vertices on which the induced subhypergraph has few hyperedges containing such pairs, allowing us to successfully use KNS-Round.

Indeed, suppose that some pair of vertices $i_0, j_0 \in V$ with inner-product $v_{i_0} \cdot v_{j_0} \geq \gamma^2/2$ participates in many hyperedges. That is, the set $S_1 = \{k \in V \mid (i, j, k) \in E\}$ is very large. In that case, we can recursively focus on the subhypergraph induced on S_1. According to our probabilistic interpretation of the SDP, we have $\Pr[x_{i_0}^* = x_{j_0}^* = 1] \geq \gamma^2/2$. Moreover, for any $k \in S_1$ the event "$x_k^* = 1$" is disjoint from the event "$x_{i_0}^* = x_{j_0}^* = 1$". Thus, if we had to repeat this recursive step due to the existence of bad pairs $(i_0, j_0), \ldots, (i_s, j_s)$, then the events "$x_{i_l}^* = x_{j_l}^* = 1$" would all be pairwise exclusive. Since each such event has probability $\Omega(\gamma^2)$, the recursion can have depth at most $O(1/\gamma^2)$, after which point there are no pairs of vertices which prevent us from using KNS-Round.

We are now ready to describe our rounding algorithm. It takes an n-vertex hypergraph H for which MAX-$\mathrm{IS}_k^{\mathrm{mix}}(H) \geq \gamma n$, where $k = \Omega(1/\gamma^2)$ and $\{v_i\}$ is the corresponding SDP solution.

H-Round$(H = (V, E), \{v_i\}, \gamma)$

1. Let $n = |V|$ and for all $i, j \in V$, let $\Gamma(i, j) \overset{\text{def}}{=} \{k \in V \mid (i, j, k) \in E\}$.
2. If for some $i, j \in V$ s.t. $v_i \cdot v_j \geq \gamma^2/2$ we have $|\Gamma(i, j)| \geq \{n^{1-v_i \cdot v_j/2}\}$, then find an ind. set using H-Round$(H|_{\Gamma(i,j)}, \{v_k \mid k \in \Gamma(i, j)\}, \gamma)$.
3. Otherwise,
   (a) Define unit vectors $\{w_i \mid i \in V\}$ s.t. for all $i, j \in V$ we have $w_i \cdot w_j = \frac{\gamma}{24}(u_i \cdot u_j)$ (outward rotation).
   (b) Let t be s.t. $N(t) = n^{-(1-\gamma^2/16)}$, and return the independent set found by KNS-Round$(H, \{w_i \mid i \in V\}, t)$.

Theorem 3. *For any constant $\gamma > 0$, given an n-vertex 3-uniform hypergraph $H = (V, E)$, and vectors $\{v_i\}$ satisfying $\mathrm{IS}^{\mathrm{mix}}_{4/\gamma^2}(H)$ and $|\,\|v_i\|^2 - \gamma| \leq 1/\log n$ (for all vertices $i \in V$), algorithm H-Round finds an independent set of size $\Omega(n^{\gamma^2/32})$ in H in time $O(n^{3+2/\gamma^2})$.*

Combining this result with Lemma 2 (applying Theorem 3 to the induced sub-hypergraph $H|_{T_l}$), we get:

Corollary 2. *For all constant $\gamma > 0$, there is a polynomial time algorithm which, given an n-vertex 3-uniform hypergraph H containing an independent set of size $\geq \gamma n$, finds an independent set of size $\tilde{\Omega}(n^{\gamma^2/32})$ in H.*

Before we prove Theorem 3, let us first see that algorithm H-Round makes only relatively few recursive calls in Step 2, and that when Step 3b is reached, the remaining hypergraph still contains a large number of vertices.

Proposition 1. *For constant $\gamma > 0$, n-vertex hypergraph $H = (V, E)$, and vectors $\{v_i\}$ as in Thereom 3:*

1. *Algorithm H-Round makes at most $2/\gamma^2$ recursive calls in Step 2.*
2. *The hypergraph in the final recursive call to H-Round contains at least $\sqrt{n}$ vertices.*

Proof. Let $(i_1, j_1), \ldots, (i_s, j_s)$ be the sequence of vertices (i, j) in the order of recursive calls to H-Round in Step 2. Let us first show that for any $s' \leq \min\{s, 2/\gamma^2\}$ we have

$$\sum_{l=1}^{s'} v_{i_l} \cdot v_{j_l} \leq 1. \tag{14}$$

Indeed, let $T = \bigcup\{i_l, j_l \mid 1 \leq l \leq s'\}$. Since vectors $\{v_i\}$ satisfy $\mathrm{IS}^{\mathrm{mix}}_{4/\gamma^2}(H)$, and $|T| \leq 2s' \leq 4/\gamma^2$, there must be some distribution on independent sets $S \subseteq T$ satisfying $\Pr[k, k' \in S] = v_k \cdot v_{k'}$ for all pairs of vertices $k, k' \in T$. Note that by choice of vertices i_l, j_l, we have $i_{l_2}, j_{l_2} \in \Gamma(i_{l_1}, j_{l_1})$ for all $l_1 < l_2$. Thus, the events "$i_l, j_l \in S$" are pairwise exclusive, and so

$$\sum_{l=1}^{s'} v_{i_l} \cdot v_{j_l} = \Pr[\exists l \leq s' : i_l, j_l \in S] \leq 1.$$

Similarly, if $s' \leq \min\{s, 2/\gamma^2 - 1\}$, then for any $k \in \bigcap_{l \leq s'} \Gamma(i_l, j_l)$ we have $\sum_{l=1}^{s'} v_{i_l} \cdot v_{j_l} + v_k \cdot v_k \leq 1$. However, by choice of i_l, j_l, we also have $\sum_{l=1}^{s'} v_{i_l} \cdot v_{j_l} + v_k \cdot v_k \geq |s'| \gamma^2/2 + \gamma - (1/\log n)$. Thus, we must have $s \leq 2/\gamma^2 - 1$, otherwise letting $k = i_{2/\gamma^2}$ above, we would derive a contradiction. This proves part 1.

For part 2, it suffices to note that the number of vertices in the final recursive call is at least $n^{\prod(1 - v_{i_l} \cdot v_{j_l}/2)}$, and that by (14) we have $\prod(1 - v_{i_l} \cdot v_{j_l}/2) \geq 1 - \sum v_{i_l} \cdot v_{j_l}/2 \geq \frac{1}{2}$. $\qquad \square$

We are now ready to prove Theorem 3.

Proof (of Theorem 3). For the sake of simplicity, let us assume that for all vertices $i \in V$, $\|v_i\|^2 = \gamma$. Violating this assumption can adversely affect the probabilities of events or sizes of sets in our analysis by at most a constant factor, whereas we will ensure that all inequalities have at least polynomial slack to absorb such errors. Thus, for any $i, j \in V$, we have

$$v_i \cdot v_j = \gamma^2 + (\gamma - \gamma^2) u_i \cdot u_j. \tag{15}$$

For brevity, we will write $v_i \cdot v_j = \theta_{ij} \gamma$ for all $i, j \in V$ (note that all $\theta_{ij} \in [0, 1]$). Moreover, we will use the notation $\alpha(e)$ introduced earlier, but this time in the context of the vector solution $\{w_i\}$:

$$\alpha(e) = \frac{1}{3} \sum_{\substack{i,j \in e \\ i < j}} w_i \cdot w_j.$$

The upper-bound on the running time follows immediately from part 1 of Proposition 1. By part 2 of Proposition 1, it suffices to show that if the condition for recursion in Step 2 of H-Round does not hold, then in Step 3b, algorithm KNS-Round finds an independent set of size $\Omega(N(t)n) = \Omega(n^{\gamma^2/16})$ (where n is the number of vertices in the current hypergraph).

Let us examine the performance of KNS-Round in this instance. Recall that for every $i \in V$, the probability that $i \in V_\zeta(t)$ is exactly $N(t)$. Thus, by linearity of expectation, the expected number of nodes in $V_\zeta(t)$ is $N(t)n$. To retain a large fraction of $V_\zeta(t)$, we must show that few vertices participate in hyperedges fully contained in this set, that is $\mathbb{E}[|\{i \in e \mid e \in E \wedge e \subseteq V_\zeta(t)\}|] = o(N(t)n)$. In fact, since every hyperedge contained in $V_\zeta(t)$ contributes at most three vertices, it suffices to show that $\mathbb{E}[|\{e \in E \mid e \subseteq V_\zeta(t)\}|] = o(N(t)n)$. We will consider separately two types of hyperedges, as we shall see.

Let us first consider hyperedges which contain some pair i, j for which $\theta_{ij} \geq \gamma/2$. We denote this set by E_+. We will assign every hyperedge in E_+ to the pair of vertices with maximum inner-product. That is, for all $i, j \in V$, define $\Gamma_+(i, j) = \{k \in \Gamma(i, j) \mid \theta_{ik}, \theta_{jk} \leq \theta_{ij}\}$. By (15), for all $i, j \in V$ and $k \in \Gamma_+(i, j)$ we have

$$\alpha(i, j, k) \leq w_i \cdot w_j = \frac{\gamma}{24}(u_i \cdot u_j) = \frac{\gamma(\theta_{ij} - \gamma)}{24(1 - \gamma)} \leq \frac{\theta_{ij} \gamma}{24}. \tag{16}$$

Now, by our assumption, the condition for recursion in Step 2 of H-Round was not met. Thus, for all $i, j \in V$ s.t. $\theta_{ij} \geq \gamma/2$, we have

$$|\Gamma_+(i,j)| \leq |\Gamma(i,j)| \leq n^{1-\theta_{ij}\gamma/2}. \tag{17}$$

By linearity of expectation, we have

$$
\begin{aligned}
\mathbb{E}[|\{e \in E_+ \mid e \subseteq V_\zeta(t)\}|] &\leq \sum_{e \in E_+} \Pr[e \subseteq V_\zeta(t)] \\
&\leq \sum_{e \in E_+} \tilde{O}(N(t)^{3/(1+2\alpha(e))}) &&\text{by Lemma 4} \\
&\leq \sum_{\substack{i,j \in V \\ \theta_{ij} \geq \gamma/2}} \sum_{k \in \Gamma_+(i,j)} \tilde{O}(N(t)^{3/(1+\frac{1}{12}\theta_{ij}\gamma)}). &&\text{by (16)}
\end{aligned}
$$

By (17), this gives

$$
\begin{aligned}
\mathbb{E}[|\{e \in E_+ \mid e \subseteq V_\zeta(t)\}|] &\leq \sum_{\substack{i,j \in V \\ \theta_{ij} \geq \gamma/2}} \tilde{O}(n^{1-\frac{1}{2}\theta_{ij}\gamma} N(t)^{3/(1+\frac{1}{12}\theta_{ij}\gamma)}) \\
&= N(t) \sum_{\substack{i,j \in V \\ \theta_{ij} \geq \gamma/2}} \tilde{O}(n^{1-\frac{1}{2}\theta_{ij}\gamma} N(t)^{(2-\frac{1}{12}\theta_{ij}\gamma)/(1+\frac{1}{12}\theta_{ij}\gamma)}) \\
&= N(t) \sum_{\substack{i,j \in V \\ \theta_{ij} \geq \gamma/2}} \tilde{O}(n^{1-\frac{1}{2}\theta_{ij}\gamma-(1-\frac{1}{16}\gamma^2)(2-\frac{1}{12}\theta_{ij}\gamma)/(1+\frac{1}{12}\theta_{ij}\gamma)}) \\
&\leq N(t) \sum_{\substack{i,j \in V \\ \theta_{ij} \geq \gamma/2}} \tilde{O}(n^{1-\frac{1}{2}\theta_{ij}\gamma-(1-\frac{1}{8}\theta_{ij}\gamma)(2-\frac{1}{12}\theta_{ij}\gamma)/(1+\frac{1}{12}\theta_{ij}\gamma)}) \\
&= N(t)\frac{1}{n} \sum_{\substack{i,j \in V \\ \theta_{ij} \geq \gamma/2}} \tilde{O}(n^{-\frac{5}{96}\theta_{ij}^2\gamma^2/(1+\frac{1}{12}\theta_{ij}\gamma)}) \\
&\leq N(t)n\tilde{O}(n^{-\frac{5}{384}\gamma^4/(1+\frac{1}{24}\gamma^2)}) = o(N(t)n).
\end{aligned}
$$

We now consider the remaining hyperedges $E_- = E \setminus E_+ = \{e \in E \mid \forall i,j \in e : \theta_{ij} \leq \gamma/2\}$. By (15), and by definition of $\{w_i\}$, we have

$$\alpha(e) \leq -\frac{\gamma^2}{48(1-\gamma)} \tag{18}$$

for every hyperedge $e \in E_-$. Thus we can bound the expected cardinality of $E_- \cap \{e \subseteq V_\zeta(t)\}$ as follows:

$$
\begin{aligned}
\mathbb{E}[|\{e \in E_- \mid e \subseteq V_\zeta(t)\}|] &\leq \sum_{e \in E_+} \Pr[e \subseteq V_\zeta(t)] \\
&\leq \sum_{e \in E_-} \tilde{O}(N(t)^{3/(1+2\alpha(e))}) &&\text{by Lemma 4}
\end{aligned}
$$

$$= N(t) \sum_{e \in E_-} \tilde{O}(N(t)^{(2-2\alpha(e))/(1+2\alpha(e))})$$

$$\leq N(t) n^3 \tilde{O}(N(t)^{(2-2\gamma+\frac{1}{24}\gamma^2)/(1-\gamma-\frac{1}{24}\gamma^2)}) \qquad \text{by (18)}$$

By our choice of t, this gives

$$\mathbb{E}[|\{e \in E_- \mid e \subseteq V_\zeta(t)\}|] \leq N(t)\tilde{O}(n^{3-(1-\frac{1}{16}\gamma^2)(2-2\gamma+\frac{1}{24}\gamma^2)/(1-\gamma-\frac{1}{24}\gamma^2)})$$

$$= N(t)\tilde{O}(n^{1-(\frac{1}{8}\gamma^3-\frac{1}{384}\gamma^4)/(1-\gamma-\frac{1}{24}\gamma^2)}) = o(N(t)n).$$

This completes the proof. □

3.4 A Further Improvement Using the Lasserre Hierarchy

Here, we present a slightly modified algorithm which takes advantage of the Lasserre hierarchy, and gives a slightly better approximation guarantee. As before, the algorithm takes an n-vertex hypergraph H for which $\text{MAX-IS}_k^{\text{Las}}(H) \geq \gamma n$, where $k = \Omega(1/\gamma^2)$ and $\{v_I\}_I$ is the corresponding SDP solution.

H-Round$^{\text{Las}}(H = (V, E), \{v_I \mid |I| \leq k\}, \gamma)$

1. Let $n = |V|$ and let $l = \gamma' \log n - 1$ be as in Lemma 2 (where $\gamma' \geq \gamma$). If $\gamma' > 2/3 + 2/\log n$, output T_l.
2. Otherwise, set $H = H|_{T_l}$, and $\gamma = \gamma'$.
3. If for some $i, j \in T_l$ s.t. $\rho_{ij} = v_i \cdot v_j \geq \gamma^2/2$ we have $|\Gamma(i,j)| \geq \{n^{1-\rho_{ij}}\}$, then find an independent set using H-Round$(H|_{\Gamma(i,j)}, \{v_I|_{x_i^*=0 \lor x_j^*=0} \mid I \subseteq \Gamma(i,j), |I| \leq k-2\}, \gamma/(1-\rho_{ij}))$.
4. Otherwise,
   (a) Define unit vectors $\{w_i \mid i \in V\}$ s.t. for all $i, j \in V$ we have $w_i \cdot w_j = \frac{\gamma}{12}(u_i \cdot u_j)$ (outward rotation).
   (b) Let t be s.t. $N(t) = n^{-(1-\gamma^2/8)}$, and return the independent set found by KNS-Round$(H, \{w_i \mid i \in V\}, t)$.

For this algorithm, we have the following guarantee:

Theorem 4. *For any constant $\gamma > 0$, given an n-vertex 3-uniform hypergraph $H = (V, E)$ for which $\text{MAX-IS}_{8/(3\gamma^2)}^{\text{Las}}(H) \geq \gamma n$ and vectors $\{v_I\}$ the corresponding solution, algorithm H-Round^{Las} finds an independent set of size $\Omega(n^{\gamma^2/8})$ in H in time $O(n^{3+8/(3\gamma^2)})$.*

We will not prove this theorem in detail, since the proof is nearly identical to that of Theorem 3. Instead, we will highlight the differences from algorithm H-Round, and the reasons for the improvement. First of all, the shortcut in step 1 (which accounts for the slightly lower level needed in the hierarchy) is valid since (as can be easily checked) constraint (3) cannot be satisfied (assuming (2) holds) when $\|v_i\|^2, \|v_j\|^2, \|v_l\|^2 > 2/3$.

The improvement in the approximation guarantee can be attributed to the following observation. Let $\{(i_1, j_1), \ldots, (i_s, j_s)\}$ be the pairs of vertices chosen for the various recursive invocations of the algorithm in Step 3. Then in the probabilistic interpretation of the SDP solution, we have carved an event of probability $\rho = \rho_{i_1 j_1} + \ldots + \rho_{i_s j_s}$ out of the sample space, and thus the SDP solution is conditioned on an event of probability $1 - \rho$. Hence, the hypergraph in the final call contains $n_\rho \geq \tilde{\Omega}(n^{1-\rho})$ vertices, and the SDP value is $\gamma_\rho n_\rho$ where $\gamma_\rho \geq \gamma/(1 - \rho)$. Thus one only needs to show that assuming the condition in Step 3 does not hold, the call to KNS-Round in Step 4b returns an independent set of size at least

$$n_\rho^{\gamma_\rho^2/8} \geq n^{\gamma^2/(8(1-\rho))} \geq n^{\gamma^2/8}.$$

The proof of this fact is identical to the proof of Theorem 3.

4   Discussion

Theorem 3, together with the integrality gap of Theorem 1, demonstrate that the hierarchy of relaxations MAX-IS$_k^{\mathrm{mix}}$ gives an infinite sequence of improved approximations for higher and higher levels k. We do not know if similar integrality gaps hold for the Lasserre hierarchy, though we know that at least the integrality gap of Theorem 1 cannot be lifted even to the second level in the Lasserre hierarchy. In light of our results, we are faced with two possible scenarios:

1. For some fixed k, the kth level of the Lasserre hierarchy gives a better approximation than MAX-IS$_l^{\mathrm{mix}}$ for any (arbitrary large constant) l, or
2. The approximation curve afforded by the kth level Lasserre relaxation gives strict improvements for infinitely many values of k.

While the second possibility is certainly the more exciting of the two, a result of either sort would provide crucial insights into the importance of lift-and-project methods for approximation algorithms. Recently Schoenebeck [27] has produced strong integrality gaps for high-level Lasserre relaxations for random 3XOR formulas, which rely on properties of the underlying 3-uniform hypergraph structure. It will be very interesting to see whether such results can be extended to confirm the second scenario, above.

Finally, we note that the existence of provably improved approximations at infinitely many constant levels of an SDP hierarchy is surprising in light of the recent work of Raghavendra [24]. One implication of that work is that if the Unique Games Conjecture [17] is true, then for every k-CSP, the kth level of a mixed hierarchy (such as MAX-IS^{mix}) suffices to get the best possible approximation (achievable in polynomial time). Our result, when combined with the work of Raghavendra [24], does not refute the Unique Games Conjecture (essentially, since the guaranteed optimality of the relaxations in [24] is only up to any arbitrary additive linear error). However, it may help shed light on the characteristics of combinatorial optimization problems which stand to benefit from the use of lift-and-project techniques.

References

1. Alekhnovich, M., Arora, S., Tourlakis, I.: Towards strong nonapproximability results in the Lovász-Schrijver hierarchy. In: Proceedings of 37th Annual ACM Symposium on Theory of Computing, pp. 294–303 (2005)
2. Arora, S., Bollobás, B., Lovász, L., Tourlakis, I.: Proving integrality gaps without knowing the linear program. Theory of Computing 2, 19–51 (2006)
3. Arora, S., Chlamtac, E., Charikar, M.: New approximation guarantee for chromatic number. In: Proceedings of 38th ACM Symposium of Theory of Computing, pp. 215–224 (2006)
4. Arora, S., Rao, S., Vazirani, U.: Expander flows, geometric embeddings and graph partitioning. In: Proceedings of the 36th Annual ACM Symposium on Theory of Computing, pp. 222–231 (2004)
5. Charikar, M., Makarychev, K., Makarychev, Y.: Integrality Gaps for Sherali-Adams Relaxations (manuscript)
6. Charikar, M., Makarychev, K., Makarychev, Y.: Near-Optimal Algorithms for Unique Games. In: Proceedings of the 38th Annual ACM Symposium on Theory of Computing, pp. 205–214 (2006)
7. Chlamtac, E.: Approximation Algorithms Using Hierarchies of Semidefinite Programming Relaxations. In: Proceedings of 48th IEEE Symposium on Foundations of Computer Science, pp. 691–701 (2007)
8. Klerk, E.D., Laurent, M., Parrilo, P.: A PTAS for the minimization of polynomials of fixed degree over the simplex. Theoretical Computer Science 361(2-3), 210–225 (2006)
9. Feige, U., Krauthgamer, R.: The Probable Value of the Lovász-Schrijver Relaxations for Maximum Independent Set. SIAM Journal on Computing 32(2), 345–370 (2003)
10. Feller, W.: An Introduction to Probability Theory and Its Applications, 3rd edn., vol. 1. John Wiley & Sons, Chichester (1968)
11. Georgiou, K., Magen, A., Pitassi, T., Tourlakis, I.: Integrality gaps of 2−o(1) for vertex cover in the Lovász-Schrijver hierarchy. In: Proceedings of 48th IEEE Symposium on Foundations of Computer Science, pp. 702–712 (2007)
12. Goemans, M., Williamson, D.: Improved approximation algorithms for maximum cut and satisfiability problems using semidefinite programming. Journal of the ACM 42(6), 1115–1145 (1995)
13. Håstad, J.: Some optimal inapproximability results. Journal of the ACM 48(4), 798–859 (2001)
14. Halperin, E., Nathaniel, R., Zwick, U.: Coloring k-colorable graphs using relatively small palettes. J. Algorithms 45(1), 72–90 (2002)
15. Karger, D., Motwani, R., Sudan, M.: Approximate graph coloring by semidefinite programming. Journal of the ACM 45(2), 246–265 (1998)
16. Karloff, H., Zwick, U.: A 7/8-approximation algorithm for max 3SAT? In: Proceedings of 38th IEEE Symposium on Foundations of Computer Science, pp. 406–415 (1997)

17. Khot, S.: On the power of unique 2-prover 1-round games. In: Proceedings of the 34th ACM Symposium on Theory of Computing, pp. 767–775 (2002)
18. Khot, S., Kindler, G., Mossel, E., O'Donnell, R.: Optimal inapproximability results for MAX-CUT and other 2-variable CSPs? SIAM Journal on Computing 37(1), 319–357 (2007); ECCC Report TR05-101 (2005)
19. Krivelevich, M., Nathaniel, R., Sudakov, B.: Approximating coloring and maximum independent sets in 3-uniform hypergraphs. Journal of Algorithms 41(1), 99–113 (2001)
20. Laurent, M.: A comparison of the Sherali-Adams, Lovász-Schrijver and Lasserre relaxations for 0-1 programming. Mathematics of Operations Research 28(3), 460–496 (2003)
21. Lasserre, J.B.: An explicit exact SDP relaxation for nonlinear 0-1 programs. In: Aardal, K., Gerards, B. (eds.) IPCO 2001. LNCS, vol. 2081, pp. 293–303. Springer, Heidelberg (2001)
22. Lovász, L., Schrijver, A.: Cones of matrices and set-functioins and 0-1 optimization. SIAM Journal on Optimization 1(2), 166–190 (1991)
23. Nie, J., Schweighofer, M.: On the complexity of Putinar's Positivstellensatz. Journal of Complexity 23(1), 135–150 (2007)
24. Raghavendra, P.: Optimal Algorithms and Inapproximability Results For Every CSP? In: Proceedings of 30th ACM Symposium on Theory of Computing, pp. 245–254 (2008)
25. Schoenebeck, G., Trevisan, L., Tulsiani, M.: Tight integrality gaps for Lovász-Schrijver LP relaxations of Vertex Cover and Max Cut. In: Proceedings of 29th ACM Symposium on Theory of Computing, pp. 302–310 (2007)
26. Sherali, H.D., Adams, W.P.: A hierarchy of relaxations between the continuous and convex hull representations for zero-one programming problems. SIAM Journal on Discrete Mathematics 3(3), 411–430 (1990)
27. Schoenebeck, G.: Personal communication
28. de la Vega, F.W., Kenyon-Mathieu, C.: Linear programming relaxations of maxcut. In: Proceedings of 18th ACM Symposium on Discrete Algorithms, pp. 53–61 (2007)
29. Zwick, U.: Computer assisted proof of optimal approximability results. In: Proceedings of 13th ACM Symposium on Discrete Algorithms, pp. 496–505 (2002)

Sweeping Points

Adrian Dumitrescu[1],[*] and Minghui Jiang[2],[**]

[1] Department of Computer Science, University of Wisconsin-Milwaukee
Milwaukee, WI 53201-0784, USA
ad@cs.uwm.edu
[2] Department of Computer Science, Utah State University
Logan, UT 84322-4205, USA
mjiang@cc.usu.edu

Abstract. Given a set of points in the plane, and a sweep-line as a tool, what is best way to move the points to a target point using a sequence of sweeps? In a sweep, the sweep-line is placed at a start position somewhere in the plane, then moved orthogonally and continuously to another parallel end position, and then lifted from the plane. The cost of a sequence of sweeps is the total length of the sweeps. Another parameter of interest is the number of sweeps. Four variants are discussed, depending whether the target is a hole or a pile, and whether the target is specified or freely selected by the algorithm. Here we present a ratio $4/\pi \approx 1.27$ approximation algorithm in the length measure, which performs at most four sweeps. We also prove that, for the two constrained variants, there are sets of n points for which any sequence of minimum cost requires $3n/2 - O(1)$ sweeps.

1 Introduction

Sweeping is a well known and widely used technique in computational geometry. In this paper we make a first study of sweeping as an operation for moving a set of points. The following question was raised by Paweł Żyliński [4]:

> There are n balls on a table. The table has a hole (at a specified point). We want to sweep all balls to the hole with a line. We can move the balls by line sweeping: all balls touched by the line are moved with the line in the direction of the sweep. The problem is to find an optimal sequence of sweeps which minimizes the total sweeping distance covered by the line.

Although the above problem is quite natural, it does not seem to have been studied before. We note an obvious application to robotics, in particular, to the automation of part feeding and to nonprehensile part manipulation [1]. Imagine a manufacturing system that produces a constant stream of small identical parts, which have to be periodically cleared out, or gathered to a collection point by a robotic arm equipped with a segment-shaped sweeping device [1]. Here we study

---

[*] Supported in part by NSF CAREER grant CCF-0444188.
[**] Supported in part by NSF grant DBI-0743670.

A. Goel et al. (Eds.): APPROX and RANDOM 2008, LNCS 5171, pp. 63–76, 2008.

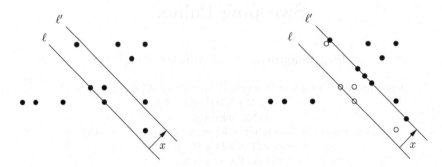

Fig. 1. A sweep of cost (length) x

an abstraction of such a scenario, when the small objects and the target are abstracted as points.

We now introduce some definitions to make the problem more precise. We refer to Figure 1. A set S of n points in the plane is given. In a *sweep*, the sweep-line is placed at a start position somewhere in the plane and is moved orthogonally and continuously to another parallel end position. All points touched by the line are moved with the line in the direction of the sweep. Then the line is lifted from the plane. Note that several points can merge during a sweep, and that merged points are subsequently treated as one point. A *sweeping sequence* for S is a sequence of sweeps that move all points in S to a target point. The *cost* of a sweeping sequence is the total length of its sweeps. As it will be evident from our Theorem 3, the sweep-line as a tool can be conveniently replaced by a finite sweep-segment of length twice the diameter of the point set.

We consider several variants of the sweeping problem, by making two distinctions on the *target*. First, the target can be either a hole or pile: if the target is a *hole*, then a point stays at the target once it reaches there, i.e., the point drops into the hole; if the target is a *pile*, then a point can still be moved away from the target after it reaches there. While it makes no difference for our algorithms whether the target is a hole or a pile (i.e., our algorithms are applicable to both variants), this subtle difference does matter when deriving lower bounds. Second, the target is either *constrained* to be a specified point or *unconstrained* (an arbitrary point freely selected by the algorithm). The four possible combinations, constrained versus unconstrained (C or U) and hole versus pile (H or P), yield thus four variants of the sweeping problem: CH, CP, UH, and UP.

Our main results are the following: although there exist sets of n points that require $\Omega(n)$ sweeps in any optimal solution (Section 3, Theorem 2), constant-factor approximations which use at most 4 sweeps can be computed in linear or nearly linear time (Section 2, Theorem 1). We also present some initial results and a conjecture for a related combinatorial question (Section 4, Theorem 3), and conclude with two open questions (Section 5).

We now introduce some preliminaries. A sweep is *canonical* if the number of points in contact with the sweep-line remains the same during the sweep. The following lemma is obvious.

Lemma 1. *Any sweep sequence can be decomposed into a sweep sequence of the same cost, consisting of only canonical sweeps. In particular, for any point set S, there is an optimal sweep sequence consisting of only canonical sweeps.*

Proof. Let $|S| = n$. A non-canonical sweep can be decomposed into a sequence of at most n canonical sweeps in the same direction and of the same total cost. □

Throughout the paper, we use the following convention: if A and B are two points, $\overleftrightarrow{AB}$ denotes the line through A and B, $\overrightarrow{AB}$ denotes the ray starting from A and going through B, AB denotes the line segment with endpoints A and B, and $|AB|$ denotes the length of the segment AB.

2   A Four-Sweep Algorithm

In this section, we present a four-sweep algorithm applicable to all four variants CH, CP, UH, and UP.

Theorem 1. *For any of the four variants CH, CP, UH, and UP of the sweeping problem (with n points in the plane),*

(I) *A ratio $\sqrt{2}$ approximation that uses at most 4 sweeps can be computed in $O(n)$ time;*

(II) *A ratio $4/\pi \approx 1.27$ approximation that uses at most 4 sweeps can be computed in $O(n \log n)$ time.*

Proof. We consider first the constrained variant, with a specified target o. Let S be the set of n points, and let $S' = S \cup \{o\}$. We next present two algorithms.

(I) **Algorithm A1.** Choose a rectilinear coordinate system xoy whose origin is o (of arbitrary orientation). Compute a minimal axis-parallel rectangle Q containing S'. Denote by w and h its width and height respectively, and assume w.l.o.g. that $h \leq w$. Perform the following (at most four) sweeps: (i) sweep from the top side of Q to the x-axis; (ii) sweep from the bottom side of Q to the x-axis; (iii) sweep from the left side of Q to the y-axis; (iv) sweep from the right side of Q to the y-axis. Figure 2 illustrates the execution of the algorithm on a small example.

Analysis. Clearly, the algorithm gives a valid solution, whose total cost is $\text{ALG} = w + h$. Let OPT be the cost of an optimal solution. We first argue that the approximation ratio of our algorithm is at most 2; we then improve this bound to $\sqrt{2}$.

We first show that $\text{OPT} \geq w$. Let p and q be the two extreme points of S' with minimum and maximum x-coordinates. Assume first that $p, q \in S$. Let p' and q' be the projection points of p and q on the x-axis throughout the execution of the sweep sequence. Put $w_1 = |p'o|$, and $w_2 = |oq'|$. Note that after the sweep sequence is complete, p' and q' coincide with the origin o. Further note that every sweep brings either p' or q' closer to o, but not both. Finally, observe that to bring p' to o requires a total sweep cost of at least w_1, and similarly, to bring

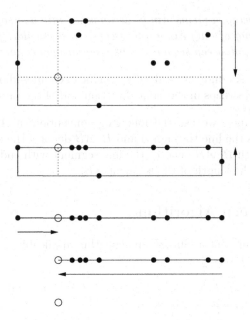

Fig. 2. Running the four-sweep algorithm

q' to o requires a total sweep cost of at least w_2. Therefore the total sweep cost is at least $w_1 + w_2 = w$, thus OPT $\geq w$. Since the total sweep cost is

$$\text{ALG} = w + h \leq 2w \leq 2 \cdot \text{OPT},$$

the ratio 2 follows when $p, q \in S$. The case when o is one of the two extreme points p and q is completely analogous.

We now argue that OPT $\geq (w + h)/\sqrt{2}$. Let X be an arbitrary sequence consisting of k sweeps which solves the given instance S. For $i = 1, \ldots, k$ let x_i be the cost of the ith sweep, and $\alpha_i \in [0, 2\pi)$ be its direction. Write $x = \sum_{i=1}^{k} x_i$. Indeed, the ith sweep of cost x_i can reduce the current semi-perimeter of Q by at most $\sqrt{2}\, x_i$. Here the points in S are considered moving, so S', and its enclosing rectangle Q change continuously as an effect of the sweeps. Since the semiperimeter of Q drops from $w + h$ to 0, by summing over all sweeps, we get that in any sweep sequence for S of total cost x,

$$\sqrt{2}\, x = \sqrt{2} \sum_{i=1}^{k} x_i \geq w + h,$$

thus

$$\text{ALG} = w + h \leq \sqrt{2} \cdot \text{OPT},$$

and the approximation ratio $\sqrt{2}$ follows.

(II) **Algorithm A2.** First compute a minimum perimeter rectangle Q_0 containing S'. This takes $O(n \log n)$ using the rotating calipers algorithm of Toussaint

[3]. Let now xoy be a rectilinear coordinate system in which Q_0 is axis-aligned. Let w and h be its width and height respectively. Then perform the four sweeps as in Algorithm A1.

Analysis. Assume w.l.o.g. that $w + h = 1$. For $\beta \in [0, \pi/2)$, let $Q(\beta)$ denote the minimum perimeter rectangle of orientation β containing S'; i.e., one of the sides of $Q(\beta)$ makes an angle β with the positive direction of the x-axis. Let $w(\beta)$ and $h(\beta)$ denote the initial values of the width and height of $Q(\beta)$ respectively. Note that $[0, \pi/2)$ covers all possible orientations β of rectangles enclosing S'.

As in the proof of the ratio $\sqrt{2}$ approximation ratio, recall that for any $i \in \{1, \ldots, k\}$, the ith sweep of cost x_i can reduce the current semi-perimeter of $Q(\beta)$ by at most $x_i\sqrt{2}$. In fact we can be more precise by taking into account the direction of the sweep: the reduction is at most

$$x_i \left(|\cos(\alpha_i - \beta)| + |\sin(\alpha_i - \beta)| \right).$$

Since X solves S, by adding up the reductions over all sweeps $i \in \{1, \ldots, k\}$, we must have—since $w(\beta) + h(\beta) \geq 1$, for every $\beta \in [0, \pi/2)$:

$$\sum_{i=1}^{k} x_i \left(|\cos(\alpha_i - \beta)| + |\sin(\alpha_i - \beta)| \right) \geq 1. \tag{1}$$

We integrate this inequality over the β-interval $[0, \pi/2]$; x_i and α_i are fixed, and each term is dealt with independently. Fix $i \in \{1, \ldots, k\}$, and write $\alpha = \alpha_i$ for simplicity. Assume first that $\alpha \in [0, \pi/2)$. A change of variables yields

$$\int_0^{\pi/2} \left(|\cos(\alpha - \beta)| + |\sin(\alpha - \beta)| \right) d\beta$$

$$= \int_\alpha^{\alpha + \pi/2} \left(|\cos\beta| + |\sin\beta| \right) d\beta$$

$$= \int_\alpha^{\pi/2} \left(\cos\beta + \sin\beta \right) d\beta + \int_{\pi/2}^{\alpha + \pi/2} \left(-\cos\beta + \sin\beta \right) d\beta$$

$$= (\sin\beta - \cos\beta) \Big|_\alpha^{\pi/2} + (-\sin\beta - \cos\beta) \Big|_{\pi/2}^{\alpha + \pi/2}$$

$$= (1 - \sin\alpha + \cos\alpha) + (-\cos\alpha + \sin\alpha + 1) = 2.$$

Let

$$G(\alpha) = \int_\alpha^{\alpha + \pi/2} \left(|\cos\beta| + |\sin\beta| \right) d\beta.$$

It is easy to verify that $G(\alpha) = G(\alpha + \pi/2)$ for any $\alpha \in [0, 2\pi)$, hence the integration gives the same result, 2, for any $\alpha_i \in [0, 2\pi)$, and for any $i \in \{1, \ldots, k\}$. Hence by integrating (1) over $[0, \pi/2]$ yields

$$2 \left(\sum_{i=0}^{k} x_i \right) \geq \frac{\pi}{2}, \quad \text{or} \quad x \geq \frac{\pi}{4}.$$

Since this holds for any valid sequence, we also have OPT $\geq \frac{\pi}{4}$. Recall that ALG $= w + h = 1$, and the approximation ratio $4/\pi$ follows.

To extend our results to the unconstrained variant requires only small changes in the proof. Instead of the minimum semi-perimeter rectangle(s) enclosing $S' = S \cup \{o\}$, consider the minimum semi-perimeter rectangle(s) enclosing S. All inequalities used in the proof of Theorem 1 remain valid. We also remark that the resulting algorithms execute only two sweeps (rather than four): from top to bottom, and left to right, with the target being the lower-right corner of the enclosing rectangle. □

2.1   A Lower Bound on the Approximation Ratio of Algorithm A2

It is likely that the approximation ratio of our four-sweep algorithm is slightly better than what we have proved: we noticed that for both cases, when h is large and when h is small relative to w, our estimates on the reduction are slightly optimistic. However, the construction we describe next, shows that the ratio of our four-sweep algorithm cannot be reduced below 1.1784 (for either variant).

Perhaps the simplest example to check first is the following. Take the three vertices of a unit (side) equilateral triangle as our point set. For the constrained variant, place the target at the triangle center: the optimal cost is at most $\sqrt{3}$ by 3 sweeps (in fact, equality holds, as shown in the proof of Theorem 3), while the four-sweep algorithm uses $1 + \sqrt{3}/2$. The ratio is about 1.077.

We now describe a better construction that gives a lower bound of about 1.1784; see Figure 3. Place n points uniformly (dense) on the thick curve $\mathcal{C}$ connecting B and C. For the constrained variant, place the target at point B. The curve is made from the two equal sides of an obtuse isosceles triangle with sharp angles $\alpha = \arctan(1/2) \approx 26.565°$, then "smoothed" around the obtuse triangle corner. $\triangle ABC$ is an isosceles triangle with sides $AB = AC = \sqrt{5}$ and $BC = 4$, with altitude $AD = 1$, and with angles $\angle ABC = \angle ACB = \alpha = \arctan(1/2)$. E and F are two points on AB and AC, respectively, such that $DE \perp AB$ and $DF \perp AC$.

The curve $\mathcal{C}$ consists of the two segments BE and CF and a curve $\mathcal{C}_0$ connecting E and F, defined as follows. For an arbitrary point G on $\mathcal{C}_0$, $\angle ADG = \beta \leq \alpha$, the length of the segment DG is

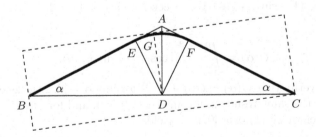

Fig. 3. A continuous convex curve that gives a lower bound of about 1.1784 on the approximation ratio of Algorithm A2

$$|DG| = f(\beta) = 4\cos\alpha + 4\sin\alpha - 4\cos\beta - 2\sin\beta.$$

Observe that $4\sin\alpha - 2\cos\alpha = 0$ holds by the definition of α, hence

$$\frac{\mathrm{d}}{\mathrm{d}\beta}f(\beta) = 4\sin\beta - 2\cos\beta \leq 4\sin\alpha - 2\cos\alpha = 0,$$

where the derivative reaches zero at E and F (when $\beta = \alpha$). So $\mathcal{C}$ is a continuous convex curve. For a rectangle that circumscribes the curve $\mathcal{C}$ with one side tangent to $\mathcal{C}_0$ at point G, its width and height are $|BC|\cos\beta$ and $|DG|+|CD|\sin\beta$, respectively. Hence its semi-perimeter is

$$
\begin{aligned}
&|BC|\cos\beta + |DG| + |CD|\sin\beta \\
&= 4\cos\beta + (4\cos\alpha + 4\sin\alpha - 4\cos\beta - 2\sin\beta) + 2\sin\beta \\
&= 4\cos\alpha + 4\sin\alpha.
\end{aligned}
$$

Therefore the semi-perimeter of a minimum rectangle with orientation β, where $0 \leq \beta \leq \alpha$, that encloses $\mathcal{C}$ is a constant: $4\cos\alpha + 4\sin\alpha$. Since the length of $\mathcal{C}_0$ is

$$
\begin{aligned}
2\int_{\beta=0}^{\alpha} f(\beta)\mathrm{d}\beta &= 2\int_{\beta=0}^{\alpha}(4\cos\alpha + 4\sin\alpha - 4\cos\beta - 2\sin\beta)\mathrm{d}\beta \\
&= 8(\cos\alpha + \sin\alpha)\alpha + 2(-4\sin\beta + 2\cos\beta)\Big|_0^{\alpha} \\
&= 8(\cos\alpha + \sin\alpha)\alpha + 2(-4\sin\alpha + 2\cos\alpha - 2) \\
&= 8(\cos\alpha + \sin\alpha)\alpha - 4,
\end{aligned}
$$

and since $|BE| = |CF| = 2\cos\alpha$, the length of $\mathcal{C}$ is $8(\cos\alpha + \sin\alpha)\alpha - 4 + 4\cos\alpha$. The ratio of the minimum semi-perimeter and the curve length is (after simplification by 4, and using the values $\cos\alpha = 2/\sqrt{5}$, $\sin\alpha = 1/\sqrt{5}$, $\alpha = \arctan(1/2)$)

$$\frac{4\cos\alpha + 4\sin\alpha}{8(\cos\alpha + \sin\alpha)\alpha - 4 + 4\cos\alpha} = \frac{3}{6\arctan(1/2) - \sqrt{5} + 2} = 1.1784\ldots$$

This gives a lower bound of 1.1784 on the approximation ratio of Algorithm A2, which holds for all four variants.

3 Point Sets for the Constrained Variants That Require Many Sweeps

In this section we show that some point sets require many sweeps in an optimal solution, i.e., the number of sweeps is not just a constant. In what follows, the target is constrained to a specified point, and may be either a hole or a pile, i.e., we refer to both constrained variants CH and CP.

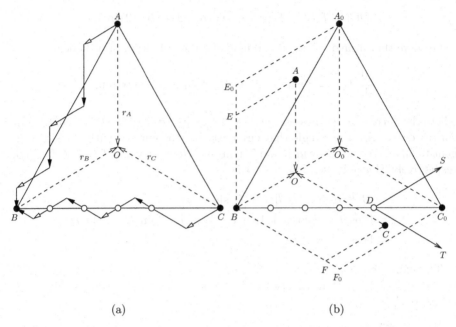

(a)   (b)

Fig. 4. A construction with three points A, B, and C (black points) forming a unit equilateral triangle and $n - 3$ arbitrary points (white points) on the edge BC. The target is at the point B. Initially: $A = A_0$, $B = B_0$, $C = C_0$. (a) An optimal sweeping sequence. (b) Some properties of optimal sweeping sequences are illustrated.

Theorem 2. *For the two constrained variants* CH *and* CP, *and for any* n, *there are sets of* n *points for which any optimal sweeping sequence consists of at least* $3n/2 - O(1)$ *sweeps.*

We now proceed with the proof of Theorem 2. We refer to Figure 4(a). Our set S consists of three points A, B, and C (black points) forming a unit equilateral triangle and $n - 3$ points (white points) arbitrary placed on the edge BC. The target is at the point B. For convenience, we place $\triangle ABC$ initially with B at the origin and $\overrightarrow{BC}$ along the x axis. In what follows, we refer to the intermediate positions of the moving points: input points from the set S (such as A, B, C, D, etc.) or other auxiliary points (such as E and F) during a sequence of sweeps. When the intermediate position of a point does not coincide with its original position, we avoid the possible ambiguity by adding a subscript 0 to the label of the original position. For example, the two labels A and A_0 in the figure refer to the intermediate and the original positions, respectively, of the same point A. Initially, we have $A = A_0$, $B = B_0$, $C = C_0$. We will show in Lemma 2 that $B = B_0$ (that is, B remains stationary) during any optimal sequence; this is evident for the CH variant, but not so for the CP variant.

Define three rays: a ray r_A from A in the $3\pi/2$ direction, a ray r_B from B in the $\pi/6$ direction, and a ray r_C from C in the $5\pi/6$ direction. The three rays from A, B, and C initially intersect at a single point $O = O_0$, the center of $\triangle A_0 B_0 C_0$. We will show below that this concurrency property is maintained throughout any optimal sweeping sequence for S. We now define six special types of sweeps:

Type A: A is moved in the direction $\overrightarrow{AO}$. B and C are not moved.
Type BC: B and C are moved together in the direction $\overrightarrow{OA}$. A is not moved.
Type B: B is moved in the direction $\overrightarrow{BO}$. A and C are not moved.
Type AC: A and C are moved together in the direction $\overrightarrow{OB}$. B is not moved.
Type C: C is moved in the direction $\overrightarrow{CO}$. A and B are not moved.
Type AB: A and B are moved together in the direction $\overrightarrow{OC}$. C is not moved.

We note that, for the CH variant, the three types involving B, namely types BC, B, and AB are in fact not used, since point B will remain at the hole throughout any sweeping sequence.

For each of the six types, each moved point (among A, B, and C) is moved for a distance equal to the sweep length, that is, the moved point is on the sweep-line during the sweep. If a sweeping sequence consists of only sweeps of the six special types, then it can be easily verified (by induction) that the three rays from A, B, and C still intersect at a single point O after each sweep; see Figure 4(b).

The three segments $A_0 O_0$, $B_0 O_0$, and $C_0 O_0$ determine two parallelograms $A_0 O_0 B_0 E_0$ and $C_0 O_0 B_0 F_0$ (each is a rhombus with two $60°$ angles), as shown in Figure 4(b). We now observe some properties of sweeps of the three types A, C, and AC. Consider how a sweep changes the two parallelograms $AOBE$ and $COBF$, initially $A_0 O_0 B_0 E_0$ and $C_0 O_0 B_0 F_0$: a sweep of type A reduces the two sides AO and BE; a sweep of type C reduces the two sides CO and BF; a sweep of type AC reduces the three sides AE, CF, and OB (note that the side OB is shared by the two parallelograms). During any sweeping sequence of the three types A, C, and AC, the point A always remains inside the rhombus $A_0 O_0 B_0 E_0$, and point C inside the rhombus $C_0 O_0 B_0 F_0$.

Lemma 2. *The optimal cost for S is $\sqrt{3}$. Moreover, any optimal sequence for S consists of only sweeps of the three special types A, C, and AC, with a subtotal cost of $\sqrt{3}/3$ for each type.*

Proof. We first show that the optimal cost for S is at most $\sqrt{3}$. We refer to Figure 4(a) for a sweeping sequence of $n-1$ alternating steps: (i) one sweep of type AC (the white arrow); (ii) two sweeps, one of type A and the other of type C (the black arrows). Each step, except the first and the last, merges C with a white point, in sequential order from right to left. The total number of sweeps in this sequence is $(3n-3)/2$ when n is odd, and is $(3n-4)/2$ when n is even. The total cost of this sequence is $|A_0 O_0| + |B_0 O_0| + |C_0 O_0| = 3 \cdot \sqrt{3}/3 = \sqrt{3}$.

We next show that the optimal cost for S is at least $\sqrt{3}$. Consider an optimal sequence for S. Assume w.l.o.g. that the sequence is canonical. We construct three paths, from the three points A_0, B_0, and C_0 to a single point, such that their total length is at most the cost of the sequence. Each sweep in the sequence that moves one or two of the three points A, B, and C corresponds to an edge in one of the three paths, with the sweep length equal to the edge length: (i) if a sweep moves only one of the three points, then the corresponding edge extends the path from that point, along the sweep direction; (ii) if a sweep moves two of the three points, then the corresponding edge extends the path from the third point, along the opposite sweep direction. We note that, for the three points A, B, and C, each three-point sweep is useless, and each two-point sweep is equivalent to a one-point sweep in the opposite direction, in the sense that the resulting triangles $\triangle ABC$ are congruent. When the three points finally meet at the target, the three paths also end at a single point (which could be different from the target).

The total length of the three paths is at least the total length of a Steiner tree for the three points A_0, B_0, and C_0. It is well known [2] that the minimum Steiner tree for the three points A_0, B_0, and C_0 is unique, and consists of exactly three edges of equal length $\sqrt{3}/3$, from the three points to the center O_0 of $\triangle A_0 B_0 C_0$. It follows that the optimal cost for S is at least $\sqrt{3}$. Together with the matching upper bound achieved by the sequence illustrated in Figure 4(a), we have shown that the optimal cost for S is exactly $\sqrt{3}$.

The uniqueness of the minimum Steiner tree for the three points A_0, B_0, and C_0 implies that every sweep in the optimal sequence must be of one of the six special types, with a subtotal cost of $|A_0 O_0| = |B_0 O_0| = |C_0 O_0| = \sqrt{3}/3$ for each of the three groups: A and BC, B and AC, and C and AB. To complete the proof, we next show that sweeps of the three types B, AB, and BC never appear in the optimal sequence. Consider the two possible cases for the target:

1. The target is a hole, that is, a point stays at the target once it reaches there. Since B is already at the target, it must stay there. So this case is obvious, as noted after our definition of the six types.
2. The target is a pile, that is, a point can be moved away from the target after it reaches there. Although B is already at the target, it can still be moved away. The only sweeps that move B are of the three types B, AB, and BC. Such sweeps all have a positive projection in the direction $\overrightarrow{BO}$, and can only move B away from the target (and cannot move it back); therefore they cannot appear in the optimal sequence.

This completes the proof of Lemma 2.   □

Let D be the rightmost white point. Figure 4(b) shows the initial position of D. Later in Lemma 4, we will prove that D remains at its initial position until it is merged with C. In Lemma 3 however, we don't make any assumption of D being at its original position. Let $\overrightarrow{DS}$ and $\overrightarrow{DT}$ be two rays from D with directions $\pi/6$ and $-\pi/6$, respectively.

Lemma 3. *Consider an optimal sweeping sequence. If C is moved above the line $\overline{DS}$ or below the line $\overline{DT}$, then C remains either above $\overline{DS}$ or below $\overline{DT}$ until either C or D coincides with the target.*

Proof. We refer to Figure 4(b). Assume w.l.o.g. that the sweeping sequence is canonical. Consider each remaining sweep in the sequence after C is at a position above $\overline{DS}$ or below $\overline{DT}$:

Type C. Consider two cases: C is above $\overline{DS}$ or below $\overline{DT}$.

1. C is above $\overline{DS}$. If D is not moved, then C is moved further above $\overline{DS}$. If both C and D are moved (when $CD \perp CO$), then they are moved for the same distance in the same direction, and C remains above $\overline{DS}$.
2. C is below $\overline{DT}$. Since $\overline{DT}$ is parallel to the sweep direction $\overrightarrow{CO}$, C remains below $\overline{DT}$,

Type AC. Consider two cases: C is above $\overline{DS}$ or below $\overline{DT}$.

1. C is above $\overline{DS}$. Since DS is parallel to the sweep direction $\overrightarrow{OB}$, C remains above $\overline{DS}$.
2. C is below $\overline{DT}$. If D is not moved, then C is moved further below $\overline{DT}$. If both C and D are moved, then they are moved for the same distance in the same direction, and C remains below $\overline{DT}$.

Type A. Note that C may be both above $\overline{DS}$ and below $\overline{DT}$. We divide the two cases in an alternative way without overlap: C is either (i) above $\overline{DS}$ and not below (i.e., above or on) $\overline{DT}$ or (ii) below $\overline{DT}$.

1. C is above $\overline{DS}$ and not below $\overline{DT}$. Then C is above D. Since C is not moved, D is not moved either. So C remains above $\overline{DS}$ and not below $\overline{DT}$.
2. C is below $\overline{DT}$. Since $\overline{DT}$ is parallel to $\overline{CO}$, D is above $\overline{CO}$. The sweep may move A down to O and correspondingly move D down until it is on the horizontal line through O, but no further. So D remains above $\overline{CO}$, and C remains below $\overline{DT}$.

$\square$

Lemma 4. *In any optimal sequence, each white point is not moved until it is merged with C, in sequential order from right to left.*

Proof. Assume w.l.o.g. that the sweeping sequence is canonical. Lemma 2 shows that the sweeps in any optimal sequence are of the three types A, C, and AC. Let σ_1 be the first sweep that moves a white point, and let D_1 be the first white point moved. If the sweep σ_1 is of type A, then A would be moved below the x axis (recall that in a sweep of type A the sweep-line always goes through A), and any subsequent sweep that moves A, of type A or AC, would move A further below the x axis and never to B. This contradicts the validity of the sequence. Therefore σ_1 must be of type C or AC.

We claim that C must be merged with the rightmost white point D before the sweep σ_1. We will prove the claim by contradiction. Suppose the contrary.

Our proof by contradiction is in two steps: In the first step, we will show that C is either above $\overline{DS}$ or below $\overline{DT}$ at the beginning of sweep σ_1. In the second step, we will show that the assumed optimal sequence is not valid.

First step. The sweep-line of σ_1 goes through D_1 during the sweep. Since σ_1 is of type C or AC, C is also on the sweep-line of σ_1. Consider two cases for the relation between D_1 and D:

1. $D_1 \neq D$ (D_1 is to the left of D on the x axis). Then every point on the sweep-line, including C, is either above $\overline{DS}$ or below $\overline{DT}$.

2. $D_1 = D$. Then every point on the sweep-line, except D, is either above $\overline{DS}$ or below $\overline{DT}$. Since C is not merged with D before σ_1, C is either above $\overline{DS}$ or below $\overline{DT}$.

In either case, C is either above $\overline{DS}$ or below $\overline{DT}$.

Second step. From Lemma 3, C remains either above $\overline{DS}$ or below $\overline{DT}$ until either C or D coincides with the target. This, as we will show in the following, implies that the sweeping sequence is not valid. Consider the two possible cases for the target as either a pile or a hole:

1. The target is a pile, that is, a point can be moved away from the target after it reaches there. Then C remains either above $\overline{DS}$ or below $\overline{DT}$ even after either C or D reaches the target. It follows that C and D never merge, and hence cannot end up together at the target. Therefore the sweeping sequence is not valid.

2. The target is a hole, that is, a point stays at the target once it reaches there. Let σ_2 be a sweep in the sequence that moves D to the target. We consider the three possible cases for the type of σ_2:

   Type AC. The sweep-line of σ_2 goes through the two points A and C. As D is moved to the point B by σ_2, both parallelograms $AOBE$ and $COBF$ shrink to the point B, that is, both A and C are moved to the target together with D. Then A, C, and D must have been merged even before the sweep σ_2. This is impossible because C is above $\overline{DS}$ or below $\overline{DT}$ until either C or D reaches the target.

   Type C. It follows by the same argument (the parallelogram $COBF$ shrinks to the point B) that C and D are merged before the sweep σ_2, which is again impossible.

   Type A. It follows by the same argument (the parallelogram $AOBE$ shrinks to the point B) that A and D are merged before the sweep σ_2, above the line $\overline{BO}$. This is impossible because D cannot be moved above $\overline{BO}$: a sweep of type AC does not change the distance from D to $\overline{BO}$; a sweep of type A can only move D further below $\overline{BO}$; a sweep of type C can move D to BO but not above $\overline{BO}$, since C itself cannot be moved above $\overline{BO}$.

   In each case, D cannot be moved to the target. Therefore the sweeping sequence is not valid.

We have shown that the sequence is not valid with the target as either a pile or a hole. By contradiction, this proves our original claim that C must be merged with D before the sweep σ_1.

As soon as C is merged with D, we can consider D as deleted. The point set now reaches a configuration similar to the original configuration: the two points B and C are on the x axis with all the (unmoved) white points between them, and A alone is above the x axis. But now we have one less white point. Repeating the argument in the preceding paragraphs inductively completes the proof of Lemma 4.   □

We are now in position to finalize the proof of Theorem 2. We have shown that in an optimal sequence, C must be merged with the white points one by one from right to left. Since the sweeps are not along the x axis, each of the $n - 3$ white point requires at least one sweep to be merged. The total number of sweeps in the sequence is at least $n - O(1)$. We obtain a tighter estimate (that matches our previous sweep sequence for S) as follows. Between two consecutive merges, C has to be moved to the left by alternating sweeps of types AC and C. Between two sweeps of type AC, since C is moved by a sweep of type C, A must also be moved by a sweep of type A, to make $AC \perp OB$ for the next sweep of type AC. Therefore each merge requires either one sweep of type AC or two sweeps of types A and C, in an alternating pattern as shown in Figure 4(a). The total number of sweeps in the sequence is at least $3n/2 - O(1)$. This completes the proof of Theorem 2.

4   A Combinatorial Question for the Unconstrained Variants

The following related question suggests itself: What is the maximum cost required for sweeping a planar point set of unit diameter to a single point? Note that the target point is unconstrained, and can be either a hole or a pile. Define

$$\rho_H = \sup_S \inf_X \text{cost}(X), \quad \text{for the variant UH,}$$

and

$$\rho_P = \sup_S \inf_X \text{cost}(X), \quad \text{for the variant UP,}$$

where S ranges over all finite planar point sets of unit diameter, and X ranges over all sweeping sequences for S. We give estimates on the two numbers ρ_H and ρ_P in the following theorem:

Theorem 3. $1.73 \approx \sqrt{3} \leq \rho_H \leq \rho_P \leq 2$.

Proof. Any sweeping sequence for the UP variant is also a sweeping sequence for the UH variant, so we have $\rho_H \leq \rho_P$. We first prove the upper bound $\rho_P \leq 2$. Let S be an arbitrary finite planar set with unit diameter. Let p and q be two points in S at unit distance. Then S is contained in a rectangle with width 1 (parallel to the line pq) and height at most 1. A sweep along the width and a sweep along the height reduce the rectangle to a single point (the pile), at a cost of at most 2.

We next prove the lower bound $\rho_H \geq \sqrt{3}$. Let T be an equilateral triangle with unit side. Let X^* be an optimal sequence of canonical sweeps for the three vertices of T. Using the same idea as in the proof for Lemma 2, we construct three paths, from the three vertices of T to a common point, such that their total length is at most the cost of X^*. It follows that the cost of X^* is at least the total length of a minimum Steiner tree for the three vertices, which is $\sqrt{3}$ [2]. Note that our analysis for this case is tight: three sweeps along the edges of the minimum Steiner tree clearly reduce the equilateral triangle T to a single point.   □

The reader can observe that a weaker lower bound $\rho_H \geq \pi/2 \approx 1.57$ follows from our result in Theorem 1 applied to a set of n points uniformly distributed on a circle (for large n). We think the upper bound in Theorem 3 is best possible, for instance, in the same case of n points uniformly distributed on a circle of unit diameter, for n going to infinity:

Conjecture 1. $\rho_H = \rho_P = 2$.

5 Concluding Remarks

Besides Conjecture 1, two interesting questions (for any of the four variants) remain open:

(1) What is the complexity of the sweeping problem? Is there a polynomial time algorithm for generating an optimal sweeping sequence?
(2) Can the number of sweeps in an optimal solution be always bounded by a polynomial in n? i.e., is there always an optimal solution with a polynomial number of sweeps?

Acknowledgment. We are grateful to Paweł Żyliński for sharing his dream problem with us.

References

1. Halperin, D., Kavraki, L., Latombe, J.-C.: Robotics. In: Goodman, J., O'Rourke, J. (eds.) Handbook of Discrete and Computational Geometry, 2nd edn., pp. 1065–1093. Chapman & Hall, Boca Raton (2004)
2. Hwang, F.K., Richards, D.S., Winter, P.: The Steiner Tree Problem. In: Annals of Discrete Mathematics, vol. 53. North-Holland, Amsterdam (1992)
3. Toussaint, G.: Solving geometric problems with the rotating calipers. In: Proceedings of Mediterranean Electrotechnical Conference (MELECON 1983), Athens (1983)
4. Żyliński, P.: personal communication (June 2007)

Constraint Satisfaction over a Non-Boolean Domain: Approximation Algorithms and Unique-Games Hardness

Venkatesan Guruswami* and Prasad Raghavendra**

Department of Computer Science & Engineering,
University of Washington,
Seattle, WA
{venkat,prasad}@cs.washington.edu

Abstract. We study the approximability of the MAX k-CSP problem over non-boolean domains, more specifically over $\{0, 1, \ldots, q-1\}$ for some integer q. We extend the techniques of Samorodnitsky and Trevisan [19] to obtain a UGC hardness result when q is a prime. More precisely, assuming the Unique Games Conjecture, we show that it is NP-hard to approximate the problem to a ratio greater than $q^2 k/q^k$. Independent of this work, Austrin and Mossel [2] obtain a more general UGC hardness result using entirely different techniques.

We also obtain an approximation algorithm that achieves a ratio of $C(q) \cdot k/q^k$ for some constant $C(q)$ depending only on q, via a subroutine for approximating the value of a semidefinite quadratic form when the variables take values on the corners of the q-dimensional simplex. This generalizes an algorithm of Nesterov [16] for the ± 1-valued variables. It has been pointed out to us [15] that a similar approximation ratio can be obtained by reducing the non-boolean case to a boolean CSP.

1   Introduction

Constraint Satisfaction Problems (CSP) capture a large variety of combinatorial optimization problems that arise in practice. In the MAX k-CSP problem, the input consists of a set of variables taking values over a domain(say $\{0, 1\}$), and a set of constraints with each acting on k of the variables. The objective is to find an assignment of values to the variables that maximizes the number of constraints satisfied. Several classic optimization problems like 3-SAT, Max Cut fall in to the general framework of CSPs. For most CSPs of interest, the problem of finding the optimal assignment turns out to be NP-hard. To cope with this intractability, the focus shifts to approximation algorithms with provable guarantees. Specifically, an algorithm $\mathcal{A}$ is said to yield an α approximation to a CSP, if on every instance Γ of the CSP, the algorithm outputs an assignment that satisfies at least α times as many constraints as the optimal assignment.

* Work done while on leave at School of Mathematics, Institute for Advanced Study, Princeton, NJ. Research supported in part by a Packard Fellowship and NSF grant CCF-0324906 to the IAS.
** Research supported by NSF CCF-0343672.

A. Goel et al. (Eds.): APPROX and RANDOM 2008, LNCS 5171, pp. 77–90, 2008.
© Springer-Verlag Berlin Heidelberg 2008

Apart from its natural appeal, the study of the MAX k-CSP problem is interesting for yet another reason. The best approximation ratio achievable for MAX k-CSP equals the optimal soundness of a PCP verifier making at most k queries. In fact, inapproximability results for MAX k-CSP have often been accompanied by corresponding developments in analysis of linearity testing.

Over the boolean domain, the problem of MAX k-CSP has been studied extensively. For a boolean predicate $P : \{0,1\}^k \to \{0,1\}$, the MAX k-CSP (P) problem is the special case of MAX k-CSP where all the constraints are of the form $P(l_1, l_2, \ldots, l_k)$ with each literal l_i being either a variable or its negation. For many natural boolean predicates P, approximation algorithms and matching NP-hardness results are known for MAX k-CSP (P)[11]. For the general MAX k-CSP problem over boolean domain, the best known algorithm yields a ratio of $\Omega(\frac{k}{2^k})$ [3], while any ratio better than $2^{\sqrt{2k}}/2^k$ is known to be NP-hard to achieve [5]. Further if one assumes the Unique Games Conjecture, then it is NP-hard to approximate MAX k-CSP problem to a factor better than $\frac{2k}{2^k}$ [19].

In this work, we study the approximability of the MAX k-CSP problem over non-boolean domains, more specifically over $\{0, 1, \ldots, q-1\}$ for some integer q, obtaining both algorithmic and hardness results (under the UGC) with almost matching approximation factors.

On the hardness side, we extend the techniques of [19] to obtain a UGC hardness result when q is a prime. More precisely, assuming the Unique Games Conjecture, we show that it is NP-hard to approximate the problem to a ratio greater than $q^2 k/q^k$. Except for constant factors depending on q, the algorithm and the UGC hardness result have the same dependence on of the arity k. Independent of this work, Austrin and Mossel [2] obtain a more general UGC hardness result using entirely different techniques. Technically, our proof extends the Gowers Uniformity based approach of Samorodnitsky and Trevisan [19] to correlations on q-ary cubes instead of the binary cube. This is related to the detection of multidimensional arithmetic progressions by a Gowers norm of appropriately large degree. Along the way, we also make a simplification to [19] and avoid the need to obtain a large *cross-influence* between two functions in a collection with a substantial Uniformity norm; instead our proof works based on large influence of just one function in the collection.

On the algorithmic side, we obtain a approximation algorithm that achieves a ratio of $C(q) \cdot k/q^k$ with $C(q) = \frac{1}{2\pi e q(q-1)^6}$. As a subroutine, we design an algorithm for maximizing a positive definite quadratic form with variables forced to take values on the corners of the q-dimensional simplex. This is a generalization of an algorithm of Nesterov [16] for maximizing positive definite quadratic form with variables forced to take $\{-1, 1\}$ values. Independent of this work, Makarychev and Makarychev [15] brought to our notice a reduction from non-boolean CSPs to the boolean case, which in conjunction with the CMM algorithm [3] yields a better approximation ratio for the MAX k-CSP problem. Using the reduction, one can deduce a $q^2(1 + o(1))k/q^k$ factor UG hardness for MAX k-CSP for arbitrary positive integers q, starting from our UG hardness result for primes q.

1.1   Related Work

The simplest algorithm for MAX k-CSP over boolean domain is to output a random assignment to the variables, thus achieving an approximation ratio of $\frac{1}{2^k}$. The first improvement over this trivial algorithm, a ratio of $\frac{2}{2^k}$ was obtained by Trevisan [20]. Hast [9] proposed an approximation algorithm with a ratio of $\Omega(\frac{k}{\log k2^k})$, which was later improved to the current best known algorithm achieving an approximation factor of $\Omega(\frac{k}{2^k})$ [3].

On the hardness side, MAX k-CSP over the boolean domain was shown to be NP-hard to approximate to a ratio greater than $\Omega(2^{2\sqrt{k}}/2^k)$ by Samorodnitsky and Trevisan [18]. The result involved an analysis of a graph-linearity test which was simplified subsequently by Håstad and Wigderson [13]. Later, using the machinery of multi-layered PCP developed in [4], the inapproximability factor was improved to $O(2^{\sqrt{2k}}/2^k)$ in [5].

A predicate P is *approximation resistant* if the best optimal approximation ratio for MAX k-CSP (P) is given by the random assignment. While no predicate over 2 variables is approximation resistant, a predicate over 3 variables is approximation resistant if and only if it is implied by the XOR of 3 variables [11,21]. Almost all predicates on 4 variables were classified with respect to approximation resistance in [10].

In recent years, several inapproximability results for MAX k-CSP problems were obtained assuming the Unique Games Conjecture. Firstly, a tight inapproximability of $\Theta\left(\frac{k}{2^k}\right)$ was shown in [19]. The proof relies on the analysis of a hypergraph linearity test using the Gowers uniformity norms. Hastad showed that if UGC is true, then as k increases, nearly every predicate P on k variables is *approximation resistant* [12].

More recently, optimal inapproximability results have been shown for large classes of CSPs assuming the Unique Games Conjecture. Under an additional conjecture, optimal inapproximability results were obtained in [1] for all boolean predicates over 2 variables. Subsequently, it was shown in [17] that for every CSP over an arbitrary finite domain, the best possible approximation ratio is equal to the integrality gap of a well known Semidefinite program. Further the same work also obtains an algorithm that achieves the best possible approximation ratio assuming UGC. Although the results of [17] apply to non-boolean domains, they do not determine the value of the approximation factor explicitly, but only show that it is equal to the integrality gap of an SDP. Further the algorithm proposed in [17] does not yield any approximation guarantee for MAX k-CSP unconditionally. Thus neither the inapproximability nor the algorithmic results of this work are subsumed by [17].

Austin and Mossel [2] obtain a sufficient condition for a predicate P to be approximation resistant. Through this sufficiency condition, they obtain strong UGC hardness results for MAX k-CSP problem over the domain $\{1, \ldots, q\}$ for arbitrary k and q. For the case when q is a prime power, their results imply a UGC hardness of $kq(q-1)/q^k$. The hardness results in this work and [2] were obtained independently and use entirely different techniques.

1.2 Organization of the Paper

We begin with background on the Unique Games conjecture, Gowers norm, and influence of variables in Section 2. In Section 3, we present a linearity test that forms the core of the UGC based hardness reduction. We prove our inapproximability result (for the case when q is a prime) by a reduction from Unique Games in Section 4. The proof uses a technical step bounding a certain expectation by an appropriate Gowers norm; this step is proved in Section 5. Finally, we state the algorithmic result in Section 6, deferring the details to the full version [6].

2 Preliminaries

In this section, we will set up notation, and review the notions of Gower's uniformity, influences, noise operators and the Unique games conjecture. Henceforth, for a positive integer n, we use the notation $[n]$ for the ring $\mathbb{Z}/(n) = \{0, 1, \ldots, n-1\}$.

2.1 Unique Games Conjecture

Definition 1. *An instance of Unique Games represented as $\Gamma = (\mathcal{X} \cup \mathcal{Y}, E, \Pi, \langle R \rangle)$, consists of a bipartite graph over node sets $\mathcal{X}, \mathcal{Y}$ with the edges E between them. Also part of the instance is a set of labels $\langle R \rangle = \{1, \ldots, R\}$, and a set of permutations $\pi_{vw} : \langle R \rangle \to \langle R \rangle$ for each edge $e = (v, w) \in E$. An assignment A of labels to vertices is said to satisfy an edge $e = (v, w)$, if $\pi_{vw}(A(v)) = A(w)$. The objective is to find an assignment A of labels that satisfies the maximum number of edges.*

For sake of convenience, we shall use the following stronger version of Unique Games Conjecture which is equivalent to the original conjecture [14].

Conjecture 1. For all constants $\delta > 0$, there exists large enough constant R such that given a bipartite unique games instance $\Gamma = (\mathcal{X} \cup \mathcal{Y}, E, \Pi = \{\pi_e : \langle R \rangle \to \langle R \rangle : e \in E\}, \langle R \rangle)$ with number of labels R, it is NP-hard to distinguish between the following two cases:

- $(1 - \delta)$-satisfiable instances: There exists an assignment A of labels such that for $1 - \delta$ fraction of vertices $v \in \mathcal{X}$, all the edges (v, w) are satisfied.
- Instances that are not δ-satisfiable: No assignment satisfies more than a δ-fraction of the edges E.

2.2 Gowers Uniformity Norm and Influence of Variables

We now recall the definition of the Gowers uniformity norm. For an integer $d \geqslant 1$ and a complex-valued function $f : G \to \mathbb{C}$ defined on an abelian group G (whose group operation we denote by $+$), the d'th uniformity norm $U_d(f)$ is defined as

$$U^d(f) := \mathop{\mathbb{E}}_{x, y_1, y_2, \ldots, y_d} \left[\prod_{\substack{S \subseteq \{1,2,\ldots,d\} \\ |S| \text{ even}}} f\left(x + \sum_{i \in S} y_i\right) \prod_{\substack{S \subseteq \{1,2,\ldots,d\} \\ |S| \text{ odd}}} \overline{f\left(x + \sum_{i \in S} y_i\right)} \right].$$

(1)

where the expectation is taken over uniform and independent choices of $x, y_0, \ldots,$ y_{d-1} from the group G. Note that $U^1(f) = \left(\mathbb{E}_x[f(x)]\right)^2$.

We will be interested in the case when the group G is $[q]^R$ for positive integers q, R, with group addition being coordinate-wise addition modulo q. G is also closed under coordinate-wise multiplication modulo q by scalars in $[q]$, and thus has a $[q]$-module structure. For technical reasons, we will restrict attention to the case when q is prime and thus our groups will be vector spaces over the field $\mathbb{F}_q$ of q elements. For a vector $\mathbf{a} \in [q]^k$, we denote by $a_1, a_2, \ldots, a_k$ its k coordinates. We will use $\mathbf{1}, \mathbf{0}$ to denote the all 1's and all 0's vectors respectively (the dimension will be clear from the context). Further denote by $\mathbf{e}_i$ the ith basis vector with 1 in the ith coordinate and 0 in the remaining coordinates. As we shall mainly be interested in functions over $[q]^R$ for a prime q, we make our further definitions in this setting. Firstly, every function $f : [q]^R \to \mathbb{C}$ has a Fourier expansion given by $f(x) = \sum_{\alpha \in [q]^R} \hat{f}_\alpha \chi_\alpha(x)$ where $\hat{f}_\alpha = \mathbb{E}_{x \in [q]^R}[f(x)\chi_\alpha(x)]$ and $\chi_\alpha(x) = \prod_{i=1}^{R} \omega^{\alpha_i x_i}$ for a qth root of unity ω.

The central lemma in the hardness reduction relates a large Gowers norm for a function f, to the existence of an influential coordinate. Towards this, we define influence of a coordinate for a function over $[q]^R$.

Definition 2. *For a function $f : [q]^R \to \mathbb{C}$ define the influence of the ith coordinate as follows:*

$$\text{Inf}_i(f) = \mathbb{E}_x[\mathbf{Var}_{x_i}[f]] .$$

The following well known result relates influences to the Fourier spectrum of the function.

Fact 1. *For a function $f : [q]^R \to \mathbb{C}$ and a coordinate $i \in \{1, 2, \ldots, R\}$,*

$$\text{Inf}_i(f) = \sum_{\alpha_i \neq 0, \alpha \in [q]^R} |\hat{f}_\alpha|^2 .$$

The following lemma is a restatement of Theorem 12 in [19].

Lemma 1. *There exists an absolute constant C such that, if $f : [q]^m \to \mathbb{C}$ is a function satisfying $|f(x)| \leqslant 1$ for every x then for every $d \geqslant 1$,*

$$U^d(f) \leqslant U^1(f) + 2^{Cd} \max_i \text{Inf}_i(f)$$

2.3   Noise Operator

Like many other UGC hardness results, one of the crucial ingredients of our reduction will be a noise operator on functions over $[q]^R$. We define the noise operator $T_{1-\varepsilon}$ formally below.

Definition 3. *For $0 \leqslant \varepsilon \leqslant 1$, define the operator $T_{1-\varepsilon}$ on functions $f : [q]^R \to \mathbb{C}$ as:*

$$T_{1-\varepsilon}f(\mathbf{x}) = \mathbb{E}_\eta[f(\mathbf{x} + \eta)]$$

where each coordinate η_i of η is 0 with probability $1 - \varepsilon$ and a random element from $[q]$ with probability ε. The Fourier expansion of $T_{1-\varepsilon}f$ is given by

$$T_{1-\varepsilon}f(\mathbf{x}) = \sum_{\alpha \in [q]^R} (1 - \varepsilon)^{|\alpha|} \hat{f}_\alpha \chi_\alpha(x)$$

Here $|\alpha|$ denotes the number of non-zero coordinates of α. Due to space constraints, we defer the proof of the following lemma(see [6]).

Lemma 2. *If a function $f : [q]^R \to \mathbb{C}$ satisfies $|f(x)| \leqslant 1$ for all x, and $g = T_{1-\varepsilon}f$ then $\sum_{i=1}^R \mathrm{Inf}_i(g) \leqslant \frac{1}{2e \ln 1/(1-\varepsilon)}$*

3   Linearity Tests and MAX k-CSP Hardness

The best approximation ratio possible for MAX k-CSP is identical to the best soundness of a PCP verifier for NP that makes k queries. This follows easily by associating the proof locations to CSP variables, and the tests of the verifier to k-ary constraints on the locations. In this light, it is natural that the hardness results of [18,5,19] are all associated with a linearity test with a strong soundness. The hardness result in this work is obtained by extending the techniques of [19] from binary to q-ary domains. In this section, we describe the test of [19] and outline the extension to it.

For the sake of simplicity, let us consider the case when $k = 2^d - 1$ for some d. In [19], the authors propose the following linearity test for functions $F : \{0,1\}^n \to \{0,1\}$.

Complete Hypergraph Test (F, d)

  – Pick $x_1, x_2, \ldots, x_d \in \{0,1\}^n$ uniformly at random.
  – Accept if for each $S \subseteq [r]$, $F(\sum_{i \in S} x_i) = \sum_{i \in S} F(x_i)$.

The test reads the value of the function F at $k = 2^d - 1$ points of a random subspace(spanned by $x_1, \ldots, x_d$) and checks that F agrees with a linear function on the subspace. Note that a random function F would pass the test with probability $2^d/2^k$, since there are 2^d different satisfying assignments to the k binary values queried by the verifier. The following result is a special case of a more general result by Samorodnitsky and Trevisan [19].

Theorem 1. *[19] If a function $F : \{0,1\}^n \to \{0,1\}$ passes the Complete Hypergraph Test with probability greater than $2^d/2^k + \gamma$, then the function $f(x) = (-1)^{F(x)}$ has a large d^{th} Gowers norm. Formally, $U^d(f) \geqslant C(\gamma, k)$ for some fixed function C of γ, k.*

Towards extending the result to the domain $[q]$, we propose a different linearity test. Again for convenience, let us assume $k = q^d$ for some d. Given a function $F : [q]^n \to [q]$, the test proceeds as follows:

Affine Subspace Test (F, d)

- Pick $\mathbf{x}, \mathbf{y_1}, \mathbf{y_2}, \ldots, \mathbf{y_d} \in [q]^n$ uniformly at random.
- Accept if for each $\mathbf{a} \subseteq [q]^d$,

$$F\left(\mathbf{x} + \sum_{i=1}^{d} a_i \mathbf{y_i}\right) = \left(1 - \sum_{i=1}^{d} a_i\right) F(\mathbf{x}) + \sum_{i=1}^{d} a_i F\left(\mathbf{x} + \mathbf{y_i}\right)$$

Essentially, the test queries the values along a randomly chosen affine subspace, and tests if the function F agrees with an affine function on the subspace. Let ω denote a q'th root of unity. From Theorem 4 presented in Section 5, the following result can be shown:

Theorem 2. *If a function $F : [q]^n \to [q]$ passes the Affine Subspace Test with probability greater than $q^{d+1}/q^k + \gamma$, then for some q'th root of unity $\omega \neq 1$, the function $f(x) = \omega^{F(x)}$ has a large dq'th Gowers norm . Formally, $U^{dq}(f) \geqslant C(\gamma, k)$ for some fixed function C of γ, k.*

The above result follows easily from Theorem 4 using techniques of [19], and the proof is ommited here. The Affine Subspace Test forms the core of the UGC based hardness reduction presented in Section 4.

4   Hardness Reduction from Unique Games

In this section, we will prove a hardness result for approximating MAX k-CSP over a domain of size q when q is prime for every $k \geqslant 2$. Let d be such that $q^{d-1} + 1 \leqslant k \leqslant q^d$. Let us consider the elements of $[q]$ to have a natural order defined by $0 < 1 < \ldots < q - 1$. This extends to a lexicographic ordering on vectors in $[q]^d$. Denote by $[q]^d_{<k}$ the set consisting of the k lexicographically smallest vectors in $[q]^d$. We shall identify the set $\{1, \ldots, k\}$ with set of vectors in $[q]^d_{<k}$. Specifically, we shall use $\{1, \ldots, k\}$ and vectors in $[q]^d_{<k}$ interchangeably as indices to the same set of variables. For a vector $\mathbf{x} \in [q]^R$ and a permutation π of $\{1, \ldots, R\}$, define $\pi(x) \in [q]^R$ defined by $(\pi(x))_i = x_{\pi(i)}$.

Let $\Gamma = (\mathcal{X} \cup \mathcal{Y}, E, \Pi = \{\pi_e : \langle R \rangle \to \langle R \rangle | e \in E\}, \langle R \rangle)$ be a bipartite unique games instance. Towards constructing a k-CSP instance Λ from Γ, we shall introduce a long code for each vertex in $\mathcal{Y}$. Specifically, the set of variables for the k-CSP Λ is indexed by $\mathcal{Y} \times [q]^R$. Thus a solution to Λ consists of a set of functions $F_w : [q]^R \to [q]$, one for each $w \in \mathcal{Y}$.

Similar to several other long code based hardness results, we shall assume that the long codes are *folded*. More precisely, we shall use *folding* to force the functions F_w to satisfy $F_w(\mathbf{x} + \mathbf{1}) = F(\mathbf{x}) + 1$ for all $\mathbf{x} \in [q]^R$. The k-ary constraints in the instance Λ are specified by the following verifier. The verifier uses an additional parameter ε that governs the level of noise in the noise operator.

- Pick a random vertex $v \in \mathcal{X}$. Pick k vertices $\{w_{\mathbf{a}} | \mathbf{a} \in [q]^d_{<k}\}$ from $N(v) \subset \mathcal{Y}$ uniformly at random independently. Let $\pi_{\mathbf{a}}$ denote the permutation on the edge $(v, w_{\mathbf{a}})$.
- Sample $\mathbf{x}, \mathbf{y_1}, \mathbf{y_2}, \dots, \mathbf{y_d} \in [q]^R$ uniformly at random. Sample vectors $\eta_{\mathbf{a}} \in [q]^R$ for each $\mathbf{a} \in [q]^d_{<k}$ from the following distribution: With probability $1 - \varepsilon$, $(\eta_{\mathbf{a}})_j = 0$ and with the remaining probability, $(\eta_{\mathbf{a}})_j$ is a uniformly random element from $[q]$.
- Query $F_{w_{\mathbf{a}}} \left(\pi_{\mathbf{a}}(\mathbf{x} + \sum_j a_j \mathbf{y_j} + \eta_{\mathbf{a}}) \right)$ for each $\mathbf{a} \in [q]^d_{<k}$. Accept if the following equality holds for each $\mathbf{a} \in [q]^d_{<k}$.

$$F_{w_{\mathbf{a}}} \left(\pi_{\mathbf{a}}(\mathbf{x} + \sum_{j=1}^{d} a_j \mathbf{y_j} + \eta_{\mathbf{a}}) \right) = \left(1 - \sum_{j=1}^{d} a_j \right) F_{w_0} \left(\pi_0(\mathbf{x} + \eta_0) \right)$$

$$+ \sum_{j=1}^{d} a_j F_{w_{\mathbf{e_j}}} \left(\pi_{\mathbf{e_j}}(\mathbf{x} + \mathbf{y_j} + \eta_{\mathbf{e_j}}) \right)$$

Theorem 3. *For all primes q, positive integers d, k satisfying $q^{d-1} < k \leqslant q^d$, and every $\gamma > 0$, there exists small enough $\delta, \varepsilon > 0$ such that*

- COMPLETENESS: *If Γ is a $(1-\delta)$-satisfiable instance of Unique Games, then there is an assignment to Λ that satisfies the verifier's tests with probability at least $(1 - \gamma)$*
- SOUNDNESS: *If Γ is not δ-satisfiable, then no assignment to Λ satisfies the verifier's tests with probability more than $\frac{q^{d+1}}{q^k} + \gamma$.*

Proof. We begin with the completeness claim, which is straightforward.

Completeness. There exists labelings to the Unique Game instance Γ such that for $1 - \delta$ fraction of the vertices $v \in \mathcal{X}$ all the edges (v, w) are satisfied. Let $A : \mathcal{X} \cup \mathcal{Y} \to \langle R \rangle$ denote one such labelling. Define an assignment to the k-CSP instance by $F_w(\mathbf{x}) = x_{A(w)}$ for all $w \in \mathcal{Y}$.

With probability at least $(1 - \delta)$, the verifier picks a vertex $v \in \mathcal{X}$ such that the assignment A satisfies all the edges $(v, w_{\mathbf{a}})$. In this case for each $\mathbf{a}$, $\pi_{\mathbf{a}}(A(v)) = A(w_{\mathbf{a}})$. Let us denote $A(v) = l$. By definition of the functions F_w, we get $F_{w_{\mathbf{a}}}(\pi_{\mathbf{a}}(x)) = (\pi_{\mathbf{a}}(x))_{A(w_{\mathbf{a}})} = x_{\pi_{\mathbf{a}}^{-1}(A(w_{\mathbf{a}}))} = x_l$ for all $x \in [q]^R$. With probability at least $(1 - \varepsilon)^k$, each of the vectors $\eta_{\mathbf{a}}$ have their l^{th} component equal to zero, i.e $(\eta_{\mathbf{a}})_l = 0$. In this case, it is easy to check that all the constraints are satisfied. In conclusion, the verifier accepts the assignment with probability at least $(1 - \delta)(1 - \varepsilon)^k$. For small enough δ, ε, this quantity is at least $(1 - \gamma)$.

Soundness. Suppose there is an assignment given by functions F_w for $w \in \mathcal{Y}$ that the verifier accepts with probability greater than $\frac{q^{d+1}}{q^k} + \gamma$.

Let $z_1, z_2, \dots, z_k$ be random variables denoting the k values read by the verifier. Thus $z_1, \dots, z_k$ take values in $[q]$. Let $P : [q]^k \to \{0, 1\}$ denote the predicate on k variables that represents the acceptance criterion of the verifier. Essentially, the value of the predicate $P(z_1, \ldots, z_k)$ is 1 if and only if $z_1, \ldots, z_k$ values are consistent with some affine function. By definition,

$$\Pr[\text{ Verifier Accepts }] = \mathop{\mathbb{E}}_{v \in \mathcal{X}} \mathop{\mathbb{E}}_{w_\mathbf{a} \in N(v)} \mathop{\mathbb{E}}_{\mathbf{x}, \mathbf{y_1}, \ldots, \mathbf{y_d}} \mathop{\mathbb{E}}_{\eta_\mathbf{a}} \left[P(z_1, \ldots, z_k) \right] \geqslant \frac{q^{d+1}}{q^k} + \gamma$$

Let ω denote a qth root of unity. The Fourier expansion of the function $P : [q]^k \to \mathbb{C}$ is given by $P(z_1, \ldots, z_k) = \sum_{\alpha \in [q]^k} \hat{P}_\alpha \chi_\alpha(z_1, \ldots, z_k)$ where $\chi_\alpha(z_1, \ldots, z_k) = \prod_{i=1}^k \omega^{\alpha_i z_i}$ and $\hat{P}_\alpha = \mathop{\mathbb{E}}_{z_1, \ldots, z_k} [P(z_1, \ldots, z_k) \chi_\alpha(z_1, \ldots, z_k)]$. Notice that for $\alpha = \mathbf{0}$, we get $\chi_\alpha(z_1, \ldots, z_k) = 1$. Further,

$$\hat{P}_\mathbf{0} = \Pr[\text{ random assignment to } z_1, z_2, \ldots, z_k \text{ satisfies } P] = \frac{q^{d+1}}{q^k}$$

Substituting the Fourier expansion of P, we get

$$\Pr[\text{ Verifier Accepts }] = \frac{q^{d+1}}{q^k} + \sum_{\alpha \neq 0} \hat{P}_\alpha \mathop{\mathbb{E}}_{v \in \mathcal{X}} \mathop{\mathbb{E}}_{w_\mathbf{a} \in N(v)} \mathop{\mathbb{E}}_{\mathbf{x}, \mathbf{y_1}, \ldots, \mathbf{y_d}} \mathop{\mathbb{E}}_{\eta_\mathbf{a}} \left[\chi_\alpha(z_1, \ldots, z_k) \right]$$

Recall that the probability of acceptance is greater than $\frac{q^{d+1}}{q^k} + \gamma$. Further $|\hat{P}_\alpha| \leqslant 1$ for all $\alpha \in [q]^k$. Thus there exists $\alpha \neq 0$ such that,

$$\left| \mathop{\mathbb{E}}_{v \in \mathcal{X}} \mathop{\mathbb{E}}_{w_\mathbf{a} \in N(v)} \mathop{\mathbb{E}}_{\mathbf{x}, \mathbf{y_1}, \ldots, \mathbf{y_d}} \mathop{\mathbb{E}}_{\eta_\mathbf{a}} \left[\chi_\alpha(z_1, \ldots, z_k) \right] \right| \geqslant \frac{\gamma}{q^k}$$

For each $w \in \mathcal{Y}, t \in [q]$, define the function $f_w^{(t)} : [q]^d \to \mathbb{C}$ as $f_w^{(t)}(x) = \omega^{t F_w(x)}$. For convenience we shall index the vector α with the set $[q]_{<k}^d$ instead of $\{1, \ldots, k\}$. In this notation,

$$\left| \mathop{\mathbb{E}}_{v \in \mathcal{X}} \mathop{\mathbb{E}}_{w_\mathbf{a} \in N(v)} \mathop{\mathbb{E}}_{\mathbf{x}, \mathbf{y_1}, \ldots, \mathbf{y_d}} \mathop{\mathbb{E}}_{\eta_\mathbf{a}} \left[\prod_{\mathbf{a} \in [q]_{<k}^d} f_{w_\mathbf{a}}^{(\alpha_\mathbf{a})} \left(\pi_\mathbf{a} (\mathbf{x} + \sum_{i=1}^d a_i \mathbf{y_i} + \eta_\mathbf{a}) \right) \right] \right| \geqslant \frac{\gamma}{q^k}$$

Let $g_w^{(t)} : [q]^d \to \mathbb{C}$ denote the *smoothened* version of function $f_w^{(t)}$. Specifically, let $g_w^{(t)}(x) = T_{1-\varepsilon} f_w^{(t)}(x) = \mathbb{E}_\eta [f_w^{(t)}(x + \eta)]$ where η is generated from ε-noise distribution. Since each $\eta_\mathbf{a}$ is independently chosen, we can rewrite the above expression,

$$\left| \mathop{\mathbb{E}}_{v \in \mathcal{X}} \mathop{\mathbb{E}}_{w_\mathbf{a} \in N(v)} \mathop{\mathbb{E}}_{\mathbf{x}, \mathbf{y_1}, \ldots, \mathbf{y_d}} \left[\prod_{\mathbf{a} \in [q]_{<k}^d} g_{w_\mathbf{a}}^{(\alpha_\mathbf{a})} \left(\pi_\mathbf{a} (\mathbf{x} + \sum_{i=1}^d a_i \mathbf{y_i}) \right) \right] \right| \geqslant \frac{\gamma}{q^k} .$$

For each $v \in \mathcal{X}, t \in [q]$, define the function $g_v^{(t)} : [q]^d \to \mathbb{C}$ as $g_v^{(t)}(x) = \mathbb{E}_{w \in N(v)} [g_w^{(t)}(\pi_{vw}(x))]$. As the vertices $w_\mathbf{a}$ are chosen independent of each other,

$$\left| \mathop{\mathbb{E}}_{v \in \mathcal{X}} \mathop{\mathbb{E}}_{\mathbf{x}, \mathbf{y_1}, \ldots, \mathbf{y_d}} \left[\prod_{\mathbf{a} \in [q]_{<k}^d} g_v^{(\alpha_\mathbf{a})} \left(\mathbf{x} + \sum_{i=1}^d a_i \mathbf{y_i} \right) \right] \right| \geqslant \frac{\gamma}{q^k} .$$

As $\alpha \neq 0$, there exists an index $\mathbf{b} \in [q]^d_{<k}$ such that $\alpha_{\mathbf{b}} \neq 0$. For convenience let us denote $c = \alpha_{\mathbf{b}}$. Define $\kappa = 2^{-Cdq}\left(\frac{\gamma}{2q^k}\right)^{2^{dq}}$ where C is the absolute constant defined in Lemma 1.

For each $v \in \mathcal{X}$, define the set of labels $L(v) = \{i \in \langle R \rangle \; : \; \mathrm{Inf}_i(g^c_v) \geqslant \kappa\}$. Similarly for each $w \in \mathcal{Y}$, let $L(w) = \{i \in \langle R \rangle \; : \; \mathrm{Inf}_i(g^c_w) \geqslant \kappa/2\}$. Obtain a labelling A to the Unique Games instance Γ as follows : For each vertex $u \in \mathcal{X} \cup \mathcal{Y}$, if $L(u) \neq \phi$ then assign a randomly chosen label from $L(u)$, else assign a uniformly random label from $\langle R \rangle$.

The functions $g^{(c)}_w$ are given by $g^{(c)}_w = T_{1-\varepsilon} f^{(c)}_w$ where $f^{(c)}_w$ is bounded in absolute value by 1. By Lemma 2, therefore, the sum of its influences is bounded by $\frac{1}{e \ln 1/(1-\varepsilon)}$. Consequently, for all $w \in \mathcal{Y}$ the size of the label set $L(w)$ is bounded by $\frac{2}{\kappa e \ln 1/(1-\varepsilon)}$. Applying a similar argument to $v \in \mathcal{X}$, $|L(v)| \leqslant \frac{1}{\kappa e \ln 1/(1-\varepsilon)}$.

For at least $\gamma/2q^k$ fraction of vertices $v \in \mathcal{X}$ we have,

$$\left| \mathop{\mathbb{E}}_{\mathbf{x}, \mathbf{y}_1, \ldots, \mathbf{y_d}} \left[\prod_{\mathbf{a} \in [q]^d_{<k}} g^{(\alpha_{\mathbf{a}})}_v \Big(\mathbf{x} + \sum_{i=1}^d a_i \mathbf{y_i}\Big) \right] \right| \geqslant \frac{\gamma}{2q^k}$$

We shall refer to these vertices as *good* vertices. Fix a *good* vertex v.

Observe that for each $u \in \mathcal{X} \cup \mathcal{Y}$ the functions $g^{(t)}_u$ satisfy $|g^{(t)}_u(x)| \leqslant 1$ for all x. Now we shall apply Theorem 4 to conclude that the functions $g^{(t)}_v$ have a large Gowers norm. Specifically, consider the collection of functions given by $f_{\mathbf{a}} = g^{(\alpha_{\mathbf{a}})}_v$ for $\mathbf{a} \in [q]^d_{<k}$, and $f_{\mathbf{a}} = 1$ for all $\mathbf{a} \notin [q]^d_{<k}$. From Theorem 4, we get

$$\min_{\mathbf{a}} U^{dq}(g^{(\alpha_{\mathbf{a}})}_v) \geqslant \left(\frac{\gamma}{2q^k}\right)^{2^{dq}}.$$

In particular, this implies $U^{dq}(g^{(c)}_v) \geqslant \left(\frac{\gamma}{2q^k}\right)^{2^{dq}}$. Now we shall use Lemma 1 to conclude that the function g_v has influential coordinates. Towards this, observe that the functions $f^{(t)}_w$ satisfy $f^{(t)}_w(x+1) = f^{(t)}_w(x) \cdot \omega^t$ due to folding. Thus for all $t \neq 0$ and all $w \in \mathcal{Y}$, $\mathbb{E}_x[f^{(t)}_w(x)] = 0$. Specifically for $c \neq 0$,

$$U^1(g^{(c)}_v) = \left(\mathbb{E}_x[g^{(c)}_v(x)]\right)^2 = \left(\mathop{\mathbb{E}}_{w \in N(v)} \mathop{\mathbb{E}}_\eta \mathbb{E}_x[f^{(c)}_w(x+\eta)]\right)^2 = 0$$

Hence it follows from Lemma 1 that there exists influential coordinates i with $\mathrm{Inf}_i(g^{(c)}_v) \geqslant 2^{-Cdq}\left(\frac{\gamma}{2q^k}\right)^{2^{dq}} = \kappa$. In other words, $L(v)$ is non-empty. Observe that, due to convexity of influences,

$$\mathrm{Inf}_i(g^{(c)}_v) = \mathrm{Inf}_i(\mathop{\mathbb{E}}_{w \in N(v)}[g^{(c)}_w]) \leqslant \mathop{\mathbb{E}}_{w \in N(v)} \mathrm{Inf}_{\pi_{vw}(i)}([g^{(c)}_w(x)]) \, .$$

If the coordinate i has influence at least κ on $g^{(c)}_v$, then the coordinate $\pi_{vw}(i)$ has an influence of at least $\kappa/2$ for at least $\kappa/2$ fraction of neighbors $w \in N(v)$. The edge π_{vw} is satisfied if i is assigned to v, and $\pi_{wv}(i)$ is assigned to w. This event happens with probability at least $\frac{1}{|L(u)||L(v)|} \geqslant (e\kappa \ln 1/(1-\varepsilon))^2/2$ for at least $\kappa/2$ fraction of the neighbors $w \in N(v)$. As there are at least $(\gamma/2q^k)$ fraction of *good* vertices v, the assignment satisfies at least $(\gamma/2q^k)(e\kappa \ln 1/(1-\varepsilon))^2\kappa/4$ fraction of the unique games constraints. By choosing δ smaller than this fraction, the proof is complete.

Since each test performed by the verifier involve k variables, by the standard connection between hardness of MAX k-CSP and k-query PCP verifiers, we get the following hardness result conditioned on the UGC.

Corollary 1. *Assuming the Unique Games conjecture, for every prime q, it is NP-hard to approximate* MAX k-CSP *over domain size q within a factor that is greater than q^2k/q^k.*

Using the reduction of [15], the above UG hardness result can be extended from primes to arbitrary composite number q.

Corollary 2. *[15] Assuming the Unique Games conjecture, for every positive integer q, it is NP-hard to approximate* MAX k-CSP *over domain size q within a factor that is greater than $q^2k(1 + o(1))/q^k$.*

5   Gowers Norm and Multidimensional Arithmetic Progressions

The following theorem forms a crucial ingredient in the soundness analysis in the proof of Theorem 3.

Theorem 4. *Let $q \geqslant 2$ be a prime and G be a $\mathbb{F}_q$-vector space. Then for all positive integers $\ell \leqslant q$ and d, and all collections $\{f_{\mathbf{a}} : G \to \mathbb{C}\}_{\mathbf{a} \in [\ell]^d}$ of ℓ^d functions satisfying $|f_{\mathbf{a}}(x)| \leqslant 1$ for every $x \in G$ and $\mathbf{a} \in [\ell]^d$, the following holds:*

$$\left| \mathop{\mathbb{E}}_{x,y_1,y_2,\dots,y_d} \left[\prod_{\mathbf{a} \in [\ell]^d} f_{\mathbf{a}}(x + a_1 y_1 + a_2 y_2 + \cdots + a_d y_d) \right] \right| \leqslant \min_{\mathbf{a} \in [\ell]^d} \left(U^{d\ell}(f_{\mathbf{a}}) \right)^{1/2^{d\ell}}$$

$$(2)$$

The proof of the above theorem is via double induction on d, ℓ. We first prove the theorem for the one-dimensional case, i.e., $d = 1$ and every $\ell, 1 \leqslant \ell < q$ (Lemma 3). This will be done through induction on ℓ. We will then prove the result for arbitrary d by induction on d.

Remark 1. Green and Tao, in their work [8] on configurations in the primes, isolate and define a property of a system of linear forms that ensures that the degree t Gowers norm is sufficient to analyze patterns corresponding to those linear forms, and called this property *complexity* (see Definition 1.5 in [8]). Gowers and Wolf [7] later coined the term Cauchy-Schwartz (CS) complexity to refer to this notion of complexity. For example, the CS-complexity of the q linear forms $x, x+y, x+2y, \ldots, x+(q-1)y$ corresponding to a q-term arithmetic progression equals $q - 2$, and the U^{q-1} norm suffices to analyze them. It can similarly be shown that the CS-complexity of the d-dimensional arithmetic progression (with q^d linear forms as in (2)) is at most $d(q-1) - 1$. In our application, we need a "multi-function" version of these statements, since we have a different function $f_{\mathbf{a}}$ for each linear form $x + \mathbf{a} \cdot \mathbf{y}$. We therefore work out a self-contained proof of Theorem 4 in this setting.

Towards proving Theorem 4, we will need the following lemma whose proof is presented in the full version[6].

Lemma 3. *Let $q \geqslant 2$ be prime and ℓ, $1 \leqslant \ell \leqslant q$, be an integer, and G be a $\mathbb{F}_q$-vector space. Let $\{h_\alpha : G \to \mathbb{C}\}_{\alpha \in [\ell]}$ be a collection of ℓ functions such that $|h_\alpha(x)| \leqslant 1$ for all $\alpha \in [\ell]$ and $x \in G$. Then*

$$\left| \mathop{\mathbb{E}}_{x, y_1} \left[\prod_{\alpha \in [\ell]} h_\alpha(x + \alpha y_1) \right] \right| \leqslant \min_{\alpha \in [\ell]} \left(U^\ell(h_\alpha) \right)^{\frac{1}{2^\ell}} . \tag{3}$$

Proof of Theorem 4: Fix an arbitrary ℓ, $1 \leqslant \ell \leqslant q$. We will prove the result by induction on d. The base case $d = 1$ is the content of Lemma 3, so it remains to consider the case $d > 1$.

By a change of variables, it suffices to upper bound the LHS of (2) by $\left(U^{d\ell}(f_{(\ell-1)\mathbf{1}}) \right)^{1/2^{d\ell}}$, and this is what we will prove. For $\alpha \in [\ell]$, and $y_2, y_3, \ldots,$ $y_d \in G$, define the function

$$g_\alpha^{y_2, \ldots, y_d}(x) = \prod_{\mathbf{b} = (b_2, b_3, \ldots, b_d) \in [\ell]^{d-1}} f_{(\alpha, \mathbf{b})}(x + b_2 y_2 + \cdots + b_d y_d) . \tag{4}$$

The LHS of (2), raised to the power $2^{d\ell}$, equals

$$\left| \mathop{\mathbb{E}}_{y_2, \ldots, y_d} \mathop{\mathbb{E}}_{x, y_1} \left[\prod_{\alpha \in [\ell]} g_\alpha^{y_2, \ldots, y_d}(x + \alpha y_1) \right] \right|^{2^{d\ell}} \leqslant \left(\mathop{\mathbb{E}}_{y_2, \ldots, y_d} \left| \mathop{\mathbb{E}}_{x, y_1} \prod_{\alpha \in [\ell]} g_\alpha^{y_2, \ldots, y_d}(x + \alpha y_1) \right|^{2^\ell} \right)^{2^{(d-1)\ell}}$$

$$\leqslant \left| \mathop{\mathbb{E}}_{y_2, \ldots, y_d} U^\ell(g_{\ell-1}^{y_2, \ldots, y_d}) \right|^{2^{(d-1)\ell}} \qquad \text{(using Lemma 3)}$$

$$= \left| \mathop{\mathbb{E}}_{y_2, \ldots, y_d} \mathop{\mathbb{E}}_{x, z_1, \ldots, z_\ell} \left[\prod_{S \subseteq \{1, 2, \ldots, \ell\}} g_{\ell-1}^{y_2, \ldots, y_d}\left(x + \sum_{i \in S} z_i\right) \right] \right|^{2^{(d-1)\ell}}$$

Defining the function

$$H_{\mathbf{b}}^{z_1, \ldots, z_\ell}(t) := \prod_{S \subseteq \{1, 2, \ldots, \ell\}} f_{(\ell-1, \mathbf{b})}\left(t + \sum_{i \in S} z_i\right) \tag{5}$$

for every $\mathbf{b} \in [\ell]^{d-1}$ and $z_1, \ldots, z_\ell \in G$, the last expression equals

$$\left| \underset{z_1,\ldots,z_\ell}{\mathbb{E}} \underset{x,y_2,\ldots,y_d}{\mathbb{E}} \left[\prod_{\mathbf{b}=(b_2,\ldots,b_d)\in[\ell]^{d-1}} H_{\mathbf{b}}^{z_1,\ldots,z_\ell} \left(x + b_2 y_2 + \cdots + b_d y_d \right) \right] \right|^{2^{(d-1)\ell}}$$

which is at most

$$\underset{z_1,\ldots,z_\ell}{\mathbb{E}} \left[\left| \underset{x,y_2,\ldots,y_d}{\mathbb{E}} \left[\prod_{\mathbf{b}=(b_2,\ldots,b_d)\in[\ell]^{d-1}} H_{\mathbf{b}}^{z_1,\ldots,z_\ell} \left(x + b_2 y_2 + \cdots + b_d y_d \right) \right] \right|^{2^{(d-1)\ell}} \right].$$

$$(6)$$

By the induction hypothesis, (6) is at most $\underset{z_1,\ldots,z_\ell}{\mathbb{E}} \left[U^{(d-1)\ell} \left(H_{(\ell-1)\mathbf{1}}^{z_1,\ldots,z_\ell} \right) \right]$. Recalling the definition of $H_{\mathbf{b}}^{z_1,\ldots,z_\ell}$ from (5), the above expectation equals

$$\underset{z_1,\ldots,z_\ell}{\mathbb{E}} \underset{\substack{x,\{z_j'\} \\ 1\leqslant j\leqslant (d-1)\ell}}{\mathbb{E}} \left[\prod_{\substack{S\subseteq\{1,2,\ldots,\ell\} \\ T\subseteq\{1,2,\ldots,(d-1)\ell\}}} f_{(\ell-1)\mathbf{1}} \left(x + \sum_{i\in S} z_i + \sum_{j\in T} z_j' \right) \right]$$

which clearly equals $U^{d\ell}(f_{(\ell-1)\mathbf{1}})$.

6 Approximation Algorithm for MAX k-CSP

On the algorithmic side, we show the following result:

Theorem 5. *There is a polynomial time algorithm that computes a* $\frac{1}{2\pi eq(q-1)^6} \cdot \frac{k}{q^k}$ *factor approximation for the* MAX k-CSP *problem over a domain of size* q.

The algorithm proceeds along the lines of [3], by formulating MAX k-CSP as a quadratic program, solving a SDP relaxation and rounding the resulting solution. The variables in the quadratic program are constrained to the vertices of the q-dimensional simplex. Hence, as a subroutine, we obtain an efficient procedure to optimize positive definite quadratic forms with the variables forced to take values on the q-dimensional simplex. Let Δ_q denote the q-dimensional simplex, and let $\mathsf{Vert}(\Delta_q)$ denote the vertices of the simplex. Formally,

Theorem 6. *Let* $A = (a_{ij}^{(k)(l)})$ *be a positive definite matrix where* $k, l \in [q]$ *and* $1 \leqslant i, j \leqslant n$. *For the quadratic program* Γ, *there exists an efficient algorithm that finds an assignment whose value is at least* $\frac{2}{\pi(q-1)^4}$ *of the optimum.*

QuadraticProgram Γ

    Maximize $\qquad \sum_{ij} a_{ij}^{(k)(l)} x_i^{(k)} \cdot x_j^{(l)}$

    Subject to $\qquad \mathbf{x}_i = (x_i^{(0)}, x_i^{(1)}, \ldots, x_i^{(q-1)}) \in \mathsf{Vert}(\Delta_q) \; 1 \leqslant i \leqslant n$

The details of the algorithm are presented in the full version[6]. It has been pointed out to us that a $\Omega(q^2 k/q^k)$-approximation for MAX k-CSP can be obtained by reducing from the non-boolean to the boolean case [15].

References

1. Austrin, P.: Towards sharp inapproximability for any 2-csp. In: FOCS: IEEE Symposium on Foundations of Computer Science, pp. 307–317 (2007)
2. Austrin, P., Mossel, E.: Approximation resistant predicates from pairwise independence. Electronic Colloqium on Computational Complexity, TR08-009 (2008)
3. Charikar, M., Makarychev, K., Makarychev, Y.: Near-optimal algorithms for maximum constraint satisfaction problems. In: Proceedings of the 18th Annual ACM-SIAM Symposium on Discrete Algorithms, pp. 62–68 (2007)
4. Dinur, I., Guruswami, V., Khot, S., Regev, O.: A new multilayered PCP and the hardness of hypergraph vertex cover. SIAM J. Computing 34(5), 1129–1146 (2005)
5. Engebretsen, L., Holmerin, J.: More efficient queries in PCPs for NP and improved approximation hardness of maximum CSP. In: Diekert, V., Durand, B. (eds.) STACS 2005. LNCS, vol. 3404, pp. 194–205. Springer, Heidelberg (2005)
6. Guruswami, V., Raghavendra, P.: Constraint Satisfaction over the Non-Boolean Domain: Approximation algorithms and Unique Games hardness. ECCC: Electronic Colloqium on Computational Complexity, TR08-008 (2008)
7. Gowers, W.T., Wolf, J.: The true complexity of a system of linear equations. arXiv:math.NT/0711.0185 (2007)
8. Green, B., Tao, T.: Linear equations in primes. arXiv:math.NT/0606088v1 (2006)
9. Hast, G.: Approximating MAX kCSP – outperforming a random assignment with almost a linear factor. In: ICALP: Annual International Colloquium on Automata, Languages and Programming (2005)
10. Hast, G.: Beating a random assignment - approximating constraint satisfaction problems. Phd Thesis, Royal Institute of Technology (2005)
11. Håstad, J.: Some optimal inapproximability results. Journal of the ACM 48(4), 798–859 (2001)
12. Håstad, J.: On the approximation resistance of a random predicate. In: APPROX-RANDOM, pp. 149–163 (2007)
13. Håstad, J., Wigderson, A.: Simple analysis of graph tests for linearity and PCP. Random Struct. Algorithms 22(2), 139–160 (2003)
14. Khot, S., Regev, O.: Vertex cover might be hard to approximate to within /spl epsi/. In: Annual IEEE Conference on Computational Complexity (formerly Annual Conference on Structure in Complexity Theory), vol. 18 (2003)
15. Makarychev, K., Makarychev, Y.: Personal communication (2008)
16. Nesterov, Y.: Quality of semidefinite relaxation for nonconvex quadratic optimization. CORE Discussion Paper 9719 (1997)
17. Raghavendra, P.: Optimal algorithm and inapproximability results for every csp? In: STOC: ACM Symposium on Theory of Computing, pp. 245–254 (2008)
18. Samorodnitsky, A., Trevisan, L.: A PCP characterization of NP with optimal amortized query complexity. In: STOC: ACM Symposium on Theory of Computing, pp. 191–199 (2000)
19. Samorodnitsky, A., Trevisan, L.: Gowers uniformity, influence of variables, and PCPs. In: STOC: ACM Symposium on Theory of Computing (2006)
20. Trevisan, L.: Parallel approximation algorithms by positive linear programming. Algorithmica 21(1), 72–88 (1998)
21. Zwick, U.: Approximation algorithms for constraint satisfaction problems involving at most three variables per constraint. In: Proceedings of the Ninth Annual ACM-SIAM Symposium on Discrete Algorithms, pp. 201–210 (1998)

Fully Polynomial Time Approximation Schemes for Time-Cost Tradeoff Problems in Series-Parallel Project Networks

Nir Halman[1], Chung-Lun Li[2], and David Simchi-Levi[3]

[1] Institute for Advanced Study, Princeton, NJ, and Massachusetts Institute of Technology, Cambridge, MA
halman@mit.edu
[2] The Hong Kong Polytechnic University, Hung Hom, Kowloon, Hong Kong
lgtclli@polyu.edu.hk
[3] Massachusetts Institute of Technology, Cambridge, MA
dslevi@mit.edu

Abstract. We consider the deadline problem and budget problem of the nonlinear time-cost tradeoff project scheduling model in a series-parallel activity network. We develop fully polynomial time approximation schemes for both problems using K-approximation sets and functions, together with series and parallel reductions.

Keywords: Project management, time-cost tradeoff, approximation algorithms.

1 Introduction

Project scheduling with time-cost tradeoff decisions plays a significant role in project management. In particular, discrete time-cost tradeoff models with deadline or budget constraints are important tools for project managers to perform time planning and budgeting for their projects. As a result, efficient and effective solution procedures for such models are highly attractive to those practitioners. Unfortunately, these models are computationally intractable, and constructing near-optimal polynomial-time heuristics for them is highly challenging. In this paper, we develop fully polynomial time approximation schemes (FPTASs) for an important class of time-cost tradeoff problems in which the underlying project network is series-parallel (see Section 4 for a discussion of how our results can be applied to problems with "near-series-parallel" networks).

Time-cost tradeoff problems in series-parallel networks have applications not only in project management. Rothfarb *et al.* [11] and Frank *et al.* [5] have applied the time-cost tradeoff model to natural-gas pipeline system design and centralized computer network design, respectively. In their applications, the underlying network is a tree network, which is a special kind of series-parallel network, and they proposed an (exponential time) enumeration method for their problems.

Consider the following time-cost tradeoff model for project scheduling: There is a (directed acyclic) project network of n activities in activity-on-arc representation. Associated with each activity i are two nonincreasing functions

A. Goel et al. (Eds.): APPROX and RANDOM 2008, LNCS 5171, pp. 91–103, 2008.
© Springer-Verlag Berlin Heidelberg 2008

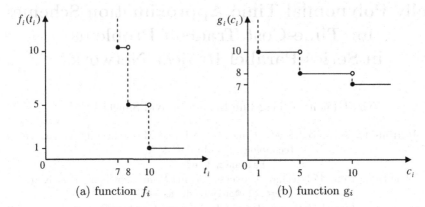

(a) function f_i (b) function g_i

Fig. 1. An activity time-cost tradeoff example

$f_i : T_i \rightarrow Z^+$ and $g_i : C_i \rightarrow Z^+$, where $f_i(t_i)$ is the cost incurred when the activity time is t_i, $g_i(c_i)$ is the activity time when an amount c_i of monetary resource is spent on the activity, $T_i = \{\underline{t}_i, \underline{t}_i+1, \ldots, \bar{t}_i\} \subset Z^+$ is the set of all possible time duration of activity i, $C_i = \{\underline{c}_i, \underline{c}_i+1, \ldots, \bar{c}_i\} \subset Z^+$ is the set of all possible cost consumption of activity i, and Z^+ is the set of all nonnegative integers. For example, if f_i is the function depicted in Figure 1(a), then g_i is the function depicted in Figure 1(b). Here, we assume that all activity times and costs are integer-valued.

Denote the activities as $1, 2, \ldots, n$. Let $\phi(t_1, t_2, \ldots, t_n)$ denote the total duration of the project (i.e., the length of the longest path in the network) when the time duration of activity i is t_i for $i = 1, 2, \ldots, n$. We are interested in two different variants of the problem: (i) given a deadline d, determine $t_1, t_2, \ldots, t_n$ so that $\phi(t_1, t_2, \ldots, t_n) \leq d$ and that $f_1(t_1) + f_2(t_2) + \cdots + f_n(t_n)$ is minimized, and (ii) given a budget b, determine $c_1, c_2, \ldots, c_n$ so that $c_1 + c_2 + \cdots + c_n \leq b$ and that $\phi(g_1(c_1), g_2(c_2), \ldots, g_n(c_n))$ is minimized. We refer to the first problem as the *deadline problem* and the second problem as the *budget problem*. In the deadline problem, we assume, for simplicity, that for each activity i, function f_i can be evaluated in constant time (i.e., for any given $t \in T_i$, $f_i(t)$ can be determined in constant time). In the budget problem, we assume, for simplicity, that for each activity i, function g_i can be evaluated in constant time. However, our FPTASs remain valid as long as f_i and g_i can be evaluated in an amount of time which is polynomial in the input size of the problems.

Note that in our model the time-cost tradeoff function of an activity can be any nonincreasing function (with nonnegative integer domain and range). Thus, our model is a generalization of the traditional "discrete" time-cost tradeoff model, which is defined in such a way that every activity i has $m(i)$ alternatives, of which alternative j requires $t(i, j) \in Z^+$ time units and $c(i, j) \in Z^+$ cost units ($j = 1, 2, \ldots, m(i)$). De et al. [3] have shown that both the deadline problem and the budget problem are NP-hard in the strong sense for the discrete time-cost tradeoff model when the underlying project network is a general directed acyclic network. This implies that both the deadline and budget problems of our model are strongly NP-hard as well. Thus, it is unlikely that there exists an FPTAS for either problem. In fact, developing polynomial-time approximation algorithms for the discrete time-cost tradeoff model is a challenging task. Skutella [12] has developed a polynomial-time algorithm for the budget problem with performance guarantee $O(\log l)$, where l is the ratio of the maximum duration and minimum nonzero duration of any activity. However, as pointed out by Deineko and Woeginger [4], unless P=NP, the budget problem does not have a polynomial-time approximation algorithm with performance guarantee strictly less than $\frac{3}{2}$.

When the underlying network is series-parallel, the deadline problem and the budget problem become "more tractable." Although the deadline and budget problems in a series-parallel project network remain NP-hard in the ordinary sense [3], they can be solved in pseudo-polynomial time by dynamic programming [3,9]. However, to the best of our knowledge, no known polynomial-time approximation scheme has been developed for these problems. Note that a series-parallel network can be reduced to a single-arc network efficiently via a sequence of simple series and parallel reduction operations [13]. In what follows, we will make use of series and parallel reductions, together with the K-approximation sets and functions introduced by Halman et al. [7], to develop FPTASs for the deadline and budget problems in series-parallel networks.

To simplify the discussion, we only consider the case where the problem is feasible. Note that it is easy to detect feasibility of the problem. The budget problem is feasible if and only if $\sum_i g_i(\bar{c}_i) \leq b$. The feasibility of the deadline problem can be detected by setting all activity times to their lower limits, solving the problem by the standard critical path method, and comparing the resulting project completion time with the deadline d.

To simplify our analysis, we expand the domains of functions f_i and g_i to $\{0, 1, \ldots, U\}$ for each activity i, where $U = \max_i\{\max\{\bar{t}_i, \bar{c}_i\}\}$. We can do so by defining $f_i(t) = M$ for $t = 0, 1, \ldots, \underline{t}_i - 1$, defining $f_i(t) = f_i(\bar{t}_i)$ for $t = \bar{t}_i + 1, \bar{t}_i + 2, \ldots, U$, defining $g_i(c) = M$ for $c = 0, 1, \ldots, \underline{c}_i - 1$, and defining $g_i(c) = g_i(\bar{c}_i)$ for $c = \bar{c}_i + 1, \bar{c}_i + 2, \ldots, U$, where M is a large integer. (Note: It suffices to set $M = \max\{\sum_i f_i(\underline{t}_i), \sum_i g_i(\underline{c}_i)\} + 1$.)

Throughout the paper, all logarithms are base 2 unless otherwise stated.

2 K-Approximation Sets and Functions

Halman et al. [7] have introduced K-approximation sets and functions, and used them to develop an FPTAS for a stochastic inventory control problem. Halman et al. [6] have applied these tools to develop a general framework for constructing FPTASs for stochastic dynamic programs. In this section we provide an overview of K-approximation sets and functions. In the next section we will use them to construct FPTASs for our time-cost tradeoff problems. To simplify the discussion, we modify Halman et al.'s definition of the K-approximation function by restricting it to integer-valued functions.

Let $K \geq 1$, and let $\psi : \{0, 1, \ldots, U\} \to Z^+$ be an arbitrary function. We say that $\hat{\psi} : \{0, 1, \ldots, U\} \to Z^+$ is a K-*approximation function* of ψ if $\psi(x) \leq \hat{\psi}(x) \leq K\psi(x)$ for all $x = 0, 1, \ldots, U$. The following property of K-approximation functions is extracted from Proposition 4.1 of [6], which provides a set of general computational rules of K-approximation functions. Its validity follows directly from the definition of the K-approximation function.

Property 1. For $i = 1, 2$, let $K_i \geq 1$, let $\psi_i : \{0, 1, \ldots, U\} \to Z^+$ be an arbitrary function, let $\tilde{\psi}_i : \{0, 1, \ldots, U\} \to Z^+$ be a K_i-approximation function of ψ_i, and let $\alpha, \beta \in Z^+$. The following properties hold:

Summation of approximation: $\alpha\tilde{\psi}_1 + \beta\tilde{\psi}_2$ is a $\max\{K_1, K_2\}$-approximation function of $\alpha\psi_1 + \beta\psi_2$.

Approximation of approximation: If $\psi_2 = \tilde{\psi}_1$ then $\tilde{\psi}_2$ is a $K_1 K_2$-approximation function of ψ_1.

Let $K > 1$. Let $\varphi : \{0, 1, \ldots, U\} \to Z^+$ be a nonincreasing function and $S = (k_1, k_2, \ldots, k_r)$ be an ordered subset of $\{0, 1, \ldots, U\}$, where $0 = k_1 < k_2 < \cdots < k_r = U$. We say that S is a K-*approximation set* of φ if $\varphi(k_j) \leq K\varphi(k_{j+1})$ for each $j = 1, 2, \ldots, r - 1$ that satisfies $k_{j+1} - k_j > 1$. (The term used in [6] is *weak K-approximation set of φ.*) Given φ, there exists a K-approximation set of φ with cardinality $O(\log_K \bar{U})$, where $\bar{U}$ is any constant upper bound of $\max_{x=0,1,\ldots,U}\{\varphi(x)\}$. Furthermore, this set can be constructed in $O((1 + \tau(\varphi))\log_K \bar{U} \log U)$ time, where $\tau(\varphi)$ is the amount of time required to evaluate φ (see Lemma 3.1 of [6]).

Given φ and a K-approximation set $S = (k_1, k_2, \ldots, k_r)$ of φ, a K-approximation function of φ can be obtained easily as follows (Definition 3.4 of [6]): Define $\hat{\varphi} : \{0, 1, \ldots, U\} \to Z^+$ such that

$$\hat{\varphi}(x) = \varphi(k_j) \text{ for } k_j \leq x < k_{j+1} \text{ and } j = 1, 2, \ldots, r - 1,$$

and that

$$\hat{\varphi}(k_r) = \varphi(k_r).$$

Note that $\varphi(x) \leq \hat{\varphi}(x) \leq K\varphi(x)$ for $x = 0, 1, \ldots, U$. Therefore, $\hat{\varphi}$ is a nonincreasing K-approximation function of φ. We say that $\hat{\varphi}$ is the K-*approximation function of φ corresponding to* S.

3   Series and Parallel Reductions

Two-terminal edge series-parallel networks (or simply "series-parallel networks") are defined recursively as follows [13]: (i) A directed network consisting of two vertices (i.e., a "source" and a "sink") joined by a single arc is series-parallel. (ii) If two directed networks G_1 and G_2 are series-parallel, then so are the networks constructed by each of the following operations: (a) Two-terminal

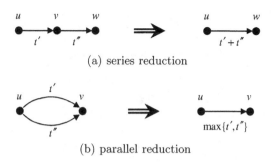

(a) series reduction

(b) parallel reduction

Fig. 2. Series and parallel reductions

series composition: Identify the sink of G_1 with the source of G_2. (b) Two-terminal parallel composition: Identify the source of G_1 with the source of G_2 and the sink of G_1 with the sink of G_2.

As mentioned in Section 1, a series-parallel network can be reduced to a single-arc network via a sequence of series and parallel reduction operations. A *series reduction* is an operation that replaces two series arcs by a single arc, while a *parallel reduction* is an operation that replaces two parallel arcs by a single arc (see Figure 2). In a project network, a reduction of two series activities with time duration t' and t'' will result in a single activity with time duration $t' + t''$, while a reduction of two parallel activities with time duration t' and t'' will result in a single activity with time duration $\max\{t', t''\}$. For example, given a series-parallel activity network depicted in Figure 3(a), we can perform a sequence of series/parallel reductions as shown in Figure 3(b). The resulting network consists of a single activity with duration 20, which is equal to the minimum project completion time of the original activity network. Thus, for a given series-parallel project network of n activities, it takes only $n-1$ series/parallel reduction operations to reduce it to a single-activity network. However, when there are time-cost tradeoff decisions for the activities, the integration of the two time-cost tradeoff functions during a series/parallel reduction operation becomes a challenge if we want to perform the computation efficiently. In the following subsections, we explain how to apply series and parallel reductions, together with K-approximation sets and functions, to develop FPTASs for the deadline and budget problems.

Note that series-parallel graphs have tree-width 2 (see [10], where "tree-width" was first introduced). It is known that many optimization problems on low tree-width graphs admit dynamic programs, which often lead to efficient exact/approximation algorithms that are unlikely to exist if the graphs were general [1]. Our paper goes along this line of research.

3.1 The Deadline Problem

For a given error tolerance $\epsilon \in (0, 1]$, our approximation algorithm for the deadline problem can be described as follows:

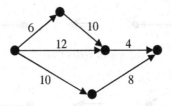

(a) Given activity network

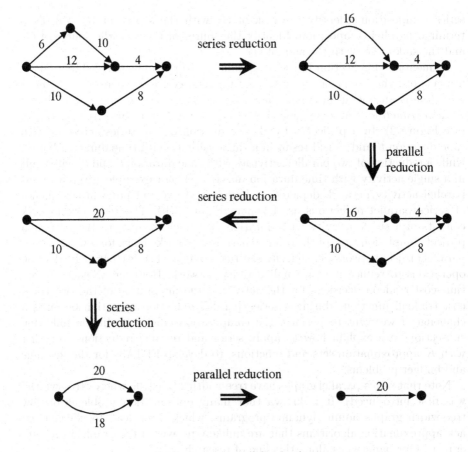

(b) Series and parallel reductions

Fig. 3. An example

Step 1: Let $K = 1 + \frac{\epsilon}{2n}$.

Step 2: For each activity i, obtain a K-approximation set S_i of f_i, and obtain the K-approximation function $\hat{f}_i$ of f_i corresponding to S_i.

Step 3: Select any pair of series or parallel activities i_1 and i_2.

Case (a): If i_1 and i_2 are series activities, then perform a series reduction to replace these two activities by an activity i. Obtain a K-approximation set $\bar{S}_i$ of $\bar{f}_i$, where

$$\bar{f}_i(t) = \min_{t' \in \{0,1,\dots,t\} \cap (S_{i_1} \cup \{t-x \,|\, x \in S_{i_2}\})} \{\hat{f}_{i_1}(t') + \hat{f}_{i_2}(t - t')\}. \tag{1}$$

Obtain the K-approximation function $\hat{f}_i$ of $\bar{f}_i$ corresponding to $\bar{S}_i$ (i.e., obtain and store the values of $\{\hat{f}_i(t) \mid t \in \bar{S}_i\}$ in an array arranged in ascending order of t).

Case (b): If i_1 and i_2 are parallel activities, then perform a parallel reduction to replace these two activities by an activity i. Obtain a K-approximation set $\bar{S}_i$ of $\bar{f}_i$, where

$$\bar{f}_i(t) = \hat{f}_{i_1}(t) + \hat{f}_{i_2}(t). \tag{2}$$

Obtain the K-approximation function $\hat{f}_i$ of $\bar{f}_i$ corresponding to $\bar{S}_i$.

Step 4: If the project network contains only one activity i_0, then the approximated solution value is given by $\hat{f}_{i_0}(d)$. Otherwise, return to Step 3.

We first discuss Case (a) of Step 3. Suppose that we allocate t time units to a pair of series activities i_1 (along arc $u \to v$) and i_2 (along arc $v \to w$); that is, we allow these two activities to spend no more than a total of t time units. Then, the merged activity i (along with merged arc $u \to w$, as shown in Figure 2(a)), which has a duration of t, will incur a cost of

$$f_i(t) = \min_{t'=0,1,\dots,t} \{f_{i_1}(t') + f_{i_2}(t - t')\}, \tag{3}$$

where $f_{i_1}(t')$ and $f_{i_2}(t - t')$ are the costs of the original activities i_1 and i_2 if they are allocated t' and $t - t'$ time units, respectively. Suppose we do not know the exact time-cost tradeoff functions f_{i_1} and f_{i_2} of these two activities, but instead we have: (i) a nonincreasing K^{k-1}-approximation function $\bar{f}_{i_1}$ of f_{i_1} and a nonincreasing $K^{\ell-1}$-approximation function $\bar{f}_{i_2}$ of f_{i_2}, where k and ℓ are positive integers, and (ii) a K-approximation set S_{i_j} of $\bar{f}_{i_j}$ and the K-approximation function $\hat{f}_{i_j}$ of $\bar{f}_{i_j}$ corresponding to S_{i_j} for $j = 1, 2$. Then, we obtain $\bar{f}_i$ using equation (1). We first show that $\bar{f}_i$ is a nonincreasing function.

Property 2. $\bar{f}_i$ defined in (1) is a nonincreasing function.

Proof: Consider any $t \in \{0, 1, \dots, U-1\}$. Then $\bar{f}_i(t) = \hat{f}_{i_1}(t^*) + \hat{f}_{i_2}(t - t^*)$ for some $t^* \in \{0, 1, \dots, t\} \cap (S_{i_1} \cup \{t-x \mid x \in S_{i_2}\})$. We have $t^* \in S_{i_1}$ or $t - t^* \in S_{i_2}$ (or both). If $t^* \in S_{i_1}$, then $t^* \in \{0, 1, \dots, t, t+1\} \cap (S_{i_1} \cup \{t+1-x \mid x \in S_{i_2}\})$, which implies that

$$\bar{f}_i(t + 1) \le \hat{f}_{i_1}(t^*) + \hat{f}_{i_2}(t + 1 - t^*) \le \hat{f}_{i_1}(t^*) + \hat{f}_{i_2}(t - t^*) = \bar{f}_i(t).$$

If $t - t^* \in S_{i_2}$, then $t^* + 1 \in \{t + 1 - x \mid x \in S_{i_2}\} \subseteq \{0, 1, \ldots, t, t+1\} \cap (S_{i_1} \cup \{t+1-x \mid x \in S_{i_2}\})$, which implies that

$$\bar{f}_i(t+1) \leq \hat{f}_{i_1}(t^* + 1) + \hat{f}_{i_2}(t - t^*) \leq \hat{f}_{i_1}(t^*) + \hat{f}_{i_2}(t - t^*) = \bar{f}_i(t).$$

Therefore, $\bar{f}_i$ is nonincreasing.   □

The following property is modified from Theorem 4.1 of [6].

Property 3. Let f_i and $\bar{f}_i$ be the functions defined in (3) and (1), respectively. Then, $\bar{f}_i$ is a $K^{\max\{k,\ell\}}$-approximation function of f_i.

Proof: Consider any fixed $t \in \{0, 1, \ldots, U\}$. Let

$$t^* = \arg \min_{t' = 0, 1, \ldots, t} \left\{ f_{i_1}(t') + f_{i_2}(t - t') \right\}$$

(with ties broken arbitrarily). Let

$$t^{**} = \arg \min_{t' \in \{0, 1, \ldots, t\} \cap (S_{i_1} \cup \{t - x \mid x \in S_{i_2}\})} \left\{ \hat{f}_{i_1}(t') + \hat{f}_{i_2}(t - t') \right\}$$

(with ties broken arbitrarily). We have

$$\bar{f}_i(t) = \hat{f}_{i_1}(t^{**}) + \hat{f}_{i_2}(t - t^{**}) \geq f_{i_1}(t^{**}) + f_{i_2}(t - t^{**}) \geq f_{i_1}(t^*) + f_{i_2}(t - t^*) = f_i(t). \tag{4}$$

Because $\hat{f}_{i_1}$ is the K-approximation function of $\bar{f}_{i_1}$ corresponding to S_{i_1}, there exists $t_0 \in S_{i_1}$ such that $t_0 \leq t^*$ and $\hat{f}_{i_1}(t_0) = \hat{f}_{i_1}(t^*)$. This implies that $\hat{f}_{i_1}(t_0) \leq K \bar{f}_{i_1}(t^*) \leq K^k f_{i_1}(t^*)$. Note that $\hat{f}_{i_2}(t - t_0) \leq \hat{f}_{i_2}(t - t^*) \leq K \bar{f}_{i_2}(t - t^*) \leq K^\ell f_{i_2}(t - t^*)$. Thus,

$$\begin{aligned} \bar{f}_i(t) &= \hat{f}_{i_1}(t^{**}) + \hat{f}_{i_2}(t - t^{**}) \leq \hat{f}_{i_1}(t_0) + \hat{f}_{i_2}(t - t_0) \\ &\leq K^k f_{i_1}(t^*) + K^\ell f_{i_2}(t - t^*) \leq K^{\max\{k,\ell\}} f_i(t). \end{aligned} \tag{5}$$

Combining (4) and (5) yields the desired result.   □

In Case (a) of Step 3, $\bar{S}_i$ is a K-approximation set of $\bar{f}_i$. Due to Property 2, $\bar{S}_i$ is well defined. Function $\hat{f}_i$ is the (nonincreasing) K-approximation function of $\bar{f}_i$ corresponding to $\bar{S}_i$. By approximation of approximation (Property 1), $\hat{f}_i$ is a nonincreasing $K^{\max\{k,\ell\}+1}$-approximation function of f_i. The amount of time required to evaluate $\bar{f}_i(t)$ for each t is

$$\tau(\bar{f}_i) = O\big((|S_{i_1}| + |S_{i_2}|)(\tau(\hat{f}_{i_1}) + \tau(\hat{f}_{i_2}))\big).$$

Note that

$$|S_{i_1}| = O(\log_K \bar{U}),$$
$$|S_{i_2}| = O(\log_K \bar{U}),$$

and

$$\tau(\hat{f}_{i_j}) = O(\log |S_{i_j}|)$$

for $j = 1, 2$ (because the values of $\{\hat{f}_{i_j}(t) \mid t \in S_{i_j}\}$ are stored in an array arranged in ascending order of t, for any $t = 0, 1, \ldots, U$, it takes only $O(\log |S_{i_j}|)$ time to search for the value of $\hat{f}_{i_j}(t)$). Thus, $\tau(\bar{f}_i) \leq O(\log_K \bar{U} \log \log_K \bar{U})$, and therefore the time required for constructing $\bar{S}_i$ is $O\big((1 + \tau(\bar{f}_i)) \log_K \bar{U} \log U\big) \leq O(\log_K^2 \bar{U} \log U \log \log_K \bar{U})$.

Next, we discuss Case (b) of Step 3. Suppose that we allocate t time units to a pair of parallel activities i_1 and i_2; that is, we allow each of these two activities to spend no more than t time units. Then, the merged activity, which has a maximum duration of t, will incur a cost of

$$f_i(t) = f_{i_1}(t) + f_{i_2}(t), \tag{6}$$

where $f_{i_1}(t)$ and $f_{i_2}(t)$ are the costs of the original activities i_1 and i_2, respectively. Suppose we do not know the exact time-cost tradeoff functions f_{i_1} and f_{i_2}, but instead we have: (i) a nonincreasing K^{k-1}-approximation function $\bar{f}_{i_1}$ of f_{i_1} and a nonincreasing $K^{\ell-1}$-approximation function $\bar{f}_{i_2}$ of f_{i_2}, where k and ℓ are positive integers, and (ii) a K-approximation set S_{i_j} of $\bar{f}_{i_j}$ and the K-approximation function $\hat{f}_{i_j}$ of $\bar{f}_{i_j}$ corresponding to S_{i_j} for $j = 1, 2$. Then, $\hat{f}_{i_1}$ is a K^k-approximation function of f_{i_1}, and $\hat{f}_{i_2}$ is a K^ℓ-approximation function of f_{i_2}.

By summation of approximation (Property 1), $\bar{f}_i$ defined in (2) is a $K^{\max\{k,\ell\}}$-approximation function of f_i. Clearly, $\bar{f}_i$ is nonincreasing. Let $\bar{S}_i$ be a K-approximation set of $\bar{f}_i$, and $\hat{f}_i$ be the (nonincreasing) K-approximation function of $\bar{f}_i$ corresponding to $\bar{S}_i$. By approximation of approximation (Property 1), $\hat{f}_i$ is a $K^{\max\{k,\ell\}+1}$-approximation function of f_i. The amount of time required to evaluate $\bar{f}_i$ is

$$\tau(\bar{f}_i) = O\big(\tau(\hat{f}_{i_1}) + \tau(\hat{f}_{i_2})\big) = O(\log |S_{i_1}| + \log |S_{i_2}|) \leq O(\log \log_K \bar{U}).$$

The amount of time required to construct $\bar{S}_i$ is $O\big((1 + \tau(\bar{f}_i)) \log_K \bar{U} \log U\big)$, which is dominated by the running time for constructing $\bar{S}_i$ in the series reduction case.

Let $f^*(d)$ denote the optimal total cost of the project for a given deadline d. We now analyze how close $\hat{f}_{i_0}(d)$ is to $f^*(d)$. Note that after performing r series/parallel reduction operations ($0 \leq r \leq n-1$), the project network has $n-r$ activities, namely $i_1, i_2, \ldots, i_{n-r}$. Associated with each activity i_j is a function $\hat{f}_{i_j}$, which is a K^{β_j}-approximation function of f_{i_j} for some positive integer β_j. We define $\sum_{j=1}^{n-r} \beta_j$ as the *approximation level* of this project.

Before performing any series/parallel reduction, the project has an approximation level n. Since $\max\{k, \ell\} + 1 \leq k + \ell$, neither a series reduction operation nor a parallel reduction operation will increase the approximation level of the project. Hence, at the end of the solution procedure, the approximation level of the project is at most n, which implies that $\hat{f}_{i_0}$ is a K^n-approximation of f^*. Recall that $K = 1 + \frac{\epsilon}{2n}$. Because $(1 + \frac{\epsilon}{2n})^n \leq 1 + \epsilon$, we conclude that $\hat{f}_{i_0}(d)$ is a $(1 + \epsilon)$-approximation solution to the deadline problem.

Finally, we analyze the running time of the approximation algorithm. Step 2 obtains a K-approximation set and function for each activity. The running time of this step is dominated by that of the series/parallel reduction operations in Step 3. The construction of $\bar{S}_i$ in each series/parallel reduction takes $O(\log_K^2 \bar{U} \log U \log \log_K \bar{U})$ time. Thus, the running time of the entire solution procedure is $O(n \log_K^2 \bar{U} \log U \log \log_K \bar{U})$. Since $\log_K \bar{U} \le \frac{1}{K-1} \log_2 \bar{U}$ (because $\log_2 K \ge K-1$ for $1 < K < 2$), the running time is $O(\frac{n^3}{\epsilon^2} \log^2 \bar{U} \log U \log(\frac{n}{\epsilon} \log \bar{U}))$. Therefore, our solution scheme is an FPTAS.

3.2 The Budget Problem

We now consider the budget problem. Let $g^*(b)$ denote the optimal duration of the project for a given budget b. Suppose we allocate c units of monetary resources to a pair of series activities i_1 (along arc $u \to v$) and i_2 (along arc $v \to w$). Then, the merged activity i (along the merged arc $u \to w$), which has a budget of c, will have a duration of

$$g_i(c) = \min_{c'=0,1,\ldots,c} \{g_{i_1}(c') + g_{i_2}(c - c')\}, \tag{7}$$

where $g_{i_1}(c')$ and $g_{i_2}(c-c')$ are the activity times of the original activities i_1 and i_2 if they are allocated monetary resources of c' and $c-c'$, respectively. Let $\bar{g}_{i_1}$ be a nonincreasing K^{k-1}-approximation function of g_{i_1}, and $\bar{g}_{i_2}$ be a nonincreasing $K^{\ell-1}$-approximation function of g_{i_2}. Let S_{i_j} be a K-approximation set of $\bar{g}_{i_j}$, and $\hat{g}_{i_j}$ be the K-approximation function of $\bar{g}_{i_j}$ corresponding to S_{i_j} ($j = 1, 2$). Then, $\hat{g}_{i_1}$ is a K^k-approximation function of g_{i_1}, and $\hat{g}_{i_2}$ is a K^ℓ-approximation function of g_{i_2}. Following the same argument as in Section 3.1, we define function $\bar{g}_i$ such that for $t = 0, 1, \ldots, U$,

$$\bar{g}_i(c) = \min_{c' \in \{0,1,\ldots,c\} \cap (S_{i_1} \cup \{c-x \mid x \in S_{i_2}\})} \{\hat{g}_{i_1}(c') + \hat{g}_{i_2}(c - c')\}.$$

By Properties 2 and 3, $\bar{g}_i$ is a nonincreasing $K^{\max\{k,\ell\}}$-approximation function of g_i. Let $\bar{S}_i$ be a K-approximation set of $\bar{g}_i$, and $\hat{g}_i$ be the (nonincreasing) K-approximation function of $\bar{g}_i$ corresponding to $\bar{S}_i$. Then, $\hat{g}_i$ is a nonincreasing $K^{\max\{k,\ell\}+1}$-approximation function of g_i, and $\bar{S}_i$ can be constructed in $O(\log_K^2 \bar{U} \log U \log \log_K \bar{U})$ time.

Now, suppose that we allocate c units of monetary resources to a pair of parallel activities i_1 and i_2. Then, the merged activity will have an activity time of

$$g_i(c) = \min_{c'=0,1,\ldots,c} \left\{ \max \{g_{i_1}(c'), g_{i_2}(c - c')\} \right\}. \tag{8}$$

We define function $\bar{g}_i$ such that for $t = 0, 1, \ldots, U$,

$$\bar{g}_i(c) = \min_{c' \in \{0,1,\ldots,c\} \cap (S_{i_1} \cup \{c-x \mid x \in S_{i_2}\})} \left\{ \max \{\hat{g}_{i_1}(c'), \hat{g}_{i_2}(c - c')\} \right\},$$

with S_{i_1}, S_{i_1}, $\hat{g}_{i_1}$, and $\hat{g}_{i_2}$ having the same definitions as before. Using the same argument as in the proofs of Properties 2 and 3, we can show that $\bar{g}_i$ is a nonincreasing $K^{\max\{k,\ell\}}$-approximation function of g_i. Let $\bar{S}_i$ be a K-approximation set of $\bar{g}_i$, and $\hat{g}_i$ be the K-approximation function of $\bar{g}_i$ corresponding to $\bar{S}_i$. Then, $\hat{g}_i$ is a $K^{\max\{k,\ell\}+1}$-approximation function of g_i, and $\bar{S}_i$ can be constructed in $O(\log_K^2 \bar{U} \log U \log \log_K \bar{U})$ time.

Similar to the deadline problem, we determine an approximation solution to the budget problem by first obtaining a K-approximation set S_i and the K-approximation function of f_i corresponding to S_i for each activity i, and then applying series and parallel reductions recursively until the project is reduced to a single activity i_0. The solution value is given by $\hat{g}_{i_0}(b)$, which is a K^n-approximation of $\hat{g}^*(b)$. Let $K = 1 + \frac{\epsilon}{2n}$, where $0 < \epsilon \le 1$. Then, $\hat{g}_{i_0}(b)$ is a $(1+\epsilon)$-approximation solution to the budget problem, and the running time of the solution procedure is $O\left(\frac{n^3}{\epsilon^2} \log^2 \bar{U} \log U \log(\frac{n}{\epsilon} \log \bar{U})\right)$. Therefore, our solution scheme is an FPTAS.

4   Concluding Remarks

We have developed FPTASs for both the deadline and budget problems. Note that although these FPTASs generate solutions with relative errors bounded by ϵ, the actual relative error of a solution is affected by the sequence of series and parallel reduction operations. For example, consider the deadline problem with only four activities i_1, i_2, i_3, i_4 arranged in series, where i_j is the immediate predecessor of i_{j+1} ($j = 1, 2, 3$). At the beginning of the solution procedure, we obtain a K-approximation set and a K-approximation function for each of these activities. Suppose we perform series reductions in the following sequence: (i) merge i_1 and i_2 to form a new activity i_{12}; (ii) merge i_{12} and i_3 to form a new activity i_{123}; and (iii) merge i_{123} and i_4 to form a network with a single activity i_0. Then, step (i) generates a K^2-approximation function $\hat{f}_{i_{12}}$ of $f_{i_{12}}$. Step (ii) generates a K^3-approximation function $\hat{f}_{i_{123}}$ of $f_{i_{123}}$. Step (iii) generates a K^4-approximation function $\hat{f}_{i_0}$ of f_{i_0}.

Now, suppose we perform the series reductions in another sequence: (i) merge i_1 and i_2 to form a new activity i_{12}; (ii) merge i_3 and i_4 to form a new activity i_{34}; and (iii) merge i_{12} and i_{34} to form a network with a single activity i_0. Then, step (i) generates a K^2-approximation function $\hat{f}_{i_{12}}$ of $f_{i_{12}}$. Step (ii) generates a K^2-approximation function $\hat{f}_{i_{34}}$ of $f_{i_{34}}$. Step (iii) generates a K^3-approximation function $\hat{f}_{i_0}$ of f_{i_0}. Hence, this sequence of series reduction operations yields a better approximation than the previous one.

Our FPTAS for the deadline problem uses only the "primal" dynamic program in (3) and (6). It not only approximates the value of the optimal solution $f^*(d)$ for the deadline problem, but also stores an approximation of the function f^* over the entire domain $\{0, 1, \ldots, d\}$ in a sorted array of size $O(\frac{n}{\epsilon} \log \bar{U})$. Therefore, for any integer $x \in \{0, 1, \ldots, d\}$, only $O(\log(\frac{n}{\epsilon} \log \bar{U}))$ additional time is needed to determine the approximated value of $f^*(x)$.

We note that it is also possible to approximate the deadline and budget problems using the traditional "scaling and rounding the data" approach. On one hand, for doing so one needs to use the "dual" dynamic program (e.g., recursions (7) and (8) for the deadline problem). On the other hand, by applying the elegant technique of Hassin [8], it is possible to reduce the $\log \bar{U}$ term in the running time to $\log \log \bar{U}$. This is done by performing binary search in the log domain and rounding/scaling $g_i(c)$ in (7) and (8) for every value c where these functions are computed. Unlike our approach, approximating $f^*(x)$ for any additional x will require the same running time.

Our solution method can be extended to non-series-parallel project networks. However, the running time of the approximation algorithm will no longer be polynomial. To tackle non-series-parallel project networks, besides series and parallel reductions, we also make use of node reduction. Any two-terminal directed acyclic network can be reduced to a single arc via series, parallel, and node reductions (see [2]). A node reduction operation can be applied when the node concerned has either in-degree 1 or out-degree 1. Suppose node v has in-degree 1. Let $u \to v$ be the arc into v, and $v \to w_1$, $v \to w_2$, ..., $v \to w_k$ be the arcs out of v. Then a node reduction at v is to replace these $k+1$ arcs by arcs $u \to w_1$, $u \to w_2$, ..., $u \to w_k$. The case where v has out-degree 1 is defined symmetrically. In our deadline and budget problems, such a node reduction implies a decomposition of the problem into $m(i)$ separate problems, where $m(i)$ is the number of time-cost alternatives of the activity i corresponding to arc $u \to v$. In each decomposed problem, we obtain the time-cost tradeoff functions for arcs $u \to w_1$, $u \to w_2$, ..., $u \to w_k$ by adding the time duration and activity cost of $u \to v$ to the time-cost tradeoff functions of $v \to w_1$, $v \to w_2$, ..., $v \to w_k$, respectively. Bein *et al.* [2] have developed an efficient method for determining the minimum number of node reductions in order to reduce the given project network to a single activity. They refer to this minimum number of node reductions as reduction complexity. Therefore, a discrete time-cost tradeoff problem in a non-series-parallel project network can be decomposed into $\bar{m}^h$ time-cost tradeoff problems with series-parallel networks, where $\bar{m} = \max_i \{m(i)\}$ and h is the reduction complexity. If h is bounded by a constant (i.e., the network is near-series-parallel) and $\bar{m}$ is bounded by a polynomial of the problem input size, then making such a decomposition and applying the algorithms presented in Section 3 will give us an FPTAS for the problem.

Note that the computational complexity of this decomposition method increases exponentially as the reduction complexity increases. Hence, this method is practical only if h is small. As mentioned in Section 1, for general non-series-parallel project networks, it is very difficult to obtain an ϵ-approximation algorithm for the time-cost tradeoff problem (for example, the budget problem does not even have a polynomial-time approximation algorithm with performance guarantee better than $\frac{3}{2}$ unless P=NP).

Acknowledgment

The authors would like to thank three anonymous reviewers for their helpful comments and suggestions, including the one about the possibility to decrease the $\log \bar{U}$ term in the running time to $\log \log \bar{U}$.

References

1. Baker, B.S.: Approximation algorithms for NP-complete problems on planar graphs. Journal of the Association for Computing Machinery 41, 153–180 (1994)
2. Bein, W.W., Kamburowski, J., Stallmann, M.F.M.: Optimal reduction of two-terminal directed acyclic graphs. SIAM Journal on Computing 21, 1112–1129 (1992)
3. De, P., Dunne, E.J., Ghosh, J.B., Wells, C.E.: Complexity of the discrete time-cost tradeoff problem for project networks. Operations Research 45, 302–306 (1997)
4. Deineko, V.G., Woeginger, G.J.: Hardness of approximation of the discrete time-cost tradeoff problem. Operations Research Letters 29, 207–210 (2001)
5. Frank, H., Frisch, I.T., Van Slyke, R., Chou, W.S.: Optimal design of centralized computer networks. Networks 1, 43–58 (1971)
6. Halman, N., Klabjan, D., Li, C.-L., Orlin, J., Simchi-Levi, D.: Fully polynomial time approximation schemes for stochastic dynamic programs. In: Proceedings of the Nineteenth ACM-SIAM Symposium on Discrete Algorithms (SODA), pp. 700–709 (2008)
7. Halman, N., Klabjan, D., Mostagir, M., Orlin, J., Simchi-Levi, D.: A fully polynomial time approximation scheme for single-item stochastic inventory control with discrete demand. Working paper submitted for publication. Massachusetts Institute of Technology, Cambridge (2006)
8. Hassin, R.: Approximation schemes for the restricted shortest path problem. Mathematics of Operations Research 17, 36–42 (1992)
9. Hindelang, T.J., Muth, J.F.: A dynamic programming algorithm for decision CPM networks. Operations Research 27, 225–241 (1979)
10. Robertson, N., Seymour, P.D.: Graph minors. II. Algorithmic aspects of tree-width. Journal of Algorithms 7, 309–322 (1986)
11. Rothfarb, B., Frank, H., Rosebbaum, D.M., Steiglitz, K., Kleitman, D.J.: Optimal design of offshore natural-gas pipeline systems. Operations Research 18, 992–1020 (1970)
12. Skutella, M.: Approximation algorithms for the discrete time-cost tradeoff problem. Mathematics of Operations Research 23, 909–929 (1998)
13. Valdes, J., Tarjan, R.E., Lawler, E.L.: The recognition of series-parallel digraphs. SIAM Journal on Computing 11, 298–313 (1982)

Efficient Algorithms for Fixed-Precision
Instances of Bin Packing and Euclidean TSP

David R. Karger and Jacob Scott*

Computer Science and Artificial Intelligence Laboratory
Massachusetts Institute of Technology
Cambridge, MA 02139, USA
{karger,jhscott}@csail.mit.edu

Abstract. This paper presents new, polynomial time algorithms for Bin
Packing and Euclidean TSP under *fixed precision*. In this model, integers
are encoded as floating point numbers, each with a mantissa and an exponent. Thus, an integer i with $i = a_i 2^{t_i}$ has mantissa a_i and exponent
t_i. This natural representation is the norm in real-world optimization. A
set of integers I has *L-bit precision* if $\max_{i \in I} a_i < 2^L$. In this framework,
we show an exact algorithm for Bin Packing and an FPTAS for Euclid-
ean TSP which run in time $\text{poly}(n)$ and $\text{poly}(n + \log 1/\epsilon)$, respectively,
when L is a fixed constant. Our algorithm for the later problem is exact
when distances are given by the L_1 norm. In contrast, both problems
are strongly NP-Hard (and yield PTASs) when precision is unbounded.
These algorithms serve as evidence of the significance of the *class* of fixed
precision polynomial time solvable problems. Taken together with algo-
rithms for the Knapsack and $Pm||C_{\max}$ problems introduced by Orlin
et al. [10], we see that fixed precision defines a class incomparable to
polynomial time approximation schemes, covering at least four distinct
natural NP-hard problems.

1   Introduction

When faced with an NP-complete problem, algorithm designers must either settle for approximation algorithms, accept superpolynomial runtimes, or identify
natural restrictions of the given problem that are tractable. In the last cate-
gory we find numerous *weakly NP-hard problems*, which have *pseudo-polynomial
algorithms* that solve them in time polynomial in the problem size *and the magnitude of the numbers in the problem instance*, and are thus polynomial in the
problem size when the number magnitudes are polynomial in the problem size.
A more recent development is that of *fixed parameter tractability*, another way
of responding to the hardness of specific problem instances. In this paper, we
explore *fixed precision tractability*. While pseudo-polynomial algorithms aim at
problems whose numbers are *integers* of bounded size, we consider the case of
floating point numbers that have bounded mantissas but arbitrary exponents.
We consider such an exploration natural for a variety of reasons:

---

* Supported by National Science Foundation grant CCF-0635286 and by an NDSEG
  Fellowship.

A. Goel et al. (Eds.): APPROX and RANDOM 2008, LNCS 5171, pp. 104–117, 2008.

1. Floating point numbers are the norm in real-world optimization. They reflect the fact that practitioners seem to need effectively unlimited ranges for their numbers but recognize that they must tolerate limited precision. It is thus worth developing a complexity theory aligned with these types of inputs.
2. Conversely, from a complexity perspective, it remains unclear exactly *why* certain problems become easy in the pseudo-polynomial sense. Is it because the *magnitude* of those numbers is limited, or because their *precision* is? The pseudo-polynomial characterization bounds both, so it is not possible to distinguish.
3. As observed by Orlin et al. [10], a fixed-precision-tractable algorithm yields a *inverse approximation algorithm* for the problem, taking an arbitrary input and yielding an *optimal* solution to an instance in which the input numbers are perturbed by a small *relative* factor (namely, by rounding to fixed precision). This is arguably a more meaningful approximation than a PTAS or FPTAS that perturbs the value of the output.

Orlin et al., while introducing the notion of fixed precision tractability, gave fixed precision algorithms for Knapsack and Three Partition. But this left open the question of how general fixed-precision tractability might be. Knapsack and Three Partition are some of the "easiest" NP-complete problems, both exhibiting trivial fully polynomial approximation schemes. It was thus conceivable that fixed-precision tractability was a fluke arising from these problems' simplicity.

In this paper we bring evidence as to the significance of the fixed precision class, by showing that Bin Packing and Euclidean TSP are both fixed precision tractable. This is interesting because neither of them is known to have a fully polynomial approximation scheme. Orlin et al. showed that, conversely, there are problems with FPTASs that are not fixed precision tractable. We therefore see that fixed precision defines a class incomparable to FPTASs, covering at least 4 distinct natural NP-hard problems.

In order to demonstrate these results it is necessary for us to introduce some new, natural solution techniques. If all our input numbers have the same exponent, then we can concentrate on the mantissas of those numbers (which will be bounded integers) and apply techniques from pseudo-polynomial algorithms. The question is how to handle the varying exponents. We develop dynamic programs that scale through increasing values of the exponents, and argue that once we have reached a certain exponent, numbers with much smaller exponents can be safely ignored. This would be trivial if we were seeking approximate solutions but takes some work as we are seeking exact solutions. We believe that our approach is a general one to developing fixed-precision algorithms.

2   Background

In this section we give an overview of generally related previous work. Then we review the L-bit precision model to which our algorithms apply. Finally, we present relevant work on Bin Packing and Euclidean TSP, the problems for which we give algorithms in Sections 3 and 4.

2.1   Approximation Algorithms, Fixed Parameter Tractability, and Inverse Optimization

When faced with an NP-Hard optimization problem, the first option is most often the development of an approximation algorithm. If the best solution for an instance I of some minimization problem P has objective value $OPT(I)$, then an α-approximation algorithm for P is guaranteed for each instance I to return a solution with an objective value of no more than $\alpha \cdot OPT(I)$. Previous work in this area is vast, and we refer the interested reader to Vazirani [12] for a wide overview of the field. The best one can hope for in this context is a *polynomial time approximation scheme*, a polynomial time algorithm which can provide a $(1 + \epsilon)$-approximation for P, for any ϵ. If the running time of such an algorithm is polynomial in $1/\epsilon$, it is referred to as *fully* polynomial, and we say that P has an *FPTAS*. If the running time is polynomial only when ϵ is fixed, we say that P has an *PTAS*.

Another way in which NP-Hard problems can be approached is through fixed parameter tractability [6]. Here, a problem may have an algorithm that is exponential, but only in the size of a particular input parameter. Given an instance I and a parameter k, a problem P is fixed parameter tractable with respect to k if there is an (exact) algorithm for P with running time $O(f(|k|) \cdot |I|^c)$, where $|x|$ gives the length of x, f depends only on $|k|$, and c is a constant. Such problems can be tractable for small values of k. For example, Balasubramanian et al. [3] show it is possible to determine if a graph $G = (V, E)$ has a vertex cover of size k in $O(|V|k + (1.324718)^k k^2)$ time. The problem is thus tractable when the size of the vertex cover is given as a fixed parameter.

Inverse optimization, introduced by Ahuja and Orlin [1], is closer to the L-bit precision regime we work in. Consider a general minimization problem over some vector space X, $\min\{cx : x \in X\}$. If this problem is NP-Hard, an α-approximation algorithm can return a solution x^* such that $\frac{cx^* - cx}{cx} \leq \alpha$. Inverse optimization instead modifies the cost vector. That is, it searches for a solution x' and cost vector c' close to c (in, for example, L_∞ distance) so that $\min\{c'x : x \in X\} = x'$. The tightness of this approximation is then measured based on the distance from c' to c. This analysis can be more natural for some problems. In Bin Packing, for example, a standard approximation algorithm requires more bins than the optimal algorithm, while an inverse approximation algorithm requires an equal number of *larger* bins. All fixed precision tractable problems can be translated into this framework via rounding.

2.2 L-Bit Precision

Our algorithms apply to the L-bit precision model introduced by Orlin et al. [10]. Problems in this model have their integer inputs represented in a nonstandard encoding. Each integer i is encoded as the unique pair (m, x) such that $i = m \cdot 2^x$, and m is odd. Following terminology for floating point numbers, we refer to m as the *mantissa* and x as the *exponent* of i. A problem instance has L-bit precision with respect to some subset $M \subseteq I$ of its integer inputs (encoded as above) if, $\forall (m, x) \in M, m < 2^L$. Each $(m, x) \in M$ can then be represented with

$O(L + \log x)$ bits. An algorithms runtime is then computed in relation to the total size T of its input, with numbers encoded appropriately. Two settings for L are considered. When L is constant, instances have *fixed precision*, and when $L = O(\log T)$, they have *logarithmic precision*.

Orlin et al. show polynomial time algorithms for fixed and logarithmic precision instances of the Knapsack problem. They show the same for $Pm||C_{\max}$, the problem of scheduling jobs on m identical machines so as to minimize the maximum completion time. The precision here is with respect to item values and job lengths, respectively. The algorithms work by reducing the problem to finding the shortest path on an appropriate graph. Such paths can take exponential time to find in the general case, but can be computed efficiently under L-bit precision. Our algorithms use similar techniques. We note that both of these problems are known to be NP-Complete, and also to support FPTASs.

The paper also demonstrates a polynomial time algorithm for the 3-partition problem when all numbers to be partitioned have fixed precision.[1] These instances can be reduced to integer programs with a linear number of constraints and a fixed number of variables, which in turn can be solved in linear time. Finally, a variant of the Knapsack problem is shown that supports an FPTAS, but is NP-Complete for even 1-bit precision. The problem, *Group Knapsack*, allows items to have affinities for other items, so that item x can be placed in the Knapsack if and only if item y is also packed. Thus, items with different exponents can express affinity for each other, essentially reconstructing items of arbitrary precision.

2.3   Bin Packing

Bin packing is a classic problem in combinatorial optimization. The problem dates back to the 1970s and there are an enormous number of variants. We restrict ourselves to the original one-dimensional version. Here we are given a set of n items $\{x_1, x_2, \ldots x_n\}$, each with a size $s(x_i) \in [0, 1]$. We must pack each item into a unit capacity bin B_k without overflowing it – that is, $\forall k, \sum_{x_i \in B_k} s(x_i) \leq 1$. Our goal is to pack the items so that the number of bins used, K, is minimized.

Bin packing is NP-Complete, and extensive research has been done on approximation algorithms for the problem in various settings (see Coffman et al. [5] for a survey). The culmination of this work is the well-known asymptotic PTAS of de la Vega and Lueker [7], which handles large items with a combination of rounding and brute force search, and then packs small items into the first bin in which they fit. There is a key difference between this and all other worst case approximation algorithms for Bin Packing and the algorithm we show in Section 3. While both run in polynomial time, the former guarantee that they can pack any set of items into a number of bins that is almost optimal. Our algorithm guarantees that for a specific class of instances, namely those whose items have fixed-precision sizes, it can pack items into *exactly* the optimal number of bins.

---

[1] Note that this is the problem of grouping $3n$ numbers into n 3-tuples with identical sums, not splitting them into three groups having the same sum, which is a special case of $Pm||C_{\max}$.

2.4   Euclidean TSP

The Euclidean TSP is a special case of the well-known Traveling Salesman Problem. The goal of the TSP is to find the minimum cost Hamiltonian cycle on a weighted graph $G = (V, E)$. That is, to find a minimum weight tour of the graph (starting from an arbitrary vertex) that visits every other vertex exactly once before returning to the starting vertex. Determining whether a Hamiltonian cycle exists in a graph is one of the original twenty-one problems that Karp shows to be NP-Complete [8], and it follows immediately that TSP is NP-Complete. Indeed, this reduction implies that no polynomial time algorithm can approximate general instances TSP to within any polynomial factor unless P=NP.

One special case of the TSP that can be approximated is metric TSP. Here there is an edge between each pair of vertices, and edge weights are required to obey the triangle inequality. A simple 2-approximation for metric TSP can be shown using a tour that "shortcuts" the cycle given by doubling every edge in a minimum spanning tree and using the resulting Eulerian tour. The best known approximation ratio for metric TSP is $3/2$, and comes from Christofides [4], who augments the minimum spanning tree more wisely than doubling each edge.

Euclidean TSP, finally, is then a further restriction of metric TSP. Vertices are taken to be points in Euclidean space (throughout this paper, we restrict ourselves to the plane), and there is an edge between each pair of vertices with weight equal to their Euclidean distance. Despite this restriction, the problem remains strongly NP-Hard [11]. Arora [2] gives what is then the best possible result (unless $P = NP$), a PTAS, for Euclidean TSP. His algorithm combines dynamic programming with the use of a newly introduced data structure, the randomly shifted quadtree, which has since found wide use in computational geometry. Our algorithm for fixed-precision Euclidean TSP uses a similar combination of recursion and brute force, but takes advantage of structural properties implied by the nature of our narrower class of inputs.

3   Bin Packing

In this section we present an $n^{O(4^L)}$ time algorithm for L-bit precision instances of Bin Packing.

3.1   Preliminaries

We consider a version of Bin Packing that is slightly modified from that discussed in Section 2. We take the following *decision* problem: given a set of n items $S = \{u_1, u_2, \ldots u_n\}$, each with a size $s(u_i) \in \mathbb{N}$, and integers K and C, can the set S be packed into K bins of capacity C? In the standard way, one can use an algorithm for this decision problem to solve its optimization analog by performing a binary search for the minimum feasible value of K. We find it convenient to consider S to be a set of item sizes, rather than items themselves, and refer to it in this way throughout the remainder of this section.

We consider the precision of an instance with respect to the set of items S. Thus, we consider an instance of Bin Packing to have L-bit precision if for all $u = m2^x \in S, m < 2^L$.

3.2 The Algorithm

Here, in broad strokes, is a nondeterministic algorithm $\mathcal{A}$ (that is, $\mathcal{A}$ can magically make correct guesses) for this problem. $\mathcal{A}$ runs in rounds, and is only allowed to pack items that are currently in a special *reservoir* R. It also maintains bin occupancies in a set O, such that $o_k \in O$ gives the current occupancy of (that is, the total size of the items in) the kth bin. Rounds proceed as follows. At the start of each round, $\mathcal{A}$ moves some items from S into R. Then, it guesses which items in R to pack, and which bins to pack them into. Finally, $\mathcal{A}$ updates the occupancy of bins that received items in the round, and performs certain bookkeeping on R. After the final round, $\mathcal{A}$ reports success if and only if R is empty and all bins have occupancy less than C.

We now describe these steps in detail. Afterwards, we demonstrate how they can be implemented efficiently. We initialize R to be empty, and all bins to have occupancy zero. In round x, we move the subset of items $S_x \subseteq S$ from S to R, where S_x contains all items in S with exponent x. In the packing step, for each size mantissa $m \in \{1, 3, \ldots 2^L - 1\}$, we guess which bins will be packed with an odd number of items of size $m2^x$. We place exactly one item of size $m2^x$ into each such bin, removing it from R and updating O.

We note the following property of the algorithm as presented so far.

Property 1. After round x, for all m, the remaining number of items of size $m2^x$ that should be packed into any bin k is even.

In light of Property 1, our bookkeeping procedure is to merge each pair of items of size $m2^x$ in R, creating a single item of size $m2^{x+1}$. Because we know that all of these items will be packed in pairs and that the size of the merged item is the sum of the sizes of its two constituents, this merging will not effect feasibility. By induction, this will mean that after round x there are no items of exponent less than x remaining in either S or R. After this step, $\mathcal{A}$ proceeds to round $x + 1$. The final round occurs when both S and R are empty, at which point all items have been packed. The following lemma bounds the number of rounds.

Lemma 1. *If a round x is skipped when both S_x and R are empty, then $\mathcal{A}$ runs at most n rounds.*

3.3 Analysis

We start by showing that the state of $\mathcal{A}$ can be represented compactly. This state can be described completely by the current round (which also dictates the items remaining in S), the items in R, and the bin occupancy O, so we represent such a state as $\{R, O, x\}$. We have already seen that x has at most n possible values. By design, every item in R at the start of round x has exponent exactly

x. Thus, we can represent R by storing a count of the number of objects with each mantissa. Expressed in this way, R has a total of $O(n^{2^{L-1}})$ possible values. The following lemma shows that O can also be represented compactly.

Lemma 2. *Consider a bin occupancy o_k at the start of round x. Then rounding o_k up to the next integer multiple of 2^x does not affect the feasible executions of $\mathcal{A}$.*

Proof. Let $o_{k'}$ be equal to o_k rounded up to the next integer multiple of 2^x, and consider two bins with these occupancies. Let $e_k = C - o_k$ and $e_{k'} = C - o_{k'}$ be the space remaining in each bin. Then we have that

$$\left\lfloor \frac{e_k}{2^x} \right\rfloor = \left\lfloor \frac{e_{k'}}{2^x} \right\rfloor$$

We conclude the proof by noting that all items remaining to be packed (in both R and S) have exponent at least x, so that any set of remaining items can be packed into bin k if and only if it can be packed into bin k'.

At the start of round x, each bin is occupied by at most one item of each size that has exponent less than x. At this point, we thus have for each k, $o_k \le \sum_{y=0}^{x} \sum_{w=1}^{2^L-1} w2^y \le 2^{2L+x}$. This fact coupled with Lemma 2 yields that O can also be represented by counting the number of bins of size $a \cdot 2^x$, for $a \in [0, 2^{2L})$. As there are at most n bins, the number of distinct O can then be bounded by $O(n^{4^L})$.

Now, consider a graph with nodes for each possible state $\{R, O, x\}$. Let us place a directed edge between $\{R, O, x\}$ and $\{R', O', x+1\}$ if it possible for $\mathcal{A}$ to begin round $x+1$ with reservoir R' and bin occupancy O' having started round x with reservoir R and occupancy O.[2] Then we can implement $\mathcal{A}$ by searching for a path between $\{\{\}, \{\}, 0\}$ and $\{\{\}, O^*, x_{\max}\}$ in this graph (where O^* satisfies all $o_k \le C$ for all $o_k \in O^*$). The following lemma states that we can test if an edge should be inserted efficiently.

Lemma 3. *Given two algorithm states $S_1 = \{R, O, x\}$ and $S_2 = \{R', O', x+1\}$, it is possible to determine if $\mathcal{A}$ can transition from S_1 to S_2 in time $O\left(n \cdot n^{4 \cdot 4^L}\right)$.*

Proof. Consider the following packing problem: given a set of n items Y, and a set of w bins with occupancies U and capacities Q (we allow bins with different capacities), is there a feasible packing of Y into the partially occupied U? We note that if all bin capacities are upper bounded by v, then this problem can be easily solved by dynamic programming in time $O(nw^{2v})$.

It should be clear that asking if a transition is possible between S_1 and S_2 can be represented as an instance of this problem. We first take $Y = R \cup S_x - R'$, noting that we may need to 'unmerge' some items in R' in order to subtract them, but that this is easy to do. Initial bin occupancies U are then given by O. Bin capacities Q are given by O', as this set describes the occupancy bins should have at the start of round $x + 1$.

---

[2] We may also need some edges between states $\{R, O, x\}$ and $\{R', O', x'\}$, where $x' \ne x + 1$. However, this only happens when rounds are skipped, and these edges are trivial to calculate.

We have one caveat here. As described above, the items in Y all have exponent 2^x, but the occupancies in O' are represented as multiples of 2^{x+1}. Thus, an occupancy $o'_k = (2i - 1)2^{x+1} \in O$ considered as a capacity $q_w \in Q$ may need to be incremented by 2^x. Letting $C[x]$ be the xth bit of our original capacity C, we can simply let $q_w = o'_k + C[x]2^x$.

Finally, we would like our capacities to have a small upper bound, so that the dynamic programming is efficient. Since all of the numbers here are multiples of 2^x, we scale them down by this factor before running the dynamic programming. This gives us a maximum bin capacity of $v = 2^{2L+2}$. Thus we can test whether there should be an edge between S_1 and S_2 in time $O\left(nw^{2^{2L+2}}\right) = O\left(n \cdot n^{4 \cdot + 4^L}\right)$.

We can now state our main theorem.

Theorem 1. *There is an $n^{O(4^L)}$ time algorithm for L-bit instances of Bin Packing.*

Proof. The Running time of $\mathcal{A}$ is dominated by constructing the edges in the state space graph. There are $n^{O(4^L)}$ of these edges, and each takes $n^{O(4^L)}$ time to test for inclusion.

4 Euclidean TSP

In this section we present an FPTAS for L-bit instances of Euclidean TSP, which provides an $(1 + \epsilon)$-approximation in time $n^{O(4^L)} \cdot \log 1/\epsilon$.

4.1 Preliminaries

The version of the Euclidean TSP we consider is the following. Given a set of vertices $V = \{(\bar{x}, y)\}$ on the plane,[3] what is the shortest tour that visits each point exactly once, when edge lengths are given by the L_2 distance between two endpoints? We consider the precision of an instance with respect to the coordinates of its points in the following manner. We say an instance of Euclidean TSP has precision L if for every $((m_{\bar{x}}, x_{\bar{x}}), (m_y, x_y)) \in V$, $\max\{m_{\bar{x}}, m_y\} < 2^L$, and both $m_{\bar{x}}$ and m_y are odd. Equivalently, all points in V must lie along a series of axis-aligned squares SQ_z centered on the origin, with side length $z = i \cdot 2^j$, and all $i \leq 2^L$. See Figure 1 for an example.

While we would like to show an exact algorithm for fixed-precision instances of Euclidean TSP, that goal is frustrated by the fact that the problem is not known to be in NP. Specifically, there are no known polynomial time algorithms for comparing the magnitude of sums of square roots. Thus, given two tours with square-root edge lengths, we have no efficient way to decide which is shorter. Instead, we provide an algorithm that achieves a $(1 + \epsilon)$-approximation with a

---

[3] We use $\bar{x}$ to denote the x-axis of the plane in order to avoid confusion with the exponent of a fixed precision integer.

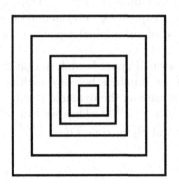

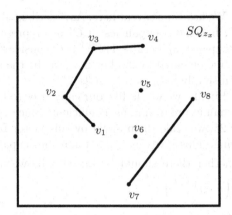

Fig. 1. The initial squares $SQ_1, SQ_2,$ SQ_3, SQ_4, SQ_6 and SQ_8 on which points of a 2-bit precision Euclidean TSP instance can lie

Fig. 2. If the edges of W_x are given as above, then $D_x = \{0 : \{v_5, v_6\}, 1 : \{(v_1, v_4), (v_7, v_8)\}\}$. All remaining vertices (v_2 and v_3) then have degree two.

runtime dependence on ϵ that is only $O(\log 1/\epsilon)$. We do this by rounding edge lengths to the nearest multiple of $2^{\lfloor \log \epsilon \rfloor}$. Going forward we assume that this rounding has been done.

For simplicity, we rescale the edge lengths and present our algorithm as an exact algorithm taking integral edge lengths in addition to the set of vertices V. We also give our bounds assuming that we can do operations on edge lengths in constant time. At the end of the section, we reintroduce the approximate nature of our result and the $O(\log 1/\epsilon)$ cost of doing operations on edge lengths. Note that this problem with radicals does not arise when edge lengths are given by the L_1 norm, and in this case the above rounding and approximation is unnecessary.

4.2   The Algorithm

Again, we begin by giving an overview of an algorithm $\mathcal{B}$ which solves the problem, and follow with an analysis of its efficiency. Like $\mathcal{A}$ discussed in the previous section, $\mathcal{B}$ is nondeterministic and runs in rounds. In each round, $\mathcal{B}$ considers points lying on a single SQ_z (as described above), in increasing order of side length. The algorithm maintains an edge set W, and for each vertex v considered in a round, the algorithm guesses all edges in an optimal tour connecting v to previously processed vertices. After the last round, when all points lying on the outermost square are processed, W will contain edges that form an optimal tour of V. The intuition underlying our algorithm is similar to that of Arora [2]. Because points lie along a series of SQ_z, and z is fixed precision, we expect few edges to cross these squares, as cheaper tours may be constructed by 'shortcutting' such in-out edges. This allows a dynamic programming formulation that keeps as its state the entry and exit points of these squares, and gathers the (rare) crossing edges.

More formally, let T be an edge set that forms an optimal tour of V. We define $T_z \subseteq T$ to be the subset of edges in T with both endpoints on or inside SQ_z. Assume that the points in V lie on a series of squares with side length $z_1 < z_2 < \ldots < z_{\max}$ (clearly, there are at most n such squares). Then in round x, $\mathcal{B}$ will add the edges in $T_{z_x} - T_{z_{x-1}}$ to W.

We use the following construction to implement $\mathcal{B}$. We define $D_x = d(W_x)$ to be an augmented degree sequence of the vertices in V induced by the edges present in W at the start of round x. We note that T_{z_x} is composed of disjoint paths; our augmentation is to record in D_x the endpoints of these paths (in pairs). See Figure 2 for an example. This information is necessary to check for the presence of premature cycles, as we will see later. We then create a graph G_{deg} with all possible D_x as vertices. We add a directed *transition edge* with weight w between D_x and D_{x+1} if there is an edge set S_x with cumulative weight w such that $D_x + d(S_x) = D_{x+1}$. We further require that adding S_x to D_x does not create any cycles, save when $x+1 = x_{\max}$. For each pair (D_x, D_{x+1}), we keep only the lightest such transition edge, and annotate it with the set S_x. From this graph, we can recover T by taking the union of the S_x attached to a shortest path between $D_0 = (0, 0, \ldots, 0)$ and $D_{x_{\max}} = (2, 2, \ldots, 2)$.

4.3 Analysis

We now show that the number of vertices in G_{deg} is bounded, and that we can efficiently construct its edge set. We begin with the following lemma, whose proof appears in Appendix A.

Lemma 4. *Consider a L-bit precision set of vertices V on the plane. Then for all z, any optimal tour of V will cross SQ_z at most $O(4^L)$ times.*

We now consider the number of distinct feasible values of D_x, for a fixed x. Clearly, all points outside of SQ_z have degree zero. By Lemma 7, we know that at most $O(4^L)$ of the remaining vertices have a degree that is less than two. Thus D_x can be represented compactly by listing the vertices in this set, their degree, and all path endpoints (that is, pairings of degree one vertices). This gives a bound of $n^{O(4^L)}$ on the number of distinct values.

What remains is to demonstrate that for each pair of degree sequences (D_x, D_{x+1}), the minimum weight edge set S_x whose addition to D_x yields D_{x+1} (and no cycles) can be computed in polynomial time. We note that all edges in S_x have at least one endpoint on SQ_{z_x}. We refer to edges with exactly one endpoint on SQ_{z_x} as *single* edges, and those with both endpoints on SQ_{z_x} as *double* edges. For each point v strictly inside SQ_{z_x}, we must add $D_{x+1}(v) - D_x(v)$ single edges incident to v. We enumerate all $n^{O(4^L)}$ possible non-cycle-inducing sets S_x^s of single edges, and claim that given such a set, the minimum weight non-cycle-inducing set of double edges S_x^d can be computed in polynomial time.

Lemma 5. *Let D_x and D_{x+1} be degree sequences as described above. Let S_x^s be a non cycle inducing set of single edges. Then the minimum weight non-cycle-inducing set of double edges S_x^d satisfying $D_x + d(S_x^s) + d(S_x^d) = D_{x+1}$ can be constructed in time $n^{O(4^L)}$.*

Proof. All possible S_x^d can be enumerated as follows. Consider any point v along SQ_{z_x} with $D_{x+1}[v] = D_x[v] + d(S_x^s)[v] + 1$. The edges in S_x^d that v can reach form a path consisting of connected segments of the perimeter of SQ_{z_x}. This path cannot cross itself (for if this is the case, then there is a less expensive tour that does not include this crossing), so there can be at most one segment on each side of SQ_{z_x}. We can enumerate the starting and ending vertex of each segment, giving the path starting from a single vertex $O(n^8)$ possibilities. By Lemma 7, there are at most $n^{O(4^L)}$ possible starting vertices. Finally, for each set of paths, we spend $O(n)$ time to check that no cycles have been created and that the degree sequence induced is correct.

We can now give our main lemma.

Lemma 6. *There is an* $n^{O(4^L)}$ *time exact algorithm for L-bit precision instances of Euclidean TSP with integral edge lengths.*

Proof. Our runtime is dominated by the time to construct G_{deg}. By Lemma 7 we have at most $n^{O(4^L)}$ possible edges in G_{deg}. By Lemma 5, the cost to check each edge is $n^{O(4^L)}$.

We conclude by relaxing our constraints on edge lengths.

Theorem 2. *There is an FPTAS for L-bit instances of Euclidean TSP that achieves an* $(1 + \epsilon)$-*approximation in time* $n^{O(4^L)} \log 1/\epsilon$.

5   Conclusion

We have given algorithms demonstrating that two natural problems, bin-packing and Euclidean TSP, are polynomial time solvable when their input numbers are given in fixed precision. We have therefore offered further evidence that the class of fixed-precision tractable NP-hard problems is a meaningful class worthy of further study. Several additional observations arise regarding this class:

1. As argued in the introduction, fixed-precision tractability seems *orthogonal* to pseudo-polynomial tractability. There are problems, such as Knapsack, that have both pseudo-polynomial and fixed precision algorithms. Others, such as Group Knapsack, have pseudo-polynomial algorithms but do not have fixed-precision algorithms. Conversely, while bin-packing and Euclidean TSP do not have fully polynomial approximation schemes, our algorithms show them to be fixed precision tractable, but only when the number of mantissa bits is *fixed* (note that if we could handle a logarithmic number of bits of precision, we would have a pseudo-polynomial algorithm).

2. Just how large is this class of fixed-precision tractable problems? Fixed precision tractability seems fragile, in a way the pseudo-polynomial algorithms are not. Add some polynomial size integers and you get another; add two low-precision numbers with different exponents and you suddenly have a high precision number. Does this fragility mean that few problems are tractable in this way?

3. Work by Korte and Schrader [9] shows a tight connection between pseudo polynomial algorithms and fully polynomial approximation schemes— problems that have one tend to have the other, in a formalizable sense. There is a similar coupling between algorithms with (non-fully) polynomial approximation schemes, and algorithms that can be solved by brute force enumeration when they are limited to a few input "types." Is there a similar connection between fixed-precision tractability and, for example, inverse approximation as defined by Orlin? One direction is obvious (as is the case for FPTAS and pseudo-polynomial algorithms) but the other is not clear.

References

1. Ahuja, R.K., Orlin, J.B.: Inverse Optimization. Operations Research 49(5), 771–783 (2001)
2. Arora, S.: Polynomial time approximation schemes for Euclidean traveling salesman and other geometric problems. Journal of the ACM (JACM) 45(5), 753–782 (1998)
3. Balasubramanian, R., Fellows, M.R., Raman, V.: An improved fixed-parameter algorithm for vertex cover. Information Processing Letters 65(3), 163–168 (1998)
4. Christofides, N.: Worst-case analysis of a new heuristic for the traveling salesman problem. In: Symposium on new directions and recent results in algorithms and complexity, page 441 (1976)
5. Coffman Jr., E.G., Garey, M.R., Johnson, D.S.: Approximation algorithms for bin packing: a survey. In: Approximation algorithms for NP-hard problems, pp. 46–93 (1996)
6. Downey, R.G., Fellows, M.R.: Fixed-Parameter Tractability and Completeness I: Basic Results. SIAM J. Comput. 24(4), 873–921 (1995)
7. Fernandez de la Vega, W., Lueker, G.S.: Bin packing can be solved within $1+\varepsilon$ in linear time. Combinatorica 1(4), 349–355 (1981)
8. Karp, R.M.: Reducibility among combinatorial problems. Complexity of Computer Computations 43, 85–103 (1972)
9. Korte, B., Schrader, R.: On the Existence of fast Approximation Schemes (1982)
10. Orlin, J.B., Schulz, A.S., Sengupta, S.: ϵ-optimization schemes and L-bit precision: alternative perspectives in combinatorial optimization (extended abstract). In: ACM Symposium on Theory of Computing, pp. 565–572 (2000)
11. Papadimitriou, C.H.: Euclidean TSP is NP-complete. Theoretical Computer Science 4, 237–244 (1977)
12. Vazirani, V.V.: Approximation Algorithms. Springer, Heidelberg (2001)

A   Geometric Proofs

Lemma 7. *Consider a L-bit precision set of vertices V on the plane. Then for all z, any optimal tour of V will cross SQ_z at most $O(4^L)$ times.*

Proof. An edge crosses SQ_z if it has exactly one endpoint strictly outside SQ_z. We refer to the set of crossing edges as Q, breaking it into two distinct subsets Q_1 and Q_2 such that $Q_1 \cup Q_2 = Q$. Edges in Q_1 have an endpoint on SQ_{z_a},

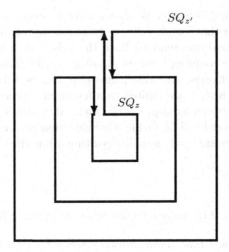

Fig. 3. This is an example 'brute force' tour that will cover all points in Q_1. We take $z' \leq 3z$. Note that the horizontal distance between the vertical edges can be arbitrarily small.

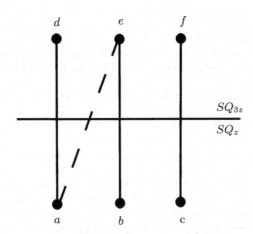

Fig. 4. The solid lines between points exist in Q_2. The dashed line indicates that there is a path from a to e that does not go through b, c, d, or f. This edge is guaranteed (modulo permutations on vertex labels) to exist, because there are no other diagonal edges between these vertices in Q_2, by definition.

$z < z_a \leq 3z$, and edges in Q_2 have an endpoint on SQ_{z_b}, $3z < z_b$. In what follows we show that $|Q_1| \leq 9 \cdot 4^L$ and $|Q_2| \leq 192$, and the claim follows immediately.

We can construct a tour that replaces edges in Q_1 with (shortcutting) the construction shown in Figure 3. Each SQ_{z_a} is toured via edges with total weight $6z_a$, while each edge in Q_1 with an endpoint on SQ_{z_a} has length at least $z_a/2^L$. Since there are at most $3 \cdot 2^{L-1}$ squares between SQ_z and SQ_{3z}, this gives an upper bound of $9 \cdot 4^L$ on the size of Q_1.

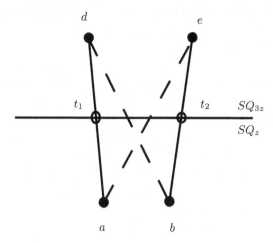

Fig. 5. We have that $\text{len}(\overline{ab}) \le \sqrt{2}z$ and $\text{len}(\overline{t_1 t_2}) \le z/2$. Also, both $\text{len}(\overline{t_1 a})$ and $\text{len}(\overline{t_2 b})$ are at least $2z$. Thus we can replace $\{(a, d), (b, d)\}$ with $\{(a, b), (d, t_1), (t_1, t_2), (t_2, e)\}$, and end up with a shorter tour without adding any cycles.

Now, assume that $|Q_2| > 192$. Then there must be a segment along the perimeter of SQ_{3z} of length $z/2$ that is intersected by 8 edges in Q_2. Then there are distinct vertices $H = \{a, b, c, d, e, f\}$ as in Figure 4 such that the set $\{(a, e)(a, f), (b, d), (b, f), (c, d), (c, e)\}$ intersected with Q_2 is empty. Thus, there must be $v_1 \in (a, b, c)$ and $v_2 \in (d, e, f)$ such that there is a path from v_1 to v_2 is neither direct nor passes through any other vertices in H. This in turn induces a graph like that found in Figure 5, where without loss of generality $v_1 = a$ and $v_2 = e$. Here, we can replace the edges $\{(a, d), (b, e)\}$ with edges $\{(a, b), (d, t_1), (t_1, t_2), (t_2, e)\}$ for a shorter tour, contradicting the assertion that $|Q_2| > 192$.

Approximating Maximum Subgraphs without Short Cycles

Guy Kortsarz[1,*], Michael Langberg[2], and Zeev Nutov[3]

[1] Rutgers University, Camden
Currently visiting IBM Research Yorktown Heights
guyk@camden.rutgers.edu
[2] The Open University of Israel
mikel@openu.ac.il
[3] The Open University of Israel
nutov@openu.ac.il

Abstract. We study approximation algorithms, integrality gaps, and hardness of approximation, of two problems related to cycles of "small" length k in a given graph. The instance for these problems is a graph $G = (V, E)$ and an integer k. The k-Cycle Transversal problem is to find a minimum edge subset of E that intersects every k-cycle. The k-Cycle-Free Subgraph problem is to find a maximum edge subset of E without k-cycles.

The 3-Cycle Transversal problem (covering all triangles) was studied by Krivelevich [Discrete Mathematics, 1995], where an LP-based 2-approximation algorithm was presented. The integrality gap of the underlying LP was posed as an open problem in the work of Krivelevich. We resolve this problem by showing a sequence of graphs with integrality gap approaching 2. In addition, we show that if 3-Cycle Transversal admits a $(2 - \varepsilon)$-approximation algorithm, then so does the Vertex-Cover problem, and thus improving the ratio 2 is unlikely. We also show that k-Cycle Transversal admits a $(k - 1)$-approximation algorithm, which extends the result of Krivelevich from $k = 3$ to any k. Based on this, for odd k we give an algorithm for k-Cycle-Free Subgraph with ratio $\frac{k-1}{2k-3} = \frac{1}{2} + \frac{1}{4k-6}$; this improves over the trivial ratio of $1/2$.

Our main result however is for the k-Cycle-Free Subgraph problem with even values of k. For any $k = 2r$, we give an $\Omega\left(n^{-\frac{1}{r} + \frac{1}{r(2r-1)} - \varepsilon}\right)$-approximation scheme with running time $\varepsilon^{-\Omega(1/\varepsilon)}\text{poly}(n)$. This improves over the ratio $\Omega(n^{-1/r})$ that can be deduced from extremal graph theory. In particular, for $k = 4$ the improvement is from $\Omega(n^{-1/2})$ to $\Omega(1/n^{-1/3-\varepsilon})$.

Similar results are shown for the problem of covering cycles of length $\leq k$ or finding a maximum subgraph without cycles of length $\leq k$.

1 Introduction

In this work, we study approximation algorithms, integrality gaps, and hardness of approximation, of two problems related to cycles of a given "small" length k

* Partially supported by NSF Award Grant number 072887.

A. Goel et al. (Eds.): APPROX and RANDOM 2008, LNCS 5171, pp. 118–131, 2008.
© Springer-Verlag Berlin Heidelberg 2008

(henceforth k-cycles) in a graph. The instance for each one of these problems is an undirected graph $G = (V, E)$ and an integer k. The goal is:

k-Cycle Transversal:
Find a minimum edge subset of E that intersects every k-cycle.

k-Cycle Free Subgraph:
Find a maximum edge subset of E without k-cycles.

Note that k-Cycle Transversal and k-Cycle-Free Subgraph are complementary problems, as the sum of their optimal values equals $|E| = m$; hence they are equivalent with respect to their optimal solutions. However, they differ substantially when considering approximate solutions. Also note that for $k = O(\log n)$ the number of k cycles in a graph can be computed in polynomial time, c.f., [3], and that it is polynomial for any fixed k. The k-Cycle Transversal problem is sometimes referred to as the "k-cycle cover" problem (as one seeks to cover k-cycles by edges). We adapt an alternative name, to avoid any mixup with an additional problem that has the same name – the problem of covering the edges of a given graph with a minimum family of k-cycles.

We will also consider problems of covering cycles of length $\leq k$ or finding a maximum subgraph without cycles of length $\leq k$. We will elaborate on the relation of these problems to our problems later. Most of our results extend to the case when edges have weights, but for simplicity of exposition, we consider unweighted and simple graphs only. We will also assume w.l.o.g. that G is connected.

1.1 Previous and Related Work

Problems related to k-cycles are among the most fundamental in the fields of Extremal Combinatorics, Combinatorial Optimization, and Approximation Algorithms, and they were studied extensively for various values of k. See for example [5,1,2,17,4,8,10,12,11,13,14,16,15,6] for only a small sample of papers on the topic. 3-Cycle Transversal was studied by Krivelevich [12]. Erdös et al. [6] considered 3-Cycle Transversal and 3-Cycle-Free Subgraph and their connections to related problems. Pevzner et al. [18] studied the problem of finding a maximum subgraph without cyles of lengt $\leq k$ in the context of computational biology, and suggested some heuristics for the problem, without analyzing their approximation ratio. However, most of the related papers studied k-Cycle-Free Subgraph in the context of extremal graph theory, and address the maximum number of edges in a graph without k-cycles (or without cycles of length $\leq k$). This is essentially the k-Cycle-Free Subgraph problem on complete graphs. In this work we initiate the study of k-Cycle-Free Subgraph in the context of approximation algorithms on general graphs.

As the state of the art differs substantially for odd and even values of k, we consider these cases separately. But for both odd and even k, note that k-Cycle Transversal is a particular case of the problem of finding a minimum transversal in a k-uniform hypergraph (which is exactly the Hitting-Set problem). Thus a simple greedy algorithm which repeatedly removes a k-cycle until no k-cycles remain, has approximation ratio k.

Odd k: For k-Cycle Transversal, an improvement over the trivial ratio of k was obtained for $k = 3$ by Krivelevich [12]. Let $\mathcal{C}_k(G)$ denote the set of k cycles in G, and let $\tau^*(G)$ denote the optimal value of the following LP-relaxation for k-Cycle Transversal:

$$\min \sum_{e \in E} x_e \tag{1}$$
$$\text{s.t. } \sum_{e \in C} x_e \geq 1 \ \forall C \in \mathcal{C}_k(G)$$
$$x_e \geq 0 \qquad \forall e \in E$$

Theorem 1 (Krivelevich [12]). 3-Cycle Transversal *admits a 2-approximation algorithm, that computes a solution of size at most* $2\tau^*(G)$.

For odd values of k, k-Cycle-Free Subgraph admits an easy $1/2$-approximation algorithm, as it is well known that any graph G has a subgraph without odd cycles (namely, a bipartite subgraph) containing at least half of the edges (such a subgraph can be computed in polynomial time). In fact, the problem of computing a maximum bipartite subgraph is exactly the Max-Cut problem, for which Goemans and Williamson [9] gave an 0.878-approximation algorithm. Note however that the solution found by the Goemans-Williamson algorithm has size at least 0.878 times the size of an optimal subgraph without odd cycles at all, and the latter can be much smaller than the optimal subgraph without k-cycles only.

Even k: For k-Cycle Transversal with even values of k we are not aware of any improvements over the trivial ratio of k. For k-Cycle-Free Subgraph with even k, it is no longer the case that G has a k-cycle free subgraph containing at least half of the edges. The maximum number $\text{ex}(n, C_{2r})$ of edges in a graph with n nodes and without cycles of length $k = 2r$ has been extensively studied. This is essentially the $2r$-Cycle-Free Subgraph problem on complete graphs. This line of research in extremal graph theory was initiated by Erdös [5]. The first major result is known as the "Even Circuit Theorem", due to Bondy and Simonovits [4], states that any undirected graph without even cycles of length $\leq 2r$ has at most $O(rn^{1+1/r})$ edges. This bound was subsequently improved. To the best of our knowledge, the currently best known upper bound on $\text{ex}(n, C_{2r})$ due to Lam and Verstraëte [15] is $\frac{1}{2}n^{1+1/r} + 2r^2 n$. We note that the best lower bounds on $\text{ex}(n, C_{2r})$ are as follows. For $r = 2, 3, 5$ it holds that $\text{ex}(n, C_{2r}) = \Theta(n^{1+1/r})$. For other values of r, the existence of a $2r$-cycle-free graph with $\Theta(n^{1+1/r})$ has not been established, and the best lower bound known is $\text{ex}(n, C_{2r}) = \Omega\left(n^{1+\frac{2}{3k-3+\varepsilon}}\right)$ where $\varepsilon = 0$ if r is odd and $\varepsilon = 1$ if r is even; we refer the reader to [16] for a summary of results of this type. All this implies that on complete graphs (a case which was studied extensively), the best known ratios for $2r$-Cycle-Free Subgraph are: constant for $r = 2, 3, 5$, and $\Omega\left(n^{-\frac{1}{r}+\frac{2}{6r-3+\varepsilon}}\right)$ otherwise. For general graphs, the bound $\text{ex}(n, C_{2r}) = O\left(n^{1+1/r}\right)$ implies an $\Omega(n^{-1/r})$-approximation by taking a spanning tree of G as a solution. In particular, for $k = 4$, the approximation ratio is $\Omega(1/\sqrt{n})$, and no better approximation ratio was known for this case.

1.2 Our Results

Our main result is for the k-Cycle-Free Subgraph problem on even values of k. It can be summarized by the following theorem:

Theorem 2. *For $k = 2r$, k-Cycle-Free Subgraph admits an $\Omega\left(n^{-\frac{1}{r}+\frac{1}{r(2r-1)}-\varepsilon}\right)$-approximation scheme with running time $\varepsilon^{-\Omega(1/\varepsilon)}\mathrm{poly}(n)$. In particular, 4-Cycle-Free Subgraph admits an $\Omega(1/n^{-1/3-\varepsilon})$-approximation scheme.*

For dense graphs, we obtain better ratios that are close to the ones known for complete graphs. Proof of the following statement will appear in the full version of this paper.

Theorem 3. *Let $G = (V, E)$ be a graph with n nodes and at least εn^2 edges. Then G contains a $2r$-cycle-free subgraph with at least $\varepsilon \cdot \mathrm{ex}(n, C_{2r})$ edges.*

On the negative side, the only hardness of approximation result we obtain (again proof will appear in the full version of this paper) is APX-hardness. Thus for even values of k there is a large gap between the upper and lower bounds we present. Resolving this large gap is an intriguing question left open in our work.

Our next results are for odd k. Krivelevevich [12] posed as an open question if his (upper) bound of 2 on the integrality gap of LP (1) is tight for $k = 3$. We resolve this question, and in addition show that the ratio 2 achieved by Krivelevich for $k = 3$ is essentially the best possible.

Theorem 4

(i) *If 3-Cycle Transversal admits a $2 - \varepsilon$ approximation ratio for some positive universal constant $\varepsilon < 1/2$, then so does the Vertex-Cover problem.*
(ii) *For any $\varepsilon > 0$ there exist infinitely many undirected graphs G for which the integrality gap of LP (1) with $k = 3$ is at least $2 - \varepsilon$.*

We note that Theorem 4 holds also for any $k \geq 4$. We also extend the 2-approximation algorithm of Krivelevich [12] for 3-Cycle Transversal to arbitrary k which is odd, and use it to improve the trivial ratio of $1/2$ for k-Cycle-Free Subgraph.

Theorem 5. *For any odd k the following holds:*

(i) *k-Cycle Transversal admits a $(k - 1)$-approximation algorithm.*
(ii) *k-Cycle-Free Subgraph admits a $\left(\frac{1}{2} + \frac{1}{4k-6}\right)$-approximation algorithm.*

Some remarks are in place: Theorem 5 is valid also for digraphs, for any value of k. Our results can be used to give approximation algorithms for the problem of covering cycles of length $\leq k$, or finding a maximum subgraph without cycles of length $\leq k$. For $k = 3$ we have for both problems the same ratios as in Theorem 5. For $k \geq 4$, the problem of covering cycles of length $\leq k$ admits a k-approximation algorithm (via the trivial reduction to the Hitting Set problem). For the problem of finding a maximum subgraph without cycles of length $\leq k$, we can show the ratio $\Omega(n^{-1/3-\varepsilon})$ for any k. For $k \geq 6$ this follows from extremal graph theory results mentioned, while for $k = 4, 5$ this is achieved by first computing a bipartite subgraph G' of G with at least $|E|/2$ edges, and then applying on G' the algorithm from Theorem 2 for 4-cycles.

1.3 Techniques

The proof of Theorem 2 is the main technical contribution of this paper. Our algorithm for k-Cycle-Free Subgraph with $k = 2r$ consists of two steps. In the first step we identify in G a subgraph G' which is an *almost* regular bipartite graph with the property that G and G' have approximately the same optimal values. The construction of G' can be viewed as a preprocessing step of our algorithm and may be of independent interest for other optimization problems as well. In the second step of our algorithm, we use the special structure of G' to analyze the simple procedure that first removes edges at random from G' until only few k-cycles remain in G', and then continues to remove edges from G' deterministically (one edge per cycle) until G' becomes k-cycle free.

The proof of Theorem 4(i) gives an approximation ratio preserving reduction from Vertex-Cover on triangle free graphs to 3-Cycle Transversal. It is well known that breaking the ratio of 2 for Vertex-Cover on triangle free graphs is as hard as breaking the ratio of 2 on general graphs. The proof of Theorem 4(ii) uses the same reduction on graphs G that on one hand are triangle free, but on the other have a minimum vertex-cover of size $(1 - o(1))n$. Such graphs exist, and appear in several places in the literature; see for example [7].

The proof of part (i) of Theorem 5 is a natural extension of the proof of Krivelevich [12] of Theorem 1. Part (ii) simply follows from part (i).

Theorems 2, 4, and 5, are proved in Sections 2, 3, and 4, respectively.

2 Proof of Theorem 2

In what follows let $\mathsf{opt}(G)$ be the optimal value of the k-Cycle-Free Subgraph problem on G. We start by a simple reduction which shows that we may assume that our input graph G is bipartite, at the price of loosing only a constant in the approximation ratio. Fix an optimal solution G^* to k-Cycle Free Subgraph. Partition the vertex set V of G randomly into two subsets, A and B, each of size $n/2$, and remove edges internal to A or B. In expectation, the fraction of edges in G^* that remain after this process is $1/2$. With probability at least $1/3$ the fraction of edges in G^* that remain is at least $1/4$; here we apply the Markov inequality on the fraction of edges inside A and B.

Assuming that the input graph G is bipartite, our algorithm has two steps. In the first step, we extract from G a family $\mathcal{G}$ of subgraphs $G_i = (A_i + B_i, E_i)$, so that either: one of these subgraphs has a "θ-semi-regularity" property (see Definition 1 below) and a k-cycle-free subgraph of size *close* to $\mathsf{opt}(G)$ or we conclude that $\mathsf{opt}(G)$ is *small*. In the latter case, we just return a spanning tree in G. In the former case, it will suffice to approximate k-Cycle-Free Subgraph on $G_i \in \mathcal{G}$, which is precisely what we do in the second step of the algorithm.

Definition 1. *A subset A of nodes in a graph is θ-semi-regular if $\Delta_A \leq \theta \cdot d_A$ where Δ_A and d_A denote the maximum and the average degree of a node in A, respectively. A bipartite graph with sides A, B is θ-semi-regular if each of A, B is θ-semi-regular.*

We will prove the following two statements that imply Theorem 2.

Lemma 1. *Let $k = 2r$. For any bipartite instance G of k-Cycle-Free Subgraph there exists an algorithm that in $\varepsilon^{-O(1/\varepsilon)}\mathrm{poly}(n)$ time finds a family $\mathcal{G}$ of at most $2\varepsilon^{-2/\varepsilon}$ subgraphs of G so that at least one of the following holds:*

(i) *$\mathcal{G}$ contains an $n^{2\varepsilon}$-semi-regular bipartite subgraph G_i of G so that $\mathrm{opt}(G_i) = \Omega(\varepsilon^{2/\varepsilon})\mathrm{opt}(G)$.*
(ii) *$\mathrm{opt}(G) = O\left(n\varepsilon^{-2/\varepsilon}\right)$.*

Lemma 2. *k-Cycle-Free Subgraph on bipartite θ-semi-regular instances $G = (A + B, E)$ and $k = 2r$ admits an $\Omega\left(\left(\theta r(|A||B|)^{\frac{r-1}{r(2r-1)}}\right)^{-1}\right)$-approximation ratio in (randomized) polynomial time.*

Let us show that Lemmas 1 and 2 imply Theorem 2 for bipartite graphs. We first compute the family $\mathcal{G}$ as in Lemma 1. Then, for each $G_i \in \mathcal{G}$ we compute a k-cycle-free subgraph H_i of G_i using the algorithm from Lemma 2, with $\theta = n^{2\varepsilon}$. Let H be the largest among the subgraphs H_i computed. If H has more than n edges, we output H. Else, we return a spanning tree in G.

2.1   Reduction to θ-Semi-Regular Graphs (Proof of Lemma 1)

Let $G = (A+B, E)$ be a bipartite connected graph, let $\varepsilon > 0$ be a small constant, let $n = |A| + |B|$, and let $\theta = n^\varepsilon$. For simplicity of exposition we will assume that θ and $\ell = 1/\varepsilon$ are integers.

We define an iterative process which partitions a subgraph $G' = (A' + B', E')$ of G with $A' \subseteq A$ and $B' \subseteq B$ into at most $\ell = 1/\varepsilon$ subgraphs so that at least one of the sides in each subgraph is θ-semi-regular. Specifically, the family $\mathcal{F}(G', A)$ is defined as follows. Partition the nodes in A' into at most ℓ sets A_j, where A_j consists of nodes in A' of degree in the range $[\theta^j, \theta^{j+1})$. The family $\mathcal{F}(G', A)$ consists of the graphs $G_j = G' - (A' - A_j)$ (namely, G_j is the induced subgraph of G' with sides A_j and B'). Note that A_j is a θ-semi-regular node set in G_j, but G_j may not be θ-semi-regular. In a similar way, the family $\mathcal{F}(G', B)$ is defined. Since the the union of the subgraphs in $\mathcal{F}(G', A)$ is G', and since $|\mathcal{F}(G', A)| = 1/\varepsilon$, there exists $G'' \in \mathcal{F}(G', A)$ so that $\mathrm{opt}(G'') \geq \varepsilon \cdot \mathrm{opt}(G')$; a similar statement holds for $\mathcal{F}(G', B)$. For a family $\mathcal{G}$ of subgraphs of G let $\mathcal{F}(\mathcal{G}, A) = \bigcup\{\mathcal{F}(G', A) : G' \in \mathcal{G}\}$ and $\mathcal{F}(\mathcal{G}, B) = \bigcup\{\mathcal{F}(G', B) : G' \in \mathcal{G}\}$.

Define a sequence of families of subgraphs of G as follows. $\mathcal{G}_0 = \{G\}$, $\mathcal{G}_1 = \mathcal{F}(\mathcal{G}_0, A)$, $\mathcal{G}_2 = \mathcal{F}(\mathcal{G}_1, B)$, and so on. Namely, $\mathcal{G}_i = \mathcal{F}(\mathcal{G}_{i-1}, A)$ if i is odd and $\mathcal{G}_i = \mathcal{F}(\mathcal{G}_{i-1}, B)$ if i is even. The following statement is immediate.

Claim. There exists a sequence of graphs $\{G_i = (A_i + B_i, E_i)\}_{i=0}^{2\ell}$ so that for every i: $G_i \in \mathcal{G}_i$, $G_i \subseteq G_{i-1}$, and $\mathrm{opt}(G_i) \geq \varepsilon \cdot \mathrm{opt}(G_{i-1})$.

We now study the structure of the graphs G_i. We show that the average degree in G_i is rapidly decreasing when i is increasing, until one of the G_i's is θ^2-semi-regular.

Claim. For every i, either G_{i+2} is θ^2-semi-regular, or at least one of the following holds:
- if i is even then $d_{A_{i+2}} < d_{A_{i+1}}/\theta$, where d_{A_i} is the average degree of A_i in G_i;
- if i is odd then $d_{B_{i+2}} < d_{B_{i+1}}/\theta$, where d_{B_i} is the average degree of B_i in G_i.

Proof. Suppose that i is even; the proof of the case when i is odd is similar. In $G_{i+1} \in \mathcal{G}_{i+1}$, the maximum degree $\Delta_{A_{i+1}}$ of A_{i+1} is at most θ times the average degree $d_{A_{i+1}}$ of A_{i+1}. If G_{i+2} is not θ^2 regular, then $\Delta_{A_{i+2}} \geq \theta^2 \cdot d_{A_{i+2}}$. However, the maximum degree in A_{i+2} is $\Delta_{A_{i+2}} \leq \Delta_{A_{i+1}} \leq \theta d_{A_{i+1}}$. This implies that $d_{A_{i+2}} \leq d_{A_{i+1}}/\theta$.

All in all, we conclude that for some $i \leq 2/\varepsilon$, G_i is θ^2-semi-regular and satisfies $\mathsf{opt}(G_i) \geq \varepsilon^i \mathsf{opt}(G)$; or $G_{2/\varepsilon}$ has constant average degree and satisfies $\mathsf{opt}(G_{2/\varepsilon}) \geq \varepsilon^{2/\varepsilon} \mathsf{opt}(G)$. The latter implies that $\mathsf{opt}(G) = O(\varepsilon^{-2/\varepsilon} n)$.

2.2   Algorithm for θ-Semi-Regular Graphs (Proof of Lemma 2)

Let $G = (A + B, E)$ be a bipartite θ-semi-regular graph. Let d_A be the average degree of nodes in A, and d_B be the average degree of nodes in B. Let $m = d_A|A| = d_B|B| = \sqrt{d_A d_B |A||B|}$ be the number of edges in G. Our algorithm builds on the following two results (the first is by A. Naor and Verstraëte [17]).

Theorem 6 ([17]). *The maximum number of edges in a bipartite graph $G = (A + B, E)$ without cycles of length $k = 2r$ is:*

$$(2r - 3)\left[(|A||B|)^{\frac{r+1}{2r}} + |A| + |B| \right] \quad \text{if r is odd}$$

$$(2r - 3)\left[|A|^{\frac{1}{2}}|B|^{\frac{r+2}{2r}} + |A| + |B| \right] \quad \text{if r is even}$$

Lemma 3. *The number of k-cycles in G is at most $m\theta^{2r-1}d_A^{r-1}d_B^{r-1}$.*

Proof. Consider picking $k = 2r$ distinct nodes in G, r from A and r from B, uniformly at random. Denote the nodes $a_1, a_2, \ldots, a_r \in A$ and $b_1, \ldots, b_r \in B$. We analyze the probability that $(a_1, b_1, a_2, b_2, \ldots, a_r, b_r, a_1)$ is a k cycle in G. In our analysis, our random choices are made according to the order of the cycle at hand, i.e., we first pick a_1, then b_1, then a_2, and so on. As a_1 has degree at most θd_A, the probability that b_1 is adjacent to a_1 is at most $\theta d_A/|B|$. Similarly, as b_1 has degree at most θd_B, the probability that a_2 is adjacent to b_1 is at most $\theta d_B/|A|$. Continuing this line of argument, it is not hard to verify that the probability that $(a_1, b_1, a_2, b_2, \ldots, a_r, b_r, a_1)$ is a k cycle in G is at most

$$\theta^{2r-1} \frac{d_A^r d_B^{r-1}}{|A|^{r-1}|B|^r} \ .$$

The number of k-tuples $(a_1, b_1, a_2, b_2, \ldots, a_r, b_r)$ in G is bounded by $|A|^r|B|^r$. Thus the number of k-cycles in G is at most $\theta^{2r-1}d_A^r d_B^{r-1}|A| = m\theta^{2r-1}d_A^{r-1}d_B^{r-1}$.

We now present our algorithm for k-Cycle Free Subgraph. In our analysis, we assume w.l.o.g. that $|A| \geq |B|$. We also assume that $|A|$ and $|B|$ are sufficiently large with respect to θ. Namely we assume that $|A||B| \geq (16\theta)^2$. Otherwise, the subgraph consisting of a single edge adjacent to v for each node $v \in A$, will suffice to yield an approximation ratio of $\Omega(1/\theta)$ which will equal $\Omega(n^{-2\varepsilon})$ in our final setting of parameters. Theorem 6 implies that

$$\mathsf{opt}(G) \leq 4r((|A||B|)^{\frac{r+1}{2r}} + |A|)$$

for any r. We now consider two cases: the case in which $(|A||B|)^{\frac{r+1}{2r}} \geq |A|$ and thus $\mathsf{opt}(G) \leq 8r(|A||B|)^{\frac{r+1}{2r}}$; and the case in which $(|A||B|)^{\frac{r+1}{2r}} \leq |A|$ and thus $\mathsf{opt}(G) \leq 8r|A|$. In the later case, the subgraph consisting of a single edge adjacent to v for each node $v \in A$ will suffice to yield an approximation ratio of $\Omega(1/r)$. We now continue to study the case in which $\mathsf{opt}(G) \leq 8r(|A||B|)^{\frac{r+1}{2r}}$.

Consider the following random process in which we remove edges from G. Each edge will be removed from G independently with probability p to be defined later. Denote the resulting graph by H. Denote by $q = 1 - p$ the probability that an edge is not removed.

Claim. As long as $mq \geq 16$, with probability at least $\frac{1}{2}$ the subgraph H satisfies:

- The number of edges in H is at least $mq/2$.
- The number of k cycles in H is at most $4q^{2r}m\theta^{2r-1}d_A^{r-1}d_B^{r-1}$.

Proof. The expected number of edges in H is $mq \geq 16$. Thus, using the Chernoff bound, the number of edges in H is at least half the expected value with probability $\geq 3/4$. In expectation, the number of k-cycles in H is at most $q^{2r}m\theta^{2r-1}d_A^{r-1}d_B^{r-1}$. With probability at least $3/4$ (Markov) the number of k-cycles in H will not exceed 4 times this expected value.

We now set q such that the number of k-cycles in H is at most $\frac{1}{2}$ the number of edges in H. Namely, we set q to satisfy $4q^{2r}m\theta^{2r-1}d_A^{r-1}d_B^{r-1} \leq mq/4$. Then:

$$q^{-1} = 16^{\frac{1}{2r-1}}\theta(d_A d_B)^{\frac{r-1}{2r-1}}.$$

With this setting of parameters and our assumption that $|A||B| \geq 16\theta^2$, we have that $mq \geq 16$ and Claim 2.2 holds. Thus, we may remove an additional single edge from each remaining k-cycle in H to obtain a k-cycle-free subgraph with at least $mq/4$ edges. This is the graph our algorithm will return. To conclude our proof, we now analyze the quality of our algorithm.

We consider 2 cases. Primarily, consider the case that $m \leq 8r(|A||B|)^{\frac{r+1}{2r}}$. This implies that $(|A||B|d_A d_B)^{\frac{1}{2}} \leq 8r(|A||B|)^{\frac{r+1}{2r}}$, which in turn implies that $d_A d_B \leq 64r^2(|A||B|)^{\frac{1}{r}}$. Using the fact that $\mathsf{opt}(G) \leq m$ we obtain in this case an approximation ratio of

$$\frac{mq}{4\mathsf{opt}(G)} \geq \frac{q}{4} = \Omega\left(\frac{1}{\theta(d_A d_B)^{\frac{r-1}{2r-1}}}\right) \geq \Omega\left(\frac{1}{\theta(64r^2|A||B|)^{\frac{r-1}{r(2r-1)}}}\right)$$

$$= \Omega\left(\frac{1}{\theta(|A||B|)^{\frac{r-1}{r(2r-1)}}}\right).$$

The second case is analyzed similarly. Assuming $m \geq 8r(|A||B|)^{\frac{r+1}{2r}}$ we get that $d_A d_B \geq 64r^2(|A||B|)^{\frac{1}{r}}$. Using the fact that $\mathrm{opt}(G) \leq 8r(|A||B|)^{\frac{r+1}{2r}}$ we obtain in this case an approximation ratio of

$$\frac{mq}{4\mathrm{opt}(G)} \geq \frac{(|A||B|d_A d_B)^{\frac{1}{2}}}{32r(|A||B|)^{\frac{r+1}{2r}} \cdot 16^{\frac{1}{2r-1}}\theta(d_A d_B)^{\frac{r-1}{2r-1}}} = \Omega\left(\frac{(d_A d_B)^{\frac{1}{2(2r-1)}}}{\theta r(|A||B|)^{\frac{1}{2r}}}\right)$$

$$= \Omega\left(\frac{1}{\theta r(|A||B|)^{\frac{r-1}{r(2r-1)}}}\right).$$

3   Proof of Theorem 4

Given an instance $J = (V_J, E_J)$ of Vertex-Cover, construct a graph $G = (V, E)$ for the 3-Cycle Transversal instance by adding to J a new node s and the edges $\{sv : v \in V_J\}$. Clearly, every edge $uv \in E_J$ corresponds to the 3-cycle $C_{uv} = \{us, sv, uv\}$ in G.

Suppose that J is 3-cycle-free. Then the set of 3-cycles of G is exactly $\{C_{uv} : uv \in E_J\}$. The following statement implies that w.l.o.g. we may consider only 3-cycle transversals that consist from edges incident to s.

Claim. Suppose that J is 3-cycle-free. Let F be a 3-cycle transversal in G and let $uv \in F \cap E_J$. Then $F - uv + su$ is also a 3-cycle transversal in G. Thus there exists a 3-cycle transversal $F' \subseteq \{sv : v \in V_J\}$ in G with $|F'| \leq |F|$.

Proof. The only 3-cycle in G that is covered by the edge uv is C_{uv}. This cycle is also covered by the edge su.

Claim. Suppose that J is 3-cycle-free. Then $U \subseteq V_J$ is a vertex-cover in J if, and only if, the edge set $F_U = \{su : u \in U\}$ is a k-cycle transversal in G.

Proof. We show that if $U \subseteq V_J$ is a vertex-cover in J then F_U is a 3-cycle transversal in G. Let C_{uv} be a 3-cycle in G. As U is a vertex-cover, $u \in U$ or $v \in U$. Thus $su \in F_U$ or $sv \in F_U$. In both cases, $C_{uv} \cap F_U \neq \emptyset$.

We now show that if F_U is a 3-cycle transversal in G, then U is a vertex-cover in J. Let $uv \in E_J$. Then C_{uv} is a 3-cycle in G, and thus $su \in F_U$ or $sv \in F_U$. This implies that $u \in U$ or $v \in U$, namely, the edge uv is covered by U.

From the claims above it follows that an α-approximation for 3-Cycle Transversal on G implies an α-approximation for Vertex-Cover on 3-cycle-free graphs J. Now we prove (for completeness, as we did not find an appropriate reference):

Claim. Any approximation algorithm with ratio $\alpha \geq 3/2$ for Vertex-Cover on 3-cycle-free graphs implies an α-approximation algorithm for Vertex-Cover (on general graphs).

Proof. Suppose that there is an α-approximation algorithm for Vertex-Cover on 3-cycle-free graphs. Let J be a general graph, and let $\mathsf{opt}(J)$ be the size of its minimum vertex cover. Consider the following two phase algorithm. Phase 1 starts with an empty cover F_1, and repeatedly, for every 3-cycle C in J, adds the nodes of C to F_1 and deletes them from J. Note that any vertex-cover contains at least two nodes of C, which implies a "local ratio" of $2/3$. Let J_2 be the triangle free graph obtained after Phase 1. In Phase 2 use the α-approximation algorithm (for 3-cycle-free graphs) to compute a vertex-cover F_2 of J_2. The statement follows since: $\mathsf{opt}(J) \geq \frac{2}{3}|F_1| + \mathsf{opt}(J_2) \geq \frac{2}{3}|F_1| + \frac{|F_2|}{\alpha} \geq \frac{|F_1|+|F_2|}{\alpha}$.

We now prove part (ii) of the theorem, namely, that for $k = 3$ the integrality gap of (1) is at least $2 - \varepsilon$. We will use the fact that for any $\varepsilon > 0$, there exist infinitely many graphs $J = (V_J, E_J)$ which are 3-cycle-free and have minimum vertex-cover of size at least $|V_J|(1 - \frac{\varepsilon}{2})$. Such graphs appear in various places in the literature. For example see Theorem 1.2 in [7] in which 3-cycle-free graphs J with independence number at most $\frac{\varepsilon}{2}|V_J|$ are presented. For such graph J, the minimum k-cycle cover in the corresponding graph G has size at least $|V_J|(1 - \frac{\varepsilon}{2})$. On the other hand, the solution $x_e = 1/2$ if e is incident to s and $x_e = 0$ otherwise is a feasible solution to LP (1) on G with value $|V_J|/2$. Hence the integrality gap is at least $\frac{(1 - \frac{\varepsilon}{2})}{1/2} = 2 - \varepsilon$.

Theorem 4 easily extends to arbitrary $k \geq 4$. We use the same construction as for the case $k = 3$, but in addition subdivide every edge of J by $k - 3$ nodes (and do not make any assumptions on J). Hence every edge $uv \in E_J$ is replaced by a path P_{uv} of the length $k - 2$, and $C_{uv} = P_{uv} + su + sv$ is a k-cycle in G. Since $k \geq 4$, G has no other k-cycles, namely, the set of k-cycles in G is $\{C_{uv} = P_{uv} + su + sv : uv \in E_J\}$. The rest of the proof of this case is identical to the case $k = 3$, and thus is omitted.

4   Proof of Theorem 5

To prove Theorem 5, we prove two theorems that consider a more general setting of a family $\mathcal{F}$ of subgraphs of G which are not necessarily k-cycles, nevertheless each subgraph $C \in \mathcal{F}$ is of size $\leq k$. We need some definitions. Let G be a graph and let $\mathcal{F}$ be a family of subgraphs (edge subsets) of G. For a subgraph H of G, let $\mathcal{F}(H)$ be the restriction of $\mathcal{F}$ to subgraphs of H; H is $\mathcal{F}$-*free* if $\mathcal{F}(H) = \emptyset$. An edge set F that intersects every member of $\mathcal{F}$ is an $\mathcal{F}$-*transversal*. We consider the following two problems, that generalize the problems k-Cycle-Free Subgraph and k-Cycle Transversal. The instance of the problems is a graph $G = (V, E)$ and a family $\mathcal{F}$ of subgraphs of G. The goal is:

$\mathcal{F}$-Transversal: Find a minimum size $\mathcal{F}$-transversal.

$\mathcal{F}$-Free Subgraph: Find a maximum size $\mathcal{F}$-free subgraph of G.

For $\mathcal{F} = \mathcal{C}_k(G)$, we get the problems k-Cycle Transversal and k-Cycle Free Subgraph, respectively. Let $\tau_{\mathcal{F}}^*(H)$ denote the optimal value of the following LP-relaxation for $\mathcal{F}$-Transversal on H:

$$\min \sum_{e \in E(H)} x_e \qquad (2)$$
$$\text{s.t. } \sum_{e \in C} x_e \geq 1 \ \forall C \in \mathcal{F}(H)$$
$$x_e \geq 0 \qquad \forall e \in E(H)$$

An edge of H is $\mathcal{F}$-*redundant* if no member of $\mathcal{F}(H)$ contains it; e.g., if $\mathcal{F} = C_k(G)$, then an edge of H is $\mathcal{F}$-redundant if it is not contained in any k-cycle of H. We prove:

Theorem 7. *Suppose that any subgraph H of G admits a polynomial time algorithm that:* (i) *Solves LP* (2) *for H;* (ii) *Finds $\mathcal{F}$-redundant edges of H;* (iii) *Finds an $\mathcal{F}(H)$-transversal of size at most $|E(H)| \cdot (k-1)/k$. Then there exist a polynomial time algorithm that finds an $\mathcal{F}(G)$-transversal of size $\leq (k-1) \cdot \tau_{\mathcal{F}}^*(G)$.*

To prove Theorem 5(ii) we connect the approximation of $\mathcal{F}$-Free Subgraph and $\mathcal{F}$-Transversal by the following theorem:

Theorem 8. *Suppose that for any graph G with m edges there exist a polynomial algorithm that finds an $\mathcal{F}(G)$-free subgraph of size $\geq \beta m$, and that $\mathcal{F}$-Transversal admits an α-approximation algorithm. Then k-Cycle-Free Subgraph admits an $\alpha\beta/(\alpha+\beta-1)$-approximation algorithm.*

Let us now show that Theorem 7 implies Theorem 5(i) and that Theorem 8 implies Theorem 5(ii). Let G be a graph with m edges. As was mentioned, it is not hard to find in G a subgraph with at least $m/2$ edges and without odd cycles. For Theorem 5(i), it is easy to see that this setting obeys the conditions of Theorem 7, hence we obtain a $(k-1)$-approximation for $\mathcal{F}$-Transversal in this case. For Theorem 5(ii), we apply Theorem 8 with $\beta = 1/2$ and $\alpha = k-1$. The ratio obtained is $\alpha\beta/(\alpha+\beta-1) = (k-1)/(2k-3) = \frac{1}{2} + \frac{1}{4k-6}$. We now prove Theorems 7 and 8 (in Sections 4.1 and 4.2, respectively).

4.1 Proof of Theorem 7

The algorithm is as follows:

Initialization: $H \leftarrow G$; $F_1 \leftarrow \emptyset$.
Phase 1:
While for an optimal solution x to (2) $x_e \geq 1/(k-1)$ for some $e \in E(H)$ do:
     $F_1 \leftarrow F_1 + e$; $H \leftarrow H - e$.
EndWhile
Phase 2:
- Remove all $\mathcal{F}(H)$-redundant edges from H. Denote the resulting graph by H_2.
- Compute an $\mathcal{F}(H_2)$-transversal F_2 of size at most $|E(H_2)| \cdot (k-1)/k$.
Return $F_1 \cup F_2$.

Under the assumptions of the Theorem, all steps can be implemented in polynomial time. It is also easy to see that the algorithm returns a feasible solution. We now analyze the approximation ratio. We start with a simple claim followed by our key Lemma.

Claim. Let H be the graph obtained after Phase 1 of our algorithm and let x_e be an optimal solution to *LP (2)* on H. Then $x_e = 0$ for every $\mathcal{F}(H)$-redundant edge e in H. Thus the restriction of x to H_2 is also an optimal solution to LP (2) on H_2.

Proof. Let e be an $\mathcal{F}(H)$-redundant edge. Assume for sake of contradiction that $x_e > 0$. We can now reduce the value of the LP solution by zeroing out x_e. The new solution is still valid, as e is $\mathcal{F}(H)$-redundant and thus does not appear in the first family of constraints of (2).

Let H_2 be obtained from H by removing all $\mathcal{F}(H)$-redundant edges. Then the restriction of x to H_2 is an optimal solution to (2) since any LP solution for H_2 can be extended to one for H by setting $x_e = 0$ for every $\mathcal{F}(H)$-redundant edge e. ∎

Using the claim above, we may assume that the subgraph H_2 has an optimal solution x to (2) in which $x_e < 1/(k-1)$ (for all $e \in E(H_2)$).

Lemma 4. *Let H_2 be a subgraph of G without $\mathcal{F}$-redundant edges and let x be an optimal solution to LP (2). If $x_e < 1/(k-1)$ for every $e \in E(H_2)$ then $\tau_{\mathcal{F}}^*(H_2) \geq |E(H_2)|/k$.*

Proof. Let $\nu_{\mathcal{F}}^*(H_2) = \tau_{\mathcal{F}}^*(H_2)$ denote the optimal value of the dual LP:

$$\max \sum_{C \in \mathcal{F}} y_C \tag{3}$$
$$\text{s.t.} \sum_{C \ni e} y_C \leq 1 \ \forall e \in E(H_2)$$
$$y_C \geq 0 \qquad \forall C \in \mathcal{F}(H_2)$$

Let x and y be optimal solutions to (2) and to (3), respectively. Consider two cases, after noting that the primal complementary slackness condition is:

$$x_e > 0 \implies \sum_{C \ni e} y_C = 1 \tag{4}$$

Case 1: $x_e > 0$ for every $e \in E(H_2)$.
In this case $\tau_{\mathcal{F}}^*(H) \geq |E(H_2)|/k$, since from (4) we get:

$$|E(H_2)| = \sum_{e \in E(H_2)} 1 = \sum_{e \in E} \sum_{C \ni e} y_C = \sum_{C \in \mathcal{F}(H_2)} |C| y_C \leq \sum_{C \in \mathcal{F}(H_2)} k y_C = k \nu_{\mathcal{F}}^*(H_2) = k \tau_{\mathcal{F}}^*(H_2).$$

Case 2: $x_f = 0$ for some $f \in E(H_2)$.
Since H_2 has no $\mathcal{F}$-redundant edges, there is $C \in \mathcal{F}(H_2)$ so that $f \in C$. Since $x_f = 0$, we have $\sum_{e \in C - f} x_e \geq 1$. Since $|C - f| \leq k - 1$, there exists $e \in C - f$ so that $x_e \geq 1/(k-1)$. A contradiction. ∎

We now bound the value of $|F_1|$ and $|F_2|$ with respect to $\tau_{\mathcal{F}}^*(G)$. We start with some notation. Let $H^0 = G$ be the starting point of our algorithm. Let H^1 be graph obtained from H^0 by the removal of e_1 after the first round of Phase 1. Similarly, for the i'th round of Phase 1, let H^i be the graph obtained from H^{i-1} by the removal of e_i. Let $H = H^\ell$ be the graph obtained after Phase 1 of our algorithm (here ℓ denotes the number of rounds in Phase 1). It is not hard to verify that $\tau_{\mathcal{F}}^*(H^{i-1}) \geq \tau_{\mathcal{F}}^*(H^i) + x_{e_i}$. Here x_{e_i} is obtained from the optimal solution to H^{i-1}. This implies that $\tau_{\mathcal{F}}^*(G) \geq \tau_{\mathcal{F}}^*(H) + \sum_{i=1}^{\ell-1} x_{e_i}$.

Now to bound $|F_1|$ and $|F_2|$. First notice that $|F_1| \leq (k-1)\sum_{i=1}^{\ell-1} x_{e_i}$. Recall that H_2 is the graph obtained in Phase 2 from H by removing all $\mathcal{F}(H)$-redundant edges. It also holds that, $|F_2| \leq |E(H_2)| \cdot (k-1)/k$. By Lemma 4, $\tau_{\mathcal{F}}^*(H_2) \geq |E(H_2)|/k$. Hence

$$\frac{|F_2|}{\tau_{\mathcal{F}}^*(H_2)} \leq \frac{|E(H_2)| \cdot (k-1)/k}{|E(H_2)|/k} = k-1 \ .$$

As by Claim 4.1, $\tau_{\mathcal{F}}^*(H) = \tau_{\mathcal{F}}^*(H_2)$ we have that

$$|F_1| + |F_2| \leq (k-1)(\tau_{\mathcal{F}}^*(H) + \sum_{i=1}^{\ell-1} x_{e_i}) \leq (k-1)\tau_{\mathcal{F}}^*(G) \ ,$$

which concludes our proof.

4.2 Proof of Theorem 8

In what follows let opt be the optimal solution value of the $\mathcal{F}$-Free Subgraph problem on G. We choose the better result F from the following two algorithms:

Algorithm 1: Find an $\mathcal{F}(G)$-free subgraph of size $\geq \beta m$.
Algorithm 2: Find an $\mathcal{F}(G)$-transversal I of size $\leq \alpha$ times an optimal $\mathcal{F}(G)$-transversal, and return $G - I$.

Algorithm 1 computes a solution of size $\geq \beta m$. Algorithm 2 computes a solution of size $\geq m - \alpha(m - \text{opt})$. The worse case is when these lower bounds coincide: $\beta m = m - \alpha(m - \text{opt})$ which implies $\text{opt} = m(\alpha + \beta - 1)/\alpha$. This gives the ratio $\frac{\beta m}{m(\alpha+\beta-1)/\alpha} = \frac{\alpha\beta}{\alpha+\beta-1}$. Formally, $|F| \geq \max\{\beta m, m - \alpha(m - \text{opt})\}$. Consider two cases:

Case 1: $\beta m \geq m - \alpha(m - \text{opt})$, so $\text{opt} \leq m(\alpha + \beta - 1)/\alpha$. Then

$$\frac{|F|}{\text{opt}} \geq \frac{\beta m}{\text{opt}} \geq \frac{\beta m}{(\alpha+\beta-1)/\alpha} = \frac{\alpha\beta}{\alpha+\beta-1} \ .$$

Case 2: $m - \alpha(m - \text{opt}) \geq \beta m$, so $m/\text{opt} \leq \alpha/(\alpha + \beta - 1)$. Then

$$\frac{|F|}{\text{opt}} \geq \frac{m - \alpha(m - \text{opt})}{\text{opt}} = \alpha - (\alpha-1) \cdot \frac{m}{\text{opt}} \geq \alpha - (\alpha-1) \cdot \frac{\alpha}{\alpha+\beta-1} = \frac{\alpha\beta}{\alpha+\beta-1} \ .$$

In both cases the ratio is bounded by $\frac{\alpha\beta}{\alpha+\beta-1}$, which concludes our proof.

5 Open Problems

For k-Cycle Transversal, we have ratios $k-1$ for odd values of k and k for even values of k. However, the best approximation threshold we have is 2. Closing this gap (even for $k = 4, 5$) is left open.

For k-Cycle-Free Subgraph, we have ratios $2/3$ for $k = 3$ and $n^{-1/3-\varepsilon}$ for $k = 4$. The best approximation threshold we have is APX-hardness. Hence, we do not even know if our ratio of $2/3$ for $k = 3$ is tight. Our result for $k = 3$ actually establishes a lower bound of $2/3$ on the integrality gap for the natural LP for 3-Cycle-Free Subgraph, but the best upper bound we have is only $3/4$. Finally, in our opinion, the most challenging open question is closing the huge gap for the case $k = 4$.

References

1. Alon, N.: Bipartite subgraphs. Combinatorica 16, 301–311 (1996)
2. Alon, N., Bollobás, B., Krivelevich, M., Sudakov, B.: Maximum cuts and judicious partitions in graphs without short cycles. J. of Comb. Th. B 88(2), 329–346 (2003)
3. Alon, N., Yuster, R., Zwick, U.: Finding and counting given length cycles. Algorithmica 17(3), 209–223 (1997)
4. Bondy, J.A., Simonovits, M.: Cycles of even lengths in graphs. J. Comb. Th. B 16, 97–105 (1974)
5. Erdös, P.: Extremal problems in graph theory. In: Fiedler, M. (ed.) Theory of Graphs and Its Applications. Academic Press, New York (1965)
6. Erdös, P., Gallai, T., Tuza, Z.: Covering and independence in triangle structures. Discrete Mathematics 150, 89–101 (1996)
7. Feige, U., Langberg, M., Schechtman, G.: Graphs with tiny vector chromatic numbers and huge chromatic numbers. SIAM J. Comput. 33(6), 1338–1368 (2004)
8. Furedi, Z., Naor, A., Verstraëte, J.: On the Turan number of the hexagon. Advances in Mathematics 203(2), 476–496 (2006)
9. Goemans, M.X., Williamson, D.P.: Improved approximation algorithms for maximum cut and satisfiability problems using semidefinite programming. J. ACM 42(6), 1115–1145 (1995)
10. Hoory, S.: The size of bipartite graphs with a given girth. Journal of Combinatorial Theory Series B 86(2), 215–220 (2002)
11. Komlós, J.: Covering odd cycles. Combinatorica 17(3), 393–400 (1997)
12. Krivelevich, M.: On a conjecture of Tuza about packing and covering of triangles. Discrete Mathematics 142(1-3), 281–286 (1995)
13. Kühn, D., Osthus, D.: Four-cycles in graphs without a given even cycle. Journal of Graph Theory 48(2), 147–156 (2005)
14. Lam, T.: A result on $2k$-cycle free bipartite graphs. Australasian Journal of Combinatorics 32, 163–170 (2005)
15. Lam, T., Verstraëte, J.: A note on graphs without short even cycles. The Electronic Journal of Combinatorics 12(1,N5) (2005)
16. Lazebnik, F., Ustimenko, V.A., Woldar, A.J.: Polarities and $2k$-cycle free graphs. Discrete Math. 197/198, 503–513 (1999)
17. Naor, A., Verstraëte, J.: A note on bipartite graphs without $2k$-cycles. Probability, Combinatorics and Computing 14(5-6), 845–849 (2005)
18. Pevzner, P.A., Tang, H., Tesler, G.: De novo repeat classification and fragment assembly. Genome Research 14(9), 1786–1796 (2004)

Deterministic 7/8-Approximation for the Metric Maximum TSP
(Extended Abstract)

Łukasz Kowalik and Marcin Mucha*

Institute of Informatics, University of Warsaw, Poland
{kowalik,mucha}@mimuw.edu.pl

Abstract. We present the first 7/8-approximation algorithm for the maximum traveling salesman problem with triangle inequality. Our algorithm is deterministic. This improves over both the randomized algorithm of Hassin and Rubinstein [2] with expected approximation ratio of $7/8 - O(n^{-1/2})$ and the deterministic $(7/8 - O(n^{-1/3}))$-approximation algorithm of Chen and Nagoya [1].

In the new algorithm, we extend the approach of processing local configurations using so-called loose-ends, which we introduced in [4].

1 Introduction

The Traveling Salesman Problem and its variants are among the most intensively researched problems in computer science and arise in a variety of applications. In its classical version, given a set of vertices V and a symmetric weight function $w : V^2 \to \mathbb{R}_{\geq 0}$ satisfying the triangle inequality one has to find a Hamiltonian cycle of minimum weight.

There are several variants of TSP, e.g. one can look for a Hamiltonian cycle of minimum or maximum weight (MAX-TSP), the weight function can be symmetric or asymmetric, it can satisfy the triangle inequality or not, etc.

In this paper, we are concerned with the MAX-TSP variant, where the weight function is symmetric and satisfies the triangle inequality. This variant is often called *the metric MAX-TSP*.

MAX-TSP (not necessarily metric) was first considered by Serdyukov in [5], where he gives a $\frac{3}{4}$-approximation. Next, a $\frac{5}{6}$-approximation algorithm for the metric case was given by Kostochka and Serdyukov [3]. Hassin and Rubinstein [2] used these two algorithms together with new ideas to achieve a randomized approximation algorithm with expected approximation ratio of $(\frac{7}{8} - O(n^{-1/2}))$. This algorithm has later been derandomized by Chen and Nagoya [1], at a cost of a slightly worse approximation factor of $(\frac{7}{8} - O(n^{-1/3}))$.

In this paper, we give a deterministic $\frac{7}{8}$-approximation algorithm for metric MAX-TSP. Our algorithm builds on the ideas of Serdyukov and Kostochka, but

---

* This research is partially supported by a grant from the Polish Ministry of Science and Higher Education, project N206 005 32/0807.

A. Goel et al. (Eds.): APPROX and RANDOM 2008, LNCS 5171, pp. 132–145, 2008.

is completely different from that of Hassin and Rubinstein. We apply techniques similar to those used earlier in [4] for the directed version of MAX-TSP with triangle inequality.

1.1 Closer Look at Previous Results

Classic undirected MAX-TSP algorithm of Serdyukov [5] starts by constructing two sets of edges of the input graph G: a maximum weight cycle cover $\mathcal{C}$ and a maximum weight matching M, and then removing a single edge from each cycle of $\mathcal{C}$ and adding it to M. It can be shown that we can avoid creating cycles in M, so in the end we get two sets of paths: $\mathcal{C}'$ and M'. These sets can be extended to Hamiltonian cycles arbitrarily. Since we started with a maximum weight cycle cover and a maximum weight matching, we have $w(\mathcal{C}') + w(M') \geq w(\mathcal{C}) + w(M) \geq \frac{3}{2}\text{OPT}$. It follows that the better of the two cycles has weight at least $\frac{3}{4}\text{OPT}$. Here, we used two standard inequalities: $w(\mathcal{C}) \geq \text{OPT}$ and $w(M) \geq \frac{1}{2}\text{OPT}$. The latter only holds for graphs with even number of vertices. The case of odd number of vertices needs separate treatment.

Serdyukov's algorithm works for any undirected graph, with weight function not necessarily satisfying the triangle inequality. However, if this inequality is satisfied, we can get a much better algorithm. Kostochka and Serdykov observed the following useful fact (see e.g. [2] for a proof).

Lemma 1 (Kostochka, Serdyukov [3]). *Let $G = (V, E)$ be a weighted complete graph with a weight function $w : E \to \mathbb{R}_{\geq 0}$ satisfying the triangle inequality. Let $\mathcal{C}$ be a cycle cover in G and let $Q = \{e_1, \dots, e_{|\mathcal{C}|}\}$ be a set of edges with exactly one edge from each cycle of $\mathcal{C}$. Then the collection of paths $\mathcal{C} \setminus Q$ can be extended in polynomial time to a Hamiltonian cycle H with*

$$w(H) \geq w(\mathcal{C}) - \sum_{i=1}^{|\mathcal{C}|} w(e_i)/2.$$

Kostochka and Serdyukov [3] propose an algorithm which starts by finding a maximum weight cycle cover $\mathcal{C}$ and then applies the above lemma with Q consisting of the lightest edges of cycles in $\mathcal{C}$. Since all cycles have length at least 3, the weight of the removed edges amounts to at most $\frac{1}{3}w(\mathcal{C})$, so we regain at least $\frac{1}{6}w(\mathcal{C})$, which leads to $\frac{5}{6}$-approximation. (Note that if it happens that all the cycles in $\mathcal{C}$ have length at least 4 we get $\frac{7}{8}$-approximation).

2 Our Approach

Similarly to Serdykov's algorithm (as well as that of Hassin and Rubinstein), our algorithm starts by constructing a maximum weight cycle cover $\mathcal{C}$ and maximum weight matching M. In our reasoning we need the inequality $w(M) \geq \frac{1}{2}\text{OPT}$, which holds only for graphs with even number of vertices. In the remainder of this paper we only consider such graphs. Our results can be extended to graphs with odd number of vertices, we defer the details to the full version of the paper.

In all previous algorithms edges are moved from the cycle cover $\mathcal{C}$ to the matching M. We do not follow this approach. Instead, we remove some edges from $\mathcal{C}$ and add some edges to M. The edges added to M are not necessarily the edges removed from $\mathcal{C}$. In fact, they might not even be cycle edges in $\mathcal{C}$. All we need to guarantee is that their total weight is sufficiently large compared to the weight loss in $\mathcal{C}$.

Here is how it works. Let $\min(C_i)$ be the lightest edge of a cycle $C_i \in \mathcal{C}$. Since removing a single edge from each C_i and then joining the resulting paths using Lemma 1 results in the weight loss equal to half the weight of the removed edges, it should be clear that we should remove $\min(C_i)$ from each C_i. The weight loss is then $\sum_i w(\min(C_i))/2$.

We are going to describe an iterative process of adding edges to a collection of paths P, initially equal to M. Edges will be added in *phases*, each phase corresponds to a single cycle $C_i \in \mathcal{C}$. After finishing the phase corresponding to C_i we will call C_i *processed*. The edges added in the phase corresponding to C_i will usually, but not necessarily belong to C_i or at least connect vertices of C_i. Their total weight will also be directly related to $w(C_i)$ and $w(\min(C_i))$. Let $(\alpha, \beta) \star C_i = \alpha w(C_i) + \beta w(\min(C_i))$. The following Lemma shows why this is a useful definition:

Lemma 2. *If during processing the cycles in $\mathcal{C}$, we can add edges of total weight at least $\sum_{C_i \in \mathcal{C}} (\alpha, 1/2) \star C_i$ to M, then we get a $(3/4 + \alpha/2)$-approximation algorithm.*

Proof. Let H_1 be the Hamiltonian cycle obtained from $\mathcal{C}$ by using Lemma 1, and let H_2 be the cycle obtained from M by processing all cycles of $\mathcal{C}$ and patching the resulting collection of paths into a Hamiltonian cycle. Then

$$w(H_1) + w(H_2) \geq \left[w(\mathcal{C}) - \sum_i w(\min(C_i))/2 \right] +$$

$$+ \left[w(M) + \alpha w(\mathcal{C}) + \sum_i w(\min(C_i))/2 \right] \geq (3/2 + \alpha)\mathrm{OPT},$$

so the heavier of the two cycles is a $(3/4 + \alpha/2)$-approximation.

In the remainder of the paper, we show that this can be done for $\alpha = 1/4$, yielding a 7/8-approximation.

2.1   Skeleton of the Algorithm

A graph P is sub-Hamiltonian if it is a family of disjoint paths or a Hamiltonian cycle (i.e. it can be extended to a Hamiltonian cycle). Let P be a family of disjoint paths. We say that set of edges S is *allowed* w.r.t. P, if S is disjoint from P and the edge sum of P and S is sub-Hamiltonian. We call an edge e *allowed* w.r.t P if $\{e\}$ is allowed w.r.t. P. If an edge is not allowed, we call it *forbidden*.

In the algorithm presented below, we maintain a sub-Hamiltonian graph P satisfying the following invariant.

Invariant 1. *For any vertex v, if $\deg_P(v) = 2$ then the cycle v belongs to has been already processed.*

Consider a phase of our algorithm and let C be the cycle that is still unprocessed. In this situation a set S of edges will be called a *support* of C if S is allowed w.r.t. P, and after adding S to P (and thus making C processed) Invariant 1 is satisfied.

The following is the skeleton of the algorithm, that we will develop in the remainder of the paper.

Algorithm 2.1. MAIN ALGORITHM

1: Let M be a heaviest matching and $\mathcal{C}$ a heaviest cycle cover in G.
2: Let H_1 be the Hamiltonian cycle obtained from $\mathcal{C}$ by using Lemma 1.
3: $P := M$
4: Mark all cycles in $\mathcal{C}$ as *unprocessed*.
5: **for** each unprocessed cycle C in $\mathcal{C}$ **do**
6:     Find S, a support of C of large weight.
7:     $P := P \cup S$
8:     Mark C as processed.
9: Arbitrarily patch P to a Hamiltonian cycle H_2.
10: Return the heavier of H_1 and H_2.

2.2   Loose-Ends

When considering a cycle C_i, we are going to extend P by adding some edges connecting the vertices of C_i. Ideally we would like to add $n_i/2$ new edges, where n_i is the length of C_i. However, this is not always possible, because some of the cycles have odd length and $n_i/2$ is not an integer. Instead we are going to use the idea of loose-ends introduced in [4].

A *loose-end* is a vertex v, for which $\deg_P(v) = 1$ even though the cycle it belongs to is already processed. A vertex v of cycle $C \in \mathcal{C}$ becomes a loose-end if no edge adjacent to v is added to P when C is processed. This vertex can be connected with some other vertex at a later stage and cease being a loose-end.

Consider two odd-length cycles C_1 and C_2, say both of length 5. When we process C_1, we can only add 2 edges to M, and some vertex $v \in C_1$ is not an endpoint of any of these edges, so it becomes a loose-end. Later, when we process C_2, we can add 3 edges to M, by connecting one of C_2's vertices with v. Using the triangle inequality, we can guarantee that this edge has large weight. So in this case we get a little less weight from C_1 and a little more weight from C_2. It is important to process cycles in order that guarantees that the weight lost when processing the earlier cycles (the ones that give loose-ends) is dominated by the weight gained when processing the later cycles (the ones that use loose-ends). We will show that the algorithm can determine this order.

Let S be a support of C in some phase of the algorithm. We will say that S is a k-support if after adding it to P (and thus processing cycle C) the number of loose-ends increases by at least k (k could be negative here).

In the following section we describe in detail how the cycles are processed in our algorithm. For even-length cycles we construct heavy 0-supports, and for odd-length cycles we construct both (-1)-supports and $(+1)$-supports.

When constructing (-1)-supports, we need to assume that at least one loose-end is available. Unfortunately, just one loose-end may be insufficient to guarantee the existence of a (-1)-support. This could happen if the loose-end u is connected to C, the cycle being processed, by a path in P. In that case, adding an edge between u and a vertex of C to P may create a cycle in P. This is acceptable only if that cycle is Hamiltonian (in particular, C would have to be the last cycle processed). Luckily, it turns out that two loose-ends are always sufficient to avoid creating such short cycles. Thus, when describing a (-1)-support for each odd cycle we will consider two situations: when there are two loose-ends, and when there is exactly one loose-end but the algorithm is in the last (i.e. $|\mathcal{C}|$-th) phase.

3   Processing Cycles

In this section we consider an arbitrary phase of the algorithm and we describe supports of unprocessed cycles. The construction of a support of such a cycle C may depend on the number of loose-ends and the way the collection P of paths constructed so far interacts with C, in particular on which edges of C are forbidden etc.

The following observations will be used in many of our proofs.

Observation 1. *Let C be an unprocessed cycle and let $M \subset E(C)$ be a matching. Let $\tilde{C}$ be any cycle in $P \cup M$. Then if $\tilde{C}$ contains an allowed edge of M, it contains at least two allowed edges. Also, if $\tilde{C}$ contains a forbidden edge of M, it contains exactly one edge of M.*   □

Observation 2. *In any phase of the algorithm and for any unprocessed cycle C, forbidden edges with both endpoints in C form a matching.*   □

Consider an unprocessed cycle C. A set of edges S will be called a *semi-support* of C when $P \cup S$ contains vertices of degree at most 2, and after adding S to P (and thus making C processed) Invariant 1 is satisfied. If after adding S to P the number of loose-ends increases by k be will also call S a k-semi-support (k may be negative).

Note that the only difference between a semi-support and a support is that after adding a semi-support to P we may get a non-Hamiltonian cycle in P. The following lemma, similar to the Kostochka-Serdyukov technique, will be used to convert a semi-support M to a support S without losing much weight. The weight loss in this process depends on how the weight of M is distributed between allowed and forbidden edges, on the weight of allowed edges of M that belong to cycles in $P \cup M$, etc.

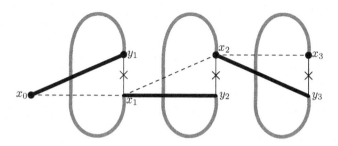

Fig. 1. Breaking the cycles in the proof of Lemma 3. Dashed edges are lighter than the corresponding solid edges. Crossed-out edges are the edges removed from the cycles.

Lemma 3. *Consider any phase of the algorithm and let C be an unprocessed cycle. Let M be a k-semi-support of C. Assume there is a vertex $x_0 \notin V(M)$, such that x_0 is a loose-end or $x_0 \in V(C)$. Moreover, assume $P \cup M$ contains cycles (possibly of length 2) $C_1, \ldots, C_q$. For each i, $1 \leq i \leq q$, let e_i be any edge in $M \cap C_i$. Let $Q = \{e_1, \ldots, e_q\}$ and let $D = \bigcup_i C_i$. Finally, let us partition edges in M into two sets: F containing forbidden edges, and A containing allowed edges. Then one can find S, a k-support of C, such that*

(i) $w(S) \geq w(M \setminus Q) + \frac{1}{2}w(Q)$,
(ii) $w(S) \geq w(A \setminus D) + \frac{3}{4}w(A \cap D) + \frac{1}{2}w(F)$.

Proof. Denote the ends of e_1 by x_1 and y_1 in such a way that $x_0 y_1$ is heavier than $x_0 x_1$. Note that $w(x_0 y_1) = \max\{w(x_0 x_1), w(x_0 y_1)\} \geq \frac{1}{2}(w(x_0 x_1) + w(x_0 y_1)) \geq \frac{1}{2}w(e_1)$, where the last step follows from the triangle inequality. Moreover, by replacing e_1 by $x_0 y_1$ we break the cycle C_1 and x_1 becomes a loose-end. We can proceed in this way for all cycles, i.e., for every $i = 1, \ldots, q$ the ends of e_i are labelled x_i and y_i so that

$$w(x_{i-1} y_i) \geq \tfrac{1}{2}w(e_i). \tag{1}$$

Let $S = M \setminus \{e_i \mid i = 1, \ldots, q\} \cup \{x_{i-1} y_i \mid i = 1, \ldots, q\}$. Clearly, $P \cup S$ does not contain cycles hence it is sub-Hamiltonian. Also, observe that there are only 2 vertices, namely x_0 and x_q whose degrees differ in graphs $P \cup M$ and $P \cup S$. Since $\deg_{P \cup S} x_0 = 2$ and $\deg_{P \cup S} x_q = 1$, after adding S to P (and thus processing C) Invariant 1 is still satisfied, and so S is a support. Also note that x_0 is a loose-end in $P \cup M$ and it is not a loose-end in $P \cup S$, while x_q is not a loose-end in $P \cup M$ and it is a loose-end in $P \cup S$. It follows that S is a k-support.

Now let us bound the weight of S. By (1), $w(S) \geq w(M \setminus Q) + \frac{1}{2}w(Q)$, which is claim (i). To prove (ii), in each cycle C_i we choose the lightest edge e_i in $M \cap C_i$ and we assume Q consists of these edges. Notice that $F \subseteq Q$ (by Observation 1) and also $A \setminus D \subseteq M \setminus Q$, so by (i) we have,

$$w(S) \geq w(M \setminus Q) + w(Q) \geq w(A \setminus D) + w((A \cap D) \setminus Q) + \tfrac{1}{2}w(A \cap Q) + \tfrac{1}{2}w(F). \tag{2}$$

By Observation 1, and since Q consists of the lightest edges in cycles, $w((A \cap D) \setminus Q) \geq \frac{1}{2}w(A \cap D)$. Then $w((A \cap D) \setminus Q) + \frac{1}{2}w(A \cap Q) = w((A \cap D) \setminus Q) +$

$\frac{1}{2}w((A \cap D) \cap Q) = \frac{1}{2}w((A \cap D) \setminus Q) + \frac{1}{2}w(A \cap D) \geq \frac{3}{4}w(A \cap D)$. By plugging it into (2) we get (ii).

3.1 Even Cycles

Lemma 4. *Let C be an unprocessed 4-cycle and assume that there is at least one loose-end. Then there is a 0-support of C of weight $\geq (\frac{1}{4}, \frac{1}{2}) \star C$.*

Proof. We consider two cases:

Case 1. $E(C)$ has at most one forbidden edge. We partition $E(C)$ into two matchings, M_1 and M_2. W.l.o.g. assume M_1 does not contain forbidden edges. Let S_1 and S_2 be the supports corresponding to M_1 and M_2 by Lemma 3 and let S be the heavier of them. Following the notation from Lemma 3, define A_1, A_2 (F_1, F_2) as the sets of allowed (resp. forbidden) edges of M_1, M_2. Let D_1, D_2 be the sets of edges of $E(C)$ that belong to cycles in $P \cup M_1$ or $P \cup M_2$ respectively. Also let $A = A_1 \cup A_2$, $F = F_1 \cup F_2$ and $D = D_1 \cup D_2$.

Notice that by inequality (ii) of Lemma 3 applied to M_i, $i = 1, 2$ we get $w(S_i) \geq w(A_i \setminus D_i) + \frac{3}{4}w(A_i \cap D_i) + \frac{1}{2}w(F_i)$. Summing up the two inequalities yields

$$w(S) \geq \frac{1}{2}(w(S_1) + w(S_2)) \geq \frac{1}{2}w(A \setminus D) + \frac{3}{8}w(A \cap D) + \frac{1}{4}w(F). \qquad (3)$$

Let us first assume that $P \cup M_1$ contains a cycle $\tilde{C}$. By Observation 1 both allowed edges of M_1 are in $\tilde{C}$. So either both chords of C are forbidden or both edges of M_2 are. Since we assumed that $E(C)$ has at most one forbidden edge, it is the chords of C that are forbidden. It now follows from Observation 2 that both edges of M_2 are allowed, so $A = C$. From (3) we get $w(S) \geq \frac{3}{8}w(A) = \frac{3}{8}w(C) \geq (\frac{1}{4}, \frac{1}{2}) \star C$.

Hence, we may assume that $P \cup M_1$ contains no cycle. It follows that $D_1 = \emptyset$, so $|A \setminus D| \geq 2$. From (3) we get $w(S) \geq \frac{1}{2}w(A \setminus D) + \frac{3}{8}w(A \cap D) + \frac{1}{4}w(F) \geq \frac{1}{4}(w(A \setminus D) + w(A \cap D) + w(F)) + \frac{1}{4}w(A \setminus D) \geq \frac{1}{4}w(C) + \frac{1}{4}w(A \setminus D) \geq (\frac{1}{4}, \frac{1}{2}) \star C$, where the last inequality follows from $|A \setminus D| \geq 2$.

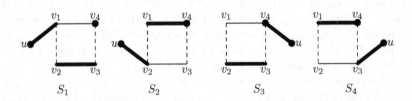

Fig. 2. Supports in Case 2 of the proof of Lemma 4

Case 2. $E(C)$ has two forbidden edges. Denote the vertices of C by $v_1, \ldots, v_4$ in the order they appear on C and assume w.l.o.g. that $v_1 v_2$ and $v_3 v_4$ are forbidden. Let u be a loose-end. Consider four edge sets $S_1 = \{uv_1, v_2 v_3\}$, $S_2 = \{uv_2, v_1 v_4\}$,

$S_3 = \{uv_4, v_2v_3\}$, and $S_4 = \{uv_3, v_1v_4\}$. Note that these sets are allowed since for any i, edges of S_i belong to a single path in $P \cup S_i$ (ending in v_4, v_3, v_1 and v_2 respectively). It follows that all S_i are supports and we choose S, the heaviest of them. Then $w(S) \geq \frac{1}{4}\sum_{i=1}^{4} w(S_i) \geq \frac{1}{4}[2w(v_2v_3) + 2w(v_1v_4) + (w(uv_1) + w(uv_2)) + (w(uv_3) + w(uv_4))] \geq \frac{1}{4}[2w(v_2v_3) + 2w(v_1v_4) + w(v_1v_2) + w(v_3v_4)]$, where the last step follows from triangle inequality. Hence $w(S) \geq \frac{1}{4}w(C) + \frac{1}{4}[w(v_2v_3) + w(v_1v_4)] \geq (\frac{1}{4}, \frac{1}{2}) \star C$.

Lemma 5. *Let C be an unprocessed even-length cycle, $|C| \geq 6$, and assume that there is at least one loose-end. Then there is a 0-support of C of weight at least $(\frac{1}{4}, \frac{1}{2}) \star C$.*

Proof. We partition $E(C)$ into two matchings, M_1 and M_2, let S_1 and S_2 be the supports corresponding to M_1 and M_2 by Lemma 3, and let S be the heavier of these supports. We follow all the definitions from the beginning of the proof of the previous lemma to obtain inequality (3).

From that inequality we get $w(S) \geq \frac{3}{8}w(A) + \frac{1}{4}w(F) = \frac{1}{4}w(C) + \frac{1}{8}w(A)$. It follows that $w(S) \geq (\frac{1}{4}, \frac{1}{2}) \star C$ if $|A| \geq 4$.

Since by Observation 2 we have $|A| \geq |C|/2$, the only case we need to consider is that of $|C| = 6$ and $|A| = 3$. W.l.o.g. $M_1 = A$ and $M_2 = F$. Let Q bet the set of the lightest edges from each cycle in $P \cup M_1$ or $P \cup M_2$, one edge from each cycle. There is at most one such cycle in $P \cup M_1$, since by Observation 1 each such cycle has to contain at least two edges. It follows that $|A \setminus Q| \geq 2$. By inequality (i) in Lemma 3 we get $w(S) \geq \frac{1}{2}(w(S_1) + w(S_2)) \geq \frac{1}{4}w(E(C) \setminus Q) + \frac{1}{4}w(Q) = \frac{1}{4}w(E(C) \setminus Q) + \frac{1}{4}w(C) = \frac{1}{4}w(A \setminus Q) + \frac{1}{4}w(C) \geq (\frac{1}{4}, \frac{1}{2}) \star C$, as required.

3.2   Triangles

For any cycle C, by $\max(C)$ we denote the heaviest edge in C.

Lemma 6. *For any unprocessed triangle C, there is a $(+1)$-support of C of weight at least $(\frac{1}{4}, \frac{1}{2}) \star C - \frac{1}{4}w(\max(C))$.*

Proof. Let x, y, z be the vertices of C and assume w.l.o.g. that both xz and yz are allowed. Let S consist of the heavier of the edges xz, yz. Clearly, S is a support and $w(S) \geq \frac{1}{2}(w(xz) + w(yz)) \geq \frac{1}{4}w(C) + \frac{1}{4}(w(xz) + w(yz)) - \frac{1}{4}w(xy) \geq (\frac{1}{4}, \frac{1}{2}) \star C - \frac{1}{4}w(xy) \geq (\frac{1}{4}, \frac{1}{2}) \star C - \frac{1}{4}w(\max(C))$.

Lemma 7. *Let C be an unprocessed triangle and assume that there are two loose-ends. Then there is a (-1)-support of C of weight at least $(\frac{1}{4}, \frac{1}{2}) \star C + \frac{1}{4}w(\max(C))$.*

Proof. Let x, y, z bet the vertices of C and let u and v be the loose-ends. We consider 2 cases:

Case 1. Both loose-ends are connected to C by paths, say u is connected to x and v to y. Note that in this case all edges of C are allowed. Let $S_1 = \{xy, zv\}$ and $S_2 = \{zy, xv\}$. Note that after adding any of these sets to P, both added

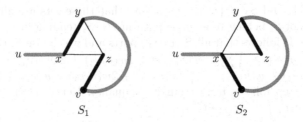

Fig. 3. Supports in Case 1 of the proof of Lemma 7. Gray lines denote the paths connecting loose-ends with C.

edges lie on a single path that ends in u (see Figure 3), so P remains sub-Hamiltonian. Hence both S_1 and S_2 are supports of C. The heavier of them has weight $\max\{w(xy) + w(zv), w(zy) + w(xv)\} \geq \frac{1}{2}(w(xy) + w(zy) + w(zv) + w(xv)) \geq \frac{1}{2}(w(xy) + w(zy) + w(xz)) \geq \frac{1}{4}w(C) + \frac{1}{2}w(\min(C)) + \frac{1}{4}w(\max(C)) = (\frac{1}{4}, \frac{1}{2}) \star C + \frac{1}{4}w(\max(C))$.

Case 2. At least one loose-end, say u, is not connected to C by a path in P. W.l.o.g. assume that both xz are yz allowed. Let $S_1 = \{xz, yu\}$ and $S_2 = \{yz, xu\}$. Note that adding S_1 to P does not create a cycle. Indeed, yu does not belong to a cycle because yu belongs to a path that ends in a vertex different from x, y or z. Also xz does not belong to a cycle because it was allowed before adding it to P. Similar reasoning shows that adding S_2 to P does not create a cycle. Hence both S_1 and S_2 are supports. Similarly to the previous case we get $\max\{w(S_1), w(S_2)\} \geq \frac{1}{2}(w(xz) + w(yu) + w(yz) + w(xu)) \geq (\frac{1}{4}, \frac{1}{2}) \star C + \frac{1}{4}w(\max(C))$.

Observation 3. *Let C be an unprocessed odd cycle in the last phase of the algorithm and assume that there is exactly one loose-end u. Then u is connected by a path in P to a vertex $z \in C$ and $E(C)$ contains exactly $\lfloor |E(C)|/2 \rfloor$ forbidden edges and none of them is adjacent to z.* □

Lemma 8. *Let C be an unprocessed triangle in the last phase of the algorithm and assume that there is exactly one loose-end u. Then there is a (-1)-support of C of weight at least $(\frac{1}{4}, \frac{1}{2}) \star C + \frac{1}{4}w(\max(C))$.*

Proof. Let x, y, z denote the vertices of C. By Observation 3 cycle C contains a forbidden edge — assume w.l.o.g. it is xy — and u is connected in P by a path to z. Let $S_1 = \{xz, yu\}$ and $S_2 = \{yz, xu\}$. Clearly, xz and yu are in the same cycle in $P \cup S_1$ and it is a Hamiltonian cycle. Hence, S_1 is a support of C, and similarly S_2. We pick the heavier of these cycle (its weight can be estimated similarly as in the proof of Lemma 7).

3.3 5-Cycles

Lemma 9. *Let C be an unprocessed 5-cycle with at most one forbidden edge. Then there is a $(+1)$-support of weight at least $(1/4, 1/2) \star C$.*

Proof. Let $v_1, \ldots, v_5$ be the vertices of C in the order they appear on C and assume w.l.o.g. that $v_1 v_5$ is the lightest edge in $E(C)$.

Let $M_1 = \{v_1 v_2, v_3 v_4\}$ and $M_2 = \{v_2 v_3, v_4 v_5\}$. Let S_1 and S_2 be the supports corresponding to M_1 and M_2 by Lemma 3 and let S be the heavier of them. Also, assume all definitions leading to inequality (3) in the proof of Lemma 4.

We consider three cases:

Case 1. $v_1 v_5$ is forbidden. Then $v_1 v_2$ belongs to a path in $P \cup M_1$ (ending in v_5), hence $v_1 v_2 \notin D$. By Observation 1, then also $v_3 v_4 \notin D$, so $M_1 \cap D = \emptyset$. By symmetry, also $M_2 \cap D = \emptyset$. Hence $A \setminus D = A$. By inequality (ii) in Lemma 3 we get $w(S) \geq \frac{1}{2}(w(S_1) + w(S_2)) \geq \frac{1}{2}w(A) \geq \frac{1}{2} \cdot \frac{4}{5}w(C) = \frac{2}{5}w(C) \geq \frac{1}{4}w(C) + \frac{3}{4}\min(C) \geq (\frac{1}{4}, \frac{1}{2}) \star C$.

Case 2. One of the matchings, say M_1, contains a forbidden edge. Hence the other edge of M_1 is allowed and by Observation 1 it does not belong to D. Also note that at least one of the edges e of M_2 has a vertex in common with the forbidden edge from M_1. It follows that e does not lie on a cycle in $M_2 \cup P$, because it lies on a path that ends with the forbidden edge from M_1. By Observation 1, the other edge of M_2 cannot lie on a cycle either. Altogether, this gives $|A \setminus D| \geq 3$.

Using inequality (3) we get $w(S) \geq \frac{1}{2}w(A \setminus D) + \frac{3}{8}w(A \cap D) + \frac{1}{4}w(F) \geq \frac{1}{4}w(C \setminus \{v_1 v_5\}) + \frac{1}{4}w(A \setminus D) + \frac{1}{8}w(A \cap D) \geq \frac{1}{4}w(C \setminus \{v_1 v_5\}) + \frac{1}{2}w(v_1 v_5) = (\frac{1}{4}, \frac{1}{2}) \star C$.

Case 3. There are no forbidden edges in $E(C)$. Suppose $P \cup M_1$ contains a cycle. Then the chords $v_1 v_3$ and $v_2 v_4$ are forbidden. It follows that the edges of M_2 belong to a path in $P \cup M_2$ (one ending in v_1), so they cannot lie on a cycle in $P \cup M_2$. We conclude that at least one of $P \cup M_1$ and $P \cup M_2$ does not contain cycles, and so $|A \setminus D| \geq 2$.

Using inequality (3) we get $w(S) \geq \frac{1}{2}w(A \setminus D) + \frac{3}{8}w(A \cap D) = \frac{3}{8}w(A) + \frac{1}{8}w(A \setminus D) \geq \frac{3}{8} \cdot \frac{4}{5}w(C) + \frac{1}{4}\min(C) = \frac{1}{4}w(C) + \frac{1}{20}w(C) + \frac{1}{4}\min(C) \geq (\frac{1}{4}, \frac{1}{2}) \star C$.

Lemma 10. *Let C be an unprocessed 5-cycle with two forbidden edges. Let e be any of the two forbidden edges of C. Then there is a $(+1)$-support of C of weight at least $(\frac{1}{4}, \frac{1}{2}) \star C - \frac{1}{4}w(e)$.*

Proof. Let $v_1, \ldots, v_5$ be the vertices of C in the order they appear on C and assume w.l.o.g. that $v_1 v_5$ and $v_2 v_3$ the forbidden edges of C and $e = v_1 v_5$. Let $M_1 = \{v_1 v_2, v_3 v_4\}$ and $M_2 = \{v_2 v_3, v_4 v_5\}$ and assume the notation from the proof of the previous lemma.

Note that the edges of M_1 belong to a path in $P \cup M_1$ ending in v_5, hence $M_1 \cap D = \emptyset$. It follows that $|A \setminus D| \geq 2$. Using inequality (3) we get $w(S) \geq \frac{1}{2}w(A \setminus D) + \frac{3}{8}w(A \cap D) + \frac{1}{4}w(F) \geq \frac{1}{4}(w(A \setminus D) + w(A \cap D) + w(F)) + \frac{1}{4}w(A \setminus D) = \frac{1}{4}w(C \setminus \{e\}) + \frac{1}{4}w(A \setminus D) \geq \frac{1}{4}w(C \setminus \{e\}) + \frac{1}{2}\min(C) = (\frac{1}{4}, \frac{1}{2}) \star C - \frac{1}{4}w(e)$.

Lemma 11. *Let C be an unprocessed 5-cycle with two forbidden edges and assume that there are two loose-ends. Let e denote any of the two forbidden edges of C. Then there is a (-1)-support of C of weight at least $(\frac{1}{4}, \frac{1}{2}) \star C + \frac{1}{4}w(e)$.*

Proof. Label the vertices of C as in the proof of the previous lemma. Observe that since there are at least two loose-ends, at least one of them, call it u, is not connected by a path to C in P.

Let $M_1 = \{v_1v_2, v_3v_4, v_5u\}$ and $M_2 = \{uv_1, v_2v_3, v_4v_5\}$, let S_1 and S_2 be the supports corresponding to M_1 and M_2 by Lemma 3, and let S be the heavier of them.

Note that the edges of M_1 belong to a path in $P \cup M_1$ (the one ending in u), hence $P \cup M_1$ does not contain cycles and we have $S_1 = M_1$. Also, neither uv_1 nor v_4v_5 belong to a cycle in $P \cup M_2$. Of course v_2v_3 belongs to a 2-cycle in $P \cup M_2$.

By inequality (i) in Lemma 3 we get $w(S) \geq \frac{1}{2}(w(S_1)+w(S_2)) \geq \frac{1}{2}[w(v_1v_2)+w(v_3v_4)+w(v_5u)+w(uv_1)+w(v_4v_5)]+\frac{1}{4}w(v_2v_3)$. Using the triangle inequality gives $w(S) \geq \frac{1}{2}[w(v_1v_2)+w(v_3v_4)+w(v_1v_5)+w(v_4v_5)]+\frac{1}{4}w(v_2v_3) \geq \frac{1}{4}w(C)+\frac{3}{4}\min(C)+\frac{1}{4}w(v_1v_5) \geq (\frac{1}{4},\frac{1}{2}) \star C + \frac{1}{4}w(e)$.

Lemma 12. *Let C be an unprocessed 5-cycle in the last phase of the algorithm and assume that there is exactly one loose-end u. Let e be any of the two forbidden edges of $E(C)$. Then there is a (-1)-support of C of weight at least $(\frac{1}{4},\frac{1}{2}) \star C + \frac{1}{4}w(e)$.*

Proof. Label the vertices of C as in Lemma 10. By Observation 3, u is connected in P to v_4 by a path.

Let $S_1 = \{v_1v_2, v_3v_4, v_5u\}$, $S_2 = \{uv_1, v_2v_4, v_3v_5\}$ and $S_3 = \{uv_1, v_2v_5, v_3v_4\}$. One may check that for any $i = 1, 2, 3$, S_i is a support and in particular $P \cup S_i$ is a Hamiltonian cycle. Let S be the heaviest of these supports.

Denote $w(v_2v_4)+w(v_3v_5)+w(v_2v_5)+w(v_3v_4)$ by X. Then $w(S) \geq \frac{1}{2}w(S_1)+\frac{1}{4}w(S_2)+\frac{1}{4}w(S_3) = \frac{1}{2}(w(v_1v_2)+w(v_3v_4)+w(v_5u)+w(uv_1))+\frac{1}{4}X$.

By triangle inequality (used twice), $X \geq 2w(v_2v_3)$. By symmetry, $X \geq 2w(v_4v_5)$. Hence, $X \geq w(v_2v_3)+w(v_4v_5)$. Let us apply triangle inequality one more time: $w(v_5u)+w(uv_1) \geq w(v_1v_5)$.

Putting it all together we get $w(S) \geq \frac{1}{2}(w(v_1v_2)+w(v_3v_4)+w(v_1v_5))+\frac{1}{4}(w(v_2v_3)+w(v_4v_5)) \geq (\frac{1}{4},\frac{1}{2}) \star C + \frac{1}{4}w(e)$.

3.4   Odd Cycles of Length at Least 7

Lemma 13. *Let C be an unprocessed odd cycle of length at least 7. Then there is a $(+1)$-support of weight at least $(\frac{1}{4},\frac{1}{2}) \star C$.*

Proof. Let $|C| = 2k + 1$, $k \geq 3$. We enumerate vertices in $V(C)$ so that $C = v_0v_1v_2 \ldots v_{2k-1}v_{2k}v_0$, both v_0v_1 and v_0v_{2k} are allowed and $w(v_0v_1) \geq w(v_0v_{2k})$. Consider two subsets of $E(C)$: $M_1 = \{v_{2i}v_{2i+1} \mid 0 \leq i \leq k-1\}$ and $M_2 = \{v_{2i+1}v_{2i+2} \mid 0 \leq i \leq k-1\}$. In other words we partition $E(C) \setminus \{v_0v_{2k}\}$ into two matchings.

Let $C_1, \ldots, C_p$ be all cycles in $P \cup M_1$ and Let $C_{p+1}, \ldots, C_q$ be all cycles in $P \cup M_2$. Similarly as in Lemma 3, let $D = \bigcup_{i=1}^{q} C_i$ and we partition edges in $M_1 \cup M_2$ into two sets: F containing forbidden edges, and A containing allowed edges.

Further, let us choose for each cycle C_i, $i = 1, \ldots, q$, some edge e_i in $C_i \cap E(C)$ and let $Q = \{e_1, \ldots, e_q\}$. Since by Observation 1 each cycle C_i that contains v_0v_1 contains also another edge from A, we assume w.l.o.g. that $v_0v_1 \notin Q$.

Using Lemma 3 we obtain supports S_1, S_2. Let S be the heavier of these supports. Then $w(S) \geq \frac{1}{2}(w(S_1) + w(S_2))$. Using Lemma 3 we obtain supports S_1, S_2. Let S be the heavier of these supports. Then $w(S) \geq \frac{1}{2}(w(S_1) + w(S_2))$.

By inequality (i) in Lemma 3, $w(S) \geq \frac{1}{2}w((M_1 \cup M_2) \setminus Q) + \frac{1}{4}w(Q) = \frac{1}{4}w(E(C) \setminus \{v_0v_{2k}\}) + \frac{1}{4}w((M_1 \cup M_2) \setminus Q)$. Since $v_0v_1 \notin Q$ and $w(v_0v_1) \geq w(v_0v_{2k})$, $w(S) \geq \frac{1}{4}w(E(C)) + \frac{1}{4}w((M_1 \cup M_2) \setminus (Q \cup \{v_0v_1\}))$. As $F \subseteq Q$, $(M_1 \cup M_2) \setminus (Q \cup \{v_0v_1\}) = (A \setminus \{v_0v_1\}) \setminus Q$ and hence

$$w(S) \geq \frac{1}{4}w(E(C)) + \frac{1}{4}w((A \setminus \{v_0v_1\}) \setminus Q). \tag{4}$$

It follows that $|(A \setminus \{v_0v_1\}) \setminus Q| \geq 2$ implies $w(S) \geq (1/4, 1/2) \star C$.

First assume there are k forbidden edges in $E(C)$. Then one of the matchings, say M_1, contains only allowed edges (and the other matching contains all the forbidden edges of C). Note that in $P \cup M_1$ all edges of M_1 belong to a path with one end in v_{2k}. It follows that $M_1 = S_1$ and $S_1 \cap Q = \emptyset$. It follows that $A \cap Q = \emptyset$ and hence $(A \setminus \{v_0v_1\}) \setminus Q$ contains at least $k - 1 \geq 2$ edges, as required.

Now assume there are at most $k-1$ forbidden edges in $E(C)$. Then $|A| \geq k+1$. By Observation 1, $|A \setminus Q| \geq \lceil \frac{|A|}{2} \rceil$. It follows that $|(A \setminus \{v_0v_1\}) \setminus Q| \geq \lceil \frac{|A|}{2} \rceil - 1$. For $|A| \geq 5$, we get $\lceil \frac{|A|}{2} \rceil - 1 \geq 2$.

Hence we are left with the case $|A| \leq 4$. Since $|A| \geq k + 1$, $k \leq 3$. So $k = 3$, $|A| = 4$ and $|F| = 2$. We consider two subcases.

Case 1. v_5v_6 is forbidden. Then v_4v_5 is allowed and after adding the matching containing v_4v_5 to P, v_4v_5 is on a path ending in v_6, hence v_4v_5 does not belong to any C_i. Hence the three remainig edges in A belong at most one cycle C_i, so $|A \cap Q| \leq 1$ and further $|(A \setminus \{v_0v_1\}) \setminus Q| \geq 2$, as required.

Case 2. v_5v_6 is allowed. If $F = \{v_2v_3, v_4v_5\}$, one of the matchings, namely M_2, contains only allowed edges. Moreover, these edges belong to a path in $P \cup M_2$ (ending in v_6), so $M_2 = S_2$ and $S_2 \cap Q = \emptyset$. There is just one allowed edge in M_1 and hence it cannot belong to a cycle C_i. It follows that $Q = F$ and hence $|(A \setminus \{v_0v_1\}) \setminus Q| \geq 3$. The case $F = \{v_1v_2, v_3v_4\}$ is symmetric. Finally, assume $F = \{v_1v_2, v_4v_5\}$. By Observation 1, in $P \cup M_1$ and $P \cup M_2$ there are at most 2 cycles with edges from A. If $P \cup M_1$ contains such cycle, then v_0v_3 is forbidden. However, then $P \cup M_2$ contains no such cycle. Hence $|A \cap Q| \leq 1$ and $|(A \setminus \{v_0v_1\}) \setminus Q| \geq 2$, as required.

4 Ordering the Cycles

4.1 Basic Setup

Based on the results from the previous section, we can see that every cycle C belongs to one of three categories:

even cycles: C has a 0-support of weight $(\frac{1}{4}, \frac{1}{2}) \star C$, if there exists at least one loose-end.

good odd cycles: C has a $(+1)$-support of weight at least $(\frac{1}{4}, \frac{1}{2}) \star C$ — that is the case if C is an odd cycle of length ≥ 7 or a 5-cycle with at most one forbidden edge.

bad odd cycles: C has a $(+1)$-support of weight smaller than $(\frac{1}{4}, \frac{1}{2}) \star C$, and it also has a (-1)-support of weight greater than $(\frac{1}{4}, \frac{1}{2}) \star C$, but only if there exist at least two loose-end or it is the last cycle processed — that is the case for all 3-cycles and for 5-cycles with two forbidden edges.

Remark 1. Notice that a good odd cycle might become bad when other cycles are processed, if it is initially a 5-cycle with zero (or one) forbidden edges and two (one, resp.) of its allowed edges becomes forbidden.

We say that a cycle C is k-processed, if it is processed using a k-support. The general order of processing the cycles consists of 4 stages:

(1) as long as there exists a good odd cycle, $(+1)$-process it,
(2) $(+1)$-process bad odd cycles until the number of loose-ends is greater or equal to the number of remaining bad odd cycles,
(3) 0-process even cycles,
(4) (-1)-process the remaining odd cycles.

When we use the above processing order all the assumptions of previous section's lemmas are satisfied. In particular in stage 3, there exists at least one loose-end, so we can process the even cycles. This is because we can assume that $\mathcal{C}$ contains at least one triange, otherwise already the Kostochka-Serdyukov algorithm gives 7/8-approximation.

It is clear that we are getting enough weight from cycles processed in stages 1 and 3. We also gain some extra weight in stage 2 and lose weight in stage 4. We want to select the cycles to be processed in stage 2 in such a way that the overall weight of edges added during stages 2 and 4 is at least $\sum_i (\frac{1}{4}, \frac{1}{2}) \star C_i$, where the sum is over all cycles processed in these stages.

4.2   Ordering Bad Odd Cycles

Let us first define certain useful notions. For any bad odd cycle C, let $B_{-1}(C)$ $(B_{+1}(C))$ be the lower bound on the weight of the (-1)-support $((+1)$-support), as guaranteed by the appropriate lemma in the previous section. Suppose that $\mathcal{C}_i$ is the set of bad odd cycles processed in stage i, $i = 2, 4$. If we use previous section's lemmas to lowerbound the weight of all edges added in stages 2 and 4, we are going to get

$$\sum_{C \in \mathcal{C}_2} B_{+1}(C) + \sum_{C \in \mathcal{C}_4} B_{-1}(C),$$

and we need to show that $\mathcal{C}_2$ and $\mathcal{C}_4$ can be chosen so that the value of this expression is at least

$$\sum_{C \in \mathcal{C}_2 \cup \mathcal{C}_4} (\tfrac{1}{4}, \tfrac{1}{2}) \star C.$$

For every bad odd cycle C there exists a non-negative number, which we call the *loose-end value for* C and denote $\text{LEV}(C)$ such that

$$B_{+1}(C) \geq (\tfrac{1}{4}, \tfrac{1}{2}) \star C - \text{LEV}(C) \quad \text{and} \quad B_{-1}(C) \geq (\tfrac{1}{4}, \tfrac{1}{2}) \star C + \text{LEV}(C).$$

Note, that this number is equal to $\frac{1}{4}w(e)$, where e is the heaviest edge of C if C is a triangle, or the heavier of the two forbidden edges of C if C is a bad 5-cycle.

The reason why we call this number the loose-end value for C is that it is essentially the price at which C should be willing to buy/sell a loose-end. In this economic analogy, the cycles that are $(+1)$-processed are selling loose-ends to the cycles that are (-1)-processed. If we can make every cycle trade a loose-end at a preferred price (LEV or better), the weight of a support of any cycle C together with its profit/loss coming from trading a loose-end adds up to at least $(\tfrac{1}{4}, \tfrac{1}{2}) \star C$. But it is obvious how to make every cycle trade a loose-end at a preferred price! It is enough to make the cycles with smallest LEV sell loose-ends (process them in stage 2), and make the remaining cycles buy loose-ends (process them in stage 4).

Note here, that some bad odd cycles will get loose-ends for free from good odd cycles processed in stage 1. Since we assume that the total number of vertices in the graph is even, the number of the remaining bad odd cycles is also even, and so they can be divided evenly into sellers and buyers.

Using Lemma 2 we get

Theorem 1. *Metric MAX-TSP problem can be 7/8-approximated for graphs with even number of vertices.*

This can be extended to graphs with odd number of vertices, at a cost of increasing the running time by a factor of $O(n^4)$, we omit the details in this extended abstract.

References

1. Chen, Z.-Z., Nagoya, T.: Improved approximation algorithms for metric max TSP. In: Brodal, G.S., Leonardi, S. (eds.) ESA 2005. LNCS, vol. 3669, pp. 179–190. Springer, Heidelberg (2005)
2. Hassin, R., Rubinstein, S.: A 7/8-approximation algorithm for metric Max TSP. Inf. Process. Lett. 81(5), 247–251 (2002)
3. Kostochka, A.V., Serdyukov, A.I.: Polynomial algorithms with the estimates 3/4 and 5/6 for the traveling salesman problem of the maximum (in Russian). Upravlyaemye Sistemy 26, 55–59 (1985)
4. Kowalik, Ł., Mucha, M.: 35/44-approximation for asymmetric maximum TSP with triangle inequality. In: Dehne, F., Sack, J.-R., Zeh, N. (eds.) WADS 2007. LNCS, vol. 4619, pp. 589–600. Springer, Heidelberg (2007)
5. Serdyukov, A.I.: The traveling salesman problem of the maximum (in Russian). Upravlyaemye Sistemy 25, 80–86 (1984)

Inapproximability of Survivable Networks[*]

Yuval Lando[1] and Zeev Nutov[2]

[1] Ben-Gurion University of the Negev
ylando@hotmail.com
[2] The Open University of Israel
nutov@openu.ac.il

Abstract. In the Survivable Network Design Problem (SNDP) one seeks to find a minimum cost subgraph that satisfies prescribed node-connectivity requirements. We give a novel approximation ratio preserving reduction from Directed SNDP to Undirected SNDP. Our reduction extends and widely generalizes as well as significantly simplifies the main results of [6]. Using it, we derive some new hardness of approximation results, as follows. We show that directed and undirected variants of SNDP and of k-Connected Subgraph are equivalent w.r.t. approximation, and that a ρ-approximation for Undirected Rooted SNDP implies a ρ-approximation for Directed Steiner Tree.

1 Introduction

Let $\kappa_H(u, v)$ (possibly $u = v$) denote the maximum number of pairwise internally-disjoint uv-paths in a graph H. Let $\kappa(H) = \min\{\kappa_H(u, v) : (u, v) \in V \times V, u \neq v\}$ be the *connectivity* of H. The following is a fundamental problem in Network Design:

Survivable Network Design Problem (SNDP)
Instance: A graph $G = (V, E)$, edge costs $\{c(e) : e \in E\}$, and requirements $r(u, v)$ on $V \times V$.
Objective: Find a minimum cost spanning subgraph $H = (V, I)$ of G so that

$$\kappa_H(u, v) \geq r(u, v) \quad \text{for all} \quad u, v \in V. \tag{1}$$

This formulation includes well known problems such as Steiner Tree/Forest, Min-Cost k-Flow, and others. If $r(u, v) = k$ for all $u, v \in V$ then we get the k-Connected Subgraph problem, which seeks a minimum cost spanning subgraph H with $\kappa(H) \geq k$. In the Rooted SNDP there is $s \in V$ so that if $r(u, v) > 0$ then: $u = s$ for directed graphs, and $u = s$ or $v = s$ for undirected graphs. In $\{0, k\}$-SNDP, requirements are either 0 or k; $\{0, 1\}$-SNDP is the Steiner Forest Problem, and Rooted $\{0, 1\}$-SNDP is the Steiner Tree Problem. See a survey in [7] for various types of SNDP problems. The following known statement (c.f., [7]) shows that undirected SNDP problems cannot be much harder to approximate than the directed ones.

---

[*] This research was supported by The Open University of Israel's Research Fund (grant no. 100685).

A. Goel et al. (Eds.): APPROX and RANDOM 2008, LNCS 5171, pp. 146–152, 2008.

Proposition 1. *A ρ-approximation algorithm for* Directed SNDP *implies a 2ρ-approximation algorithm for* Undirected SNDP.

The reduction in Proposition 1 is very simple: just apply the ρ-approximation algorithm on the "bidirected" instance, and return the underlying graph of the directed solution computed.

Following the classification of cost types of [7], we assume that the input graph G to an SNDP instance is complete. The case of $\{0, 1\}$-costs gives the augmentation problems when we seek to augment a graph G_0 (formed by edges of cost 0 in G) by a minimum size edge-set F (any edge is allowed) so that $G_0 + F$ satisfies the requirements. The case of $\{1, \infty\}$-costs gives the min-size subgraph problems: edges in G have unit costs, while edges not in G have cost ∞.

Most undirected variants of $\{0, 1\}$-SNDP are substantially easier to approximate than the directed ones. For example, Undirected Steiner Tree/Forest admits a constant ratio approximation algorithm, while the directed variants are not known to admit even a polylogarithmic approximation ratio. Specifically, Dodis and Khanna [2] showed that Directed Steiner Forest is at least as hard to approximate as Label-Cover. By extending the construction of [2], Kortsarz, Krauthgamer, and Lee [6] showed a similar hardness result for Undirected $\{0, k\}$-SNDP; the same hardness is valid even for $\{0, 1\}$-costs, see [9]. The currently best known approximation lower bound for Directed Steiner Tree is $\log^{2-\varepsilon} n$ [5], while the best known approximation ratio is $n^{\varepsilon}/\varepsilon^3$ [1].

So far, there was no unifying hardness result indicating that the inverse to Proposition 1 is also true, namely, that Undirected SNDP is at least as hard to approximate as Directed SNDP. We will give such a reduction, which looks surprisingly simple, after it is found. Our reduction transforms a directed instance on n nodes with costs in the range C into an undirected instance with $n' = 2n$ nodes and costs in the range $C' = C \cup \{0\}$; hence if the range C includes 0 costs, we have $C = C'$. Every requirement $r(u, v)$ transforms into the requirement $r'(u, v') = r(u, v) + n = r(u, v) + n'/2$. We note that the reduction does *not* preserves *metric costs* (because we add edges of cost zero), and transforms small requirements into large requirements. However, in several cases, it has the advantage of preserving the "type" of requirements and costs; see [7] for a classification of SNDP problems w.r.t. costs and requirements. In particular, we obtain the following results:

Theorem 1. *The following holds for any range of costs that includes the zero costs:*

(i) *A ρ-approximation for* Undirected SNDP *implies a ρ-approximation for* Directed SNDP.

(ii) *A ρ-approximation for* Undirected k-Connected Subgraph *implies a ρ-approximation for* Directed k-Connected Subgraph.

(iii) *A ρ-approximation for* Undirected $\{0, k\}$-SNDP *implies a ρ-approximation for* Directed Steiner Forest.

(iv) *A ρ-approximation for* Undirected Rooted $\{0, k\}$-SNDP *implies a ρ-approximation for* Directed Steiner Tree.

To illustrate the power and the limitations of our result, we list and discuss some specific consequences. Note that Theorem 1 is true for $\{0,1\}$-costs. This fact is however redundant for the problems in parts (ii) and (iii), since for $\{0,1\}$-costs the directed versions in parts (ii) and (iii) are known to be "easier" than the undirected ones. Specifically, *for* $\{0,1\}$-*costs* we have:

- Directed k-Connected Subgraph with $\{0,1\}$-costs is in P [4], while the complexity status of the undirected variant is unknown.
- Directed Steiner Forest with $\{0,1\}$-costs admits an $O(\log n)$-approximation [8], while the undirected variant is unlikely to admit a polylogarithmic approximation [9].

Dodis and Khanna [2] showed by a relatively simple proof that Directed Steiner Forest cannot be approximated within $O(2^{\log^{1-\varepsilon} n})$ for any fixed $\varepsilon > 0$, unless $NP \subseteq$ quasi-P. Thus part (iii) immediately implies the main result of [6]:

Corollary 1 ([6]). Undirected $\{0,k\}$-SNDP *does not admit an* $O(2^{\log^{1-\varepsilon} n})$ *approximation for any fixed* $\varepsilon > 0$, *unless* $NP \subseteq$ quasi-P.

In [9] it is proved that the same hardness result holds even for $\{0,1\}$-costs, for both directed and undirected graphs (for large values of k). It seems that this result of [9] cannot be deduced from our work, as the proof of the directed case is essentially the same as that of the undirected one.

It was already observed by A. Frank [3] long time ago by an easy proof that Directed Rooted $\{0,1\}$-SNDP with $\{0,1\}$-costs is at least as hard as the Set-Cover problem. Hence from part (iv) we obtain the following hardness result, which proof required considerable effort in [6] and in [10].

Corollary 2 ([6,10]). Undirected Rooted $\{0,k\}$-SNDP *with* $\{0,1\}$-*costs cannot be approximated within* $C \ln n$ *for some* $C > 0$, *unless* P=NP.

Now we give two new results. Combining part (ii) with Proposition 1, we obtain:

Corollary 3. *Directed and undirected variants of* SNDP *and of* k-Connected Subgraph *are equivalent (up to a constant factor) w.r.t. approximation.*

Finally, we can combine part (iv) with the hardness result of Halperin and Krauthgamer [5] for Directed Steiner Tree to obtain:

Corollary 4. *There exists a constant* $C > 0$ *so that* Undirected Rooted $\{0,k\}$-SNDP *does not admit a* $C \log^{2-\varepsilon} n$ *approximation for any fixed* $\varepsilon > 0$, *unless* NP *has quasi-polynomial Las-Vegas algorithms.*

The hardness in part (iv) however seems "stronger" than the one in Corollary 4, as currently no polylogarithmic approximation is known for Directed Steiner Tree. We also note that the statements in Corollaries 1 – 4 are valid even for instances when we are only interested to increase the connectivity by 1 between pairs with requirement k, namely, when G contains a subgraph G_0 of cost 0 with $\kappa_{G_0}(u,v) = k - 1$ for all $u,v \in V$ with $r(u,v) = k$.

2   The Reduction

Definition 1. *Let $H = (V, I)$ be a directed graph. The* bipartite *(undirected) graph of H is obtained by adding a copy V' of V and replacing every directed edge $ab \in I$ by the undirected edge ab', where b' denotes the copy of b in V'.*

The key observation is the following.

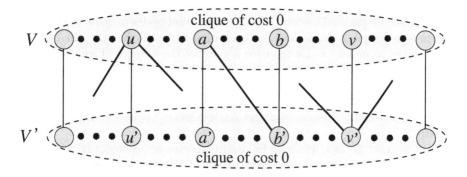

Fig. 1. The construction in Lemma 1; edges in M are shown by thin lines

Lemma 1. *Let $H' = (V + V', I')$ be an undirected graph obtained by adding to the bipartite graph of a directed graph $H = (V, I)$ edge sets of cliques on each of V and V', and the matching $M = \{aa' : v \in V\}$ (see Figure 1). Then:*

$$\kappa_{H'}(u, v') = \kappa_H(u, v) + n \quad \forall u, v \in V . \tag{2}$$

Proof. Let $k = \kappa_H(u, v)$ and $k' = \kappa_{H'}(u, v')$. We may assume that $uv \notin I$; otherwise the same proof applies on $G - uv$. Note that then $n \geq k + 2$ if $u \neq v$.

We prove that $k' \geq k+n$ by showing that H' contains $k+n$ pairwise internally-disjoint uv'-paths. Assuming $u \neq v$, we will show $2k + 2$ paths of the length 2 each, and the rest $n - k - 2$ paths of the length 3 each. Consider a set Π of k pairwise internally-disjoint uv-paths in H.

The length 2 paths are as follows:

- The two paths $u - u' - v'$ and $u - v - v'$.
- The k paths $u - w' - v'$ for any edge uw belonging to some path in Π.
- The k paths $u - w - v'$ for any edge wv belonging to some path in Π.

The length 3 paths are as follows:

- A path $u - a - b' - v'$ for every edge ab belonging to some path in Π and not incident to u, v.
- A path $u - a - a' - v'$ for every node a not belonging to any path in Π.

It is easy to see that these paths are pairwise internally-disjoint, and we now count their number. Excluding u, v', every node of H' appears as an internal node in exactly one of these paths. The number of internal nodes in the paths of length 2 is $2k + 2$. Hence the number of internal nodes in the length 3 paths is $(2n - 2) - (2k + 2) = 2(n - k - 2)$. As each of the length 3 paths has exactly 2 internal nodes, their number is $n - k - 2$. Hence the total number of paths is $(2k + 2) + (n - k - 2) = n + k$, as claimed.

Now consider the case $u = v$. In this case, we have only $2k$ paths of the length 2 each, but there is one path of the length 1, namely, the edge vv'. So we have a total of $2k + 1$ paths. The total number of internal nodes in these paths is $2k$. We can form length 3 paths that use as internal nodes all the other $2n - 2 - 2k$ nodes, in the same way as for the case $u \neq v$. So, the number of length 3 paths is $n - 1 - k$. This gives a total of $(2k + 1) + (n - k - 1) = n + k$ paths, as claimed.

To prove that $k \geq k' - n$ we show that H contains $k' - n$ pairwise internally-disjoint uv-paths. A uv'-path of length 3 in H' is an M-*path* if its internal edge belongs to M; in the case $u = v$, the single edge vv' is also considered as an M-path.

Consider a set Π' of k' pairwise internally-disjoint uv'-paths in H' with maximum number of M-paths. From the structure of H', we may assume that every path in Π' has length 2 or 3, or that it is the edge vv' in the case $u = v$. Note that every node of H' belongs to some path in Π'. Otherwise, if say $a \in V$ does not belong to some path in Π', then by replacing the path in Π' containing a' by the M-path $u - a - a' - v'$ the number of M-paths in Π' increases by 1; a similar argument applies if there is $a' \in V'$ that does not belong to some path in Π'.

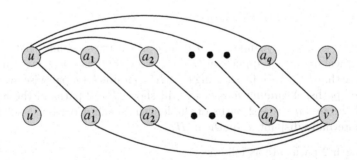

Fig. 2. Illustration to the proof of Lemma 1

We claim that there exists a sequence $a_1, \ldots, a_q$ of nodes in V satisfying (see Figure 2):

(i) $u - a_1' - v'$ and $u - a_q - v'$ belong to Π';
(ii) $u - a_{i-1} - a_i' - v'$ belongs to Π' for every $i = 2, \ldots, q$;
(iii) $ua_i' \notin I$ and $a_{i-1}v' \notin I$ for every $i = 2, \ldots, q$.

Such a sequence can be constructed as follows. Note that u has at least $k' - n$ neighbors in $V' - u'$ in the paths in Π', since u has n neighbors in $V + u'$. We choose one such neighbor a_1' of u. Since every node belongs to some path in Π', there must be a non M-path containing a_1. If this paths has length 2, namely, if it is $u - a_1 - v'$, we are done. Otherwise, this path has length 3, say $u - a_1 - a_2' - v'$. Note that a_2' cannot be a neighbor of u, since otherwise we could replace these two paths by the two paths $u - a_2' - v$ and $u - a_1 - a_1' - v'$ to increase the number of M-paths by 1. So, we can continue this process until at some iteration q a path $u - a_q - v'$ of length 2 is found.

Now we observe that a sequence above defines the uv-path $u - a_1 - a_2 - \cdots - a_q$ in H. As u has at least $k' - n$ neighbors in $V' - u'$, we can form $k' - n$ such sequences, which gives a set of $k' - n$ pairwise internally-disjoint paths in H.

Corollary 5. $\kappa(H') = \kappa(H) + n$ for H, H' as in Lemma 1.

Proof. Let $k = \kappa(H)$. By Lemma 1, it is sufficient to show that:

- $\kappa_{H'}(u, v), \kappa_{H'}(u', v') \geq k + n$ for all $u, v \in V$.
- $\kappa_{H'}(v, v') \geq k + n$ for all $v \in V$.

We prove that $\kappa_{H'}(u, v) \geq k + n$; the proof that $\kappa_{H'}(u', v') \geq k + n$ is identical. We may again assume that $uv \notin I$. A set of $k + n$ pairwise internally-disjoint uv paths in H' is as follows. There are $n - 1$ internally-disjoint uv-paths in $H' - V'$. Another path is $u - u' - v' - v$. We now show additional k paths using nodes from $V' - \{u', v'\}$ only. Let A be the set of neighbors of u and B the set of neighbors of v in $V' - \{u', v'\}$. We have $|A|, |B| \geq k$, since H is k-connected, and since $uv \notin I$. There are $|A \cap B|$ uv-paths of the length 2 each through $A \cap B$, and $\min\{|A - B|, |B - A|\}$ uv-paths of the length 3 each through $(A \cup B) - (A \cap B)$. Hence we have $|A \cap B| + \min\{|A - B|, |B - A|\} \geq k$ pairwise internally-disjoint uv-paths through $V' - \{u', v'\}$, as claimed.

We prove that $\kappa_{H'}(v, v') \geq k + n$. The key point here is that $\kappa_H(v, v) \geq k$, namely, that in H there are at least k pairwise internally-disjoint paths from v to itself. Otherwise, by Menger's Theorem, there is a set C with $|C| = k - 1$ so that $H - C$ has no path from v to itself. This implies that either $V = C + v$ so $|V| = k$, or that $H - C$ is *not* strongly connected. In both cases, we obtain the contradiction $\kappa(H) \leq k - 1$. Thus by Lemma 1 we have $\kappa_{H'}(v, v') = \kappa_H(v, v) + n \geq k + n$, as claimed.

Given an instance $\mathcal{I} = (G = (V, E), c, r)$ of Directed SNDP with $n = |V|$, construct an instance $\mathcal{I}' = (G' = (V + V', E'), c', r')$ of Undirected SNDP as follows. G' is as in Lemma 1, keeping costs of edges in E, and with zero costs of other edges in E'. The requirements are $r'(u, v') = n + r(u, v)$ for $(u, v) \in V \times V$, and $r'(u, v)$ or $r'(u', v')$ can be any integer between 0 and n otherwise. Every directed edge in E has a (unique) appearance in E', so we identify a directed edge in E with the corresponding undirected edge in E'. This establishes a bijective correspondence between edge subsets $I \subseteq E$ and edge subsets $I' \subseteq E'$ containing $E' - E$, namely, $I' = I + (E' - E)$. From Lemma 1 and Corollary 5 we get the following statement, which implies Theorem 1.

Corollary 6. (1) *holds for* $H = (V, I), r$ *if, and only if,* (1) *holds for* $H' = (V + V', I'), r'$. *Furthermore,* $\kappa(H') = \kappa(H) + n$, *namely,* H *is* k-*connected if, and only if,* H' *is* $(k + n)$-*connected.*

Corollary 6 implies that $H = (V, I)$ is a feasible solution to instance $\mathcal{I}$ if, and only if, $H' = (V, I')$ is a feasible solution to the constructed instance $\mathcal{I}'$; furthermore, $c(I) = c'(I')$, since I' is obtained from I by adding edges of cost 0. In particular, we obtain that the optimal solution values for $\mathcal{I}$ and for $\mathcal{I}'$ coincide. This is so both for SNDP and for k-Connected Subgraph. Thus if $H' = (V', I')$ is a ρ-approximate solution for $\mathcal{I}'$, then $H = (V, I)$ is a ρ-approximate solution for $\mathcal{I}$. Even if the approximation ratio ρ is given in terms of n (and is non-decreasing in n), then we get an $O(\rho)$-approximation for $\mathcal{I}$, since $|V'| = 2|V| \geq |V|$.

As when transforming $\mathcal{I}$ to $\mathcal{I}'$, the requirement are shifted by the additive term $n = n'/2$, Directed Steiner Forest is transformed into Undirected $\{0, k\}$-SNDP, while Directed Steiner Tree is transformed into Undirected Rooted $\{0, k\}$-SNDP, where $k = n'/2 + 1$. Furthermore, an instance for k-Connected Subgraph is transformed into an instance of k'-Connected Subgraph, where $k' = k + n/2$.

Finally, note that since the only change in the range of the costs when transforming $\mathcal{I}$ to $\mathcal{I}'$ was adding edges of cost 0, then the ranges of costs of $\mathcal{I}$ and $\mathcal{I}'$ coincide, provided 0 costs are in the range of $\mathcal{I}$.

The proof of Theorem 1 is now complete.

References

1. Charikar, M., Chekuri, C., Cheung, T., Dai, Z., Goel, A., Guha, S., Li, M.: Approximation algorithms for directed Steiner problems. Journal of Algorithms 33, 73–91 (1999)
2. Dodis, Y., Khanna, S.: Design networks with bounded pairwise distance. In: STOC, pp. 750–759 (2003)
3. Frank, A.: Augmenting graphs to meet edge-connectivity requirements. SIAM Journal on Discrete Math. 5(1), 25–53 (1992)
4. Frank, A., Jordán, T.: Minimal edge-coverings of pairs of sets. J. on Comb. Theory B 65, 73–110 (1995)
5. Halperin, E., Krauthgamer, R.: Polylogarithmic inapproximability. In: STOC, pp. 585–594 (2003)
6. Kortsarz, G., Krauthgamer, R., Lee, J.R.: Hardness of approximation for vertex-connectivity network design problems. SIAM Journal on Computing 33(3), 704–720 (2004)
7. Kortsarz, G., Nutov, Z.: Approximating minimum-cost connectivity problems. In: Gonzalez, T.F. (ed.) Approximation Algorithms and Metaheuristics. Chapman & Hall/CRC, Boca Raton (2007)
8. Kortsarz, G., Nutov, Z.: Tight approximation algorithm for connectivity augmentation problems. Journal of Computer and System Sciences 74, 662–670 (2008)
9. Nutov, Z.: Approximating connectivity augmentation problems. In: SODA, pp. 176–185 (2005)
10. Nutov, Z.: Approximating rooted connectivity augmentation problems. Algorithmica 44, 213–231 (2006)

Approximating Single Machine Scheduling with Scenarios

Monaldo Mastrolilli, Nikolaus Mutsanas, and Ola Svensson

IDSIA - Lugano, Switzerland
{monaldo,nikolaus,ola}@idsia.ch

Abstract. In the field of robust optimization, the goal is to provide solutions to combinatorial problems that hedge against variations of the numerical parameters. This constitutes an effort to design algorithms that are applicable in the presence of uncertainty in the definition of the instance. We study the single machine scheduling problem with the objective to minimize the weighted sum of completion times. We model uncertainty by replacing the vector of numerical values in the description of the instance by a set of possible vectors, called *scenarios*. The goal is to find the schedule with minimum value in the worst-case scenario.

We first show that the general problem is intractable by proving that it cannot be approximated within $O(\log^{1-\varepsilon} n)$ for any $\varepsilon > 0$, unless NP has quasi-polynomial algorithms. We then study more tractable special cases and obtain an LP based 2-approximation algorithm for the unweighted case. We show that our analysis is tight by providing a matching lower bound on the integrality gap of the LP. Moreover, we prove that the unweighted version is NP-hard to approximate within a factor less than $6/5$. We conclude by presenting a polynomial time algorithm based on dynamic programming for the case when the number of scenarios and the values of the instance are bounded by some constant.

1 Introduction

In classical optimization problems, it is often assumed that the parameters of the instances are precisely defined numerical values. In many cases, however, such a precise definition is impossible due to inadequate knowledge on the side of the decision maker. The necessity to provide algorithms for minimizing the cost in uncertain environments lead to the fields of stochastic and robust optimization.

In *stochastic optimization* [4], it is assumed that we have knowledge of the probability distribution of the data and the goal is to find a solution that minimizes the expected cost. *Robust optimization* [3,15] can be considered as the worst-case counterpart of the stochastic optimization. In a robust optimization problem, we have a set of possible configurations of the numerical parameters of the problem, and the goal is to find a solution that minimizes the cost in a worst-case scenario for the given solution. In the following we will focus on this latter approach.

Within robust optimization, two common ways of modeling uncertainty are *interval data* and *discrete scenarios*. In the case of interval data the vector of

A. Goel et al. (Eds.): APPROX and RANDOM 2008, LNCS 5171, pp. 153–164, 2008.

numerical parameters in the description of the instance is replaced by a vector
of intervals, one for each parameter. On the other hand, in the case of discrete
scenarios the vector of numerical parameters is replaced by a set of vectors, each
of them corresponding to a different scenario. An advantage of this model is
that, whereas in the case of interval data the fluctuations of the different numerical parameters are implicitly assumed to be independent, the use of discrete
scenarios allows the implementation of dependencies among parameters.

Several objective functions for robust minimization[1] problems have been proposed in literature (see e.g. the book by Kouvelis & Yu [15]). In the *absolute
robustness* approach, the goal is to minimize the maximum among all feasible
solutions and all scenarios. This is often referred to as the "min-max" version of
the problem. In the *robust deviation* approach, the goal is to minimize the maximum deviation from optimality among all feasible solutions and all scenarios.
Recent examples of these two families of approaches can be found in [1,12,9].

In this paper we investigate the min-max version of the following classical
scheduling problem. There is a set $N = \{1, \ldots, n\}$ of n jobs to be scheduled on a
single machine. The machine can process at most one job at a time. Each job j
is specified by its length p_j and its weight w_j, where p_j and w_j are nonnegative
integers. Jobs must be processed for p_j time units without interruptions on the
machine. The goal is to find a schedule (i.e. permutation $\pi : N \to \{1, \ldots, n\}$)
such that the sum $\sum_{j=1}^{n} w_j C_j$, where C_j is the time at which job j completes in
the given schedule, is minimized. In standard scheduling notation (see e.g. Gra-
ham et al. [10]), this problem is known as $1 || \sum w_j C_j$. Smith [22] gave a simple
polynomial time algorithm for this problem, by showing that scheduling jobs in
non-decreasing order of the ratio of their processing time to their weight is optimal: given a set of n jobs with weights w_j and processing times p_j, $1 \leq j \leq n$,
schedule the jobs such that $\pi(i) < \pi(j)$ if and only if $p_i/w_i \leq p_j/w_j$. When there
are precedence constraints among jobs, then the problem becomes NP-hard [16].
Several 2-approximation algorithms are known for this variant [19,11,6,5,17], as
observed in [7], all of them can be seen as obtained by rounding a linear relaxation of an integer program formulation ILP due to Potts [18]. The integrality gap
of ILP is known [5] to be 2, and understanding if a better than 2-approximation
algorithm exists is considered an outstanding open problem in scheduling theory (see e.g. [21]). In this paper we consider the robust version of this classical
scheduling problem, as defined below.

Definition 1. *In the robust scheduling problem, we are given a set of jobs
$N = \{1, \ldots, n\}$ and a set of scenarios $S = \{s_1, \ldots, s_m\}$ where $s_i = (p_1^{s_i}, \ldots, p_n^{s_i}, w_1^{s_i}, \ldots, w_n^{s_i})$ for $s_i \in S$. A feasible schedule is a permutation π
of the jobs and the problem is to find a permutation π^* of the jobs such that*

$$\pi^* = \min_{\pi} \max_{s_i \in S} \left(\sum_{j \in N} w_j^{s_i} C_j^{s_i}(\pi) \right),$$

where $C_j^{s_i}(\pi) = \sum_{j' \in N, \pi(j') \leq \pi(j)} p_{j'}^{s_i}$.

---

[1] The definition for robust maximization problems are analogous.

Whereas $1||\sum w_j C_j$ is polynomial time solvable in the case of a single scenario, Kouvelis & Yu [15] prove that the robust version is weakly NP-complete even for the case of two scenarios and unit processing times.

In this paper we take on the task of studying the approximability of the robust variant. We show that, unless NP has quasi-polynomial algorithms, it cannot be approximated within factor $O(\log^{1-\varepsilon} n)$ in polynomial time, for any $\varepsilon > 0$. Moreover, under $P \neq NP$, we show that it remains hard to approximate within $6/5$ even if we assume that processing times, or alternatively weights, are equal to one and do not vary across the scenarios.

Then, we study the natural generalization of the ILP due to Potts [18] for the robust version. We provide a lower bound on the integrality gap and a matching upper bound for the special case where processing times or, alternatively, weights do not vary across the scenarios. Interestingly, the upper bound can be extended to include precedence constraints, and we obtain the same performance guarantee, namely a 2-approximation, as for the single scenario case. Proving good hardness of approximation results for $1|prec|\sum w_j C_j$ is a long standing open problem in scheduling theory. In contrast, for the robust variant, we show that it is NP-hard to approximate within a factor less than $6/5$.

We conclude by presenting a polynomial time algorithm based on dynamic programming for the case that the number of scenarios and the values of the instance are bounded by some constant.

2   Hardness of the Robust Scheduling Problem

2.1   Inapproximability Result for the General Problem

Here, we show that the general problem with non-constant number of scenarios has no $O(\log^{1-\varepsilon} n)$-approximation algorithm for any $\varepsilon > 0$, unless NP has quasi-polynomial algorithms. The hardness result is obtained by reducing the following version of the Label Cover problem to the scheduling problem.

Definition 2. *The Label Cover problem* $\mathcal{L}(V, W, E, [R], \{\sigma_{v,w}\}_{(v,w)\in E})$ *is defined as follows. We are given a regular bipartite graph with left side vertices V, right side vertices W, and set of edges $E \subseteq V \times W$. In addition, for every edge $(v, w) \in E$ we are given a map $\sigma_{v,w} : [R] \to [R]$. A labeling of the instance is a function ℓ assigning a set of labels to each vertex of the graph, namely $\ell : V \cup W \to 2^{[R]}$. A labeling ℓ satisfies an edge (v, w) if*

$$\exists a \in \ell(v), \exists b \in \ell(w) : \sigma_{v,w}(a) = b.$$

A total-labeling is a labeling that satisfies all edges. The value of a Label Cover instance, denoted $val(\mathcal{L})$, is defined to be the minimum, over all total-labelings, of $\max_{x \in V \cup W} |\ell(x)|$.

Observe that the variant of the Label Cover problem that is considered assumes that an edge is covered if, among the chosen labels, there *exists* a satisfying pair of labels. The following hardness result easily follows from the hardness result for the max version by using the "weak duality" relationship between the two versions (see e.g. [2]).

Theorem 1. *There exists a constant $\gamma > 0$ so that for any language L in NP, any input w and any $R > 0$, one can construct a labeling instance $\mathcal{L}$, with $|w|^{O(\log R)}$ vertices, and label set of size R, so that: If $w \in L, val(\mathcal{L}) = 1$ and otherwise $val(\mathcal{L}) > R^{\gamma}$. Furthermore, $\mathcal{L}$ can be constructed in time polynomial in its size.*

We prove the following theorem by presenting a reduction from the label cover problem.

Theorem 2. *There exists a constant $\gamma > 0$ so that for any language L in NP, any input w, any $R > 0$ and for $g \leq R^{\gamma}$, one can, in time $O(|w|^{O(g \log R)} \cdot R^{O(g)})$, construct a robust scheduling instance that has optimal value $1 + o(1)$ if $w \in L$ and optimal value g otherwise.*

Proof. Given a Label Cover instance $\mathcal{L}(V, W, E, [R], \{\sigma_{v,w}\}_{(v,w) \in E})$, we construct a robust scheduling instance I. Before giving a more formal definition of the reduction, we first give the intuition behind it.

For $x \in V \cup W$ let $R_x \subseteq [R]$ be the possible labels of x. For each $(v, w) \in E$, let $R_{v,w} \subseteq R_v \times R_w$ contain all pairs of labels of v and w that satisfy the map $\sigma_{v,w}$, i.e., $R_{v,w} = \{(a, b) \in R_v \times R_w : b = \sigma_{v,w}(a)\}$.

Clearly, for any feasible label cover ℓ there is at least one pair (a, b) from $R_{v,w}$ such that $a \in \ell(v)$ and $b \in \ell(w)$, and we say that (a, b) *covers* (v, w). In order to have a "corresponding" situation in the scheduling instance I, we define for each $(v, w) \in E$ a set $\mathcal{J}^{(v,w)} = \{J_1^{(v,w)}, J_2^{(v,w)}, \ldots, J_{n_{v,w}}^{(v,w)}\}$ of $n_{v,w} = |R_{v,w}|$ jobs. Let us consider some total ordering $r_{v,w} : R_{v,w} \to \{1, \ldots, n_{v,w}\}$ of the pairs in $R_{v,w}$. In any feasible schedule of the jobs from $\mathcal{J}^{(v,w)}$ there exists an $i = 1, \ldots, n_{v,w}$, such that $J_{i+1}^{(v,w)}$ is scheduled before $J_i^{(v,w)}$ (assume $i + 1$ equal to 1 when $i = n_{v,w}$), otherwise we would have a cycle in that schedule. The reduction that we are going to present will associate this situation (job $J_{i+1}^{(v,w)}$ scheduled before $J_i^{(v,w)}$) to the case where the ith pair in $R_{v,w}$ is in the label cover, i.e. $r_{v,w}^{-1}(i)$ covers edge (v, w). Then, for each $x \in V \cup W$, a set of scenarios is defined such that the maximum value of them counts (up to g) the number of different labels of x. A precise description of the reduction is given below.

Jobs. The jobs of instance I are the union of all jobs $\bigcup_{(v,w) \in E} \mathcal{J}^{(v,w)}$.

Ordering Scenarios. Let $m = |E|$ and let $\pi : E \to \{1, \ldots, m\}$ be some order of the edges. For each $i : 1 \leq i < m$, we have a scenario that sets the weights of the jobs in $\mathcal{J}^{\pi^{-1}(i)}$ to m and the processing time of the jobs in $\bigcup_{j>i} \mathcal{J}^{\pi^{-1}(j)}$ to m. The purpose of these scenarios is to ensure that any optimal schedule will schedule the jobs in the order

$$\mathcal{J}^{\pi^{-1}(1)} \prec \mathcal{J}^{\pi^{-1}(2)} \prec \cdots \prec \mathcal{J}^{\pi^{-1}(m)}. \tag{1}$$

Counting Scenarios. For each $v \in V$, let $E_v \subseteq E$ denote the set of edges incident to v. For each tuple $((v, w_1), \ldots, (v, w_g)) \in E_v \times \cdots \times E_v$ of pairwise different edges, for each tuple $(a_1, \ldots, a_g) \in R_v \times \cdots \times R_v$ of pairwise different labels, and for each tuple $(b_1, \ldots, b_g) \in R_{w_1} \times \cdots \times R_{w_g}$ so that $\sigma_{(v,w_i)}(a_i) = b_i$ for $i = 1, \ldots, g$, we have a different scenario $\mathcal{S}^{(v,w_1), \ldots, (v,w_g)}_{(a_1,b_1), \ldots, (a_g,b_g)}$. Each scenario $\mathcal{S}^{(v,w_1), \ldots, (v,w_g)}_{(a_1,b_1), \ldots, (a_g,b_g)}$ represents the situation in which label (a_i, b_i) covers edge (v, w_i) and the number of different labels of v is at least g. This label cover (partial) solution corresponds to the scheduling solutions $\sigma^{(v,w_1), \ldots, (v,w_g)}_{(a_1,b_1), \ldots, (a_g,b_g)}$ that schedule job $J^{(v,w_i)}_{h+1}$ before $J^{(v,w_i)}_h$, where $h = r_{v,w_i}(a_i, b_i)$, for each $i = 1, \ldots, g$. The value of these schedules is made larger than g by setting the processing time of $J^{(v,w_i)}_{h+1}$ equal to $m^{2\pi(v,w_i)}$ and the weight of $J^{(v,w_i)}_h$ equal to $1/m^{2\pi(v,w_i)}$, for each $i = 1, \ldots, g$, and zero all the others. Observe that the processing times and weights have been picked in such a way that jobs in $\mathcal{J}^{\pi^{-1}(i)}$ only contribute a negligible amount to the weighted completion time of jobs in $\mathcal{J}^{\pi^{-1}(j)}$ for $i < j$. This defines weights and processing times of scenarios for every $v \in V$. In a symmetric way we define scenarios $\mathcal{S}^{(v_1,w), \ldots, (v_g,w)}_{(a_1,b_1), \ldots, (a_g,b_g)}$, for every $w \in W$, to count the number of labels that are assigned to w.

The total number of scenarios is at most $|E| - 1 + 2|E|^g \cdot R^g \cdot R^g$ and the total number of jobs is at most $|E| \cdot R^2$. As $|E| = |w|^{O(\log R)}$, the total size of the robust scheduling instance is $O(|w|^{O(g \log R)} \cdot R^{O(g)})$.

Completeness Analysis. By Theorem 1, there exists a feasible labeling of $\mathcal{L}$ that assigns one label to each vertex. Let ℓ be such a labeling and consider a schedule σ of I that respects (1) and such that, for each element $(v, w) \in E$, the jobs in $\mathcal{J}^{(v,w)}$ are scheduled as follows: for $h = 1, \ldots, n_{v,w}$, if $h = r_{v,w}(\ell(v), \ell(w))$ then job $J^{(v,w)}_{h+1}$ is scheduled before $J^{(v,w)}_h$, otherwise $J^{(v,w)}_h$ is before $J^{(v,w)}_{h+1}$. This gives a feasible schedule. Moreover, since only one label is assigned to each vertex, it is easy to see that the value of any scenario is at most $1 + o(1)$.

Soundness Analysis. Consider a schedule σ of I. Define a labeling ℓ as follows:

$$\begin{cases} \ell(v) = \{a : \text{if } J^{(v,w)}_{h+1} \prec J^{(v,w)}_h \text{ for some } h = r_{v,w}(a, b), w \in W \text{ and } b \in R_w\} \\ \ell(w) = \{b : \text{if } J^{(v,w)}_{h+1} \prec J^{(v,w)}_h \text{ for some } h = r_{v,w}(a, b), v \in V \text{ and } a \in R_v\} \end{cases}$$

As at least one scenario for each edge will have value 1, ℓ is a feasible labeling of $\mathcal{L}$. Furthermore, by Theorem 1, there exists a vertex $x \in V \cup W$ so that $|\ell(x)| \geq g$, and this implies that there is a scenario of value g. Indeed, if $x \in V$ let $\ell(x) = \{a_1, \ldots, a_g\}$ be the set of g labels assigned to x, and let $(b_1, \ldots, b_g)$ and $(w_1, \ldots, w_g)$ be such that $J^{(v,w)}_{h+1} \prec J^{(x,w)}_h$ with $h = r_{x,w_i}(a_i, b_i)$. Then scenario $\mathcal{S}^{(x,w_1), \ldots, (x,w_g)}_{(a_1,b_1), \ldots, (a_g,b_g)}$ has been constructed to have value g according to this schedule. The same holds when $x \in W$.

By setting $g = O(\log^c n)$ (and $R = O(\log^{O(c)} n)$), where $|w| = n$ and $c \geq 1$ any large constant, we obtain that the input size is equal to $s = n^{O(g \log R)} \cdot R^{O(g)} = n^{O(\log^c n \cdot \log \log n)} \cdot (\log n)^{O(\log^c n)} = n^{O(\log^{c+\delta} n)} = 2^{O(\log^{c+1+\delta} n)}$, for any arbitrarily small $\delta > 0$. It follows that $g = O(\log s)^{\frac{c}{c+1+\delta}} = O(\log s)^{1-\varepsilon}$, for any arbitrarily small $\varepsilon > 0$.

Theorem 3. *For every $\varepsilon > 0$, the robust scheduling problem cannot be approximated within ratio $O(\log^{1-\varepsilon} s)$, where s is the input size, unless NP has quasi-polynomial algorithms.*

2.2   Inapproximability for Unit-Time/Unweighted Case

We now restrict the above problem to the case where the processing times do not vary across scenarios. We note that this case is symmetric to the one where the processing times may vary across scenarios while the weights are common. We show that, if the number of scenarios is unbounded, the robust scheduling problem is not approximable within $6/5$ even for the special case that all processing times are equal to one.

Our reduction is from the E3-Vertex-Cover problem, defined as follows. Given a 3-uniform hypergraph $G = (V, E)$ (each edge has size 3), the E3-Vertex-Cover problem is to find a subset $S \subseteq V$ that "hits" every edge in G, i.e. such that for all $e \in E$, $e \cap S \neq \emptyset$. Dinur et al. [8] showed that it is NP-hard to distinguish whether a k-uniform hypergraph has a vertex cover of weight $(\frac{1}{k-1} + \varepsilon)n$ from those whose minimum vertex cover has weight at least $(1 - \varepsilon)n$ for an arbitrarily small $\varepsilon > 0$.

Given a 3-uniform hypergraph $G(V, E)$, we construct a robust scheduling instance as follows.

- For every vertex $v_i \in V$ we create a job $i \in N$ with $p_i = 1$.
- For every hyperedge $e = \{v_1^e, v_2^e, v_3^e\} \in E$ we create a scenario s_e defined by

$$w_i^{s_e} = \begin{cases} 1 \text{, if } v_i \in \{v_1^e, v_2^e, v_3^e\} \\ 0 \text{, otherwise.} \end{cases}$$

Given the size of a minimum vertex cover c, one can calculate upper and lower bounds on the optimal value of the corresponding scheduling instance, as follows: given a schedule, i.e., a permutation π of the jobs, we can define a vertex cover solution VC by letting

$$VC = \{v_i \mid v_i \text{ covers an edge not covered by } \{v_j | \pi(j) < \pi(i)\}\}.$$

Let $v_j \in VC$ be the vertex in VC that is scheduled last, i.e., any $v_i \in VC$ with $i \neq j$ satisfies $\pi(i) < \pi(j)$. As v_j was added to VC, it covers an edge, say $e = \{v_j, v_k, v_l\}$, with $\pi(j) < \pi(k)$ and $\pi(j) < \pi(l)$. Furthermore, since $|VC| \geq c$ we have that $\pi(j) \geq c$ and hence $\pi(k) + \pi(l) \geq (c+1) + (c+2)$. It follows that there is an $s \in S$ with value at least

$$LB(c) = c + (c+1) + (c+2) > 3c$$

which is thus also a lower bound on $\min_{\pi} \max_{s \in S}(\text{val}(\pi, s))$.

For the upper bound, consider the schedule where we schedule c jobs corresponding to a minimum vertex cover first. Observe that a scenario in which the last of these c jobs has weight one takes its maximal value if the other two jobs of the corresponding edge are scheduled last, yielding

$$UB(c) = c + (n - 1) + n < c + 2 \cdot n.$$

Using the inapproximability results of Dinur et al. [8] we get the following gap for the robust scheduling problem:

$$\frac{LB((1 - \varepsilon)n)}{UB((\frac{1}{3-1} + \varepsilon)n)} > \frac{(1 - \varepsilon)n \cdot 3}{(\frac{1}{2} + \varepsilon)n + 2 \cdot n} = \frac{6}{5} - \varepsilon'$$

for some $\varepsilon' > 0$ that can be made arbitrarily small. As the unit-time and unweighted robust scheduling problem are symmetric, this yields the following theorem.

Theorem 4. *It is NP-hard to approximate the unit-time/unweighted robust scheduling problem within a factor less than 6/5.*

Assuming the *Unique Games Conjecture* [13], the inapproximability result for Ek-uniform Vertex Cover improves to a gap of $k - \varepsilon$ [14]. A similar reduction from 2-uniform hypergraphs (i.e. graphs) using the same bounds as above yields an inapproximability gap of 4/3.

Finally, we note that an easy numerical analysis shows that, in both cases, the inapproximability results cannot be improved by changing the uniformity of the hypergraphs in the vertex cover problems considered.

3   An LP-Based Approximation Algorithm and Integrality Gap

In this section, we consider the special case that processing times do not vary among scenarios, i.e. for every $i \in N$ we have $p_i^{s_1} = \ldots = p_i^{s_m} = p_i$. Note that this is symmetric to the case that processing times may vary across scenarios while weights are common. Inspired by Potts [18] integer linear program (ILP) formulation of $1|prec| \sum w_j C_j$, we formulate the robust scheduling problem with common processing times as follows:

$$\min \quad t$$
$$\sum_{j \in N} p_j w_j^{s_k} + \sum_{(i,j) \in N^2} \delta_{ij} \cdot p_i w_j^{s_k} \leq t \qquad 1 \leq k \leq m$$
$$\delta_{ij} + \delta_{ji} = 1 \qquad (i, j) \in N^2$$
$$\delta_{ij} + \delta_{jk} + \delta_{ki} \geq 1 \qquad (i, j, k) \in N^3$$
$$\delta_{ij} \in \{0, 1\} \qquad (i, j) \in N^2$$

The variables, δ_{ij} for $(i, j) \in N^2$, are called ordering variables with the natural meaning that job i is scheduled before job j if and only if $\delta_{ij} = 1$. The LP relaxation of the above ILP is obtained by relaxing the constraint $\delta_{ij} \in \{0,1\}$ to $\delta_{ij} \geq 0$. We will show that the resulting LP has an integrality gap of 2.

Consider the following family of instances, consisting of n jobs and an equal number of scenarios. The (scenario-independent) processing times are set to $p_j = 1$, $j \in N$. The weights of the jobs in scenario s_k are defined as follows:

$$w_j^{s_k} = \begin{cases} 1 \, , \text{ if } j = k \\ 0 \, , \text{ otherwise} \end{cases}, \ j \in N.$$

It is easy to see that setting

$$\delta_{ij} = 1/2, \quad 1 \leq i, j \leq n, \ i \neq j$$

yields a feasible solution. For this solution, all scenarios assume the same objective value

$$p_j + \sum_{i \neq j} \delta_{ij} p_i = 1 + (n-1) \cdot \frac{1}{2} = \frac{n+1}{2}$$

which therefore equals the objective value of this solution. This gives an upper bound on the value of the optimal solution.

On the other hand, for any feasible integral solution, there is a scenario s_k for which the job j is scheduled last. This scenario has value $w_j^k \cdot C_j = n$. Thus the integrality gap of the above presented LP with n scenarios is at least $2n/(n+1)$, which tends to 2 as n tends to infinity.

We now provide a 2-approximation algorithm based on the above LP-relaxation, thus showing that the analysis of the integrality gap is tight.

Given a solution of the LP, let

$$\tilde{C}_j = p_j + \sum_{i \neq j} \delta_{ij} p_i$$

be the fractional completion time of job j. Assume, without loss of generality, that $\tilde{C}_1 \leq \ldots \leq \tilde{C}_n$. We will use the following property to derive a 2-approximation algorithm.

Lemma 1 (Schulz [20]). *Given a solution of the above LP, with $\tilde{C}_1 \leq \ldots \leq \tilde{C}_n$ the following inequality holds*

$$\tilde{C}_j \geq \frac{1}{2} \sum_{i=1}^{j} p_i$$

This property can be used to derive a simple 2-approximation algorithm: schedule the jobs in non-decreasing order of $\tilde{C}_j$. The integral completion time is

$$C_j = \sum_{i=1}^{j} p_i \leq 2 \cdot \tilde{C}_j.$$

Since every completion time increases by at most a factor of 2, we have a 2-approximate solution.

It is worth noting that the above analysis holds also for the case that there are precedence constraints among the jobs, a significant generalization of this problem. For instance, in the single scenario case, $1|prec| \sum w_j C_j$ is NP-complete whereas $1|| \sum w_j C_j$ is polynomial time solvable. We summarize with the following theorem.

Theorem 5. *The robust version of $1|prec| \sum w_i C_i$ has a polynomial time 2-approximation algorithm when the processing times or, alternatively, the weights of the jobs do not vary among the scenarios.*

4   A Polynomial Time Algorithm for Constant Number of Scenarios and Constant Values

In this section we assume that the number of scenarios m as well as the weights and processing times are bounded by some constant. Given an instance I of the robust scheduling problem, let W be the maximum weight and P the maximum processing time occurring in the description of I. We present a polynomial time algorithm that solves this problem. In fact, we are going to solve the related multi-criteria scheduling problem. This result carries over to our problem by use of Theorem 1 in Aissi et. al. [1].

In the context of multi-criteria optimization, given two vectors $v, w \in \mathbb{N}^k$, $v \neq w$, $k > 0$, we say that v *dominates* w, if $v_i \leq w_i$ for all $1 \leq i \leq k$. A vector that is not dominated is called *efficient*. Analogously, given a set of vectors S, a subset $S' \subseteq S$ is called an *efficient set* if there is no pair $(v, v'), v \in S, v' \in S'$ such that v dominates v'. The goal in multi-criteria optimization is to find a maximal efficient set of solutions.

For a fixed set of scenarios $S = \{s_1, \ldots, s_m\}$, we define the *multivalue* of a schedule π by $\text{val}(\pi) = (\text{val}(\pi, s_1), \ldots, \text{val}(\pi, s_m))$. Furthermore, we call $\alpha = ((w_1, p_1), \ldots, (w_m, p_m))$ with $1 \leq w_i \leq W$, $1 \leq p_i \leq W$ a *job profile*, and let $p(\alpha) = (p_1, \ldots, p_m)$ and similarly $w(\alpha) = (w_1, \ldots, w_m)$. Note that, since we assumed that P, W and m are all bounded by a constant, there can only be a constant number of different job profiles. Let $\alpha_1, \ldots, \alpha_k$ be the different job profiles that occur in instance I. We can now identify I by the tuple $((\alpha_1, \ldots, \alpha_k), (n_1, \ldots, n_k))$ where n_i is the number of jobs with profile α_i occurring in I. We will present a dynamic programming approach for solving the min-max scheduling problem with a constant number of scenarios and constant values in polynomial time.

4.1   Polynomial Time Algorithm

Consider a k-dimensional dynamic programming table DPT of size $(n_1 + 1) \times (n_2 + 1) \times \ldots \times (n_k + 1)$. Each cell of this table represents a subinstance I' of I, where the coordinates of the cell encode the number of jobs of the corresponding profile that are present in I' (for instance, the cell $(1, 0, 4)$ represents the subinstance of I that contains one job of type α_1 and four jobs of type α_3). We denote the number of jobs in an instance represented by a cell $c = (c_1, \ldots, c_k)$ by $n(c) = \sum_{i=1}^{k} c_i$. Each of these cells will accommodate an efficient set M_c of multivalues of schedules in which only the jobs of the subinstance are considered (note that since the maximum value in any scenario is bounded, there can only be a polynomial number of different efficient vectors). Since the cell $(n_1, \ldots, n_k)$ represents the whole instance, filling in the last cell of the table would allow us to solve the multi-criteria scheduling problem, and thus also the min-max scheduling problem.

We initialize the table by filling in the cells whose coordinates sum up to one, i.e. the cells $c = (c_1, \ldots, c_k)$ with $n(c) = 1$, as follows: for $c_t = 1$ add to M_c the multivalue of the schedule consisting of a single job with profile α_t. We continue filling in the rest of the cells in order of increasing $n(c)$ in the following manner.

Consider the cell c with coordinates $(c_1, \ldots, c_k)$ and let $T_c = \{(c_1', \ldots, c_k') \mid n(c') = n(c) - 1, \ c_i' \geq c_i - 1\}$. In other words, T_c contains those cells representing subinstances that result by removing one job from I_c. Note that, since we fill in the table in order of increasing $n(c)$, all cells in T_c have been filled in at this point. For each $c' \in T_c$ with $c_t - c_t' = 1$, add to the set M_c the multivalues of the schedules that result from the schedules in $M_{c'}$ by adding a job of profile α_t in the end of the schedule. More formally, for each π' with $\mathrm{val}(\pi') \in M_{c'}$, add $\mathrm{val}(\pi)$ to M_c, with π defined as follows:

$$\pi(j) = \pi'(j) \text{ for } 1 \leq j \leq n(c') \text{ and } \pi(n(c)) = \alpha_t.$$

Given $\mathrm{val}(\pi')$, the multivalues of these schedules can easily be computed by:

$$\mathrm{val}(\pi) = \mathrm{val}(\pi') + w(\alpha_t) \cdot \sum_{i=1}^{k} c_i \cdot p(\alpha_i)$$

Note that only the multivalue of π' is needed in the above calculations, not π' itself.

We conclude the computation for cell c by replacing M_c by $\mathrm{Red}(M_c)$, which retains only the efficient elements of M_c.

Lemma 2. *For every cell c of the table DPT, the set M_c is a maximal efficient subset of the set of all multivalues achieved by scheduling the jobs of I_c.*

Proof. We need to show that for every cell c of the table DPT and every multivalue $\mathrm{val}(\pi)$, where π is a schedule of I_c, either

- $\mathrm{val}(\pi) \in M_c$, or
- $\exists v \in M_c$, such that $v \leq \mathrm{val}(\pi)$

Suppose, towards contradiction, that this is not the case, and let c be a cell with minimal $n(c)$ that does not satisfy the above condition. Thus, there is a schedule π of the instance I_c with $\mathrm{val}(\pi) \notin M_c$ and for any $v \in M_c$ there is an $l \in \{1, \ldots, k\}$ with $\mathrm{val}(\pi)_l < v_l$. Clearly, this can only happen for $n(c) \geq 2$. Let α_f be the profile of the job scheduled last in π and let c' be the cell with coordinates $(c_1, c_2, \ldots, c_{f-1}, c_f - 1, c_{f+1}, \ldots, c_k)$. Furthermore, let π' be the schedule derived from π by omitting the last job. The multivalue of π' is $\mathrm{val}(\pi) - w(\alpha_f) \cdot \sum_{i=1}^{k} c_i \cdot p(\alpha_i)$. If there were a $v \in M_c$ such that $\mathrm{val}(v) \leq \mathrm{val}(\pi')$, then $\mathrm{val}(\pi)$ would be dominated by $v + w(\alpha_f) \cdot \sum_{i=1}^{k} c_i \cdot p(\alpha_i)$. Thus, for every $v \in M_{c'}$, there is an $l \in \{1, \ldots, k\}$ such that $v_l > \mathrm{val}(\pi')_l$ and thus c' does not satisfy the above property either. Since $n(c') < n(c)$, this contradicts the minimality of c.

It is easy to see that the initialization of the table, as well as the computations of $\mathrm{val}(\pi)$ can be done in polynomial time. Furthermore, since $(n^2 \cdot P \cdot W)^2$ is an upper bound on the value of any schedule in any scenario, there can be at most $(n^2 \cdot P \cdot W)^{2m}$ efficient vectors in any stage of the computation. The size of the dynamic programming table is bounded by n^k and for each computation of a cell, at most k cells need to be considered. Moreover, the operator Red can be implemented in time $(n^2 \cdot P \cdot W)^{4m}$ by exhaustive comparison. Thus, a single cell can be filled-in in time $k(n^2 \cdot P \cdot W)^{2m} + (n^2 \cdot P \cdot W)^{4m}$, and the whole table in time $n^k \cdot (k \cdot (n^2 \cdot P \cdot W)^{2m} + (n^2 \cdot P \cdot W)^{4m})$. The number of different profiles k is bounded by $(P \cdot W)^m$, which is a constant. Thus our algorithm runs in time $O(n^{8m+W^m P^m})$, i.e. polynomial in n.

Acknowledgements

This research is supported by Swiss National Science Foundation project 200021-104017/1, "Power Aware Computing", by the Swiss National Science Foundation project 200020-109854, "Approximation Algorithms for Machine scheduling Through Theory and Experiments II", and by the Swiss National Science Foundation project PBTI2-120966, "Scheduling with Precedence Constraints".

References

1. Aissi, H., Bazgan, C., Vanderpooten, D.: Approximating min-max (regret) versions of some polynomial problems. In: Chen, D.Z., Lee, D.T. (eds.) COCOON 2006. LNCS, vol. 4112, pp. 428–438. Springer, Heidelberg (2006)
2. Arora, S., Lund, C.: Hardness of approximations. In: Hochbaum, D.S. (ed.) Approximation Algorithms for NP-Hard Problems. PWS (1995)
3. Bertsimas, D., Sim, M.: Robust discrete optimization and network flows. Programming Series B 98, 49–71 (2002)
4. Birge, J., Louveaux, F.: Introduction to stochastic programming. Springer, Heidelberg (1997)
5. Chekuri, C., Motwani, R.: Precedence constrained scheduling to minimize sum of weighted completion times on a single machine. Discrete Applied Mathematics 98(1-2), 29–38 (1999)
6. Chudak, F.A., Hochbaum, D.S.: A half-integral linear programming relaxation for scheduling precedence-constrained jobs on a single machine. Operations Research Letters 25, 199–204 (1999)

7. Correa, J.R., Schulz, A.S.: Single machine scheduling with precedence constraints. Mathematics of Operations Research 30(4), 1005–1021 (2005); Extended abstract in Proceedings of the 10th Conference on Integer Programming and Combinatorial Optimization (IPCO 2004), pp. 283–297 (2004)
8. Dinur, I., Guruswami, V., Khot, S., Regev, O.: A new multilayered pcp and the hardness of hypergraph vertex cover. In: STOC 2003: Proceedings of the thirty-fifth annual ACM symposium on Theory of computing, pp. 595–601. ACM, New York (2003)
9. Feige, U., Jain, K., Mahdian, M., Mirrokni, V.S.: Robust combinatorial optimization with exponential scenarios. In: Fischetti, M., Williamson, D.P. (eds.) IPCO 2007. LNCS, vol. 4513, pp. 439–453. Springer, Heidelberg (2007)
10. Graham, R., Lawler, E., Lenstra, J.K., Rinnooy Kan, A.H.G.: Optimization and approximation in deterministic sequencing and scheduling: A survey. Annals of Discrete Mathematics 5, 287–326 (1979)
11. Hall, L.A., Schulz, A.S., Shmoys, D.B., Wein, J.: Scheduling to minimize average completion time: off-line and on-line algorithms. Mathematics of Operations Research 22, 513–544 (1997)
12. Kasperski, A., Zieliński, P.: On the existence of an fptas for minmax regret combinatorial optimization problems with interval data. Operations Research Letters 35(4), 525–532 (2007)
13. Khot, S.: On the power of unique 2-prover 1-round games. In: IEEE Conference on Computational Complexity, p. 25 (2002)
14. Khot, S., Regev, O.: Vertex cover might be hard to approximate to within $2 - \epsilon$. In: Proc. of 18th IEEE Annual Conference on Computational Complexity (CCC), pp. 379–386 (2003)
15. Kouvelis, P., Yu, G.: Robust Discrete Optimization and Its Applications. Kluwer Academic Publishers, Dordrecht (1997)
16. Lawler, E.L.: Sequencing jobs to minimize total weighted completion time subject to precedence constraints. Annals of Discrete Mathematics 2, 75–90 (1978)
17. Margot, F., Queyranne, M., Wang, Y.: Decompositions, network flows and a precedence constrained single machine scheduling problem. Operations Research 51(6), 981–992 (2003)
18. Potts, C.N.: An algorithm for the single machine sequencing problem with precedence constraints. Mathematical Programming Study 13, 78–87 (1980)
19. Schulz, A.S.: Scheduling to minimize total weighted completion time: Performance guarantees of LP-based heuristics and lower bounds. In: Cunningham, W.H., Queyranne, M., McCormick, S.T. (eds.) IPCO 1996. LNCS, vol. 1084, pp. 301–315. Springer, Heidelberg (1996)
20. Schulz, A.S.: Scheduling to minimize total weighted completion time: performance guarantees of LP-based heuristics and lower bounds. In: Cunningham, W.H., Queyranne, M., McCormick, S.T. (eds.) IPCO 1996. LNCS, vol. 1084, pp. 301–315. Springer, Heidelberg (1996)
21. Schuurman, P., Woeginger, G.J.: Polynomial time approximation algorithms for machine scheduling: ten open problems. Journal of Scheduling 2(5), 203–213 (1999)
22. Smith, W.E.: Various optimizers for single-stage production. Naval Research Logistics Quarterly 3, 59–66 (1956)

Streaming Algorithms for k-Center Clustering with Outliers and with Anonymity

Richard Matthew McCutchen* and Samir Khuller**

University of Maryland
{rmccutch,samir}@cs.umd.edu

Abstract. Clustering is a common problem in the analysis of large data sets. *Streaming* algorithms, which make a single pass over the data set using small working memory and produce a clustering comparable in cost to the optimal offline solution, are especially useful. We develop the first streaming algorithms achieving a constant-factor approximation to the cluster radius for two variations of the k-center clustering problem. We give a streaming $(4+\epsilon)$-approximation algorithm using $O(\epsilon^{-1}kz)$ memory for the problem with *outliers*, in which the clustering is allowed to drop up to z of the input points; previous work used a random sampling approach which yields only a bicriteria approximation. We also give a streaming $(6 + \epsilon)$-approximation algorithm using $O(\epsilon^{-1}\ln(\epsilon^{-1})k + k^2)$ memory for a variation motivated by anonymity considerations in which each cluster must contain at least a certain number of input points.

Keywords: clustering, k-center, streaming, outliers, anonymity.

1   Introduction

Clustering is a common problem arising in the analysis of large data sets. For many applications in document and image classification [3,9,13,15,16] and data mining, clustering plays a central role [5]. In a typical clustering problem, we have a set of n input points from an arbitrary metric space (with a distance function satisfying the triangle inequality) and wish to partition the points into k clusters. We select a center point for each cluster and consider the distance from each point to the center of the cluster to which it belongs. In the k-center problem, we wish to minimize the maximum of these distances, while in the k-median problem, we wish to minimize their sum. In this paper we focus on k-center clustering, since it is an important problem for which a variety of approaches have been presented. Hochbaum and Shmoys [14] and Gonzalez [10] developed algorithms that achieve a factor 2 approximation in the cluster radius. This is the best possible since one can show by a reduction from the dominating set problem that it is *NP*-hard to approximate k-center with factor $2 - \epsilon$ for any $\epsilon > 0$.

* Supported by an NSF REU Supplement to Award CCF-0430650.
** Supported by NSF CCF-0430650 and CCF-0728839.

A. Goel et al. (Eds.): APPROX and RANDOM 2008, LNCS 5171, pp. 165–178, 2008.
© Springer-Verlag Berlin Heidelberg 2008

In the analysis of extremely large data sets, it is not possible to hold the entire input in memory at once. Thus, we consider the *streaming* or *incremental* model in which the algorithm reads input points one by one and maintains a valid clustering of the input seen so far using a small amount of working memory. (We contrast such algorithms with *offline* algorithms that use memory polynomial in the size of the input.)

Charikar, Chekuri, Feder and Motwani [5] introduced the incremental model for the k-center problem and gave a very elegant "Doubling Algorithm" that achieves a factor 8 approximation using only $O(k)$ memory. The result is slightly surprising, since it is not obvious at all how to do this incrementally. The key idea is to maintain a lower bound on the radius of an optimal solution. For example, after $k + 1$ input points have been presented, examining the closest pair of points gives us an obvious lower bound on the optimal radius, since at least two of these points must belong to the same cluster.

The key focus of this paper is to deal with outliers, an issue originally raised in [7]. Data is often noisy and a very small number of outliers can dramatically affect the quality of the solution if not taken into account, especially under the k-center objective function, which is extremely sensitive to the existence of points far from cluster centers. The formal definition of the problem is as follows: group *all but z points* into k clusters, minimizing the radius of the largest cluster. An offline factor 3 approximation for the outlier version was developed [7]; it greedily chooses clusters of a certain radius so as to cover as many new input points with each cluster as possible. The factor 3 assumes that we can enumerate all center points in the metric space that the optimal clustering is allowed to use. If not, the algorithm is easily modified to produce a clustering that uses only input points as centers but has radius at most 4 times that of an optimal clustering with unrestricted centers.[1] The same paper also considered the k-median objective function and developed a bicriteria algorithm: if there exists a solution of cost C that drops z outliers, it finds one of cost at most $O(C)$ that drops at most $O(z)$ outliers. More recently, a polynomial time algorithm has been developed for k-medians that delivers a solution of cost $O(C)$ while dropping only z outliers [8].

The offline algorithm for k-center clustering with outliers is not easily adapted to the streaming model because it relies on the ability to count the input points that would be covered by a potential cluster, which is difficult to implement without having the entire data set in memory. In general, dealing with outliers in the streaming model is quite tricky because we have no way to know, as points arrive, which should be clustered and which are outliers. This problem was first considered by Charikar, O'Callaghan and Panigrahy [6], who developed a streaming-model bicriteria approximation for the k-center problem (see also [12]). Their approach is based on taking a random sample of the data set that is small enough to fit in memory and running the offline algorithm [7] on the sample. They then prove that, with high probability, the set of clusters found for the sample is also a good solution for the entire data set. This construction preserves the radius approximation factor of the underlying offline algorithm

---

[1] We simply expand the disks G_i to radius $2r$ and the disks E_i to radius $4r$.

(3 or 4) but increases the number of outliers to $(1 + \epsilon)^2 z$. The sample has size roughly $O(\epsilon^{-2}kn/z)$, where n is the data set size. Therefore, the sampling approach is good when z is linear in n and a slight increase in the number of outliers is acceptable; otherwise, it requires an unreasonable amount of memory.

We present a streaming algorithm for k-center clustering with outliers that is in several ways complementary to that of [6]. Our deterministic algorithm is based on the Doubling Algorithm [5] and also uses the offline algorithm for outliers [7] as a subroutine. It increases the radius approximation factor to $3 + \epsilon$ or $4 + \epsilon$ but meets the outlier bound z exactly; as far as we are aware, it is the first streaming constant-factor approximation with the latter property. Our algorithm uses $O(\epsilon^{-1}kz)$ memory, so it is suitable when z is *small* and an additional slight increase in the *cluster radius* is acceptable.

Agarwal et al. [1] present an algorithm for shape-fitting with outliers that may be applicable to k-center clustering, and Bădoiu et al. [4] present a sampling-based streaming k-center algorithm that uses coresets. However, both techniques work only in Euclidean spaces $\mathbb{R}^d$; furthermore, the first requires multiple passes over the input and the second has running time exponential in k.

Other recent applications of k-center clustering (with and without outliers) for the purposes of anonymity are considered in [2], but the algorithms given there do not work in the streaming model. We present a streaming $(6 + \epsilon)$-approximation algorithm for the k-center clustering problem with a lower bound b on the number of points per cluster. The precise requirement is that it must be possible to allocate each input point to a center within the appropriate radius so that each center gets at least b points, i.e., centers cannot meet the bound by sharing points.

2   Improving Streaming Algorithms by Parallelization

In this section, we develop a parallelization construction that improves the approximation factor of the Doubling Algorithm to $2 + \epsilon$ while increasing the running time and memory usage by a factor of $O(\epsilon^{-1} \ln(\epsilon^{-1}))$.[2] We first generalize the Doubling Algorithm to a "Scaling Algorithm" based on a parameter $\alpha > 1$ that maintains a lower bound r on the radius of the optimal cluster and raises it by a factor of exactly α at a time. As it reads points, this algorithm keeps centers separated by at least $2\alpha r$ and ensures that every input point seen so far is within $\eta r = \left(2\alpha^2/(\alpha - 1)\right)r$ of a center. Let r^* denote the optimal radius, and let r_0 be half the least distance between two of the first $k + 1$ distinct input points, which is used to initialize r.

Naively, the Scaling Algorithm is an η-approximation because it gives us a solution with radius within a factor η of its own lower bound r, and we minimize $\eta = 8$ by choosing $\alpha = 2$. But observe that if $r^* = 1.9r_0$, we get lucky: the algorithm cannot raise r to $2r_0$ because $2r_0$ is not a lower bound on r^*, so it is

---

[2] Sudipto Guha independently discovered a similar construction [11] based on Gonzalez's algorithm [10]; it also yields a streaming $(2 + \epsilon)$-approximation algorithm for k-center clustering.

obliged to return a solution with radius at most $8r_0$, which is only a factor of about 4.2 from optimal.

To ensure that we always get lucky in this way, we run m instances of the Scaling Algorithm in parallel (feeding each input point to each instance) with interleaved sequences of r values. Specifically, we initialize the r value of the ith instance ($i = 1$, $\ldots, m$) to $\alpha^{(i/m)-1}r_0$ so that the instance takes on the r values $\alpha^{t+(i/m)-1}r_0$, where $t = 0, 1, \ldots$. Consequently, any desired r of the form $\alpha^{(j/m)-1}r_0$ for a positive integer j will eventually be taken on by some instance. Letting j be the smallest integer greater than $m\log_\alpha(r^*/r_0)$, we have $\alpha^{j/m}r_0 > r^*$, so the instance that takes $r = \alpha^{(j/m)-1}r_0$ will be unable to raise r again and thus will return a solution whose radius R is at most $\eta\alpha^{(j/m)-1}r_0$. And by our choice of j, $\alpha^{(j-1)/m}r_0 \leq r^*$, so $R \leq \eta\alpha^{(1/m)-1}r^*$. Therefore, by taking the best solution produced by any of the m instances, we achieve a factor $(\eta/\alpha)\sqrt[m]{\alpha}$ approximation.

Substituting the expression for η, the approximation factor of the parallelized algorithm becomes $2\big(1+1/(\alpha-1)\big)\sqrt[m]{\alpha}$. Now, we want α large to make $1/(\alpha-1)$ small; intuitively, with larger α, accounting for previous rounds across an increase of r costs less in the Scaling Algorithm's approximation factor. We also want m large to keep $\sqrt[m]{\alpha}$ close to 1. Letting $\alpha = O(\epsilon^{-1})$ and $m = O(\epsilon^{-1}\ln(\epsilon^{-1}))$ gives a factor of $2 + \epsilon$. This approximation factor is essentially the best we can hope for since the best offline algorithms [14,10] are 2-approximations, but there may be a better construction that uses less time and memory. We will apply the same parallelization construction to the streaming-model clustering algorithms described in the following sections.

2.1 Suitability of Parallelized Algorithms

The original model of Charikar et al. [5] requires that a clustering algorithm maintain a single clustering of the input points read so far and modify it only by merging clusters. This model has the advantage that a forest describing the merges can be incrementally written to secondary storage; the forest can later be traversed to enumerate the input points in any desired output cluster without a second pass over the entire input. Parallelized algorithms do not fit this model because they maintain many clusterings and do not choose one until the end. (This is why they do not contradict the lower bound of $1+\sqrt{2}$ on the approximation factor in [5].) Writing out a forest for each of the many partial clusterings under consideration may be impractical. However, parallelized algorithms are still useful when the goal is only to produce statistics for each output cluster (along with the centers themselves) because the statistics can be maintained independently for each partial clustering.

3   Clustering with Outliers

In this section, we develop a streaming algorithm for k-center clustering with z outliers that achieves a constant factor approximation to the cluster radius using $O(kz)$ memory. We then parallelize it to a $(4+\epsilon)$-approximation using $O(\epsilon^{-1}kz)$

memory. The essential difficulty in designing a deterministic streaming algorithm for clustering *with outliers* is that it is dangerous to designate an input point as a cluster center and start forgetting nearby points because they could all be outliers and the center might be needed to cover points elsewhere. Our algorithm overcomes the difficulty by delaying the decision as long as necessary. Specifically, it accumulates input points (remembering all of them) until it sees $z + 1$ points close together. These cannot all be outliers, so it creates a cluster for them and only then can safely forget any later points that fall in that cluster.

The algorithm's state consists of:

- some number $\ell \leq k$ of stored cluster centers, each of which carries a list of $z + 1$ nearby "support points" from which it was originally formed;
- some "free points" that do not fall into existing clusters but cannot yet be made into new clusters because they might be outliers; and
- a lower bound r on the optimal radius, as in the Doubling Algorithm.

The algorithm ensures that clusters of radius ηr at the ℓ stored centers cover all forgotten points, and it checks after processing each input point that it can cover all but at most z of the free points with $k - \ell$ additional clusters of radius ηr. Thus, whenever the algorithm encounters the end of the input, it can produce a solution with radius ηr. The algorithm is based on parameters α, β, and η, which we will choose later to optimize the approximation factor; for the proof of correctness to hold, these parameters must satisfy some constraints that we will state as they arise.

The algorithm is designed so that, whenever its partial solution with radius ηr becomes invalid, it can establish a new lower bound αr on the optimal radius, raise r by a factor of α, and adapt the partial solution to the new value of r; this process is repeated until the validity of the partial solution is restored. Furthermore, we will show that the algorithm will never store more than $O(kz)$ free points at a time, establishing the memory requirement.

As in the Doubling Algorithm [5], we need a certain separation between centers in order to raise r. To this end, we say that two distinct centers *conflict* if some support point of the first is within distance $2\alpha r$ of some support point of the second.

Algorithm 3.1 (Clustering with outliers). Peek at the first $k+z+1$ distinct input points, initialize r to half the least distance between any two of those points, and start with no cluster centers and no free points. Then read the input points in batches. Batches of size kz appear to give the best trade-off between running time and memory usage, but a different size can be used if desired. For each batch, add the points as free points and then perform the following procedure:

1. Drop any free points that are within distance ηr of cluster centers.
2. If some free point p has at least $z+1$ free points within distance βr (including itself), then add p as a cluster center, choosing any $z + 1$ of the free points within distance βr as its support points, and repeat from step 1. If no such p exists, proceed to the next step.

3. Let ℓ be the number of stored cluster centers. Check that $\ell \leq k$ and that at most $(k - \ell)z + z$ free points are stored. Run the 4-approximation offline algorithm for k-center clustering with outliers (see the Introduction) to attempt to cover all but at most z of the free points using $k - \ell$ clusters of radius ηr. If the checks and the offline algorithm both succeed, processing of the current input batch is complete. Otherwise, set $r \leftarrow \alpha r$ and continue to the next step.

4. Unmark all the stored centers and then process them as follows: while there exists an unmarked center c, mark c and drop any other centers that conflict with c with respect to the new value of r. When a center is dropped, its support points are forgotten. (Note that once a center c is marked, it cannot later be dropped on account of another center c' because c' would already have been dropped on account of c.) Repeat from step 1.

When the end of the input is reached, return clusters of radius ηr at the stored centers plus the clusters found by the last run of the offline algorithm.   □

Figure 1 shows an intermediate state of the algorithm on a data set with $k = 3$ and $z = 4$. The algorithm is storing $\ell = 2$ cluster centers c_1 and c_2, and each center has $z + 1 = 5$ support points (including itself), which are within βr of it. Several other input points within distance ηr of the centers have been forgotten. The algorithm

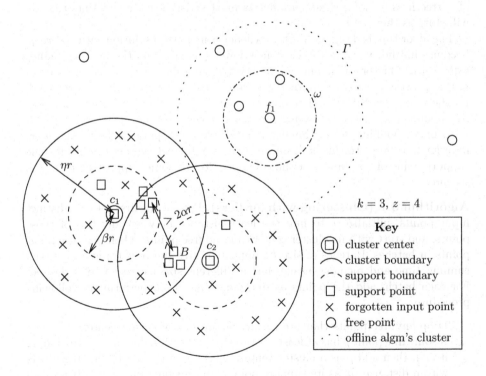

Fig. 1. Example of clustering with outliers

is also storing seven free points, including f_1, which would be converted to a cluster center if there were just one more free point inside circle ω; but as it stands, the algorithm cannot rule out the possibility that all four of the points in ω are outliers. The offline algorithm found the cluster Γ (centered at f_1), which covers all but $2 \le z$ of the free points; if we combine it with the stored centers, we have a valid clustering of radius ηr for the input points seen so far.

Notice that the three support points A are just far enough from the support points B to avoid a conflict. If they were any closer, then the optimal solution could conceivably cover all six points with a single cluster of radius αr and leave the remaining four support points as outliers, and the proof of correctness of the algorithm's decision to set $r \leftarrow \alpha r$ (see Lemma 3.2(e) below) would fail.

Suppose several free points arrive inside Γ but outside ω. The current clustering covers these points, but if the algorithm allowed them to accumulate indefinitely, it would violate the $O(kz)$ memory bound. Thus, when the number of free points exceeds $(k - \ell)z + z = 8$, the algorithm raises r on the following logic: in the optimal solution, two clusters are busy covering support points of the stored centers, and there is no way a third cluster of radius αr containing at most $z = 4$ points can cover all the free points with at most 4 outliers. (If there were a potential third cluster of more than 4 points, the algorithm would already have recognized it in step 2.) Once r is raised, the support points A and B conflict, so one of the centers c_1, c_2 subsumes the other in step 4.

Lemma 3.2. *The algorithm maintains the following invariants:*

(a) *Every time step 1 completes, the remaining free points are at least distance ηr from cluster centers.*

(b) *Each stored center has $z + 1$ support points within distance βr of it.*

(c) *No two stored cluster centers conflict.*

(d) *Every input point the algorithm has read so far either is a free point or is covered by a cluster of radius ηr at a stored center.*

(e) *The optimal clustering for the input points the algorithm has read so far requires a radius of at least r.*

Proof. (a) is obvious. (b) is checked when a center is added and remains true when r increases.

To prove (c), we place the constraint $\eta \geq 2\alpha + \beta$ on our later choice of the parameters. With this constraint, addition of a center c in step 2 preserves the invariant. For if s_1 is a support point of an existing center c_1, then s_1 is within βr of c_1, which is at least ηr from c's support points (since they were previously free points). By the triangle inequality, s_1 is at least distance $\eta r - \beta r \geq 2\alpha r$ from c's support points, so no conflict results. Furthermore, temporary conflicts created by an increase in r are removed in step 4.

Each point the algorithm reads is initially a free point, so the algorithm endangers invariant (d) only when it drops free points or centers. Free points dropped in step 1 are covered by stored centers, so they do not break the invariant. Steps 3 and 4 effectively drop some clusters while expanding the remaining ones to radius $\eta \alpha r$; we must show that any input point that was covered by a dropped cluster before the change is covered by an expanded cluster afterwards. To this end, we constrain $\eta + 2\alpha^2 + 2\beta \leq \eta\alpha$. Let r_0 and $r_1 = \alpha r_0$ denote the old and new values of r, respectively. Consider an input point p that was covered by a dropped center c, meaning that it was within distance ηr_0 of c. c was dropped when a conflicting center c' was marked. The support points causing the conflict were within distance $2\alpha r_1$ of each other and distance βr_0 of their respective centers, so the distance from p to c' is at most

$$\eta r_0 + \beta r_0 + 2\alpha r_1 + \beta r_0 = (\eta + 2\alpha^2 + 2\beta)r_0 \leq \eta\alpha r_0 = \eta r_1.$$

Thus, p is covered by c', and the invariant holds.

Invariant (e) is established by the initial setting of r because one of the k clusters of the optimal solution must cover two of the first $k + z + 1$ distinct input points. To show that increases in r maintain the invariant, we will show that, if step 3 is reached and the optimal clustering C^* for the input read so far has radius *less than* αr, then the algorithm does *not* set $r \leftarrow \alpha r$.

Let c be a stored cluster center. C^* cannot designate all $z + 1$ of c's support points as outliers, so some cluster $c^* \in C^*$ must cover one of c's support points; we say that c^* *bites* c. No two stored cluster centers conflict, so no cluster of C^* (having diameter less than $2\alpha r$) can bite two of them; thus, each stored cluster center is bitten by a different cluster of C^*. In particular, this means $\ell \leq k$. Similarly, by invariant (a) and our assumption that $\eta \geq 2\alpha + \beta$, no cluster of C^* can both bite a stored cluster center and cover a free point. Finally, we constrain $\beta \geq 2\alpha$; then no cluster of C^* can cover $z + 1$ or more free points because, if it did, each of those free points would be within distance βr of all the others and they would have become the support points of a cluster center in step 2.

Now, at least ℓ of the clusters of C^* are devoted to biting stored cluster centers, so at most $k - \ell$ clusters can cover free points; let F^* be the set of these clusters. In order for C^* to be a valid solution for the input read so far, F^* must be a valid clustering of all but at most z of the free points. But we showed that each of the at most $k - \ell$ clusters in F^* covers at most z free points, so there can be at most $(k - \ell)z + z$ free points in total. Finally, the offline algorithm is a 4-approximation, so if we assume that $\eta \geq 4\alpha$, the existence of F^* with radius less than αr guarantees that the offline algorithm will find a clustering of radius ηr. The result is that r is not raised, as desired.   □

Theorem 3.3. *The algorithm produces a valid clustering of radius ηr using $O(kz)$ memory and $O(kzn + (kz)^2 \log P)$ time, where P is the ratio of the optimal radius to the shortest distance between any two distinct input points.*

Proof. Validity of the clustering: The first set of ℓ clusters covers all input points except the free points by Lemma 3.2(d), and the second set of $k - \ell$ clusters covers all but at most z of the free points. Thus, together, the k clusters cover all but at most z of the input points.

Memory usage: At any time, the algorithm remembers $\ell(z + 1)$ support points (including the centers), at most $(k - \ell)z + z$ free points from before the current batch, and at most kz free points from the current batch. This is a total of at most $(2k + 1)(z + 1)$ points, and the working storage needed to carry out the steps is constant per point.

Running time: At the beginning of a batch, we perform step 1 exhaustively in $O(k^2 z)$ time. We identify potential centers in step 2 by maintaining a count for each free point p of the free points within distance βr of p. Each time we add or drop a free point, which happens at most $O(kz)$ times per batch, we perform a scan of the other free points to update their counts (this takes $O(kz)$ time). When we convert a free point c to a center, we identify its support points and the free points to be dropped in step 1 on the same scan that drops c itself as a free point. The offline algorithm in step 3 runs in $O((kz)^2)$ time using its own set of distance-$\eta r/2$ counts; we charge a successful run of the offline algorithm, which ends a batch, to that batch and a failed run to the resulting increase in r. In step 4, we have $O(k)$ centers (k from the previous batch and at most one per $z + 1$ of the $O(kz)$ free points), so we test each of the $O(k^2)$ pairs of centers for a conflict in $O(z^2)$ time; this takes $O((kz)^2)$ time, which we charge to the increase in r. Now, there are $O(n/kz)$ batches and $O(\log P)$ increases in r, and each batch or increase is charged $O((kz)^2)$ time, giving the desired bound.   □

The construction in Section 2 yields an m-instance parallelized algorithm with approximation factor $(\eta/\alpha)\sqrt[m]{\alpha}$. We wish to choose the parameters to minimize this factor. We have the constraints:

$$\eta \geq 2\alpha + \beta \tag{1}$$
$$\eta\alpha \geq \eta + 2\alpha^2 + 2\beta \tag{2}$$
$$\beta \geq 2\alpha \tag{3}$$
$$\eta \geq 4\alpha \tag{4}$$

Setting $\alpha = 4$, $\beta = 8$, and $\eta = 16$ satisfies the constraints and gives an approximation factor of $4^{1+(1/m)}$, so we can achieve a $(4 + \epsilon)$-approximation with $m = O(\epsilon^{-1})$. The memory usage and running time of the parallelized algorithm increase by a factor of m to $O(\epsilon^{-1}kz)$ and $O(\epsilon^{-1}(kzn + (kz)^2 \log P))$. Note that two things limit the approximation performance: that of the offline algorithm via (4), and the constraints (3) and (1) that limit what an optimal cluster can do. Thus, an improvement in the approximation factor of the offline algorithm will not carry through to the streaming algorithm unless it comes with a correspondingly better way to analyze optimal clusters.

3.1 Improvement Using a Center-Finding Oracle

There is a $(3+\epsilon)$-approximation version of the streaming algorithm, corresponding to the 3-approximation offline algorithm, when the metric space comes with a *center-finding* oracle. Given a positive integer j, a distance x, and a point set S, the oracle returns a point p having at least j points of S within distance x or announces that no such p exists in the metric space. Such an oracle may be impractical to implement in high-dimensional spaces, but when one is available, we can use it to improve the algorithm.

In step 2, instead of looking for potential centers among the free points, we invoke the oracle with $x = \beta r$, $j = z + 1$, and S being the current set of free points, and we add the resulting point (if any) as a center. Now, when the oracle fails, we know there is no cluster of radius βr centered *anywhere* that covers more than z free points, so we can relax constraint (3) to $\beta \geq \alpha$. In step 3, we substitute the 3-approximation offline algorithm, choosing centers using the oracle, and hence relax constraint (4) to $\eta \geq 3\alpha$. With the modified constraints, we choose $\alpha = \beta = 5$ and $\eta = 15$ to achieve a $(3 + \epsilon)$-approximation with the same $O(\epsilon^{-1}kz)$ memory usage; the running time depends on that of the oracle.

4   Clustering with Anonymity

For the problem of k-center clustering with a lower bound b on the number of points per cluster, we present a construction based on the parallelized Scaling Algorithm of Section 2 that achieves a $(6 + \epsilon)$-approximation. Applications of this problem for anonymity are considered by Aggarwal et al. [2].

Algorithm 4.1 (Clustering with anonymity). Let $\delta = \epsilon/2$. First run the m-instance parallelized Scaling Algorithm with m chosen to achieve a $(2 + \delta)$-approximation, but modify it to keep a count of how many input points "belong" to each center under an assignment of each point to a center within distance $(2 + \delta)r$ of it. (The algorithm does not store this assignment explicitly, but we use it in the proof of correctness.) When an existing center catches a new input point, the center's count is incremented, and when centers are merged, their counts are added. The Scaling Algorithm returns a lower bound r on the radius of the optimal k-center clustering of the input, a list of k *preliminary* centers $c_1, \ldots, c_k$, and the number n_i of input points belonging to each preliminary center c_i.

If $n_i \geq b$ for all i, the preliminary centers c_i constitute a solution within factor $2 + \delta$ of the optimal and we are done. Otherwise, we merge some centers using a scheme resembling the offline algorithm for k-center clustering with anonymity [2]. Given a merging radius R, the scheme works as follows. Initialize all preliminary centers to inactive; then, while there exists a preliminary center c that has no active center within distance $2R$, activate c. Next, attempt to allocate each input point p (belonging to a preliminary center c) to an active center within distance $2R + (2 + \delta)r$ of c in such a way that each active center gets at least b input points. To do this, construct a bipartite graph on the sets P of preliminary centers and A of currently active centers with an edge of infinite capacity connecting a node $x \in P$ to a node $y \in A$ if their distance is at most $2R + (2 + \delta)r$. Add a source s with an edge of capacity n_i to each $c_i \in P$ and a sink t with an edge of capacity b from each $c_i \in A$, and compute a max flow from s to t. If this flow saturates all edges entering t, it represents a valid allocation of the input points, which the merging scheme returns.

We attempt the merging scheme for various values of R in a binary search (which need only consider values of the form $d/2$ and $(d - (2 + \delta)r)/2$ for intercenter distances d) and keep the successful allocation with the smallest value of R. The algorithm returns a clustering consisting of the active centers under this allocation with radius $(4 + 2\delta)r + 2R$. □

Theorem 4.2. *The algorithm produces a clustering with at least b points per cluster whose radius is at most $6 + \epsilon = 6 + 2\delta$ times that of the optimal such clustering.*

Proof. Every input point p belongs to a preliminary center c within distance $(2+\delta)r$ of it and is allocated to an active center c' within distance $2R + (2+\delta)r$ of c, so p is within distance $(4 + 2\delta)r + 2R$ of c'. The algorithm's clustering consists of the active centers, so the clustering covers every input point at radius $(4 + 2\delta)r + 2R$ by virtue of the active center to which the point is allocated. Furthermore, each active center is allocated b points within distance $(4+2\delta)r+2R$ of it. Therefore, the algorithm's clustering is valid. We must show that it is a $(6 + 2\delta)$-approximation.

Let r^* be the radius of the optimal clustering, and consider an execution of the merging scheme with $R \geq r^*$. Active centers are separated by more than $2R \geq 2r^*$ by construction, so each lies in a different optimal cluster. We now claim that there exists an allocation of the form sought by the merging scheme, namely the allocation A that gives each input point to the unique active center (if any) lying in its optimal cluster. Let p be an input point; since optimal clusters have diameter $2r^*$, A gives p to an active center c within distance $2r^*$ of it. At the end of the Scaling Algorithm, p belonged to a center c' within distance $(2 + \delta)r$, so the distance between c and c' is at most $2r^* + (2+\delta)r \leq 2R + (2+\delta)r$. Thus, the merging scheme could legally allocate p (as counted by c') to c. This is true of every input point p, so the claim is established. Consequently, the merging scheme must succeed whenever $R \geq r^*$.

Thus, when the algorithm takes the smallest R for which the merging scheme succeeds, it will take an $R \leq r^*$. (The algorithm might consider not r^* itself but a slightly smaller value of R for which the merging scheme makes all the same decisions and therefore still must succeed.) The Scaling Algorithm ensures that r is a lower bound on r^*, i.e., $r \leq r^*$. Combining these two inequalities, the radius $(4 + 2\delta)r + 2R$ of the algorithm's clustering is at most $6 + 2\delta$ times the optimal radius r^*, as desired. □

Theorem 4.3. *The algorithm runs in $O(m(kn + k^2 \log P) + k^3 \log k)$ time using $O(mk+k^2)$ memory, where $m = O(\epsilon^{-1} \ln(\epsilon^{-1}))$ and P is the ratio of the optimal radius to the shortest distance between any two distinct input points.*

Proof. We use the simple $O(k)$-memory implementation of the Scaling Algorithm that stores only the centers; it performs each of the $O(\log P)$ scalings in $O(k^2)$ time and otherwise processes each point in $O(k)$ time for a total running time of $O(kn + k^2 \log P)$. Parallelization multiplies these bounds by m. The running time of the second phase is dominated by the max flow computation, which is done $O(\log k)$ times because there are $O(k^2)$ possible values for R. Using the relabel-to-front algorithm, each max flow invocation takes $O(k^3)$ time and $O(k^2)$ memory. The desired bounds follow. □

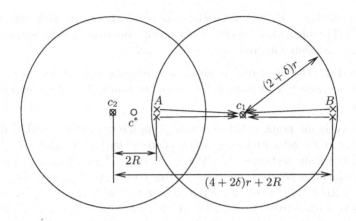

Fig. 2. Example of clustering with anonymity, with $b = 3$

The example in Figure 2 should help clarify the argument and motivate the final radius of $(4 + 2\delta)r + 2R$. We have two preliminary clusters c_1 and c_2 of radius $(2 + \delta)r$ with $n_1 = 5$ and $n_2 = 1$. The Scaling Algorithm decided to make the points A belong to c_1 even though they are actually much closer to c_2, perhaps because c_2 was not created until after they were read. Suppose we activate c_2 in the merging scheme. All we know about it is that it is in some optimal cluster of diameter $2r^*$ that contains b input points. For example, suppose $b = 3$ and there is an optimal cluster centered at c^* (not an input point) that contains c_2 and the points A. In order to guarantee that we can successfully allocate three points to c_2 whenever $R \geq r^*$, we must make all input points within distance $2R$ of c_2 (here the points A) available for allocation to it. But these points could belong to a different center (here c_1) that is another $(2+\delta)r$ away, so to be sure of catching them, we must allow c_2 to take points belonging to centers up to $2R + (2 + \delta)r$ away. However, the algorithm knows only that the five points belonging to c_1 are within $(2 + \delta)r$ of it; it knows nothing else about where they lie. In allowing c_2 to take points from c_1 to ensure that it has access to the points A, we are also opening the possibility of it taking the points B, which are $(4 + 2\delta)r + 2R$ away; there is no obvious way to avoid this. Thus, we set the radius of the final clustering to $(4 + 2\delta)r + 2R$ to make sure the clustering is valid. (For example, if $b = 6$, the algorithm might return a clustering consisting of a single cluster centered at c_2 containing all six points, which would need radius $(4+2\delta)r+2R$.)

The best-known offline algorithm [2], which essentially performs the allocation phase without the perturbation caused by the initial Scaling Algorithm phase, achieves an approximation factor of 2. Whether there is a streaming algorithm with a factor closer to 2 is an open problem.

4.1   Computing Per-cluster Statistics

Suppose that we wish to compute some statistics about the points allocated to each cluster as well as the center itself. In most cases, the algorithm can be extended to compute the statistics without a second pass over the input. For example, consider a medical application in which each input point represents a person who may or may not have a certain disease, and suppose we want to know what percentage of the people in each cluster have the disease. The first phase already maintains a count of people belonging to each center, and we can maintain a second count of people with the disease in the same way. When we allocate the people belonging to a preliminary center in the second phase, we simply allocate the people with the disease in the same proportion. For example, suppose 100 people belong to a preliminary center c_1 and 11 of them have the disease; if we allocate 30 of these 100 people to an active center c_2, we assume that 3.3 of them have the disease. In effect, we are allocating to c_2 30% of each individual who belongs to c_1. The fractionality of the allocation may appear silly but does not really harm the statistics.

In the same way, if we want the average height of the people in each cluster, we can maintain a "total height" value for each center, allocate height values in proportion to people, and then divide the total height allocated to a cluster by the number of people allocated to it. We can even compute several statistics on the same run. In full generality, if each input point comes with a vector of real-number weights, we can compute a total-weight vector for each cluster and divide by the number of points if we desire averages.

5   Conclusions

It is probably possible to combine our techniques for clustering with outliers and with anonymity to obtain an algorithm for the problem with both outliers and anonymity (albeit with a worse approximation factor), but we have not investigated this. One obvious open problem is to find an algorithm for the outlier problem with better running time and memory usage than our approach or the sampling approach of [6], particularly for the case where neither z nor n/z is small, or to prove a lower bound on the amount of memory needed.

If we are allowed multiple passes over the input, we can use a scaling-style algorithm to determine the optimal radius up to a constant factor on the first pass and then bound it more tightly on each subsequent pass by testing multiple guesses in parallel. By spreading the work across passes, we achieve the same approximation factor with a much smaller number of parallel instances. (The basic Hochbaum-Shmoys method [14] works naturally for guess-checking in the streaming model, but the offline algorithm for outliers [7] does not; one could instead use a cut-down guess-checking version of our outlier algorithm.) Developing a better algorithm that fully exploits multiple passes to achieve the same approximation factor using even less memory is another open problem.

Acknowledgment. The authors are grateful to Sudipto Guha for useful discussions.

References

1. Agarwal, P., Har-Peled, S., Yu, H.: Robust shape fitting via peeling and grating coresets. In: Proc. of ACM Symp. on Discrete Algorithms (SODA), pp. 182–191 (2006)
2. Aggarwal, G., Feder, T., Kenthapadi, K., Khuller, S., Panigrahy, R., Thomas, D., Zhu, A.: Achieving anonymity via clustering. In: Proc. of ACM Principles of Database Systems (PODS), pp. 153–162 (2006)
3. Aldenderfer, M.S., Blashfield, R.K.: Cluster Analysis. Sage, Beverly Hills (1984)
4. Bădoiu, M., Har-Peled, S., Indyk, P.: Approximate clustering via core-sets. In: Proc. of ACM Symp. on Theory of Computing (STOC), pp. 250–257 (2002)
5. Charikar, M., Chekuri, C., Feder, T., Motwani, R.: Incremental clustering and dynamic infomation retrieval. In: Proc. of ACM Symp. on Theory of Computing (STOC), pp. 626–635 (1997)
6. Charikar, M., O'Callaghan, L., Panigrahy, R.: Better streaming algorithms for clustering problems. In: Proc. of ACM Symp. on Theory of Computing (STOC), pp. 30–39 (2003)
7. Charikar, M., Khuller, S., Mount, D., Narasimhan, G.: Algorithms for facility location problems with outliers. In: Proc. of ACM-SIAM Symp. on Discrete Algorithms (SODA), pp. 642–651 (2001)
8. Chen, K.: A constant factor approximation algorithm for k-median clustering with outliers. In: Proc. of ACM-SIAM Symp. on Discrete Algorithms (SODA), pp. 826–835 (2008)
9. Everitt, B.: Cluster Analysis. Heinemann Educational, London (1974)
10. Gonzalez, T.: Clustering to minimize the maximum inter-cluster distance. Theoretical Computer Science 38, 293–306 (1985)
11. Guha, S.: The k-center karma of a data stream (unpublished manuscript) (2007)
12. Guha, S., Mishra, N., Motwani, R., O'Callaghan, L.: Clustering data streams. In: Proc. of IEEE Foundations of Computer Science (FOCS), pp. 359–366 (2000)
13. Hartigan, J.A.: Clustering Algorithms. Wiley, Chichester (1975)
14. Hochbaum, D., Shmoys, D.B.: A best possible approximation algorithm for the k-center problem. Math. of Operations Research 10, 180–184 (1985)
15. Jain, A.K., Dubes, R.C.: Algorithms for clustering data. Prentice Hall, NJ (1988)
16. Rasmussen, E.: Clustering algorithms. In: Frakes, W., Baeza-Yates, R. (eds.) Information Retrieval: Data Structures and Algorithms. Prentice Hall, Englewood Cliffs (1992)

A General Framework for Designing Approximation Schemes for Combinatorial Optimization Problems with Many Objectives Combined into One

Shashi Mittal and Andreas S. Schulz

Operations Research Center and Sloan School of Management
Massachusetts Institute of Technology
Cambridge, MA 02139
{mshashi,schulz}@mit.edu

Abstract. In this paper, we propose a general framework for designing fully polynomial time approximation schemes for combinatorial optimization problems, in which more than one objective function are combined into one using any norm. The main idea is to exploit the approximate Pareto-optimal frontier for multi-criteria optimization problems. Using this approach, we obtain an FPTAS for a novel resource allocation problem, for the problem of scheduling jobs on unrelated parallel machines, and for the Santa Claus problem, when the number of agents/machines is fixed, for any norm, including the l_∞-norm. Moreover, either FPTAS can be implemented in a manner so that the space requirements are polynomial in all input parameters. We also give approximation algorithms and hardness results for the resource allocation problem when the number of agents is not fixed.

1   Introduction

Consider the following resource allocation problem. There are m agents, and n resources, which are to be distributed among the agents. Each resource is assumed to be unsplittable; that is, a resource can be allocated to only one of the agents. However, agents may need to access resources assigned to other agents as well. The cost incurred by agent i, if it needs to access resource k from agent j, is c_{ij}^k. We assume that the c_{ij}^k are non-negative integers, and that $c_{ii}^k = 0$. The goal is to have a fair allocation of the resources among the agents; in other words, the maximum cost of an agent is to be minimized.

A practical setting where such a resource allocation problem can arise is page sharing in a distributed shared memory multiprocessor architecture [1]. In this architecture, the shared memory is distributed among different processors (also referred to as nodes), and each node contains a part of the shared memory locally. Typically, accessing the local memory is faster than accessing the remote memory. Every physical page in this architecture is allocated to a fixed node, which is referred to as the home node of the page. Also, there cannot be more

A. Goel et al. (Eds.): APPROX and RANDOM 2008, LNCS 5171, pp. 179–192, 2008.
© Springer-Verlag Berlin Heidelberg 2008

than one copy of a page in the system. Suppose each node knows in advance the number of accesses it will need to make to a page. The total delay, or latency, faced by a node is the sum of latencies over all the pages it needs to access. Suppose the latency between node i and j is t_{ij}, and the number of times node i needs to access page k is a_{ik}. If page k is stored in node j, then the cost of accessing page k for node i will be $c_{ij}^k = t_{ij}a_{ik}$. The performance of the system is governed by the node having maximum total latency. Thus, the objective is to allocate pages among the nodes in an offline fashion so that the maximum total latency over all the nodes is minimized.

We present an FPTAS for this resource allocation problem when the number of agents is fixed. There are many standard techniques for obtaining approximation schemes for combinatorial optimization problems. They include rounding of the input parameters (e.g. [2,3,4]), and shrinking the state space of dynamic programs [5]. We propose a novel framework for designing approximation schemes. The idea behind the new procedure is to treat the cost of each agent as a separate objective function, and to find an approximate Pareto-optimal frontier corresponding to this multi-objective optimization problem. Safer et al. [6][1] give necessary and sufficient conditions for the existence of fully polynomial time approximation schemes in multi-criteria combinatorial optimization. Papadimitriou and Yannakakis [9] propose an efficient procedure to construct an approximate Pareto-optimal frontier for discrete multi-objective optimization problems, and we use their procedure in constructing the approximation scheme for the resource allocation problem.

A closely related problem is the Santa Claus problem [10,11,12]. In this problem, each agent has a utility corresponding to each resource allocated to it, and the objective is to allocate the resources among the agents so that the minimum utility over all the agents is maximized. Our problem is different from the Santa Claus problem in that there is a cost associated with accessing each resource an agent does not get, instead of having a utility for each resource it gets. Using the above framework, we obtain the first FPTAS for the Santa Claus problem with a fixed number of agents.

Another closely related problem is scheduling jobs on unrelated parallel machines to minimize the makespan, also referred to in the literature as the $Rm||C_{\max}$ problem. There are m machines and n jobs, and each job is to be scheduled on one of the machines. The processing time of job k on machine i is p_{ik}. The objective is to minimize the makespan, that is the time at which the last job finishes its execution. Our procedure yields the first FPTAS for this problem that has space requirements that are polynomial in all the input parameters.

The resource allocation problem is NP-hard even when there are only two agents, and strongly NP-hard when the number of agents is variable (see the proof of NP-hardness in the Appendix). It remains strongly NP-hard for the special case of *uniform costs*, in which for each agent i and each resource k, $c_{ij}^k = c_i^k$ for all agents $j \neq i$. In this paper, we give a 2-approximation algorithm for

---

[1] This paper is a combined version of two earlier working papers by Safer and Orlin [7,8].

the uniform cost case. The algorithm makes use of the well-known technique of parametric linear programming and rounding, which has been successfully used in obtaining approximation algorithm for scheduling problems in the past [3]. Our rounding procedure, however, differs from the one given in [3]; it is more similar to the one used by Bezàkovà and Dani [12] for the Santa Claus problem.

Our results: The results in this paper can be summarized as follows.

1. *Approximation schemes.* We present a general framework for designing approximation schemes for problems with multiple objective functions combined into one using norms or other functions. We illustrate the versatility of this scheme by applying it to the resource allocation problem, the $Rm||C_{\max}$ problem, and the Santa Claus problem. An interesting byproduct is that, by a careful implementation of the FPTAS, the space requirements can be made polynomial in all the input parameters. Previously, all FPTASes for the $Rm||C_{\max}$ problem had space complexity exponential in the number of machines. This settles an open question raised by Lenstra et al. [3].

2. *A 2-approximation algorithm.* We propose a 2-approximation algorithm for the resource allocation problem with an arbitrary number of agents, for the special case of uniform costs, in which each agent incurs the same cost to access a resource from another agent, irrespective of the agent the resource is allocated to. This is achieved by solving a linear programming relaxation of the problem, and then rounding the fractional solution.

3. *Hardness of approximation.* We show that the general resource allocation problem cannot be approximated within a factor better than $3/2$ in polynomial time, unless P=NP. We achieve this by giving an approximation preserving reduction from the $R||C_{\max}$ problem to the resource allocation problem. In [3], it had been shown that the former problem cannot be approximated better than $3/2$ in polynomial time, unless P=NP, hence a similar result holds for the resource allocation problem, too. This reduction also establishes a direct connection between the resource allocation problem and the $R||C_{\max}$ scheduling problem.

Related work: Lenstra et al. [3] presented a 2-approximation algorithm for the $R||C_{\max}$ problem, based on a linear programming relaxation and rounding. For the case of a fixed number of machines, Horowitz and Sahni [2] gave the first FPTAS, which, however, has exponential space requirements. Lenstra et al. [3] derived a PTAS for this problem, which has better space complexity. In their paper, the authors mentioned that, "An interesting open question is whether this result can be strengthened to give a *fully* polynomial approximation scheme for fixed values of m, where the space required is bounded by a polynomial in the input size, m, and $1/\epsilon$ (or, even better, $\log(1/\epsilon)$)." We settle this open question in the affirmative in this paper. Azar et al. [13] gave an FPTAS for this problem for fixed m for any l_p-norm, but they do not analyze the space complexity of their approximation scheme.

The Santa Claus problem was first studied by Lipton et al. [11]. Bezàkovà and Dani [12] proposed a linear factor approximation algorithm for this problem, which is based on a linear programming relaxation and rounding; our rounding procedure is similar to the rounding procedure used in their paper. Bansal and

Sviridenko [10] obtained a tighter approximation algorithm for the restricted assignment version of the problem, where each resource can be allocated to only a subset of the agents, and each such agent has the same utility for that resource. As of now, no FPTAS has been proposed for the Santa Claus problem with a fixed number of agents.

The focus of this paper will be mainly on the resource allocation problem, since this problem was our original motivation for taking up this study. We will refer to the $Rm||C_{\max}$ problem and the Santa Claus problem whenever our techniques for the resource allocation problem also apply to these two problems. We begin by giving an integer programming formulation of the resource allocation problem. Let x_{ik} be a variable which is 1 if the kth resource is given to agent i, otherwise it is 0. Then the total cost incurred by agent i is $\sum_{k=1}^{n} \sum_{j=1}^{m} c_{ij}^{k} x_{jk}$. An integer programming formulation of the resource allocation problem is given by

$$\min \quad S$$

$$\text{s.t.} \quad \sum_{k=1}^{n} \sum_{j=1}^{m} c_{ij}^{k} x_{jk} \leq S \qquad \text{for } i = 1, \ldots, m,$$

$$\sum_{i=1}^{m} x_{ik} = 1 \qquad \text{for } k = 1, \ldots n,$$

$$x_{ik} \in \{0, 1\} \qquad \text{for } i = 1, \ldots, m, \quad k = 1, \ldots, n.$$

2   An FPTAS for a Fixed Number of Agents

In this section, we give an FPTAS for the resource allocation problem with a fixed number of agents. We first discuss a polynomial-time procedure to compute an approximate Pareto-optimal frontier for general multi-objective optimization problems. We then show that using the approximate Pareto-optimal frontier, we can get an approximate solution for the resource allocation problem. Subsequently, we use this technique for obtaining an FPTAS for the $Rm||C_{\max}$ problem and the Santa Claus problem as well, and then extend it to the case of general l_p-norms, other norms, and beyond.

2.1   Formulation of the FPTAS

An instance π of a multi-objective optimization problem Π is given by a set of m functions $f_1, \ldots, f_m$. Each $f_i : X \to \mathbb{R}_+$ is defined over the same set of feasible solutions, X. Let $|\pi|$ denote the binary-encoding size of the instance π. Assume that each f_i takes values in the range $[2^{-p(|\pi|)}, 2^{p(|\pi|)}]$ for some polynomial p. We first define the Pareto-optimal frontier for multi-objective optimization problems.

Definition 1. *Let π be an instance of a multi-objective optimization problem. A* Pareto-optimal frontier *(with respect to minimization), denoted by $P(\pi)$, is a set of solutions $x \in X$, such that there is no $x' \in X$ such that $f_i(x') \leq f_i(x)$ for all i with strict inequality for at least one i.*

In other words, $P(\pi)$ consists of all undominated solutions. In many cases, it may not be tractable to compute $P(\pi)$ (e.g., determining whether a point belongs to the Pareto-optimal frontier for the two-objective shortest path problem is NP-hard), or the number of undominated solutions can be exponential in $|\pi|$ (e.g., for the two-objective shortest path problem [14]). One way of getting around this problem is to look at an approximate Pareto-optimal frontier, which is defined below.

Definition 2. *Let π be an instance of a multi-objective optimization problem. For $\epsilon > 0$, an ϵ-approximate Pareto-optimal frontier, denoted by $P_\epsilon(\pi)$, is a set of solutions, such that for all $x \in X$, there is $x' \in P_\epsilon(\pi)$ such that $f_i(x') \leq (1 + \epsilon) f_i(x)$, for all i.*

In the rest of the paper, whenever we refer to an (approximate) Pareto-optimal frontier, we mutually refer to both its set of solutions and their vectors of objective function values.

Papadimitriou and Yannakakis [9] showed that whenever m is fixed, there is always an approximate Pareto-optimal frontier that has polynomially many elements.

Theorem 1 (Papadimitriou and Yannakakis [9]). *Let π be an instance of a multi-objective optimization problem. For any $\epsilon > 0$ and for fixed m, there is an ϵ-approximate Pareto-optimal frontier $P_\epsilon(\pi)$ whose cardinality is bounded by a polynomial in $|\pi|$ and $1/\epsilon$.*

Let us consider the following optimization problem:

$$\text{minimize } g(x) = \max_{i=1,\ldots,m} f_i(x), \qquad x \in X. \tag{1}$$

We show that if an approximate Pareto curve can be constructed in polynomial time, then there is an FPTAS to solve this min-max problem.

Lemma 1. *There is at least one optimal solution x^* to (1) such that $x^* \in P(\pi)$.*

Proof. Let $\hat{x}$ be an optimal solution of (1). Suppose $f_k(\hat{x})$ is the maximum among all function values for $\hat{x}$; that is, $f_k(\hat{x}) \geq f_i(\hat{x})$ for all $i = 1, \ldots, m$. Suppose $\hat{x} \notin P(\pi)$. Then there exists $x' \in P(\pi)$ such that $f_i(x') \leq f_i(\hat{x})$ for $i = 1, \ldots, m$. Therefore, $f_i(x') \leq f_k(\hat{x})$ for all i, that is $\max_{i=1,\ldots,m} f_i(x') \leq f_k(\hat{x})$, or $g(x') \leq g(\hat{x})$. Thus x' minimizes the function g and is in $P(\pi)$. $\square$

Lemma 2. *Let $\hat{x}$ be a solution in $P_\epsilon(\pi)$ that minimizes $g(x)$ over all points $x \in P_\epsilon(\pi)$. Then $\hat{x}$ is a $(1 + \epsilon)$-approximate solution of (1); that is, $g(\hat{x})$ is at most $(1 + \epsilon)$ times the value of an optimal solution to (1).*

Proof. Let x^* be an optimal solution of (1) that is in $P(\pi)$. By the definition of ϵ-approximate Pareto-optimal frontier, there exists $x' \in P_\epsilon(\pi)$ such that $f_i(x') \leq (1 + \epsilon) f_i(x^*)$, for all $i = 1, \ldots, m$. Therefore $g(x') \leq (1 + \epsilon) g(x^*)$. Since $\hat{x}$ is a minimizer of $g(x)$ over all solutions in $P_\epsilon(\pi)$, $g(\hat{x}) \leq g(x') \leq (1 + \epsilon) g(x^*)$. $\square$

From these two lemmas, we get the following theorem regarding the existence of an FPTAS for solving (1).

Theorem 2. *Suppose there is an algorithm that computes $P_\epsilon(\pi)$ in time polynomial in $|\pi|$ and $1/\epsilon$ for a fixed value of m. Then there is an FPTAS for solving the min-max optimization problem (1).*

Thus, the only thing we are left with is to find a polynomial-time algorithm for computing an approximate Pareto-optimal frontier. Papadimitriou and Yannakakis [9] give a necessary and sufficient condition under which such a polynomial-time algorithm exists.

Theorem 3 (Papadimitriou and Yannakakis [9]). *Let m be fixed, and let $\epsilon, \epsilon' > 0$ be such that $(1 - \epsilon')(1 + \epsilon) = 1$. One can determine a $P_\epsilon(\pi)$ in time polynomial in $|\pi|$ and $1/\epsilon$ if and only if the following 'gap problem' can be solved in polynomial-time: Given an m-vector of values $(v_1, \dots, v_m)$, either*
(i) return a solution $x \in X$ such that $f_i(x) \le v_i$ for all $i = 1, \dots, m$, or
(ii) assert that there is no $x \in X$ such that $f_i(x) \le (1 - \epsilon')v_i$ for all $i = 1, \dots, m$.

We sketch the proof because our approximation schemes are based on it.

Proof. Suppose we can solve the gap problem in polynomial time. An approximate Pareto-optimal frontier can then be constructed as follows. Consider the box in $\mathbb{R}^m$ of possible function values given by $\{(v_1, \dots, v_m) : 2^{-p(|\pi|)} \le v_i \le 2^{p(|\pi|)}$ for all $i\}$. We divide this box into smaller boxes, such that in each dimension, the ratio of successive divisions is equal to $1 + \epsilon''$, where $\epsilon'' = \sqrt{1 + \epsilon} - 1$. For each corner point of all such smaller boxes, we call the gap problem. Among all solutions returned by solving the gap problems, we keep only those solutions that are not Pareto-dominated by any other solution. This is the required $P_\epsilon(\pi)$. Since there are $O((p(|\pi|)/\epsilon)^m)$ many smaller boxes, this can be done in polynomial time.

Conversely, suppose we can construct $P_\epsilon(\pi)$ in polynomial time. To solve the gap problem for a given m-vector $(v_1, \dots, v_m)$, if there is a solution point $(f_1(x), \dots, f_m(x))$ in $P_\epsilon(\pi)$ such that $f_i(x) \le v_i$ for all i, then we return x. Otherwise we assert that there is no $x \in X$ such that $f_i(x) \le (1 - \epsilon')v_i$ for all $i = 1, \dots, m$. □

Thus, we only need to solve the gap problem to get a $(1+\epsilon)$-approximate solution for the min-max problem. This is accomplished in a manner similar to that given in [9]. Our description here is with respect to minimization problems; a similar description for maximization problems can be found in [9].

We restrict our attention to the case when $X \subseteq \{0, 1\}^d$, since many combinatorial optimization problems can be framed as 0/1-integer programming problems. Further, we consider linear objective functions; that is, $f_i(x) = \sum_{j=1}^d a_{ij} x_j$, and each a_{ij} is a non-negative integer. Suppose we want to solve the gap problem for the m-vector $(v_1, \dots, v_m)$. Let $r = \lceil d/\epsilon \rceil$. We first define a "truncated" objective function. For all $j = 1, \dots, d$, if for some i, $a_{ij} > v_i$, we set $x_j = 0$, and drop the variable x_j from each of the objective functions. Let V be the index set of the remaining variables. Thus, the coefficients in each objective function are now less than or equal to v_i. Next, we define a new objective function $f_i'(x) = \sum_{j \in V} a_{ij}' x_j$, where $a_{ij}' = \lceil a_{ij} r / v_i \rceil$. In the new objective function, the maximum value of a coefficient is now r. For $x \in X$, the following two statements hold.

- If $f_i'(x) \leq r$, then $f_i(x) \leq v_i$.
- If $f_i(x) \leq v_i(1 - \epsilon')$, then $f_i'(x) \leq r$.

Therefore, to solve the gap problem, it suffices to find an $x \in X$ such that $f_i'(x) \leq r$, for $i = 1, \ldots, m$, or assert that no such x exists. Since all the coefficients of $f_i'(x)$ are non-negative integers, there are $r + 1$ ways in which $f_i'(x) \leq r$ can be satisfied. Hence there are $(r+1)^m$ ways overall in which all inequalities $f_i'(x) \leq r$ can be simultaneously satisfied. Suppose we want to find if there is an $x \in X$ such that $f_i'(x) = b_i$ for $i = 1, \ldots, m$. This is equivalent to finding an x such that $\sum_{i=1}^m M^{i-1} f_i'(x) = \sum_{i=1}^m M^{i-1} b_i$, where $M = dr + 1$ is a number greater than the maximum value that $f_i'(x)$ can take.

Given an instance π of a multi-objective linear optimization problem over a discrete set X, the exact version of the problem is: Given a non-negative integer C and a vector $(c_1, \ldots, c_d) \in \mathbb{Z}_+^d$, does there exist a solution $x \in X$ such that $\sum_{j=1}^d c_j x_j = C$?

Theorem 4. *Suppose we can solve the exact version of the problem in pseudo-polynomial time, then there is an FPTAS for solving* (1).

Proof. The gap problem can be solved by making at most $(r + 1)^m$ calls to the pseudo-polynomial time algorithm, and the input to each call has numerical values of order $O((d^2/\epsilon)^{m+1})$. Therefore, all calls to the algorithm take polynomial time, hence the gap problem can be solved in polynomial time. The theorem now follows from Theorems 2 and 3.   □

Now we give a pseudo-polynomial time algorithm for solving the exact version of the resource allocation problem for a fixed number of agents. The exact version for resource allocation is this: Given an integer C, does there exist a 0/1-vector x such that $\sum_{k=1}^n \sum_{j=1}^m c_{jk} x_{jk} = C$, subject to the constraints that $\sum_{j=1}^m x_{jk} = 1$ for $k = 1, \ldots, n$, and $x_{jk} \in \{0, 1\}$? The exact problem can be viewed as a reachability problem in a directed graph. The graph is an $(n+1)$-partite directed graph; let us denote the partitions of this digraph by $V_0, \ldots, V_n$. The partition V_0 has only one node, labeled as $v_{0,0}$ (the source node), all other partitions have $C + 1$ nodes. The nodes in V_i for $1 \leq i \leq n$ are labeled as $v_{i,0}, \ldots, v_{i,C}$. The arcs in the digraph are from nodes in V_i to nodes in V_{i+1} only, for $0 \leq i \leq n - 1$. For all $c \in \{c_{1,i+1}, \ldots, c_{m,i+1}\}$, there is an arc from $v_{i,j}$ to $v_{i+1,j+c}$, if $j + c \leq C$. Then there is a solution to the exact version if and only if there is a directed path from the source node $v_{0,0}$ to the node $v_{n,C}$. Finding such a path can be accomplished by doing a depth-first search from the node $v_{0,0}$. The corresponding solution for the exact problem (if it exists) can be obtained using the path found by the depth-first search algorithm.

Thus, the above pseudo-polynomial algorithm implies the following theorem.

Theorem 5. *There is an FPTAS for the resource allocation problem with a fixed number of agents.*

2.2   Space Complexity of the FPTAS

A straightforward implementation of the above algorithm will have substantial storage requirements. The bottleneck for space requirements appears at two places: one is storing the approximate Pareto-optimal frontier, and the other is in solving the exact problem. However, by a careful implementation of the algorithm, the storage requirements can be reduced significantly. We give an outline below for a space-efficient implementation of the above algorithm.

1. We do not need to store all the corner points of the smaller boxes into which the region of possible objective function values has been divided. By simply iterating over the corner points using loops, we can cover all the corner points.

2. We also do not need to store the approximate Pareto-optimal frontier, as it is sufficient to store the current best solution obtained after solving each gap problem.

3. When solving the exact problem using the depth-first search algorithm, we do not need to generate the whole graph explicitly. The only data we need to store in the execution of the depth-first search algorithm are the stack corresponding to the path traversed in the graph so far (the path length is at most n), and the coefficients of the modified objective function. There are mn coefficients that need to be stored, and the maximum magnitude of each coefficient is $O((m^2 n^2 / \epsilon)^{m+1})$, thus the space complexity of the FPTAS is $O(m^2 n \log (mn/\epsilon))$.

Thus, we have the following theorem.

Theorem 6. *There is an FPTAS for the resource allocation problem whose space requirements are polynomial in m, n and $\log (1/\epsilon)$.*

2.3   An FPTAS for Scheduling on Unrelated Parallel Machines and the Santa Claus Problem

Recall the $Rm||C_{\max}$ scheduling problem defined in the introduction. There are m machines and n jobs, and the processing time of job k on machine i is p_{ik}. The objective is to schedule the jobs to minimize the makespan. The m objective functions in this case are given by $f_i(x) = \sum_{k=1}^{n} p_{ik} x_{ik}$, and the set X is given by $\sum_{i=1}^{m} x_{ik} = 1$ for each $k = 1, \ldots, n$, and $x_{ik} \in \{0, 1\}$. The Santa Claus problem is similar to this scheduling problem, except that the objective here is to maximize the minimum execution time over all the machines.

The exact version of the $Rm||C_{\max}$ problem and the Santa Claus problem is the same as that for the resource allocation problem, and hence we get an FPTAS for either problem for fixed m. For the $Rm||C_{\max}$ problem, we obtain the first FPTAS that has space requirements which are polynomial in m, n

and $\log(1/\epsilon)$, whereas all the previously obtained FPTASes for this problem had space complexity exponential in m. For the Santa Claus problem, we give the first FPTAS for a fixed number of agents. We therefore have the following theorem.

Theorem 7. *There are FPTASes for the $Rm||C_{\max}$ problem and the Santa Claus problem with a fixed number of agents whose space requirements are polynomial in n, m, and $\log(1/\epsilon)$.*

2.4   FPTAS for Any Norm

The above technique for obtaining an FPTAS in fact can be extended to include any norm used for combining the different objective functions. More generally, let $h : \mathbb{R}_+^m \to \mathbb{R}_+$ be any function that satisfies

(i) $h(y) \leq h(y')$ for all $y, y' \in \mathbb{R}_+^m$ such that $y_i \leq y_i'$ for all $i = 1, \ldots, m$, and
(ii) $h(\lambda y) \leq \lambda h(y)$ for all $y \in \mathbb{R}_+^m$ and $\lambda > 1$.

Consider the following generalization of the optimization problem given by (1):

$$\text{minimize } g(x) = h(f(x)), \qquad x \in X. \tag{2}$$

Then Lemma 1 and 2 can be easily generalized as follows.

Lemma 3. *There is at least one optimal solution x^* to (2) such that $x^* \in P(\pi)$.*

Lemma 4. *Let $\hat{x}$ be a solution in $P_\epsilon(\pi)$ that minimizes $g(x)$. Then $\hat{x}$ is a $(1+\epsilon)$-approximate solution of (2); that is, $g(\hat{x})$ is at most $(1 + \epsilon)$ times the optimal value of (2).*

These two lemmata then imply that the technique given in this section can be used to obtain an FPTAS for (2). The only difference is in selecting the solution from the approximate Pareto-optimal frontier: we have to choose the solution which is the best according to the given h. Thus we have the following theorem.

Theorem 8. *There is an FPTAS for the resource allocation problem, the problem of scheduling jobs on unrelated parallel machines, and the Santa Claus problem with fixed m when the objectives for the different agents/machines are combined into one using a function h that satisfies (i) and (ii). Moreover, this algorithm can be made to run with space requirements that are polynomial in m, n, and $\log(1/\epsilon)$.*

3   A 2-Approximation Algorithm for the Uniform Cost Case

Recall that in the case of the resource allocation problem with uniform costs, for each agent i and each resource k, $c_{ij}^k = c_i^k$ for all $j \neq i$, and $c_{ii}^k = 0$. Let $A_k(s)$

denote the set of all agents such that if resource k is allocated to an agent in this set, the cost that any other agent will have to pay to access resource k is no more than s.

We will consider a parametric linear programming relaxation of the problem, in which we have the constraint that no agent has a cost of more than s in the relaxed solution. For each resource k, we consider only the agents in the set $A_k(s)$ as possible candidates for allocating that resource. We show that if this parametric linear program has a feasible solution, then an extreme point of the feasible set of the linear program can be rounded to an integer solution in which each agent has cost no more than $2s$.

Theorem 9. *For $s \in \mathbb{Z}_+$, consider the following set of linear inequalities, which we denote by $LP(s)$:*

$$\sum_{k=1}^{n} \sum_{j \in A_k(s)} c_{ij}^k x_{jk} \leq s \qquad for \ i = 1, \ldots, m, \tag{3a}$$

$$\sum_{i \in A_k(s)} x_{ik} = 1 \qquad for \ k = 1, \ldots, n, \tag{3b}$$

$$x_{ik} \geq 0 \qquad for \ k = 1, \ldots, n, \quad i \in A_k(s). \tag{3c}$$

Suppose $LP(s)$ has a feasible solution, then, for the case of uniform costs, one can find $x_{ik}^R \in \{0, 1\}$ in polynomial time such that

$$\sum_{k=1}^{n} \sum_{j \in A_k(s)} c_{ij}^k x_{jk}^R \leq 2s \qquad for \ i = 1, \ldots, m, \tag{4a}$$

$$\sum_{i \in A_k(s)} x_{ik}^R = 1 \qquad for \ k = 1, \ldots, n. \tag{4b}$$

Proof. Let x^{LP} be an extreme point of the non-empty polytope defined by the inequalities of $LP(s)$. Let v be the total number of variables defining the system $LP(s)$. There are $v + m + n$ inequalities in $LP(s)$. Since $LP(s)$ has v variables, at any extreme point of this polytope, at least v linearly independent inequalities will be satisfied with equality. Hence, at most $m + n$ inequalities will not be satisfied with equality. Therefore, it follows from (3c) that at most $m + n$ variables will have a non-zero value.

Consider the bipartite graph G in which one of the partitions has nodes corresponding to each agent, and the other partition has nodes corresponding to each resource. There is an edge between agent i and resource k in G if $x_{ik}^{LP} > 0$. In this graph, the number of edges is less than or equal to the number of nodes. For the $R||C_{\max}$ problem, which has a similar integer programming formulation, Lenstra et al. [3] showed that each connected component of G also has the property that the number of edges is less than or equal to the number of nodes. This result holds here as well.

We now construct an integral solution x^R by rounding the fractional solution. Let G' be a connected component of G. The rounding is performed in two stages. In the first stage, the following two operations are performed on G' repeatedly, in the given order.

1. For all resource nodes k such that in G', exactly one edge, say (i, k), is incident to it, we set $x_{ik}^R = 1$, and remove all such resource nodes and the edges incident to these nodes from G'.
2. For all agent nodes i such that there is exactly one edge, say (i, k), incident to it, we set $x_{ik}^R = 0$, and remove all such agent nodes and all the edges incident to these nodes from G'.

The first stage of rounding ends when the above two operations can no longer be performed. Let the resulting subgraph after the first stage of rounding be G''. Note that in the first stage, whenever we are deleting a node, we are also deleting at least one edge from the graph. Hence after the first stage, the number of edges is still less than or equal to the number of nodes in G''. For the second stage, there are three possibilities.

1. There are no nodes corresponding to resources in G''. This means that all resources in this subgraph have already been allocated to some agent. In this case we are done for G''.
2. There are some nodes corresponding to resources in G'', but there are no edges incident to these resource nodes. That is, some of the resources in G' have not yet been assigned to any of the agents. In this case, each such resource is assigned to one of the agents to which it was incident before the starting of the rounding procedure.
3. If both the above cases do not hold, then each node in G'' has at least two edges incident to it. Since the number of edges is less than or equal to the number of nodes, this component is actually a cycle, and the number of agent nodes is the same as the number of resource nodes. In this component, there is now a perfect matching between the agent nodes and the resource nodes. We find any perfect matching in this component, and for each matching edge (i, k) we set $x_{ik}^R = 1$. All the remaining variables corresponding to G' whose values have not been determined yet, are assigned the value zero.

This rounding procedure is performed on each connected component of G to get a 0/1-solution x^R. Note that x_{ik}^R satisfies the constraint (4b), since each resource is allocated to exactly one of the agents. Also, for each agent i, there is at most one resource, say $r(i)$, for which the LP solution was fractional, and in the integral solution that resource was not allocated to i, but was instead allocated to agent $i' \in A_{r(i)}(s)$. This is because in the first stage of rounding, an agent node is deleted only when there is just one resource node in the graph to which it remains incident to, and hence it does not get that resource. And in the second stage, in the third case, there will be exactly one resource to which an agent is incident to, but that resource is not allocated to the agent.

For an agent i, define a partition of the resources into $R^i_{=0}$ and $R^i_{>0}$ as follows: $R^i_{=0} = \{k : x^{LP}_{ik} = 0\}$, and $R^i_{>0} = \{k : x^{LP}_{ik} > 0\}$. For all $i \in \{1, \ldots, m\}$,

$$
\sum_{k=1}^{n} \sum_{j \in A_k(s)} c^k_{ij} x^R_{jk} = \sum_{k \in R^i_{=0}} \sum_{j \in A_k(s)} c^k_i x^R_{jk} + \sum_{k \in R^i_{>0}} \sum_{j \in A_k(s)} c^k_{ij} x^R_{jk}
$$

$$
= \sum_{k \in R^i_{=0}} \sum_{j \in A_k(s)} c^k_i x^{LP}_{jk} + \sum_{k \in R^i_{>0}} \sum_{j \in A_k(s)} c^k_{ij} x^R_{jk} \tag{5a}
$$

$$
\leq \sum_{k \in R^i_{=0}} \sum_{j \in A_k(s)} c^k_i x^{LP}_{jk} + \sum_{k \in R^i_{>0}} \sum_{j \in A_k(s)} c^k_{ij} x^{LP}_{jk} + c^{r(i)}_{ii'} \tag{5b}
$$

$$
= \sum_{k=1}^{n} \sum_{j \in A_k(s)} c^k_{ij} x^{LP}_{jk} + c^{r(i)}_{ii'}
$$

$$
\leq s + s = 2s. \tag{5c}
$$

The equality in (5a) follows from the fact that for each resource k, $\sum_{j \in A_k(s)} x^{LP}_{jk} = \sum_{j \in A_k(s)} x^R_{jk} = 1$, and also because we are dealing with the case of uniform costs. The inequality in (5b) holds because for each agent i, there is at most one resource $r(i)$ such that $x^{LP}_{i,r(i)} > 0$, but $x^R_{i,r(i)} = 0$. And finally, the inequality in (5c) is true by the definition of the set $A_{r(i)}(s)$, $c^{r(i)}_i \leq s$, and (3a). □

To get a 2-approximation algorithm for the problem that runs in polynomial time, one starts by choosing a trivial lower and upper bound on the optimum value of the objective function. The lower bound can be $\min\{c^k_{ij}\}$, and the upper bound can be $mn \max\{c^k_{ij}\}$. Then, by adopting a binary search procedure, one can find the minimum integer value of s, say s^*, for which $LP(s)$ is feasible, and get a corresponding vertex x^{LP} of the non-empty polytope in polynomial time by using the ellipsoid algorithm [15]. Clearly, s^* is a lower bound on the optimal objective function value of the resource allocation problem. Using the above rounding procedure, one can obtain a rounded solution whose value is at most $2s^*$. We therefore obtain a 2-approximation algorithm for the resource allocation problem with uniform costs.

4   Hardness of Approximation

In this section, we give a hardness of approximation result for the resource allocation problem with general costs.

Theorem 10. *There is no polynomial-time algorithm that yields an approximation ratio smaller than 3/2 for the resource allocation problem, unless P=NP.*

Proof. We prove this by a reduction from the problem of scheduling jobs on unrelated parallel machines $(R||C_{\max})$, which cannot have a better than 3/2-approximation algorithm, unless P=NP [3].

Consider an instance of the $R||C_{max}$ problem with m machines and n jobs, where the processing time of job k on machine i is p_{ik}. Let $p_{max} = \max\{p_{ik}\}$. We construct a corresponding instance of the resource allocation problem as follows. There are $2m$ agents and n resources. For $i, j \in \{1, \ldots, m\}, i \neq j$, let $c_{ij}^k = np_{max} + 1$, and $c_{i,m+i}^k = p_{ik}$. All other cost coefficients are zero. Then, in any optimal allocation of resources in the resource allocation problem, all the resources will be distributed among the agents $m + 1, \ldots, 2m$. It is easy to see that if there is an optimal solution of the $R||C_{max}$ instance in which job k is allocated to machine $m(k)$, there is a corresponding optimal solution for the resource allocation problem in which resource k is allocated to agent $m + m(k)$, and vice-versa. Also, the optimal objective function value of both instances will be the same.

Thus, if the resource allocation problem could be approximated better than $3/2$ in polynomial time, then so can the $R||C_{max}$ problem, which is impossible, unless P=NP [3].   □

Acknowledgments

The first author thanks Mainak Chaudhuri and Maunendra Sankar Desarkar for fruitful discussions on framing the resource allocation problem. The authors also thank Retsef Levi for pointing out the work by Bansal and Sviridenko [10]. This research was supported by NSF awards #0426686 and #0700044, and by ONR grant N00014-08-1-0029.

References

1. Laudon, J., Lenoski, D.: The SGI origin: A ccNUMA highly scalable server. In: Proceedings of the 24th International Symposium on Computer Architecture, Boulder, CO, pp. 241–251 (1997)
2. Horowitz, E., Sahni, S.: Exact and approximate algorithms for scheduling nonidentical processors. Journal of the ACM 23, 317–327 (1976)
3. Lenstra, J.K., Shmoys, D.B., Tardos, É.: Approximation algorithms for scheduling unrelated parallel machines. Mathematical Programming 46, 259–271 (1990)
4. Hochbaum, D.S., Shmoys, D.B.: Using dual approximation algorithms for scheduling problems: Theoretical and practical results. Journal of the ACM 34, 144–162 (1987)
5. Woeginger, G.J.: When does a dynamic programming formulation guarantee the existence of a fully polynomial time approximation scheme (FPTAS)? INFORMS Journal on Computing 12, 57–74 (2000)
6. Safer, H.M., Orlin, J.B., Dror, M.: Fully polynomial approximation in multi-criteria combinatorial optimization. Technical report, Operations Research Center, MIT (2004)
7. Safer, H.M., Orlin, J.B.: Fast approximation schemes for multi-criteria combinatorial optimzation. Technical report, Operations Research Center, MIT (1995)
8. Safer, H.M., Orlin, J.B.: Fast approximation schemes for multi-criteria flow, knapsack, and scheduling problems. Technical report, Operations Research Center, MIT (1995)

9. Papadimitriou, C.H., Yannakakis, M.: On the approximability of trade-offs and optimal access of web sources. In: Proceedings of the 41st Annual IEEE Symposium on Foundations of Computer Science, Redondo Beach, CA, pp. 86–92 (2000)
10. Bansal, N., Sviridenko, M.: The Santa Claus problem. In: Proceedings of the 38th Annual ACM Symposium on Theory of Computing, Seattle, WA, pp. 31–40 (2006)
11. Lipton, R.J., Markakis, E., Mossel, E., Saberi, A.: On approximately fair allocations of indivisible goods. In: Proceedings of the 5th ACM Conference on Electronic Commerce, New York, NY, pp. 125–131 (2004)
12. Bezàkovà, I., Dani, V.: Allocating indivisible goods. ACM SIGecom Exchanges 5, 11–18 (2005)
13. Azar, Y., Epstein, L., Richter, Y., Woeginger, G.J.: All-norm approximation algorithms. Journal of Algorithms 52, 120–133 (2004)
14. Hansen, P.: Bicriterion path problems. In: Fandel, G., Gál, T. (eds.) Multiple Criteria Decision Making: Theory and Application. Lecture Notes in Economics and Mathematical Systems, vol. 177, pp. 109–127. Springer, Heidelberg (1980)
15. Grötschel, M., Lovász, L., Schrijver, A.: Geometric Algorithms and Combinatorial Optimization. Springer, Berlin (1988)
16. Garey, M.R., Johnson, D.S.: Computers and Intractability: A Guide to the Theory of NP-completeness. W. H. Freeman, New York (1979)

Appendix

Lemma 5. *The resource allocation problem with uniform costs is NP-hard for two agents, and strongly NP-hard in general.*

Proof. The proof of NP-hardness for the two-agents case is by reduction from PARTITION [16]. Consider an instance of PARTITION given by a set A of n elements, where element $a \in A$ has size $s(a) \in \mathbb{Z}_+$. We construct an instance of the resource allocation problem with two agents and n resources as follows: $c_{12}^a = c_{21}^a = s(a)$ for each $a \in A$, and $c_{ii}^a = 0$ for $i = 1, 2$. Then, A can be partitioned into two sets of equal size if and only if the optimal solution for the given resource allocation problem has cost $\sum_{a \in A} s(a)/2$.

The strong NP-hardness proof for the general case is by a reduction from 3-PARTITION [16]. Let an instance of this problem be given by the set $A = \{a_1, \ldots, a_{3m}\}$, with $\sum_{a \in A} s(a) = mB$. The corresponding instance of the resource allocation problem is constructed as follows: There are m agents, and $3m$ resources. For each agent i, $c_{ij}^k = s(a_k)$ for $k = 1, \ldots, 3m; j = 1, \ldots, m, i \neq j$, and $c_{ii}^k = 0$. Then the answer to the 3-PARTITION instance is "Yes" if and only if the optimal solution to the given resource allocation problem has cost $(m-1)B$.   □

The Directed Minimum Latency Problem[*]

Viswanath Nagarajan and R. Ravi

Tepper School of Business, Carnegie Mellon University, Pittsburgh USA
{viswa,ravi}@cmu.edu

Abstract. We study the *directed minimum latency problem*: given an n-vertex asymmetric metric (V, d) with a root vertex $r \in V$, find a spanning path originating at r that minimizes the sum of latencies at all vertices (the latency of any vertex $v \in V$ is the distance from r to v along the path). This problem has been well-studied on symmetric metrics, and the best known approximation guarantee is 3.59 [3]. For any $\frac{1}{\log n} < \epsilon < 1$, we give an $n^{O(1/\epsilon)}$ time algorithm for directed latency that achieves an approximation ratio of $O(\rho \cdot \frac{n^\epsilon}{\epsilon^3})$, where ρ is the integrality gap of an LP relaxation for the *asymmetric traveling salesman path* problem [13,5]. We prove an upper bound $\rho = O(\sqrt{n})$, which implies (for any fixed $\epsilon > 0$) a polynomial time $O(n^{1/2+\epsilon})$-approximation algorithm for directed latency.

In the special case of metrics induced by shortest-paths in an unweighted directed graph, we give an $O(\log^2 n)$ approximation algorithm. As a consequence, we also obtain an $O(\log^2 n)$ approximation algorithm for minimizing the weighted completion time in *no-wait permutation flowshop scheduling*. We note that even in unweighted directed graphs, the directed latency problem is at least as hard to approximate as the well-studied *asymmetric traveling salesman problem*, for which the best known approximation guarantee is $O(\log n)$.

1 Introduction

The minimum latency problem [17,6,14,2] is a variant of the basic traveling salesman problem, where there is a metric with a specified root vertex r, and the goal is to find a spanning path starting from r that minimizes the sum of arrival times at all vertices (it is also known as the *deliveryman problem* or *traveling repairman problem*). This problem can model the traveling salesman problem, and hence is NP-complete. To the best of our knowledge, all previous work has focused on symmetric metrics– the first constant-factor approximation algorithm was in Blum et al. [2], and the currently best known approximation ratio is 3.59 due to Chaudhuri et al. [3]. In this paper, we consider the minimum latency problem on *asymmetric metrics*.

Network design problems on directed graphs are often much harder to approximate than their undirected counterparts– the traveling salesman and Steiner tree

---

[*] Supported by NSF grant CCF-0728841.

A. Goel et al. (Eds.): APPROX and RANDOM 2008, LNCS 5171, pp. 193–206, 2008.

problems are well known examples. The currently best known approximation ratio for the asymmetric traveling salesman problem (ATSP) is $O(\log n)$ [9,7], and improving this bound is an important open question. On the other hand, there is a 1.5-approximation algorithm for the symmetric TSP.

The *orienteering* problem is closely related to the minimum latency problem that we consider– given a metric with a length bound, the goal is to find a bounded-length path between two specified vertices that visits the maximum number of vertices. Blum et al. [1] gave the first constant factor approximation for the undirected version of this problem. Recently, Chekuri et al. [4] and the authors [15] independently gave $O(\log^2 n)$ approximation algorithms for the directed orienteering problem.

1.1   Problem Definition

We represent an asymmetric metric by (V, d), where V is the vertex set (with $|V| = n$) and $d : V \times V \to \mathbb{R}_+$ is a distance function satisfying the triangle inequality. For a directed path (or tour) π and vertices u, v, $d^\pi(u, v)$ denotes the distance from u to v along π; if v is not reachable from u along π, then $d^\pi(u, v) = \infty$. The directed minimum latency problem is defined as follows: given an asymmetric metric (V, d) and a root vertex $r \in V$, find a spanning *path* π originating at r that minimizes $\sum_{v \in V} d^\pi(r, v)$; the quantity $d^\pi(r, v)$ is the *latency* of vertex v in path π. Another possible definition of this problem would require a *tour* covering all vertices, where the latency of the root r is defined to be the distance required to return to r (i.e. the total tour length); note that in the previous definition of directed latency, the latency of r is zero. The approximability of both these versions of directed latency are related as below (the proof is deferred to the full version).

Theorem 1. *The approximability of the path-version and tour-version of directed latency are within a factor 4 of each other.*

In this paper, we work with the path version of directed latency.

For a directed graph $G = (V, E)$ and any $S \subseteq V$, we denote by $\delta^+(S) = \{(u, v) \in E \mid u \in S, v \notin S\}$ the arcs leaving set S, and $\delta^-(S) = \{(u, v) \in E \mid u \notin S, v \in S\}$ the arcs entering set S. When dealing with asymmetric metrics, the edge set E is assumed to be $V \times V$ unless mentioned otherwise. Given an asymmetric metric and a special vertex r, an r-path (resp. r-tour) is any directed path (resp. tour) originating at r.

Asymmetric Traveling Salesman Path (ATSP-path). The following problem is closely related to the directed latency problem. In ATSP-path, we are given a directed metric (V, d) and specified start and end vertices $s, t \in V$. The goal is to compute the minimum length $s - t$ path that visits all the vertices. It is easy to see that this problem is at least as hard to approximate as the ATSP (tour-version, where $s = t$). Lam and Newmann [13] were the first to consider this problem, and they gave an $O(\sqrt{n})$ approximation based on the Frieze et al. [9] algorithm for ATSP. This was improved to $O(\log n)$ in Chekuri and Pal [5], which extended the algorithm of Kleinberg and Williamson [12] for ATSP. Subsequently Feige and Singh [7] showed that the approximability of ATSP-tour and ATSP-path are within a constant factor of each other. We are concerned with the following LP relaxation of the ATSP-path problem.

$$\min \ \sum_e d_e \cdot x_e$$

$$s.t.$$

$$
\begin{aligned}
& x(\delta^+(u)) = x(\delta^-(u)) && \forall u \in V - \{s,t\} \\
& x(\delta^+(s)) = x(\delta^-(t)) = 1 \\
(ATSP-path) \quad & x(\delta^-(s)) = x(\delta^+(t)) = 0 \\
& x(\delta^-(S)) \geq \tfrac{2}{3} && \forall \{u\} \subseteq S \subseteq V \setminus \{s\}, \quad \forall u \in V \\
& x_e \geq 0 && \forall \text{ arcs } e
\end{aligned}
$$

The most natural LP relaxation for ATSP-path would have a 1 in the right-hand-side of the cut constraints, instead of $\frac{2}{3}$ as above. The above LP further relaxes the cut-constraints, and is still a valid relaxation of the problem. The precise value in the right-hand-side of the cut constraints is not important: we only require it to be some constant strictly between $\frac{1}{2}$ and 1.

1.2   Results and Paper Outline

Our main result is a reduction from the directed latency problem to the *asymmetric traveling salesman path* problem (ATSP-path) [13,5], where the approximation ratio for directed latency depends on the integrality gap of an LP relaxation for ATSP-path. We give an $n^{O(1/\epsilon)}$ time algorithm for the directed latency problem that achieves an approximation ratio of $O(\rho \cdot \frac{n^\epsilon}{\epsilon^3})$ (for any $\frac{1}{\log n} < \epsilon < 1$), where ρ is the integrality gap of an LP relaxation for the ATSP-path problem. The best upper bound we obtain is $\rho = O(\sqrt{n})$ (Section 3); however we conjecture that $\rho = O(\log n)$. In particular, our result implies a polynomial time $O(n^{1/2+\epsilon})$-approximation algorithm (any fixed $\epsilon > 0$) for directed latency. We study the LP relaxation for ATSP-path in Section 3, and present the algorithm for latency in Section 2. Our algorithm for latency first guesses a sequence of break-points (based on distances along the optimal path) and uses a linear program to obtain an assignment of vertices to segments (the portions between consecutive break-points), then it obtains local paths servicing each segment, and finally stitches these paths across all segments.

We also consider the special case of metrics given by shortest paths in an underlying unweighted directed graph, and obtain an $O(\log^2 n)$ approximation for minimum latency in this case (Section 4). This algorithm is essentially based on using the directed orienteering algorithm [15,4] within the framework for undirected latency [10]. On the hardness side, we observe that the directed latency problem (even in this 'unweighted' special case) is at least as hard to approximate as ATSP, for which the best known ratio is $O(\log n)$.

We note that ideas from the 'unweighted' case, also imply an $O(\log^2 n)$ approximation algorithm for minimizing weighted completion time in the no-wait permutation flowshop scheduling problem [20,18]– this can be cast as the latency problem in a special directed metric. We are not aware of any previous results on this problem.

2   The Directed Latency Algorithm

For a given instance of directed latency, let π denote an optimal latency path, $L = d(\pi)$ its length, and Opt its total latency. For any two vertices $u, v \in V$, recall that $d^\pi(u, v)$ denotes the length along path π from u to v; note that $d^\pi(u, v)$ is finite only if u appears before v on path π. The algorithm first guesses the length L (within factor 2) and $l = \lceil \frac{1}{\epsilon} \rceil$ vertices as follows: for each $i = 1, \cdots, l$, v_i is the last vertex on π with $d^\pi(r, v_i) \leq n^{i\epsilon} \frac{L}{n}$. We set $v_0 = r$ and note that v_l is the last vertex visited by π. Let $F = \{v_0, v_1, \cdots, v_l\}$. Consider now the following linear program (MLP):

$$\min \ \sum_{i=0}^{l-1} n^{(i+1)\epsilon} \frac{L}{n} \left(\sum_{u \notin F} y_u^i \right)$$

s.t.

$$
\begin{array}{ll}
z^i(\delta^+(u)) = z^i(\delta^-(u)) & \forall u \in V \setminus \{v_i, v_{i+1}\}, \quad \forall i = 0, \cdots, l-1 \\
z^i(\delta^+(v_i)) = z^i(\delta^-(v_{i+1})) = 1 & \forall i = 0, \cdots, l-1 \\
z^i(\delta^-(v_i)) = z^i(\delta^+(v_{i+1})) = 0 & \forall i = 0, \cdots, l-1 \\
z^i(\delta^-(S)) \geq y_u^i & \forall \{u\} \subseteq S \subseteq V \setminus \{v_i\}, \quad \forall u \in V \setminus F, \\
& \forall i = 0, \cdots, l-1 \\
\sum_e d_e \cdot z^i(e) \leq n^{(i+1)\epsilon} \cdot \frac{L}{n} & \forall i = 0, \cdots, l-1 \\
\sum_{i=0}^{l-1} y_u^i \geq 1 & \forall u \in V \setminus F \\
z^i(e) \geq 0 & \forall \text{ arcs e}, \quad \forall i = 0, \cdots, l-1 \\
y_u^i \geq 0 & \forall u \in V \setminus F, \quad \forall i = 0, \cdots, l-1
\end{array}
$$

Basically this LP requires one unit of flow to be sent from v_i to v_{i+1} (for all $0 \leq i \leq l-1$) such that the total extent to which each vertex u is covered (over all these flows) is at least 1. In addition, the i-th flow is required to have total cost (under the length function d) at most $n^{(i+1)\epsilon} \cdot \frac{L}{n}$. It is easy to see that this LP can be solved in polynomial time for any guess $\{v_i\}_{i=1}^l$. Furthermore the number of possible guesses is $O(n^{1/\epsilon})$, hence we can obtain the optimal solution of (MLP) over all guesses, in $n^{O(1/\epsilon)}$ time.

Claim 1. *The minimum value of (MLP) over all possible guesses of $\{v_i\}_{i=0}^l$ is at most $2n^\epsilon \cdot$ Opt.*

Proof: This claim is straightforward, based on the guesses from an optimal path. Recall that π is the optimal latency path for the given instance. One of the guesses of the vertices $\{v_i\}_{i=0}^l$ satisfies the condition desired of them, namely each v_i (for $i = 1, \cdots, l$) is the last vertex on π with $d^\pi(s, v_i) \leq n^{i\epsilon} \frac{L}{n}$. For each $i = 0, \cdots, l-1$, define O_i to be the set of vertices that are visited between v_i and v_{i+1} in path π. Let z^i denote the (integral) edge values corresponding to path π restricted to the vertices $O_i \cup \{v_i, v_{i+1}\}$; note that the cost of this flow $d \cdot z^i \leq d^\pi(r, v_{i+1}) \leq n^{(i+1)\epsilon} \frac{L}{n}$. Also set $y_u^i = 1$ for $u \in O_i$ and 0 otherwise, for all $i = 0, \cdots, l-1$. Note that each vertex in $V \setminus \{v_i\}_{i=0}^l$ appears in some set O_i, and each z^i supports unit flow from v_i to all vertices in O_i; hence this (integral) solution $\{z^i, y^i\}_{i=0}^{l-1}$ is feasible for (MLP). The cost of this solution is $\sum_{i=0}^{l-1} n^{(i+1)\epsilon} \frac{L}{n} \cdot |O_i| \le n^\epsilon L + n^\epsilon \sum_{i=1}^{l-1} n^{i\epsilon} \frac{L}{n} \cdot |O_i| \le 2n^\epsilon \cdot \mathsf{Opt}$, since $|O_0| \le n$, $L \le \mathsf{Opt}$, and each vertex $u \in O_i$ (for $i = 1, \cdots, l-1$) has $d^\pi(r, u) > n^{i\epsilon} \frac{L}{n}$.

We now assume that we have an optimal fractional solution $\{z^i, y^i\}_{i=0}^{l-1}$ to (MLP) over all guesses (with objective value as in Claim 1), and show how to round it to obtain $v_i - v_{i+1}$ paths for each $i = 0, \cdots, l-1$, which when stitched give rise to *one* r-path having a small latency objective. We say that a vertex u is *well-covered* by flow z^i if $y_u^i \ge \frac{1}{4l}$. We partition the vertices $V \setminus F$ into two parts: V_1 consists of those vertices that are well-covered for *at least two* values of $i \in [0, l]$, and V_2 consists of all other vertices. Note that each vertex in V_2 is covered by some flow z^i to the extent at least $\frac{3}{4}$. We first show how to service each of V_1 and V_2 separately using local paths, and then stitch these into a single r-path.

Splitting off: A directed graph is called *Eulerian* if the in-degree equals the out-degree at each vertex. In our proofs, we make use of the following 'splitting-off' theorem for Eulerian digraphs.

Theorem 2 (Frank [8] (Theorem 4.3) and Jackson [11]). *Let $D = (U + r, A)$ be an Eulerian directed multi-graph. For each arc $f = (r, v) \in A$ there exists an arc $e = (u, r) \in A$ so that after replacing arcs e and f by arc (u, v), the directed connectivity between every pair of vertices in U is preserved.*

Note that any vector $\tilde{x}$ of rational edge-capacities that is Eulerian (namely $\tilde{x}(\delta^-(v)) = \tilde{x}(\delta^+(v))$ at all vertices v) corresponds to an Eulerian multi-graph by means of a (sufficiently large) uniform scaling of all arcs. Based on this correspondence, one can use the above splitting-off theorem directly on fractional edge-capacities that are Eulerian.

2.1 Servicing Vertices V_1

We partition V_1 into l parts as follows: U_i (for $i = 0, \cdots, l-1$) consists of those vertices of V_1 that are well-covered by z^i but *not* well-covered by any flow z^j for $j > i$. Each set U_i is serviced separately by means of a suitable ATSP solution on $U_i \cup \{v_i\}$ (see Lemma 1): this step requires a bound on the length of back-arcs from U_i-vertices to v_i, which is ensured by the next claim.

Claim 2. *For each vertex $w \in U_i$, $d(w, v_i) \le 8l \cdot n^{i\epsilon} \frac{L}{n}$.*

Proof: Let $j \le i - 1$ be such that $y_w^j \ge \frac{1}{4l}$; such an index exists by the definition of V_1 and U_i. In other words, arc-capacities z^j support at least $\frac{1}{4l}$ flow from w to v_{j+1}; so $4l \cdot z^j$ supports a unit flow from w to v_{j+1}. Thus $d(w, v_{j+1}) \le 4l(d \cdot z^j) \le 4l \cdot n^{(j+1)\epsilon} \frac{L}{n}$. Note that for any $0 \le k \le l$, z^k supports a unit flow from v_k to v_{k+1}; hence $d(v_k, v_{k+1}) \le d \cdot z^k \le n^{(k+1)\epsilon} \frac{L}{n}$. Now, $d(w, v_i) \le d(w, v_{j+1}) + \sum_{k=j+1}^{i-1} d(v_k, v_{k+1}) \le 4l \frac{L}{n} \sum_{k=j}^{i-1} n^{(k+1)\epsilon} \le 8l \cdot n^{i\epsilon} \frac{L}{n}$.

We now show how all vertices in U_i can be covered by a v_i-tour.

Lemma 1. *For each $i = 0, \cdots, l-1$, there is a poly-time computable v_i-tour covering vertices U_i, of length $O(\frac{1}{\epsilon^2} n^{(i+1)\epsilon} \log n \cdot \frac{L}{n})$.*

Proof: Fix an $i \in \{0, \cdots, l-1\}$; note that the arc capacities z^i are Eulerian at all vertices except v_i and v_{i+1}. Although applying splitting-off (Theorem 2) requires an Eulerian graph, we can apply it to z^i after adding a dummy (v_{i+1}, v_i) arc of capacity 1, and observing that flows from v_i or flows into v_{i+1} do not use the dummy arc. So using Theorem 2 on vertices $V \setminus (U_i \cup \{v_i, v_{i+1}\})$ and triangle inequality, we obtain arc capacities α on the arcs induced by $U_i \cup \{v_i, v_{i+1}\}$ such that: $d \cdot \alpha \leq d \cdot z^i \leq n^{(i+1)\epsilon} \cdot \frac{L}{n}$ and α supports $y_u^i \geq \frac{1}{4l}$ flow from v_i to u and from u to v_{i+1}, for every $u \in U_i$. Below we use B to denote the quantity $n^{(i+1)\epsilon} \cdot \frac{L}{n}$. Consider adding a dummy arc from v_{i+1} to v_i of length B in the induced metric on $U_i \cup \{v_i, v_{i+1}\}$, and set the arc capacity $\alpha(v_{i+1}, v_i)$ on this arc to be 1. Note that α is Eulerian, has total cost at most $2B$, and every non-trivial cut has value at least $\min\{y_u^i : u \in U_i\} \geq \frac{1}{4l}$. So scaling α uniformly by $4l$, we obtain a fractional feasible solution to ATSP on the vertices $U_i \cup \{v_i, v_{i+1}\}$ (in the modified metric), having cost at most $8l \cdot B$. Since the Frieze et al. [9] algorithm computes an integral tour of length at most $O(\log n)$ times any fractional feasible solution (see Williamson [19]), we obtain a v_i-tour τ on the modified metric of length at most $O(l \log n) \cdot B$. Since the dummy (v_{i+1}, v_i) arc has length B, it may be used at most $O(l \log n)$ times in τ. So removing all occurrences of this dummy arc gives a set of $O(l \log n)$ $v_i - v_{i+1}$ paths in the original metric, that together cover U_i. Ignoring vertex v_{i+1} and inserting the direct arc to v_i from the last U_i vertex in each of these paths gives us the desired v_i-tour covering U_i. Finally note that each of the arcs to v_i inserted above has length $O(l \cdot n^{i\epsilon}) \frac{L}{n} = O(l) \cdot B$ (from Claim 2), and the number of arcs inserted is $O(l \log n)$. So the length of this v_i-tour is at most $O(l \log n) \cdot B + O(l^2 \log n) \cdot B = O(\frac{1}{\epsilon^2} n^{(i+1)\epsilon} \log n \cdot \frac{L}{n})$.

2.2 Servicing Vertices V_2

We partition vertices in V_2 into $W_0, \cdots, W_{l-1}$, where each W_i contains the vertices in V_2 that are well-covered by z^i. As noted earlier, each vertex $u \in W_i$ in fact has $y_u^i \geq \frac{3}{4} > \frac{2}{3}$. We now consider any particular W_i and obtain a $v_i - v_{i+1}$ path covering the vertices of W_i. Vertices in W_i are covered by a fractional $v_i - v_{i+1}$ path as follows. Splitting off vertices $V \setminus (W_i \cup \{v_i, v_{i+1}\})$ in the fractional solution z^i gives us edge capacities β in the metric induced on $W_i \cup \{v_i, v_{i+1}\}$, such that: β supports at least $\frac{2}{3}$ flow from v_i to u and from u to v_{i+1} for all $u \in W_i$, and $d \cdot \beta \leq d \cdot z^i$ (this is similar to how arc-capacities α were obtained in Lemma 2.1). Note that β is a fractional feasible solution to the LP relaxation $(ATSP - path)$ for the ATSP-path instance on the metric induced by $W_i \cup \{v_i, v_{i+1}\}$ with start-vertex v_i and end-vertex v_{i+1}. So if ρ denotes the (constructive) integrality gap of $(ATSP - LP)$, we can obtain an integral v_i-v_{i+1} path that spans W_i of length at most $\rho(d \cdot \beta) \leq \rho(d \cdot z^i) \leq \rho n^{(i+1)\epsilon} \frac{L}{n}$. This requires a polynomial time algorithm that computes an integral path of length at most ρ times the LP value; However even a non-constructive proof of integrality gap ρ' implies a constructive integrality gap $\rho = O(\rho' \log n)$, using the algorithm in Chekuri and Pal [5]. So we obtain:

Lemma 2. *For each $i = 0, \cdots, l-1$, there is a poly-time computable v_i-v_{i+1} path covering W_i of length at most $\rho \cdot n^{(i+1)\epsilon} \frac{L}{n}$.*

2.3   Stitching the Local Paths

We now stitch the v_i-tours that service V_1 (Lemma 1) and the $v_i - v_{i+1}$ paths that service V_2 (Lemma 2), to obtain a single r-path that covers all vertices. For each $i = 0, \cdots, l-1$, let π_i denote the v_i-tour servicing U_i, and let τ_i denote the $v_i - v_{i+1}$ path servicing W_i. The final r-path that the algorithm outputs is the concatenation $\tau^* = \pi_0 \cdot \tau_0 \cdot \pi_1 \cdots \pi_{l-1} \cdot \tau_{l-1}$. From Lemmas 1 and 2, it follows that for all $0 \le i \le l-1$, $d(\pi_i) \le O(\frac{1}{\epsilon^2} \log n) \cdot n^{(i+1)\epsilon} \frac{L}{n}$ and $d(\tau_i) \le O(\rho) \cdot n^{(i+1)\epsilon} \frac{L}{n}$. So the length of τ^* from r until all vertices of $U_i \cup W_i$ are covered is at most $O(\rho + \frac{1}{\epsilon^2} \log n) \cdot n^{(i+1)\epsilon} \frac{L}{n}$ (as $\epsilon \ge \Omega(\frac{1}{\log n})$). This implies that the total latency of vertices in $U_i \cup W_i$ along path τ^* is at most $O(\rho + \frac{1}{\epsilon^2} \log n) \cdot n^{(i+1)\epsilon} \frac{L}{n} \cdot (|W_i| + |U_i|)$.

Moreover, the contribution of each vertex in U_i (resp., W_i) to the LP objective is at least $\frac{1}{4l} \cdot n^{(i+1)\epsilon} \frac{L}{n}$ (resp., $\frac{3}{4} \cdot n^{(i+1)\epsilon} \frac{L}{n}$). Thus the contribution of $U_i \cup W_i$ to the LP objective is at least $\frac{1}{4l} \cdot n^{(i+1)\epsilon} \frac{L}{n} \cdot (|W_i| + |U_i|)$. Using the upper bound on the latency along τ^* for $U_i \cup W_i$, and summing over all i, we obtain that the total latency along τ^* is at most $O(\frac{1}{\epsilon} \rho + \frac{1}{\epsilon^3} \log n)$ times the optimal value of (MLP). From Claim 1, it now follows that the latency of τ^* is $O(\frac{1}{\epsilon} \rho + \frac{1}{\epsilon^3} \log n) n^\epsilon \cdot \mathsf{Opt}$.

Theorem 3. *For any $\Omega(\frac{1}{\log n}) < \epsilon < 1$, there is an $O(\frac{\rho + \log n}{\epsilon^3} \cdot n^\epsilon)$-approximation algorithm for directed latency, that runs in time $n^{O(1/\epsilon)}$, where ρ is the integrality gap of the LP $(ATSP - path)$. Using $\rho = O(\sqrt{n})$, we have a polynomial time $O(n^{\frac{1}{2}+\epsilon})$ approximation algorithm for any fixed $\epsilon > 0$.*

We prove the bound $\rho = O(\sqrt{n})$ in the next section. A bound of $\rho = O(\log n)$ on the integrality gap of $(ATSP - path)$ would imply that this algorithm is a quasi-polynomial time $O(\log^4 n)$ approximation for directed latency.

Remark: The $(ATSP - path)$ rounding algorithm in Section 3 can be modified slightly to obtain (for any $0 < \delta < 1$), an $(O(n^\delta \log n), \lfloor \frac{1}{\delta} \rfloor)$ bi-criteria approximation for ATSP-path. This implies the following generalization of Theorem 3.

Corollary 1. *For any $\Omega(\frac{1}{\log n}) < \epsilon < 1$ and $0 < \delta < 1$, there is an $n^{O(1/\epsilon)}$ time algorithm for directed latency, that computes $\lfloor \frac{1}{\delta} \rfloor$ paths covering all vertices, having a total latency of $O(\frac{\log n}{\epsilon^3} \cdot n^{\epsilon+\delta}) \cdot \mathsf{Opt}$, where Opt is the minimum latency of a single path covering all the vertices.*

3   Bounding the Integrality Gap of ATSP-Path

We prove an upper bound of $O(\sqrt{n})$ on the integrality gap ρ of the linear relaxation $(ATSP - path)$ (c.f. Section 1.1). Even for the seemingly stronger LP with 1 in the right-hand-side of the cut constraints, the best bound on the integrality gap we can obtain is $O(\sqrt{n})$: this follows from the cycle-cover based algorithm of Lam and Newmann [13]. As mentioned in Chekuri and Pal [5], it is unclear whether their $O(\log n)$-approximation can be used to bound the integrality gap of such a linear program. In this section, we present a rounding algorithm for the weaker LP $(ATSP - path)$, which shows $\rho = O(\sqrt{n})$. Our algorithm is similar to the ATSP-path algorithm of Lam and Newmann [13] and the ATSP algorithm of Frieze et al. [9]; but it needs some more work as we compare the algorithm's solution against a fractional solution to $(ATSP - path)$.

Let x be any feasible solution to $(ATSP - path)$. We now show how x can be rounded to obtain an integral path spanning all vertices, of total length $O(\sqrt{n})(d \cdot x)$. Let N denote the network corresponding to the directed metric with the *cost* of each arc equal to its metric length, and an extra (t, s) arc of cost 0. The *capacity* of this extra (t, s) arc is set to 3, and all other capacities are set to ∞. The rounding algorithm for x is as follows.

1. Initialize the set of representatives $R \leftarrow V \setminus \{s, t\}$, and the current integral solution $\sigma = \emptyset$.
2. While $R \neq \emptyset$, do:
   (a) Compute a minimum cost circulation $\mathcal{C}$ in $N[R \cup \{s, t\}]$ that sends at least 2 units of flow through each vertex in R (note: $\mathcal{C}$ can be expressed as a sum of cycles).
   (b) Repeatedly extract from $\mathcal{C}$ all cycles that do not use the extra arc (t, s), to obtain circulation $\mathcal{A} \subseteq \mathcal{C}$. Let $R' \subseteq R$ be the set of R-vertices that have degree at least 1 in $\mathcal{A}$.
   (c) Let $\mathcal{B} = \mathcal{C} \setminus \mathcal{A}$; note that $\mathcal{B}$ is Eulerian and each cycle in it uses arc (t, s).
   (d) If $|R'| \geq \sqrt{n}$, do:
      i. Set $\sigma \leftarrow \sigma \cup \mathcal{A}$.
      ii. *Modify R by dropping all but one R'-vertex from each strong component of $\mathcal{A}$.*
   (e) If $|R'| < \sqrt{n}$, do:
      i. Take an Euler tour on $\mathcal{B}$ and remove all (at most 3) occurrences of arc (t, s) to obtain s-t paths P_1, P_2, P_3.
      ii. Restrict each path P_1, P_2, P_3 to vertices in $R \setminus R'$ by short-cutting over R'-vertices, to obtain paths $\tilde{P}_1, \tilde{P}_2, \tilde{P}_3$.
      iii. Take a topological ordering $s = w_1, w_2, \cdots, w_h = t$ of vertices $(R \setminus R') \cup \{s, t\}$ relative to the arcs $\tilde{P}_1 \cup \tilde{P}_2 \cup \tilde{P}_3$.
      iv. Set $\sigma \leftarrow \sigma \cup \{(w_j, w_{j+1}) : 1 \leq j \leq h - 1\}$.
      v. Repeat for each vertex $u \in R'$: find an arc $(w, w') \in \sigma$ such that x supports $\frac{1}{6}$ flow from w to u and from u to w', and modify $\sigma \leftarrow (\sigma \setminus (w, w')) \cup \{(w, u), (u, w')\}$.
      vi. Set $R \leftarrow \emptyset$.
3. Output any spanning s-t walk in σ.

We now show the correctness and performance guarantee of the rounding algorithm. We first bound the cost of the circulation obtained in Step 2a during any iteration.

Claim 3. *For any $R \subseteq V \setminus \{s, t\}$, the minimum cost circulation $\mathcal{C}$ computed in step 2a has cost at most $3(d \cdot x)$.*

Proof: The arc values x define a fractional $s - t$ path in network N. Extend x to be a (fractional) circulation by setting $x(t, s) = 1$. We can now apply splitting-off (Theorem 2) on each vertex in $V \setminus R$, to obtain capacities x' in network $N[R \cup \{s, t\}]$, such that every pairwise connectivity is preserved and (by triangle inequality) $d \cdot x' \leq d \cdot x$. Note that the extra (t, s) arc is not modified in the splitting-off steps. So x' supports $\frac{2}{3}$ flow from s to each vertex in R; this implies that $3x'$ is a feasible fractional solution to the circulation instance solved in step 2a (note that $x'(t, s)$ remains 1, so solution $3x'$ satisfies the capacity of arc (t, s)). Finally, note that the linear relaxation for circulation is integral (c.f. Nemhauser and Wolsey [16]). So the minimum cost (integral) circulation computed in step 2a has cost at most $3d \cdot x' \leq 3d \cdot x$.

Note that each time step 2d is executed, $|R|$ decreases by at least $\sqrt{n}/2$ (each strong component in $\mathcal{A}$ has at least 2 vertices); so there are at most $O(\sqrt{n})$ such iterations and the cost of σ due to additions in this step is $O(\sqrt{n})(d \cdot x)$ (using Claim 3). Step 2e is executed at most once (at the end); the next claim shows that this step is well defined and bounds the cost incurred.

Claim 4. *In step 2(e)iii, there exists a topological ordering $w_1, \cdots, w_h$ of $(R \setminus R') \cup \{s, t\}$ w.r.t. arcs $\tilde{P}_1 \cup \tilde{P}_2 \cup \tilde{P}_3$. Furthermore, $\{(w_j, w_{j+1}) : 1 \leq j \leq h - 1\} \subseteq \tilde{P}_1 \cup \tilde{P}_2 \cup \tilde{P}_3$.*

Proof: Note that any cycle in $P_1 \cup P_2 \cup P_3$ is a cycle in $\mathcal{B}$ that does not use arc (t, s), which is not possible by the definition of $\mathcal{B}$ (every cycle in $\mathcal{B}$ uses arc (t, s)); so $P_1 \cup P_2 \cup P_3$ is acyclic. It is clear that if $\tilde{P}_1 \cup \tilde{P}_2 \cup \tilde{P}_3$ contains a cycle, so does $P_1 \cup P_2 \cup P_3$ (each path $\tilde{P}_i$ is obtained by short-cutting the corresponding path P_i). Hence $\tilde{P}_1 \cup \tilde{P}_2 \cup \tilde{P}_3$ is also acyclic, and there is a topological ordering of $(R \setminus R') \cup \{s, t\}$ relative to arcs $\tilde{P}_1 \cup \tilde{P}_2 \cup \tilde{P}_3$. We now prove the second part of the claim. In circulation $\mathcal{C}$, each vertex of R has at least 2 units of flow through it; but vertices $R \setminus R'$ are not covered (even to an extent 1) in the circulation $\mathcal{A}$. So each vertex of $R \setminus R'$ is covered to extent at least 2 in circulation $\mathcal{B}$, and hence in $P_1 \cup P_2 \cup P_3$. In other words, each vertex of $R \setminus R'$ appears on at least two of the three $s - t$ paths P_1, P_2, P_3. This also implies that (after the short-cutting) each $R \setminus R'$ vertex appears on at least two of the three $s - t$ paths $\tilde{P}_1, \tilde{P}_2, \tilde{P}_3$. Now observe that for each consecutive pair (w_j, w_{j+1}) $(1 \leq j \leq h - 1)$ in the topological order, there is a common path $\tilde{P}_k$ (for some $k = 1, 2, 3$) that contains both w_j and w_{j+1}. Furthermore, in $\tilde{P}_k$, w_j and w_{j+1} are consecutive in that order (otherwise, the topological order would contain a back arc!). Thus each arc (w_j, w_{j+1}) (for $1 \leq j \leq h - 1$) is present in $\tilde{P}_1 \cup \tilde{P}_2 \cup \tilde{P}_3$, and we obtain the claim.

We also need the following claim to bound the cost of insertions in step 2(e)v.

Claim 5. *For any two vertices $u', u'' \in V$, if $\lambda(u', u''; x)$ (resp. $\lambda(u'', u'; x)$) denotes the maximum flow supported by x from u' to u'' (resp. u'' to u'), then $\lambda(u', u''; x) + \lambda(u'', u'; x) \geq \frac{1}{3}$.*

Proof: If either u' or u'' is in $\{s, t\}$, the claim is obvious since for every vertex v, x supports $\frac{2}{3}$ flow from s to v and from v to t. Otherwise $\{s, t, u', u''\}$ are distinct, and define capacities $\hat{x}$ as:

$$\hat{x}(v_1, v_2) = \begin{cases} x(v_1, v_2) & \text{for arcs } (v_1, v_2) \neq (t, s) \\ 1 & \text{for arc } (v_1, v_2) = (t, s) \end{cases}$$

Observe that $\hat{x}$ is Eulerian; now apply Theorem 2 to $\hat{x}$ and split-off all vertices of V except $T = \{s, t, u', u''\}$, and obtain capacities y on the arcs induced on T. We have $\lambda(t_1, t_2; y) = \lambda(t_1, t_2; \hat{x})$ for all $t_1, t_2 \in T$. Note that since neither t nor s is split-off, their degrees in y are unchanged from $\hat{x}$, and also $y(t, s) \geq \hat{x}(t, s) = 1$. Since the out-degree of t in $\hat{x}$ (hence in y) is 1 and $y_{t,s} \geq 1$, we have $y(t, u') = y(t, u'') = 0$ and $y(t, s) = 1$. The capacities y support at least $\frac{2}{3}$ flow from s to u'; so $y(s, u') + y(u'', u') \geq \frac{2}{3}$. Similarly for u'', we have $y(s, u'') + y(u', u'') \geq \frac{2}{3}$, and adding these two inequalities we get $y(u', u'') + y(u'', u') + (y(s, u') + y(s, u'')) \geq \frac{4}{3}$. Note that $y(s, u') + y(s, u'') \leq y(\delta^+(s)) = \hat{x}(\delta^+(s)) = 1$ (the degree of s is unchanged in the splitting-off). So $y(u', u'') + y(u'', u') \geq \frac{1}{3}$. Since y is obtained from $\hat{x}$ by a sequence of splitting-off operations, it follows that $\hat{x}$ supports flows corresponding to all edges in y *simultaneously*. In particular, the following flows are supported disjointly in $\hat{x}$: $\mathcal{F}_1$ that sends $y(u', u'')$ units from u' to u'', $\mathcal{F}_2$ that sends $y(u'', u')$ units from u'' to u', and $\mathcal{F}_3$ that sends $y(t, s) = 1$ unit from t to s. Hence the flows $\mathcal{F}_1$ and $\mathcal{F}_2$ are each supported by $\hat{x}$ and *do not* use the extra (t, s) arc (since $\hat{x}(\delta^+(t)) = \hat{x}(t, s) = 1$). This implies that the flows $\mathcal{F}_1$ and $\mathcal{F}_2$ are both supported by the original capacities x (where $x(t, s) = 0$). Hence $\lambda(u', u''; x) + \lambda(u'', u'; x) \geq y(u', u'') + y(u'', u') \geq \frac{1}{3}$.

From Claim 4, we obtain that the cost addition in step 2e(iv) is at most $d(\tilde{P}_1) + d(\tilde{P}_2) + d(\tilde{P}_3) \leq d(P_1) + d(P_2) + d(P_3) \leq 3(d \cdot x)$ (from Claim 3). We now consider the cost addition to σ in step 2(e)v. Claim 5 implies that for any pair of vertices $u', u'' \in V$, x supports $\frac{1}{6}$ flow either from u' to u'' or from u'' to u'. Also for every vertex u, x supports $\frac{2}{3}$ flow from s to u and from u to t. Since σ always contains an $s - t$ path in step 2(e)v, there is always some position along this $s - t$ path to insert any vertex $u \in R'$ as required in step 2(e)v. Furthermore, the cost increase in any such insertion step is at most $12(d \cdot x)$. Hence the total cost for inserting all the vertices R' into σ is at most $12|R'|(d \cdot x) = O(\sqrt{n})(d \cdot x)$. Thus the total cost of σ at the end of the algorithm is $O(\sqrt{n})(d \cdot x)$. Finally note that σ is connected (in the undirected sense), Eulerian at all vertices in $V \setminus \{s, t\}$ and has outdegree 1 at s. This implies that σ corresponds to a spanning $s - t$ walk. This completes the proof of the following.

Theorem 4. *The integrality gap of* $(ATSP - path)$ *is at most* $O(\sqrt{n})$.

4   Unweighted Directed Metrics

In the special case where the metric is induced by shortest paths in an unweighted directed graph, we obtain an improved approximation guarantee for the minimum latency problem. This draws on ideas from the undirected latency problem, and the $O(\log^2 n)$ approximation ratio for directed orienteering ([15] and [4]). The directed orienteering problem is as follows: given a starting vertex r in an asymmetric metric and length bound L, find an r-path of length at most L covering the maximum number of vertices. We note that the reduction from ATSP to directed latency also holds in unweighted directed metrics, and the best known approximation ratio for ATSP even on this special class is $O(\log n)$. Here we show the following.

Theorem 5. *An α-approximation algorithm for directed orienteering implies an $O(\alpha+\gamma)$ approximation algorithm for the directed latency problem on unweighted digraphs, where γ is the best approximation ratio for ATSP. In particular there is an $O(\log^2 n)$ approximation.*

Let $G = (V, A)$ denote the underlying digraph that induces the given metric, and r the root vertex. We first argue (Section 4.1) that if G is strongly connected, then there is an $O(\alpha)$-approximation algorithm. Then we show (Section 4.2) how this can be extended to the case when G is not strongly connected.

4.1 G Is Strongly Connected

In this case, the distance from any vertex to the root r is at most $n = |V|$. The algorithm and analysis for this case are identical to those for the undirected latency problem [2,10,3]. Details are deferred to the full version.

Remark: This 'greedy' approach does not work in the general directed case since it is unclear how to bound the length of back-arcs to the root r (which is required to stitch the paths that are computed greedily). In the undirected case, back-arcs can be easily bounded by the forward length, and this approach results in a constant approximation algorithm. In the unweighted strongly-connected case (considered above), the total length of back-arcs used by the algorithm could be bounded by roughly n^2 (which is also a lower bound for the latency problem). By an identical analysis, it also follows that there is an $O(\alpha)$-approximation for the directed latency problem on metrics (V, d) with the following property: for every vertex $v \in V$, the back-arc length to r is within a constant factor of the forward-arc length from r, i.e. $d(v, r) \leq O(1) \cdot d(r, v)$. As a consequence, we obtain an $O(\alpha) = O(\log^2 n)$ approximation for *no-wait flowshop scheduling* with the *weighted completion time* objective (n is the number of jobs in the given instance); this seems to be the first approximation ratio for the problem. The no-wait flowshop problem can be modeled as a minimum latency problem in an appropriate directed metric [20,18], with the property that all back-arcs to the root r have length 0; hence the above greedy approach applies.

4.2 G Is Not Strongly Connected

In this case, we show an $O(\gamma + \beta)$-approximation algorithm, where γ is the approximation guarantee for ATSP and β is the approximation guarantee for the minimum latency problem on unweighted strongly-connected digraphs. From

Section 4.1, $\beta = O(\alpha)$, where α is the approximation ratio for directed orienteering. Consider the strong components of G, which form a directed acyclic graph. If the instance is feasible, there is a Hamilton path in G from r; so we can order the strong components of G as $C_1, \cdots, C_l$ such that $r \in C_1$ and any spanning path from r visits the strong components in that order. For each $1 \le i \le l$, let $n_i = |C_i|$, and pick an arbitrary vertex $s_i \in C_i$ as root for each strong component (setting $s_1 = r$).

Lemma 3. *There exists a spanning r-path τ^* having latency objective at most $7 \cdot \mathsf{Opt}$ such that $\tau^* = \tau_1 \cdot (s_1, s_2) \cdot \tau_2 \cdot (s_2, s_3) \cdots (s_{l-1}, s_l) \cdot \tau_l$, where each τ_i (for $1 \le i \le l$) is an s_i-tour covering all vertices in C_i.*

Proof: Consider the optimal latency r-path P^*: this is a concatenation $P_1 \cdot P_2 \cdots P_l$ of paths in each strong component (P_i is a spanning path on C_i). For each $1 \le i \le l$, let $\mathsf{Lat}(P_i)$ denote the latency of vertices C_i just along path P_i, and $D_i = \sum_{j=1}^{i-1} d(P_j)$ be the distance traversed by P^* before P_i. Then the total latency along P^* is $\mathsf{Opt} = \sum_{i=1}^{l} (n_i \cdot D_i + \mathsf{Lat}(P_i))$.

For each $1 \le i \le l$, let τ_i denote a spanning tour on C_i, obtained by adding to P_i the direct arcs: from s_i to its first vertex and from its last vertex to s_i. Each of these extra arcs in τ_i has length at most $n_i - 1$ (since C_i is strongly connected), and $d(P_i) \ge n_i - 1$ (it is spanning on C_i); so $d(\tau_i) \le 3d(P_i)$. Let $\mathsf{Lat}(\tau_i)$ denote the latency of vertices C_i along τ_i; from the above observation we have $\mathsf{Lat}(\tau_i) \le n_i \cdot (n_i - 1) + \mathsf{Lat}(P_i)$. Now we obtain τ^* as the concatenation $\tau_1 \cdot (s_1, s_2) \cdot \tau_2 \cdots (s_{t-1}, s_l) \cdot \tau_l$. Note also that for any $1 \le i \le l-1$, $d(s_i, s_{i+1}) \le n_i + n_{i+1}$. So the latency in τ^* of vertices C_i is:

$$
\begin{aligned}
& n_i \cdot \textstyle\sum_{j=1}^{i-1} (d(\tau_j) + d(s_j, s_{j+1})) + \mathsf{Lat}(\tau_i) \\
\le\ & n_i \cdot \textstyle\sum_{j=1}^{i-1} (3d(P_j) + n_j + n_{j+1}) + n_i \cdot (n_i - 1) + \mathsf{Lat}(P_i) \\
\le\ & n_i \cdot \textstyle\sum_{j=1}^{i-1} (3d(P_j) + 2n_j) + n_i^2 + n_i \cdot (n_i - 1) + \mathsf{Lat}(P_i) \\
\le\ & n_i \cdot \textstyle\sum_{j=1}^{i-1} 7d(P_j) + n_i^2 + n_i \cdot (n_i - 1) + \mathsf{Lat}(P_i) \\
\le\ & 7n_i \cdot D_i + 2n_i^2 + \mathsf{Lat}(P_i) \\
\le\ & 7n_i \cdot D_i + 5 \cdot \mathsf{Lat}(P_i)
\end{aligned}
$$

The last inequality follows from the fact that $\mathsf{Lat}(P_i) \ge n_i^2/2$ (P_i is a path on n_i vertices in an unweighted metric). So the total latency of τ^* is at most $7 \sum_{i=1}^{l} (n_i \cdot D_i + \mathsf{Lat}(P_i)) = 7 \cdot \mathsf{Opt}$.

The algorithm for directed latency in this case computes an approximately minimum latency s_i-path for each C_i separately (using the algorithm in Section 4.1); by adding the direct arc from the last vertex back to s_i, we obtain C_i-spanning tours $\{\sigma_i\}_{i=1}^{l}$. We now use the following claim from [2] to bound the length of each tour σ_i.

Claim 6 ([2]). *Given C_i-spanning tours σ_i and π_i, there exists a poly-time computable tour σ_i' on C_i of length at most $3 \cdot d(\pi_i)$ and latency at most thrice that of σ_i.*

Proof: Tour σ_i' is constructed as follows: starting at s_i, traverse tour σ_i until a length of $d(\pi_i)$, then traverse tour π_i from the current vertex to visit all remaining vertices and then return to s_i. Note that tour π_i will have to be traversed at most twice, and so the length of σ_i' is at most $3d(\pi_i)$. Furthermore, the total latency along σ_i' for vertices visited in the σ_i part is at most $\mathsf{Lat}(\sigma_i)$ (the latency along σ_i). Also the latency along σ_i' of each vertex v visited in the π_i part is at most $3d(\pi_i)$, which is at most thrice its latency in σ_i. Hence the total latency along σ_i' is at most $3 \cdot \mathsf{Lat}(\sigma_i)$.

This implies that by truncating σ_i with a γ-approximate TSP on C_i, we obtain another spanning tour σ_i' of length $3\gamma \cdot L_i$ and latency $3 \cdot \mathsf{Lat}(\sigma_i)$ (where L_i is length of the minimum TSP on C_i). The final r-path is the concatenation of these local tours, $\pi = \sigma_1' \cdot (s_1, s_2) \cdot \sigma_2' \cdots (s_{l-1}, s_l) \cdot \sigma_l'$.

Claim 7. *The latency of r-path π is at most $O(\gamma + \beta) \cdot \mathsf{Opt}$.*

Proof: Consider the near-optimal r-path τ^* given by Lemma 3. For $1 \leq i \leq l$, let Opt_i denote the latency of the C_i-spanning tour τ_i, and $\tilde{D}_i = \sum_{j=1}^{i-1}(d(\tau_j) + d(s_j, s_{j+1}))$ denote the length of τ^* before C_i. Then the total latency of τ^* can be written as $\sum_{i=1}^{l}(n_i \cdot \tilde{D}_i + \mathsf{Opt}_i) \leq 7 \cdot \mathsf{Opt}$.

Now consider the r-path π output by the algorithm. The s_i-tour τ_i is a feasible solution to the minimum latency instance on C_i; so the latency of tour σ_i is at most $\beta \cdot \mathsf{Opt}_i$, since we use a β-approximation for each such instance. So for each $1 \leq i \leq l$, the truncated tour σ_i' has latency $\mathsf{Lat}(\sigma_i') \leq 3\beta \cdot \mathsf{Opt}_i$, and length $d(\sigma_i') \leq 3\gamma L_i$. Again, the latency of π can be written as $\sum_{i=1}^{l}(n_i \cdot D_i' + \mathsf{Lat}(\sigma_i'))$, where $D_i' = \sum_{j=1}^{i-1}(d(\sigma_j') + d(s_j, s_{j+1}))$ is the length of π before C_i. So the latency of vertices C_i in π is:

$$n_i \cdot \sum_{j=1}^{i-1}(d(\sigma_j') + d(s_j, s_{j+1})) + \mathsf{Lat}(\sigma_i')$$
$$\leq n_i \cdot \sum_{j=1}^{i-1}(3\gamma \cdot L_j + d(s_j, s_{j+1})) + 3\beta \cdot \mathsf{Opt}_i$$
$$\leq n_i \cdot \sum_{j=1}^{i-1}(3\gamma \cdot d(\tau_j) + d(s_j, s_{j+1})) + 3\beta\mathsf{Opt}_i$$
$$\leq 3\gamma n_i \cdot \sum_{j=1}^{i-1}(d(\tau_j) + d(s_j, s_{j+1})) + 3\beta\mathsf{Opt}_i$$
$$= 3\gamma n_i \cdot \tilde{D}_i + 3\beta\mathsf{Opt}_i$$
$$\leq 3(\gamma + \beta)(n_i \cdot \tilde{D}_i + \mathsf{Opt}_i)$$

So the total latency of π is at most $3(\gamma+\beta)\sum_{i=1}^{l}(n_i \cdot \tilde{D}_i + \mathsf{Opt}_i) \leq O(\gamma+\beta) \cdot \mathsf{Opt}$.

Theorem 5 now follows.

References

1. Blum, A., Chawla, S., Karger, D.R., Lane, T., Meyerson, A., Minkoff, M.: Approximation Algorithms for Orienteering and Discounted-Reward TSP. In: FOCS, pp. 46–55 (2003)
2. Blum, A., Chalasani, P., Coppersmith, D., Pulleyblank, W.R., Raghavan, P., Sudan, M.: The minimum latency problem. In: STOC, pp. 163–171 (1994)

3. Chaudhuri, K., Godfrey, B., Rao, S., Talwar, K.: Paths, Trees, and Minimum Latency Tours. In: FOCS, pp. 36–45 (2003)
4. Chekuri, C., Korula, N., Pal, M.: Improved Algorithms for Orienteering and Related Problems. In: SODA, pp. 661–670 (2008)
5. Chekuri, C., Pal, M.: An $O(\log n)$ Approximation Ratio for the Asymmetric Traveling Salesman Path Problem. Theory of Computing 3, 197–209 (2007)
6. Papadimitriou, C.H., Papageorgiou, G., Papakostantinou, N., Afrati, F., Cosmadakis, S.: The complexity of the traveling repairman problem. Informatique Theorique et Applications 20, 79–87 (1986)
7. Feige, U., Singh, M.: Improved Approximation Ratios for Traveling Salesperson Tours and Paths in Directed Graphs. In: APPROX-RANDOM, pp. 104–118 (2007)
8. Frank, A.: On Connectivity properties of Eulerian digraphs. Annals of Discrete Mathematics 41, 179–194 (1989)
9. Frieze, A., Galbiati, G., Maffioli, F.: On the worst-case performance of some algorithms for the asymmetric travelling salesman problem. Networks 12, 23–39 (1982)
10. Goemans, M., Kleinberg, J.: An improved approximation ratio for the minimum latency problem. Mathematical Programming 82, 111–124 (1998)
11. Jackson, B.: Some remarks on arc-connectivity, vertex splitting, and orientation in digraphs. Journal of Graph Theory 12(3), 429–436 (1988)
12. Kleinberg, J., Williamson, D.: Unpublished note (1998)
13. Lam, F., Newman, A.: Traveling salesman path problems. Mathematical Programming (2006) (online)
14. Minieka, E.: The delivery man problem on a tree network. Annals of Operations Research 18, 261–266 (1989)
15. Nagarajan, V., Ravi, R.: Poly-logarithmic approximation algorithms for directed vehicle routing problems. In: APPROX-RANDOM, pp. 257–270 (2007)
16. Nemhauser, G.L., Wolsey, L.A.: Integer and Combinatorial Optimization (1999)
17. Sahni, S., Gonzalez, T.: P-complete approximation problems. Journal of the ACM 23, 555–565 (1976)
18. Sviridenko, M.: Makespan Minimization in No-Wait Flow Shops: A Polynomial Time Approximation Scheme. SIAM J. Discret. Math. 16(2), 313–322 (2003)
19. Williamson, D.: Analysis of the held-karp heuristic for the traveling salesman problem. Master's thesis, MIT Computer Science (1990)
20. Wismer, D.A.: Solution of the flowshop sheduling problem with no intermediate queues. Operations Research 20, 689–697 (1972)

A Simple LP Relaxation for the Asymmetric Traveling Salesman Problem

Thành Nguyen[*]

Cornell University,
Center for Applies Mathematics
657 Rhodes Hall, Ithaca, NY, 14853, USA
thanh@cs.cornell.edu

Abstract. A long-standing conjecture in Combinatorial Optimization is that the integrality gap of the Held-Karp LP relaxation for the Asymmetric Traveling Salesman Problem (ATSP) is a constant. In this paper, we give a simpler LP relaxation for the ASTP. The integrality gaps of this relaxation and of the Held-Karp relaxation are within a constant factor of each other. Our LP is simpler in the sense that its extreme solutions have at most $2n - 2$ non-zero variables, improving the bound $3n - 2$ proved by Vempala and Yannakakis for the extreme solutions of the Held-Karp LP relaxation. Moreover, more than half of these non-zero variables can be rounded to integers while the total cost only increases by a constant factor.

We also show that given a partially rounded solution, in an extreme solution of the corresponding LP relaxation, at least one positive variable is greater or equal to 1/2.

Keywords: ATSP, LP relaxation.

1 Introduction

The Traveling Salesman Problem (TSP) is a classical problem in Combinatorial Optimization. In this problem, we are given an undirected or directed graph with nonnegative costs on the edges, and we need to find a Hamiltonian cycle of minimum cost. A Hamiltonian cycle is a simple cycle that covers all the nodes of the graph. It is well known that the problem is in-approximable for both undirected and directed graphs. A more tractable version of the problem is to allow the solution to visit a vertex/edge more than once if necessary. The problem in this version is equivalent to the case when the underlying graph is a complete graph, and the edge costs satisfy the triangle inequality. This problem is called the metric-TSP, more specifically Symmetric-TSP (STSP) or Asymmetric-TSP (ATSP) when the graph is undirected or directed, respectively. In this paper, we consider the ATSP.

---

[*] Part of this work was done while the author visited the Egerváry Research Group on Combinatorial Optimization (EGRES), Budapest, Hungary.

A. Goel et al. (Eds.): APPROX and RANDOM 2008, LNCS 5171, pp. 207–218, 2008.
© Springer-Verlag Berlin Heidelberg 2008

Notation. In the rest of the paper, we need the following notation. Given a directed graph $G = (V, E)$ and a set $S \subset V$, we denote the set of edges going in and out of S by $\delta^+(S)$ and $\delta^-(S)$, respectively. Let x be a nonnegative vector on the edges of the graph G, the in-degree or out-degree of S (with respect to x) is the sum of the value of x on $\delta^+(S)$ and $\delta^-(S)$. We denote them by $x(\delta^+(S))$ and $x(\delta^-(S))$.

An LP relaxation of the ATSP was introduced by Held and Karp [9] in 1970. It is usually called the Held-Karp relaxation. Since then it has been an open problem to show whether this relaxation has a constant integrality gap. The Held-Karp LP relaxation can have many equivalent forms, one of which requires a solution $x \in \mathbb{R}_+^{|E|}$ to satisfy the following two conditions: *i)* the in-degree and the out-degree of every vertex are at least 1 and equal to each other, and *ii)* the out-degree of every subset $S \subset V - \{r\}$ is at least 1, where r is an arbitrary node picked as a root. Fractional solutions of this LP relaxation are found to be hard to round because of the combination of the degree conditions on each vertex and the connectivity condition. A natural question is to relax these conditions to get an LP whose solutions are easier to round. In fact, when these conditions are considered separately, their LP forms integral polytopes, thus the optimal solution can be found in polynomial time. However, the integrality gap of these LPs with respect to the integral solutions of the ATSP can be arbitrarily large. Another attempt is to keep the connectivity condition and relax the degree condition on each vertex. It is shown recently by Lau et al. [15] that one can find an integral solution whose cost is at most a constant times the cost of the LP described above, furthermore it satisfies the connectivity condition and violates the degree condition at most a constant. The solution found does not satisfy the balance condition on the vertices, and such a solution can be very far from a solution of the ATSP.

Generally speaking, there is a trade-off in writing an LP relaxation for a discrete optimization problem: between having "simple enough" LP to round and a "strong enough" one to prove an approximation guarantee. It is a major open problem to show how strong the Held-Karp relaxation is. And, as discussed above, it seems that all the simpler relaxations can have arbitrarily big integrality gaps. In this paper, we introduce a new LP relaxation of the ATSP which is as strong as the Help-Karp relaxation up to a constant factor, and is simpler. Our LP is simpler in the sense that an extreme solution of this LP has at most $2n - 2$ non-zero variables, improving the bound $3n - 2$ on the extreme solutions of the Held-Karp relaxation. Moreover, out of such $2n - 2$ variables, at least n can be rounded to integers. This result shows that the integrality gap of the Held-Karp relaxation is a constant if and only if our *simpler* LP also has a constant gap.

The new LP. The idea behind our LP formulation is the following. Consider the Held-Karp relaxation in one of its equivalent forms:

$$\min \ c_e x_e$$

$$\text{Sbjt:} \quad x(\delta^+(S)) \geq 1 \ \forall S \subset V - \{r\} \quad (\textit{Connectivity condition})$$
$$x(\delta^+(v)) = x(\delta^-(v)) \ \forall v \in V \quad (\textit{Balance condition}) \tag{1}$$
$$x_e \geq 0.$$

Our observation is that because of the balance condition in the LP above, the in-degree $x(\delta^+(S))$ is equal to the out-degree $x(\delta^-(S))$ for every set S . If one can guarantee that the ratio between $x(\delta^+(S))$ and $x(\delta^-(S))$ is bounded by a constant, then using a theorem of A. J. Hoffman [10] about the condition for the existence of a circulation in a network, we can still get a solution satisfying the balance condition for every node with only a constant factor loss in the total cost. The interesting fact is that when allowed to relax the balance condition, we can combine it with the connectivity condition in a single constraint. More precisely, consider the following fact. Given a set $S \subset V - \{r\}$, the balance condition implies $x(\delta^+(S)) - x(\delta^-(S)) = 0$, and the connectivity condition is $x(\delta^+(S)) \geq 1$. Adding up these two conditions, we have:

$$2x(\delta^+(S)) - x(\delta^-(S)) \geq 1.$$

Thus we can have a valid LP consisting of these inequalities for all $S \subset V - \{r\}$ and two conditions on the in-degree and out-degree of r. Observe that given a vector $x \geq 0$, the function $f(S) = 2x(\delta^+(S)) - x(\delta^-(S))$ is a submodular function, therefore, we can apply the uncrossing technique as in [11] to investigate the structure of an extreme solution. We introduce the following LP:

$$
\begin{aligned}
\min \ & c_e x_e \\
\text{Subject to: } & 2x(\delta^+(S)) - x(\delta^-(S)) \geq 1 \ \forall S \subset V - \{r\} \\
& x(\delta^+(r)) = x(\delta^-(r)) = 1 \\
& x_e \geq 0.
\end{aligned}
\tag{2}
$$

This LP has exponentially many constraints. But because $2x(\delta^+(S)) - x(\delta^-(S))$ is a submodular function, the LP can be solved in polynomial time via the ellipsoid method and a subroutine to minimize submodular setfunctions.

Our results. It is not hard to see that our new LP (2) is weaker than the Held-Karp relaxation (1). In this paper, we prove the following result in the reverse direction. Given a feasible solution x of (2), in polynomial time we can find a solution y feasible to (1) on the support of x such that the cost of y is at most a constant factor of the cost of x. Furthermore, if x is integral then y can be chosen to be integral as well. Thus, given an integral solution of (2) we can find a Hamiltonian cycle of a constant approximate cost. This also shows that the integrality gaps of these two LPs are within a constant factor of each other. In section 3, we show that our new LP is simpler than the Held-Karp relaxation. In particular, we prove that an extreme solution of the new LP has at most $2n - 2$ non-zero variables, improving the bound $3n - 2$ proved by Vempala and Yannakakis [17] for the extreme solutions of the Held-Karp relaxation. We then show how to round at least n variables of a fractional solution of (2) to integers. And finally, we prove the existence of a big fractional variable in an extreme point of our LP in a partially rounded instance.

Note that one can have a more general LP relaxation by adding the Balance Condition and the Connectivity Condition in (1) with some positive coefficient

(a, b) to get: $(a + b)x(\delta^+(S)) - bx(\delta^-(S)) \geq a$. All the results will follow, except that the constants in these results depend on a and b. One can try to find a and b to minimize these constants. But, to keep this extended abstract simple, we only consider the case where $a = b = 1$.

Related Work. The Asymmetric TSP is an important problem in Combinatorial Optimization. There is a large amount of literature on the problem and its variants. See the books [8], [16] for references and details. A natural LP relaxation was introduced by Held-Karp [9] in 1970, and since then many works have investigated this LP in many aspects. See [8] for more details. Vempala and Yannakakis [17] show a sparse property of an extreme solution of the Held-Karp relaxation. Carr and Vempala [4] investigated the connection between the Symmetric TSP (STSP) and the ATSP. They proved that if a certain conjecture on STSP is true then the integrality gap of this LP is bounded by 4/3. Charikar et al. [3] later refuted this conjecture by showing a lower bound of 2 for the integrality gap of the Held-Karp LP, this is currently the best known lower bound. On the algorithmic side, a $\log_2 n$ approximation algorithm for the ATSP was first proved by Frieze et al. [6]. This ratio is improved slightly in [2], [12]. The best ratio currently known is $0.842 \log_2 n$ [12].

Some proofs of our results are based on the uncrossing technique, which was first used first by László Lovász [5] in a mathematical competition for university students in Hungary. The technique was later used successfully in Combinatorial Optimization. See the book [13] for more details. In Approximation Algorithms, the uncrossing technique was applied to the the generalized Steiner network problem by Kamal Jain [11]. And it is recently shown to be a useful technique in many other settings [7,14,15].

2   The Integrality Gaps of the New LP and of the Held-Karp Relaxation Are Essentially the Same

In this section, we prove that our LP and the Held-Karp relaxation have integrality gaps within a constant factor of each other. We also show that given an integral solution of the new LP (2), one can find a Hamilton cycle while only increasing the cost by a constant factor.

Theorem 1. *Given a feasible solution of the Held-Karp relaxation (1), we can find a feasible solution of (2) with no greater cost. Conversely, if x is a solution of (2) then there is a feasible solution y of (1) on the support of x, whose cost is at most a constant times the cost of x. Moreover, such a y can be found in polynomial time, and if x is integral then y can also be chosen to be an integral vector.*

The second part of this theorem is the technical one. As discussed in the introduction, at the heart of our result is the following theorem of Alan Hoffman [10] about the condition for the existence of a circulation in a network.

Lemma 1 (Hoffman). *Consider the LP relaxation of a circulation problem on a directed graph $G = (V, E)$ with lower and an upper bounds $l_e \leq u_e$ on each edge $e \in E$:*

$$x(\delta^+(v)) = x(\delta^-(v)) \;\; \forall v \in V$$
$$l_e \leq x_e \leq u_e \;\; \forall e \in E. \tag{3}$$

The LP is solvable if and only if for every set S:

$$\sum_{e \in \delta^+(S)} l_e \leq \sum_{e \in \delta^-(S)} u_e.$$

Furthermore, if l_e, u_e are integers, then the solution can be chosen to be integral.

$\square$

Given a solution x of our new LP, we use it to set up a circulation problem. Then, using Lemma 1, we prove that there exists a solution y to this circulation problem. And this vector is a feasible solution of the Help-Karp relaxation. Before proving Theorem 1, we need the following lemmas:

Lemma 2. *For a solution x of (2) and every set $S \subsetneq V$, the in-degree $x(\delta^+(S))$ and the out-degree $x(\delta^-(S))$ are at least $\frac{1}{3}$.*

Proof. Because of symmetry, we assume that $r \notin S$. Since x is a solution of (2), we have: $2x(\delta^+(S)) - x(\delta^-(S)) \geq 1$. This implies $x(\delta^+(S)) \geq \frac{1}{2} + \frac{1}{2}x(\delta^-(S)) > \frac{1}{3}$.

We now prove that $x(\delta^-(S)) \geq \frac{1}{3}$. If $S = V - \{r\}$, then because of the LP (2) the out-degree of S is 1, which is of course greater than $\frac{1}{3}$. Now, assume S is a real subset of $V - \{r\}$, let $T = V - \{r\} - S \neq \emptyset$.

To make the formula easy to follow, we use the following notation. Let α, β be the total value of the edges going from S to r and T respectively. See Figure 1. Thus, the out-degree of S is $x(\delta^-(S)) = \alpha + \beta$. We denote the total value of the edges going from r to S by a, and the total value of edges from T to S by b. Due to (2), the in-degree and out-degree of r is 1, therefore the total value of edges from r to T is $1 - a$ and from T to r is $1 - \alpha$.

Now, from $2x(\delta^+(T)) - x(\delta^-(T)) \geq 1$, we have $2(1 - a + \beta) - (1 - \alpha + b)$. Therefore

$$2\beta + \alpha \geq 2a + b.$$

And $2x(\delta^+(S)) - x(\delta^-(S)) \geq 1$ is equivalent to $2(a + b) - (\alpha + \beta) \geq 1$, which implies

$$(a + b) \geq \frac{(\alpha + \beta) + 1}{2}.$$

Combine these two inequalities:

$$2\beta + \alpha \geq 2a + b \geq a + b \geq \frac{\alpha + \beta + 1}{2}.$$

Thus we have $2\beta + \alpha \geq \frac{\alpha + \beta + 1}{2}$. From this, $4\beta + 2\alpha \geq \alpha + \beta + 1$ and $3\beta + \alpha \geq 1$. Hence, $3(\beta + \alpha) \geq 3\beta + \alpha \geq 1$. Therefore

$$\alpha + \beta \geq \frac{1}{3}.$$

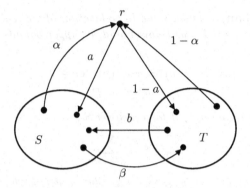

Fig. 1. Out-degree and in-degree of the set S

This inequality is what we need to prove.   □

The next lemma shows that for any S, the ratio between its out-degree and in-degree is bounded by a constant.

Lemma 3. *Given a solution x of (2), for any $S \subsetneq V$,*

$$\frac{1}{8}x(\delta^-(S)) \leq x(\delta^+(S)) \leq 8x(\delta^-(S)).$$

Proof. Because of symmetry, we can assume that $r \notin S$. From the inequality $2x(\delta^+(S)) - x(\delta^-(S)) \geq 1$ we have:

$$x(\delta^-(X)) < 2x(\delta^+(X)). \text{ Therefore } \frac{1}{8}x(\delta^-(S)) \leq x(\delta^+(S)).$$

To show the second inequality, we observe that when $S = V - \{r\}$, its out-degree is equal to its in-degree, thus we can assume that S is a real subset of $V - \{r\}$. As in the previous lemma, let $T = V - S - \{r\} \neq \emptyset$. We then apply the inequality $2x(\delta^+(T)) - x(\delta^-(T)) \geq 1$ to get the desired inequality.

First, observe that S and T are almost complements of each other, except that there is a node r with in and out degrees of 1 outside S and T. Thus, the out-degree of S is almost the same as the in-degree of T and vice versa. More precisely, using the same notation as in the previous lemma, one has $x(\delta^+(S)) - x(\delta^-(T)) = a - (1 - \alpha) \leq 1$. Therefore $x(\delta^+(S)) \leq x(\delta^-(T)) + 1$.

By symmetry, we also have: $x(\delta^+(T)) \leq x(\delta^-(S)) + 1$.

Now, $2x(\delta^+(T)) - x(\delta^-(T)) \geq 1$ implies:

$$1 + x(\delta^-(T)) \leq 2x(\delta^+(T)).$$

Using the relations between the in/out-degrees of S and T, we have the following:

$$x(\delta^+(S)) \leq 1 + x(\delta^-(T)) \leq 2x(\delta^+(T)) \leq 2(x(\delta^-(S)) + 1).$$

But because of the previous lemma, $x(\delta^-(S)) \geq \frac{1}{3}$. Therefore

$$x(\delta^+(S)) \leq 2(x(\delta^-(S)) + 1) \leq 8x(\delta^-(S)).$$

This is indeed what we need to prove.   □

Note: We believe the constant in this lemma can be reduced if we use a more careful analysis.

We are now ready to prove our main theorem:

Proof (Proof of Theorem 1). First, given a solution y, if $y(\delta^+(r)) = y(\delta^-(r)) = 1$, then y is also a feasible solution of (2). When $y(\delta^+(r)) = y(\delta^-(r)) \geq 1$, we can short-cut the fractional tour to get the solution satisfying the degree constraint on r: $y(\delta^+(r)) = y(\delta^-(r)) = 1$ without increasing the cost. This solution is a feasible solution of (2).

We now prove the second part of the theorem. Given a solution x of (2), consider the following circulation problem:

$$\min\ c_e y_e$$
$$sbt.\ \ y(\delta^+(v)) = y(\delta^-(v)) \forall v \in V$$
$$3x_e \leq y_e \leq 24x_e.$$

For every set $S \subset V$, Lemma 2 states that the ratio between its in-degree and its out-degree is bounded by 8. Therefore

$$\sum_{e \in \delta^+(S)} 3x_e \leq \sum_{e \in \delta^-(S)} 24x_e.$$

Using Lemma (1), the above LP has a solution y, and y can be chosen to be integral if x is integral. We need to show that y is a feasible solution of the Held-Karp relaxation. y satisfies the Balance Constraint on every node, thus we only need to show the Connectivity Condition. Because $y \geq 3x$, for every cut S we have :

$$y(\delta^+(S)) \geq 3x(\delta^+(S)) \geq 1.$$

The last inequality comes from Lemma 2. We have shown that given a feasible solution x of the new LP, there exists a feasible solution of the Held-Karp relaxation 1 whose cost is at most 24 times the cost of x. This completes the proof of our theorem.   □

3   Rounding an Extreme Solution of the New LP

In this section, we show that an extreme solution of our LP contains at most $2n - 2$ non-zero variables (Theorem 2). And at least n variables of this solution can be rounded to integers (Theorem 3). Finally, given a partially rounded solution, let x be an extreme solution of the new LP for this instance. We show that among the other positive variables, there is at least one with a value greater or equal to $1/2$ (Theorem 4).

Theorem 2. *The LP (2) can be solved in polynomial time, and an extreme solution has at most $2n - 2$ non-zero variables.*

Proof. First, observe that given a vector $x \geq 0$, $f_x(S) = 2x(\delta^+(S)) - x(\delta^-(S))$ is a submodular function. To prove this, one needs to check that $f_x(S) + f_x(T) \geq f_x(S \cup T) + f_x(S \cap T)$. Or more intuitively: $f_x(S) = x(\delta^+(S)) + (x(\delta^+(S)) - x(\delta^-(S)))$ is a sum of two submodular functions, thus f_x is also a submodular function.

The constraints in our LP is $f_x(S) \geq 1 \forall S \subset V - \{r\}$ and $x(\delta^+(r)) = x(\delta^-(r)) = 1$. Thus with a subroutine to minimize a submodular function, we can decide whether a vector x is feasible to our LP, and therefore the LP can be solved in polynomial time by the ellipsoid method.

Now, assume x is an extreme solution. Let S, T be two tight sets, i.e., $f_x(S) = f_x(T) = 1$. Then, it is not hard to see that if $S \cup T \neq \emptyset$ then both $S \cup T$ and $S \cap T$ are tight. Furthermore, the constraint vectors corresponding to $S, T, S \cup T, S \cap T$ are dependent. Now, among all the tight sets, take the maximal laminar set family. The constraints corresponding to these sets span all the other tight constraints. Thus x is defined by 2 constraints for the root node r and the constraints corresponding to a laminar family of sets on $n - 1$ nodes, which contains at most $2(n-1) - 1$ sets. However, the constraint corresponding to the set $V - r$ is dependent on the two constraints of the node r, therefore we have at most $2 + 2(n-1) - 1 - 1 = 2n - 2$ independent constraints. This shows that x has at most $2n - 2$ non-zero variables. □

We prove the next theorem about rounding at least n variables of a fractional solution of our new LP (2).

Theorem 3. *Given an extreme solution x of (2), we can find a solution $\tilde{x}$ on the support of x. Thus $\tilde{x}$ contains at most $2n - 2$ non-zero edges such that it satisfies the constraint $2\tilde{x}(\delta^+(S)) - \tilde{x}(\delta^-(S)) \geq 1 \quad \forall S \subset V - \{r\}$, and it has at least n non-zero integral variables. Furthermore, the cost of $\tilde{x}$ is at most a constant times the cost of x.*

Proof. x is a solution of (2). Due to Theorem 1, on the support of x, there exists a solution y of (1) whose cost is at most a constant times the cost of x. Because y satisfies $y(\delta^+(v)) = y(\delta^-(v)) \geq 1$ for every $v \in V$, y is a fractional cycle cover on the support of x. Recall that a cycle cover on directed graph is a Eulerian subgraph (possibly with parallel edges) covering all the vertices. However, we can find an integral cycle cover in a directed graph whose cost is at most the cost of a fractional solution. Let z be such an integral solution. Clearly, z has at least n non-zero variables, and the cost of z is at most the cost of y which is at most a constant times the cost of x.

Next consider the solution $w = x + \frac{3}{2}z$. For every edge e where $z_e > 0$, we have $w_e = x_e + \frac{3}{2}z_e > \frac{3}{2}$. Round w_e to the closest integer to get the solution $\tilde{x}$. Clearly, $\tilde{x}$ has at most $2n - 2$ non-zero variables and at least n non-zero integral variables. We will show that the cost of $\tilde{x}$ is at most a constant times the cost of x, and that $\tilde{x}$ satisfies $2\tilde{x}(\delta^+(S)) - \tilde{x}(\delta^-(S)) \geq 1 \forall S \subset V - \{r\}$.

Rounding each w_e to the closest integer will sometimes cause an increase in w_e of at most $1/2$. But, because we only round the value w_e when the corresponding $z_e \geq 1$, and note that z is an integral vector, the total increase is at most half the cost of z which is at most a constant times the cost of x.

Consider a set $S \subset V - \{r\}$. We have $2x(\delta^+(S)) - x(\delta^-(S)) \geq 1$. Let k be the total value of the edges of z going out from S, that is $k = z(\delta^+(S)) = z(\delta^-(S))$. This is true because z is a cycle cover. Hence, when adding $w := x + \frac{3}{2}z$, we have:

$$2w(\delta^+(S)) - w(\delta^-(S)) = 2x(\delta^+(S)) - x(\delta^-(S)) + \frac{3}{2}(2z(\delta^+(S)) - z(\delta^-(S))).$$

Therefore,

$$2w(\delta^+(S)) - w(\delta^-(S)) \geq 1 + \frac{3}{2}k. \tag{4}$$

Now, $\tilde{x}$ is a rounded vector of w on the edges where z is positive. For the set S, there are at most $2k$ such edges, at most k edges going out and k edges coming in. Rounding each one to the closest integer will sometimes cause a change at most $\frac{1}{2}$ on each edge, and thus causes the change of $2w(\delta^+(S)) - w(\delta^-(S))$ in at most $k(2 \cdot \frac{1}{2} - (-\frac{1}{2})) = \frac{3}{2}k$. But, because of (4), we have :

$$2\tilde{x}(\delta^+(S)) - \tilde{x}(\delta^-(S)) \geq 1$$

which is what we need to show.   □

Our last theorem shows that there always exists a "large" variable in an extreme solution in which some variables are assigned fixed integers.

Theorem 4. *Consider the following LP which is the corresponding LP of (2) when some variables $x_e, e \in F$ are assigned fixed integral values. $x_e = a_e \in \mathbb{N}$ for $e \in F$.*

$$\begin{aligned}
\min \quad & c_e x_e \\
\text{sbt.} \quad & 2x(\delta^+(S)) - x(\delta^-(S)) \geq 1 \; \forall S : r \notin S. \\
& x(\delta^+(r)) = r_1 \\
& x(\delta^-(r)) = r_2 \quad (r_1, r_2 \in \mathbb{N}) \\
& x_e = a_e \; \forall e \in F \\
& x_e \geq 0.
\end{aligned} \tag{5}$$

Given an extreme solution x of this LP, let $H = \{e \in E - F | x_e > 0\}$. If $H \neq \emptyset$, then there exists an $e \in H$ such that $x_e \geq \frac{1}{2}$.

Proof. Let $\mathcal{L}$ be the laminar set family whose corresponding constraints together with two constraints on the root node r determine the value of $\{x_e | e \in H\}$ uniquely. As we have seen in the proof of Theorem 2, one can see that such an $\mathcal{L}$ exists, and $|\mathcal{L}|$ is at least $|H|$. Assume all the values in $\{x_e | e \in H\}$ is less than a half. We assign one token to each edge in H. If we can redistribute these tokens to the sets in $\mathcal{L}$ and the constraints on the root r such that each constraint gets at least 1 token, but at the end there are still some tokens remaining, we will get a contradiction to prove the theorem.

We apply the technique used in [11] and the other recent results [14], [15], [1]. For each $e \in H$, we distribute a fraction $1 - 2x_e$ of the token to the head of the edge, a fraction x_e of the token to the tail and the remaining x_e to the edge itself. See Figure 2. Because $0 < x_e < \frac{1}{2}$, all of these values are positive. Given a set S in $\mathcal{L}$, we now describe the set of tokens assigned to this set. First, we use the following notation: for a set T, let $E(T)$ be the set of edges in $\{x_e | e \in E - F\}$ that have both endings in T. Now, let $S_1, .., S_k \in \mathcal{L}$ be the maximal sets which are real subsets of S. The set of tokens that S gets is all the tokens on the edges in $E(S) - (E(S_1) \cup ... \cup E(S_k))$ plus the tokens on the vertices in $S - (S_1 \cup ... \cup S_k)$. Clearly, no tokens are assigned to more than one set. The constraint on the in-degree of r gets all the tokens on the heads of edges going into r, and the constraint on the in-degree of r gets all the tokens on the tails of edges going out from r.

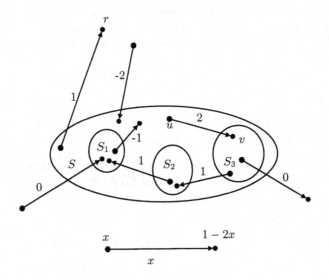

Fig. 2. Tokens distributed to S

Consider now the equalities corresponding to the set $S, S_1, ..., S_k$. If we add the equalities of $S_1, S_2, ..., S_k$ together and subtract the equality on the set S we will get an linear equality on the variables $\{x_e | e \in H\}$:

$$\sum_{e \in H} \alpha_e x_e = \text{an integer number.}$$

It is not hard to calculate α_e for each type of e. For example, if e connects S_i and S_j , $i \neq j$ then $\alpha_e = 1$, if e connects from vertex outside S to a vertex in $S - (S_1 \cup ... \cup S_k)$ then $\alpha_e = -2$, etc. See Figure 2 for all other cases.

On the other hand, if we calculate the amount of tokens assigned to the set S, it also has a linear formula on $\{x_e | e \in H\}$:

$$\sum_{e \in H} \beta_e x_e + \text{an integer number.}$$

We can also calculate the coefficient β_e for every e. For example, if e connects S_i and S_j , $i \neq j$ then the edge e is the only one that gives an amount of tokens which is a function of x_e, and it is exactly x_e. Thus $\beta_e = 1$. Consider another example, $e = u \to v$ where $u \in S - (S_1 \cup ... \cup S_k)$ and $v \in S_3$. See Figure 2. Then only the amounts of tokens on the edge uv and the node u depend on x_e. On the edge uv, it is x_e and, on the node u, it is x_e plus a value not depending on x_e. Thus $\beta_e = 2$ in this case.

It is not hard to see that the coefficient $\alpha_e = \beta_e \forall e \in H$. Thus the amount of tokens S gets is an integer number, and it is positive, thus it is at least 1. Similarly, one can show that this fact also holds for the constraints on the root node r.

We now show that there are some tokens that were not assigned to any set. Consider the biggest set in the laminar set family $\mathcal{L}$, it has some non-zero edges going in or out but the tokens on this edge is not assigned to any constraint. This completes the proof.   □

Acknowledgment. The author thanks Tamás Király, Éva Tardos and László Végh for numerous discussions.

References

1. Bansal, N., Khandekar, R., Nagarajan, V.: Additive Guarantees for Degree Bounded Directed Network Design. In: STOC 2008 (2008)
2. Bläser, M.: A new Approximation Algorithm for the Asymmetric TSP with Triangle Inequality. In: SODA 2002, pp. 638–645 (2002)
3. Charikar, M., Goemans, M.X., Karloff, H.J.: On the Integrality Ratio for Asymmetric TSP. In: FOCS 2004, pp. 101–107 (2004)
4. Carr, R., Vempala, S.: On the Held-Karp relaxation for the asymmetric and symmetric TSPs. Mathematical Programming 100(3), 569–587 (2004)
5. Frank, A.: Personal communication
6. Frieze, A., Galbiati, G., Maffioli, M.: On the worst-case performance of some algorithms for the asymmetric traveling salesman problem. Networks 12 (1982)
7. Goemans, M.X.: Minimum Bounded Degree Spanning Trees. In: FOCS 2006, pp. 273–282 (2006)
8. Gutin, G., Punnen, A.P. (eds.): Traveling Salesman Problem and Its Variations. Springer, Berlin (2002)
9. Held, M., Karp, R.M.: The traveling salesman problem and minimum spanning trees. Operation Research 18, 1138–1162 (1970)
10. Hoffman, A.J.: Some recent applications of the theory of linear inequalities to extremal combinatorial analysis. In: Proc. Symp. in Applied Mathematics, Amer. Math. Soc., pp. 113–127 (1960)
11. Jain, K.: A factor 2 approximation for the generalized Steiner network problem. Combinatorica 21, 39–60 (2001)
12. Kaplan, H., Lewenstein, M., Shafir, N., Sviridenko, M.: Approximation Algorithms for Asymmetric TSP by Decomposing Directed Regular Multidigraphs. In: Proc. of IEEE FOCS, pp. 56–67 (2003)
13. Schrijver, A.: Combinatorial Optimization - Polyhedra and Efficiency. Springer, Berlin (2003)

14. Lau, L.C., Naor, J., Salavatipour, M.R., Singh, M.: Survivable network design with degree or order constraints. In: STOC 2007, pp. 651–660 (2007)
15. Lau, L.C., Singh, M.: Approximating minimum bounded degree spanning trees to within one of optimal. In: STOC 2007, pp. 661–670 (2007)
16. Lawler, E., Lenstra, J.K., Rinnooy Kan, A.H.G., Shmoys, D. (eds.): The Traveling Salesman Problem: A Guided Tour of Combinatorial Optimization. John Wiley & Sons Ltd., Chichester (1985)
17. Vempala, S., Yannakakis, M.: A Convex Relaxation for the Asymmetric TSP. In: SODA 1999, pp. 975–976 (1999)

Approximating Directed Weighted-Degree Constrained Networks

Zeev Nutov

The Open University of Israel, Raanana, Israel
nutov@openu.ac.il

Abstract. Given a graph $H = (V, F)$ with edge weights $\{w(e) : e \in F\}$, the *weighted degree* of a node v in H is $\sum\{w(vu) : vu \in F\}$. We give bicriteria approximation algorithms for problems that seek to find a minimum cost *directed* graph that satisfies both *intersecting supermodular* connectivity requirements and *weighted degree* constraints. The input to such problems is a directed graph $G = (V, E)$, edge-costs $\{c(e) : e \in E\}$, edge-weights $\{w(e) : e \in E\}$, an intersecting supermodular set-function f on V, and degree bounds $\{b(v) : v \in V\}$. The goal is to find a minimum cost f-connected subgraph $H = (V, F)$ (namely, at least $f(S)$ edges in F enter every $S \subseteq V$) of G with weighted degrees $\leq b(v)$. Our algorithm computes a solution of cost $\leq 2 \cdot \mathsf{opt}$, so that the weighted degree of every $v \in V$ is at most: $7b(v)$ for arbitrary f and $5b(v)$ for a $0, 1$-valued f; $2b(v) + 4$ for arbitrary f and $2b(v) + 2$ for a $0, 1$-valued f in the case of unit weights. Another algorithm computes a solution of cost $\leq 3 \cdot \mathsf{opt}$ and weighted degrees $\leq 6b(v)$. We obtain similar results when there are both indegree and outdegree constraints, and better results when there are indegree constraints only: a $(1, 4)$-approximation algorithm for arbitrary weights and a polynomial time algorithm for unit weights. Finally, we consider the problem of packing maximum number k of edge-disjoint arborescences so that their union satisfies weighted degree constraints, and give an algorithm that computes a solution of value at least $\lfloor k/36 \rfloor$.

1 Introduction

1.1 Problem Definition

In many Network Design problems one seeks to find a low-cost subgraph H of a given graph G that satisfies prescribed connectivity requirements. Such problems are vastly studied in Combinatorial Optimization and Approximation Algorithms. Known examples are Min-Cost k-Flow, b-Edge-Cover, Min-Cost Spanning Tree, Traveling Salesperson, directed/undirected Steiner Tree, Steiner Forest, k-Edge/Node-Connected Spanning Subgraph, and many others. See, e.g., surveys in [16,4,8,10,12].

In Degree Constrained Network Design problems, one seeks the cheapest subgraph H of a given graph G that satisfies both prescribed connectivity requirements and degree constraints. One such type of problems are the matching/ edge-cover problems, which are solvable in polynomial time, c.f., [16]. For other

A. Goel et al. (Eds.): APPROX and RANDOM 2008, LNCS 5171, pp. 219–232, 2008.

degree constrained problems, even checking whether there exists a feasible solution is NP-complete, hence one considers bicriteria approximation when the degree constraints are relaxed.

The connectivity requirements can be specified by a set function f on V, as follows.

Definition 1. *For an edge set or a graph H and node set S let $\delta_H(S)$ $(\delta_H^{in}(S))$ denote the set of edges in H leaving (entering) S. Given a set-function f on subsets of V and a graph $H = (V, F)$, we say that H is f-connected if*

$$|\delta_H^{in}(S)| \geq f(S) \quad \text{for all } S \subseteq V. \tag{1}$$

Several types of f are considered in the literature, among them the following known one:

Definition 2. *A set function f on V is* intersecting supermodular *if for any $X, Y \subseteq V$, $X \cap Y \neq \emptyset$*

$$f(X) + f(Y) \leq f(X \cap Y) + f(X \cup Y) . \tag{2}$$

We consider *directed* network design problems with *weighted-degree* constraints. For simplicity of exposition, we will consider mainly out-degree constraints, but our results easily extend to the case when there are also in-degree constraints, see Section 6. The problem we consider is:

Directed Weighted Degree Constrained Network (DWDCN)
Instance: A directed graph $G = (V, E)$, edge-costs $\{c(e) : e \in E\}$, edge-weights $\{w(e) : e \in E\}$, set-function f on V, and degree bounds $\{b(v) : v \in V\}$.
Objective: Find a minimum cost f-connected subgraph $H = (V, F)$ of G that satisfies the weighted degree constraints

$$w(\delta_H(v)) \leq b(v) \quad \text{for all } v \in V . \tag{3}$$

We assume that f admits a polynomial time evaluation oracle. Since for most functions f even checking whether DWDCN has a feasible solution is NP-complete, we consider bicriteria approximation algorithms. Assuming that the problem has a feasible solution, an (α, β)-approximation algorithm for DWDCN either computes an f-connected subgraph $H = (V, F)$ of G of cost $\leq \alpha \cdot$ opt that satisfies $w(\delta_H(v)) \leq \beta \cdot b(v)$ for all $v \in V$, or correctly determines that the problem has no feasible solution. Note that even if the problem does not have a feasible solution, the algorithm may still return a subgraph that violates the degree constraints (3) by a factor of β.

A graph H is k-*edge-outconnected from* r if it has k-edge-disjoint paths from r to any other node. DWDCN includes as a special case the Weighted Degree Constrained k-Outconnected Subgraph problem, by setting $f(S) = k$ for all $\emptyset \neq S \subseteq V - r$, and $f(S) = 0$ otherwise. For $k = 1$ we get the Weighted Degree Constrained Arborescence problem. We also consider the problem of packing maximum number k of edge-disjoint arborescences rooted at r so that their union H satisfies (3). By Edmond's Theorem, this is equivalent to requiring that H is k-edge-otconnected from r and satisfies (3). This gives the following problem:

Weighted Degree Constrained Maximum Arborescence Packing (WDCMAP)
Instance: A directed graph $G = (V, E)$, edge-weights $\{w(e) : e \in E\}$, degree bounds $\{b(v) : v \in V\}$, and $r \in V$.
Objective: Find a k-edge-outconnected from r spanning subgraph $H = (V, F)$ of G that satisfies the degree constraints (3) so that k is maximum.

1.2   Our Results

Our main results are summarized in the following theorem. For an edge set I, let $x(I) = \sum_{e \in I} x(e)$. Let opt denote the optimal value of the following natural LP-relaxation for DWDCN that seeks to minimize $c \cdot x$ over the following polytope P_f:

$$x(\delta_E^{in}(S)) \geq f(S) \qquad \text{for all } \emptyset \neq S \subset V$$

$$\sum_{e \in \delta_E(v)} x(e)w(e) \leq b(v) \qquad \text{for all } v \in V$$

$$0 \leq x(e) \leq 1 \qquad \text{for all } e \in E$$

Theorem 1. DWDCN *with intersecting supermodular f admits a polynomial time algorithm that computes an f-connected graph of cost $\leq 2\mathsf{opt}$ so that the weighted degree of every $v \in V$ is at most: $7b(v)$ for arbitrary f and $5b(v)$ for a $0, 1$-valued f; for unit weights, the degree of every $v \in V$ is at most $2b(v) + 4$ for arbitrary f and $2b(v) + 2$ for a $0, 1$-valued f. The problem also admits a $(3, 6)$-approximation algorithm for arbitrary weights and arbitrary intersecting supermodular f.*

Interistingly, we can show a much better result for the case of indegree constraints only (for the case of both indegree and outdegree constraints see Section 6).

Theorem 2. DWDCN *with intersecting supermodular f and with indegree constraints only, admits a $(1, 4)$-approximation algorithm for arbitrary weights, and a polynomial time algorithm for unit weights.*

Theorem 1 has several applications. Bang-Jensen, Thomassé, and Yeo [1] conjectured that every k-edge-connected directed graph $G = (V, E)$ contains a spanning arborescence H so that $|\delta_H(v)| \leq |\delta_G(v)|/k + 1$ for every $v \in V$. Bansal, Khandekar, and Nagarajan [2] proved that even if G is only k-edge-outconnected from r, then G contains such H so that $|\delta_H(v)| \leq |\delta_G(v)|/k + 2$. We prove that for any $\ell \leq k$, G contains an ℓ-outconnected from r spanning subgraph H which cost and weighted degrees are not much larger than the "expected" values $c(G) \cdot (\ell/k)$ and $w_G(v) \cdot (\ell/k)$. In particular, one can find an arborescence with both low weighted degrees and low cost.

Corollary 1. *Let $H_k = (V, F)$ be a k-outconnected from r directed graph with costs $\{c(e) : e \in F\}$ and weights $\{w(e) : e \in F\}$. Then for any $\ell \leq k$ the graph H_k contains an ℓ-outconnected from r subgraph H_ℓ so that $c(H_\ell) \leq c(H_k) \cdot (2\ell/k)$ and so that for all $v \in V$: $w(\delta_{H_\ell}(v)) \leq w(\delta_{H_k}(v)) \cdot (7\ell/k)$, and $w(\delta_{H_\ell}(v)) \leq w(\delta_{H_k}(v)) \cdot (5/k)$ for $\ell = 1$; for unit weights, $|\delta_{H_\ell}(v)| \leq |\delta_{H_k}(v)| \cdot (2\ell/k) + 2$. There also exists H_ℓ so that $c(H_\ell) \leq c(H_k) \cdot (3\ell/k)$ and $w(\delta_{H_\ell}(v)) \leq w(\delta_{H_k}(v)) \cdot (6\ell/k)$ for all $v \in V$.*

Proof. Consider the Weighted Degree Constrained ℓ-Outconnected Subgraph problem on H_k with degree bounds $b(v) = w(\delta_{H_k}(v)) \cdot (\ell/k)$. Clearly, $x(e) = \ell/k$ for every $e \in F$ is a feasible solution of cost $c(H_k) \cdot (\ell/k)$ to the LP-relaxation $\min\{c \cdot x : x \in P_f\}$ where $f(S) = \ell$ for all $\emptyset \neq S \subseteq V - r$, and $f(S) = 0$ otherwise. By Theorem 1, our algorithm computes a subgraph H_ℓ as required.

Another application is for the WDCMAP problem. Ignoring costs, Theorem 1 implies a "pseudo-approximation" algorithm for WDCMAP that computes the maximum number k of packed arborescences, but violates the weighted degrees. E.g., using the $(3, 6)$-approximation algorithm from Theorem 1, we can compute a k-outconnected H that violates the weighted degree bounds by a factor of 6, where k is the optimal value to WDCMAP. Note that assuming P$\neq$NP, WDCMAP cannot achieve a $1/\rho$-approximation algorithm for any $\rho > 0$, since deciding whether $k \geq 1$ is equivalent to the Degree Constrained Arborescence problem, which is NP-complete. We can however show that if the optimal value k is not too small, then the problem does admit a constant ratio approximation.

Theorem 3. WDCMAP *admits a polynomial time algorithm that computes a feasible solution H that satisfies (3) so that H is $\lfloor k/36 \rfloor$-outconnected from r.*

Proof. The algorithm is very simple. We set $b'(v) \leftarrow b(v)/6$ for all $v \in V$ and apply the $(3, 6)$-approximation algorithm from Theorem 1. The degree of every node v in the subgraph computed is at most $6b'(v) \leq b(v)$, hence the solution is feasible. All we need to prove is that if the original instance admits a packing of size k, then the new instance admits a packing of size $\lfloor k/36 \rfloor$. Let H_k be an optimal solution to WDCMAP. Substituting $\ell = \lfloor k/36 \rfloor$ in the last statement of Corollary 1 and ignoring the costs we obtain that H_k contains a subgraph H_ℓ which is ℓ-outconnected from r so that $w(\delta_{H_\ell}(v)) \leq w(\delta_{H_k}(v)) \cdot (6\ell/k) \leq w(\delta_{H_k}(v))/6 \leq b(v)/6$ for all $v \in V$, as claimed.

We note that Theorem 3 easily extends to the case when edges have costs; the cost of the subgraph H computed is at most the minimum cost of a feasible k-outconnected subgraph.

1.3   Previous and Related Work

Fürer and Raghavachari [6] considered the problem of finding a spanning tree with maximum degree $\leq \Delta$, and gave an algorithm that computes a spanning tree of maximum degree $\leq \Delta + 1$. This is essentially the best possible since computing the optimum is NP-hard. A variety of techniques were developed in attempt to generalize this result to the minimum-cost case – the Minimum Degree Spanning Tree problem, c.f., [15,11,3]. Goemans [7] presented an algorithm that computes a spanning tree of cost $\leq$ opt and with degrees at most $b(v) + 2$ for all $v \in V$, where $b(v)$ is the degree bound of v. An optimal result was obtained by Singh and Lau [17]; their algorithm computes a spanning tree of cost $\leq$ opt and with degrees at most $b(v) + 1$ for all $v \in V$. The algorithm of Singh and Lau [17] uses the method of *iterative rounding*. This method was initiated in a seminal paper of Jain [9] that gave a 2-approximation algorithm for the Steiner Network problem. Without degree constraints, this method is as follows: given an optimal basic solution to an LP-relaxation for the problem, round at least one entry, and recurse on the residual instance. The algorithm of Singh and Lau [17] for the Minimum Bounded Degree Spanning Tree problem is a surprisingly simple extension – either round at least one entry, or remove a degree constraint from some node v. The non-trivial part usually is to prove that basic fractional solution have certain "sparse" properties.

For unit weights, the following results were obtained recently. Lau, Naor, Salvatipour, and Singh [13] were the first to consider general connectivity requirements. They gave a $(2, 2b(v)+3)$-approximation for undirected graphs in the case when f is skew-supermodular. For directed graphs, they gave a $(4\text{opt}, 4b(v)+6)$-approximation for intersecting supermodular f, and $(8\text{opt}, 8b(v)+6)$-approximation for *crossing supermodular* f (when (2) holds for any X, Y that cross). Recently, in the full version of [13], these ratios were improved to $(3\text{opt}, 3b(v)+5)$ for crossing supermodular f, and $(2\text{opt}, 2b(v) + 2)$ for $0, 1$-valued intersecting supermodular f. For the latter case we obtain the same ratio, but our proof is simpler than the one in the full version of [13].

Bansal, Khandekar, and Nagarajan [2] gave for intersecting supermodular f a $(\frac{1}{\varepsilon} \cdot \text{opt}, \lceil \frac{b(v)}{1-\varepsilon} \rceil + 4)$-approximation scheme, $0 \leq \varepsilon \leq 1/2$. They also showed, that this ratio cannot be much improved based on the standard LP-relaxation. For crossing supermodular f [2] gave a $(\frac{2}{\varepsilon} \cdot \text{opt}, \lceil \frac{b(v)}{1-\varepsilon} \rceil + 4 + f_{\max})$-approximation scheme. For the degree constrained arborescence problem (without costs) [2] give an algorithm that computes an arborescence H with $|\delta_H(v)| \leq b(v) + 2$ for all $v \in V$. Some additional results for related problems can also be found in [2].

For weighted degrees, Fukunaga and Nagamochi [5] considered *undirected* network design problems and gave a $(1, 4)$-approximation for minimum spanning trees and a $(2, 7)$-approximation algorithm for arbitrary weakly supermodular set-function f.

2   Proof of Theorem 1

During the algorithm, F denotes the partial solution, I are the edges to add to F, and B is the set of nodes on which the outdegree bounds constraints are still present. The algorithm starts with $F = \emptyset$, $B = V$ and performs iterations. In any iteration, we work with the "residual problem" polytope $P_f(I, F, B)$ ($\alpha \geq 1$ is a fixed parameter):

$$x(\delta_I^{in}(S)) \geq f(S) - |\delta_F^{in}(S)| \qquad \text{for all } \emptyset \neq S \subset V$$

$$\sum_{e \in \delta_I(v)} x(e)w(e) \leq b(v) - w(\delta_F(v))/\alpha \qquad \text{for all } v \in B$$

$$0 \leq x(e) \leq 1 \qquad \text{for all } e \in I$$

Recall some facts from polyhedral theory. Let x belong to a polytope $P \subseteq R^m$ defined by a system of linear inequalities; an inequality is *tight* (for x) if it holds as equality for x. $x \in P$ is a *basic solution* for (the system defining) P if there exist a set of m tight inequalities in the system defining P such that x is the unique solution for the corresponding equation system; that is, the corresponding m tight equations are linearly independent. It is well known that if $\min\{c \cdot x : x \in P\}$ has an optimal solution, then it has an optimal solution which is basic, and that a basic optimal solution for $\{c \cdot x : x \in P_f(I, F, B)\}$ can be computed in polynomial time, c.f., [13].

Note that if $x \in P_f(I, F, B)$ is a basic solution so that $0 < x(e) < 1$ for all $e \in I$, then every tight equation is induced by either:

- *cut constraint* $x(\delta_I^{in}(S)) \geq f(S) - |\delta_F^{in}(S)|$ defined by some set $\emptyset \neq S \subset V$ with $f(S) - |\delta_F^{in}(S)| \geq 1$.

- *degree constraint* $\sum_{e \in \delta_I(v)} x(e)w(e) \leq b(v) - w(\delta_F(v))/\alpha$ defined by some node $v \in B$.

A family $\mathcal{F}$ of sets is *laminar* if for every $S, S' \in \mathcal{F}$, either $S \cap S' = \emptyset$, or $S \subset S'$, or $S' \subset S$. We use the following statement observed in [13] for unit weights, which also holds in our setting.

Lemma 1. *For any basic solution x to $P_f(I, F, B)$ with $0 < x(e) < 1$ for all $e \in I$, there exist a laminar family $\mathcal{L}$ on V and $T \subseteq B$ such that x is the unique solution to the linear equation system:*

$$x(\delta_I^{in}(S)) = f(S) - |\delta_F^{in}(S)| \qquad \text{for all } S \in \mathcal{L}$$

$$\sum_{e \in \delta_I(v)} x(e)w(e) = b(v) - w(\delta_F(v))/\alpha \qquad \text{for all } v \in T$$

where $f(S) - |\delta_F^{in}(S)| \geq 1$ for all $S \in \mathcal{L}$. In particular, $|\mathcal{L}| + |T| = |I|$ and the characteristic vectors of $\delta_I^{in}(S)$ for all $S \in \mathcal{L}$ are linearly independent.

Proof. Let $\mathcal{F} = \{\emptyset \neq S \subset V : x(\delta_E^{in}(S)) = f(S) - |\delta_F^{in}(S)| \geq 1\}$, (i.e., the tight sets) and $T = \{v \in B : \sum_{e \in \delta_I(v)} x(e)w(e) = b(v) - w(\delta_F(v))/\alpha\}$ (i.e., the tight nodes in B). For $\mathcal{F}' \subseteq \mathcal{F}$ let $span(\mathcal{F}')$ denote the linear space generated by the characteristic vectors of $\delta_I^{in}(S)$, $S \in \mathcal{F}'$. Similarly, $span(T')$ is the linear space generated by the weight vectors of $\delta_I(v)$, $v \in T'$. In [9] (see also [14]) it is proved that a maximal laminar subfamily $\mathcal{L}$ of $\mathcal{F}$ satisfies $span(\mathcal{L}) = span(\mathcal{F})$. Since $x \in P_f(I, F, B)$ is a basic solution, and $0 < x(e) < 1$ for all $e \in I$, $|I|$ is at most the dimension of $span(\mathcal{F}) \cup span(T) = span(\mathcal{L}) \cup span(T)$. Hence repeatedly removing from T a node v so that $span(\mathcal{L}) \cup span(T - v) = span(\mathcal{L}) \cup span(T)$ results in $\mathcal{L}$ and T as required.

Definition 3. *The polytope $P_f(I, F, B)$ is (α, Δ)-sparse for integers $\alpha, \Delta \geq 1$ if any basic solution $x \in P_f(I, F, B)$ has an edge $e \in I$ with $x(e) = 0$, or satisfies at least one of the following:*

$$x(e) \geq 1/\alpha \quad \text{for some } e \in I \tag{4}$$

$$|\delta_I(v)| \leq \Delta \quad \text{for some } v \in B \tag{5}$$

We prove the following two general statements that imply Theorem 1:

Theorem 4. *If for any I, F the polytope $P_f(I, F, B)$ is (α, Δ)-sparse (if non-empty), then* DWDCN *admits an $(\alpha, \alpha + \Delta)$-approximation algorithm; for unit weights the algorithm computes a solution F so that $c(F) \leq \alpha \cdot \mathsf{opt}$ and $|\delta_F(v)| \leq \alpha b(v) + \Delta - 1$ for all $v \in V$.*

Theorem 5. *$P_f(I, F, B)$ is $(2, 5)$-sparse and $(3, 3)$-sparse for intersecting supermodular f; if f is $0, 1$-valued, then $P_f(I, F, B)$ is $(2, 3)$-sparse.*

3   The Algorithm (Proof of Theorem 4)

The algorithm perform iterations. Every iteration either removes at least one edges from I or at least one node from B. In the case of unit weights we assume that all the degree bounds are integers.

Algorithm for DWDCN **with intersecting supermodular** f
Initialization: $F \leftarrow \emptyset$, $B \leftarrow V$, $I \leftarrow E - \{vu \in E : w(vu) > b(v)\}$.
If $P_f(I, F, B) = \emptyset$, then return "UNFEASIBLE" and STOP.
While $I \neq \emptyset$ do:
    1. Find a basic solution $x \in P_f(I, F, B)$.
    2. Remove from I all edges with $x(e) = 0$.
    3. Add to F and remove from I all edges with $x(e) \geq 1/\alpha$.
    4. Remove from B every $v \in B$ with $|\delta_I(v)| \leq \Delta$.
EndWhile

Lemma 2. DWDCN *admits an $(\alpha, \alpha + \Delta)$-approximation algorithm if every polytope $P_f(I, F, B)$ constructed during the algorithm is (α, Δ)-sparse; furthermore, for unit weights, the algorithm computes a solution F so that $c(F) \leq \alpha \cdot \mathsf{opt}$ and $|\delta_F(v)| \leq \alpha b(v) + \Delta - 1$ for all $v \in V$.*

Proof. Clearly, if $P_f(I, F, B) = \emptyset$ at the beginning of the algorithm, then the problem has no feasible solution, and the algorithm indeed outputs "INFEA-SIBLE". It is also easy to see that if $P_f(I, F, B) \neq \emptyset$ at the beginning of the algorithm, then $P_f(I, F, B) \neq \emptyset$ throughout the subsequent iterations. Hence if the problem has a feasible solution, the algorithm returns an f-connected graph, and we need only to prove the approximation ratio. As for every edge added we have $x(e) \geq 1/\alpha$, the algorithm indeed computes a solution of cost $\leq \alpha \cdot \mathsf{opt}$.

Now we prove the approximability of the degrees. Consider a node $v \in V$. Let F' be the set of edges in $\delta_F(v)$ added to F while $v \in B$, and let F'' be the set of edges in I leaving v at Step 3 when v was excluded from B. Clearly, $\delta_F(v) \subseteq F' \cup F''$. Note that at the moment when v was excluded from B we had

$$w(F') \leq \alpha \left(b(v) - \sum_{e \in F''} x(e)w(e) \right)$$

In particular, $w(F') \leq \alpha b(v)$. Also, $|F''| \leq \Delta$ and thus $w(F'') \leq |F''| \cdot b(v) \leq \Delta b(v)$. Consequently, $w(\delta_F(v)) \leq w(F') + w(F'') \leq \alpha b(v) + \Delta b(v) = (\alpha + \Delta)b(v)$.

Now consider the case of unit weights. We had $|F'| \leq \alpha \left(b(v) - \sum_{e \in F''} x(e) \right)$ when v was excluded from B. Moreover, we had $x(e) > 0$ for all $e \in F''$, since edges with $x(e) = 0$ were removed at Step 2, before v was excluded from B. Hence if $F'' \neq \emptyset$ then $|F'| < \alpha b(v)$, and thus $|F| \leq |F'| + |F''| < \alpha b(v) + \Delta$. Since all numbers are integers, this implies $|F| \leq \alpha b(v) + \Delta - 1$. If $F'' = \emptyset$, then $|F| = |F'| \leq \alpha b(v) \leq \alpha b(v) + \Delta - 1$. Consequently, in both cases $|F| \leq \alpha b(v) + \Delta - 1$, as claimed.

4   Sparseness of $P_f(I, F, B)$ (Proof of Theorem 5)

Let $\mathcal{L}$ and T be as in Lemma 1. Define a child-parent relation on the members of $\mathcal{L} + T$ as follows. For $S \in \mathcal{L}$ or $v \in T$, its parent is the inclusion minimal member of $\mathcal{L}$ properly containing it, if any. Note that if $v \in T$ and $\{v\} \in \mathcal{L}$, then $\{v\}$ is the parent of v, and that no members of T has a child. For every edge $uv \in I$ assign one *tail-token* to u and one *head-token* to v, so every edge contributes exactly 2 tokens. The number of tokens is thus $2|I|$.

Definition 4. *A token contained in S is an S-token if it is not a tail-token of an edge vu leaving S so that $v \notin T$ (so a tail-token of an edge vu leaving S is an S-token if, and only if, $v \in T$).*

Recall that we need to prove that if $x \in P_f(I, F, B)$ is a basic solution so that $0 < x(e) < 1$ for all $e \in I$, then there exists $e \in I$ with $x(e) \geq 1/\alpha$ or there exists $v \in B$ with $|\delta_I(v)| \leq \Delta$. Assuming this is not so, we have:

The Negation Assumption:
- $|\delta_I^{in}(S)| \geq \alpha + 1$ for all $S \in \mathcal{L}$;
- $|\delta_I(v)| \geq \Delta + 1$ for all $v \in T$.

We obtain the contradiction $|I| > |\mathcal{L}| + |T|$ by showing that for any $S \in \mathcal{L}$ we can assign the S-tokens so that every proper descendant of S in $\mathcal{L} + T$ gets 2 S-tokens and S gets at least 3 S-tokens. Except the proof of $(2,3)$-sparseness of $0, 1$-valued f, our assignment scheme will be:

The $(2, \alpha + 1)$-Scheme:
- every proper descendant of S in $\mathcal{L} + T$ gets 2 S-tokens;
- S gets $\alpha + 1$ S-tokens.

Initial assignment:
For every $v \in T$, assign the $|\delta_I(v)|$ tail-tokens of the edges in $\delta_I(v)$.

The rest of the proof is by induction on the number of descendants of S in $\mathcal{L}$. If S has no children/descendants in $\mathcal{L}$, it has at least $|\delta_I^{in}(S)| \geq \alpha + 1$ head-tokens of the edges in $\delta_I^{in}(S)$. We therefore assume that S has in $\mathcal{L}$ at least one child. Given $S \in \mathcal{L}$ with at least one child in $\mathcal{L}$, let C be the set of edges entering some child of S, J the set of edges entering S or a child of S but not both, and D the set of edges that enter a child of S and their tail is in $T \cap S$ but not in a child of S. Formally:

$$C = \bigcup \{\delta_I^{in}(R) : R \text{ is a child in } \mathcal{L} \text{ of } S\}$$
$$J = (\delta_I^{in}(S) - C) \cup (C - \delta_I^{in}(S))$$
$$D = \{e = vu \in C - \delta_I^{in}(S) : v \in T\} .$$

Lemma 3. *Let $S \in \mathcal{L}$ and suppose that $0 < x(e) < 1$ for all $e \in E$. Then $|J| \geq 2$, and every edge $e \in J - D$ has an endnode that owns an S-token that is not an R-token of any child R of S in $\mathcal{L}$.*

Proof. $C = \delta_I^{in}(S)$ contradicts linear independence, hence one of the sets $\delta_I^{in}(S) - C, C - \delta_I^{in}(S)$ is nonempty. If one of these sets is empty, say $\delta_I^{in}(S) - C = \emptyset$, then $x(C) - x(\delta_I^{in}(S))$ must be a *positive* integer. Thus $|C - \delta_I^{in}(S)| \geq 2$, as otherwise there is an edge $e \in C - \delta_I^{in}(S)$ with $x(e) = 1$. The proof of the case $C - \delta_I^{in}(S) = \emptyset$ is identical. The second statement is straightforward.

4.1 Arbitrary Intersecting Supermodular f

For $(2,5)$-sparseness the Negation Assumption is $|\delta_I^{in}(S)| \geq 3$ for all $S \in \mathcal{L}$, and $|\delta_I(v)| \geq 6$ for all $v \in T$. We prove that then the $(2,3)$-Scheme is feasible. First, for every $v \in T$, we reassign the $|\delta_I(v)|$ tail-tokens assigned to v as follows:
- 3 tokens to v;
- 1/2 token to every edge in $\delta_I(v)$ (this is feasible since $|\delta_I(v)| \geq 6$).

Claim. If S has at least 3 children in $\mathcal{L}$, then the $(2,3)$-Scheme is feasible.

Proof. By moving one token from each child of S to S we get an assignment as required.

Claim. If S has exactly 2 children in $\mathcal{L}$ then the $(2,3)$-Scheme is feasible.

Proof. S can get 2 tokens by taking one token from each child, and needs 1 more token. If there is $e \in J - D$ then S can get 1 token from an endnode of e, by Lemma 3. Else, $|D| = |J| \geq 2$. As every edge in D owns 1/2 token, S can collect 1 token from edges in D.

Claim. If S has exactly 1 child in $\mathcal{L}$, say R, then the $(2,3)$-Scheme is feasible.

Proof. S gets 1 token from R, and needs 2 more tokens. We can collect $|J - D| + |D|/2 + |T \cap (S - R)|$ S-tokens that are not R-tokens, from edges in J and from the children of S in T, by Lemma 3 and our assignment scheme. We claim that $|J - D| + |D|/2 + |T \cap (S - R)| \geq 2$. This follows from the observation that if $|J - D| \leq 1$ then $|T \cap (S - R)| \geq 1$, and if $|J - D| = 0$ then $|D| = |J| \geq 2$, by Lemma 3.

It is easy to see that during our distribution procedure no token was assigned twice. For "node" tokens this is obvious. For $1/2$ tokens on the edges, this follows from the fact that each time we assigned a $1/2$ token of an edge, both endnodes of this edge were inside S, as this edge was connecting the two children of S.

For $(3,3)$-sparseness the Negation Assumption is $|\delta_I^{in}(S)| \geq 4$ for all $S \in \mathcal{L}$ and $|\delta_I(v)| \geq 4$ for all $v \in T$. In this case we can easily prove that the $(2,4)$-Scheme is feasible. If S has at least 2 children in $\mathcal{L}$, then by moving 2 tokens from each child to S we get an assignment as required. If S has exactly 1 child in $\mathcal{L}$, say R, then S gets 2 tokens from R, and needs 2 more tokens. If $D = \emptyset$ then S can get 2 tokens from endnodes of the edges in J. Else, S has a child in T, and can get 2 tokens from this child.

4.2   Improved Sparseness for $0, 1$-Valued f

Here the Negation Assumption is $|\delta_I^{in}(S)| \geq 3$ for all $S \in \mathcal{L}$ and $|\delta_I(v)| \geq 4$ for all $v \in T$. Assign colors to members of $\mathcal{L} + T$ as follows. All nodes in T are black; $S \in \mathcal{L}$ is black if $S \cap T \neq \emptyset$, and S is white otherwise. We show that given $S \in \mathcal{L}$, we can assign the S-tokens so that:

The $(2, 3, 4)$-Scheme
- every proper descendant of S gets 2 S-tokens;
- S gets at least 3 S-tokens, and S gets 4 S-tokens if S is black.

As in the other cases, the proof is by induction on the number of descendants of S in $\mathcal{L}$. If S has no descendants in $\mathcal{L}$, then S gets $|\delta_I^{in}(S)| \geq 3$ head tokens of the edges in $\delta_I^{in}(S)$; if S is black, then S has a child in T and S gets 1 more token from this child.

Lemma 4. *If $J = D$ then S has a child in T or at least 2 black children in $\mathcal{L}$.*

Proof. Otherwise, all edges in J must have tails in $T \cap R$ for some child R of S, and every edge that enters S also enters some child of S. Thus $\delta_I^{in}(R) \subseteq \delta_I^{in}(S)$, and since $x(\delta_I^{in}(R)) = x(\delta_I^{in}(S)) = 1$, we must have $\delta_I^{in}(R) = \delta_I^{in}(S)$. This contradicts linear independence.

Claim. If S has in $\mathcal{L} + T$ at least 3 children, then the $(2, 3, 4)$-Scheme is feasible.

Proof. S gets 3 tokens by taking 1 token from each child; if S is black, then one of these children is black, and S can get 1 more token.

Claim. If S has in $\mathcal{L}$ exactly 2 children, say R, R', then the $(2, 3, 4)$-Scheme is feasible.

Proof. If S has a child $v \in T$, then we are in the case of Claim 4.2. If both R, R' are black, then S gets 4 tokens, 2 from each of R, R'. Thus we assume that S has no children in T, and that at least one of R, R' is white, say R' is white. In particular, S is black if, and only if, R is black. Thus S only lacks 1 token, that does not come directly from R, R'. By Lemma 4 there is $e \in J - D$, and S can get a token from an endnode of e, by Lemma 3.

Claim. If S has in $\mathcal{L}$ exactly one child, say R, then the $(2, 3, 4)$-Scheme is feasible.

Proof. Suppose that $T \cap (S - R) = \emptyset$. Then S is black if, and only if, R is black. Thus S needs 2 S-tokens not from R. As every edge in D has tail in $T \cap (S - R)$ and head in R, $D = \emptyset$ so $|J - D| = |J| \geq 2$, and thus S can get 2 S-tokens from endnodes of the edges in J, by Lemma 3.

If there is $v \in T \cap (S - R)$, then S can get 1 token from R, 2 tokens from v, and needs 1 more token. We claim that there is $e \in \delta_I^{in}(S) - \delta_I^{in}(R)$, and thus S can get the head-token of e. Otherwise, $\delta_I^{in}(S) \subseteq \delta_I^{in}(R)$, and since $x(\delta_I^{in}(S)) = x(\delta_I^{in}(R)) = 1$, we obtain $\delta_I^{in}(S) = \delta_I^{in}(R)$, contradicting linear independence.

This finishes the proof of Theorem 5, and thus also the proof of Theorem 1 is complete.

5 Indegree Constraints only (Proof of Theorem 2)

Here we prove Theorem 2. Consider the following polytope $P_f^{in}(I, F, B)$:

$$x(\delta_I^{in}(S)) \geq f(S) - |\delta_F^{in}(S)| \qquad \text{for all } \emptyset \neq S \subset V$$
$$\sum_{e \in \delta_I^{in}(v)} x(e)w(e) \leq b(v) - w(\delta_F^{in}(v)) \qquad \text{for all } v \in B$$
$$0 \leq x(e) \leq 1 \qquad \text{for all } e \in I$$

Theorem 6. $P_f^{in}(I, F, B)$ *is* $(1, 3)$*-sparse for intersecting supermodular* f. *For unit weights and integral indegree bounds, any basic solution of* $P_f^{in}(I, F, B)$ *always has an edge* e *with* $x(e) = 1$.

In Lemma 1, we have a set T^{in} of nodes corresponding to tight in-degree constraints. We prove that if $x \in P_f^{in}(I, F, B)$ is a basic solution so that $x(e) > 0$ for all $e \in I$, then there exists $e \in I$ with $x(e) = 1$ or there exists $v \in T^{in}$ with $|\delta_I^{in}(v)| \leq 3$. Otherwise, we must have:

The Negation Assumption:
- $|\delta_I^{in}(S)| \geq 2$ for all $S \in \mathcal{L}$;
- $|\delta_I^{in}(v)| \geq 4$ for all $v \in T^{in}$.

Assuming Theorem 5 is not true, we show that given $S \in \mathcal{L}$, we can assign the S-tokens so that (here token is an S-token if it is not a tail-token of an edge leaving S):

The $(2, 2)$-Scheme:
S and every proper descendant of S in $\mathcal{L} + T$ gets 2 S-tokens.

The contradiction $|I| > |\mathcal{L}| + |T^{in}|$ is obtained by observing that if S is an inclusion maximal set in $\mathcal{L}$, then there are at least 2 edges entering S, and their tail-tokens are not assigned, since they are not S'-tokens for any $S' \in \mathcal{L}$.

Initial assignment:
For every $v \in T$, we assign the 4 tail-tokens of some edges in $\delta_I^{in}(v)$.

The rest of the proof is by induction on the number of descendants of S, as before. If S has no children/descendants, it contains at least $|\delta_I^{in}(S)| \geq 2$ head-tokens, as claimed. If S has in $\mathcal{L} + T^{in}$ at least one child $v \in T^{in}$, then S gets 2 tokens from this child.

Thus we may assume that S has at least 1 child in $\mathcal{L}$ and no children in T^{in}. Let J be as in Lemma 3, so $|J| \geq 2$. One can easily verify that S can collect 1 S-token from an endnode of every edge in J. Thus the $(2, 2)$-Scheme is feasible.

For the case of unit weights (and integral degree bounds), we can prove that any basic solution to $P_f^{in}(I, F, B)$ has an edge e with $x(e) = 1$. This follows by the same proof as above, after observing that if $v \in T^{in}$ is a child of $S \in \mathcal{L}$, then $\delta_I^{in}(v) \neq \delta_I^{in}(S)$, as otherwise we obtain a contradiction to the linear independence in Lemma 1. Thus assuming that there are at least 2 edges in I entering any member of $\mathcal{L} + T^{in}$, we obtain a contradiction in the same way as before, by showing that the $(2, 2)$-Scheme is feasible. Initially, every minimal member of $\mathcal{L} + T^{in}$ gets 2 tail-tokens of some edges entering it. In the induction step, any $S \in \mathcal{L}$ can collect at least 2 S-tokens that are not tokens of its children, by Lemma 3.

Remark: Note that we also showed the well known fact (c.f., [16]), that if there are no degree constraints at all, then there is an edge $e \in I$ with $x(e) = 1$.

6   The Case of Both Indegree and Outdegree Constraints

Here we describe the slight modifications required to handle the case when there are both indegree and outdegree constraints. In this case, in Lemma 1, we have sets T and T^{in} of nodes corresponding to tight out-degree and in-degree constraints, respectively. Let $S \in \mathcal{L}$ and suppose that S has in $\mathcal{L} + T + T^{in}$ a unique child $v \in T^{in}$ (possibly $S = \{v\}$).

Arbitrary weights: For arbitrary weights, we can show that an appropriate polytope has sparseness $(\alpha, \Delta, \Delta^{in}) = (2, 5, 4)$, in the same way as in Section 4.1. The Negation Assumption for $v \in T^{in}$ is $|\delta_I^{in}| \geq 5$, and we do not put any tokens on the edges leaving v (unless their tail is in T). Even if $\delta_I^{in}(S) = \delta_I^{in}(v)$ (note that in the case of arbitrary weights this may not contradict linear independence), the head-tokens of at least 5 edges entering v suffice to assign 2 tokens for v and 3 tokens to S. Hence in this case the approximation ratio is $(\alpha, \alpha + \Delta, \alpha + \Delta^{in}) = (2, 7, 6)$. In a similar way we can also show the sparseness $(\alpha, \Delta, \Delta^{in}) = (3, 3, 4)$, and in this case the ratio is $(3, 6, 7)$.

Unit weights: In the case of unit weights, we must have $\delta_I^{in}(S) \neq \delta_I^{in}(v)$, as otherwise the equations of S and v are linearly dependent. Hence in this case, it is sufficient to require $|\delta_I^{in}| \geq 4$, and the sparseness is $(\alpha, \Delta, \Delta^{in}) = (2, 5, 3)$. Consequently, the approximation is $(\alpha \cdot \mathsf{opt}, \alpha b(v) + \Delta - 1, \alpha b^{in}(v) + \Delta^{in} - 1) = (2 \cdot \mathsf{opt}, 2b(v) + 4, 2b^{in}(v) + 2)$.

$0, 1$-*valued* f: In the case of $0, 1$-valued f, we can show that the corresponding polytope has sparseness $(\alpha, \Delta, \Delta^{in}) = (2, 3, 4)$, in the same way as in Section 4.2. The negation assumption for a node $v \in T^{in}$ is $|\delta_I^{in}| \geq 5$; a member in $\mathcal{L}$ containing a node from T^{in} only is *not* black, unless it also contains a node from T. Hence in this case the approximation ratio is $(\alpha, \alpha + \Delta, \alpha + \Delta^{in}) = (2, 5, 6)$. If we have also unit weights, then $\delta_I^{in}(S) \neq \delta_I^{in}(v)$, as otherwise we obtain a contradiction to the linear independence; hence for unit weights we can obtain sparseness $(\alpha, \Delta, \Delta^{in}) = (2, 3, 3)$, and the ratio $(\alpha \cdot \mathsf{opt}, \alpha b(v) + \Delta - 1, \alpha b^{in}(v) + \Delta^{in} - 1) = (2 \cdot \mathsf{opt}, 2b(v) + 2, 2b^{in}(v) + 2)$.

Summarizing, we obtain the following result:

Theorem 7. DWDCN *with intersecting supermodular* f *admits a polynomial time algorithm that computes an* f-*connected graph* H *of cost* $\leq 2 \cdot \mathsf{opt}$ *so that the weighted* (*degree,indegree*) *of every* $v \in V$ *is at most* $(7b(v), 6b^{in}(v))$ *for arbitrary* f, *and* $(5b(v), 6b^{in}(v))$ *for* $0, 1$-*valued* f. *Furthermore, for unit weights, the* (*degree,indegree*) *of every* $v \in V$ *is at most* $(2b(v)+4, 2b^{in}(v)+2)$ *for arbitrary* f, *and* $(2b(v) + 2, 2b^{in}(v) + 2)$ *for a* $0, 1$-*valued* f.

References

1. Bang-Jensen, J., Thomassé, S., Yeo, A.: Small degree out-branchings. J. of Graph Theory 42(4), 287–307 (2003)
2. Bansal, N., Khandekar, R., Nagarajan, V.: Additive gurantees for degree bounded directed network design. In: STOC 2008, pp. 769–778 (2008)
3. Chaudhuri, K., Rao, S., Riesenfeld, S., Talwar, K.: A push-relabel algorithm for approximating degree bounded MSTs. In: Bugliesi, M., Preneel, B., Sassone, V., Wegener, I. (eds.) ICALP 2006. LNCS, vol. 4051, pp. 191–201. Springer, Heidelberg (2006)
4. Frank, A.: Connectivity and network flows. In: Graham, R.L., Grötschel, M., Lovász, L. (eds.) Handbook of Combinatorics, ch. 2, pp. 111–177. Elsevier, Amsterdam (1995)
5. Fukunaga, T., Nagamochi, H.: Network design with weighted degree constraints. TR 2008-005, Kyoto University (manuscript, 2008)
6. Furer, M., Raghavachari, B.: Approximating the minimum-degree steiner tree to within one of optimal. Journal of Algorithms 17(3), 409–423 (1994)
7. Goemans, M.X.: Minimum bounded degree spanning trees. In: FOCS, pp. 273–282 (2006)
8. Goemans, M.X., Williamson, D.P.: The primal-dual method in approximation algorithms and its applications to network design problems. In: Hochbaum, D.S. (ed.) Approximation Algorithms For NP-hard Problems, ch. 4. PWS (1997)
9. Jain, K.: A factor 2 approximation algorithm for the generalized Steiner network problem. Combinatorica 21(1), 39–60 (2001)
10. Khuller, S.: Approximation algorithm for finding highly connected subgraphs. In: Hochbaum, D.S. (ed.) Approximation Algorithms For NP-hard Problems, ch. 6. PWS (1997)
11. Könemann, J., Ravi, R.: A matter of degree: Improved approximation algorithms for degree bounded minimum spanning trees. SIAM Journal on Computing 31(3), 1783–1793 (2002)

12. Kortsarz, G., Nutov, Z.: Approximating minimum-cost connectivity problems. In: Gonzalez, T.F. (ed.) Approximation Algorithms and Metaheuristics, ch. 58. Chapman & Hall/CRC, Boca Raton (2007)
13. Lau, L.C., Naor, J., Salavatipour, M.R., Singh, M.: Survivable network design with degree or order constraints. In: STOC, pp. 651–660 (2007)
14. Melkonian, V., Tardos, E.: Algorithms for a network design problem with crossing supermodular demands. Networks 43(4), 256–265 (2004)
15. Ravi, R., Marathe, M.V., Ravi, S.S., Rosenkrantz, D.J., Hunt III, H.B.: Many birds with one stone: Multi objective approximation algorithms. In: STOC, pp. 438–447 (1993)
16. Schrijver, A.: Combinatorial Optimization, Polyhedra and Efficiency. Springer, Heidelberg (2004)
17. Singh, M., Lau, L.C.: Approximating minimum bounded degree spanning trees to within one of optimal. In: STOC, pp. 661–670 (2007)

A Constant Factor Approximation for Minimum λ-Edge-Connected k-Subgraph with Metric Costs

MohammadAli Safari* and Mohammad R. Salavatipour**

Department of Computing Science
University of Alberta
Edmonton, Alberta T6G 2E8, Canada
{msafarig,mreza}@cs.ualberta.ca

Abstract. In the (k, λ)-subgraph problem, we are given an undirected graph $G = (V, E)$ with edge costs and two parameters k and λ and the goal is to find a minimum cost λ-edge-connected subgraph of G with at least k nodes. This generalizes several classical problems, such as the minimum cost k-Spanning Tree problem or k-MST (which is a $(k, 1)$-subgraph), and minimum cost λ-edge-connected spanning subgraph (which is a $(|V(G)|, \lambda)$-subgraph). The only previously known results on this problem [12,5] show that the $(k, 2)$-subgraph problem has an $O(\log^2 n)$-approximation (even for 2-node-connectivity) and that the (k, λ)-subgraph problem in general is almost as hard as the densest k-subgraph problem [12]. In this paper we show that if the edge costs are metric (i.e. satisfy triangle inequality), like in the k-MST problem, then there is an $O(1)$-approximation algorithm for (k, λ)-subgraph problem. This essentially generalizes the k-MST constant factor approximability to higher connectivity.

1 Introduction

Network design is a central topic in combinatorial optimization, approximation algorithms, and operations research. A fundamental problem in network design is to find a minimum cost subgraph satisfying some given connectivity requirements between vertices. Here by a network we mean an undirected graph together with non-negative costs on the edges. For example, with a connectivity requirement $\lambda = 1$ between all the vertices, we have the classical minimum spanning tree problem. For larger values of λ, i.e. finding minimum cost λ-edge-connected spanning subgraphs, the problem is APX-hard. These are special cases of the more general problem of survivable network design problem (SNDP), in which we have a connectivity requirement of r_{uv} between every pair u, v of vertices. Even for this general setting there is a 2-approximation algorithm by Jain [11].

A major line of research in this area has focused on problems with connectivity requirements where one has another parameter k, and the goal is to find a subgraph satisfying the connectivity requirements with a lower bound k on the total number of vertices. The most well-studied problem in this class is the minimum k-spanning tree

---

* Supported by Alberta Ingenuity.
** Supported by NSERC and an Alberta Ingenuity New Faculty award.

A. Goel et al. (Eds.): APPROX and RANDOM 2008, LNCS 5171, pp. 233–246, 2008.

problem, a.k.a. k-MST. In this problem, we have to find a minimum cost connected subgraph spanning at least k-vertices. The approximation factor for this problem was improved from $\sqrt{k}$ and $O(\log^2 k)$ in [14,1] down to $O(\log n)$ in [13] and to a constant in [3,9] and recently to 2 [10]. The algorithm of [10] can be used to obtain a constant approximation for the slightly more general setting in which we have a set of nodes T, called terminals, and the goal is to find a minimum cost connected subgraph containing at least k terminals. This is known as the k-Steiner tree problem. The problem of k-TSP, in which one has to find a minimum cost TSP tour containing at least k nodes, can be approximated using very similar technique. We note that in all these problems, the input graph is assumed to be complete and the edge cost function is metric, i.e. satisfies triangle inequality. Most of these problems are motivated from their applications in vehicle routing or profit maximization with respect to a given fixed budget. For example, suppose that we have a battery operated robot and the goal is to find the minimum battery charge required to travel a sequence of at least k nodes in a given graph such that the total length of the tour can be travelled in a single battery charge. See [2,6] for similar problems.

Recently, Lau et al. [12] considered a very natural common generalization of both the k-MST problem and minimum cost λ-edge-connected spanning subgraph problem, which they called the (k, λ)-subgraph problem. In this problem, we are given a graph $G = (V, E)$ with a (not necessarily metric) cost function $c : E \rightarrow \mathbb{R}^+$ on the edges, a positive integers k, and a connectivity requirement $\lambda \geq 1$; the goal is to find a minimum cost λ-edge-connected subgraph of G with at least k vertices. We should point out that we are not allowed to take more copies of an edge than are present in the graph. Otherwise, a 4-approximate solution can be computed by taking a 2-approximate k-MST solution T, and then taking λ copies of T. It is easy to observe that the (k, λ)-subgraph problem contains, as special cases, the minimum cost λ-edge-connected spanning subgraph problem (it becomes the $(|V(G)|, \lambda)$-subgraph problem), and the k-MST problem (which becomes the $(k, 1)$-subgraph problem). Lau et al. [12] present an $O(\log^2 n)$-approximation for $(k, 2)$-subgraph and show that for arbitrary values of λ, (k, λ)-subgraph is almost as hard as the k-densest subgraph problem[1]. In the k-densest subgraph problem, one has to find a subgraph with k nodes in a given graph G that has the largest number of edges. Despite considerable attempts, the best known approximation algorithm for this problem has ratio $O(n^{\frac{1}{3}-\epsilon})$ for some fixed $\epsilon > 0$ [8]. Chekuri and Korula [5] have recently (and independently of [12]) shown that an algorithm similar to the one in [12] yields an $O(\log^2 n)$-approximation for the $(k, 2)$-subgraph problem even if we want a 2-node-connectivity requirement in the solution.

In light of the result of [12] on the hardness of (k, λ)-subgraph for arbitrary values of λ and general cost functions, it is natural to try to obtain good approximation algorithms for the class of graphs where the edge cost function is metric, i.e. satisfies triangle inequality. Remember that the constant factor approximation algorithms for k-MST and k-TSP are on graphs with metric cost function. Our main result of this paper is the following theorem:

---

[1] The extended abstract version claimed an $O(\log^3 n)$-approximation but the proof was inaccurate. The full version has the improved result with a complete proof.

Theorem 1. *Given a (complete) graph G with metric costs on the edges and two parameters k, λ, there is an $O(1)$-approximation algorithm for finding a (k, λ)-subgraph in G.*

Our algorithm is combinatorial and uses ideas from [4] for metric-cost subset node-connectivity problem as well as the algorithm for k-Steiner tree [7,10,14]. The constant factor we obtain is relatively large (between 400 and 500), however, most of our efforts have been to show that the problem has a constant factor approximation rather than trying to obtain the best possible ratio.

The organization of the paper is as follows. We start by some definitions and preliminary bounds used throughout paper in the next section. For the ease of exposition, we first present an algorithm that finds a λ-edge-connected subgraph with at least $k - \lambda/7$ nodes whose cost is at most $O(\text{OPT})$. In Section 3 we show how to extend this solution to a feasible solution to the (k, λ)-subgraph problem while keeping the total cost still bounded by $O(\text{OPT})$. We finish the paper with some concluding remarks.

2 Preliminaries

As mentioned earlier, we assume we are given a (complete) graph $G = (V, E)$, with a cost function $c : E \to \mathbb{R}^+$ on the edges that satisfies triangle inequality, and two positive integers k and $\lambda \geq 1$. For every subgraph $F \subseteq G$, we use $c(F)$ to denote the total cost of the edges in F. Throughout, $G^* \subseteq G$ denotes the optimum solution and $\text{OPT} = c(G^*)$ denotes the cost of the optimum solution. We will use two lower bounds on OPT in the analysis of our algorithm. These lower bounds were used earlier in [4] for the problem of minimum cost subset k-node-connectivity. The first lower bound comes from the cost of a minimum spanning tree of the G^*, which we call it T^*. Considering the cut-constraint IP formulation of MST, it is easy to see that $\frac{\lambda}{2} \sum_{e \in T^*} c_e \leq \text{OPT}$. The second lower bound comes from the minimum cost subgraph that has minimum degree at least λ. Note that in a λ-edge-connected subgraph, every vertex has degree at least λ. For any λ-edge-connected subgraph $F \subseteq G$ and any vertex $u \in F$ we let $S_u(F)$ to be the set of λ nearest neighbors of u in F and $s_u(F)$ be $\sum_{v \in S_u(F)} c_{uv}$. Clearly, for any λ-edge-connected subgraph $F \subseteq G$ and any vertex $u \in F$: $s_u(G) \leq s_u(F)$. We often use S_u and s_u instead of $S_u(G)$ and $s_u(G)$, respectively, unless the graph is different from G. It is easy to see that $\frac{1}{2} \sum_{u \in G^*} s_u \leq \frac{1}{2} \sum_{u \in G^*} s_u(G^*) \leq \text{OPT}$. Thus, if T^* is a minimum spanning tree of G^*, then we obtain the following two lower bounds for OPT: (i) $\frac{1}{2} \sum_{u \in T^*} s_u \leq \text{OPT}$, and (ii) $\frac{\lambda}{2} \sum_{e \in T^*} c_e \leq \text{OPT}$, and in particular:

$$\frac{1}{2} \sum_{u \in T^*} s_u + \frac{\lambda}{2} \sum_{e \in T^*} c_e \leq 2\text{OPT}. \tag{1}$$

In our algorithm we will use these two lower bounds frequently, often without referring to them.

We present an algorithm for a modified version of the problem in which along with G, k, and λ, we are also given a vertex $r \in G$ as the root which we are told belongs to the optimum solution G^* and among all the vertices in G^* it has the smallest value s_u. Clearly if we can solve this rooted version, then we can try every vertex as the root and return the minimum solution among all as the final answer.

Ravi et al [14] showed that any α-approximation for k-MST implies a 2α-approximation for k-Steiner tree. Therefore, together with Garg's algorithm [10], we have a 4-approximation for k-Steiner tree. In fact, we can have a 4-approximation algorithm for the rooted version of the problem, in which a specific vertex r is given as the root and the goal is to find a minimum cost rooted at r Steiner tree containing at least k terminals. Our algorithm will use the best known approximation algorithm for finding a minimum cost rooted k-Steiner tree problem; let us denote the approximation ratio of this algorithm by ρ (by the argument above, we know that $\rho \leq 4$). We denote this approximation algorithm by ST-Alg.

3 Obtaining a Low Cost $(k - O(\lambda), \lambda)$-Subgraph

Observe that to have λ-edge-connectivity, we must have $k \geq \lambda + 1$. We start by presenting an algorithm that returns a λ-edge-connected subgraph (containing root r) that has at least $k - \lambda/7$ nodes and whose cost is within constant factor of OPT. Our algorithm is influenced by the work of Cheriyan and Vetta [4] for minimum cost subset k-node-connectivity.

3.1 Overview of the Algorithm

The main idea of the algorithm is as follows. We create a new graph $G'(V \cup V', E')$ from G by creating a new (dummy) vertex u' (in V') for each vertex $u \in G$ and $E' = E \cup \{uu'|u \in V\}$. Each edge $uu' \in E'$ has weight s_u. For every other edge in G' (that also exists in G) we multiply its weight by λ. Suppose that T^* is an optimum (rooted at r) k-Steiner tree of G' with terminal set V'. We show that $c(T^*) \leq 4\text{OPT}$. We can obtain an approximation of T^*, call it T', by running the ST-Alg. Let us assume that $\tilde{T} \subset G$ is the tree obtained from T' by deleting the dummy vertices and for ease of exposition, suppose that T^* and $\tilde{T}$ are binary trees. For simplicity, suppose that all these sets S_u (for $u \in \tilde{T}$) are disjoint and let us assume that $v_1, \ldots, v_p$ (with $p = k/(\lambda+1)$) have the smallest s_u values among all the nodes in $\tilde{T}$. Our next steps would be to obtain a λ-edge-connected subgraph by selecting v_i and S_{v_i} (for $1 \leq i \leq p$) and forming a $(\lambda + 1)$-clique on each to get λ-edge connectivity among themselves. The cost of each of these cliques will be at most λs_{v_i}, for $1 \leq i \leq p$, and since $p \approx k/\lambda$ the total cost of all the cliques is at most $X = \sum_{u \in \tilde{T}} s_u$. Considering each of these cliques as a big "blub", we need to establish λ-edge connectivity among these blubs. For that we need to find a tree with the blubs being the nodes and for each edge in the tree between two blubs we add about λ edges between the cliques corresponding to the blubs to maintain λ-edge-connectivity. We can use the structure of $\tilde{T}$ itself to establish a tree over these blubs. Roughly speaking, the total cost of all the λ edges between the blubs will be at most $O(\lambda)$ times the edges in $\tilde{T}$ (using triangle inequality) which is $Y = O(\lambda \sum_{e \in \tilde{T}} c_e)$. Noting that $X + Y \leq O(c(T'))$ implies that we will have a solution within constant factor of the optimum. The main difficulty here will be the possible (or lack of) intersections between S_u and S_v for two vertices $u, v \in T$. A lot of details have been skipped over in this overview and are explained in the next subsection.

3.2 Details of the Algorithm

We build the graph G' from G as described above and compute an approximate (rooted) k-Steiner tree with terminal set $V' = \{v'|v \in V(G)\}$ and root r' (copy of r) using the ST-Alg. Let's call this tree T'.

Lemma 1. $c(T') \leq 4\rho\text{OPT}$.

Proof. Consider the optimal solution G^* to the (k, λ)-subgraph problem on G and let T^* be a MST of G^* (we assume that $r \in T^*$). Then $\tilde{T} = T^* \cup \{uu' \in G'|u \in T^*\}$ is clearly a Steiner tree in G' containing at least k terminals with total cost at most 4OPT (using the bound in (1) for T^*). The lemma follows by observing that the ST-Alg (for k-Steiner tree) has approximation ratio ρ.

Without loss of generality, we can assume that T' has exactly k terminals, as if it has more we can safely delete them. Let $T_0 \subset G$ be the tree obtained from $T' \subset G'$ by deleting the dummy nodes (i.e. the nodes in V') and V_0 be the vertex set of T_0. Note that by Lemma 1:

$$\lambda \sum_{e \in T_0} c_e + \sum_{u:u' \in T'} s_u \leq 4\rho\text{OPT}. \tag{2}$$

We should also point out that V_0 might have some vertices $v \in V$ (and therefore $v \in T'$) but $v' \notin T'$. We obtain another tree $T_1 = (V_1, E_1) \subset G$ from T_0 with the following properties: (i) $V_1 \subseteq V_0$, (ii) $c(T_1) \leq 2c(T_0)$, and (iii) for every vertex $v \in V_1$, the corresponding vertex $v' \in G'$ belongs to T'. To do this, we duplicate every edge of T_0 and do an Eulerian walk of T_0; now shortcut over every vertex $v \in T_0$ with $v' \notin T'$. It is easy to see that we are left with a tree T_1 whose cost is at most $2c(T_0)$ and every vertex $v \in T_1$ has its copy v' in T'. Also, property (iii) implies that T_1 has exactly k vertices. Thus:

Lemma 2. $V_1 \subseteq V_0$, with $|V_1| = k$ and $c(T_1) \leq 2c(T_0)$.

Suppose that we have an ordering of the vertices of T_1, say $v_1 = r, v_2, \ldots, v_k$, such that $s_{v_2} \leq s_{v_3} \leq \cdots \leq s_{v_k}$. Note that although r has the smallest s_u value among all vertices $u \in G^*$, it is not necessarily the case in T_1. For each $1 \leq i \leq k$, let $\mu_i = \frac{s_{v_i}}{\lambda}$. We call S_{v_i} the ball of v_i and the core of S_{v_i}, denote by B_{v_i}, is the set of nodes in S_{v_i} with distance at most $2\mu_i$ to v_i. By a simple averaging argument, one can easily show that $|B_{v_i}| \geq \lambda/2$. We partition the nodes of T_1 into two sets of *active* and *inactive* nodes using the following procedure to cluster the balls. Initially, all the nodes of T_1 are active and we have $\mathcal{S} = \emptyset$ ($\mathcal{S}$ will contain the centers of active balls). For each $1 \leq i \leq k$, if v_i is active and there is no $v_j \in \mathcal{S}$ (with $j < i$) such that $c_{ij} \leq 4\mu_i + 2\mu_j$ then add v_i to $\mathcal{S}$ and make all the nodes in S_{v_i} inactive (except v_i itself). Note that S_{v_i} might include some vertices not in T_1. So at the end, for every two active nodes $v_i, v_j \in \mathcal{S}$ (with $j < i$) we have $c_{ij} > 4\mu_i + 2\mu_j$ and $B_{v_i} \cap B_{v_j} = \emptyset$. Now for every value of $1 \leq i \leq k$ such that v_i is active but $v_i \notin \mathcal{S}$, there exists a $j < i$ such that $v_j \in \mathcal{S}$ and $c_{ij} \leq 4\mu_i + 2\mu_j$. Let j^* be the smallest such index and define $p(i) = j^*$, meaning that v_i is assigned to ball $S_{v_{j^*}}$. So each active node v_i is either the center of an active ball (and it belongs to $\mathcal{S}$) or is assigned to a ball $S_{p(i)}$ with $p(i) \in \mathcal{S}$, and all the remaining nodes (that are inside the balls with centers in $\mathcal{S}$) are inactive. Thus:

Lemma 3. *Every core B_{v_i}, for $v_i \in S$, is disjoint from any other B_{v_j} (with $v_j \in S$, $j \neq i$) and $|B_{v_i}| > \frac{\lambda}{2}$.*

For every value of i, consider the union of active nodes v_j and their ball S_{v_j} (if $v_j \in S$), for all $j \leq i$, and define this set of vertices U_i, i.e.

$$U_i = \{\text{active nodes } v_j \text{ with } j \leq i\} \cup \bigcup_{\substack{\text{active } v_j, j \leq i, \\ v_j \in S}} S_{v_j}. \tag{3}$$

Let $i^* \leq k$ be the smallest index such that $|U_{i^*}| \geq k - \lambda/7$. It is easy to see that from the definition of U_i and the choice of i^*:

Lemma 4. $k - \frac{\lambda}{7} \leq |U_{i^*}| \leq k + \frac{6\lambda}{7}$ *and if $|U_{i^*}| > k$ then $v_{i^*} \in S$ and $S_{v_{i^*}}$ has at least $\frac{\lambda}{7} + 2$ vertices not in U_{i^*-1}.*

The solution of our algorithm will be a graph on vertex set U_{i^*}.

Let V_2 be the set of active nodes in S with index at most i^*. We compute a tree $T_2 = (V_2, E_2)$ starting from T_0 as follows. Duplicate every edge in T_0 and find an Eulerian tour. Shortcut all the edges that go through vertices that were deleted while computing T_1 from T_0 or those vertices of T_1 that are not in V_2. The cost of T_2 is clearly at most $2c(T_0)$, using triangle inequality. Also, it only contains a subset of vertices of T_1, namely V_2. Thus:

Lemma 5. $V_2 \subseteq V_1$ *and* $c(T_2) \leq 2c(T_0)$.

Note that T_2 is in fact a path, so the maximum degree of every vertex in T_2 is at most 2. The next steps of the algorithm would be to make a $(\lambda + 1)$-clique over $S_{v_i} \cup \{v_i\}$, for each $v_i \in T_2$ which are precisely those $v_i \in S$ with $i \leq i^*$. For each active node $v_i \notin S$, we connect v_i to all the λ vertices in $S_{p(i)}$. It is easy to observe that each ball S_{v_i} with $v_i \in S$ together with all the active nodes assigned to it will form a λ-edge-connected subgraph. The final step is to make good connectivity between these balls. For that, we look at every edge $v_i v_j \in T_2$; note that both $v_i, v_j \in S$. Let $a = |S_{v_i} \cap S_{v_j}|$. We add an arbitrary matching (of size $\lambda - a$) between the $\lambda - a$ vertices in $S_{v_i} - S_{v_j}$ and $S_{v_j} - S_{v_i}$. The full description of the algorithm, is given in Figure 1, and Figure 2 illustrates the approximate Steiner tree computed and the balls around the active nodes and some of the edges added to make the graph λ-edge-connected.

3.3 Analysis of Algorithm

It is easy to see that H contains exactly those active nodes v_i with $i \leq i^*$ as well as all the nodes in $\bigcup_{v_i \leq i^* \in S} S_{v_i}$; which is exactly set U_{i^*}. Thus, by Lemma 4:

Lemma 6. $k - \frac{\lambda}{7} \leq |H| \leq k + \frac{6\lambda}{7}$ *and if $|H| > k$ then at the iteration in which v_{i^*} is added to H, $S_{v_{i^*}} \cup \{v_{i^*}\}$ adds at least $\frac{\lambda}{7} + 2$ new vertices to H.*

In the remaining of this subsection we show that H has the required connectivity while its cost is bounded by $O(\text{OPT})$.

---

1 Build graph $G' = (V' \cup V, E')$ by starting from G and creating a new vertex u' for every $u \in G$ and adding the edge uu' to G'; define $c_{u'u} = s_u$ and the cost of every other edge in G' (that also belongs to G) is multiplied by λ.

2 Compute an approximate k-Steiner tree T' (using ST-Alg) with terminals V'.

3 Compute tree $T_1 = (V_1, E_1)$ from T' (as described) which only consist of vertices v s.t. $v' \in T'$.

4 Compute the clustering of active balls $\mathcal{S}$; let i^* be the first index such that the union of active nodes v_i (and their ball S_{v_i} if it belongs to $\mathcal{S}$), for all $i \leq i^*$, is at least $k - \lambda/7$, and let $V_2 = \{v_i \in \mathcal{S} | i \leq i^*\}$.

5 Compute a tree $T_2 = (V_2, E_2)$ out of T_0 s.t. $c(T_2) \leq 2c(T_0)$; we first duplicate edges in T_0, find a Eulerian tour, and shortcut all the edges that go through vertices not in V_2.

6 Let H be an empty graph on vertex set consisting of union of all active nodes v_i with $i \leq i^*$. and $\bigcup_{v_i \in \mathcal{S}} B_{v_i}$.

7 **foreach** *active node v_i with $i \leq i^*$* **do**

8   **if** $v_i \in \mathcal{S}$ **then**

9 | Add v_i and every $u \in S_{v_i}$ to H as well as every edge uv, with $u, v \in S_{v_i} \cup \{v_i\}$

10   **else**

11 | Add v_i to H and every edge uv_i, with $u \in S_{p(i)}$

12   **end**

13 **end**

14 **foreach** *edge $v_i v_j \in T_2$* **do**

15 | Add an arbitrary matching of size $\lambda - |S_{v_i} \cap S_{v_j}|$ from $S_{v_i} - S_{v_j}$ to $S_{v_j} - S_{v_i}$ in H

16 **end**

17 **return** H

Fig. 1. Algorithm 1, which is an approximation algorithm for low cost $(k - \lambda/7, \lambda)$-subgraph

Lemma 7. *Solution H returned by the algorithm is λ-edge-connected.*

Proof. For every $v \in H$, let us define the hub for v, denoted by $h(v)$, to be (i) v itself if $v \in \mathcal{S}$, (ii) $p(i)$ if $v = v_i$ is an active node but not in $\mathcal{S}$, and (iii) $v_\ell \in \mathcal{S}$ if v_ℓ is the first vertex added to $\mathcal{S}$ with $v \in S_{v_\ell}$. Observe that the set of hub nodes are precisely the nodes in $\mathcal{S}$ with index at most i^*, which is the same as the set of nodes of T_2. First it is easy to see that each v has λ-edge connectivity to its hub (for case (iii) we have made a clique out of $h(v)$ and all the vertices in its ball including v, and for case (ii) v is adjacent to λ vertices in the clique made from the ball of $h(v)$). So it is enough to show that we have λ-edge-connectivity between the hub vertices. For any two adjacent vertices $v_i, v_j \in T_2$, the matching edges added between the balls of v_i and v_j (together with possible nodes in $S_{v_i} \cap S_{v_j}$) establish λ-edge-connectivity between v_i and v_j. By transitivity, we have λ-edge-connectivity between any pair of nodes $v_i, v_j \in T_2$.

Lemma 8. *The cost of edges of H added in line 15 is at most $8\rho\mathrm{OPT}$.*

Proof. Let $v_i v_j$ be an edge in T_2, thus $v_i, v_j \in \mathcal{S}$. For any edge xy with $x \in S_{v_i}$ and $y \in S_{v_j}$ that we add in line 15: $c_{xy} \leq c_{xv_i} + c_{v_i v_j} + c_{v_j y}$. Since the matching added between the balls of v_i and v_j has size at most λ, the cost of this matching is at most $\lambda c_{v_i v_j} + s_{v_i} + s_{v_j}$. Noting that the degree of each vertex $v_i \in T_2$ is at most 2, there

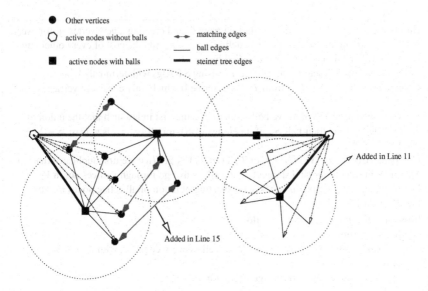

Fig. 2. A sample showing how the edges between the balls of active nodes are added and how the active nodes without a ball are connected to the balls of other active nodes

are at most two vertices like v_j with $v_i v_j \in T_2$ for which we have to add a matching between the balls of S_{v_i} and S_{v_j}. So the total cost of all the edges added in line 15 is at most: $\lambda \sum_{e \in T_2} c_e + 2 \sum_{v_i \in T_2} s_{v_i}$. Using Equation (2) and noting that $V_2 \subseteq V_0$ and $c(T_2) \leq 2c(T_0)$ (by Lemma 5), the total cost of the edges added in line 15 is at most $2\lambda \sum_{e \in T_0} c_e + 2 \sum_{v_i \in T_0} s_{v_i} \leq 8\rho\text{OPT}$.

In order to bound the cost of the edges added in lines 9 and 11 we need the following lemma.

Lemma 9. *For every $v_i \in S$, with $i \geq 2$, (that is every node in S except the root $r = v_1$) and every node $v_j \in T_1$ with $c_{v_i v_j} \leq 2\mu_i$ such that v_j became inactive once we added v_i to S: $\mu_i \leq 2\mu_j$.*

Proof. If $i < j$ (i.e. v_i was considered before v_j) then clearly $s_{v_i} \leq s_{v_j}$ and therefore $\mu_i \leq \mu_j$. Now suppose that $i > j$. It means that v_j was an active node but not in S at the time v_i was examined. This can happen only if there is $\ell < j$ with $v_\ell \in S$ and

$$c_{v_j v_\ell} \leq 4\mu_j + 2\mu_\ell. \tag{4}$$

On the other hand, since v_i was not inactivated by v_ℓ:

$$c_{v_i v_\ell} > 4\mu_i + 2\mu_\ell \tag{5}$$

Using triangle inequality:

$$c_{v_i v_\ell} \leq c_{v_i v_j} + c_{v_j v_\ell}. \tag{6}$$

Combining (4), (5), and (6), together with the assumption that $c_{v_i v_j} \leq 2\mu_i$ implies that: $4\mu_i + 2\mu_\ell < c_{v_i v_\ell} \leq c_{v_i v_j} + c_{v_j v_\ell} \leq 2\mu_i + 4\mu_j + 2\mu_\ell$; therefore $\mu_i \leq 2\mu_j$ as wanted.

Lemma 10. *The cost of edges of H added in lines 9 and 11 is at most $(2 + 28\rho)$OPT.*

Proof. We will charge the cost of the edges added to H to the vertices incident to them (at the time those edges and vertices are added) and show that the total charge is bounded by $O(\text{OPT})$. To achieve this goal, we make sure that for every vertex $v_i \in H \cap T_1$, the charge assigned to v_i is no more than $7s_{v_i}$ and for every vertex $u \in H - T_1$ (which implies it is added by adding S_{v_i} for some $v_{i \leq i^*} \in S$), the charge assigned to u is no more than $2s_{v_i}$. We consider the following cases.

Case 1: First consider an active node $v_{i \leq i^*} \notin S$; so $i \geq 2$ and we have added vertex v_i to H plus every edge uv_i with $u \in S_{p(i)}$ in line 11. Let us assume $p(i) = j^*$. Note that $c_{uv_i} \leq c_{uv_{j^*}} + c_{v_i v_{j^*}}$. So the total cost of edges added at line 11 (for adding vertex v_i to H) is at most $\lambda c_{v_i v_{j^*}} + s_{v_{j^*}}$. By definition of j^*: $c_{v_i v_{j^*}} \leq 4\mu_i + 2\mu_{j^*}$. Noting that $s_{v_{j^*}} \leq s_{v_i}$ (and therefore $\mu_{j^*} \leq \mu_i$), the total cost of edges added for v_i is at most $6\lambda\mu_i + s_{v_i} \leq 7s_{v_i}$. We charge this cost to v_i.

Case 2: Now consider an active node $v_i \in S$ for which we add all the vertices in $S_{v_i} \cup \{v_i\}$ to H and make a $(\lambda+1)$-clique on these vertices in line 9. For any two vertices $x, y \in S_{v_i} \cup \{v_i\}$: $c_{xy} \leq c_{v_i x} + c_{v_i y}$. Since each vertex x is incident with λ edges in this clique, the total cost of the edges of the clique is at most $\lambda \sum_{y \in S_{v_i}} c_{v_i y} = \lambda s_{v_i}$. Now we show how to pay for this cost by charging the vertices in $S_{v_i} \cup \{v_i\}$.

Sub-case 2a: In this sub-case we assume $i = 1$, i.e. the case of $v_1 = r$. In this case, we are adding λ new vertices in S_{v_1} in line 9 at a cost of at most λs_{v_1}. Assume that $v'_1, v'_2, v'_3, \ldots, v'_k$ are the vertices of the optimum solution where $v'_1 = v_1 = r$. Without loss of generality, and using the assumption that $v'_1 = r$ has the smallest $s_{v'_i}$ value among all the nodes in the optimum solution we assume that $s_r \leq s_{v'_2} \leq \ldots \leq s_{v'_k}$. Using the first lower bound given for OPT in the previous section: $\sum_{1 \leq i \leq k} s_{v'_i} \leq 2\text{OPT}$. Thus, using the fact that $k \geq \lambda + 1$: $\lambda s_r < \sum_{1 \leq i \leq k} s_{v'_i} \leq 2\text{OPT}$. So if we charge the root by 2OPT, we can pay for the cost of edges added for the ball of the root in line 9.

Sub-case 2b: In this sub-case we consider other active nodes $v_i \in S$ with $i \geq 2$ that are added (in line 9). As mentioned earlier, the total cost of the edges of the clique added in line 9 is at most λs_{v_i}. We will charge this cost to the vertices in B_{v_i}. Using Lemma 3, there are at least $\lambda/2$ nodes in B_{v_i} (the core of S_{v_i}) that do not belong to any B_{v_j} for $v_j \in S$ with $j \neq i$. So the vertices in B_{v_i} that we charge the total cost of λs_{v_i} to, are not charged any cost in a different core. We can pay for this (at most) λs_{v_i} cost if we charge every node in B_{v_i} by $2s_{v_i}$ and that is what we do. Remember that we want to ensure that for every vertex $v_j \in H \cap T_1$, the total charge for it is no more than $7s_{v_j}$. For every vertex $v_j \in B_{v_i} \cap T_1$ that we charge $2s_{v_i}$, if $j \geq i$ then clearly the charge $2s_{v_i}$ assigned to v_j is no more than $7s_{v_j}$. But if $j < i$, it means that v_j was de-activated when adding v_i to S; in this case the charge of $2s_{v_i}$ which is assigned to it is upper bounded by $4s_{v_j}$ for the following reason. Note that $c_{v_j v_i} \leq 2\mu_i$ (by definition of core B_{v_i}); so using Lemma 9: $\mu_i \leq 2\mu_j$, which implies $s_{v_i} \leq 2s_{v_j}$. Thus the total charge assigned to v_j is bounded by $2s_{v_i} \leq 4s_{v_j}$.

Hence, the charge for every node $v_i \in H \cap T_1$ $(i \geq 2)$ analyzed in Cases (1) and (2b) is at most $7s_{v_i}$ and the charge of every node $u \in H - T_1$, which means $u \in S_{v_i}$ for some active node $v_{i \leq i^*} \in S$, is at most $2s_{v_i}$; i.e. the property we required at the beginning of the lemma holds. We show the total charge is at most $O(\text{OPT})$, which together with the charge of root, which is 2OPT, is still $O(\text{OPT})$.

Let $\tilde{H}$ be $H - (S_{v_1} \cup \{v_1\}) - (S_{v_{i^*}} \cup \{v_{i^*}\})$, and $|\tilde{H}| = \ell$. By definition of i^*: $\ell < k - \frac{\lambda}{7} - 1$. We define a one-to-one function π from the vertices of $\tilde{H}$ to vertices $v_2, \ldots, v_{\ell+1}$ in T_1 in the following way: every vertex $v_i \in \tilde{H} \cap T_1$ is mapped to itself (therefore the charge assigned to v_i is at most $7s_{\pi(v_i)}$). Every other vertex $u \in \tilde{H} - T_1$, which is added in line 9 by adding $S_{v_i} \cup \{v_i\}$ (for some $v_{i \leq i^* - 1} \in S$), is mapped to a vertex $v_{j \geq i} \in T_1$ to which no other vertex of $\tilde{H}$ is mapped to already (in this case the charge assigned to u is at most $2s_{v_i}$, which is at most $2s_{\pi(u)}$). Thus, the total charge assigned to the vertices in $\tilde{H}$ is at most:

$$7 \sum_{2 \leq i \leq \ell+1} s_{v_i} \leq 7 \sum_{2 \leq i < k - \lambda/7} s_{v_i} \tag{7}$$

where the inequality follows from the fact that $\ell < k - \frac{\lambda}{7} - 1$. Noting that $i^* \leq k - \frac{\lambda}{7}$, the total charge of the nodes in $S_{v_{i^*}} \cup \{v_{i^*}\}$ is at most

$$\lambda s_{v_{i^*}} \leq \lambda s_{v_{k-\lambda/7}} \leq 7 \sum_{k-\lambda/7 \leq i \leq k} s_{v_i}. \tag{8}$$

Using Equations (7) and (8), together with the bound of 2OPT for the charge of v_1 in Sub-case 2b, the total charge of the nodes in H is at most

$$2\text{OPT} + 7 \sum_{2 \leq i < k - \lambda/7} s_{v_i} + 7 \sum_{k-\lambda/7 \leq i \leq k} s_{v_i} \leq 2\text{OPT} + 7 \sum_{1 \leq i \leq k} s_{v_i}.$$

Using Equation (2) and the fact $V_1 \subseteq V_0$: $7 \sum_{1 \leq i \leq k} s_{v_i} \leq 28\rho\text{OPT}$. Thus, the total charge of the nodes in H is at most $(2 + 28\rho)\text{OPT}$.

Theorem 2. *Algorithm 1 (in Figure 1) returns a graph of size at least $k - \frac{\lambda}{7}$ which is λ-edge-connected and has cost at most $(2 + 36\rho)\text{OPT}$.*

Proof. By Lemma 6 and Lemma 7, H is a λ-edge-connected graph with at least $k - \frac{\lambda}{7}$ nodes. Using Lemmas 8 and 10: $c(H) \leq 8\rho\text{OPT} + (2 + 28\rho)\text{OPT} = (2 + 36\rho)\text{OPT}$.

4   From Size $k - O(\lambda)$ to Size k

As mentioned in the previous section, graph H computed by Algorithm 1 has at least $k - \lambda/7$ vertices and has non-empty intersection with G^* (at least root r belongs to both). If $|H| \geq k$ then we are done. Otherwise, we in this section show how to augment H to have at least k vertices without loosing its edge-connectivity. For every vertex $u \in G \backslash H$, let the distance of u to H, denoted by $d(u, H)$, be the cost of the cheapest edge from u to a vertex in H. We compute two different graphs $H_1 \supseteq H$ and $H_2 \supseteq H$

---

1 If $|H| < k$ then compute H_1 and H_2 as described below. If $H_1 = H$ then return H_2.
    Otherwise, return the one with the smallest cost among H_1 and H_2.
2 H_1:
3   Start with $H_1 = H$.
4   If there is a vertex $u \in G \backslash H$ such that S_u contains at least $\lambda/7$ vertices in $G \backslash H$
        then:
5   Find such a vertex $u \in G \backslash H$ with the smallest $s_u + d(u, H)$ value.
6   If $S_u \cap H = \emptyset$ then find the cheapest edge from u to a vertex in H, say v. Add
            all the edges from u to the λ nearest neighbors of v in H_1.
7   Add all the vertices in $S_u \cup \{u\}$ to H_1 (if they do not already belong to H_1)
            and make $S_u \cup \{u\}$ a clique by adding all the necessary edges to H_1.
8 H_2:
9   Start with $H_2 = H$.
10   Find a minimum weight matching M of size $k - |H|$ between $G \backslash H$ and H, let Y
        be the set of vertices in $G \backslash H$ that participate in this matching.
11   Add each $y \in Y$ to H_2 and all the edges between y to the λ closest neighbors of
        $M(y)$ in H.

Fig. 3. Algorithm 2, which augments H (the result of Algorithm 1) to have at least k vertices

that are λ-edge-connected and return whichever has at least k nodes and the least cost. The description of the algorithm is given in Figure 3.

Figure 4 shows how graphs H_1 and H_2 are built from expanding H. To perform line 10 of the algorithm, we can use one of the known minimum weighted (bipartite) matching algorithms or a minimum cost flow algorithms (see [15]). In what follows we show that both H_1 and H_2 are λ-edge-connected and at least one of them has at least k vertices and cost at most $O(\text{OPT})$.

Lemma 11. *If $H_1 \neq H$ then it is λ-edge connected and has at least k vertices. Also, H_2 is λ-edge-connected and has at least k vertices.*

Proof. By the description of Algorithm 2 (in Figure 3), if $H_1 \neq H$ then we have added at least $\lambda/7$ new vertices (belonging to $S_u \cup \{u\}$) to H, so the size will be at least k (given that $|H| \geq k - \lambda/7$). For H_2, we add the vertices of Y and $|Y| = k - |H|$, so $|H_2| = k$. For λ-edge-connectivity, note that H was originally λ-edge-connected. If $H_1 \neq H$ then we have added a vertex u together with S_u. The vertices in $S_u \cup \{u\}$ form a clique, so are λ-edge-connected among themselves. Also, if $S_u \cap H = \emptyset$, u is λ-edge connected to some vertex $v \in H$ by the λ edges added between u and the λ nearest neighbors of v (in H). By transitivity, this implies the λ-edge connectivity of H_1. For the connectivity of H_2, every new vertex $y \in Y$ is connected to at least λ vertices in H which makes it λ-edge connected to all the vertices in H.

Lemma 12. *If H is the solution of Algorithm 1, then the solution of Algorithm 2 has cost at most $\max\{12\text{OPT} + 3c(H), 13\text{OPT} + 2c(H)\}$.*

Proof. We prove this by considering the following two cases.

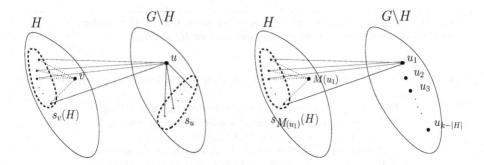

Fig. 4. Constructing H_1 (in the left picture) and H_2 (in the right picture)

Case 1: $|G^* \backslash H| \leq \lambda/3$

In this case we show that cost of H_2 is at most $O(\text{OPT})$. Every vertex $u \in G^* \backslash H$ is connected, in G^*, to at least $2\lambda/3$ vertices in $G^* \cap H$ and (by a simple averaging argument) the distance of u to at least $\lambda/3$ of them is at most $3s_u(G^*)/\lambda$ (recall that $s_u(G^*)$ is the sum of distances from u to its λ closest neighbors in G^*). Therefore there is a matching $\tilde{M}$ between $G^* \backslash H$ and $G^* \cap H$ such that $d(u, \tilde{M}(u)) \leq 3s_u(G^*)/\lambda$ for every $u \in G^* \backslash H$, and $|\tilde{M}| = |G^* \backslash H| \geq k - |H| = |M|$, where M is the matching we find in the algorithm. Since M is a minimum weight matching:

$$c(M) \leq c(\tilde{M}) \leq \frac{3}{\lambda} \sum_{u \in G^* \backslash H} s_u(G^*) \leq \frac{6\text{OPT}}{\lambda}, \tag{9}$$

where we use the lower bound of $\sum_{u \in G^* \backslash H} s_u(G^*) \leq \sum_{u \in G^*} s_u(G^*) \leq 2\text{OPT}$. Connecting u to the λ nearest neighbors of $M(u)$ costs at most $\lambda \cdot d(u, M(u)) + s_{M(u)}(H)$, by triangle inequality. Thus, if $c(M)$ and $c(H)$ denote the cost of edges of matching M and graph H, respectively, the total cost of the edges added in line 11 is at most:

$$\sum_{u \in G^* \backslash H} (\lambda \cdot d(u, M(u)) + s_{M(u)}(H)) \leq \lambda \cdot c(M) + 2c(H)$$

$$\leq 6\text{OPT} + 2c(H)$$

where the first inequality follows from the fact that $\cup_{u \in G^* \backslash H} S_{M(u)}(H)$ counts every edge of H at most twice and, therefore, its cost is at most twice as much as the cost of H. Thus, $c(H_2) \leq 3c(H) + 6\text{OPT}$.

Case 2: $|G^* \backslash H| > \lambda/3$

In this case (again by an averaging argument) there is a set Y of $\lambda/6$ vertices in $G^* \backslash H$ such that $s_y \leq 12\text{OPT}/\lambda$ for each $y \in Y$ (Otherwise, the remaining at least $\lambda/6$ vertices would have total s value more than $\lambda/6 \times 12\text{OPT}/\lambda \geq 2OPT$ which is a contradiction). If every vertex in Y is connected, in G^*, to at least $2\lambda/6$ vertices in $G^* \cap H$ then, with an argument similar to the previous case, we can upper bound the cost of H_2 by:

$$c(H_2) \leq c(H) + \lambda \sum_{y \in Y} d(y, M(y)) + 2c(H)$$

$$\leq \lambda \sum_{y \in Y} \frac{6s_y}{\lambda} + 3c(H)$$

$$= 6 \sum_{y \in Y} s_y + 3c(H)$$

$$\leq 6 \cdot \frac{\lambda}{6} \cdot \frac{12\text{OPT}}{\lambda} + 3c(H)$$

$$= 12\text{OPT} + 3c(H),$$

where the 2nd inequality follows from the fact that Y has $\lambda/6$ nodes and each $y \in Y$ has at least $\lambda/3$ neighbors in $G^* \cap H$.

Otherwise, let $y \in Y$ be a vertex such that S_y has more than $4\lambda/6 > \lambda/7$ vertices in $G^* \backslash H$. In this case we show that $H_1 \neq H$ and $c(H_1) = O(\text{OPT})$. First we claim that the clique on vertices $S_y \cup \{y\}$ costs at most $\lambda \cdot s_y \leq 12\text{OPT}$. The reason is, for each pair $u, v \in S_y$: $c_{uv} \leq c_{uy} + c_{yv}$. Since edge uy participates in λ such inequalities, we get that the total cost of the clique is at most λs_y, and because each vertex $y \in Y$ has $s_y \leq 12\text{OPT}/\lambda$, we get the upper bound of 12OPT. Furthermore, since $H \cap G^*$ is non-empty (at least $r \in H$), y must have distance at most OPT/λ to some vertex $v \in H$ (because there are at least λ edge disjoint paths between y and v and the cost of each is at least c_{yv} by the triangle inequality). Thus, with an argument similar to the previous case, it costs at most $\lambda c_{yv} + s_v(H) \leq \text{OPT} + c(H)$ to connect y to the λ nearest neighbors of v in H. So the cost of building the clique on $S_y \cup \{y\}$ and connecting y to the λ nearest neighbors of v in H is at most $13\text{OPT} + c(H)$. This implies that H_1 will have at least k vertices and costs at most $13\text{OPT} + 2c(H)$

Combining the two Algorithms 1 (in Figure 1) and 2 (in Figure 3), and using Lemmas 12 and 11, and Theorem 2 we have an algorithm that returns a λ-edge connected subgraph on at least k vertices with cost at most $\max\{12\text{OPT} + 3c(H), 13\text{OPT} + 2c(H)\} \leq 3(2 + 36\rho)\text{OPT} + 12\text{OPT} = (18 + 108\rho)\text{OPT}$. Thus, we have the following theorem which is essentially Theorem 1:

Theorem 3. *There is a polynomial time algorithm for the (k, λ)-subgraph problem on graphs with metric edge costs which has approximation factor at most $18 + 108\rho$, with $\rho \leq 4$ being the best approximation factor for the k-Steiner tree problem.*

5   Concluding Remarks

In this paper, we proved that the (k, λ)-subgraph problem with metric costs has a polynomial time $O(1)$-approximation algorithm. However, the approximation ratio of our algorithm is relatively large (between 400 and 500). Although it is very likely that one can achieve an approximation ratio close to 100 using the same algorithm by fine tuning the parameters, getting a small constant factor approximation seems challenging, for general values of λ.

For general cost functions, the only known results on this problem (that we are aware of) are the papers [12,5] which prove that the $(k, 2)$-subgraph problem (on general graphs) has $O(\log^2 n)$-approximation, even if we require 2-node-connectivity in the solution (instead of 2-edge-connectivity). Even for the special case of $\lambda = 3$, there is no known non-trivial approximation algorithm or lower bound (hardness result).

Acknowledgments. The second author thanks Joseph Cheriyan for some initial discussions on the problem.

References

1. Awerbuch, B., Azar, Y., Blum, A., Vempala, S.: New approximation guarantees for minimum-weight k-trees and prize-collecting salesmen. SIAM J. Computing 28(1), 254–262 (1999)
2. Blum, A., Chawla, S., Karger, D., Lane, T., Meyerson, A., Minkoff, M.: Approximation Algorithms for Orienteering and Discounted-Reward TSP. SIAM J. on Computing 28(1), 254–262 (1999); Earlier version in Proc of STOC 1995
3. Blum, A., Ravi, R., Vempala, S.: A constant-factor approximation algorithm for the k-MST problem. J. Comput. Syst. Sci. 58(1), 101–108 (1999); Earlier in Proceedings of the 28th Annual ACM Symposium on the Theory of Computing (STOC 1996), pp. 442–448
4. Cheriyan, J., Vetta, A.: Approximation algorithms for network design with metric costs. In: Proceedings of the thirty-seventh annual ACM symposium on Theory of computing (STOC), pp. 167–175 (2005)
5. Chekuri, C., Korula, N.: Min-Cost 2-Connected Subgraphs with k Terminals (manuscript, 2008), http://arxiv.org/abs/0802.2528
6. Chekuri, C., Korula, N., Pál, M.: Improved Algorithms for Orienteering and Related Problems. In: Proc of ACM-SIAM SODA (2008)
7. Chudak, F., Roughgarden, T., Williamson, D.P.: Approximate MSTs and Steiner trees via the primal-dual method and Lagrangean relaxation. Math. Program. 100(2), 411–421 (2004)
8. Feige, U., Kortsarz, G., Peleg, D.: The dense k-subgraph problem. Algorithmica 29(3), 410–421 (2001); Preliminary version in the Proc. 34-th IEEE Symp. on Foundations of Computer Science (FOCS) pp. 692–701 (1993)
9. Garg, N.: A 3-Approximation for the minim tree spanning k vertices. In: Proceedings of the 37th Annual Symposium on Foundations of Computer Science (FOCS), pp. 302–309 (1996)
10. Garg, N.: Saving an epsilon: a 2-approximation for the k-MST problem in graphs. In: Proceedings of the thirty-seventh annual ACM symposium on Theory of computing (STOC), pp. 396–402 (2005)
11. Jain, K.: A factor 2 approximation algorithm for the generalized Steiner network problem. Combinatorica 21, 39–60 (2001)
12. Lau, L., Naor, S., Salavatipour, M., Singh, M.: Survivable Network Design with Degree or Order Constraints. SIAM J. on Computing (submitted); Earlier version in Proceedings of the thirty-ninth annual ACM symposium on Theory of computing (STOC) (2007)
13. Rajagopalan, S., Vazirani, V.: Logarithmic approximation of minimum weight k trees (unpublished manuscript) (1995)
14. Ravi, R., Sundaram, R., Marathe, M.V., Rosenkrants, D.J., Ravi, S.S.: Spanning trees short or small. SIAM Journal on Discrete Mathematics 9(2), 178–200 (1996)
15. Schrijver, A.: Combinatorial Optimization. Springer, Heidelberg (2003)

Budgeted Allocations in the Full-Information Setting[*]

Aravind Srinivasan

Dept. of Computer Science and Institute for Advanced Computer Studies,
University of Maryland, College Park, MD 20742

Abstract. We build on the work of Andelman & Mansour and Azar, Birnbaum, Karlin, Mathieu & Thach Nguyen to show that the full-information (i.e., offline) budgeted-allocation problem can be approximated to within $4/3$: we conduct a rounding of the natural LP relaxation, for which our algorithm matches the known lower-bound on the integrality gap.

1 Introduction

Sponsored-search auctions are a key driver of advertising, and are a topic of much current research (Lahaie, Pennock, Saberi & Vohra [10]). A fundamental problem here is *online budgeted allocation*, formulated and investigated by Mehta, Saberi, Vazirani & Vazirani [12]. Recent work has also focused on the offline version of this basic allocation problem; we improve on the known results, demonstrating a rounding approach for a natural LP relaxation that yields a $4/3$-approximation, matching the known integrality gap. We also show that in the natural scenario where bidders' individual bids are much smaller than their budgets, our algorithm solves the problem almost to optimality.

Our problem is as follows. We are given a set U of *bidders* and a set V of *keywords*. Each bidder i is willing to pay an amount $b_{i,j}$ for keyword j to be allocated to them; each bidder i also has a budget B_i at which their total payment is capped. Our goal is to assign each keyword to at most one bidder, in order to maximize the total payment obtained. This models the problem of deciding which bidder (if any) gets to be listed for each keyword, in order to maximize the total revenue obtained by, say, a search engine. That is, we want to solve the following integer linear program (ILP), where $x_{i,j}$ is the indicator variable for keyword j getting assigned to bidder i: maximize $\sum_{i \in U} \min\{B_i, \sum_{j \in V} b_{i,j} x_{i,j}\}$, subject to $\sum_i x_{i,j} \le 1$ for each j, and $x_{i,j} \in \{0,1\}$ for all (i,j). (It is easy to see that the "min" term can be appropriately rewritten in order to express this as a standard ILP.)

Known results. This NP-hard problem has been studied by Garg, Kumar & Pandit, who presented an $(1 + \sqrt{5})/2 \sim 1.618$-approximation algorithm for

---

[*] Research supported in part by NSF ITR Award CNS-0426683 and NSF Award CNS-0626636. Part of this work was done while the author was on sabbatical at the Network Dynamics and Simulation Science Laboratory of the Virginia Bioinformatics Institute, Virginia Tech.

A. Goel et al. (Eds.): APPROX and RANDOM 2008, LNCS 5171, pp. 247–253, 2008.
© Springer-Verlag Berlin Heidelberg 2008

the problem [8]. (As usual, for our maximization problem, a ρ-approximation algorithm, for $\rho \geq 1$, is a polynomial-time algorithm that always presents a solution of value at least $1/\rho$ times optimal; in the case of randomized algorithms, the *expected* solution-value should be at least $1/\rho$ times optimal.) In addition to other results, Lehmann, Lehmann & Nisan [11] have developed a greedy 2-approximation algorithm for this problem. Now, the natural LP relaxation for the problem is obtained by relaxing each $x_{i,j}$ to lie in $[0, 1]$, in the above ILP. Andelman & Mansour [2] presented a rounding algorithm for this LP that achieves an approximation of $e/(e - 1) \sim 1.582$; this was improved – for a more general problem – by Feige & Vondrak to $e/(e - 1) - \epsilon$, for an ϵ that is about 10^{-4} [6]. More recently, Azar, Birnbaum, Karlin, Mathieu & Thach Nguyen [3] have improved the approximation ratio to $3/2$. There are also two interesting special cases of the problem: the *uniform* case, where each j has a price p_j such that $b_{i,j} \in \{0, p_j\}$ for all i, and the case where all the budgets B_i are the same. Two additional results are obtained in [2]: that the integrality gap of the above LP-relaxation is at least $4/3$ even for the first (i.e., uniform) special case, and that the second special case can be approximated to within 1.39. See, e.g., [12, 4, 9] for online versions of the problem.

Our results. We build on the work of [2, 3] and show how to round the LP to obtain an approximation of $4/3$: note from the previous paragraph that this meets the integrality gap. Anna Karlin (personal communication, March 2008) has informed us that Chakrabarty & Goel have independently obtained this approximation ratio, as well as improved hardness-of-approximation results – a preprint of this work is available [5]. We also present two extensions in Section 3: (a) the important special case where each bidder's bids are much smaller than their budget [12, 4] can be solved near-optimally: if, for some $\epsilon \in [0, 1]$, $b_{i,j} \leq \epsilon \cdot B_i$ for all (i, j), our algorithm's approximation ratio is $4/(4 - \epsilon)$; and (b) suppose that for some $\lambda \geq 1$, we have for all (i, j, j') that if $b_{i,j}$ and $b_{i,j'}$ are nonzero, then $b_{i,j} \leq \lambda \cdot b_{i,j'}$. For this case, our algorithm yields a better-than-$4/3$ approximation if $\lambda < 2$. In particular, if $\lambda = 1$, our algorithm has an approximation ratio of $(\sqrt{2} + 1)/2 \sim 1.207$.

2 The Algorithm and Analysis

We will round the natural LP-relaxation mentioned in Section 1. Our algorithm is randomized, and can be derandomized using the method of conditional probabilities.

Observe that for the original (integral) problem, setting

$$b_{i,j} := \min\{b_{i,j}, B_i\} \tag{1}$$

keeps the problem unchanged. Thus, we will assume

$$\forall (i, j), \ b_{i,j} \leq B_i. \tag{2}$$

Notation. When we refer to the *load* on a bidder i w.r.t. some (fractional) allocation x, we mean the sum $\sum_j b_{i,j} x_{i,j}$; note that we do not truncate at B_i in this definition.

Suppose we are given some feasible fractional allocation x; of course, the case of interest is where this is an optimal solution to the LP, but we do not require it. It is also immediate that the following assumption is without loss of generality:

$$\text{if } b_{i,j} = 0, \text{ then } x_{i,j} = 0. \tag{3}$$

As in [3], we may assume that the bipartite graph (with (U, V) as the partition) induced by those $x_{i,j}$ that lie in $(0, 1)$, is a *forest* F. This can be effected by an efficient algorithm, such that the resulting fractional objective-function value equals that of the original value that we started with [3]. This forest F is the structure that we start with; we show how to round those $x_{i,j}$ in F. We are motivated by the approaches of [1, 13, 7]; however, our method is different, especially in step **(P2)** below. Each iteration is described next.

2.1 Iteration s, $s \geq 1$

Remove all (i, j) that have already been rounded to 0 or 1; let F be the current forest consisting of those $x_{i,j}$ that lie in $(0, 1)$. Choose any **maximal** path $P = (w_0, w_1, \ldots, w_k)$ in F; we will now probabilistically round at least one of the edges in P to 0 or 1. For notational simplicity, let the current x value of edge $e_t = (w_{t-1}, w_t)$ in P be denoted y_t; note that all the y_t lie in $(0, 1)$. We will next choose values $z_1, z_2, \ldots, z_k$ probabilistically, and update the x value of each edge $e_t = (w_{t-1}, w_t)$ to $y_t + z_t$. Suppose we initialize some value for z_1, and that we have chosen the increments $z_1, z_2, \ldots, z_t$, for some $t \geq 1$. Then, the value z_{t+1} (corresponding to edge $e_{t+1} = (w_t, w_{t+1})$) is chosen as follows:

(P1) If $w_t \in V$ (i.e., is a keyword), then $z_{t+1} = -z_t$ (i.e., we retain the total assignment value of w_t);

(P2) if $w_t \in U$ (i.e., is a bidder), then we choose z_{t+1} so that the load on w_t remains unchanged (recall that in computing the load, we do not truncate at B_{w_t}); i.e., we set $z_{t+1} = -b_{w_t, w_{t-1}} z_t / b_{w_t, w_{t+1}}$, which ensures that the incremental load $b_{w_t, w_{t-1}} z_t + b_{w_t, w_{t+1}} z_{t+1}$ is zero. (Since $x_{w_t, w_{t+1}}$ is nonzero by the definition of F, $b_{w_t, w_{t+1}}$ is also nonzero by (3); therefore, dividing by $b_{w_t, w_{t+1}}$ is admissible.)

Observe that the vector $z = (z_1, z_2, \ldots, z_k)$ is completely determined by z_1, the path P, and the matrix of bids b; more precisely, there exist reals $c_1, c_2, \ldots, c_k$ that depend only on the path P and the matrix b, such that

$$\forall t, \ z_t = c_t z_1. \tag{4}$$

We will denote this resultant vector z by $f(z_1)$.

Now let μ be the smallest positive value such that if we set $z_1 := \mu$, then all the x values (after incrementing by the vector z as mentioned above) stay in $[0, 1]$, and at least one of them becomes 0 or 1. Similarly, let γ be the smallest positive value such that if we set $z_1 := -\gamma$, then this "rounding-progress" property holds. (It is easy to see that μ and γ are strictly positive, since all the y_i lie in $(0,1)$.) We now choose the vector z as follows:

(R1) with probability $\gamma/(\mu+\gamma)$, let $z = f(\mu)$;
(R2) with the complementary probability of $\mu/(\mu+\gamma)$, let $z = f(-\gamma)$.

2.2 Analysis

If $Z = (Z_1, Z_2, \ldots, Z_k)$ denotes the random vector z chosen in steps (R1) and (R2), the choice of probabilities in (R1) and (R2) ensures that $\mathbf{E}[Z_1] = 0$. So, we have from (4) that
$$\forall t, \ \mathbf{E}[Z_t] = 0. \tag{5}$$

The algorithm clearly rounds at least one edge permanently in each iteration (and removes all such edges from the forest F), and therefore terminates in polynomial time. We now analyze the expected revenue obtained from each bidder i, and prove that it is not too small.

Let $L_i^{(s)}$ denote the load on bidder i at the end of iteration s; the values $L_i^{(0)}$ refer to the initial values obtained by running the subroutine of [3] that obtains the forest F. Property (P2) shows that as long as i has degree at least two in the forest F, $L_i^{(s)}$ stays at its initial value $L_i^{(0)}$ with probability 1. (Recall that whenever we refer to F etc., we always refer to its subgraph containing those edges with x values in $(0,1)$; edges that get rounded to 0 or 1 are removed from F.) In particular, if i never had degree one at the end of any iteration, then its final load equals $L_i^{(0)}$ with probability one, so the expected approximation ratio for i is one. So, suppose the degree of i came down to one at the end of some iteration s. Let the corresponding unique neighbor of i be j, let $\beta = b_{i,j}$, and suppose, at the end of iteration s, the total already-rounded load on i and the value of $x_{i,j}$ are $\alpha \geq 0$ and $p \in (0,1)$ respectively. Note that j, α, β as well as p are all random variables, and that $L_i^{(s)} = \alpha + \beta p$; so,

$$\Pr[\alpha + \beta p = L_i^{(0)}] = 1.$$

Fix any j, α, β and p that satisfy $\alpha + \beta p = L_i^{(0)}$; all calculations from now on will be conditional on this fixed choice, and on all random choices made up to the end of iteration s. Property (5) and induction on the iterations show that the final load on i (which is now a random variable that is a function of the random choices made from iteration $s+1$ onward) is:

$$\alpha, \text{ with probability } 1 - p; \text{ and } \alpha + \beta, \text{ with probability } p. \tag{6}$$

Let $B = B_i$ for brevity. Thus, the final expected revenue from i is $(1-p) \cdot \min\{\alpha, B\} + p \cdot \min\{\alpha + \beta, B\}$; the revenue obtained from i in the LP solution is

$\min\{\alpha + \beta p, B\}$. So, by the linearity of expectation, the expected approximation ratio is the maximum possible value of

$$\frac{\min\{\alpha + \beta p, B\}}{(1 - p) \cdot \min\{\alpha, B\} + p \cdot \min\{\alpha + \beta, B\}}.$$

It is easily seen that this ratio is 1 if $\alpha > B$ or if $\alpha + \beta < B$. Also note from (2) that $\beta \leq B$. Thus, we want the minimum possible value of the reciprocal of the approximation ratio:

$$r = \frac{(1 - p)\alpha + pB}{\min\{\alpha + \beta p, B\}}, \tag{7}$$

subject to the constraints

$$p \in [0, 1]; \ \alpha, \beta \leq B; \ \alpha + \beta \geq B. \tag{8}$$

(Of course, we assume the denominator of (7) is nonzero. In the case where it is zero, it is easy to see that so is the numerator, in which case it follows trivially that $(1 - p)\alpha + pB \geq (3/4) \cdot \min\{\alpha + \beta p, B\}$.)

We consider two cases, based on which term in the denominator of r is smaller:

Case I: $\alpha + \beta p \leq B$. Here, we want to minimize

$$r = \frac{(1 - p)\alpha + pB}{\alpha + \beta p}. \tag{9}$$

Keeping all other variables fixed and viewing α as a variable, r is minimized when α takes one of its extreme values, since r is a non-negative rational function of α. From our constraints, we have $B - \beta \leq \alpha \leq B - \beta p$. Thus, r is minimized at one of these two extreme values of α. If $\alpha + \beta = B$, then $r = 1$. Suppose $\alpha = B - \beta p$. Then,

$$r = \frac{(1 - p)\alpha + pB}{B}. \tag{10}$$

Since

$$\alpha = B - \beta p \geq B(1 - p), \tag{11}$$

we have

$$r = \frac{(1 - p)\alpha + pB}{B} \geq (1 - p)^2 + p,$$

which attains a minimum value of $3/4$ when $p = 1/2$.

Case II: $\alpha + \beta p \geq B$. We once again fix all other variables and vary α; the extreme values for α now are $\alpha = B - Bp$ (with $\beta = B$) and $\alpha = B$. In the former case, the argument of Case I shows that $r \geq 3/4$; in the latter case, r is easily seen to be 1.

This completes the proof that our expected approximation ratio is at most $4/3$. Also, it is easy to derandomize the algorithm by picking one of the two possible updates in each iteration using the method of conditional probabilities; we will describe this in the full version. Thus we have the following theorem:

Theorem 1. *Given any feasible fractional solution to the LP-relaxation of the offline budgeted-allocation problem with the truncations (1) done without loss of generality, it can be rounded to a feasible integer solution with at least 3/4-th the value of the fractional solution in deterministic polynomial time. Therefore, the offline budgeted-allocation problem can be approximated to within 4/3 in deterministic polynomial time.*

3   Extensions

The following two extensions hold.

3.1   The Case of Bids Being Small w.r.t. Budgets

Here we consider the case where for some $\epsilon \in [0,1]$, we have for all i,j that $b_{i,j} \leq \epsilon B_i$. The only modification needed to the analysis of Section 2.2 is that (11) now becomes "$\alpha = B - \beta p \geq B(1 - \epsilon p)$", and that the function to minimize is $(1-p)\cdot(1-\epsilon p)+p$ instead of $(1-p)^2+p$. This is again minimized at $p = 1/2$, giving $r \geq 1-\epsilon/4$. Thus, the approximation ratio in this case is at most $4/(4-\epsilon)$.

3.2   The Case of Similar Bids for Any Given Bidder

We now study the case where for each i, all its nonzero bids $b_{i,j}$ are within some factor λ of each other, where $1 \leq \lambda \leq 2$. Note that different bidders can still have widely-differing bid values.

Consider the analysis of Section 2.2. In the trivial case where $\alpha = 0$, it easily follows from (6) that the approximation ratio for machine i is 1. So suppose $\alpha > 0$; then the additional constraint that

$$\beta \leq \alpha\lambda \qquad (12)$$

must hold, by our assumption about the bid-values.

By a tedious proof along the lines of Section 2.2, it can be shown that we get a better-than-4/3 approximation if $\lambda < 2$. We will present the calculation-details in the full version. For now, we just focus on the case where $\lambda = 1$. Recall that we aim to minimize r from (7), subject to (8) and the constraint (12), i.e., $\alpha \geq \beta$. Let us first argue that if the minimum value of r is smaller than 1, then $\alpha = \beta$ at any minimizing point. To see this, assume for a contradiction that there is a minimizing pair (α, β) with $\alpha > \beta$, and observe that we may make the following three sets of assumptions w.l.o.g.: (i) if $\alpha = 0$ or $\alpha + \beta = B$, then $r = 1$: so, we may assume that $\alpha > 0$ and $\alpha + \beta > B$; (ii) if $\beta = B$, then $\alpha \geq \beta = B = \beta$ and we are done, so we can assume $\beta < B$; (iii) if $p = 0$, then $r = 1$, so we can take $p > 0$. Now, if we perturb as $\alpha := \alpha - \delta$ and $\beta := \beta + \delta/p$ for some tiny positive δ, then we stay in the feasible region and get a smaller value for r from (7), a contradiction. So, we can take $\alpha = \beta$, and have from (8) that $\alpha = \beta \geq B/2$.

We repeat the case analysis of Section 2.2. In Case I, the extreme value $\alpha = B/2$ gives $r = 1$. The other extreme value is $\alpha = B - \beta p = B - \alpha p$, i.e., $\alpha = B/(1+p)$. So, the r of (10) becomes $(1-p)/(1+p)+p$, whose minimum value is $2(\sqrt{2}-1)$. Similarly in Case II. Thus, $r \geq 2(\sqrt{2}-1)$, and taking the reciprocal, we see that the approximation ratio is $(\sqrt{2}+1)/2 \sim 1.207$.

Acknowledgment. I thank Anna Karlin for informing me about related work, and the APPROX 2008 referees for their helpful suggestions.

References

[1] Ageev, A.A., Sviridenko, M.: Pipage rounding: A new method of constructing algorithms with proven performance guarantee. J. Comb. Optim. 8(3), 307–328 (2004)

[2] Andelman, N., Mansour, Y.: Auctions with budget constraints. In: Hagerup, T., Katajainen, J. (eds.) SWAT 2004. LNCS, vol. 3111, pp. 26–38. Springer, Heidelberg (2004)

[3] Azar, Y., Birnbaum, B., Karlin, A.R., Mathieu, C., Nguyen, C.T.: Improved approximation algorithms for budgeted allocations. In: Aceto, L., Damgård, I., Goldberg, L.A., Halldórsson, M.M., Ingolfsdóttir, A., Walukiewicz, I. (eds.) ICALP. LNCS, vol. 5125, pp. 186–197. Springer, Heidelberg (2008)

[4] Buchbinder, N., Jain, K., Naor, J.: Online primal-dual algorithms for maximizing ad-auctions revenue. In: Arge, L., Hoffmann, M., Welzl, E. (eds.) ESA 2007. LNCS, vol. 4698, pp. 253–264. Springer, Heidelberg (2007)

[5] Chakrabarty, D., Goel, G.: On the approximability of budgeted allocations and improved lower bounds for submodular welfare maximization and GAP (manuscript)

[6] Feige, U., Vondrák, J.: Approximation algorithms for allocation problems: Improving the factor of 1 - 1/e. In: FOCS, pp. 667–676 (2006)

[7] Gandhi, R., Khuller, S., Parthasarathy, S., Srinivasan, A.: Dependent rounding and its applications to approximation algorithms. J. ACM 53(3), 324–360 (2006)

[8] Garg, R., Kumar, V., Pandit, V.: Approximation algorithms for budget-constrained auctions. In: RANDOM-APPROX, pp. 102–113 (2001)

[9] Goel, G., Mehta, A.: Online budgeted matching in random input models with applications to Adwords. In: SODA, pp. 982–991 (2008)

[10] Lahaie, S., Pennock, D.M., Saberi, A., Vohra, R.V.: Sponsored search auctions. In: Nisan, N., Roughgarden, T., Tardos, É., Vazirani, V.V. (eds.) Algorithmic Game Theory, ch. 28, pp. 699–716. Cambridge University Press, Cambridge (2007)

[11] Lehmann, B., Lehmann, D., Nisan, N.: Combinatorial auctions with decreasing marginal utilities. Games and Economic Behavior 55(2), 270–296 (2006)

[12] Mehta, A., Saberi, A., Vazirani, U.V., Vazirani, V.V.: Adwords and generalized online matching. J. ACM 54(5) (2007)

[13] Srinivasan, A.: Distributions on level-sets with applications to approximation algorithms. In: FOCS, pp. 588–597 (2001)

Optimal Random Matchings on Trees and Applications

Jeff Abrahamson[1], Béla Csaba[2,*], and Ali Shokoufandeh[1,**]

[1] Dept. of Computer Science
Drexel University
Philadelphia, PA
{jeffa,ashokouf}@cs.drexel.edu
[2] Dept. of Mathematics
Western Kentucky University
Bowling Green, KY
bela.csaba@wku.edu

Abstract. In this paper we will consider tight upper and lower bounds on the weight of the optimal matching for random point sets distributed among the leaves of a tree, as a function of its cardinality. Specifically, given two n sets of points $R = \{r_1, ..., r_n\}$ and $B = \{b_1, ..., b_n\}$ distributed uniformly and randomly on the m leaves of λ-Hierarchically Separated Trees with branching factor b such that each of its leaves is at depth δ, we will prove that the expected weight of optimal matching between R and B is $\Theta(\sqrt{nb}\sum_{k=1}^{h}(\sqrt{b}\lambda)^k)$, for $h = \min(\delta, \log_b n)$. Using a simple embedding algorithm from $\mathbb{R}^d$ to HSTs, we are able to reproduce the results concerning the expected optimal transportation cost in $[0,1]^d$, except for $d = 2$. We also show that giving random weights to the points does not affect the expected matching weight by more than a constant factor. Finally, we prove upper bounds on several sets for which showing reasonable matching results would previously have been intractable, e.g., the Cantor set, and various fractals.

Keywords: Random Matching, Hierarchically Separated Trees, Supremum Bounds.

1 Introduction

The problem of computing a large similar common subset of two point sets arises in many areas of computer science, ranging from computer vision and pattern recognition, to bio-informatics [2,12,4]. Most of the recent related work concerns the design of efficient algorithms to compute rigid transformations for establishing correspondences between two point sets in $\mathbb{R}^d$ subject to minimization of a

* Part of this research was done while the author worked at the Analysis and Stochastics Research Group at the University of Szeged, Hungary. Partially supported by OTKA T049398.
** Partially supported by NSF grant IIS-0456001 and by ONR grant ONR-N000140410363.

A. Goel et al. (Eds.): APPROX and RANDOM 2008, LNCS 5171, pp. 254–265, 2008.
© Springer-Verlag Berlin Heidelberg 2008

distance measure. In comparison, less attention has been devoted to extremal matching problems related to random point sets, such as: *Presented with two random point sets, how do we expect matching weight to vary with data set size?*

Perhaps the most seminal work in extremal random matching is the 1984 paper of Ajtai, Komlós and Tusnády [1] presenting a very deep and important result which has found many applications since then. Considering two sets of points X_n and Y_n chosen uniformly at random in $[0,1]^2$, with $|X_n| = |Y_n| = n$, they determined (asymptotic) bounds on the sequence $\{\mathbf{E}M\}$, where M is the optimal matching weight, or *transportation cost* between X_n and Y_n: $M = \min_\sigma \sum_i ||X_i - Y_{\sigma(i)}||_2$ where σ runs through all the possible permutations on $[n]$. Shortly after Ajtai et al., Leighton and Shor [8] addressed the problem of 2-dimensional grid matching, analyzing the maximum cost of any edge in the matching instead of the sum. Shor and Yukich [14] extended this minimax grid matching result to dimensions greater than two. Shor [13] applied the AKT result to obtain bounds on the average case analysis of several algorithms. Talagrand [15] introduced the notion of majorizing measures and as an illustration of this powerful technique derived the theorem of Ajtai et al. Rhee and Talagrand [9] have explored upward matching (in $[0,1]^2$): the case where points from X must be matched to points of Y that have greater x- and y-coordinates. They have also explored a similar problem in the cube [10]. In [16] Talagrand gave insight to exact behavior of expected matching weight for dimensions $d \geq 3$ for arbitrary norms.

In this paper we will introduce the random matching problem on *hierarchically separated trees*. The notion of a hierarchically (well-)separated tree (HST) was introduced by Bartal [3]. A λ-HST is a rooted weighted tree with two additional properties: (1) edge weights between nodes and their children are the same for any given parent node, and (2) the ratio of incident edge weights along a root-leaf path is λ (so edges get lighter by a factor of λ as one approaches leaves). We primarily consider balanced trees, i.e., trees in which the branching factor of all nodes other than the leaves is an integer b, and in which every leaf is at depth δ. Using the notion of balanced λ-HST, we can state the first contribution of this manuscript on the expected transportation cost of optimal matching $\mathbf{E}M_T(R,B)$:

Theorem 1. *Let $T = T(b,\delta,\lambda)$ be a balanced HST, and R and B two randomly chosen n-element submultisets of the set of leaves of T and define $h = \min(\delta, \log_b n)$. Then there exist positive constants K_1 and K_2 such that*

$$K_1 \sqrt{bn} \sum_{k=1}^{h} (\sqrt{b}\lambda)^k \leq \mathbf{E}M_T(R,B) \leq K_2 \sqrt{bn} \sum_{k=1}^{h} (\sqrt{b}\lambda)^k.$$

Theorem 1 will also allow us easily to approach and duplicate the upper-bound results of optimal matching for point sets distributed in $[0,1]^d$ found in the literature (see [7]), with a slightly loose result in the single (and most interesting) case of $d = 2$. Since we use crude approximations of $[0,1]^d$ by HSTs, we cannot expect much more.

On the other hand, this method is general enough to attack the randomized matching problem in general for finite metric spaces. It can always give upper bounds (by using Theorem 2 or Corollary 1). If the metric space is sufficiently symmetric (e.g., fractals), one can get reasonable lower bounds by applying the theorem of Fakcharoenphol et al. ([5]) on approximating a finite metric space by HSTs. We further extend the upper bound of the transportation cost to the case of weighted point sets. This model is commonly used in texture mapping in computer vision (see [11]).

The final application of the newly developed machinery will include extending upper-bound matching results to finite approximations of certain fractals. We generalize Theorem 1 for non-uniformly distributed point sets and for subtrees of balanced trees as well.

2   The Upper and Lower Bounds for Matching on HSTs

In this section, our modus operandi will be to prove upper- and lower bounds for the weight of the matching problems on HSTs. The trees considered in this paper are a somewhat restricted variation of HSTs defined as follows:

Definition 1. *Let b, δ be positive integers and $0 < \lambda < 1$ be a real number. We call a rooted tree T a balanced (b, δ, λ)-HST, if every edge incident on the root has unit weight, every edge not incident on the root has weight λ times the weight of the edge immediately closer to the root, every non-leaf node has the same number of children (which we will call the branching factor b), and every leaf has the same depth δ.*

We remark that having the same depth δ for every leaf of T can be assumed without loss of generality, and as we will see, in several cases the branching factor is naturally bounded.

Given a balanced HST T, let $R = \{r_1, ..., r_n\}$ and $B = \{b_1, ..., b_n\}$ respectively denote the multisets of n red and n blue points chosen among the leaves of T. We define a *matching* between R and B as a one-to-one mapping σ between them. The weight of the optimal matching (optimal transportation cost) with respect to T will be defined as $M_T(R, B) = \min_\sigma \left(\sum_{1 \leq i \leq n} d_T(r_i, b_{\sigma(i)}) \right)$, where $d_T(r, b)$ is the length of the path between leaves containing points r and b in T. Note that $M_T(R, B)$ is the Earth Mover's Distance of R and B on the metric defined by T. For a pair of points (r, b) matched under a mapping σ and belonging to distinct leaves u_r and u_b in T, we will say the matched pair (r, b) results in a *transit* at vertex v, if v is an ancestor of both u_r and u_b and the path between u_r and u_b passes through v. We will also use τ_v to denote the total number of transits at vertex v in an optimal matching between R and B. Any red-blue pair that is mapped under a matching σ at a leaf of T contributes no weight to the transportation cost. For a vertex v let $\delta(v)$ denote its level in the tree, that is, the number of edges on the path from r to v. Observe that the weight of the optimal matching can be restated as follows:

$$M_T(R, B) = \sum_{k=0}^{\delta-1} \sum_{v:\delta(v)=k} \tau_v S(k, \delta - 1), \tag{1}$$

where $S(i,j) = 2(\lambda^i + \lambda^{i+1} + \ldots + \lambda^j) \leq C_\lambda \lambda^i$ for $0 \leq i \leq j \leq \delta - 1$, and $C_\lambda = \sum_{j \geq 0} \lambda^j$.

Our goal is to estimate tight bounds on the expected optimal transportation cost $\mathbf{E}M_T(R, B)$ for randomly chosen R and B. Throughout the paper we will denote the standard deviation of a random variable X by $\mathbf{D}X$. The following pair of observations will be useful in the proof of Theorem 1:

Observation 1. *Given a balanced (b, δ, λ)-HST tree T, and a multiset R of n red points and a multiset B of m blue points distributed among the leaves of T, we have $M_T(R, B) \leq \min(n, m)S(1, \delta)$.*

Lemma 1. *Let X be the sum of a finite number of independent bounded random variables. Then $\mathbf{E}|X - \mathbf{E}X| = \Theta(\mathbf{D}X)$.*

We omit the details, but comment that to show the upper bound of Lemma 1, one can repeatedly use Chebyshev's inequality, while the lower bound is the consequence of Hölder's inequality.

The process of randomly and uniformly choosing the leaves of a balanced HST T with branching factor b to host the points in R and B can be stated as follows: starting from the root, choose a child of the current vertex uniformly at random among its b children; if the new vertex is not a leaf, repeat this random selection process. Otherwise, this leaf is our random choice. We will distribute the "random" sets R and B among the leaves of T by repeating this procedure independently for every point of $R \cup B$. It is obvious that this procedure results in two random submultisets of the set of leaves of T. For an arbitrary vertex $v \in T$, let R_v and B_v, respectively, denote the cardinality of the set of red (respectively, blue) points that when distributed reach their host leaves in T on a path from the root through v. In particular, R_l is the number of red points assigned to the leaf l, and B_l is the number of blue points assigned to l.

Next, we will estimate the number of transits, τ_r, at the root r of a star (HST-) tree T with b leaves $L = \{u_1, u_2, \ldots, u_b\}$ when n red and n blue points are distributed randomly among the elements of L. Let $X_u = R_u - B_u$ for the leaf u. Then $\sum_{u \in L} X_u = 0$ and

$$\tau_r = \sum_{u \in L} \max\{X_u, 0\} = -\sum_{u \in L} \min\{X_u, 0\}.$$

It follows that $\sum_{u \in L} |X_u| = 2\sum_{u \in L} \max\{X_u, 0\}$, and hence

$$\mathbf{E}\tau_r = \frac{1}{2} \sum_{u \in L} \mathbf{E}|X_u|.$$

Observe that X_u is the combination of $2n$ independent indicator random variables

$$X_u = \sum_{j=1}^{n} R_u(j) - \sum_{j=1}^{n} B_u(j),$$

where $R_u(j) = 1$ if and only if the jth red points reaches leaf u; we define $B_u(j)$ similarly. Hence, $\mathbf{E}X_u = 0$, and Lemma 1 can be applied. Setting $\beta = (1/b - 1/b^2)$, it is an easy exercise to verify that $\mathbf{D}R_u(j) = \mathbf{D}B_u(j) = \sqrt{\beta}$ for every $1 \leq j \leq n$, and hence $\mathbf{D}X = \sqrt{2n\beta}$. In summary, we have

Lemma 2. *There exist positive constants c_1 and c_2 such that $c_1 b \sqrt{n\beta} \leq \mathbf{E}\tau_r \leq c_2 b \sqrt{n\beta}$ for a star T with root r on b leaves, when n red and n blue points are distributed randomly and uniformly among its leaves.*

We note that Lemma 2 proves Theorem 1 when $\delta = 1$. We need a generalization of the above, when $\sum_{u \in L} (R_u - B_u) = \Delta \neq 0$. In this case there will be $|\Delta|$ points which will remain unmatched in the star tree. The number of transits at r is easily seen to be

$$\tau_r = \frac{1}{2} \left(\sum_{u \in L} |X_u| - |\Delta| \right),$$

whereby we get

Lemma 3. *The expected number of transits at root r of a star with leaf set L is*

$$\mathbf{E}\tau_r = \frac{1}{2} \mathbf{E} \left(\sum_{u \in L} |X_u| - \left| \sum_{u \in L} (R_u - B_u) \right| \right).$$

Next, we present the proof of Theorem 1 for two randomly chosen n-element submultisets R and B among the leaves of a balanced (b, δ, λ)-HST T. The following simple combinatorial lemma is crucial for the proof.

Lemma 4. *Let R, B and T be as above, and let $k \geq 1$. Then $\mathcal{T}_{k-1}$, the total number of transits at level $k - 1$ is*

$$\mathcal{T}_{k-1} = \sum_{\delta(v)=k-1} \tau_v = \frac{1}{2} \sum_{\delta(u)=k} |X_u| - \frac{1}{2} \sum_{\delta(u')=k-1} |X_{u'}|.$$

The lemma follows easily by induction on the depth of the tree; we omit the details.   □

Now we are ready to prove our main result.

Proof of Theorem 1. Since $M_T(R, B)$ is a finite sum (see Equation 1), we can restate it as the sum of the expectation at each level of the tree T, i.e.,

$$\mathbf{E}M_T(R, B) = \sum_{k=0}^{\delta-1} \sum_{v:\delta(v)=k} \mathbf{E}\tau_v \Theta(C_\lambda \lambda^k).$$

Applying Lemma 4 we get

$$\mathbf{E}M_T(R, B) = \sum_{k=0}^{\delta-1} \Theta(C_\lambda \lambda^k) \mathbf{E} \left(\sum_{\delta(u)=k} |X_u| - \sum_{\delta(u')=k-1} |X_{u'}| \right).$$

Therefore, it suffices to compute $\mathbf{E}|X_u|$ for every $u \in T$. Notice that we are in a situation very similar to that of the star tree. At level k we have b^k vertices, hence the expected number of transits at level k is $b^k \mathbf{E}|X_u|$, where u is an arbitrary vertex at level k. Let $\beta_k = \frac{1}{b^k}(1 - \frac{1}{b^k})$. Applying Lemma 1 we get that the expected number of transits at level k is of order

$$T_k = b^{k-1}\sqrt{n}(b\sqrt{\beta_k} - \sqrt{\beta_{k-1}}).$$

Simple calculation shows that

$$b^{k/2}\frac{\sqrt{nb}}{2} \leq T_k \leq 2b^{k/2}\sqrt{nb}.$$

This will allow us to conclude that

$$K_1\sqrt{bn}\sum_{k=0}^{\delta-1}(\lambda\sqrt{b})^k \leq \mathbf{E}M_T(R, B) \leq K_2\sqrt{bn}\sum_{k=0}^{\delta-1}(\lambda\sqrt{b})^k. \tag{2}$$

If $\delta \leq \log_b n$ then the above proves Theorem 1. So, assume that $\delta > \log_b n$. In this case there is at least one vertex w such that $\delta(w) = \log_b n$. Then using Observation 1, the expected transportation cost of the matching for the subtree T_w rooted at w can be bounded as

$$M_{T_w}(R_w, B_w) \leq \min(|R_w|, |B_w|)S(\delta(w), \delta).$$

Therefore, we get the following upper bound for the expected matching length in T_w:

$$\mathbf{E}M_{T_w}(R_w, B_w) \leq C_\lambda\lambda^{\log_b n}\sum_{k=0}^{n}\Pr(R_w = k) \times k$$

$$= C_\lambda\lambda^{\log_b n}\sum_{k=0}^{n}\frac{k}{k!}$$

$$\leq eC_\lambda\lambda^{\log_b n}.$$

Here we used the fact that if $\delta(w) = \log_b n$ then R_w has a Poisson distribution. Observing that there are b^k vertices at level k, we have

$$\mathbf{E}M_T(R, B) \leq K_2\sqrt{bn}\sum_{k=0}^{\log_b n-1}(\lambda\sqrt{b})^k + eC_\lambda n\lambda^{\log_b n}.$$

Now we are in a position to estimate the precise upper bound on $\mathbf{E}M_T(R, B)$. We will consider three distinct cases, depending on the value of $\sqrt{b}\lambda$:

Case I: If $\sqrt{b}\lambda < 1$ then $\lambda^{\log_b n} < n^{-1/2}$, and hence, $neC_\lambda\lambda^{\log_b n} < eC_\lambda\sqrt{n}$. Since in this case $\sum_{k=0}(\lambda\sqrt{b})^k$ is a constant, we get the desired upper bound.

Case II: If $\sqrt{b}\lambda = 1$ we will have

$$\sum_{k=0}^{\log_b n-1}(\lambda\sqrt{b})^k = \log_b n,$$

and $\lambda^{\log_b n} = n^{-1/2}$, which again gives us the upper bound of the theorem.

Case III: If $\sqrt{b}\lambda > 1$ then

$$\sqrt{n}\sum_{k=0}^{\log_b n - 1}(\lambda\sqrt{b})^k = \sqrt{n}\frac{(\lambda\sqrt{b})^{\log_b n} - 1}{\lambda\sqrt{b} - 1} = O(n\lambda^{\log_b n}),$$

which implies the desired upper bound.

The lower bound of Theorem 1 follows trivially from the fact that truncating the lower bound of the sum in (2) (which has only non-negative elements) at the $\log_b n$-th term will result in the desired lower bound. □

An important generalization emerges when the points of R and B are not necessarily uniformly distributed among the leaves of T. Given any non-leaf vertex of T we can distribute the red and blue points among its children according to an arbitrary probability distribution. Conversely, it is easy to see that given any probability distribution on the leaves one can find appropriate probabilities for every non-leaf vertex of T in order to arrive at the desired distribution of the red and blue points at the leaf-level. This gives rise to the following theorem:

Theorem 2. *Let $T = T(b, \delta, \lambda)$ be a balanced HST, and $\mathbf{P}$ a probability distribution on the leaves of T. Let R and B be two n-element submultisets of the set of leaves of T chosen randomly and independently from $\mathbf{P}$. Then there exists a positive constant K_3 (depending only on λ) such that*

$$\mathbf{E}M_T(R, B) \leq K_3\sqrt{bn}\sum_{k=0}^{\delta-1}(\sqrt{b}\lambda)^k.$$

Sketch of the proof. The proof follows the same line of argument as Theorem 1, except that in addition we use the following elementary inequality: if $a_1, a_2, \ldots, a_t \in [0, 1]$, $\sum a_i \leq 1$ then

$$\sum_{1\leq i\leq t}\sqrt{a_i(1 - a_i)} \leq t\sqrt{\sum a_i/t(1 - \sum a_i/t)}.$$

Applying the above inequality we can perform the following balancing algorithm: First, we make the probability of choosing an arbitrary child of the root equal to the reciprocal of the number of its children; that is, we choose uniformly among the children of the root. Then we repeat the above for all the subtrees originating from these children. Proceeding top-down, at the end we achieve that every leaf of the tree has the same chance to be chosen, moreover, we never decreased the expected number of transitions at any intermediate vertex. This implies the theorem. □

We also note the following consequence of Theorem 2, which follows by choosing certain edge probabilities to be 0.

Corollary 1. *If T' is an arbitrary subtree of a balanced (b, δ, λ)-HST T, then the expected optimal transportation cost on T' is upper bounded by the expected optimal transportation cost on T.*

Observe that one cannot expect any reasonable general lower bound in the case of a non-uniform distribution or for subtrees: if the subtree T' is a path, the transportation cost is zero.

3   The Case of Matching in $[0,1]^d$

As a first application of the theory we developed in Section 2, we reproduce the results of [7] concerning the expected optimal transportation cost in the d-dimensional unit cube. We remark that for finding nearest neighbors in Euclidean space, Indyk and Thaper [6] used similar ideas for approximating the d-dimensional unit cube with HSTs.

We begin by presenting the general idea for approximating $[0,1]^d$ by a balanced HST. The number of iterations of this process will be the depth δ of the tree. In the jth step we construct a grid G_j with 2^{jd} cells, each cell having an edge length of 2^{-j}. G_j is a refinement of G_{j-1} for every j: we obtain the cells of G_j by dividing each cell of G_{j-1} into 2^d subcells of equal volume. We stop when $j = \delta$. The tree is going to have 2^{jd} vertices at level j, with each vertex corresponding to a cell of G_j. A vertex v at level j will be adjacent to a vertex w at level $j+1$ if and only if the cell of v in G_j contains the cell of w in G_{j+1}. The weight of edge (v,w) will be 2^{1-j}. Clearly, the construction will result in a balanced $(2^d, \delta, 1/2)$-HST. Moreover, the resulting HST will dominate the distances of the lattice points of G_δ: the Euclidean distance of any two lattice points is no greater than their distance in the HST. Finally, we will approximate a set of points in $[0,1]^d$ by $discretizing$ the point set: we assign every point to the available lattice point that is closest to it.

We will first consider the case of the unit interval, i.e., $d = 1$:

Proposition 1. *Given n red points and n blue points distributed uniformly at random on $[0,1]$, the expected weight of an optimal matching is $O(\sqrt{n})$.*

Proof: We approximate the $[0,1]$ interval with an equidistant set of $O(n^2)$ lattice points, as is described above. We will approximate this metric space by a balanced $(2, 2\log n, 1/2)$-HST T whose leaves are the lattice points. The discretization overhead associated with approximating the red and blue points with the leaves is no more than the cost of moving each point to the nearest leaf, i.e., $2n \cdot 1/(2n) = 1$. Applying Theorem 1 with parameters $b = 2$ and $\lambda = 1/2$, results in the desired bound.   □

In the plane our HST technique offers loose results. Ajtai et al. [1] showed that the expected weight of the optimal matching in $[0,1]^2$ is $\Theta(\sqrt{n \log n})$. In Proposition 2, we use Theorem 1 to obtain the bound of $O(\sqrt{n} \log n)$.

Proposition 2. *Let $B = \{b_i\}_{i=1}^n$ and $R = \{r_i\}_{i=1}^n$ be sets of blue and red points distributed uniformly at random in $[0,1]^2$ and let M_n be the expected weight of an optimal matching of B against R. Then $\mathbf{E}M_n = O(\sqrt{n} \log n)$.*

Proof: As discussed in the general process, we construct the 2-dimensional grid, then the balanced $(4, 1/2, 2\log_4 n)$-HST T. The discretization overhead associated with approximating the red and blue points with the leaves of T is again negligible for $\delta \geq 2\log_4 n$. Applying Theorem 1 with parameters $b = 4$ and $\lambda = 1/2$, we get the upper bound of $O(\sqrt{n}\log n)$. □

We now jump to real dimension 3 and above, showing (Proposition 3) that expected weight of optimal matching is $O(n^{(d-1)/d})$.

Proposition 3. *Let $B = \{b_i\}_{i=1}^n$ and $R = \{r_i\}_{i=1}^n$ be sets of blue and red points distributed uniformly at random in $[0,1]^d$, $d \geq 3$, and let M_n be the expected weight of an optimal matching of B against R. Then $\mathbf{E}M_n = O(n^{(d-1)/d})$.*

Proof: As before, we construct a sufficiently dense grid. Its lattice points will be used to aproximate the real vectors of $[0,1]^d$. The finite metric space of the lattice points will be dominated by a balanced HST $T = T(2^d, 3\log_{2^d} n, 1/2)$. We are now in a position to apply Theorem 1 with parameters $b = 2^d$ and $\lambda = 1/2$, from which will follow the bound of $O(n^{(d-1)/d})$. □

Observe that for $d \neq 2$, our seemingly crude approximations by HSTs result in tight bounds up to a constant factor (see e.g., [1]).

4   Optimal Matching for Weighted Point Sets

In this section we will estimate the expected weight of the optimal weighted matching for point sets $R = \{r_1, ..., r_n\}$ and $B = \{b_1, ..., b_n\}$ distributed uniformly and at random among the leaves of an HST T. We assume that every leaf u of T is associated with a randomly and independently chosen mass $m(u) \in [0,1]$. Then the total transportation cost is defined to be

$$M_{T,m}(R,B) = \min_\sigma \left(\sum_{1 \leq i \leq n} d_T(r_i, b_{\sigma(i)}) \min\{m(r_i), m(b_{\sigma(i)})\} \right)$$

We will use the following folklore result: if x and y are chosen randomly and independently from $[0,1]$, their expected distance $\mathbf{E}|x-y|$ is $1/3$.

Theorem 3. *Let $T = T(b, \delta, \lambda)$ be a balanced HST with set of leaves L, and R and B two randomly chosen n-element submultisets of L. Let $m : L \to [0,1]$ be a function with values drawn randomly and independently, and define $h = \min(\delta, \log_b n)$. Then there exist positive constants K_4 and K_5 such that*

$$K_4\sqrt{bn} \sum_{k=1}^h (\sqrt{b}\lambda)^k \leq \mathbf{E}M_{T,m}(R,B) \leq K_5\sqrt{bn} \sum_{k=1}^h (\sqrt{b}\lambda)^k.$$

Sketch of the proof. The proof follows the same line of arguments as Theorem 1, except that when computing the expected transportation cost, one has to multiply the number of transitions not only by the edge weight of T but also by the expected mass which is to be moved. Since this latter number is $1/3$ on the average and was chosen independently from the distribution of the points, the theorem is proved. □

5   The Case of Finite Approximation of Fractals

The machinery developed in Section 2 is general enough to use for matching on a finite approximation of a self-similar set. The notion of a finite approximation of fractals is best explained through an example. Recall that the Cantor set is formed by repeatedly removing the open middle third of each line segment in a set of line segments, starting with $[0, 1]$. If we stop this process after α iterations, we will refer to the resulting set as the α-approximation of the Cantor set.

Next, consider sets $R = \{r_1, ..., r_n\}$ and $B = \{b_1, ..., b_n\}$ of red and blue points respectively, distributed uniformly at random on the δ-approximation of the Cantor set, with $\delta \geq 2 \log n$. We are interested in the expected weight of an optimal matching between R and B. We can think of the δ-approximation of the Cantor set as being embedded into a balanced $(2, \delta, 1/3)$-HST T over the unit interval. We have $b = 2$, since at every step we double the number of subintervals, and $\lambda = 1/3$, because the length of these subintervals shrink by a factor of $1/3$. The discretization overhead associated with approximating the red and blue points with the leaves is no more than the cost of moving each point to the nearest leaf $2n \cdot 1/(2n) = 1$. We can apply Theorem 1 with parameters $b = 2$ and $\lambda = 1/3$, and conclude that $\mathbf{E}M_T(R, B) = O(\sqrt{n})$.

The tree metric of $T(2, \delta, 1/3)$ dominates the Euclidean metric on the Cantor set. Therefore, the expected optimal matching weight of n blue and n red points distributed randomly on the Cantor set is no heavier than the same points distributed on $[0, 1]$ itself. We have proved the following

Theorem 4. *Let $R = \{r_1, ..., r_n\}$ and $B = \{b_1, ..., b_n\}$ be sets of red and blue points distributed uniformly and at random in the δ-approximation of the Cantor set process with $\delta \geq 2 \log n$. Then the expected weight of an optimal matching between R and B is $O(\sqrt{n})$.*

Next, we consider the $\log_3 n$-approximation of a Sierpinski triangle. Here a balanced HST with branching factor $b = 3$, $\lambda = 1/2$, and depth δ ($\delta \geq 2 \log_3 n$) dominates the Euclidean metric, and provides a good approximation after discretization. As with the Cantor set, we can prove the following about the Sierpinski triangle:

Theorem 5. *Let $R = \{r_1, ..., r_n\}$ and $B = \{b_1, ..., b_n\}$ be sets of red and blue points distributed uniformly and at random in the interior of the δ-approximation of a Sierpinski triangle for δ large enough. Then the expected weight of an optimal matching between R and B is $O(\sqrt{n})$.*

Note the lack of the $\log n$ factor in the upper bound. The expected optimal matching weight in a triangle would be $O(\sqrt{n \log n})$ by the result of Ajtai et al.

As a final example for the application of Theorem 1 on fractals, we will consider the *Menger sponge*. A Menger sponge results from recursively dividing the unit cube into $3^3 = 27$ sub-cubes, removing the middle cube on each face and the cube in the center, then recursing on each sub-cube. To find an upper bound on the expected weight of matchings on the Menger sponge, consider a balanced

HST T with $\lambda = 1/3$ (the diameter decreases by a factor of $1/3$ at every recursion step), branching factor $b = 20$, and depth $\delta \geq 3 \log_{20} n$ (this depth is sufficiently large to provide good approximation in the discretization). Because T is dominating, we can state the following upper bound:

Theorem 6. *Let $R = \{r_1, ..., r_n\}$ and $B = \{b_1, ..., b_n\}$ be sets of red and blue points distributed uniformly and at random in the interior of the $3 \log_{20} n$- approximation of a Menger sponge. Then the expected weight of an optimal matching between R and B is $O(n^{1-\log_{20} 3})$.*

6   Conclusions

In this paper we presented a tight bound on the expected weight of transportation cost for matching of points on balanced HSTs. We extended our upper bounds for subtrees of balanced HSTs, and for non-uniform distributions. Using low-distortion embedding of $\mathbb{R}^d$ to HSTs, we reproduce the results concerning the expected optimal transportation cost in the $[0, 1]^d$, except for the case of $d = 2$ for which we have a discrepancy of a factor of $\sqrt{\log n}$. We also proved upper bounds on several sets for which showing reasonable matching results would previously have been intractable. By existing approximation theorems for finite metric spaces, we could give bounds on the expected transportation cost in any finite metric space. We plan to consider the analogues of other related matching problems, such as upright matchings.

References

1. Ajtai, M., Komlós, J., Tusnády, G.: On optimal matchings. Combinatorica 4(4), 259–264 (1984)
2. Alt, H., Guibas, L.: Discrete geometric shapes: Matching, interpolation, and approximation. In: Sack, J., Urrutia, J. (eds.) Handbook of Comput. Geom., pp. 121–153. Elsevier Science Publishers, Amsterdam (1996)
3. Bartal, Y.: Probabilistic approximation of metric spaces and its algorithmic applications. In: FOCS 1996: 37th Annual Symposium on Foundations of Computer Science, pp. 184–193 (1996)
4. Cabello, S., Giannopoulos, P., Knauer, C., Rote, G.: Matching point sets with respect to the earth mover's distance. Comput. Geom. Theory Appl. 39(2), 118–133 (2008)
5. Fakcharoenphol, J., Rao, S., Talwar, K.: A tight bound on approximating arbitrary metrics by tree metrics. In: STOC 2003: Proceedings of the thirty-fifth annual ACM symposium on Theory of computing, pp. 448–455 (2003)
6. Indyk, P., Thaper, N.: Fast image retrieval via embeddings. In: The 3rd International Workshop on Statistical and Computational Theories of Vision (2003)
7. Karp, R.M., Luby, M., Marchetti-Spaccamela, A.: A probabilistic analysis of multidimensional bin packing problems. In: STOC 1984: Proceedings of the sixteenth annual ACM symposium on Theory of computing, pp. 289–298 (1984)
8. Leighton, T., Shor, P.W.: Tight bounds for minimax grid matching, with applications to the average case analysis of algorithms. In: STOC 1986: Proceedings of the Eighteenth Annual ACM Symposium on Theory of Computing, pp. 91–103 (1986)

9. Rhee, W.T., Talagrand, M.: Exact bounds for the stochastic upward matching problem. Transactions of the American Mathematical Society 307(1), 109–125 (1988)
10. Rhee, W.T., Talagrand, M.: Matching random subsets of the cube with a tight control on one coordinate. The Annals of Applied Probability 2(3), 695–713 (1992)
11. Rubner, Y., Tomasi, C.: Texture based image retrieval without segmentation. In: IEEE International Conference on Computer Vision, pp. 1018–1024 (1999)
12. Rubner, Y., Tomasi, C., Guibas, L.: The earth movers distance as a metric for image retrieval. International Journal of Computer Vision 40(2), 99–122 (2000)
13. Shor, P.W.: The average-case analysis of some on-line algorithms for bin packing. In: FOCS 1984: 25th Annual Symposium on Foundations of Computer Science, pp. 193–200 (1986)
14. Shor, P.W., Yukich, J.E.: Minimax grid matching and empirical measures, vol. 19(3), pp. 1338–1348 (1991)
15. Talagrand, M.: Matching random samples in many dimensions. Annals of Applied Probability 2(4), 846–856 (1992)
16. Talagrand, M.: Matching theorems and empirical discrepancy computations using majorizing measures. J. ACM 7, 455–537 (1994)

Small Sample Spaces Cannot Fool Low Degree Polynomials

Noga Alon[1,*], Ido Ben-Eliezer[2], and Michael Krivelevich[3,**]

[1] School of Mathematics, Institute for Advanced Study, Princeton, NJ 08540, USA,
and Schools of Mathematics and Computer Science, Raymond and Beverly Sackler
Faculty of Exact Sciences, Tel Aviv University, Tel Aviv 69978, Israel
nogaa@tau.ac.il
[2] School of Computer Science, Raymond and Beverly Sackler Faculty of Exact
Sciences, Tel Aviv University, Tel Aviv 69978, Israel
idobene@post.tau.ac.il
[3] School of Mathematical Sciences, Raymond and Beverly Sackler Faculty of Exact
Sciences, Tel Aviv University, Tel Aviv 69978, Israel
krivelev@post.tau.ac.il

Abstract. A distribution D on a set $S \subset \mathbb{Z}_p^N$ ϵ-fools polynomials of
degree at most d in N variables over $\mathbb{Z}_p$ if for any such polynomial P,
the distribution of $P(x)$ when x is chosen according to D differs from
the distribution when x is chosen uniformly by at most ϵ in the ℓ_1 norm.
Distributions of this type generalize the notion of ϵ-biased spaces and
have been studied in several recent papers. We establish tight bounds
on the minimum possible size of the support S of such a distribution,
showing that any such S satisfies

$$|S| \geq c_1 \cdot \left(\frac{(\frac{N}{2d})^d \cdot \log p}{\epsilon^2 \log(\frac{1}{\epsilon})} + p \right).$$

This is nearly optimal as there is such an S of size at most

$$c_2 \cdot \frac{(\frac{3N}{d})^d \cdot \log p + p}{\epsilon^2}.$$

1 Introduction

Let P be a polynomial in N variables over $\mathbb{Z}_p$ of degree at most d. Let D be
a distribution over a set S of vectors from $\mathbb{Z}_p^N$, and denote by U_N the uniform
distribution on $\mathbb{Z}_p^N$. The distribution D is an ϵ-approximation of U_N with respect
to P if

$$\sum_{a \in \mathbb{Z}_p} \left| \Pr_{x \sim D}[P(x) = a] - \Pr_{x \sim U_N}[P(x) = a] \right| \leq \epsilon.$$

* Research supported in part by the Israel Science Foundation, by a USA-Israeli BSF
grant and by the Hermann Minkowski Minerva Center for Geometry at Tel Aviv
University.
** Research supported in part by USA-Israel BSF Grant 2006322, and by grant 526/05
from the Israel Science Foundation.

A. Goel et al. (Eds.): APPROX and RANDOM 2008, LNCS 5171, pp. 266–275, 2008.
© Springer-Verlag Berlin Heidelberg 2008

We say that S (with the distribution D) is an (ϵ, N, d)-*biased space* if it is an ϵ-approximation with respect to any polynomial on N variables of degree at most d. Note that D is not necessarily a uniform distribution over its support S.

The case $d = 1$ is known as ϵ-*biased spaces*. Many works deal with such spaces, including efficient constructions, lower bounds and applications (see, for example, [3,4,5,7,16,17,19] and their references).

Luby et al. [15] gave an explicit construction for the general case, but the size of their sample space S is $2^{2^{O(\sqrt{\log(N/\epsilon)})}}$ even for the case $d = 2$. They used it to construct a deterministic approximation algorithm to the probability that a given depth-2 circuit outputs a certain value on a random input.

Bogdanov [8] gave better constructions that work for fields of size at least $poly(d, \log N, \frac{1}{\epsilon})$. Bogdanov and Viola [10] suggested a construction for general fields. The construction is the sum of d copies of ϵ'-biased spaces, and the sample size is $N^d \cdot f(\epsilon, d, p)$ for some function f. However, the analysis of their construction relies on the so called "Inverse Gowers Conjecture" which was recently shown to be false [14]. Lovett [13] proved unconditionally for $p = 2$ that the sum of 2^d copies of ϵ'-biased spaces fools polynomials of degree d (where ϵ' is exponentially small in ϵ), thus giving an explicit construction of size $(\frac{N}{\epsilon})^{2^{O(d)}}$. Later, Viola [20] proved that the sum of d copies is sufficient. This yields an explicit construction of size $\frac{N^d}{\epsilon^{O(d \cdot 2^d)}}$ using the best known constructions of ϵ-biased spaces. Recently, Bogdanov et. al. [9] showed how to fool width-2 branching programs using such distributions.

Here we study the minimum possible size of (ϵ, N, d)-biased spaces. Bogdanov and Viola [10] observed that for $p = 2$ and $\epsilon < 2^{-d}$ every such space is of size at least $\binom{N}{d}$. Their argument is very simple: The set of polynomials of degree at most d forms a linear space of dimension $\sum_{i=0}^{d} \binom{N}{i} > \binom{N}{d}$. If S is of size less than $\binom{N}{d}$ then there is a non-zero polynomial P such that $P(x) = 0$ for every $x \in S$, and since every non zero polynomial is not zero with probability at least 2^{-d} (as follows, for example, by considering the minimal distance of the Reed-Muller code of order d) we get the desired bound. However, their bound doesn't depend on ϵ and, for small values of ϵ, is far from optimal and also from the known bound for ϵ-biased space, which is nearly optimal for $d = 1$. Our main contribution is a nearly tight lower bound on the size of such spaces as a function of all four parameters ϵ, N, d and p. Note that as spaces of this type can be useful in derandomization, where the running time of the resulting algorithms is proportional to the size of the space, it is interesting to get a tight bound for their smallest possible size.

Theorem 1. *There exists an absolute constant $c_1 > 0$ such that for every $d \leq \frac{N}{10}$ and $\epsilon \geq d \cdot p^{-\frac{N}{2d}}$, every (ϵ, N, d)-biased space over $\mathbb{Z}_p$ has size at least*

$$\max \left\{ c_1 \cdot \frac{(\frac{N}{2d})^d \log p}{\epsilon^2 \log(\frac{1}{\epsilon})}, \, p(1 - \epsilon) \right\}.$$

We also observe that this bound is nearly tight by proving the following simple statement:

Proposition 1. *There is an absolute constant $c_2 > 0$ so that for every $d \leq \frac{N}{10}$ there is an (ϵ, N, d)-biased space over $\mathbb{Z}_p$ of size at most $c_2 \cdot \frac{(\frac{3N}{d})^d \log p + p}{\epsilon^2}$.*

The proofs are described in the next section; for completeness, we include some of the details in the appendix. The final section contains some concluding remarks. Throughout the proofs we omit all floor and ceiling signs whenever these are not crucial.

2   Proofs

In this section we present the proofs of our results. The proof of our main result, Theorem 1, lower bounding the size of an (ϵ, N, d)-biased set, is given in Section 2.1. The proof of the upper bound (Proposition 1) is in Section 2.2.

2.1   Lower Bound

First we observe that a bound of $p(1-\epsilon)$ follows easily as otherwise the distribution doesn't fool every polynomial P for which $P(x)$ is the uniform distribution (for example, all the linear polynomials). Let N be the number of variables and let d be the degree of the polynomial. Assume for simplicity that $N = nd$, where n is an integer. For every $i \geq 1$ define the set of variables $S_i = \{x_{i,1}, ..., x_{i,n}\}$. A monomial over $\mathbb{Z}_p$ is called d-partite if it has the form $\prod_{1 \leq i \leq d} x_{i,j_i}$, and a polynomial over $\mathbb{Z}_p$ is called d-partite if it is a sum of d-partite monomials. Note that d-partite polynomials are homogeneous polynomials of degree d.

Let $P_{n,d}$ be the uniform distribution on the set of d-partite polynomials. A random element in $P_{n,d}$ is a sum of d-partite monomials, where every one of the possible n^d monomials has a random coefficient selected uniformly and independently from $\mathbb{Z}_p$.

An assignment to the variables $\{x_i\}$ is non-trivial if there is an i such that $x_i \neq 0$. Similarly, if $v_1, v_2, ..., v_n \in V$ for some vector space V, a linear combination $\sum_i \alpha_i v_i$ is non-trivial if there is i such that $\alpha_i \neq 0$. For a prime p, a polynomial P over $\mathbb{Z}_p$ is δ-balanced if

$$\sum_{a \in \mathbb{Z}_p} \left| \frac{|\{x : P(x) = a\}|}{p^N} - \frac{1}{p} \right| \leq \delta.$$

A polynomial is *balanced* if it is 0-balanced. We have the following key lemma:

Lemma 1. *The probability $\phi(n,d)$ that a random element from $P_{n,d}$ is $d \cdot p^{-\frac{n}{2}}$-balanced is at least*

$$1 - p^{-(\frac{n}{2})^d + 2(\frac{n}{2})^{d-1} + \sum_{i=0}^{d-3}(\frac{n}{2})^i(\frac{n^2}{4}+n)} \geq 1 - p^{-(\frac{n}{2})^d + 4(\frac{n}{2})^{d-1}}.$$

Proof: We apply induction on d. For $d = 1$, as every non-trivial linear polynomial is balanced, we have $\phi(n, 1) = 1 - p^{-n}$, and the statement holds. Assuming that the statement is valid for d, we prove it for $d + 1$. A random $(d + 1)$-partite polynomial P can be represented as $\sum_{i=1}^{n} x_{1,i} P_i$, where for every i, P_i is a random polynomial (distributed uniformly and independently over $P_{n,d}$) over the sets of variables $S_2, S_3, ..., S_{d+1}$. Denote the set $\{P_i\}$ of polynomials by $\mathcal{P}$. We use the following claim:

Claim 1. *With probability at least $1 - p^{-(\frac{n}{2})^{d+1} + 2(\frac{n}{2})^d + \sum_{i=0}^{d-1}(\frac{n}{2})^i(\frac{n^2}{4} + n)}$ over the choice of polynomials in $\mathcal{P}$, there is a subset $B \subseteq \mathcal{P}$ of size at least $\frac{n}{2}$ such that for any non-trivial choice of $\{\alpha_i\}$, the polynomial $\sum_{P_i \in B} \alpha_i P_i$ is $d \cdot p^{-\frac{n}{2}}$-balanced.*

Proof: Let $B_0 := \emptyset$. In the i'th step, we consider the polynomial P_i. If P_i as well as all its combinations with elements from B_{i-1} are $d \cdot p^{-\frac{n}{2}}$-balanced, we set $B_i := B_{i-1} \bigcup \{P_i\}$, otherwise we call the step *bad* and let $B_i := B_{i-1}$. After the last polynomial, set $B := B_n$. We want to bound the probability that there are more than $\frac{n}{2}$ bad steps. Consider a certain step i and assume that $|B_{i-1}| < \frac{n}{2}$. Since P_i is a random polynomial, the sum of P_i with every fixed polynomial is uniformly distributed over the set $P_{n,d}$. By the induction hypothesis, it is $d \cdot p^{-\frac{n}{2}}$-balanced with probability at least $1 - p^{-(\frac{n}{2})^d + 2(\frac{n}{2})^{d-1} + \sum_{i=0}^{d-2}(\frac{n}{2})^i(\frac{n^2}{4} + n)}$. By the union bound, the probability that the step is bad is at most

$$p^{n/2} \cdot p^{-(\frac{n}{2})^d + 2(\frac{n}{2})^{d-1} + \sum_{i=0}^{d-2}(\frac{n}{2})^i(\frac{n^2}{4} + n)}.$$

We bound the probability that there are more than $\frac{n}{2}$ bad steps. For $d = 2$ the probability is at most

$$\binom{n}{\frac{n}{2}} \left(p^{n/2} \cdot p^{-n}\right)^{n/2} \leq p^{-(\frac{n}{2})^2 + n}.$$

For $d \geq 3$, we have:

$$\binom{n}{\frac{n}{2}} \left(p^{n/2} \cdot p^{-(\frac{n}{2})^d + 2((\frac{n}{2})^{d-1}) + \sum_{i=0}^{d-2}(\frac{n}{2})^i(\frac{n^2}{4} + n)}\right)^{n/2}$$

$$\leq p^{-(\frac{n}{2})^{d+1} + 2((\frac{n}{2})^d) + \sum_{i=0}^{d-1}(\frac{n}{2})^i(\frac{n^2}{4} + n)}.$$

The claim follows. ■

Assume that the condition of the claim holds, and without loss of generality assume that $\{P_1, P_2, ..., P_{n/2}\} \subseteq B$. Let $P' = \sum_{i=1}^{n/2} x_{1,i} P_i$. By Claim 1, for every non-trivial assignment of the variables $\{x_{1,i}\}$, the obtained polynomial is $d \cdot p^{-\frac{n}{2}}$-balanced. The probability that the assignment of the variables $\{x_{1,i}\}$ is trivial is $p^{-\frac{n}{2}}$. Therefore, P' is δ-balanced, where

$$\delta \leq p^{-\frac{n}{2}} + d \cdot p^{-\frac{n}{2}} = (d + 1) \cdot p^{-\frac{n}{2}}. \tag{1}$$

We use this fact to prove that the polynomial P is $(d + 1) \cdot p^{-\frac{n}{2}}$-balanced. For every assignment of the variables from $\bigcup_{2 \leq i \leq d+1} S_i$, P reduces to a linear polynomial, which depends only on the variables from S_1. Denote by $\mu(P)$ (respectively, $\mu(P')$) the probability over the assignments of $\bigcup_{2 \leq i \leq d+1} S_i$ that P (respectively, P') reduces to a trivial linear polynomial. Clearly $\mu \leq \mu'$ and μ is an upper bound on the imbalance of P. Therefore, it is sufficient to prove that μ' is bounded by $(d+1) \cdot p^{-\frac{n}{2}}$. To this end, note that whenever P' is reduced to a constant polynomial it is actually reduced to the zero polynomial. Therefore, as the bias of P' is bounded by $(d+1) \cdot p^{-\frac{n}{2}}$, the lemma follows.   ∎

We construct a set of polynomials Q as follows. Let

$$r = \log_p \left(\frac{1}{1 - \phi(n,d)} \right) - 1 \geq \left(\frac{N}{2d} \right)^d - 4 \left(\frac{N}{2d} \right)^{d-1} - 1.$$

For every $1 \leq i \leq r$ let q_i be a polynomial distributed uniformly and independently over $P_{n,d}$. Denote by Q the set of all non-trivial combinations of $\{q_1, ..., q_r\}$.

By the union bound and by Lemma 1, with positive probability all the elements of Q are $d \cdot p^{-\frac{n}{2}}$-balanced. Fix Q to be such a set. It follows also that the vectors $q_1, q_2, ..., q_r$ are linearly independent (otherwise Q contains the zero vector, which is not $d \cdot p^{-\frac{n}{2}}$-balanced). Therefore, $|Q| \geq p^{(\frac{N}{2d})^d - 4(\frac{N}{2d})^{d-1} - 1} - 1$.

The following lemma is due to Alon [2]:

Lemma 2 ([2]). *There exists an absolute positive constant c so that the following holds. Let B be an n by n real matrix with $b_{i,i} \geq \frac{1}{2}$ for all i and $|b_{i,j}| \leq \epsilon$ for all $i \neq j$ where $\frac{1}{2\sqrt{n}} \leq \epsilon \leq \frac{1}{4}$. Then the rank of B satisfies*

$$rank(B) \geq \frac{c \log n}{\epsilon^2 \log \left(\frac{1}{\epsilon} \right)}.$$

Here we need the following complex variant of the lemma:

Lemma 3. *There exists an absolute positive constant c so that the following holds. Let C be an n by n complex matrix with $|c_{i,i}| \geq \frac{1}{2}$ for all i and $|c_{i,j}| \leq \epsilon$ for all $i \neq j$ where $\frac{1}{2\sqrt{n}} \leq \epsilon \leq \frac{1}{4}$. Then the rank of C satisfies*

$$rank(C) \geq \frac{c \log n}{\epsilon^2 \log \left(\frac{1}{\epsilon} \right)}.$$

We give the proof of this lemma in the appendix. For completeness we also reproduce there the proof of Lemma 2.

We are now ready to prove Theorem 1:

Proof of Theorem 1. Suppose that W is an (ϵ, N, d)-biased space, and that $W = \{w_1, w_2, ..., w_m\}$, $\Pr[w_i] = t_i$. Define a $|Q|$-by-m complex matrix U whose rows are indexed by the elements of Q and whose columns are indexed by the elements of W. Set $U_{q,w_i} = (\xi_p)^{q(w_i)} \sqrt{t_i}$, where ξ_p is a primitive root of unity of order p and the value of $q(w_i)$ is computed over $\mathbb{Z}_p$. Note that by our choice of Q and the definition of an (ϵ, N, d)-biased space, for every $q \in Q$:

$$\left| \sum_{i=1}^{m} (\xi_p)^{q(w_i)} \cdot t_i \right| \leq \epsilon + d \cdot p^{-\frac{n}{2}} \leq 2\epsilon.$$

Also, obviously:

$$\sum_{i=1}^{m} t_i = 1.$$

For every two distinct polynomials $q_1, q_2 \in Q$, the polynomial $q_1 - q_2$ is also in Q, and for every w_i we have

$$(\xi_p)^{(q_1 - q_2)(w_i)} = (\xi_p)^{q_1(w_i)} \cdot (\xi_p)^{-q_2(w_i)}.$$

Set $A = UU^*$. For every distinct $q_1, q_2 \in Q$ we have:

$$|A_{q_1,q_2}| = |\sum_{i=1}^{m} (\xi_p)^{(q_1-q_2)(w_i)} \cdot t_i| \le 2\epsilon.$$

All the diagonal entries in A are 1. Since the rank of U is at most m the rank of A is also at most m. By Lemma 3:

$$m \ge rank(A) \ge c' \cdot \frac{\log |Q|}{\epsilon^2 \log\left(\frac{1}{\epsilon}\right)} \ge c_1 \cdot \frac{(\frac{N}{2d})^d \cdot \log p}{\epsilon^2 \log\left(\frac{1}{\epsilon}\right)}.$$

The desired result follows. ∎

2.2 Upper Bound

Here we prove the simple upper bound:

Proof of Propostion 1. Let $R \subseteq \mathbb{Z}_p^N$ be a random set of size $m = 2 \cdot \frac{(\frac{3N}{d})^d \log(p)+p}{\epsilon^2}$. We bound the probability that for a given polynomial P, the uniform distribution on R is not an ϵ-approximation with respect to P.

Let $L \subset \mathbb{Z}_p$, and let $\mu_L = m \sum_{a \in L} \Pr_{x \in U_n} [p(x) = a]$ be the expected number of vectors from R such that P evaluates to elements from L. By the Chernoff bounds (see, e.g., [6], Appendix A), we have:

$$\Pr_{R} \left[\Pr_{x \in U_n} [P(x) \in L] - \Pr_{x \in R} [P(x) \in L] > \epsilon \right] \le e^{-\mu_L \cdot (\frac{\epsilon m}{\mu_L})^2 / 2} \le e^{-m\epsilon^2 / 2}.$$

By the union bound over all 2^p possible sets L, the probability that the uniform distribution on R is not an ϵ-approximation is at most $e^{-m\epsilon^2/2+p}$.

The number of normalized monomials of degree at most d is exactly the number of ways to put d identical balls in $N + 1$ distinct bins, and is bounded by

$$\binom{d+N}{d} \le \left(\frac{e(N+d)}{d}\right)^d \le \left(\frac{3N}{d}\right)^d.$$

Therefore the total number of polynomials of degree at most d is at most

$$p^{(\frac{3N}{d})^d} = 2^{(\frac{3N}{d})^d \cdot \log p}.$$

By applying the union bound, with high probability the uniform distribution on R is an ϵ-approximation with respect to any polynomial on N variables with degree at most d, and the theorem follows. ∎

3   Concluding Remarks

For $p \ll \binom{n}{d}$, the ratio between the lower and upper bounds is $c \cdot (2e)^d \log\left(\frac{1}{\epsilon}\right)$ for some constant c. In particular, for fixed d the ratio is $\Theta(\log\left(\frac{1}{\epsilon}\right))$. This matches the ratio between the best known upper and lower bounds in the case $d = 1$ that corresponds to ϵ-biased spaces.

Our bound is valid only for $\epsilon \geq d \cdot p^{-\frac{N}{2d}}$. As noted in [2], for $\epsilon \leq p^{-\frac{N}{2}}$ every ϵ-biased space must be essentially the whole space (even for $d = 1$). It may be interesting to close the gap between $p^{-\frac{N}{2}}$ and $d \cdot p^{-\frac{N}{2d}}$. In a recent joint work with Tali Kaufman, we could actually replace the probabilistic construction with an explicit set of polynomials with smaller bias. Using this construction, we can extend our result for every value of ϵ. The details will appear in the final version of this paper.

Recently, Schechtman and Shraibman [18] proved a strengthening of Lemma 2. They showed that under the conditions of Lemma 2, if A is also positive-semidefinite then we need only an upper bound on the values of non-diagonal entries, instead of an upper bound on their absolute values. In our case, for $p = 2$ the matrix A is positive semidefinite, and we can thus relax the conditions and establish a similar lower bound for the size of the support of any distribution in which no polynomial attains the value zero with probability bigger by $\epsilon/2$ than the probability it attains it in the uniform distribution. That is, for $p = 2$ the lower bound for the size of the distribution holds, even if there is no lower bound on the probability that each polynomial attains the value zero.

Lemma 1 can also be formulated in the language of error correcting codes. For given N and d, it states that every Reed-Muller code with parameters N and d contains a dense linear subcode in which every nontrivial codeword is balanced.

Recently, Dvir and Shpilka [12] gave an efficient encoding and decoding procedures for the construction of sum of d copies of ϵ-biased spaces.

Acknowledgements. We thank Avi Wigderson, Shachar Lovett and Tali Kaufman for fruitful discussions.

References

1. Alon, N.: Problems and results in extremal combinatorics, I. Discrete Math. 273, 31–53 (2003)
2. Alon, N.: Perturbed identity matrices have high rank: proof and applications. Combinatorics, Probability and Computing (to appear)
3. Alon, N., Andoni, A., Kaufman, T., Matulef, K., Rubinfeld, R., Xie, N.: Testing k-wise and almost k-wise independence. In: Proceedings of the 39th Annual ACM Symposium, STOC 2007, pp. 496–505 (2007)
4. Alon, N., Goldreich, O., Håstad, J., Peralta, R.: Simple constructions of almost k-wise independent random variables. Random Structures and Algorithms 3, 289–304 (1992)
5. Alon, N., Roichman, Y.: Random Cayley graphs and expanders. Random Structures and Algorithms 5, 271–284 (1994)

6. Alon, N., Spencer, J.: The Probabilistic Method, 2nd edn. Wiley, New York (2000)
7. Ben-Sasson, E., Sudan, M., Vadhan, S., Wigderson, A.: Randomness-efficient low degree tests and short PCPs via epsilon-biased sets. In: Proceedings of the 35th Annual ACM Symposium, STOC 2003, pp. 612–621 (2003)
8. Bogdanov, A.: Pseudorandom generators for low degree polynomials. In: Proceedings of the 37th Annual ACM Symposium, STOC 2005, pp. 21–30 (2005)
9. Bogdanov, A., Dvir, Z., Verbin, E., Yehudayoff, A.: Pseudorandomness for width 2 branching programs (manuscript, 2008)
10. Bogdanov, A., Viola, E.: Pseudorandom bits for polynomials. In: Proceedings of the 38th Annual Symposium on Foundations of Computer Science (FOCS), pp. 41–51 (2007)
11. Codenotti, B., Pudlák, P., Resta, G.: Some structural properties of low-rank matrices related to computational complexity. Theoret. Comput. Sci. 235, 89–107 (2000)
12. Dvir, Z., Shpilka, A.: Noisy interpolating sets for low degree polynomials. In: Proceedings of the 23th IEEE Conference on Computational Complexity (CCC), pp. 140–148 (2008)
13. Lovett, S.: Unconditional pseudorandom generators for low degree polynomials. In: Proceedings of the 40th Annual ACM Symposium, STOC 2008, pp. 557–562 (2008)
14. Lovett, S., Meshulam, R., Samorodnitsky, A.: Inverse conjecture for the Gowers norm is false. In: Proceedings of the 40th Annual ACM Symposium, STOC 2008, pp. 547–556 (2008)
15. Luby, M., Velickovic, B., Wigderson, A.: Deterministic approximate counting of depth-2 circuits. In: Proceedings of the 2nd ISTCS (Israeli Symposium on Theoretical Computer Science), pp. 18–24 (1993)
16. Motwani, R., Naor, J., Naor, M.: The probabilistic method yields deterministic parallel algorithms. JCSS 49(3), 478–516 (1994)
17. Naor, J., Naor, M.: Small bias probability spaces: efficient constructions and applications. In: Proceedings of the 22th Annual ACM Symposium, STOC 1990, pp. 213–223 (1990)
18. Schechtman, G., Shraibman, A.: Lower bounds for local versions of dimension reductions (manuscript, 2007)
19. Shpilka, A.: Constructions of low-degree and error-correcting ε-biased generators. In: Proceedings of the 21st Annual IEEE Conference on Computational Complexity (CCC), Prague, Czech Republic, pp. 33–45 (2006)
20. Viola, E.: The sum of d small-bias generators fools polynomials of degree d. In: Proceedings of the 23th IEEE Conference on Computational Complexity (CCC), pp. 124–127 (2008)

A   A Complex Variant of Lemma 2

In this section we reproduce the proof of Lemma 2 (omitting the final detailed computation) as given in [2], and also prove Lemma 3.

We start with the following lemma from which Lemma 2 will follow:

Lemma 4. *There exists an absolute positive constant c so that the following holds. Let B be an n by n real matrix with $b_{i,i} = 1$ for all i and $|b_{i,j}| \leq \epsilon$ for all $i \neq j$. If $\frac{1}{\sqrt{n}} \leq \epsilon < 1/2$, then*

$$rank(B) \geq \frac{c}{\epsilon^2 \log(1/\epsilon)} \log n.$$

We need the following well known lemma proved, among other places, in [11], [1].

Lemma 5. *Let $A = (a_{i,j})$ be an n by n real, symmetric matrix with $a_{i,i} = 1$ for all i and $|a_{i,j}| \leq \epsilon$ for all $i \neq j$. If the rank of A is d, then*

$$d \geq \frac{n}{1 + (n-1)\epsilon^2}.$$

In particular, if $\epsilon \leq \frac{1}{\sqrt{n}}$ then $d > n/2$.

Proof: Let $\lambda_1, \ldots, \lambda_n$ denote the eigenvalues of A, then their sum is the trace of A, which is n, and at most d of them are nonzero. Thus, by Cauchy-Schwartz, $\sum_{i=1}^{n} \lambda_i^2 \geq d(n/d)^2 = n^2/d$. On the other hand, this sum is the trace of $A^t A$, which is precisely $\sum_{i,j} a_{i,j}^2 \leq n + n(n-1)\epsilon^2$. Hence $n + n(n-1)\epsilon^2 \geq n^2/d$, implying the desired result. ∎

Lemma 6. *Let $B = (b_{i,j})$ be an n by n matrix of rank d, and let $P(x)$ be an arbitrary polynomial of degree k. Then the rank of the n by n matrix $(P(b_{i,j}))$ is at most $\binom{k+d}{k}$. Moreover, if $P(x) = x^k$ then the rank of $(P(b_{i,j}))$ is at most $\binom{k+d-1}{k}$.*

Proof: Let $\mathbf{v_1} = (v_{1,j})_{j=1}^n, \mathbf{v_2} = (v_{2,j})_{j=1}^n, \ldots, \mathbf{v_d} = (v_{d,j})_{j=1}^n$ be a basis of the row-space of B. Then the vectors $(v_{1,j}^{k_1} \cdot v_{2,j}^{k_2} \cdots v_{d,j}^{k_d})_{j=1}^n$, where $k_1, k_2, \ldots, k_d$ range over all non-negative integers whose sum is at most k, span the rows of the matrix $(P(b_{i,j}))$. In case $P(x) = x^k$ it suffices to take all these vectors corresponding to $k_1, k_2, \ldots, k_d$ whose sum is precisely k. ∎

Proof of Lemma 4. We may and will assume that B is symmetric, since otherwise we simply apply the result to $(B + B^t)/2$ whose rank is at most twice the rank of B. Put $d = rank(B)$. If $\epsilon \leq 1/n^\delta$ for some fixed $\delta > 0$, the result follows by applying Lemma 5 to a $\lfloor \frac{1}{\epsilon^2} \rfloor$ by $\lfloor \frac{1}{\epsilon^2} \rfloor$ principal submatrix of B. Thus we may assume that $\epsilon \geq 1/n^\delta$ for some fixed, small $\delta > 0$. Put $k = \lfloor \frac{\log n}{2\log(1/\epsilon)} \rfloor$, $n' = \lfloor \frac{1}{\epsilon^{2k}} \rfloor$ and note that $n' \leq n$ and that $\epsilon^k \leq \frac{1}{\sqrt{n'}}$. By Lemma 6 the rank of the n' by n' matrix $(b_{i,j}^k)_{i,j \leq n'}$ is at most $\binom{d+k}{k} \leq (\frac{e(k+d)}{k})^k$. On the other hand, by Lemma 5, the rank of this matrix is at least $n'/2$. Therefore

$$\left(\frac{e(k+d)}{k} \right)^k \geq \frac{n'}{2} = \frac{1}{2} \lfloor \frac{1}{\epsilon^{2k}} \rfloor,$$

and the desired result follows by some simple (though somewhat tedious) manipulation, which we omit. ∎

Proof of Lemma 2. Let $C = (c_{i,j})$ be the n by n diagonal matrix defined by $c_{i,i} = 1/b_{i,i}$ for all i. Then every diagonal entry of CB is 1 and every off-diagonal entry is of absolute value at most 2ϵ. The result thus follows from Lemma 4. ∎

Proof of Lemma 3. Let P be an n by n diagonal matrix defined by $p_{i,i} = 1/c_{i,i}$ and set $D = CP$. Then every diagonal entry of D is 1 and every off-diagonal entry is of absolute value at most 2ϵ. Set $D' = (D+D^*)/2$. Then D' is a real matrix and $rank(D') \leq 2 \cdot rank(D)$. The desired result follows by applying Lemma 4 to D'. ∎

Derandomizing the Isolation Lemma and Lower Bounds for Circuit Size

V. Arvind and Partha Mukhopadhyay

Institute of Mathematical Sciences
C.I.T Campus,Chennai 600 113, India
{arvind,partham}@imsc.res.in

Abstract. The isolation lemma of Mulmuley et al [MVV87] is an important tool in the design of randomized algorithms and has played an important role in several nontrivial complexity upper bounds. On the other hand, polynomial identity testing is a well-studied algorithmic problem with efficient randomized algorithms and the problem of obtaining efficient *deterministic* identity tests has received a lot of attention recently. The goal of this paper is to compare the isolation lemma with polynomial identity testing:

1. We show that derandomizing reasonably restricted versions of the isolation lemma implies circuit size lower bounds. We derive the circuit lower bounds by examining the connection between the isolation lemma and polynomial identity testing. We give a randomized polynomial-time identity test for non-commutative circuits of polynomial degree based on the isolation lemma. Using this result, we show that derandomizing the isolation lemma implies noncommutative circuit size lower bounds. For the commutative case, a stronger derandomization hypothesis allows us to construct an explicit multilinear polynomial that does not have subexponential size commutative circuits. The restricted versions of the isolation lemma we consider are natural and would suffice for the standard applications of the isolation lemma.

2. From the result of Klivans-Spielman [KS01] we observe that there is a randomized polynomial-time identity test for commutative circuits of polynomial degree, also based on a more general isolation lemma for linear forms. Consequently, derandomization of (a suitable version of) this isolation lemma implies that either NEXP $\not\subset$ P/poly or the Permanent over $\mathbb{Z}$ does not have polynomial-size arithmetic circuits.

1 Introduction

We recall the Isolation Lemma [MVV87]. Let $[n]$ denote the set $\{1, 2, \cdots, n\}$. Let U be a set of size n and $\mathcal{F} \subseteq 2^U$ be any family of subsets of U. Let $w : U \to \mathbb{Z}^+$ be a weight function that assigns positive integer weights to the elements of U. For $T \subseteq U$, define its weight $w(T)$ as $w(T) = \sum_{u \in T} w(u)$. Then Isolation Lemma guarantees that for any family of subsets $\mathcal{F}$ of U and for any random weight assignment $w : U \to [2n]$, with high probability there will be a unique minimum weight set in $\mathcal{F}$.

Lemma 1 (Isolation Lemma). [MVV87] Let U be an universe of size n and $\mathcal{F}$ be any family of subsets of U. Let $w : U \to [2n]$ denote a weight assignment function to elements of U. Then,

A. Goel et al. (Eds.): APPROX and RANDOM 2008, LNCS 5171, pp. 276–289, 2008.

$$\text{Prob}_w [\text{ There exists a unique minimum weight set in } \mathcal{F}] \geq \frac{1}{2},$$

where the weight function w is picked uniformly at random.

In the seminal paper [MVV87] Mulmuley et al apply the isolation lemma to give a randomized NC algorithm for computing maximum cardinality matchings for general graphs (also see [ARZ99]). Since then the isolation lemma has found several other applications. For example, it is crucially used in the proof of the result that NL $\subset$ UL/poly [AR00] and in designing randomized NC algorithms for linear representable matroid problems [NSV94]. It is also known that the isolation lemma can be used to prove the Valiant-Vazirani lemma that SAT is many-one reducible via randomized reductions to USAT.

Whether the matching problem is in deterministic NC, and whether NL $\subseteq$ UL are outstanding open problems. Thus, the question whether the isolation lemma can be derandomized is clearly important.

As noted in [Agr07], it is easy to see by a counting argument that the isolation lemma can not be derandomized, in general, because there are 2^{2^n} set systems $\mathcal{F}$. More formally, the following is observed in [Agr07].

Observation 1. [Agr07] *The Isolation Lemma can not be fully derandomized if we allow weight functions* $w : U \to [n^c]$ *for a constant c (i.e. weight functions with a polynomial range). More precisely, for any polynomially bounded collection of weight assignments* $\{w_i\}_{i \in [n^{c_1}]}$ *with weight range* $[n^c]$, *there exists a family* $\mathcal{F}$ *of* $[n]$ *such that for all* $j \in [n^{c_1}]$, *there exists two minimal weight subsets with respect to* w_j.

However that does not rule out the derandomization of any special usage of the isolation lemma. Indeed, for all applications of the isolation lemma (mentioned above, for instance) we are interested only in exponentially many set systems $\mathcal{F} \subseteq 2^U$.

We make the setting more precise by giving a general framework. Fix the universe $U = [n]$ and consider an n-input boolean circuit C where $size(C) = m$. The set 2^U of all subsets of U is in a natural 1-1 correspondence with the length n-binary strings $\{0,1\}^n$: each subset $S \subseteq U$ corresponds to its characteristic binary string $\chi_S \in \{0,1\}^n$ whose i^{th} bit is 1 iff $i \in S$. Thus the n-input boolean circuit C implicitly defines the set system

$$\mathcal{F}_C = \{S \subseteq [n] \mid C(\chi_S) = 1\}.$$

As an easy consequence of Lemma 1 we have the following.

Lemma 2. *Let U be an universe of size n and C be an n-input boolean circuit of size m. Let $\mathcal{F}_C \subseteq 2^U$ be the family of subsets of U defined by circuit C. Let $w : U \to [2n]$ denote a weight assignment function to elements of U. Then,*

$$\text{Prob}_w [\text{ There exists a unique minimum weight set in } \mathcal{F}_C] \geq \frac{1}{2},$$

where the weight function w is picked uniformly at random. Furthermore, there is a collection of weight functions $\{w_i\}_{1 \leq i \leq p(m,n)}$, *where $p(m,n)$ is a fixed polynomial, such that for each $\mathcal{F}_C$ there is a weight function w_i w.r.t. which there is a unique minimum weight set in $\mathcal{F}_C$.*

Lemma 2 allows us to formulate two natural and reasonable derandomization hypotheses for the isolation lemma.

Hypothesis 1. There is a deterministic algorithm $\mathcal{A}_1$ that takes as input (C, n), where C is an n-input boolean circuit, and outputs a collection of weight functions $w_1, w_2, \cdots, w_t$ such that $w_i : [n] \rightarrow [2n]$, with the property that for some w_i there is a unique minimum weight set in the set system $\mathcal{F}_C$. Furthermore, $\mathcal{A}_1$ runs in time subexponential in $size(C)$.

Hypothesis 2. There is a deterministic algorithm $\mathcal{A}_2$ that takes as input (m, n) in unary and outputs a collection of weight functions $w_1, w_2, \cdots, w_t$ such that $w_i : [n] \rightarrow [2n]$, with the property that for each size m boolean circuit C with n inputs there is some weight function w_i w.r.t. which $\mathcal{F}_C$ has a unique minimum weight set. Furthermore, $\mathcal{A}_2$ runs in time polynomial in m.

Clearly, Hypothesis 2 is stronger than Hypothesis 1. It demands a "black-box" derandomization in the sense that $\mathcal{A}_2$ efficiently computes a collection of weight functions that will work for *any* set system in 2^U specified by a boolean circuit of size m.

Notice that a random collection $w_1, \cdots, w_t$ of weight functions will fulfil the required property of either hypotheses with high probability. Thus, the derandomization hypotheses are plausible. Indeed, it is not hard to see that suitable standard hardness assumptions that yield pseudorandom generators for derandomizing BPP would imply these hypotheses. We do not elaborate on this here. In this paper we show the following consequences of Hypotheses 1 and 2.

1. Hypothesis 1 implies that either NEXP $\not\subset$ P/poly or the Permanent does not have polynomial size noncommutative arithmetic circuits.
2. Hypothesis 2 implies that for almost all n there is an explicit multilinear polynomial $f_n(x_1, x_2, \cdots, x_n) \in \mathbb{F}[x_1, x_2, \cdots, x_n]$ in *commuting* variables x_i (where by explicit we mean that the coefficients of the polynomial f_n are computable by a uniform algorithm in time exponential in n) that does not have commutative arithmetic circuits of size $2^{o(n)}$ (where the field $\mathbb{F}$ is either the rationals or a finite field).

The first result is a consequence of an identity testing algorithm for noncommutative circuits that is based on the isolation lemma. This algorithm is based on ideas from [AMS08] where we used automata theory to pick matrices from a suitable matrix ring and evaluate the given arithmetic circuit on these matrices. In the next section, we describe the background and then give the identity test in the following section.

Remark 1. Notice that derandomizing the isolation lemma in specific applications like the RNC algorithm for matchings [MVV87] and the containment NL $\subseteq$ UL/poly [AR00] might still be possible without implying such circuit size lower bounds.

Noncommutative Circuits

Noncommutative polynomial identity testing has been the focus of recent research [RS05, BW05, AMS08]. One reason to believe that it could be easier than the commutative case to derandomize is because lower bounds are somewhat easier to prove in the noncommutative setting as shown by Nisan [N91]. Using a rank argument Nisan has shown exponential size lower bounds for noncommutative formulas (and noncommutative algebraic branching programs) that compute the noncommutative permanent or determinant polynomials in the ring $\mathbb{F}\{x_1, \cdots, x_n\}$ where x_i are noncommuting variables. In [CS07], Chien and Sinclair further extend Nisan's idea to prove exponential size lower bounds for noncommutative formulas computing noncommutative permanent or determinant polynomial over matrix algebra, quaternion algebra and group algebra. However, no superpolynomial lower bounds are known for the size of noncommutative circuits for explicit polynomials.

Our result in this paper is similar in flavour to the Impagliazzo-Kabanets result [KI03], where for *commutative* polynomial identity testing they show that derandomizing polynomial identity testing implies circuit lower bounds. Specifically, it implies that either NEXP $\not\subset$ P/poly or the integer Permanent does not have polynomial-size arithmetic circuits.

In [AMS08] we have observed that an analogous result also holds in the noncommutative setting. I.e., if noncommutative PIT has a deterministic polynomial-time algorithm then either NEXP $\not\subset$ P/poly or the *noncommutative* Permanent function does not have polynomial-size noncommutative circuits.

The connection that we show here between derandomizing the isolation lemma and noncommutative circuit size lower bounds is based on the above observation and our noncommutative polynomial identity test based on the isolation lemma.

Commutative Circuits

As a consequence of Hypothesis 2 we are able to show that for almost all n there is an explicit multilinear polynomial $f_n(x_1, x_2, \cdots, x_n) \in \mathbb{F}[x_1, x_2, \cdots, x_n]$ in *commuting* variables x_i (where by explicit we mean that the coefficients of the polynomial f_n are computable by a uniform algorithm in time exponential in n) that does not have commutative arithmetic circuits of size $2^{o(n)}$ (where the field $\mathbb{F}$ is either the rationals or a finite field). This is a fairly easy consequence of the univariate substitution idea and the observation that for arithmetic circuits computing multilinear polynomials, we can efficiently test if a monomial has nonzero coefficient (Lemma 4).

Klivans and Spielman [KS01] apply a more general form of the isolation lemma to obtain a polynomial identity test (in the commutative) case. This lemma is stated below.

Lemma 3. [KS01, Lemma 4] *Let L be any collection of linear forms over variables $z_1, z_2, \cdots, z_n$ with integer coefficients in the range $\{0, 1, \cdots, K\}$. If each z_i is picked independently and uniformly at random from $\{0, 1, \cdots, 2Kn\}$ then with probability at least $1/2$ there is a unique linear form from C that attains minimum value at $(z_1, \cdots, z_n)$.*

We can formulate a restricted version of this lemma similar to Lemma 2 that will apply only to sets of linear forms L accepted by a boolean circuit C. More precisely, an integer vector $(\alpha_1, \cdots, \alpha_n)$ such that $\alpha_i \in \{0, \cdots, K\}$ is in L if and only if $(\alpha_1, \cdots, \alpha_n)$ is accepted by the boolean circuit C.

Thus, for this form of the isolation lemma we can formulate another derandomization hypothesis analogous to Hypothesis 2 as follows.

Hypothesis 3. There is a deterministic algorithm $\mathcal{A}_3$ that takes as input (m, n, K) and outputs a collection of weight functions $w_1, w_2, \cdots, w_t$ such that $w_i : [n] \to [2Kn]$, with the property that for any size m boolean circuit C that takes as input $(\alpha_1, \cdots, \alpha_n)$ with $\alpha_i \in \{0, \cdots, K\}$ there is some weight vector w_i for which there is a *unique* linear form $(\alpha_1, \cdots, \alpha_n)$ accepted by C which attains the minimum value $\sum_{j=1}^{n} w_i(j)\alpha_j$. Furthermore, $\mathcal{A}_3$ runs in time subexponential in $size(C)$.

We show that Hypothesis 3 yields a lower bound consequence for the integer permanent.

2   Automata Theory Background

We recall some standard automata theory [HU78]. Fix a finite automaton $A = (Q, \Sigma, \delta, q_0, q_f)$ which takes inputs in Σ^*, Σ is the alphabet, Q is the set of states, $\delta : Q \times \Sigma \to Q$ is the transition function, and q_0 and q_f are the initial and final states respectively (we only consider automata with unique accepting states). For each $b \in \Sigma$, let $\delta_b : Q \to Q$ be defined by: $\delta_b(q) = \delta(q, b)$. These functions generate a submonoid of the monoid of all functions from Q to Q. This is the transition monoid of the automaton A and is well-studied in automata theory [Str94, page 55]. We now define the 0-1 matrix $M_b \in \mathbb{F}^{|Q| \times |Q|}$ as follows:

$$M_b(q, q') = \begin{cases} 1 \text{ if } \delta_b(q) = q', \\ 0 \text{ otherwise.} \end{cases}$$

The matrix M_b is the adjacency matrix of the graph of δ_b. As M_b is a 0-1 matrix, we can consider it as a matrix over any field $\mathbb{F}$.

For a string $w = w_1 w_2 \cdots w_k \in \Sigma^*$ we define M_w to be the matrix product $M_{w_1} M_{w_2} \cdots M_{w_k}$. If w is the empty string, define M_w to be the identity matrix of dimension $|Q| \times |Q|$. Let δ_w denote the natural extension of the transition function to w; if w is the empty string, δ_w is simply the identity function. We have

$$M_w(q, q') = \begin{cases} 1 \text{ if } \delta_w(q) = q', \\ 0 \text{ otherwise.} \end{cases} \tag{1}$$

Thus, M_w is also a matrix of zeros and ones for any string w. Also, $M_w(q_0, q_f) = 1$ if and only if w is accepted by the automaton A.

2.1   Noncommutative Arithmetic Circuits and Automata

This subsection is reproduced from [AMS08] to make this paper self-contained.

Consider the ring $\mathbb{F}\{x_1, \cdots, x_n\}$ of polynomials with noncommuting variables $x_1, \cdots, x_n$ over a field $\mathbb{F}$. Let C be a noncommutative arithmetic circuit computing a polynomial $f \in \mathbb{F}\{x_1, \cdots, x_n\}$. Let d be an upper bound on the degree of f. We can consider monomials over $x_1, \cdots, x_n$ as strings over the alphabet $\Sigma = \{x_1, x_2, \cdots, x_n\}$.

Let $A = (Q, \Sigma, \delta, q_0, q_f)$ be a finite automaton over the alphabet $\Sigma = \{x_1, x_2, \cdots, x_n\}$. We have matrices $M_{x_i} \in \mathbb{F}^{|Q| \times |Q|}$ as defined in Section 2. We are interested in the output matrix obtained when the inputs x_i to the circuit C are replaced by the matrices M_{x_i}. This output matrix is defined in the obvious way: the inputs are $|Q| \times |Q|$ matrices and we do matrix additions and multiplications at the circuit's addition and multiplication gates, respectively. We define the *output of C on the automaton A* to be this output matrix M_{out}. Clearly, given circuit C and automaton A, the matrix M_{out} can be computed in time $\mathrm{poly}(|C|, |A|, n)$.

We observe the following property: the matrix output M_{out} of C on A is determined completely by the polynomial f computed by C; the structure of the circuit C is otherwise irrelevant. This is important for us, since we are only interested in f. In particular, the output is always 0 when $f \equiv 0$.

More specifically, consider what happens when C computes a polynomial with a single term, say $f(x_1, \cdots, x_n) = cx_{j_1} \cdots x_{j_k}$, with a non-zero coefficient $c \in \mathbb{F}$. In this case, the output matrix M_{out} is clearly the matrix $cM_{x_{j_1}} \cdots M_{x_{j_k}} = cM_w$, where $w = x_{j_1} \cdots x_{j_k}$. Thus, by Equation 1 above, we see that the entry $M_{out}(q_0, q_f)$ is 0 when A rejects w, and c when A accepts w. In general, suppose C computes a polynomial $f = \sum_{i=1}^{t} c_i m_i$ with t nonzero terms, where $c_i \in \mathbb{F} \setminus \{0\}$ and $m_i = \prod_{j=1}^{d_i} x_{i_j}$, where $d_i \leq d$. Let w_i denotes the string representing monomial m_i. Finally, let $S_A^f = \{i \in \{1, \cdots, t\} \mid A \text{ accepts } w_i\}$.

Theorem 2. [AMS08] *Given any arithmetic circuit C computing polynomial $f \in \mathbb{F}\{x_1, \cdots, x_n\}$ and any finite automaton $A = (Q, \Sigma, \delta, q_0, q_f)$, then the output M_{out} of C on A is such that $M_{out}(q_0, q_f) = \sum_{i \in S_A^f} c_i$.*

Proof. The proof is an easy consequence of the definitions and the properties of the matrices M_w stated in Section 2. Note that $M_{out} = f(M_{x_1}, \cdots, M_{x_n})$. But $f(M_{x_1}, \cdots, M_{x_n}) = \sum_{i=1}^{s} c_i M_{w_i}$, where w_i is the string representing monomial m_i. By Equation 1, we know that $M_{w_i}(q_0, q_f)$ is 1 if w_i is accepted by A, and 0 otherwise. Adding up, we obtain the result.

We now explain the role of the automaton A in testing if the polynomial f computed by C is identically zero. Our basic idea is to design an automaton A that accepts exactly one word among all the words that correspond to the nonzero terms in f. This would ensure that $M_{out}(q_0, q_f)$ is the nonzero coefficient of the monomial filtered out. More precisely, we will use the above theorem primarily in the following form, which we state as a corollary.

Corollary 1. [AMS08] *Given any arithmetic circuit C computing polynomial $f \in \mathbb{F}\{x_1, \cdots, x_n\}$ and any finite automaton $A = (Q, \Sigma, \delta, q_0, q_f)$, then the output M_{out} of C on A satisfies:*

(1) If A rejects every string corresponding to a monomial in f, then $M_{out}(q_0, q_f) = 0$.

(2) If A accepts exactly one string corresponding to a monomial in f, then $M_{out}(q_0, q_f)$ is the nonzero coefficient of that monomial in f.

Moreover, M_{out} can be computed in time $\mathrm{poly}(|C|, |A|, n)$.

Proof. Both points (1) and (2) are immediate consequences of the above theorem. The complexity of computing M_{out} easily follows from its definition.

Another corollary to the above theorem is the following.

Corollary 2. [AMS08] *Given any arithmetic circuit C over $\mathbb{F}\{x_1, \cdots, x_n\}$, and any monomial m of degree d_m, we can compute the coefficient of m in C in time* poly$(|C|, d_m, n)$.

Proof. Apply Corollary 1 with A being any standard automaton that accepts the string corresponding to monomial m and rejects every other string. Clearly, A can be chosen so that A has a unique accepting state and $|A| = O(nd_m)$.

In fact corollary 2 says that, given an arithmetic circuit C and a monomial m, there is an uniform way to generate a polynomial-size boolean circuit C' such that C' can decide whether m is a nonzero monomial in the polynomial computed by C. The boolean circuit C' is simply the description of the algorithm described in the proof of corollary 2.

Corollary 3. *Given an arithmetic circuit C over $\mathbb{F}x_1, \cdots, x_n$ and a monomial m of degree d, there is an uniform polynomial-time algorithm that generates a* poly$(|C|, d, n)$ *size boolean circuit C' that accepts (C, m) if and only if m is a nonzero monomial in the polynomial computed by C.*

Remark 2. Corollary 2 is very unlikely to hold in the commutative ring $\mathbb{F}[x_1, \cdots, x_n]$. For, it is easy to see that in the commutative case computing the coefficient of the monomial $\prod_{i=1}^{n} x_i$ in even a product of linear forms $\Pi_i \ell_i$ is at least as hard as computing the permanent over $\mathbb{F}$, which is #P-complete when $\mathbb{F} = \mathbb{Q}$. However, we can show the following for commutative circuits computing multilinear polynomials.

Corollary 4. *Given a commutative arithmetic circuit $\hat{C}$ over $\mathbb{F}[x_1, \cdots, x_n]$, with the promise that $\hat{C}$ computes a* multilinear *polynomial, and any monomial $m = \prod_{i \in S} x_i$ where $S \subseteq [n]$, we can compute the coefficient of m in C in time* poly$(|\hat{C}|, n)$. *Furthermore, there is a uniform polynomial-time algorithm that generates a boolean circuit C' of size* poly$(|C|, n)$ *such that C' takes as input a description of circuit C and monomial m and it decides whether the coefficient of m is nonzero in the polynomial computed by C.*

Proof. Let $m = \prod_{i \in S} x_i$ be the given monomial. The algorithm will simply substitute 1 for each x_i such that $i \in S$ and 0 for each x_i such that $i \notin S$ and evaluate the circuit $\hat{C}$ to find the coefficient of the monomial m. The boolean circuit C' is simply the description of the above algorithm. It is clear that C' can be uniformly generated.

3   Noncommutative Identity Test Based on Isolation Lemma

We now describe a new identity test for noncommutative circuits based on the isolation lemma. It is directly based on the results from [AMS08]. This is conceptually quite different from the randomized identity test of Bogdanov and Wee [BW05].

Theorem 3. *Let $f \in \mathbb{F}\{x_1, x_2, \cdots, x_n\}$ be a polynomial given by an arithmetic circuit C of size m. Let d be an upper bound on the degree of f. Then there is a randomized algorithm which runs in time $\mathrm{poly}(n, m, d)$ and can test whether $f \equiv 0$.*

Proof. Let $[d] = \{1, 2, \cdots, d\}$ and $[n] = \{1, 2, \cdots, n\}$. Consider the set of tuples $U = [d] \times [n]$. Let $v = x_{i_1} x_{i_2} \cdots x_{i_t}$ be a nonzero monomial of f. Then the monomial can be identified with the following subset S_v of U :

$$S_v = \{(1, i_1), (2, i_2), \cdots, (t, i_t)\}$$

Let $\mathcal{F}$ denotes the family of subsets of U corresponding to the nonzero monomials of f i.e,

$$\mathcal{F} = \{S_v \mid v \text{ is a nonzero monomial in } f\}$$

By the Isolation Lemma we know that if we assign random weights from $[2dn]$ to the elements of U, with probability at least $1/2$, there is a unique minimum weight set in $\mathcal{F}$. Our aim will be to construct a family of small size automatons which are indexed by weights $w \in [2nd^2]$ and $t \in [d]$, such that the automata $A_{w,t}$ will precisely accept all the strings (corresponding to the monomials) v of length t, such that the weight of S_v is w. Then from the isolation lemma we will argue that the automata corresponding to the minimum weight will precisely accept only one string (monomial). Now for $w \in [2nd^2]$, and $t \in [d]$, we describe the construction of the automaton $A_{w,t} = (Q, \Sigma, \delta, q_0, F)$ as follows: $Q = [d] \times [2nd^2] \cup \{(0,0)\}$, $\Sigma = \{x_1, x_2, \cdots, x_n\}$, $q_0 = \{(0,0)\}$ and $F = \{(t, w)\}$. We define the transition function $\delta : Q \times \Sigma \to Q$,

$$\delta((i, V), x_j) = (i + 1, V + W),$$

where W is the random weight assign to $(i + 1, j)$. Our automata family $\mathcal{A}$ is simply,

$$\mathcal{A} = \{A_{w,t} \mid w \in [2nd^2], t \in [d]\}.$$

Now for each of the automaton $A_{w,t} \in \mathcal{A}$, we mimic the run of the automaton $A_{w,t}$ on the circuit C as described in Section 2. If the output matrix corresponding to any of the automaton is nonzero, our algorithm declares $f \neq 0$, otherwise declares $f \equiv 0$.

The correctness of the algorithm follows easily from the Isolation Lemma. By the Isolation Lemma we know, on random assignment, a unique set S in $\mathcal{F}$ gets the minimum weight $w_{\min}$ with probability at least $1/2$. Let S corresponds to the monomial $x_{i_1} x_{i_2} \cdots x_{i_\ell}$. Then the automaton $A_{w_{\min}, \ell}$ accepts the string (monomial) $x_{i_1} x_{i_2} \cdots x_{i_\ell}$. Furthermore, as no other set in $\mathcal{F}$ get the same minimum weight, $A_{w_{min}, \ell}$ rejects all the other monomials. So the (q_0, q_f) entry of the output matrix M_o, that we get in running $A_{w_{\min}, \ell}$ on C is nonzero. Hence with probability at least $1/2$, our algorithm correctly decide that f is nonzero. The success probability can be boosted to any constant by standard independent repetition of the same algorithm. Finally, it is trivial to see that the algorithm always decides correctly if $f \equiv 0$.

4   Noncommutative Identity Testing and Circuit Lower Bounds

For commutative circuits, Impagliazzo and Kabanets [KI03] have shown that derandomizing PIT implies circuit lower bounds. It implies that either NEXP $\not\subset$ P/poly or the integer Permanent does not have polynomial-size arithmetic circuits.

In [AMS08] we have observed that this also holds in the noncommutative setting. That is, if noncommutative PIT has a deterministic polynomial-time algorithm then either NEXP $\not\subset$ P/poly or the *noncommutative* Permanent function does not have polynomial-size noncommutative circuits. We note here that noncommutative circuit lower bounds are sometimes easier to prove than for commutative circuits. E.g. Nisan [N91], Chien and Sinclair [CS07] have shown exponential-size lower bounds for noncommutative formula size and further results are known for pure noncommutative circuits [N91, RS05]. However, proving superpolynomial size lower bounds for general noncommutative circuits computing the Permanent has remained an open problem.

To keep this paper self contained, we briefly recall the discussion from [AMS08].

The noncommutative Permanent function over integer $Perm(x_1, \cdots, x_n) \in \mathbb{Z}\{x_1, \cdots, x_n\}$ ($\mathbb{Z}$ is the set of integer) is defined as:

$$Perm(x_1, \cdots, x_n) = \sum_{\sigma \in S_n} \prod_{i=1}^{n} x_{i,\sigma(i)}.$$

Let SUBEXP denote $\cap_{\epsilon>0}\mathrm{DTIME}(2^{n^\epsilon})$ and NSUBEXP denote $\cap_{\epsilon>0}\mathrm{NTIME}(2^{n^\epsilon})$.

Theorem 4. [AMS08] *If* PIT *for noncommutative circuits of polynomial degree* $C(x_1, \cdots, x_n) \in \mathbb{Z}\{x_1, \cdots, x_n\}$ *is in SUBEXP, then either NEXP $\not\subset$ P/poly or the noncommutative Permanent function does not have polynomial-size noncommutative circuits.*

Proof. Suppose NEXP $\subset$ P/poly. Then, by the main result of [IKW02] we have NEXP $=$ MA. Furthermore, by Toda's theorem MA $\subseteq$ P$^{Perm_{\mathbb{Z}}}$, where the oracle computes the integer permanent. Now, assuming PIT for noncommutative circuits of polynomial degree is in deterministic subexponential-time, we will show that the (noncommutative) Permanent function does not have polynomial-size noncommutative circuits. Suppose to the contrary that it does have polynomial-size noncommutative circuits. Clearly, we can use it to compute the integer permanent as well. Furthermore, as in [KI03] we notice that the noncommutative $n \times n$ Permanent is also uniquely characterized by the identities $p_1(x) \equiv x$ and $p_i(X) = \sum_{j=1}^{i} x_{1j}p_{i-1}(X_j)$ for $1 < i \leq n$, where X is a matrix of i^2 noncommuting variables and X_j is its j-th minor w.r.t. the first row. I.e. the polynomials $p_i, 1 \leq i \leq n$ satisfy these n identities over *noncommuting* variables $x_{ij}, 1 \leq i, j \leq n$ if and only if p_i computes the $i \times i$ permanent of noncommuting variables. The rest of the proof is exactly as in Impagliazzo-Kabanets [KI03]. We can easily describe an NP machine to simulate a P$^{Perm_{\mathbb{Z}}}$ computation. The NP machine guesses a polynomial-size noncommutative circuit for $Perm$ on $m \times m$ matrices, where m is a polynomial bound on the matrix size of the queries made in the computation of the P$^{Perm_{\mathbb{Z}}}$ machine. Then the NP machine verifies that the circuit computes the permanent by checking the m *noncommutative* identities it must satisfy. This can be done in SUBEXP by assumption. Finally, the NP machine uses the circuit to answer all the integer permanent queries that are made in the computation of P$^{Perm_{\mathbb{Z}}}$ machine. Putting it together, we get NEXP $\subseteq$ NSUBEXP which contradicts the nondeterministic time hierarchy theorem.

5   The Results

We are now ready to prove our first result. Suppose the derandomization Hypothesis 1 holds (as stated in the introduction): i.e. suppose there is a deterministic algorithm $\mathcal{A}_1$ that takes as input (C, n) where C is an n-input boolean circuit and in subexponential time computes a set of weight functions $w_1, w_2, \cdots, w_t$, $w_i : [n] \rightarrow [2n]$ such that the set system $\mathcal{F}_C$ defined by the circuit C has a unique minimum weight set w.r.t. at least one of the weight functions w_i.

Let $C'(x_1, x_2, \cdots, x_n)$ be a noncommutative arithmetic circuit of degree d bounded by a polynomial in $size(C')$. By Corollary 2, there is a deterministic polynomial-time algorithm that takes as input C' and a monomial m of degree at most d and accepts if and only if the monomial m has nonzero coefficient in the polynomial computed by C'. Moreover by corollary 3, we have a uniformly generated boolean circuit C of size polynomial in $size(C')$ that accepts only the monomials $x_{i_1} x_{i_2} \cdots x_{i_k}$, $k \leq d$ that have nonzero coefficients in the polynomial computed by C'. Now, as a consequence of Theorem 3 and its proof we have a *deterministic* subexponential algorithm for checking if $C' \equiv 0$, assuming algorithm $\mathcal{A}_1$ exists. Namely, we compute the boolean circuit C from C' in polynomial time. Then, invoking algorithm $\mathcal{A}_1$ with C as input we compute at most subexponentially many weight functions $w_1, \cdots, w_t$. Then, following the proof of Theorem 3 we construct the automata corresponding to these weight functions and evaluate C' on the matrices that each of these automata define in the prescribed manner. By assumption about algorithm $\mathcal{A}_1$, if $C' \not\equiv 0$ then one of these w_i will give matrix inputs for the variables $x_j, 1 \leq j \leq n$ on which C' evaluates to a nonzero matrix. We can now show the following theorem.

Theorem 5. *If the subexponential time algorithm $\mathcal{A}_1$ satisfying Hypothesis 1 exists then noncommutative identity testing is in* SUBEXP *which implies that either* NEXP $\not\subset$ P/poly *or the Permanent does not have polynomial size noncommutative circuits.*

Proof. The result is a direct consequence of the discussion preceding the theorem statement and Theorem 4.

Commutative Circuits

We now turn to the result under the *stronger* derandomization Hypothesis 2 (stated in the introduction). More precisely, suppose there is a deterministic algorithm $\mathcal{A}_2$ that takes as input (m, n) and in time polynomial in m computes a set of weight functions $w_1, w_2, \cdots, w_t$, $w_i : [n] \rightarrow [2n]$ such that for *each* n-input boolean circuit C of size m, the set system $\mathcal{F}_C$ defined by the circuit C has a unique minimum weight set w.r.t. at least one of the weight functions w_i. We show that there is an *explicit* polynomial[1] $f(x_1, \cdots, x_n)$ in commuting variables x_i that does not have subexponential size *commutative* circuits. The following theorem is similar in flavour to the Agrawal's result that a black-box derandomization of PIT for a class of arithmetic circuit via pseudorandom generator will show similar lower bound (Lemma 5.1 of [Agr05]).

---

[1] By explicit we mean that the coefficients of f are computable in time exponential in n.

Theorem 6. *Suppose there is a polynomial-time algorithm $\mathcal{A}_2$ satisfying Hypothesis 2. Then for all but finitely many n there is an explicit multilinear polynomial (where by explicit we mean that the coefficients of the polynomial f_n are computable by a uniform algorithm in time exponential in n) $f(x_1, \cdots, x_n) \in \mathbb{F}[x_1, x_2, \cdots, x_n]$ (where $\mathbb{F}$ is either $\mathbb{Q}$ or a finite field) that is computable in $2^{n^{O(1)}}$ time (by a uniform algorithm) and does not have arithmetic circuits of size $2^{o(n)}$.*

Proof. We will pick an appropriate multilinear polynomial $f \in \mathbb{F}[x_1, x_2, \cdots, x_n]$:

$$f(x_1, x_2, \cdots, x_n) = \sum_{S \subseteq [n]} c_S \prod_{i \in S} x_i,$$

where the coefficients $c_S \in \mathbb{F}$ will be determined appropriately so that the polynomial f has the claimed property.

Suppose $\mathcal{A}_2$ runs in time m^c for constant $c > 0$, where m denotes the size bound of the boolean circuit C defining set system $\mathcal{F}_C$. Notice that the number t of weight functions output by $\mathcal{A}_2$ is bounded by m^c.

The total number of coefficients c_S of f is 2^n. For each weight function w_i let $(w_{i,1}, \cdots, w_{i,n})$ be the assignments to the variables x_i. For each weight function $w_i, 1 \leq i \leq t$ we write down the following equations

$$f(y^{w_{i,1}}, y^{w_{i,2}}, \cdots, y^{w_{i,n}}) = 0.$$

Since f is of degree at most n, and the weights $w_{i,j}$ are bounded by $2n$, $f(y^{w_{i,1}}, y^{w_{i,2}}, \cdots, y^{w_{i,n}})$ is a univariate polynomial of degree at most $2n^2$ in y. Thus, each of the above equations will give rise to at most $2n^2$ linear equations in the unknowns c_S.

In all, this will actually give us a system of at most $2n^2 m^c$ linear equations over $\mathbb{F}$ in the unknown scalars c_S. Since the total number of distinct monomials is 2^n, and 2^n asymptotically exceeds m^c for $m = 2^{o(n)}$, the system of linear equations has a *nontrivial* solution in the c_S provided $m = 2^{o(n)}$. Furthermore, a nontrivial solution for c_S can be computed using Gaussian elimination in time exponential in n.

We claim that f does not have commutative circuits of size $2^{o(n)}$ over $\mathbb{F}$. Assume to the contrary that $\hat{C}(x_1, \cdots, x_n)$ is a circuit for $f(x_1, \cdots, x_n)$ of size $2^{o(n)}$. By Lemma 4 notice that we can uniformly construct a boolean circuit C of size $m = 2^{o(n)}$ that will take as input a monomial $\prod_{i \in S} x_i$ (encoded as an n bit boolean string representing S as a subset of $[n]$) and test if it is nonzero in $\hat{C}$ and hence in $f(x_1, \cdots, x_n)$.

Assuming Hypothesis 2, let $w_1, \cdots, w_t$ be the weight functions output by $\mathcal{A}_2$ for input (m, n). By Hypothesis 2, for some weight function w_i there is a unique monomial $\prod_{j \in S} x_j$ such that $\sum_{j \in S} w_{i,j}$ takes the minimum value. Clearly, the commutative circuit $\hat{C}$ must be nonzero on substituting $y^{w_{i,j}}$ for x_j (the coefficient of $y^{\sum_{j \in S} w_{i,j}}$ will be nonzero). However, f evaluates to zero on the integer assignments prescribed by all the weight functions $w_1, \cdots, w_t$. This is a contradiction to the assumption and it completes the proof.

Remark 3. We note that Hypothesis 2 also implies the existence of an explicit polynomial in noncommuting variables that does not have noncommutative circuits of subexponential size (we can obtain it as an easy consequence of the above proof).

We now show that under the derandomization Hypothesis 3 yields a different consequence (about the integer permanent rather than some explicit function).

Theorem 7. *If a subexponential-time algorithm $\mathcal{A}_3$ satisfying Hypothesis 3 exists then identity testing over $\mathbb{Z}$ is in* SUBEXP *which implies that either* NEXP $\not\subset$ P/poly *or the integer Permanent does not have polynomial size arithmetic circuits.*

Proof. Using Lemma 3 it is shown in [KS01, Theorem 5] that there is a randomized identity test for small degree polynomials in $\mathbb{Q}[x_1, \cdots, x_n]$, where the polynomial is given by an arithmetic circuit $\hat{C}$ of polynomially bounded degree d. The idea is to pick a random weight vector $w : [n] \rightarrow [2nd]$ and replace the indeterminate x_i by $y^{w(i)}$, where d is the total degree of the input polynomial. As the circuit $\hat{C}$ has small degree, after this univariate substitution the circuit can be evaluated in deterministic polynomial time to explicitly find the polynomial in y. By Lemma 3 it will be nonzero with probability $1/2$ if $\hat{C}$ computes a nonzero polynomial.

Coming to the proof of this theorem, if NEXP $\not\subset$ P/poly then we are done. So, suppose NEXP $\subset$ P/poly. Notice that given any monomial $x_1^{d_1} \cdots x_n^{d_n}$ of total degree bounded by d we can test if it is a nonzero monomial of $\hat{C}$ in exponential time (explicitly listing down the monomials of the polynomial computed by $\hat{C}$). Therefore, since NEXP $\subset$ P/poly there is a polynomial-size boolean circuit C that accepts the vector $(d_1, \cdots, d_n)$ iff $x_1^{d_1} \cdots x_n^{d_n}$ is a nonzero monomial in the given polynomial (as required for application of Hypothesis 3).

Now, we invoke the derandomization Hypothesis 3. We can apply the Klivans-Spielman polynomial identity test, explained above, to the arithmetic circuit $\hat{C}$ for each of the t weight vectors $w_1, \cdots, w_t$ generated by algorithm $\mathcal{A}_3$ to obtain a subexponential deterministic identity test for the circuit $\hat{C}$ by the properties of $\mathcal{A}_3$. Now, following the argument of Impagliazzo-Kabanets [KI03] it is easy to derive that the integer Permanent does not have polynomial size arithmetic circuits.

Remark 4. Although the permanent is a multilinear polynomial, notice that Hypothesis 2 does not seem strong enough to prove the above theorem. The reason is that the arithmetic circuit for the permanent that is nondeterministically guessed may not compute a multilinear polynomial and hence the application of Lemma 4 is not possible. There does not appear any easy way of testing if the guessed circuit computes a multilinear polynomial.

Remark 5. We can formulate both Hypothesis 1 and Hypothesis 2 more generally by letting the running time of algorithms $\mathcal{A}_1$ and $\mathcal{A}_2$ be a function $t(m, n)$. We will then obtain suitably quantified circuit lower bound results as consequence.

6 Discussion

An interesting open question is whether derandomizing similar restricted versions of the Valiant-Vazirani lemma also implies circuit lower bounds. We recall the Valiant-Vazirani lemma as stated in the original paper [VV86].

Lemma 4. *Let $S \subseteq \{0,1\}^t$. Suppose $w_i, 1 \leq i \leq t$ are picked uniformly at random from $\{0,1\}^t$. For each i, let $S_i = \{v \in S \mid v.w_j = 0, 1 \leq j \leq i\}$ and let $p_t(S)$ be the probability that $|S_i| = 1$ for some i. Then $p_t(S) \geq 1/4$.*

Analogous to our discussion in Section 1, here too we can consider the restricted version where we consider $S_C \subseteq \{0,1\}^n$ to be the set of n-bit vectors accepted by a boolean circuit C of size m. We can similarly formulate derandomization hypotheses similar to Hypotheses 1 and 2.

We do not know if there is another randomized polynomial identity test for noncommutative arithmetic circuits based on the Valiant-Vazirani lemma. The automata-theoretic technique of Section 3 does not appear to work. Specifically, given a matrix $h : \mathbb{F}_2^n \to \mathbb{F}_2^k$, there is no deterministic finite automaton of size poly(n, k) that accepts $x \in \mathbb{F}_2^n$ if and only if $h(x) = 0$.

Acknowledgements. We are grateful to Manindra Agrawal for interesting discussions and his suggestion that Theorem 6 can be obtained from the stronger hypothesis. We also thank Srikanth Srinivasan for comments and discussions.

References

[Agr05] Agrawal, M.: Proving Lower Bounds Via Pseudo-random Generators. In: Ramanujam, R., Sen, S. (eds.) FSTTCS 2005. LNCS, vol. 3821, pp. 92–105. Springer, Heidelberg (2005)

[Agr07] Agrawal, M.: Rings and Integer Lattices in Computer Science. In: Barbados Workshop on Computational Complexity, Lecture no. 9 (2007)

[AR00] Reinhardt, K., Allender, E.: Making Nondeterminism Unambiguous. SIAM J. Comput. 29(4), 1118–1131 (2000)

[ARZ99] Allender, E., Reinhardt, K., Zhou, S.: Isolation, matching and counting uniform and nonuniform upper bounds. Journal of Computer and System Sciences 59(2), 164–181 (1999)

[AMS08] Arvind, V., Mukhopadhyay, P., Srinivasan, S.: New results on Noncommutative and Commutative Polynomial Identity Testing. In: Proceedings of the 23rd IEEE Conference on Computational Complexity (to appear, June 2008); Technical report version in ECCC report TR08-025 (2008)

[BW05] Bogdanov, A., Wee, H.: More on Noncommutative Polynomial Identity Testing. In: Proc. of the 20th Annual Conference on Computational Complexity, pp. 92–99 (2005)

[CS07] Chien, S., Sinclair, A.: Algebras with Polynomial Identities and Computing the Determinants. SIAM J. of Comput. 37(1), 252–266 (2007)

[HU78] Hopcroft, J.E., Ullman, J.D.: Introduction to Automata Theory, Languages and Computation. Addison-Wesley, Reading (1979)

[IKW02] Impagliazzo, R., Kabanets, V., Wigderson, A.: In search of an easy witness: Exponential time vs. probabilistic polynomial time. Journal of Computer and System Sciences 65(4), 672–694 (2002)

[KI03] Kabanets, V., Impagliazzo, R.: Derandomization of polynomial identity tests means proving circuit lower bounds. In: Proc. of the thirty-fifth annual ACM Sym. on Theory of computing, pp. 355–364 (2003)

[KS01]   Klivans, A.R., Spielman, D.A.: Randomness Efficient Identity Testing. In: Proceedings of the 33rd Symposium on Theory of Computing (STOC), pp. 216–223 (2001)

[MVV87]   Mulmuley, K., Vazirani, U., Vazirani, V.: Matching is as easy as matrix inversion. In: Proc. of the nineteenth annual ACM conference on Theory of Computing, pp. 345–354. ACM Press, New York (1987)

[N91]   Nisan, N.: Lower bounds for non-commutative computation. In: Proc. of the 23rd annual ACM Sym. on Theory of computing, pp. 410–418 (1991)

[NSV94]   Narayanan, H., Saran, H., Vazirani, V.V.: Randomized Parallel Algorithms for Matroid Union and Intersection, With Applications to Arboresences and Edge-Disjoint Spanning Trees. SIAM J. Comput. 23(2), 387–397 (1994)

[RS05]   Raz, R., Shpilka, A.: Deterministic polynomial identity testing in non commutative models. Computational Complexity 14(1), 1–19 (2005)

[Sch80]   Schwartz, J.T.: Fast Probabilistic algorithm for verification of polynomial identities. J. ACM 27(4), 701–717 (1980)

[Str94]   Straubing, H.: Finite automata, formal logic, and circuit complexity. In: Progress in Theoretical Computer Science, Birkhuser Boston Inc., Boston (1994)

[VV86]   Valiant, L.G., Vazirani, V.V.: NP is as Easy as Detecting Unique Solutions. Theor. Comput. Sci. 47(3), 85–93 (1986)

[Zip79]   Zippel, R.: Probabilistic algorithms for sparse polynomials. In: Proc. of the Int. Sym. on Symbolic and Algebraic Computation, pp. 216–226 (1979)

Tensor Products of Weakly Smooth Codes Are Robust*

Eli Ben-Sasson and Michael Viderman

Computer Science Department
Technion — Israel Institute of Technology
Haifa, 32000, Israel
{eli,viderman}@cs.technion.ac.il

Abstract. We continue the study of *robust* tensor codes and expand the class of base codes that can be used as a starting point for the construction of locally testable codes via robust two-wise tensor products. In particular, we show that all unique-neighbor expander codes and all locally correctable codes, when tensored with any other good-distance code, are robust and hence can be used to construct locally testable codes. Previous works by [2] required stronger expansion properties to obtain locally testable codes.

Our proofs follow by defining the notion of *weakly smooth* codes that generalize the *smooth* codes of [2]. We show that weakly smooth codes are sufficient for constructing robust tensor codes. Using the weaker definition, we are able to expand the family of base codes to include the aforementioned ones.

1 Introduction

A linear code over a finite field F is a linear subspace $C \subseteq F^n$. A code is *locally testable* if given a word $x \in F^n$ one can verify whether $x \in C$ by reading only a few (randomly chosen) symbols from x. More precisely such a code has a *tester*, which is a randomized algorithm with oracle access to the received word x. The tester reads at most q symbols from x and based on this "local view" decides if $x \in C$ or not. It should accept codewords with probability one, and reject words that are "far" (in Hamming distance) with "noticeable" probability.

Locally Testable Codes (LTCs) were first explicitly studied by Goldreich and Sudan [9] and since then a few constructions of LTCs were suggested (See [8] for an extensive survey of those constructions). All known efficient constructions of LTCs, i.e. that obtain subexponential rate, rely on some form of "composition" of two (or more) codes. One of the simplest ways to compose codes for the construction of LTCs is by use of the tensor product, as suggested by Ben-Sasson

---

* Research supported in part by a European Community International Reintegration Grant, an Alon Fellowship, and grants by the Israeli Science Foundation (grant number 679/06) and by the US-Israel Binational Science Foundation (grant number 2006104).

A. Goel et al. (Eds.): APPROX and RANDOM 2008, LNCS 5171, pp. 290–302, 2008.

and Sudan [1]. They introduced the notion of *robust* LTCs: An LTC is called robust if whenever the received word is far from the code, then with noticeable probability the local view of the tester is *far* from an accepting local view (see robust definition [2]). It was shown in [1] that a code obtained by tensoring three or more codes (i.e. $C_1 \otimes C_2 \otimes C_3$) is robustly testable when the distances of the codes are big enough, and used this result to construct LTCs. Then they considered the tensor product of two codes. Given two linear codes R, C their tensor product $R \otimes C$ consists of all matrices whose rows are codewords of R and whose columns are codewords of C. If R and C are locally testable, we would like $R \otimes C$ to be locally testable. [1] suggested using the following test for the testing the tensor product $R \otimes C$ and asked whether it is robust:

Test for $R \otimes C$: Pick a random row (or column), accept iff it belongs to R (or C).

Valiant [3] showed a surprising example of two linear codes R and C for which the test above is not robust, by exhibiting a word x that is far from $R \otimes C$ but such that the rows of x are very close to R and the columns of x are very close to C. Additional examples give a codes whose tensor product with itself is not robust [4] and two good codes (with linear rate) whose tensor product is not robust [7].

Despite these examples Dinur et al. showed in [2] that the above test is robust as long as one of the base codes is *smooth*, according to a definition of the term introduced there (see Definition 5). The family of smooth codes includes locally testable codes and certain codes constructed from expander graphs with very good expansion properties. In this work we continue this line of research and enlarge the family of base codes that result in robust tensor codes and do this by working with a weaker definition of smoothness (Definition 4). Using the weaker definition, we still manage to get pretty much the same results as in [2] and do this using the same proof strategy as there. However, our weaker definition allows us to argue — in what we view as the main technical contributions of this paper (Sections 6 and 7) — that a larger family of codes is suitable for forming robust tensor codes. One notable example is that our definition allows us to argue that any expander code with unique-neighbor expansion (i.e., with expansion parameter $\gamma < 1/2$ as per Definition 3) is also weakly smooth, hence robust. We stress that unique-neighbor expansion is the minimal requirement in terms of expansion needed to argue an expander code has good (i.e., constant relative) distance, so our our work shows all "combinatorially good" expander codes[1] are robust. In comparison, the work of [2] required stronger expansion parameters ($\gamma < 1/4$) of the kind needed to ensure an expander code is not merely good in terms of its distance, but can also be decoded in linear time [10].

Another family of codes shown here to be robust under two-wise tensor products is the family of locally correctable codes (LCCs), see Definition 7.

---

[1] Clearly, there exist non-unique-neighbor expander codes with good distance. However, the distance of these codes cannot be argued merely using the combinatorial structure of the underlying parity check matrix.

We end this section by pointing out that recently, tensor codes have played a role in the combinatorial construction by Meir [6] of quasilinear length locally testable codes. Better base codes may result in LTCs with improved rate, hence the importance in broadening the class of base codes that can be used to construct robust tensor codes.

Organization of the Paper. In the following section we provide the now-standard definitions regarding robust tensor codes. In Section 3 We formally define weakly smooth codes and state our main results. In Section 4 We prove weakly smooth codes are robust. Section 5 shows the smooth codes of [2] are also weakly smooth. The last two sections prove that unique-neighbor expander codes, and locally correctable codes, respectively, are weakly smooth.

2   Preliminary Definitions

The definitions appearing here are pretty much standard in the literature on tensor-based LTCs.

Throughout this paper F is a finite field and C, R are linear codes over F. For $c \in C$ let $\text{supp}(c) = \{i|c_i \neq 0\}$ and $\text{wt}(c) = |\text{supp}(c)|$. We define the *distance* between two words $x, y \in F^n$ to be $d(x,y) = |\{i \mid x_i \neq y_i\}|$ and the relative distance to be $\delta(x,y) = \frac{d(x,y)}{n}$. The distance of a code is denoted $d(C)$ and defined to be the minimal value of $d(x,y)$ for two distinct codewords $x, y \in C$. Similarly, the relative distance of the code is denoted $\delta(C) = \frac{d(C)}{n}$. For $x \in F^n$ and $C \subseteq F^n$, let $\delta_C(x) = min_{y \in C}\{\delta(x,y)\}$ denote the relative distance of x from code C. We let $\dim(C)$ denote the dimension of C. The vector inner product between u_1 and u_2 is denoted by $\langle u_1, u_2 \rangle$. For code C let $C^\perp = \{u \in F^n \mid \forall c \in C : \langle u, c \rangle = 0\}$ be its dual code and let $C_t^\perp = \{u \in C^\perp \mid \text{wt}(u) = t\}$. In similar way we define $C_{<t}^\perp = \{u \in C^\perp \mid \text{wt}(u) < t\}$ and $C_{\leq t}^\perp = \{u \in C^\perp \mid \text{wt}(u) \leq t\}$. For $w \in F^n$ and $S \subseteq [n]$ we let $w|_S = (w_{j_1}, w_{j_2}, ..., w_{j_m})$ when $\{j_1, j_2, ..., j_m\} = S$ be the projection of w on subset S. Similarly, we let $C|_S = \{c|_S \mid c \in C\}$ to denote the projection of code C on subset S.

2.1   Tensor Product of Codes

For $x \in F^m$ and $y \in F^n$ we let $x \otimes y$ denote tensor product of x and y (i.e. the $n \times m$ matrix xy^T). Let $R \subseteq F^m$ and $C \subseteq F^n$ be linear codes. We define the tensor product code $R \otimes C$ to be the linear subspace spanned by words $r \otimes c \in F^{n \times m}$ for $r \in R$ and $c \in C$. Some immediate facts:

- The code $R \otimes C$ consists of all $n \times m$ matrices over F whose rows belong to R and whose columns belong to C.
- $\dim(R \otimes C) = \dim(R) \cdot \dim(C)$
- $\delta(R \otimes C) = \delta(R) \cdot \delta(C)$

Let $M \in F^m \otimes F^n$ and let $\delta(M) = \delta_{R \otimes C}(M)$. Let $\delta^{row}(M) = \delta_{R \otimes F^n}(M)$ denote the distance from the space of matrices whose rows are codewords of R. This is expected distance of a random row in x from R. Similarly let $\delta^{col}(M) = \delta_{F^m \otimes C}(M)$.

2.2 Robust Locally Testable Codes

Definition 1 (Robustness). *Let M be a candidate codeword for $R \otimes C$. The robustness of M is defined as $\rho(M) = (\delta^{row}(M) + \delta^{col}(M))/2$, i.e., it is the average distance of "local views" of the codeword. The code $R \otimes C$ is robustly testable if there exists a constant α such that $\frac{\rho(M)}{\delta(M)} \geq \alpha$ for every M.*

The robustness of a Tester T is defined as $\rho^T = min_{M \in R \otimes C} \frac{\rho(M)}{\delta_{R \otimes C}(M)}$.

2.3 Low Density Parity Check (LDPC) Codes

The following definition is the natural generalization of a LDPC codes to fields of size > 2.

Definition 2 (LDPC codes). *A check graph $([n], [m], E, F)$ is a bipartite graph $([n], [m], E)$ over F for a code $C \subset F^n$ where each edge $e = (i, j) \in E$ is labeled by some $e_{(i,j)} \neq 0 \in F$ and the following holds (let $N(j)$ denote the neighbors of j in the graph):*

$$x \in C \Longleftrightarrow \forall j \in [m] \sum_{i \in N(j)} x_i \cdot e_{(i,j)} = 0,$$

where the sum $\sum_{i \in N(j)} x_i \cdot e_{(i,j)}$ is computed over F.

Clearly, any linear code $C \subseteq F$ has a corresponding check graph $([n], [m], E, F)$. Moreover if $C^\perp = span(C^\perp_{\leq d})$ then without loss of generality every right hand node $j \in [m]$ has degree at most d.

Definition 3 (Expander graphs). *Let $c, d \in N$ and let $\gamma, \delta \in (0, 1)$. Define a (c, d)-regular (γ, δ)-expander to be a bipartite graph (L, R, E, F) with vertex sets L, R such that all vertices in L have degree c, and all vertices in R have degree d; and the additional property that every set of vertices $L' \subset L$, such that $|L'| \leq \delta|L|$, has at least $(1 - \gamma)c|L'|$ neighbors.*

We say that a code C is an (c, d, γ, δ)-expander code if it has a check graph that is a (c, d)-regular (γ, δ)-expander. It is well-known that if $\gamma < 1/2$ then the graph has *unique-neighbor* expansion, meaning that for every $L' \subset L$ there exists a set of unique neighbors R' on the right such that each member of R' is a neighbor of a *unique* member of L'. Thus, from here on we refer to (γ, δ)-expanders as *unique-neighbor* expanders. The following well-known proposition (the proof of which is included for the sake of completeness) shows that unique-neighbor expansion of G is sufficient to guarantee the code whose check graph is G has large distance.

Proposition 1. *If C is (c, d, γ, δ)-expander code over F and $\gamma < \frac{1}{2}$, then $\delta(C) \geq \delta$.*

Proof. We prove that every non-zero word in C must have weight more than δn. Indeed let (L, R, E, F) be check graph of C that is a (c, d)-regular (γ, δ)-expander. The proposition follows by examining the unique neighbor structure of the graph. Let $x \in C$ be such that $0 < \text{wt}(x) < \delta n$ and $L' = \text{supp}(x) \subseteq L$. But then L' has at least $(1 - \gamma)c|L'| > \frac{c}{2}|L'|$ neighbors in R. At least one of these sees only one element of L', so the check by this element (corresponding dual word) will give $x_i \cdot e_{(i,j)}$ when $x_i \neq 0, e_{(i,j)} \neq 0$ and thus $x_i \cdot e_{(i,j)} \neq 0$, violating the corresponding constraint and contradicting $x \in C$.

3   Main Results

Our first main result says that codes obtained by the tensor product of a code with constant relative distance and a unique-neighbor expander code is robust:

Theorem 1 (Unique-Neighbor Expander codes are robust). *Let $R \subseteq F^m$ be a code of distance at least $\delta_R > 0$. Let $C \subseteq F^n$ be a (c, d, γ, δ)-expander code for some $c, d \in N, \delta > 0$, and $0 < \gamma < 1/2$. Then,*

$$\rho^T \geq min\{\frac{0.5\delta \cdot \delta_R}{2d^*}, \frac{\delta_R \cdot 0.25\delta}{2}, 1/8\}.$$

Where $d^* < d^k$, $k = (log_{(0.5+\gamma)}0.05) + 1$.

The above theorem extends the result of [2] where a similar result was proved for expanders with the stronger requirement $\gamma < 1/6$. Notice the difference between $\gamma < 1/6$ and unique-neighbor expansion ($\gamma < 1/2$) is qualitative, not merely quantitative. This is because expansion $\gamma < 1/4$ is required to guarantee efficient decoding algorithms, as shown by Sipser and Spielman in [10] whereas $\gamma < 1/2$ is sufficient for claiming the code has large distance, but does not necessarily warrant efficient decoding.

Our next result extends [2] in a different direction by showing that locally correctable codes are also robust. Informally, locally correctable codes allow to recover each entry of a codeword with high probability by reading only a few entries of the codeword even if a large fraction of it is adversely corrupted (see Definition 7).

Theorem 2 (Locally correctable codes are robust). *Let $R \subseteq F^m$ be a code of distance at least $\delta_R > 0$. Let $C \subseteq F^n$ be a (ϵ, δ, q)-locally correctable code with $\epsilon > 0$. Then,*

$$\rho^T \geq min\{\frac{0.5\delta \cdot \delta_R}{2(q+1)}, 1/8\}.$$

To prove both theorems we first define *weakly smooth* codes and prove that the tensor of a weakly smooth code and another code with constant relative distance is robust. Then we show that *smooth* codes are also weakly smooth. Finally we show that all unique-neighbor expander codes (with $\gamma < 1/2$) and all locally correctable codes are weakly smooth, thus obtaining Theorems 1, 2, respectively.

3.1 Weakly Smooth Codes

We are coming now to the central definition of the paper, that of a weakly smooth code. This definition allows us to generalize the work of [2] by using pretty much the same proof as there. In particular, in Section 5 we show that every code that is *smooth* according to [2] is also weakly smooth as per Definition 4. Furthermore, using our definition we get robust tensor from a broader family of base codes.

Both the *smooth* codes of [2] and our weakly smooth codes require the code retain large distance even after a portion of its coordinates and constraints have been removed. However there are two subtle differences between the two notions.

1. In the *smooth* codes setting an adversary removes a fraction of *constraints* and then a "Good" player removes a fraction of *indices*. In our Definition 4 both the adversary and the good player remove sets of indices.
2. In the *smooth* codes work with a predefined set of low weight constraints coming from a regular bipartite graph. Our Definition 4 does not assume any graph, nor does it require any regularity of degrees. This slackness and nonregularity will be crucial in arguing that unique-neighbor expanders are weakly smooth.

Definition 4 (Weakly smooth codes). *Let* $0 \leq \alpha_1 \leq \alpha_1' < 1$, $0 < \alpha_2 < 1$, d^* *be constants. Code C is $(\alpha_1, \alpha_1', \alpha_2, d^*)$-weakly smooth if $\forall I \subseteq [n]$, $|I| < \alpha_1 n$ letting*

$$\text{Constr}_{(I)} = \{u \in C_{\leq d^*}^{\perp} \mid \text{supp}(u) \cap I = \emptyset\}$$

and $C' = (\text{Constr}_{(I)})^{\perp}$ there exists $I' \subset [n]$, $I \subseteq I'$, $|I'| < \alpha_1' n$ such that $d(C'|_{[n] \setminus I'}) \geq \alpha_2 n$.

The following is the main technical lemma used to show weakly smooth codes are robust. Its proof follows in the next section.

Lemma 1 (Main Lemma). *Let $R \subseteq F^m$ and $C \subseteq F^n$ be codes of distance δ_R and δ_C. Assume C is $(\alpha_1, \alpha_1' < \delta_C, \alpha_2, d^*)$-weakly smooth and let $M \in F^m \otimes F^n$. If $\rho(M) < min\{\frac{\alpha_1 \delta_R}{2d^*}, \frac{\delta_R \alpha_2}{2}\}$ then $\delta(M) \leq 8\rho(M)$.*

4 Weakly Smooth Codes Are Robust — Proof of Lemma 1

We pretty much follow the proof of the Main Lemma in [2], but attend to the required modifications needed to carry the proof with the weaker requirement of smoothness. The main place where we use the weakly smooth property is the Proposition 3.

Proof (Proof of Lemma 1). For row $i \in [n]$, let $r_i \in R$ denote the codeword of R closest to the ith row of M. For column $j \in [m]$, let $c^{(j)} \in C$ denote the codeword of C closest to the jth column of M. Let M_R denote the $n \times m$ matrix whose ith row is r_i, and let M_C denote the matrix whose jth column is $c^{(j)}$. Let $E = M_R - M_C$.

In what follows matrices M_R, M_C and (especially) E will be central objects of attention. We refer to E as the error matrix. Note that $\delta(M, M_R) = \delta^{row}(M)$ and $\delta(M, M_C) = \delta^{col}(M)$ and with some abuse of notation let $\mathrm{wt}(E)$ be the relative weight of E, so

$$\mathrm{wt}(E) = \delta(M_R, M_C) \le \delta(M, M_R) + \delta(M, M_C)$$
$$= \delta^{row}(M) + \delta^{col}(M) = 2\rho(M). \tag{1}$$

Our proof strategy is to show that the error matrix E is actually very structured. We do this in two steps. First we show that its columns satisfy most constraints of the column code. Then we show that E contains a large submatrix which is all zeroes. Finally using this structure of E we show that M is close to some codeword in $R \otimes C$. The following is from [2, Proposition 4], we give the proof for the sake of completeness.

Proposition 2. *Let $u \in C_d^\perp$ be a constraint of C with $\mathrm{supp}(u) = \{i_1, ..., i_d\}$. Let e_i denote the ith row of E. Suppose $\mathrm{wt}(e_{i_j}) < \delta_R/d$ for every $j \in [d]$. Then $u^T \cdot E = 0$.*

Proof. Note that $\forall c \in C: \langle c, u \rangle = 0$. Let c_i denote the i-th row of the matrix M_C. (Recall that the rows of M_C are not necessarily codewords of any nice code - it is only the columns of M_C that are codewords of C). For every column j, we have $\langle (M_C)_j, u \rangle = 0$ (since the columns of M_C are codewords of C).

Thus we conclude that $u^T \cdot M_C = 0$ as a vector. Clearly, $u^T \cdot M_R \in R$ since each one of the rows of M_R is a codeword of R. But this implies

$$u^T \cdot E = u^T \cdot (M_R - M_C) = u^T \cdot M_R - u^T \cdot M_C = u^T \cdot M_R - 0 \in R$$

Now we use the fact that the e_{i_j}s have small weight for $i_j \in [d]$. This implies that

$$\mathrm{wt}(u^T \cdot E) \le \mathrm{wt}(u) \cdot (\delta_R/d) < \delta_R.$$

But R is an error-correcting code of the minimum distance δ_R so the only word of weight less than δ_R in it is the zero codeword, yielding $u^T \cdot E = 0$. ∎

Proposition 3. *There exist subsets $U \subseteq [m]$ and $V \subseteq [n]$ with $|U|/m < \delta_R$ and $|V|/n < \delta_C$ such that letting $\bar{V} = [n] \setminus V$ and $\bar{U} = [m] \setminus U$ we have for all $i \in \bar{V}, j \in \bar{U}$ that $E(i, j) = 0$.*

Proof. Let $V_1 \subseteq [n]$ be the set of indices corresponding to rows of the error matrix E with weight more than δ_R/d^*, i.e.

$$V_1 = \{i \in [n] \mid \mathrm{wt}(e_i) \ge \delta_R/d^*\}.$$

Clearly, $|V_1| < \alpha_1 n$, since $\frac{|V_1|}{n} \cdot \frac{\delta_R}{d^*} \le \mathrm{wt}(E) \le 2\rho(M)$ and thus $\frac{|V_1|}{n} \le \frac{2\rho(M)}{\delta_R/d^*} < \alpha_1$ where the last inequality follows from the assumption $\rho(M) < \frac{\alpha_1 \delta_R}{2d^*}$. Let $\mathrm{Constr}_{(V_1)} = \{u \in C_{\le d^*}^\perp \mid \mathrm{supp}(u) \cap V_1 = \emptyset\}$ and $C' = (\mathrm{Constr}_{(V_1)})^\perp$. Proposition

2 implies that $\forall u \in \mathrm{Constr}_{(V_1)}$ we have $u^T \cdot E = 0$, i.e. every column of E, denoted by E_j, satisfies constraint u and thus $E_j \in C'$.

Recall that C is $(\alpha_1, \alpha_1' < \delta_C, \alpha_2, d^*)$-weakly smooth. Associate the set V_1 with I from Definition 4. Following this definition, there exists a set I' (let $V = I'$), $|V| = |I'| < \alpha_1' n$ such that $d(C'_{[n]\setminus I'}) = d(C_{[n]\setminus V}) \geq \alpha_2 n$. We notice that for every column of E, denoted by E_j, we have $(E_j)|_{[n]\setminus I'} \in C_{[n]\setminus V}$. Thus E_j is either zero outside V or has at least $\alpha_2 n$ non-zero elements outside V.

Let U be the set of indices corresponding to the "heavy columns" of E that have $\alpha_2 n$ or more non-zero elements in the rows outside V. We conclude that every column of E that is not zero outside V is located in U. We argue that for each $(i,j) \in \bar{V} \times \bar{U}$ we have $E(i,j) = 0$. This is true since after we remove rows from V all projected nonzero columns have weight at least $\alpha_2 n$ and thus all nonzero columns are located in U. Hence all columns of $\bar{V} \times \bar{U}$ are zero columns.

Clearly, $\frac{|U|}{m} < \delta_R$, since $\frac{|U|}{m} \cdot \alpha_2 \leq \mathrm{wt}(E) \leq 2\rho(M)$ and thus $\frac{|U|}{m} \leq \frac{2\rho(M)}{\alpha_2} < \delta_R$, where the last inequality follows from the assumption $\rho(M) < \frac{\delta_R \alpha_2}{2}$.

We now use a standard property of tensor products to claim M_R, M_C and M are close to a codeword of $R \otimes C$. Recall that $M \in F^{n \times m}$ and that $\delta(M_C, M_R) \leq 2\rho(M)$. We reproduce the following proof from [2, Proposition 6] for the sake of completeness.

Proposition 4. *Assume there exist sets $U \subseteq [m]$ and $V \subseteq [n]$, $|U|/m < \delta_R$ and $|V|/n < \delta_C$ such that $M_R(i,j) \neq M_C(i,j)$ implies $j \in U$ or $i \in V$. Then $\delta(M) \leq 8\rho(M)$.*

Proof. First we note that there exists a matrix $N \in R \otimes C$ that agrees with M_R and M_C on $\bar{V} \times \bar{U}$ (See [1, Proposition 3]). Recall also that $\delta(M, M_R) = \delta^{row} \leq 2\rho(M)$. So it suffices to show $\delta(M_R, N) \leq 6\rho(M)$. We do so in two steps. First we show that $\delta(M_R, N) \leq 2\rho(M_R)$. We then show that $\rho(M_R) \leq 3\rho(M)$ concluding the proof.

For the first part we start by noting that M_R and N agree on every row in $\bar{V}$. This is the case since both rows are codewords of R which may disagree only on entries from the columns of U, but the number of such columns is less that $\delta_R m$. Next we claim that for every column $j \in [m]$ the closest codeword of C to the $M_R(\cdot, j)$, the jth column of M_R, is $N(\cdot, j)$, the jth column of N. This is true since $M_R(i,j) \neq N(i,j)$ implies $i \in V$ and so the number of such i is less than $\delta_C n$. Thus for every j, we have $N(\cdot, j)$ is the (unique) decoding of the jth column of M_R. Averaging over j, we get that $\delta^{col}(M_R) = \delta(M_R, N)$. In turn this yields $\rho(M_R) \geq \delta(M_R)/2 = \delta(M_R, N)/2$. This yields the first of the two desired inequalities.

Now to bound $\rho(M_R)$, note that for any pair of matrices M_1 and M_2 we have $\rho(M_1) \leq \rho(M_2) + \delta(M_1, M_2)$. Indeed it is the case that $\delta^{row}(M_1) \leq \delta^{row}(M_2) + \delta(M_1, M_2)$ and $\delta^{col}(M_1) \leq \delta^{col}(M_2) + \delta(M_1, M_2)$. When the above two arguments are combined it yields $\rho(M_1) \leq \rho(M_2) + \delta(M_1, M_2)$. Applying this inequality to $M_1 = M_R$ and $M_2 = M$ we get $\rho(M_R) \leq \rho(M) + \delta(M_R, M) \leq 3\rho(M)$. This yields the second inequality and thus the proof of the proposition.

The Main Lemma 1 follows immediately from the two last propositions.

5   Smooth Codes Are also Weakly so

We now show that our Definition 4 is indeed a generalization of *smooth* codes of Dinur et al. [2]. In what follows F_2 denotes the two-element field and $C(R_0)$ is a code defined by constraints in $R \setminus R_0$ (For further information and definitions see [2].). Recall the definition of smooth code:

Definition 5 (Smooth Codes). *A code $C \subseteq F_2^n$ is $(d, \alpha, \beta, \delta)$-smooth if it has a parity check graph $B = (L, R, E)$ where all the right vertices R have degree d, the left vertices have degree $c = d|R|/|L|$, and for every set $R_0 \subseteq R$ such that $|R_0| \leq \alpha|R|$, there exist a set $L_0 \subseteq L$, $|L_0| \leq \beta|L|$ such that the code $C(R_0)|_{[n] \setminus L_0}$ has distance at least δ.*

Claim. If $C \subseteq F_2^n$ is a $(d, \alpha, \beta, \delta)$-smooth code then it is $(\alpha_1, \alpha_1', \alpha_2, d^*)$-weakly smooth with $\alpha_1 = \frac{\alpha}{d}$, $\alpha_1' = \beta$, $\alpha_2 = \delta$, $d^* = d$.

Proof. Let R be a set of constraints of degree d and let $I \subseteq [n]$, $|I| < \alpha_1 n = \frac{\alpha n}{d}$ be the index set from Definition 4. Remove all d-constraints that touch at least one index in I. Let R_0 be a set of removed constraints from R. We have left degree $c = \frac{d|R|}{n}$, so, we removed at most $c \cdot \alpha_1 n = d|R|\alpha_1 = \alpha|R|$ constraints. Let $\mathrm{Constr}_{(I)} = \{u \in C_d^{\perp} \mid \mathrm{supp}(u) \cap I = \emptyset\}$ be the set of constraints in $R \setminus R_0$ (low weight dual words). We notice that $C(R_0) = (\mathrm{Constr}_{(I)})^{\perp}$. Let $I' \subseteq [n]$, $|I'| < \beta n = \alpha_1' n$ be index set from smooth codes definition (Definition 5) that should be thrown out in order to remain with good distance, i.e. $d(C(R_0)|_{[n] \setminus I'}) \geq \delta n = \alpha_2 n$. Clearly $I \subseteq I'$ as otherwise $d(C(R_0)|_{[n] \setminus I'}) = 1$. Thus from the definition of smoothness, letting $C' = (\mathrm{Constr}_{(I)})^{\perp}$ we have $d(C'|_{[n] \setminus I'}) \geq \alpha_2 n$ which proves that C is $(\alpha_1, \alpha_1', \alpha_2, d^*)$-weakly smooth.

6   Unique-Neighbor Expander Codes Are Weakly Smooth

As explained in Section 3.1 Dinur et al. [2] showed that expander codes with $\gamma < \frac{1}{6}$ are smooth and thus result in robust tensor product. In this section we show that it is possible to obtain robust tensor codes from expander code with the weaker assumption $\gamma < \frac{1}{2}$. We first define the *gap property* (Definition 6) and prove that it implies weak smoothness. Then we show that unique-neighbor expander codes have the *gap property*.

Definition 6 (Gap property). *Code C has a (α, δ, d)-gap property if $\forall J \subseteq [n], |J| < \alpha n$ letting $\mathrm{Constr}_{(J)} = \{u \in C_{\leq d}^{\perp} \mid \mathrm{supp}(u) \cap J = \emptyset\}$ and $C' = (\mathrm{Constr}_{(J)})^{\perp}$ we have that $\forall c \in C'|_{[n] \setminus J}$ either $\mathrm{wt}(c) < 0.1\delta n$ or $\mathrm{wt}(c) > 0.8\delta n$.*

Claim. If C has (α, δ, d)-gap property then it is $(\alpha, \alpha + 0.3\delta, 0.5\delta, d)$-weakly smooth.

Proof. Clearly, C has no codewords of weight between $0.1\delta n$ and $0.8\delta n$. To see this take $J = \emptyset$ and then gap property implies that $\forall w \in F^n$ if $0.1\delta n \leq \mathrm{wt}(w) \leq 0.8\delta n$ then $\langle w, u \rangle \neq 0$ for some $u \in C_{\leq d}^{\perp}$.

Let $S = \{c \in C \mid 0 < \mathrm{wt}(c) < 0.1\delta n\}$ be a set of all non-zero low weight codewords. Let J_S be the union of supports of non-zero low weight words, i.e. $J_S = \bigcup_{c \in S} \mathrm{supp}(c)$ and for any set $A \subseteq C$ let $J_A = \bigcup_{c \in A} \mathrm{supp}(c)$. We show that $|J_S| < 0.3\delta n$.

Assume the contrary, i.e. $|J_S| \geq \delta \cdot 0.3n$. Then there exists $S' \subseteq S$, such that $0.2\delta n < |J_{S'}| < 0.3\delta n$. To see this remove low weight words one by one from S, each time decreasing S at most by $0.1\delta n$.

Consider a random linear combination of codewords from S'. The expected weight of the above is more than $0.1\delta n$ but can not exceed $0.3\delta n$, thus there exists such a linear combination of low weight codewords that produces a codeword with weight more than $0.1\delta n$ but less than $0.3\delta n$. Contradiction.

Thus for the rest of the proof we assume $|J_S| < 0.3\delta n$. We are ready to show that C is $(\alpha, \alpha + 0.3\delta n, 0.5\delta n, d)$-weakly smooth. Let $I \subset [n]$, $|I| < \alpha n$ be arbitrarily chosen set. Let $\mathrm{Constr}_{(I)} = \{u \in C^{\perp}_{\leq d} \mid \mathrm{supp}(u) \cap I = \emptyset\}$ and $C' = (\mathrm{Constr}_{(I)})^{\perp}$.

¿From the definition of the gap property and from the above it follows that $\forall c \in C'|_{[n] \setminus I}$ either $\mathrm{wt}(c) < 0.1\delta n$ and thus $\mathrm{supp}(c) \subseteq J_S$ or $\mathrm{wt}(c) > 0.8\delta n$.

Let $I' = J_S \cup I$ and then $|I'| \leq |J_S| + |I| < \alpha n + 0.3\delta n$. We claim that $d(C'|_{[n]\setminus(I \cup J_S)}) = d(C'|_{[n]\setminus(I')}) \geq 0.5\delta n$. To see this assume $c' \in C'|_{[n]\setminus I}$, $c'' = c'|_{[n]\setminus(I \cup J_S)}$, $c'' \in C'|_{[n]\setminus(I \cup J_S)}$ such that $0 < \mathrm{wt}(c'') < 0.5\delta n$ but then $0 < \mathrm{wt}(c'') \leq \mathrm{wt}(c') \leq |J_S| + \mathrm{wt}(c'') < 0.8\delta n$ and thus c' is a low weight word, i.e. $\mathrm{supp}(c') \subseteq J_S$. Hence $c'' = c'|_{[n]\setminus(I \cup J_S)} = 0$, contradicting $\mathrm{wt}(c'') > 0$.

Proposition 5. *Let C be a linear code over F. If $u_1 \in C^{\perp}_{<f}$ and $u_2 \in C^{\perp}_{<g}$ and $i \in \mathrm{supp}(u_1) \cap \mathrm{supp}(u_2)$ then exists $u_3 \in C^{\perp}_{<f+g}$ such that $\mathrm{supp}(u_3) \subseteq (\mathrm{supp}(u_1) \cup \mathrm{supp}(u_2)) \setminus \{i\}$.*

Proof. Let $a \in F$ be ith entry in u_1 and $b \in F$ be ith entry in u_2. Then $u_3 = a^{-1}u_1 + b^{-1}u_2 \in C^{\perp}_{<f+g}$ has desired properties.

Claim. Let C be a (c, d, γ, δ)-expander code over F with constant $\gamma < \frac{1}{2}$. Let $w \in F^n$ with $0 < \mathrm{wt}(w) < \delta n$ with $I = \mathrm{supp}(w)$. Then at least a 0.95-fraction of indices $i \in I$ have $u_i \in C^{\perp}_{<d^*}$ where $d^* < d^k$, $k = (log_{0.5+\gamma}(0.05)) + 1$ such that $\mathrm{supp}(u_i) \cap I = \{i\}$.

Proof. Fix set I with $|I| < \delta n$. Let (L, R, E) be a check graph of C that is a (c, d)-regular (γ, δ)-expander. The claim follows from examining the unique neighbor structure of the graph. We prove this by induction on $j = 1...k$ and show set constructions I_j satisfying

- $I_1 = I$, $I_{j+1} \subset I_j$
- $|I_{j+1}| \leq (\frac{1}{2} + \gamma)|I_j|$
- $\forall i \in I_j \setminus I_{j+1}$ exists $u_i \in C^{\perp}_{\leq d^j}$ with $\mathrm{supp}(u_i) \cap I = \{i\}$

We then conclude $(\frac{1}{2} + \gamma)^k < 0.05$ and thus from the induction follows that $I_k \subset I$, $|I_k| < 0.05 \cdot |I|$ and $\forall i \in I \setminus I_k$ exists $u_i \in C^{\perp}_{<d^k}$ with $\mathrm{supp}(u_i) \cap I = \{i\}$. And the the proof of the claim is completed.

For the base case let $I_1 = I$. Since C is an expander and $|I_1| \leq \delta n$, I_1 has at least $(1 - \gamma)c|I_1| = (\frac{c}{2} + (0.5 - \gamma)c)|I_1|$ neighbors in R. Each index $i \in I_1$ is asked by c constraints in R. And thus the number of neighbors that ask at least 2 indices from I_1 is bounded from above by $(\frac{c}{2})|I_1|$. Hence there are at least $((\frac{1}{2} - \gamma)c)|I_1|$ unique neighbors in R. Since a single index can not have more than c unique neighbors in R, the number of indices in I_1 having unique neighbor is at least $(\frac{1}{2} - \gamma)|I_1|$. I.e. at least $(\frac{1}{2} - \gamma)$-fraction of all indices in I_1 have a unique neighbor with support $d = d^1$. Let $I_2 \subset I_1$ be subset of all indices $i \in I_1$ that have no unique neighbor of weight at most d^1. We constructed set I_1, I_2 such that

- $I_1 = I$, $I_2 \subset I_1$
- $|I_2| \leq (\frac{1}{2} + \gamma)|I_1|$
- $\forall i \in I_1 \setminus I_2$ exists $u_i \in C^{\perp}_{\leq d^1}$ with $\mathrm{supp}(u_i) \cap I = \{i\}$

And this completes the base case.

Assume correctness until $j - 1$ and let us prove for j. Consider I_j, $|I_j| \leq |I_1| \leq \delta n$. By the unique neighbor expansion at least $(\frac{1}{2} - \gamma)$-fraction of indices $i \in I_j$ have bounded unique neighbor, i.e. $u_i \in C^{\perp}_d$ such that $\mathrm{supp}(u_i) \cap I_j = \{i\}$. Let $I_{j+1} \subset I_j$ be indices $i \in I_j$ that have no bounded unique neighbor and thus $|I_{j+1}| \leq (\frac{1}{2} + \gamma)|I_j|$.

Fix $i \in I_j \setminus I_{j+1}$ arbitrarily. There exists $u_i \in C^{\perp}_d$ such that $\mathrm{supp}(u_i) \cap I_j = \{i\}$. Every index $l \in \mathrm{supp}(u_i)$, $l \neq i$ is located either in $[n] \setminus I_1$ or in $I_1 \setminus I_j$. We handle all $l \in I_1 \setminus I_j$ using linear combination according to Proposition 5 to obtain a constraint $u'_i \in C^{\perp}_{\leq d^j}$ such that $\mathrm{supp}(u'_i) \cap I = \{i\}$. This is possible since every $l \in I_1 \setminus I_j$ is located in some I_f for $1 \leq f < j$ and thus from induction assumption has $u_l \in C^{\perp}_{\leq d^{j-1}}$ such that $\mathrm{supp}(u_l) \cap I = \{l\}$. Since $\mathrm{wt}(u_i) \leq d$ we obtain $u'_i \in C^{\perp}_{\leq d^{j-1} \cdot d} = C^{\perp}_{\leq d^j}$ such that $\mathrm{supp}(u'_i) \cap I = \{i\}$. So we showed

- $I_{j+1} \subset I_j$
- $|I_{j+1}| \leq (\frac{1}{2} + \gamma)|I_j|$
- $\forall i \in I_j \setminus I_{j+1}$ exists $u_i \in C^{\perp}_{\leq d^j}$ with $\mathrm{supp}(u_i) \cap I = \{i\}$

This yields the induction and the claim.

Corollary 1. *If C is (c, d, γ, δ) expander code with $\gamma < \frac{1}{2}$ then C has $(0.5\delta, 0.5\delta, d^*)$ gap property where $d^* < d^k$, $k = (\log_{(0.5 + \gamma)} 0.05) + 1$.*

Proof. Let $J \subset [n]$, $|J| < 0.5\delta$ be arbitrarily chosen. Let $\mathrm{Constr}_{(J)} = \{u \in C^{\perp}_{< d^k} \mid \mathrm{supp}(u) \cap J = \emptyset\}$ and $C' = (\mathrm{Constr}_{(J)})^{\perp}$. Assume by contradiction, there exists $w \in C'_{[n] \setminus J}$ such that $0 < 0.1 \cdot (0.5\delta)n \leq \mathrm{wt}(w) \leq 0.8 \cdot (0.5\delta)n$. And thus there is no $u \in \mathrm{Constr}_{(J)}$ such that $|\mathrm{supp}(u) \cap \mathrm{supp}(w)| = 1$.

Let $I = J \cup \mathrm{supp}(w)$, $|I| \leq |J| + \mathrm{wt}(w) < 0.5\delta n + 0.4\delta n < \delta n$. We notice that $\mathrm{supp}(w) \cap J = \emptyset$ and $|\mathrm{supp}(w)| > 0.05 \cdot |I|$. Thus Claim 6 implies that there exists $u \in C^{\perp}_{< d^k}$ such that $|\mathrm{supp}(u) \cap \mathrm{supp}(w)| = 1$ and $|\mathrm{supp}(u) \cap I| = |\mathrm{supp}(u) \cap \mathrm{supp}(w)| = 1$. Thus $u \in \mathrm{Constr}_{(J)}$ such that $|\mathrm{supp}(u) \cap \mathrm{supp}(w)| = 1$. Contradiction.

Claim. If C is (c, d, γ, δ) expander code with $\gamma < \frac{1}{2}$ then C is $(0.5\delta, 0.65\delta, 0.25\delta, d^*)$-weakly smooth where $d^* < d^k$, $k = (log_{(0.5+\gamma)}0.05) + 1$.

Proof. Follows immediately from Corollary 1 and Claim 6. Corollary 1 implies that C has $(0.5\delta, 0.5\delta, d^*)$ gap property where $d^* < d^k$, $k = (log_{(0.5+\gamma)}0.05) + 1$. Claim 6 implies that C is $(0.5\delta, 0.5\delta + 0.15\delta, 0.25\delta, d^*)$-weakly smooth .

Proof (Proof of Theorem 1). Let $R \subseteq F^m$ and $C \subseteq F^n$ be codes of distance δ_R and δ_C. Let $M \in F^m \otimes F^n$. Claim 6 implies that C is $(0.5\delta, 0.65\delta, 0.25\delta, d^*)$-weakly smooth where $d^* < d^k$, $k = (log_{(0.5+\gamma)}0.05) + 1$. Main Lemma implies that if $\rho(M) < min\{\frac{(0.5\delta)\cdot\delta_R}{2d^*}, \frac{\delta_R\cdot(0.25\delta)}{2}\}$ then $\delta(M) \leq 8\rho(M)$.

7   Locally Correctable Codes Are Weakly Smooth

Definition 7 (Locally Correctable Code). *A $[n, k, d]_{|F|}$ code C is called (q, ϵ, δ) locally correctable code if there exists a randomized decoder (D) that reads at most q entries and the following holds: $\forall c \in C$, $\forall i \in [n]$ and $\forall \hat{c} \in F^n$ such that $d(c, \hat{c}) \leq \delta n$ we have*

$$Pr[D^{\hat{c}}[i] = c_i] \geq \frac{1}{|F|} + \epsilon,$$

i.e. with probability at least $\frac{1}{|F|} + \epsilon$ entry c_i will be recovered correct.

Without loss of generality we assume that given $\hat{c} \in F^n$ the "correction" of entry i (obtaining c_i) is done by choosing random $u \in S \subseteq C^{\perp}_{\leq q+1}$ such that $i \in supp(u)$. Formally, assume the ith entry of u is u_i, let $u^{proj} = u|_{[n]\setminus\{i\}}$, $\hat{c}^{proj} = \hat{c}|_{[n]\setminus\{i\}}$ and then c_i is recovered by $D^{\hat{c}}[i] = \frac{\langle u^{proj}, \hat{c}^{proj} \rangle}{u_i}$, notice that $u_i \neq 0$.

The next claim holds for every $\epsilon > 0$ which can be arbitrarily close to 0 (e.g. $o(1)$) whereas usually locally correctable codes are defined with $\epsilon = \Omega(1)$.

Claim. If C is (ϵ, δ, q)-locally correctable code with $\epsilon > 0$ then it is $(0.5\delta, 0.5\delta, 0.5\delta, q + 1)$-weakly smooth and its relative distance is at least δ.

Proof. We first show that $\forall I \subseteq [n]$, $|I| \leq \delta n$ and $\forall i \in I$ we have $u_i \in C^{\perp}_{\leq q+1}$ with $supp(u_i) \cap I = \{i\}$. Assume the contrary and fix $I \subseteq [n]$, $|I| \leq \delta n$ and $i \in I$. So, for all $u_i \in C^{\perp}_{\leq q+1}$ with $i \in supp(u_i) \cap I$ we have $|supp(u_i) \cap I| \geq 2$. Consider an adversary that takes $c \in C$ and sets c_j to random element from F for all $j \in I$, obtaining $\hat{c}$. Clearly, c_i will be recovered with probability at most $\frac{1}{|F|}$ since for every $u^{(i)} \in C^{\perp}_{\leq q+1}$ such that $i \in supp(u^{(i)})$ the inner product $\langle (u^{(i)})|_{[n]\setminus\{i\}}, c|_{[n]\setminus\{i\}} \rangle$ will produce a uniformly random value in F.

We next show that $d(C) \geq \delta n$. To see this assume $c \in C$ such that $0 < wt(c) < \delta n$. Let $I = supp(c)$, $|I| < \delta n$ and $i \in I$. There exists $u \in C^{\perp}_{\leq q+1}$ with $supp(u) \cap supp(c) = \{i\}$ and thus $\langle u, w \rangle \neq 0$ implies $c \notin C$.

We finally show the weak smoothness of C. Let $I \subset [n]$, $|I| < 0.5\delta n$ be the adversary chosen set and let $I' = I$. Let $Constr_{(I)} = \{u \in C^{\perp}_{\leq q+1} \mid supp(u) \cap I = \emptyset\}$

and $C' = (\text{Constr}_{(I)})^{\perp}$. We claim that $d(C'|_{[n]\setminus I}) \geq 0.5\delta n$. This is true, since otherwise we have $c' \in C'$, $c'_{[n]\setminus I} \in C'|_{[n]\setminus I}$ such that $0 < \text{wt}(c'_{[n]\setminus I}) < 0.5\delta n$. But then $0 < \text{wt}(c') < 0.5\delta n + |I| \leq \delta n$ and thus exists $u \in \text{Constr}_{(I)}$ such that $|\text{supp}(u) \cap \text{supp}(c')| = 1$ which implies $\langle u, c' \rangle \neq 0$ and $c' \notin C'$. Contradiction. So, C is $(0.5\delta, 0.5\delta, 0.5\delta, q+1)$-weakly smooth.

Proof (of Theorem 2). Let $R \subseteq F^m$ and $C \subseteq F^n$ be linear codes such that $\delta(R) \geq \delta_R$. Let $M \in F^m \otimes F^n$. Claim 7 implies that C is $(0.5\delta, 0.5\delta, 0.5\delta, q+1)$-weakly smooth and $\delta(C) \geq \delta$. The Main Lemma 1 implies that if $\rho(M) < \min\{\frac{(0.5\delta)\cdot\delta_R}{2(q+1)}, \frac{\delta_R\cdot(0.5\delta)}{2}\} = \frac{(0.5\delta)\cdot\delta_R}{2(q+1)}$ then $\delta(M) \leq 8\rho(M)$.

Acknowledgements. We thank Madhu Sudan for helpful discussions.

References

1. Ben-Sasson, E., Sudan, M.: Robust locally testable codes and products of codes. In: APPROX-RANDOM, pp. 286–297 (2004) (See ECCC TR04-046, 2004)
2. Dinur, I., Sudan, M., Wigderson, A.: Robust local testability of tensor products of LDPC codes. In: APPROX-RANDOM, pp. 304–315 (2006)
3. Valiant, P.: The tensor product of two codes is not necessarily robustly testable. In: APPROX-RANDOM, pp. 472–481 (2005)
4. Copersmith, D., Rudra, A.: On the robust testability of tensor products of codes, ECCC TR05-104 (2005)
5. Meir, O.: On the rectangle method in proofs of robustness of tensor products, ECCC TR07 (2007)
6. Meir, O.: Combinatorial Construction of Locally Testable Codes. M.Sc. Thesis, Weizmann Institute of Science (2007)
7. Goldreich, O., Meir, O.: The tensor product of two good codes is not necessarily robustly testable, ECCC TR07 (2007)
8. Goldreich, O.: Short locally testable codes and proofs (survey), ECCC TR05-014 (2005)
9. Goldreich, O., Sudan, M.: Locally testable codes and PCPs of almost linear length. In: FOCS (2002), pp. 13-22 (See ECCC TR02-050 2002)
10. Spielman, D.: Linear-time encodable and decodable error-correcting codes. IEEE Trans. on Information Theory, 1723–1731 (1996)

On the Degree Sequences of Random Outerplanar and Series-Parallel Graphs

Nicla Bernasconi, Konstantinos Panagiotou*, and Angelika Steger

Institute of Theoretical Computer Science, ETH Zurich
Universitätsstrasse 6, CH-8092 Zurich
{nicla,panagiok,steger}@inf.ethz.ch

Abstract. Let $\mathcal{G}$ be a class of labeled connected graphs and let $\mathcal{B}$ be the class of biconnected graphs in $\mathcal{G}$. In this paper we develop a general framework that allows us to derive mechanically the degree distribution of random graphs with n vertices from certain 'nice' classes $\mathcal{G}$ as a function of the degree distribution of the graphs in $\mathcal{B}$ that are drawn under a specific probabilistic model, namely the Boltzmann model. We apply this framework to obtain the degree distribution of a random outerplanar graph and a random series-parallel graph. For the latter we formulate a generic concentration result that allows us to make statements that are true with high probability for a large family of variables defined on random graphs drawn according to the Boltzmann distribution.

1 Introduction and Results

One of the central questions of interest in theoretical computer science is the analysis of algorithms. Here one usually distinguishes between *worst case analysis* and *average case analysis*. From a practical point of view, an average case analysis is particularly important when the worst case analysis does not result in satisfactory quality characteristics about the given algorithm: it is possible that the algorithm behaves well in real world scenarios, although a bad worst case behavior can be mathematically proved. In order to prove qualitatively strong and meaningful results about the average case behavior of a particular algorithm, we usually require precise knowledge about properties of "typical" input instances.

The standard example of a successful average case analysis is that of the QuickSort algorithm. It relies on the fact that properties of random permutations are well understood. In the context of graph algorithms an average case analysis can be performed if we assume the uniform distribution on the set of all graphs with a given number of vertices: one can then model a "typical" input by the classical Erdős-Rényi random graph, and can thus use the wide and extensive knowledge about random graphs, see the two excellent monographs [1] and [2], to derive properties that can be used to analyze performance measures like the running time or the achieved approximation ratio of the algorithm in question.

---

* This work was partially supported by the SNF, grant nr. 20021-107880/1.

A. Goel et al. (Eds.): APPROX and RANDOM 2008, LNCS 5171, pp. 303–316, 2008.
© Springer-Verlag Berlin Heidelberg 2008

The picture changes dramatically if we are interested in *natural* graph classes where it seems hard – if not impossible – to derive suitable independence properties. A standard example that has evolved over the last decade as a reference model in this context is the class of *planar* graphs. The random planar graph R_n was first investigated in [3] by Denise, Vasconcellos and Welsh and has attracted considerable attention since then. We mention selectively a few results. McDiarmid, Steger and Welsh [4] showed the surprising fact that R_n does not share the 0-1-law known from standard random graph theory: the probability of connectedness is bounded away from 0 and 1 by positive constant values. Moreover, the situation is similar if the average degree is fixed [5]. These results relied on a (crude) counting of the number of planar graphs with n vertices. A breakthrough occurred with the recent results of Giménez and Noy [6], who not only managed to determine the asymptotic value of the number of planar graphs with n vertices, but also showed that the number of edges in R_n is asymptotically normally distributed. Moreover, they studied the number of connected and 2-connected components in R_n. The proofs of these results are based on singularity analysis of generating functions, a powerful method from analytic combinatorics that has led to many beautiful results, see the forthcoming book by Flajolet and Sedgewick [7].

Our results. In this paper we further elaborate and extend significantly an approach that was recently used in [8] to obtain the degree sequence and subgraph counts of random dissections of convex polygons. More precisely, we exploit the so-called Boltzmann sampler framework by Duchon, Flajolet, Louchard, and Schaeffer [9] to reduce the study of degree sequences to properties of sequences of *independent* and *identically distributed* random variables. Hence, we can – and do – use many tools developed in classical random graph theory to obtain extremely tight results.

Our first main contribution is a general *framework* that allows to derive mechanically the degree distribution of random graphs from certain "nice" graph classes from the degree distribution of random graphs from the 2-connected objects in the graph class in question, see Section 2 for the details. Recall that a 2-connected (or biconnected) graph is a graph that can not be disconnected by deleting any single vertex. Our framework can be readily applied to obtain the degree sequence of random graphs from "simple" classes, like Cayley trees, or graphs which have the property that their maximal 2-connected components (or equivalently, blocks) have a simple structure. We mention as examples cactus graphs, where the blocks are just cycles, and clique graphs, where the blocks are complete graphs. The second main contribution of our work are two applications of this framework.

A graph is called *series-parallel* (SP) if it does not contain a subdivision of the complete graph K_4, or equivalently if it does not contain K_4 as a minor. Hence, the class of SP graphs is a subclass of all planar graphs. Moreover, an *outerplanar* graph is a planar graph that can be embedded in the plane so that all vertices are incident to the outer face. Outerplanar graphs are characterized as those graphs that do not contain a K_4 or a $K_{2,3}$ minor. The classes of outerplanar and

SP graphs are often used as the first non-trivial test cases for results about the class of all planar graphs.

As a first important application of our framework we derive the degree distribution of random outerplanar graphs, and show that the number of vertices of degree $k = k(n)$ (where k is not allowed to grow too fast) is concentrated around a specific value with very high probability (i.e. the probability of observing a large deviation is *exponentially* small). Here we use that the degree distribution of 2-connected outerplanar graphs can be derived from a result in [8], where the class of polygon dissections was studied. In order to state our results we need to fix some notation. Let $\mathcal{O}_n$ be the class of labeled, connected outerplanar graphs with n vertices, and let O_n be a graph drawn uniformly at random from $\mathcal{O}_n$. Furthermore, for a function $G(z)$ let $[z^k]G(z)$ be the coefficient of z^k in the series expansion of $G(z)$ around zero, and for real numbers α and X let us write "$(1 \pm \alpha)X$" for the interval $((1 - \alpha)X, (1 + \alpha)X)$.

Theorem 1. *There exist constants $A, \ldots, G$, $c > 0$ and λ such that for all $\delta > 0$ and $1 \le k = k(n) \le (1 - \delta) \log_{C/D} n$ the following is true. For every $0 < \varepsilon < 1$ and sufficiently large n*

$$\mathbb{P}\left[\deg(k; O_n) \in (1 \pm \varepsilon)o_k n\right] \ge 1 - e^{-c\varepsilon^2 o_k e^{-3\sqrt{k}}n}, \tag{1}$$

where $o_k = [z^k]\lambda \cdot b(z) \cdot e^{\lambda(r(z)-1)}$, and b and r are the functions

$$r(z) = \frac{Az + Bz^2}{C - Dz} \quad \text{and} \quad b(z) = Ez + \frac{Fz^2 + Gz^3}{(C - Dz)^2}. \tag{2}$$

In fact, we do get explicit expressions for $A, \ldots, G$, and λ, cf. (12)–(14). Using these we can determine the values o_k explicitly for small k and asymptotically for large k, by evaluating the kth coefficient of the function $\lambda b(z)e^{\lambda(r(z)-1)}$ through the Cauchy integral formula.

Corollary 1. *We have $o_1 \doteq 0.13659$, $o_2 \doteq 0.28753$, $o_3 \doteq 0.24287$, $o_4 \doteq 0.15507$, $o_5 \doteq 0.08743, \ldots$, and there are analytically given constants c_1, c_2 such that for large k*

$$o_k = (1 + o(1)) \cdot c_1 \cdot (D/C)^k \cdot e^{c_2\sqrt{k}} \cdot k^{1/4}.$$

The formula in Corollary 1 (and the proof of Theorem 1) strongly indicate that the maximum degree $\Delta(O_n)$ of a random outerplanar graph O_n is roughly $\log_{C/D}(n)$. Unfortunately, our current techniques are not strong enough to prove this, although we come very close to this value. We thus formulate it as a conjecture.

Conjecture 1. $\lim_{n \to \infty} \mathbb{P}[\Delta(O_n) \in (1 + o(1)) \log_{C/D} n] = 1$.

In the last part of the paper we study the degree distribution of series-parallel graphs. With the general framework developed in Section 2 it will again be sufficient to derive the degree distribution of 2-connected series parallel graphs. This, however, was unknown and our second main contribution is an approach to derive this. The main difference (and complication) in comparison to the outerplanar graphs is that 2-connected outerplanar graphs can be constructed by an easy recursive procedure, as was shown in [8]: start with a cycle (whose length follows an appropriate distribution) and glue onto each edge except one a graph constructed recursively by the same principle. Clearly, as the name already indicates, the nature of SP graphs does not allow such an easy generation. Instead we have to consider several types of graph classes and several types of gluing operations simultaneously. In a first step we obtain the *expected number* of vertices of a certain degree, and then, most importantly, we develop a general lemma which allows us to derive the *actual* degree distribution. This generic concentration result is of its own interest and importance, and may have applications to other parameters and graphs classes as well. Unfortunately, we currently only get a little bit weaker – but still exponentially small – bounds for the tail probabilities compared to those in Theorem 1. We believe that with the same methods and some extra work one should be able to obtain similarly strong bounds as above.

Let $\mathcal{SP}$ be the class of all labeled connected series-parallel graphs, and let SP_n be a graph drawn uniformly at random from $\mathcal{SP}_n$. Our result for SP_n is then the following statement.

Theorem 2. *There exist constants $C, c > 0$ such that the following is true. For $1 \leq k = k(n) \leq C \log n$ and for every $0 < \varepsilon < 1$ and sufficiently large n*

$$\mathbb{P}\left[\deg(k; SP_n) \in (1 \pm \varepsilon)s_k n\right] \geq 1 - e^{-c\varepsilon^2 s_k n^{1/3}}, \tag{3}$$

where $s_k = [z^k]\lambda \cdot b(z) \cdot e^{\lambda(r(z)-1)}$. The constant λ and the functions b and r are given explicitly in (19), (16) and (18).

Corollary 2. *We have $s_1 \doteq 0.11021$, $s_2 \doteq 0.35637$, $s_3 \doteq 0.22335$, $s_4 \doteq 0.12576$, $s_5 \doteq 0.07172$, and there are analytically given constants c_1 and $\rho \doteq 1.33259$ such that for large k*

$$s_k = c_1 \rho^{-k} k^{-3/2} + O(\rho^{-k} k^{-5/2}).$$

We conjecture also in this case that the asymptotic value of the maximum degree of a random series parallel graph is $\log_\rho n$.

The number of vertices of a given degree in random outerplanar and series-parallel graphs is studied also in [10], independently from our work. Using different techniques, the authors show for constant k that the number of vertices of degree k is asymptotically normally distributed, with linear expectation and variance. Moreover, they provide exponential estimates for the tails of the distributions for such k.

Techniques & Outline. All graph classes considered in this paper allow a socalled *decomposition*, which is a *description* of the class in terms of general-purpose combinatorial constructions. These constructions appear frequently in modern systematic approaches to asymptotic enumeration and random sampling of several combinatorial structures. It is beyond of the scope of this work to survey these results, and we refer the reader to [7] and references therein for a detailed exposition.

One benefit of the knowledge of the decomposition is that it allows us to develop *mechanically* algorithms that sample objects from the graph class in question by using the framework of *Boltzmann samplers*. This framework was introduced by Duchon et. al. in [9], and was extended by Fusy [11] to obtain an (expected) linear time approximate-size sampler for planar graphs. Here we just present the basic ideas of this framework. Let $\mathcal{G}$ be a class of labeled graphs. In the Boltzmann model of parameter x, we assign to any object $\gamma \in \mathcal{G}$ the probability

$$\mathbb{P}_x[\gamma] = \frac{1}{G(x)} \frac{x^{|\gamma|}}{|\gamma|!}, \tag{4}$$

if the expression above is well-defined, where $G(x)$ is the exponential generating function enumerating the elements of $\mathcal{G}$. It is straightforward to see that the expected size of an object in $\mathcal{G}$ under this probability distribution is $\frac{xG'(x)}{G(x)}$. A *Boltzmann sampler* $\Gamma G(x)$ for $\mathcal{G}$ is an algorithm that generates graphs from $\mathcal{G}$ according to (4). In [9,11] several general procedures which translate common combinatorial construction rules like union, set, etc. into Boltzmann samplers are given. Notice that the probability above only depends on the choice of x and on the size of γ, such that every object of the same size has the *same* probability of being generated. This means that if we condition on the output being of a certain size n, then the Boltzmann sampler $\Gamma G(x)$ is a *uniform* sampler of the class $\mathcal{G}_n$.

In Section 2 we use Boltzmann samplers to derive the degree distribution of connected random graphs, given the degree distribution of the 2-connected graphs in suitably defined "nice" graph classes. In Section 3 we use this and a result from [8] to obtain the degree distribution of a random outerplanar graph. Finally, in Section 4 we first determine the number of vertices of degree k in a random 2-connected series-parallel (SP) graph, and then apply the result of Section 2 in order to obtain the degree distribution of a random SP graph.

Notation. Before we proceed, let us introduce some notation which will be extensively used in the next sections. Let $\mathcal{G}$ be a class of labeled graphs. We denote by $\mathcal{G}_n$ the subset of graphs in $\mathcal{G}$ which have precisely n vertices, and we write $g_n := |\mathcal{G}_n|$. Moreover, we write $G(x) = \sum_{n \geq 0} g_n \frac{x^n}{n!}$ for its corresponding *exponential generating function (egf)*.

In the following we will frequently use the *pointing* and *derivative* operators. Given a labeled class of graphs $\mathcal{G}$, we define $\mathcal{G}^\bullet$ as the class of *pointed* (or *rooted*) graphs, where a vertex is distinguished from all other vertices. The *derived* class $\mathcal{G}'_n$ is obtained by removing the label n from every object in $\mathcal{G}_n$, such that the obtained objects have $n-1$ labeled vertices, i.e., vertex n can be considered as a distinguished vertex that does not contribute to the size. Consequently, there is a bijection between the classes $\mathcal{G}'_{n-1}$ and $\mathcal{G}_n$. We set $\mathcal{G}' := \bigcup_{n \geq 0} \mathcal{G}'_n$. On generating function level, the pointing operation corresponds to taking the derivative with respect to x, and multiplying by it by x, i.e. $G^\bullet(x) = xG'(x)$. Similarly, the egf of $\mathcal{G}'$ is simply $G'(x)$. Finally, we denote by $\rho_{\mathcal{G}}$ the dominant singularity of G (in this work we are going to deal only with functions that have a unique singularity on the real axis).

2 A Framework for Nice Graph Classes

The aim of this section is to develop a general framework that will allow us to mechanically give tight bounds for the number of vertices of degree k in a random graph drawn from a graph class that satisfies certain technical assumptions. Before we state our main result formally, let us introduce some notation. We denote by $\mathcal{Z}$ the graph class consisting of one single labeled vertex. Furthermore, for two graph classes $\mathcal{X}$ and $\mathcal{Y}$, we denote by $\mathcal{X} \times \mathcal{Y}$ the *cartesian product* of $\mathcal{X}$ and $\mathcal{Y}$ followed by a relabeling step, so as to guarantee that all labels are distinct. Moreover, $\text{SET}(\mathcal{X})$ is the graph class such that each object in it is an ordered collection of graphs in $\mathcal{X}$. Finally, the class $\mathcal{X} \circ \mathcal{Y}$ consists of all graphs that are obtained from graphs from $\mathcal{X}$, where each vertex is replaced (in a unique way) by a graph from $\mathcal{Y}$. This set of combinatorial operators (cartesian product, set, and substitution) appears frequently in modern theories of combinatorial analysis [9,7,12,13] as well as in systematic approaches to random generation of combinatorial objects [9,14]. For a very detailed description of these operators and numerous applications we refer to [7].

With this notation we may now define the graph classes we are going to consider.

Definition 1. *Let $\mathcal{G}$ be a class of labeled graphs, and $\mathcal{B} = \mathcal{B}(\mathcal{G}) \subset \mathcal{G}$ the subclass of biconnected graphs in $\mathcal{G}$. We say that $\mathcal{G}$ is* nice *if it fulfills the following two conditions.*

i) $\mathcal{G}^{\bullet}$ satisfies

$$\mathcal{G}^{\bullet} = \mathcal{Z} \times \text{SET}(\mathcal{B}' \circ \mathcal{G}^{\bullet}). \tag{5}$$

ii) The egf $G^{\bullet}(x)$ of $\mathcal{G}^{\bullet}$ has a unique finite singularity $\rho_{\mathcal{G}}$ and there exist constants $c, \alpha > 0$ such that

$$g_n^{\bullet} \sim c n^{-1-\alpha} \cdot \rho_{\mathcal{G}}^{-n} \cdot n!. \tag{6}$$

We call α the critical exponent *of $\mathcal{G}$.*

This definition states that nice classes allow the following decomposition: a rooted graph can be viewed as a collection of rooted *biconnected* graphs, which are "glued" together at their roots, and every vertex in them is substituted by a rooted connected graph. In particular, every graph in a nice class is connected. The above definition is not very restrictive, as many natural graph classes are nice. Probably the most prominent examples are classes with forbidden minors, in particular connected planar, outerplanar, and series-parallel graphs, or cactus, block graphs and many kinds of trees (like Cayley trees). Note that condition (6) restricts the set of possible classes to such ones that have only a "small" number of graphs. Hence, many graph classes like the class of connected triangle-free graphs, or the class of connected k-colorable graphs, are not nice, although they satisfy (5).

The remainder of this section is structured as follows. In the next subsection we shall define an algorithm that generates pointed graphs from a nice class $\mathcal{G}$ according to the Boltzmann model for $\mathcal{G}$. Then, in Section 2.2 we shall exploit this sampler to prove our main result (Lemma 2), which translates in a general way the degree sequence of the graphs in $\mathcal{B}$ to the degree sequence of $\mathcal{G}$.

2.1   A Sampler for Nice Graph Classes

Recall that due to (5) a rooted graph from a nice class $\mathcal{G}^{\bullet}$ of graphs can be viewed as a set of rooted biconnected graphs, which are "glued" together at their roots, and every vertex in them is substituted by a rooted connected graph. A sampler for $\mathcal{G}^{\bullet}$ reverses this description: it starts with a single vertex, attaches to it a random set of biconnected graphs, and proceeds recursively to substitute every newly generated vertex by a rooted connected graph.

Let us now define formally the generic sampler. For this we need some additional notation. Let $G(x)$ and $B(x)$ be the egfs of $\mathcal{G}$ and $\mathcal{B}$ respectively, and let $\rho_{\mathcal{G}}$ and $\rho_{\mathcal{B}}$ be their singularities. Define $\lambda_{\mathcal{G}} := B'(G^{\bullet}(\rho_{\mathcal{G}}))$, and let $\Gamma B'(x)$ be a Boltzmann sampler for $\mathcal{B}'$, i.e. $\Gamma B'(x)$ samples according to the Boltzmann distribution (4) with parameter x for $\mathcal{B}'$. Then the sampler $\Gamma G^{\bullet}$ for $\mathcal{G}^{\bullet}$ is defined as follows.

> $\Gamma G^{\bullet} : \gamma \leftarrow$ a single node r
> $\quad\quad k \leftarrow \mathrm{Po}(\lambda_{\mathcal{G}})$ $\hspace{4cm}$ ($\star$)
> $\quad\quad$ **for** $(j = 1, \ldots, k)$
> $\quad\quad\quad \gamma' \leftarrow \Gamma B'(G^{\bullet}(\rho_{\mathcal{G}})),$ discard the labels of γ' ($\star\star$)
> $\quad\quad\quad \gamma \leftarrow$ merge γ and γ' at their roots
> $\quad\quad$ **foreach** vertex $v \neq r$ of γ
> $\quad\quad\quad \gamma_v \leftarrow \Gamma G^{\bullet},$ discard the labels of γ_v $\hspace{1.6cm}$ ($\ast$)
> $\quad\quad$ replace all nodes $v \neq r$ of γ with γ_v
> $\quad\quad$ label the vertices of γ uniformly at random
> $\quad\quad$ **return** γ

The following lemma is an immediate consequence of the compilation rules in [9,11].

Lemma 1. *Let* $\gamma \in \mathcal{G}^{\bullet}$. *Then* $\mathbb{P}[\Gamma G^{\bullet} = \gamma] = \frac{\rho_{\mathcal{G}}^{|\gamma|}}{|\gamma|! G^{\bullet}(\rho_{\mathcal{G}})}.$

2.2   Degree Sequence

Our goal is to analyze the execution of $\Gamma G^{\bullet}$ so as to obtain information about the degree sequence of random graphs from $\mathcal{G}_n^{\bullet}$. Before we proceed let us make a few important observations. Note that every vertex v different from the root goes through two *phases*. In the first phase, v is generated in a biconnected graph (i.e., in a call to $\Gamma B'$ in the line marked with ($\star\star$)), and has a specific degree. We will also say that v was *born* with this degree. In the second phase, when $\Gamma G^\bullet$ is recursively called, a certain number of new biconnected graphs will be attached to v, such that its degree increases by the sum of the degrees of the roots of those graphs. After this, the degree will not change anymore, such that the final degree is the sum of the degrees in the two phases. Hence, to count vertices of a given degree k, we will fix a $1 \leq \ell \leq k$ and count how many vertices are born with degree ℓ. Let B_ℓ be the number of such vertices. Then, we will compute the fraction of vertices among those B_ℓ that will receive $k - \ell$ neighbors in their second phase. Let us call this fraction $R_{k-\ell}$. The total number of vertices with degree k is then the sum of these numbers over all possible ℓ, namely $\sum_{\ell=1}^{k} B_\ell R_{k-\ell}$.

In order to determine the degree sequences of the resulting graphs it is therefore important to understand how many vertices are born with a given degree during a (random) execution of $\Gamma G^\bullet$, and what happens to the vertices in their second phase. In Lemma 2 below we describe this idea formally. In order to state it we need some additional notation. Let $\mathsf{B}_1', \mathsf{B}_2', \ldots,$ be random graphs from $\mathcal{B}'$, drawn independently according to the Boltzmann distribution with parameter $x = G^\bullet(\rho_\mathcal{G})$, and denote by $\mathsf{rd}\,(\mathsf{B}_i')$ the degree of the root vertex of B_i'. We say that a variable X is *sumRootBlock* distributed, $X \sim \mathrm{sRB}$, if it is distributed like $\sum_{i=1}^{\mathrm{Po}(\lambda_\mathcal{G})} \mathsf{rd}\,(\mathsf{B}_i')$. Moreover, let $\mathsf{deg}'(k; \mathsf{B}_i')$ be the number of vertices different from the root vertex of B_i' that have degree k.

Lemma 2. *Let $k \in \mathbb{N}$. Let $\mathcal{G}$ be a nice class of graphs with critical exponent α, and let $\mathcal{B}$ be the subclass of $\mathcal{G}$ containing all biconnected graphs in $\mathcal{G}$. Suppose that for all $0 < \varepsilon < 1$ there is a decreasing function $f(N) = f(N; \mathcal{G}, k, \varepsilon)$ such that the following holds for sufficiently large N.*

(B) Let $\mathsf{B}_1', \ldots, \mathsf{B}_N'$ be independent random graphs drawn according to the Boltzmann distribution for $\mathcal{B}'$ with parameter $x = G^\bullet(\rho_\mathcal{G})$. Then for every $1 \leq \ell \leq k$ there is a constant $b_{\mathcal{B},\ell}$ such that with probability at least $1 - f(N)$

$$\sum_{i=1}^{N} \mathsf{deg}'(\ell; \mathsf{B}_i') \in \left(1 \pm \frac{\varepsilon}{5}\right) b_{\mathcal{B},\ell} N. \tag{7}$$

Then, for every $0 < \varepsilon < 1$ and sufficiently large n there is a constant $C = C(k) > 0$ such that

$$\mathbb{P}\left[\mathsf{deg}(k; \mathsf{G}_n) \in (1 \pm \varepsilon)g_k n\right] \geq 1 - n^{\alpha+5}\left(f\left(\frac{\lambda_\mathcal{G} n}{2}\right) + e^{-C\varepsilon^2 n}\right), \tag{8}$$

where $g_k = \lambda_\mathcal{G} \sum_{\ell=1}^{k} b_{\mathcal{B},\ell} \cdot s_{\mathcal{B},k-\ell}$ is the kth coefficient of the generating function $\lambda_\mathcal{G} \cdot b(z) \cdot p(r(z))$, and $s_{\mathcal{B},k-\ell} := \mathbb{P}[\mathrm{sRB} = k - \ell]$. Here b, p, and r denote the functions

$$b(z) := \sum_{\ell \geq 1} b_{\mathcal{B},\ell} z^\ell \ , \quad p(z) := e^{\lambda_\mathcal{G}(z-1)} \ , \quad \text{and} \quad r(z) := \sum_{\ell \geq 1} \mathbb{P}[\mathsf{rd}\,(\mathsf{B}') = \ell] z^\ell,$$

where B' is drawn according to the Boltzmann distribution with parameter $x = G^\bullet(\rho_\mathcal{G})$ for $\mathcal{B}'$.

The statement of the above lemma generalizes to $k = k(n)$ with one additional technical assumption. This generalization can be found in the full version of the paper [15].

3   Outerplanar Graphs

In this section we are going to consider labeled connected outerplanar graphs, that we will simply call outerplanar graphs. Let $\mathcal{O}$ be the class of all outerplanar graphs, and $\mathcal{B}$ the class of labeled biconnected outerplanar graphs. By applying a standard decomposition of a graph into 2-connected blocks (see e.g. [16, p. 10]) we obtain the following lemma.

Lemma 3. *The classes $\mathcal{O}$ and $\mathcal{B}$ satisfy the relation $\mathcal{O}^\bullet = \mathcal{Z} \times \mathrm{SET}(\mathcal{B}' \circ \mathcal{O}^\bullet)$.*

In words, a rooted outerplanar graph is just a collection of rooted 2-connected outerplanar graphs, merged at their roots, in which every vertex may be substituted with another rooted outerplanar graph. The decomposition above translates immediately to a relation of the egf's $O^\bullet(x)$ for $\mathcal{O}^\bullet$ and $B'(x)$ for $\mathcal{B}'$. Bodirsky et al. exploited this relation in [17], where among other results they determined the singular expansion for $O^\bullet(x)$ (which yields straightforwardly an asymptotic estimate for $|\mathcal{O}_n^\bullet|$). Here we state only the result from [17] that we are going to exploit.

Lemma 4 ([17]). *The singular expansion of $O^\bullet(x)$ at its singularity $\rho_\mathcal{O}$ is*

$$O^\bullet(x) \overset{x \to \rho_\mathcal{O}}{=} O_0^\bullet - O_1^\bullet (1 - x/\rho_\mathcal{O})^{1/2} + o\left((1 - x/\rho_\mathcal{O})^{1/2}\right),$$

where $O_0^\bullet$ is the solution of the equation $xB''(x) = 1$, $\rho_\mathcal{O} = O_0^\bullet e^{-B'(O_0^\bullet)}$, and $O_1^\bullet$ is given analytically. Moreover, $|\mathcal{O}_n^\bullet| = (1 + O(n^{-1}))\frac{O_1^\bullet}{2\sqrt{\pi}} n^{-3/2} \rho_\mathcal{O}^{-n} n!$.

By combining the above two lemmas we obtain immediately the following corollary.

Corollary 3. *The class $\mathcal{O}$ of labeled connected outerplanar graphs is nice in the sense of Definition 1, and has critical exponent $\alpha = \frac{1}{2}$.*

3.1   The Degree Sequence of Random Outerplanar Graphs

In order to apply Lemma 2 we have to check that assertion (B) of that lemma is true for the class of outerplanar graphs. For this we prove the following statement; before we state it let us introduce some quantities, which are closely related to the number of vertices of degree k in a random outerplanar graph (the connection will become more explicit in Lemma 5). Let

$$c(x) := \frac{x}{2B'(x) - x}, \quad \text{and} \quad t(x) := \frac{x^2}{2B^\bullet(x)} = \frac{x}{2B'(x)},$$

and define the quantities

$$d_\ell(x) := \begin{cases} 2t(x), & \text{if } \ell = 1 \\ c(x)(1 - c(x))^{\ell-1} \left[1 - t(x) + \frac{c(x)}{1-c(x)}(\ell-1)\left(\frac{xB''(x)}{B'(x)} - 1\right)\right], & \text{if } \ell \geq 2 \end{cases},$$

(9)

and

$$r_\ell(x) := \begin{cases} 2t(x), & \text{if } \ell = 1 \\ (1 - t(x))c(x)(1 - c(x))^{\ell-1}, & \text{if } \ell \geq 2. \end{cases}$$

(10)

The next statement says that assertion (B) of Lemma 2 is true for outerplanar graphs.

Lemma 5. *Let* $B'_1, \ldots, B'_N$ *be independent random graphs drawn according to the Boltzmann distribution with parameter* $0 < x < \rho_{B'}$ *for* $\mathcal{B}'$. *Let* $\deg'(\ell; B')$ *denote the number of vertices different from the root with degree* ℓ *in* B'. *There is a* $C = C(x) > 0$ *such that for every* $0 < \varepsilon < 1$ *and* ℓ *such that* $d_\ell(x)N > \log^4 N$

$$\mathbb{P}\left[\sum_{i=1}^N \deg'(\ell; B'_i) \in (1 \pm \varepsilon)d_\ell(x)N\right] \geq 1 - e^{-C\varepsilon^2 d_\ell(x)N}.$$

(11)

Furthermore, let $\mathsf{rd}(B'_i)$ *be the degree of the root vertex of* B'_i. *Then* $\mathbb{P}[\mathsf{rd}(B'_i) = \ell] = r_\ell(x)$.

Proof (Proof of Theorem 1). We will only prove the case $k \in \mathbb{N}$, the proof for general $k = k(n)$ can be found in the full version of the paper [15]. Let us first make a technical observation. Recall that $O^\bullet(\rho_O) = O_0^\bullet$ is due to Lemma 4 the smallest solution of $xB''(x) = 1$, where $B(x)$ is the egf enumerating biconnected outerplanar graphs. As B is explicitly given (see [17]), one can easily verify that $O_0^\bullet < 3 - 2\sqrt{2} = \rho_{B'}$.

Recall that the class of outerplanar graphs is nice. Having the previous discussion in mind, by applying Lemma 5 with $x = O^\bullet(\rho_O) < \rho_{B'}$ we see that the assertions of Lemma 2 are satisfied with $b_{B,\ell} = d_\ell(O_0^\bullet)$ and $f(N) = e^{-C\varepsilon^2 N}$, for a suitably chosen $C = C(k) > 0$. By applying (8) we immediately obtain (1).

It remains to show (2). Observe that due to Lemma 4 we have $O_0^\bullet B''(O_0^\bullet) = 1$. With (9) we then obtain

$$b(z) = d_1(O_0^\bullet)z + \sum_{\ell \geq 2} d_\ell(O_0^\bullet)z^\ell = \frac{O_0^\bullet}{\lambda_O} \cdot z + \frac{Fz^2 + Gz^3}{(C - Dz)^2}, \quad \text{where} \quad \lambda_O = B'(O_0^\bullet),$$

(12)

and

$$F = O_0^\bullet \lambda_O^{-1}\left(2\lambda_O^2 - 4\lambda_O O_0^\bullet + (O_0^\bullet)^2 + O_0^\bullet\right), \quad C = O_0^\bullet - 2\lambda_O,$$
$$G = O_0^\bullet \lambda_O^{-1}\left(-2\lambda_O^2 + 4\lambda_O O_0^\bullet - 2(O_0^\bullet)^2\right), \quad D = 2(O_0^\bullet - \lambda_O).$$

(13)

To complete the proof we determine $r(z)$. By applying Lemma 5 and by exploiting (10)

$$r(z) = \sum_{\ell \geq 1} r_\ell(O_0^\bullet)z^\ell = \frac{Az + Bz^2}{C - Dz},$$

(14)

where $A = O_0^\bullet \lambda_O^{-1}(O_0^\bullet - 2\lambda_O)$ and $B = O_0^\bullet \lambda_O^{-1}(\lambda_O - O_0^\bullet)$.

4   Series-Parallel Graphs

In this section we determine the degree sequence of graphs drawn uniformly at random from the class of connected series-parallel (SP) graphs. To achieve this we will again apply Lemma 2. Before we proceed let us state some facts that we are going to need.

Let $\mathcal{SP}$ be the class of all labeled connected SP graphs, and $\mathcal{B}$ the class of labeled biconnected SP graphs. Then the following is true, similar to the case of outerplanar graphs.

Lemma 6. *The classes $\mathcal{SP}$ and $\mathcal{B}$ satisfy the relation $\mathcal{SP}^\bullet = \mathcal{Z} \times \text{SET}(\mathcal{B}' \circ \mathcal{SP}^\bullet)$.*

This decomposition translates into a relation of the egf's $SP^\bullet(x)$ for $\mathcal{SP}^\bullet$ and $B'(x)$ for $\mathcal{B}'$. Bodirsky et al. exploited this relation in [17], where they determined the singular expansion for $SP^\bullet(x)$ and the asymptotic value of $|\mathcal{SP}^\bullet_n|$.

Lemma 7 ([17]). *The singular expansion of $SP^\bullet(x)$ around its singularity $\rho_{\mathcal{SP}}$ is*

$$SP^\bullet(x) \stackrel{x \to \rho_{\mathcal{SP}}}{=} SP^\bullet_0 - SP^\bullet_1 (1 - x/\rho_{\mathcal{SP}})^{1/2} + o\left((1 - x/\rho_{\mathcal{SP}})^{1/2}\right),$$

where $\rho_{\mathcal{SP}} \doteq 0.11021$, $SP^\bullet_0 \doteq 0.12797$ and $SP^\bullet_1 \doteq 0.00453$ are implicitly given constants. Moreover, $|\mathcal{SP}^\bullet_n| = (1 + O(n^{-1}))\frac{SP^\bullet_1}{2\sqrt{\pi}}n^{-3/2}\rho_{\mathcal{SP}}^{-n}n!$.

By combining the above two lemmas we obtain immediately the following corollary.

Corollary 4. *The class $\mathcal{SP}$ is nice with critical exponent $\alpha = \frac{1}{2}$.*

In order to apply Lemma 2 we have to check that condition (B) is true for the class $\mathcal{SP}$, i.e., we have to determine the fraction of vertices of degree ℓ in a "typical" sequence of random graphs from $\mathcal{B}'$, drawn according to the Boltzmann distribution. The second ingredient needed to apply Lemma 2 is the distribution of the root degree of a random graph from $\mathcal{B}'$.

The next lemma provides us with information about the root degree of random objects from $\mathcal{B}'$. Before we state it, let us introduce an auxiliary graph class, which plays an important role in the the decomposition of 2-connected series-parallel graphs. Following [18,17], we define a *network* as a connected graph with two distinguished vertices, called the left and the right *pole*, such that adding the edge between the poles the resulting (multi)graph is 2-connected. Let $\mathcal{D}$ be the class of series-parallel networks, such that $\mathcal{D}_n$ contains all networks with n non-pole vertices. We write for brevity $\mathcal{D}_0 \equiv e$ for the network consisting of a single edge. Let $\vec{\mathcal{B}}$ be the class containing all graphs in $\mathcal{B}$ rooted at any of their edges, where the root edge is oriented. Then the classes $\mathcal{B}$ and $\mathcal{D}$ are due to the definition of $\mathcal{D}$ related as follows:

$$(\mathcal{D} + 1) \times \mathcal{Z}^2 \times e = (1 + e) \times \vec{\mathcal{B}}. \tag{15}$$

Although this decomposition can be used to obtain detailed information about the generating function enumerating $\mathcal{B}$ (see e.g. [19]), as well as the degree sequence of a "typical" graph from $\vec{\mathcal{B}}$, it turns out that it is quite involved to derive from it information about the degree sequence of a random graph from $\mathcal{B}$. This difficulty is mainly due to the fact that the number of ways to root a graph at an edge varies for graphs of the same size (w.r.t. the number of vertices), and would require to perform a very laborious integration. We attack this problem differently: we exploit a very general recent result by Fusy, Kang, and Shoilekova [20], which allows to decompose families of 2-connected graphs in a direct combinatorial way (again based on networks), but avoiding the often complicated and intractable integration steps. We will now write only the results that we have obtained, we refer the reader to the full version of this paper [15] for the proofs and all the details.

Let $\mathcal{G}$ be a class consisting of graphs that have a distinguished vertex. $\mathcal{G}$ could be for example $\mathcal{B}'$, or $\mathcal{D}$ with the left pole as distinguished vertex. Let G be a graph from $\mathcal{G}$, drawn according to the Boltzmann distribution with parameter x, and denote by $\mathrm{rd}\,(\mathrm{G})$ the degree of the distinguished vertex of G. Then we write $R_{\mathcal{G}}(z;x)$ for the probability generating function of $\mathrm{rd}\,(\mathrm{G})$, i.e.

$$R_{\mathcal{G}}(z;x) := \sum_{k\geq 0} \mathbb{P}[\mathrm{rd}\,(\mathrm{G}) = k]z^k.$$

We will use this notation throughout the paper without further reference and we will omit the parameter x if this is clear from the context. Having this, the distribution of the root-degree of random graphs from $\mathcal{B}'$ is as follows.

Lemma 8. *Let B' be a graph drawn from $\mathcal{B}'$ according to the Boltzmann distribution with parameter $0 \leq x \leq \rho_{\mathcal{B}'}$. Then*

$$\mathbb{P}[\mathrm{rd}\,(\mathrm{B}') = \ell] = [z^\ell]R_{\mathcal{B}'}(z) = [z^\ell]\frac{R_{\mathcal{D}}(z)(xD(x)^2R_{\mathcal{D}}(z) - 2)}{xD(x)^2 - 2}, \qquad (16)$$

where $D(x)$ is the egf enumerating series-parallel networks, and $R_{\mathcal{D}}(z)$ satisfies

$$R_{\mathcal{D}}(z) = \frac{1}{D(x)}\left(-1 + (1+z)\left(\frac{1+D(x)}{2}\right)^{R_{\mathcal{D}}(z)}\right). \qquad (17)$$

Next we determine the *expected* number of non-root vertices having degree ℓ in a random graph from $\mathcal{B}'$. Let $I_{\mathcal{B}'}(z)$ be the function whose kth coefficient is the expected number of non-root vertices of degree k in B'.

Lemma 9. *Let B' be a graph from $\mathcal{B}'$ drawn according to the Boltzmann distribution with parameter $0 \leq x \leq \rho_{\mathcal{B}'}$. Then*

$$\mathbb{E}[\mathrm{deg}'(\ell;\mathrm{B}')] = [z^\ell]I_{\mathcal{B}'}(z) = [z^\ell]\left(\left(\frac{xB''(x)}{B'(x)} - 1\right)R_{\mathcal{D}}(z)^2 + R_{\mathcal{B}'}(z)\right). \qquad (18)$$

Lemma 9 is unfortunately not sufficient to apply Lemma 2, as it provides us only with information about the expected number of non-root vertices with a given degree, and not with the appropriate concentration statement. Our final lemma solves this problem in a general way, with the slight disadvantage that the obtained tail probability might not be sharp.

Lemma 10. *Let* $G_1, \ldots, G_N$ *be random graphs drawn independently from a class* $\mathcal{G}$ *of graphs, according to a distribution that satisfies* $\mathbb{P}[|G_i| = s] \leq c^{-s}$ *for all* s, *and some* $c > 1$. *Let* $X : \mathcal{G} \to \mathbb{N}$ *be a function such that* $X(G) \leq |G|$ *for all* $G \in \mathcal{G}$, *and* $\mathbb{E}[X] \geq \frac{\log^2 N}{\sqrt{N}}$. *Then there is a* $C > 0$ *such that for all* $0 < \varepsilon < 1$ *and sufficiently large* N

$$\mathbb{P}\left[\sum_{i=1}^{N} X(G_i) \in (1 \pm \varepsilon)\mathbb{E}[X]N\right] \geq 1 - e^{-C\varepsilon^2\mathbb{E}[X]^{2/3}N^{1/3}}.$$

If we choose $\mathcal{G}$ to be $\mathcal{B}'$ and X as the variable counting internal vertices of degree k, all conditions of Lemma 10 are fulfilled. Then condition (B) of Lemma 2 holds for $\mathcal{SP}$, and we can apply the lemma to prove Theorem 2 with $b(z) = I_{\mathcal{B}'}(z)$, $r(z) = R_{\mathcal{B}'}(z)$ and

$$\lambda_{\mathcal{SP}} = B'(SP_0^{\bullet}). \tag{19}$$

The proof can be found in the full version of the paper [15].

References

1. Bollobás, B.: Random graphs, 2nd edn. Cambridge Studies in Advanced Mathematics, vol. 73. Cambridge University Press, Cambridge (2001)
2. Janson, S., Łuczak, T., Ruciński, A.: Random Graphs. John Wiley & Sons, Chichester (2000)
3. Denise, A., Vasconcellos, M., Welsh, D.J.A.: The random planar graph. Congr. Numer. 113, 61–79 (1996)
4. McDiarmid, C., Steger, A., Welsh, D.J.A.: Random planar graphs. J. Combin. Theory Ser. B 93(2), 187–205 (2005)
5. Gerke, S., McDiarmid, C., Steger, A., Weißl, A.: Random planar graphs with n nodes and a fixed number of edges. In: Proceedings of the Sixteenth Annual ACM-SIAM Symposium on Discrete Algorithms, pp. 999–1007. ACM, New York (2005) (electronic)
6. Giménez, O., Noy, M.: The number of planar graphs and properties of random planar graphs. In: 2005 Int. Conf. on An. of Alg. Discrete Math. Theor. Comput. Sci. Proc., AD. Assoc. Discrete Math. Theor. Comput. Sci., Nancy, pp. 147–156 (2005) (electronic)
7. Flajolet, P., Sedgewick, R.: Analytic combinatorics (Book in preparation, October 2005)
8. Bernasconi, N., Panagiotou, K., Steger, A.: On properties of random dissections and triangulations. In: Proceedings of the 19th Annual ACM-SIAM Symposium on Discrete Algorithms (SODA 2008), pp. 132–141 (2008), www.as.inf.ethz.ch/research/publications/2008/index/
9. Duchon, P., Flajolet, P., Louchard, G., Schaeffer, G.: Boltzmann samplers for the random generation of combinatorial structures. Combin. Probab. Comput. 13(4-5), 577–625 (2004)
10. Drmota, M., Giménez, O., Noy, M.: Vertices of given degree in series-parallel graphs (preprint)

11. Fusy, É.: Quadratic exact size and linear approximate size random generation of planar graphs. In: Martínez, C. (ed.) 2005 In. Conf. on An. of Al. DMTCS Proceedings. Discrete Mathematics and Theoretical Computer Science, vol. AD, pp. 125–138 (2005)
12. Bergeron, F., Labelle, G., Leroux, P.: Combinatorial species and tree-like structures. Encyclopedia of Mathematics and its Applications, vol. 67. Cambridge University Press, Cambridge (1998)
13. Stanley, R.P.: Enumerative combinatorics. Vol. 1. Cambridge Studies in Advanced Mathematics, vol. 49. Cambridge University Press, Cambridge (1997); With a foreword by Gian-Carlo Rota, Corrected reprint of the 1986 original
14. Flajolet, P., Zimmerman, P., Van Cutsem, B.: A calculus for the random generation of labelled combinatorial structures. Theoret. Comput. Sci. 132(1-2), 1–35 (1994)
15. Bernasconi, N., Panagiotou, K., Steger, A.: On the degree sequences of random outerplanar and series-parallel graphs,
www.as.inf.ethz.ch/research/publications/2008/index/
16. Harary, F., Palmer, E.: Graphical Enumeration. Academic Press, New York (1973)
17. Bodirsky, M., Giménez, O., Kang, M., Noy, M.: On the number of series parallel and outerplanar graphs. In: 2005 European Conference on Combinatorics, Graph Theory and Applications (EuroComb 2005). DMTCS Proceedings., Discrete Mathematics and Theoretical Computer Science, vol. AE, pp. 383–388 (2005)
18. Walsh, T.R.S.: Counting labelled three-connected and homeomorphically irreducible two-connected graphs. J. Combin. Theory Ser. B 32(1), 1–11 (1982)
19. Bender, E.A., Gao, Z., Wormald, N.C.: The number of labeled 2-connected planar graphs. Electron. J. Combin. 9(1) (2002); Research Paper 43, 13 pp. (electronic)
20. Fusy, E., Kang, M., Shoilekova, B.: A complete grammar for decomposing a family of graphs into 3-connected components. (submitted for publication) (2008)

Improved Bounds for Testing Juntas

Eric Blais[*]

Carnegie Mellon University
eblais@cs.cmu.edu

Abstract. We consider the problem of testing functions for the property of being a k-junta (i.e., of depending on at most k variables). Fischer, Kindler, Ron, Safra, and Samorodnitsky (*J. Comput. Sys. Sci., 2004*) showed that $\tilde{O}(k^2)/\epsilon$ queries are sufficient to test k-juntas, and conjectured that this bound is optimal for non-adaptive testing algorithms.

Our main result is a non-adaptive algorithm for testing k-juntas with $\tilde{O}(k^{3/2})/\epsilon$ queries. This algorithm disproves the conjecture of Fischer et al.

We also show that the query complexity of non-adaptive algorithms for testing juntas has a lower bound of $\min\left(\tilde{\Omega}(k/\epsilon), 2^k/k\right)$, essentially improving on the previous best lower bound of $\Omega(k)$.

1   Introduction

A function $f : \{0,1\}^n \to \{0,1\}$ is said to be a k-junta if it depends on at most k variables. Juntas provide a clean model for studying learning problems in the presence of many irrelevant features [4,6], and have consequently been of particular interest to the computational learning theory community [5,6,17,12, 16]. A problem closely related to learning juntas is the problem of *testing juntas*: given query access to a function, is it possible to efficiently determine if all but at most k of the variables in the function represent irrelevant features?

We consider the problem of testing juntas in the standard framework of property testing, as originally introduced by Rubinfeld and Sudan [19]. In this framework, we say that a function f is ϵ-*far* from being a k-junta if for every k-junta g, the functions f and g disagree on at least an ϵ fraction of inputs. A randomized algorithm $\mathcal{A}$ that makes q queries to its input function is an ϵ-*testing* algorithm for k-juntas if

1. All k-juntas are accepted by $\mathcal{A}$ with probability at least $2/3$, and
2. All functions that are ϵ-far from being k-juntas are rejected by $\mathcal{A}$ with probability at least $2/3$.

A testing algorithm $\mathcal{A}$ is *non-adaptive* if does not use the answers of some queries to determine later queries; otherwise, the algorithm $\mathcal{A}$ is *adaptive*.

In this article we consider the problem of determining the query complexity for the problem of testing juntas: given fixed $k \geq 1$ and $\epsilon > 0$, what is the minimum number $q = q(k, \epsilon)$ of queries required for any algorithm $\mathcal{A}$ to ϵ-test k-juntas?

---

[*] Supported in part by a scholarship from the Fonds québécois de recherche sur la nature et les technologies (FQRNT).

A. Goel et al. (Eds.): APPROX and RANDOM 2008, LNCS 5171, pp. 317–330, 2008.

Background. The first result related to testing juntas was obtained by Bellare, Goldreich, and Sudan [2] in the context of testing long codes. That result was generalized by Parnas, Ron, and Samorodnitsky [18] to obtain an algorithm for ϵ-testing 1-juntas with only $O(1/\epsilon)$ queries.

The next important step in testing k-juntas was taken by Fischer, Kindler, Ron, Safra, and Samorodnitsky [10], who developed multiple algorithms for testing k-juntas with $\mathrm{poly}(k)/\epsilon$ queries. Those algorithms were particularly significant for showing explicitly that testing juntas can be done with a query complexity *independent* of the total number of variables. The most query-efficient algorithms they presented require $\tilde{O}(k^2)/\epsilon$ queries[1] to ϵ-test k-juntas.

Fischer et al. [10] also gave the first non-trivial lower bound on the query complexity for the testing juntas problem. They showed that any non-adaptive algorithm for ϵ-testing k-juntas requires at least $\tilde{\Omega}(\sqrt{k})$ queries and conjectured that the true query complexity for non-adaptive algorithms is k^2/ϵ queries.

Chockler and Gutfreund [8] improved the lower bound for testing juntas by showing that all algorithms – adaptive or non-adaptive – for ϵ-testing k-juntas require $\Omega(k)$ queries. This result applies for all values of $\epsilon < 1/8$, but the bound itself does not increase as ϵ decreases.

Our results and techniques. Our main result is an improvement on the upper bound for the query complexity of the junta testing problem.

Theorem 1.1. *The property of being a k-junta can be ϵ-tested by a non-adaptive algorithm with $\tilde{O}(k^{3/2})/\epsilon$ queries.*

The new algorithm presented in this article is the first for testing juntas with a number of queries sub-quadratic in k, and disproves the lower bound conjecture of Fischer et al.

Our algorithm is based on an algorithm of Fischer et al. for testing juntas [10, §4.2]). The observation that led to the development of the new algorithm is that the algorithm of Fischer et al. can be broken up into two separate tests: a "block test" and a simple "sampling test". In this article, we generalize the sampling test, and we establish a structural Lemma for functions that are ϵ-far from being k-juntas to show how the two tests can be combined to ϵ-test k-juntas more efficiently.

Our second result is an improved lower bound on the number of queries required for testing juntas with non-adaptive algorithms. The new bound is the first lower bound for the query complexity of the junta testing problem that incorporates the accuracy parameter ϵ.

Theorem 1.2. *Any non-adaptive algorithm for ϵ-testing k-juntas must make at least* $\min\left(\Omega\left(\frac{k/\epsilon}{\log k/\epsilon}\right), \Omega\left(\frac{2^k}{k}\right)\right)$ *queries.*

We prove Theorem 1.2 via Yao's Minimax Principle [20]. The proof involves an extension of the argument of Chockler and Gutfreund [8] and an application of the Edge-Isoperimetric Inequality of Harper [13], Bernstein [3], and Hart [14].

---

[1] Here and in the rest of this article, the $\tilde{O}(\cdot)$ notation is used to hide polylog factors. (i.e., $\tilde{O}(f(x)) = O(f(x)\log^c f(x))$ and $\tilde{\Omega}(f(x)) = \Omega(\frac{f(x)}{\log^c f(x)})$ for some $c \geq 0$.)

Organization. We introduce some notation and definitions in Section 2. We present the new algorithm for ϵ-testing k-juntas and its analysis in Section 3. In Section 4, we present the proof for the lower bound on the query complexity of non-adaptive algorithms for testing juntas. Finally, we conclude with some remarks and open problems in Section 5.

2 Preliminaries

Notation. For $n \geq 1$, let $[n] = \{1, \ldots, n\}$. For a set $A \subseteq [n]$, we write $\bar{A} = [n] \backslash A$ to represent the complement of A in $[n]$. When $x, y \in \{0, 1\}^n$, we define $x_A y_{\bar{A}}$ to be the hybrid string z where $z_i = x_i$ for every $i \in A$ and $z_j = y_j$ for every $j \in \bar{A}$.

We write $\mathbf{Pr}_x[\,\cdot\,]$ (resp., $\mathbf{E}_x[\,\cdot\,]$) to denote the probability (resp., expectation) over the choice of x taken uniformly at random from $\{0, 1\}^n$. We also write $H_k = \sum_{j=1}^{k} \frac{1}{j}$ to denote the k-th harmonic number.

Variation. In the analysis of the new algorithm for testing juntas, we consider the *variation* of sets of coordinates in a function, a concept introduced by Fischer et al. [10].[2]

Definition 2.1. *The variation of the set $S \subseteq [n]$ of coordinates in the function $f : \{0, 1\}^n \to \{0, 1\}$ is*

$$\mathrm{Vr}_f(A) = \mathbf{Pr}_x\left[f(x) \neq f(x_{\bar{S}} y_S)\right].$$

We write $\mathrm{Vr}_f(i) = \mathrm{Vr}_f(\{i\})$ to represent the variation of the ith coordinate. The variation of a single coordinate is equivalent to the notion of *influence*, as defined in, e.g., [15].

Some useful properties of variation are its monotonicity, subadditivity, and submodularity.

Fact 2.2 (Fischer et al. [10]) *For any function $f : \{0, 1\}^n \to \{0, 1\}$, and any sets $A, B, C \subseteq [n]$, the following three properties hold:*

(i) Monotonicity: $\mathrm{Vr}_f(A) \leq \mathrm{Vr}_f(A \cup B)$
(ii) Subadditivity: $\mathrm{Vr}_f(A \cup B) \leq \mathrm{Vr}_f(A) + \mathrm{Vr}_f(B)$
(iii) Submodularity: $\mathrm{Vr}_f(A \cup B) - \mathrm{Vr}_f(B) \geq \mathrm{Vr}_f(A \cup B \cup C) - \mathrm{Vr}_f(B \cup C)$

The Independence Test. A function f is said to be *independent* of a set $S \subseteq [n]$ of coordinates if $\mathrm{Vr}_f(S) = 0$. The definition of variation suggests a natural test for independence:

INDEPENDENCETEST [10]: Given a function $f : \{0, 1\}^n \to \{0, 1\}$ and a set $S \subseteq [n]$, generate two inputs $x, y \in \{0, 1\}^n$ independently and uniformly at random. If $f(x) = f(x_{\bar{S}} y_S)$, then accept; otherwise, reject.

---

[2] The definition of variation used in [10] is slightly different, but is equivalent to the one used in this article up to a constant factor.

BLOCKTEST(f, k, η, δ)

Additional parameters: $s = \lceil 2k^2/\delta \rceil$, $r = \lceil 4k\ln(s/\delta) \rceil$, $m = \lceil \ln(2r/\delta)/\eta \rceil$

1. Randomly partition the coordinates in $[n]$ into s sets $I_1, \ldots, I_s$.
2. For each of r rounds,
   2.1. Pick a random subset $T \subseteq [s]$ by including each index independently with probability $1/k$.
   2.2. Define the block of coordinates $B_T = \bigcup_{j \in T} I_j$.
   2.3. If INDEPENDENCETEST(f, B_T, m) accepts, mark I_j as "variation-free" for every $j \in T$.
3. **Accept** f if at most k of the sets $I_1, \ldots, I_s$ are not marked as "variation-free"; otherwise **reject** f.

Fig. 1. The algorithm for the block test

Let us define INDEPENDENCETEST(f, S, m) to be the algorithm that runs m instances of the INDEPENDENCETEST on f and S and accepts if and only if every instance of the INDEPENDENCETEST accepts. By the definition of variation, this algorithm accepts with probability $\left(1 - \text{Vr}_f(S)\right)^m$. In particular, this test always accepts when f is independent of the set S of coordinates, and rejects with probability at least $1 - \delta$ when $\text{Vr}_f(S) \geq \ln(1/\delta)/m$.

3   The Algorithm for Testing Juntas

In this section, we present the algorithm for ϵ-testing k-juntas with $\tilde{O}(k^{3/2})/\epsilon$ queries. The algorithm has two main components: the BLOCKTEST and the SAMPLINGTEST. We introduce the BLOCKTEST in Section 3.1 and the SAMPLINGTEST in Section 3.2. Finally, in Section 3.3 we show how to combine both tests to obtain an algorithm for testing juntas.

3.1   The Block Test

The purpose of the BLOCKTEST is to accept k-juntas and reject functions that have at least $k + 1$ coordinates with "large" variation.

The BLOCKTEST first randomly partitions the coordinates in $[n]$ into s sets $I_1, \ldots, I_s$. It then applies the INDEPENDENCETEST to blocks of these sets to identify the sets of coordinates that have low variation. The test accepts if all but at most k of the sets $I_1, \ldots, I_s$ are identified as having low variation. The full algorithm is presented in Fig. 1.

The BLOCKTEST is based on Fischer et al.'s non-adaptive algorithm for testing juntas [10, §4.2], which uses a very similar test.[3] As the following two

---

[3] The principal difference between our version of the BLOCKTEST and Fischer et al.'s version of the test is that in [10], the set T is generating by including exactly k indices chosen at random from $[s]$.

Propositions show, with high probability the BLOCKTEST accepts k-juntas and rejects functions with $k + 1$ coordinates with variation at least η.

Proposition 3.1 (Completeness). *Fix $\eta > 0$, and let $f : \{0, 1\}^n \to \{0, 1\}$ be a k-junta. Then the BLOCKTEST accepts f with probability at least $1 - \delta$.*

Proof. Let I_j be a set that contains only coordinates i with variation $\mathrm{Vr}_f(i) = 0$. In a given round, the probability that I_j is included in B_T and none of the sets $I_{j'}$ that contain a coordinate with positive variation are included in B_T is at least $(1/k)(1 - 1/k)^k \geq 1/4k$ since $(1 - 1/k)^k \geq 1/4$ for all $k \geq 2$. So the probability that I_j is not marked as "variation-free" in any of the r rounds is at most $(1 - 1/4k)^r \leq e^{-r/4k} \leq \delta/s$ when $r \geq 4k \ln(s/\delta)$. By the union bound, all the sets I_j that contain only coordinates with no variation are identified as "variation-free" with probability at least $1 - s(\delta/s) = 1 - \delta$. □

Proposition 3.2 (Soundness). *Let $f : \{0, 1\}^n \to \{0, 1\}$ be a function for which there exists a set $S \subseteq [n]$ of size $|S| = k + 1$ such that every coordinate $i \in S$ has variation $\mathrm{Vr}_f(i) \geq \eta$. Then the BLOCKTEST rejects f with probability at least $1 - \delta$.*

Proof. There are two ways in which the block test can wrongly accept the input function. The first way it can do so is by mapping all the coordinates with variation at least η into at most k sets during the random partition. We can upper bound the probability of this event with the probability that any collision occurs during the mapping of the first $k + 1$ coordinates with high variation, which is at most $\frac{1}{s} + \frac{2}{s} + \cdots + \frac{k}{s} = \frac{k(k+1)}{2s} \leq \frac{k^2}{s} \leq \delta/2$.

The second way in which the block test can wrongly accept the input function is by erroneously marking one of the sets I_j that contains a coordinate with variation at least η as "variation-free". To bound the probability of this event happening, consider a given round in which B_T contains at least one of the coordinates i with variation $\mathrm{Vr}_f(i) \geq \eta$. By Fact 2.2 (i), the variation of B_T is at least η, so when $m \geq \ln(2r/\delta)/\eta$, the INDEPENDENCETEST accepts B_T with probability at most $\delta/2r$. By the union bound, the probability that one of the r rounds results in a false "variation-free" marking is at most $\delta/2$. So the total probability that the algorithm wrongly accepts the function f is at most $\delta/2 + \delta/2 = \delta$. □

The BLOCKTEST algorithm makes $2m$ queries to f in each round, so the total query complexity of the algorithm is $2rm = O(k \log^2(k/\delta)/\eta)$.

3.2 The Sampling Test

The purpose of the SAMPLINGTEST is to accept k-juntas and reject functions that have a large number of coordinates with non-zero variation.

The SAMPLINGTEST, as its name implies, uses a sampling strategy to estimate the number of coordinates with non-negligible variation in a given function f. The sampling test generates a random subset $T \subseteq [n]$ of coordinates in each

---

SAMPLINGTEST(f, k, l, η, δ)

Additional parameters: $r = \lceil 128k^2 \ln(2/\delta)/l^2 \rceil$, $m = \lceil \ln(2r/\delta)/\eta \rceil$

1. Initialize the success counter $c \leftarrow 0$.
2. For each of r rounds,
    2.1. Pick a random subset $T \subseteq [n]$ by including each coordinate independently with probability $1/k$.
    2.2. If INDEPENDENCETEST(f, T, m) accepts, set $c \leftarrow c + 1$.
3. **Accept** f if $c/r \geq (1 - 1/k)^k - l/16k$; otherwise **reject** f.

Fig. 2. The algorithm for the sampling test

round, and uses the INDEPENDENCETEST to determine if f is independent of the coordinates in T. The test accepts when the fraction of rounds that pass the independence test is not much smaller than the expected fraction of rounds that pass the test when f is a k-junta. The details of the algorithm are presented in Fig. 2.

Proposition 3.3 (Completeness). *Fix $\eta > 0$, $l \in [k]$. Let $f : \{0,1\}^n \to \{0,1\}$ be a k-junta. Then the SAMPLINGTEST accepts f with probability at least $1 - \delta$.*

Proof. When f is a k-junta, the probability that the set T in a given round contains only coordinates i with variation $\mathrm{Vr}_f(i) = 0$ is at least $(1 - 1/k)^k$. When this occurs, the set T also has variation $\mathrm{Vr}_f(T) = 0$. Let t be the number of rounds for which the set T satisfies $\mathrm{Vr}_f(T) = 0$. By Hoeffding's bound,

$$\mathbf{Pr}\left[\frac{t}{r} < (1 - 1/k)^k - \frac{l}{16k} \right] \leq e^{-2r \cdot (l/16k)^2} \leq \delta/2$$

when $r \geq 128k^2 \ln(2/\delta)/l^2$. Every set T with variation $\mathrm{Vr}_f(T) = 0$ always passes the INDEPENDENCETEST, so $c \geq t$ and the completeness claim follows. $\square$

Proposition 3.4 (Soundness 1). *Fix $\eta > 0$, $l \in [k]$. Let $f : \{0,1\}^n \to \{0,1\}$ be a function for which there is a set $S \subseteq [n]$ of size $|S| = k + l$ such that every coordinate $i \in S$ has variation $Vr_f(i) \geq \eta$. Then the SAMPLINGTEST rejects f with probability at least $1 - \delta$.*

Proof. In a given round, the probability that the random set T does not contain any of the $k + l$ coordinates with large variation is $(1 - 1/k)^{k+l}$. When $l \leq k$, $(1 - 1/k)^l \leq 1 - l/2k$, and when $k \geq 2$, $(1 - 1/k)^k \geq 1/4$. So the probability that T contains none of the $k + l$ coordinates with large varation is $(1 - 1/k)^{k+l} \leq (1 - 1/k)^k (1 - l/2k) \leq (1 - 1/k)^k - l/8k$.

Let t represent the number of rounds whose sets T contain no coordinate with variation at least η. By Hoeffding's bound,

$$\mathbf{Pr}\left[\frac{t}{r} > ((1 - 1/k)^k - l/8k) + l/16k \right] \leq e^{-2r(l/16k)^2} \leq \delta/2$$

when $r \geq 128k^2 \ln(2/\delta)/l^2$. By Fact 2.2 (i), every set T that contains one of the coordinates i with variation $\mathrm{Vr}_f(i) \geq \eta$ also has variation $\mathrm{Vr}_f(T) \geq \eta$. By our choice of m, the probability that the INDEPENDENCETEST accepts a set with variation η is at most $\delta/2r$. By the union bound, the INDEPENDENCETEST correctly rejects all the sets with variation at least η except with probability at most $\delta/2$.

The sampling test can accept f only if more than a $(1-1/k)^k - l/16k$ fraction of the random sets contain no coordinate with variation η, or if at least one of those random sets contains such a coordinate but still passes the INDEPENDENCETEST. So the probability that the sampling test erroneously accepts f is at most $\delta/2 + \delta/2 = \delta$.   □

Proposition 3.5 (Soundness 2). *Let* $\eta = \frac{\epsilon}{64H_k k}$,[4] *and let* $f : \{0,1\}^n \to \{0,1\}$ *be a function for which there exists a set* $S \subseteq [n]$ *of coordinates satisfying the following two properties:*

(i) Each coordinate $i \in S$ *has variation* $\mathrm{Vr}_f(i) < \eta$, *and*
(ii) The total variation of the set S *is* $\mathrm{Vr}_f(S) \geq \epsilon/2$.

Then when $l = k$, *the* SAMPLINGTEST *rejects* f *with probability at least* $1 - \delta$.

The proof of Proposition 3.5 follows very closely the proof of Fischer et al. [10, Lem. 4.3]. In particular, the proof uses the following Chernoff-like bound.

Lemma 3.6 (Fischer et al. [10, Prop. 3.5]). *Let* $X = \sum_{i=1}^{l} X_i$ *be a sum of non-negative independent random variables* X_i. *If every* X_i *is bounded above by* t, *then for every* $\lambda > 0$

$$\mathbf{Pr}\big[X < \lambda\mathbf{E}[X]\big] < \exp\left(\frac{\mathbf{E}[X]}{et}(\lambda e - 1)\right).$$

The proof of Proposition 3.5 also makes extensive use of Fischer et al.'s concept of *unique variation* [10].

Definition 3.7 (Fischer et al. [10]). *The* unique variation *of the coordinate* $i \in [n]$ *with respect to the set* $S \subseteq [n]$ *in the function* $f : \{0,1\}^n \to \{0,1\}$ *is*

$$\mathrm{Ur}_{f,S}(i) = \mathrm{Vr}_f([i] \cap S) - \mathrm{Vr}_f([i-1] \cap S).$$

Furthermore, the unique variation *of the set* $I \subseteq [n]$ *of coordinates with respect to* S *in* f *is* $\mathrm{Ur}_{f,S}(I) = \sum_{i \in I} \mathrm{Ur}_{f,S}(i)$.

Fact 3.8 (Fischer et al. [10]) *For any function* $f : \{0,1\}^n \to \{0,1\}$ *and sets of coordinates* $S, T \subseteq [n]$, *the following two properties hold:*

(i) $\mathrm{Ur}_{f,S}(T) \leq \mathrm{Vr}_f(T)$, *and*
(ii) $\mathrm{Ur}_{f,S}([n]) = \mathrm{Vr}_f(S)$.

We are now ready to complete the proof of Proposition 3.5.

---

[4] Recall that $H_k = \sum_{j=1}^{k} \frac{1}{j}$ is the kth harmonic number.

Proof (of Proposition 3.5). There are two ways in which the SAMPLINGTEST can accept f. The test may accept f if at least a $(1 - 1/k)^k - 1/16$ fraction of the random sets T have variation $\mathrm{Vr}_f(T) < \eta$. Alternatively, the test may also accept if some of the sets T with variation $\mathrm{Vr}_f(T) \geq \eta$ pass the INDEPENDENCETEST. By our choice of m and the union bound, this latter event happens with probability at most $\delta/2$. So the proof of Proposition 3.5 is complete if we can show that the probability of the former event happening is also at most $\delta/2$.

Let t represent the number of rounds where the random set T has variation $\mathrm{Vr}_f(T) \geq \eta$. We want to show that $\mathbf{Pr}\left[t/r \geq (1 - 1/k)^k - 1/16\right] \leq \delta/2$. In fact, since $(1-1/k)^k \geq 1/4$ for all $k \geq 2$, it suffices to show that $\mathbf{Pr}\left[t/r \geq 3/16\right] \leq \delta/2$.

In a given round, the expected unique variation of the random set T with respect to S in f is

$$\mathbf{E}[\mathrm{Ur}_{f,S}(T)] = \sum_{i \in [n]} \frac{1}{k} \mathrm{Ur}_{f,S}(i) = \frac{\mathrm{Ur}_{f,S}([n])}{k} = \frac{\mathrm{Vr}_f(S)}{k} \geq \frac{\epsilon}{2k},$$

where the third equality uses Fact 3.8 (ii). By Property (i) of the Proposition, $\mathrm{Ur}_{f,S}(T)$ is the sum of non-negative variables that are bounded above by η. So we can apply Lemma 3.6 with $\lambda = 1/32H_k$ to obtain

$$\mathbf{Pr}\left[\mathrm{Ur}_f(T) < \eta\right] < e^{\frac{\epsilon}{2ek\eta}\left(\frac{\epsilon}{32H_k}-1\right)}.$$

By Fact 3.8 (i) and the fact that $e^{\frac{\epsilon}{2ek\eta}\left(\frac{\epsilon}{32H_k}-1\right)} < 1/8$ for all $k \geq 1$, we have that

$$\mathbf{E}[t/r] = \mathbf{Pr}\left[\mathrm{Vr}_f(T) < \eta\right] < 1/8.$$

The final result follows from an application of Hoeffding's inequality and the choice of r. □

The SAMPLINGTEST algorithm makes $2m$ queries to f in each round, so the total query complexity of the algorithm is $2rm = O(k^2 \log(k/l\delta)/l^2\eta)$.

3.3   The Junta Test

In the previous two subsections, we defined two tests: the BLOCKTEST that distinguishes k-juntas from functions with $k+1$ coordinates with large variation, and the SAMPLINGTEST that distinguishes k-juntas from functions that have some variation distributed over a large number of coordinates. The following structural Lemma on functions that are ϵ-far from being k-juntas shows that these two tests are sufficient for testing juntas.

Lemma 3.9. *Let $f : \{0,1\}^n \to \{0,1\}$ be ϵ-far from being a k-junta. Then for any $t > 0$, f satisfies at least one of the following two properties:*

(i) *There exists an integer $l \in [k]$ such that there are at least $k+l$ coordinates i with variation $\mathrm{Vr}_f(i) \geq \frac{\epsilon}{tH_k l}$ in f.*

JUNTATEST(f, k, ϵ)

Additional parameters: $\delta = \frac{1}{3(\lceil \log k^{1/2} \rceil + 2)}, \quad \tau = \frac{\epsilon}{64 H_k}$

1. Run BLOCKTEST$(f, k, \tau/\lceil k^{1/2} \rceil, \delta)$.
2. For $l = \lceil k^{1/2} \rceil, \lceil 2k^{1/2} \rceil, \lceil 4k^{1/2} \rceil, \lceil 8k^{1/2} \rceil, \ldots, k$,
   2.1. Run SAMPLINGTEST$(f, k, l, \tau/2l, \delta)$.
3. Run SAMPLINGTEST$(f, k, k, \tau/k, \delta)$.
4. **Accept** f if all of the above tests accept; otherwise **reject** f.

Fig. 3. The algorithm for the junta test

(ii) The set S of coordinates $i \in [n]$ with variation $\mathrm{Vr}_f(i) < \frac{\epsilon}{tH_k k}$ has total variation $\mathrm{Vr}_f(S) \geq (1 - 1/t)\epsilon$.

Proof. Let f be a function that does not satisfy the Property (i) of the Lemma. Define $J \subseteq [n]$ to be the set of the k coordinates in f with highest variation, and let T be the set of coordinates $i \in [n] \setminus J$ with variation $\mathrm{Vr}_f(i) \geq \frac{\epsilon}{tH_k k}$. Since f does not satisfy Property (i) of the Lemma, Fact 2.2 (ii) ensures that the variation of T is bounded by $\mathrm{Vr}_f(T) \leq \frac{\epsilon}{tH_k} + \frac{\epsilon}{2tH_k} + \cdots + \frac{\epsilon}{ktH_k} = \frac{\epsilon}{t}$. Since $S \cup T \supseteq [n] \setminus J$ and any function ϵ-far from being a k-junta must satisfy $\mathrm{Vr}_f([n] \setminus J) \geq \epsilon$, a second application of Fact 2.2 (ii) shows that f must satisfy Property (ii) of the Lemma. $\square$

Lemma 3.9 naturally suggests an algorithm for testing k-juntas: use the BLOCK-TEST (with parameter $\eta = \epsilon/64H_k k$) to reject functions that satisfy Property (i) of the Lemma, and use the SAMPLINGTEST (with parameters $l = k$ and η as above) to reject the functions that satisfy Property (ii) of the Lemma. This algorithm is equivalent to the non-adaptive algorithm of Fischer et al. [10], and requires $\tilde{O}(k^2)/\epsilon$ queries.

We can improve the query complexity of the algorithm by splitting up the task of identifying functions that satisfy Property (i) of Lemma 3.9 into multiple tasks for more specific ranges of l. The result of this approach is the JUNTATEST algorithm presented in Fig. 3. With this algorithm, we are now ready to prove Theorem 1.1.

Theorem 1.1. *The property of being a k-junta can be ϵ-tested by a non-adaptive algorithm with $\tilde{O}(k^{3/2})/\epsilon$ queries.*

Proof. Let us begin by showing that the JUNTATEST is a valid algorithm for ϵ-testing k-juntas. By Propositions 3.1 and 3.3, k-juntas pass the BLOCKTEST and each of the SAMPLINGTEST instances with probability δ. So by our choice of δ and the union bound, k-juntas are accepted by the JUNTATEST with probability at least $2/3$.

Let f be any function that is ϵ-far from being a k-junta. If f satisfies Property (i) of Lemma 3.9 with parameter $t = 64$, consider the minimum integer

$l' \in [k]$ for which there is a set $S \subseteq [n]$ of size $k + l'$ such that every coordinate $i \in S$ has variation $\mathrm{Vr}_f(i) \geq \frac{\epsilon}{64 H_k l'}$. If $l' < \lceil k^{1/2} \rceil$, then by Proposition 3.2, the BLOCKTEST rejects f with probability $1 - \delta > 2/3$. If $l' \geq k^{1/2}$, then by Proposition 3.4, the SAMPLINGTEST with the parameter l that satisfies $l \leq l' \leq 2l$ rejects the function with probability $1 - \delta > 2/3$.

If f satisfies Property (ii) of Lemma 3.9, by Proposition 3.5, the last SAMPLINGTEST rejects the function with probability $1 - \delta > 2/3$. Since Lemma 3.9 guarantees that any function ϵ-far from being a k-junta must satisfy at least one of the two properties of the Lemma, this completes the proof of soundness of the JUNTATEST.

To complete the proof of Theorem 1.1, it suffices to show that the JUNTATEST is a non-adaptive algorithm and that it makes only $\tilde{O}(k^{3/2})/\epsilon$ queries to the function. The non-adaptivity of the JUNTATEST is apparent from the fact that all queries to the input function come from independent instances of the INDEPENDENCETEST. The query complexity of the JUNTATEST also follows from the observation that each instance of the BLOCKTEST or the SAMPLINGTEST in the algorithm requires $\tilde{O}(k^{3/2})/\epsilon$ queries. Since there are a total of $O(\log k)$ calls to those tests, the total query complexity of the JUNTATEST is also $\tilde{O}(k^{3/2})/\epsilon$. $\square$

4   The Lower Bound

In this section, we show that every non-adaptive algorithm for ϵ-testing k-juntas must make at least $\min \left(\tilde{\Omega}(k/\epsilon), 2^k/k \right)$ queries to the function.

To prove Theorem 1.2, we introduce two distributions, $\mathcal{D}_{yes}$ and $\mathcal{D}_{no}$, over functions that are k-juntas and functions that are ϵ-far from k-juntas with high probability, respectively. We then show that no deterministic non-adaptive algorithm can reliably distinguish between functions drawn from $\mathcal{D}_{yes}$ and functions drawn from $\mathcal{D}_{no}$. The lower bound on all non-adaptive algorithms for ϵ-testing k-juntas then follows from an application of Yao's Minimax Principle [20].

A central concept that we use extensively in the proof of Theorem 1.2 is Chockler and Gutfreund's definition of twins [8].

Definition 4.1. *Two vectors $x, y \in \{0, 1\}^n$ are called i-twins if they differ exactly in the ith coordinate (i.e., if $x_i \neq y_i$ and $x_j = y_j$ for all $j \in [n] \setminus \{i\}$). The vectors x, y are called* twins *if they are i-twins for some $i \in [n]$.*

We now define the distributions $\mathcal{D}_{yes}$ and $\mathcal{D}_{no}$. To generate a function from the distribution $\mathcal{D}_{no}$, we first define a function $g : \{0, 1\}^{k+1} \to \{0, 1\}$ by setting the value $g(x)$ for each input $x \in \{0, 1\}^{k+1}$ independently at random, with $\mathbf{Pr}[g(x) = 1] = 6\epsilon$. We then extend the function over the full domain by defining $f(x) = g(x_{[k+1]})$ for every $x \in \{0, 1\}^n$. The distribution $\mathcal{D}_{yes}$ is defined to be the uniform mixture distribution over the distributions $\mathcal{D}_{yes}^{(1)}, \mathcal{D}_{yes}^{(2)}, \ldots, \mathcal{D}_{yes}^{(k+1)}$, where the distribution $\mathcal{D}_{yes}^{(i)}$ is defined similarly to the $\mathcal{D}_{no}$ distribution, but over the set $[k+1] \setminus \{i\}$ instead of $[k+1]$.

By construction, the functions drawn from $\mathcal{D}_{yes}$ are all k-juntas. The following Lemma shows that a function drawn from $\mathcal{D}_{no}$ is ϵ-far from being a k-junta with high probability.

Lemma 4.2. *When $k/2^k < \epsilon \leq 1/12$ and $k \geq 3$, a function $f : \{0,1\}^n \to \{0,1\}$ drawn from $\mathcal{D}_{no}$ is ϵ-far from being a k-junta with probability at least $11/12$.*

Proof. A function f drawn from $\mathcal{D}_{no}$ is ϵ-far from being a k-junta iff the function $g : \{0,1\}^{k+1} \to \{0,1\}$ that was extended to form f is ϵ-far from being a k-junta. In turn, g is ϵ-far from being a k-junta iff for every coordinate $i \in [k+1]$, we must change the value of $g(x)$ on at least $\epsilon 2^{k+1}$ different inputs $x \in \{0,1\}^{k+1}$ to make the function g independent of the ith variable – which is equivalent to requiring that at least $\epsilon 2^{k+1}$ pairs of i-twins have distinct values in g.

Consider a fixed $i \in [k+1]$. Since each value $g(x)$ is generated independently and takes value $g(x) = 1$ with probability 6ϵ, each pair of i-twins has distinct values with probability $2 \cdot 6\epsilon(1 - 6\epsilon)$. Let t_i represent the number of i-twins with distinct values in g. Then when $\epsilon \leq 1/12$, $\mathbf{E}[t_i] = 12\epsilon(1 - 6\epsilon)2^k \geq 6\epsilon 2^k$, and we can apply Chernoff's bound to obtain

$$\mathbf{Pr}\left[t_i \leq \epsilon 2^{k+1}\right] \leq e^{-6\epsilon 2^k(1-1/3)^2/2} = e^{-\epsilon 2^{k+2}/3} .$$

The Lemma then follows from the union bound and the conditions that $\epsilon > k/2^k$ and $k \geq 3$. □

Consider any sequence of q queries that a deterministic non-adaptive algorithm may make to a function f. We want to show that when q is small, the responses observed by the algorithm when f is drawn from $\mathcal{D}_{yes}$ are very similar to the responses observed when f is drawn from $\mathcal{D}_{no}$. The following Lemma provides a first step toward that goal.

Lemma 4.3. *Let Q be a set of q queries containing t_i i-twins. Let $\mathcal{R}_{yes}^{(i)}$ and $\mathcal{R}_{no}$ be the distributions of the responses to the queries in Q when the input function is drawn from $\mathcal{D}_{yes}^{(i)}$ or $\mathcal{D}_{no}$, respectively. Then the statistical distance between $\mathcal{R}_{yes}^{(i)}$ and $\mathcal{R}_{no}$ is bounded above by*

$$\sum_{y \in \{0,1\}^q} \left|\mathcal{R}_{yes}^{(i)}(y) - \mathcal{R}_{no}(y)\right| \leq 24t_i\epsilon.$$

Proof. We apply a hybridization argument. Let the pairs of i-twins in Q be represented by $(\alpha_1, \beta_1), \ldots, (\alpha_{t_i}, \beta_{t_i})$. For $j \in \{0, 1, \ldots, t_i\}$, define the response distribution $\mathcal{H}_j$ to be the distribution where each response is independent and 6ϵ-biased, except for the responses $\beta_1, \ldots, \beta_j$, which are constrained to satisfy $\alpha_1 = \beta_1, \ldots, \alpha_j = \beta_j$. Note that $\mathcal{H}_0 = \mathcal{R}_{no}$ and $\mathcal{H}_{t_i} = \mathcal{R}_{yes}^{(i)}$, so

$$\sum_y \left|\mathcal{R}_{yes}^{(i)}(y) - \mathcal{R}_{no}(y)\right| = \sum_y \left|\mathcal{H}_{t_i}(y) - \mathcal{H}_0(y)\right| \leq \sum_{j=1}^{t_i} \sum_y \left|\mathcal{H}_j(y) - \mathcal{H}_{j-1}(y)\right|.$$

The distributions $\mathcal{H}_j$ and $\mathcal{H}_{j-1}$ are nearly identical. The only difference between the two distributions is that β_j is constrained to take the value α_j in $\mathcal{H}_j$, while it is an independent 6ϵ-biased random variable in $\mathcal{H}_{j-1}$. So the statistical distance between $\mathcal{H}_j$ and $\mathcal{H}_{j-1}$ is twice the probability that $\beta_j \neq \alpha_j$ in $\mathcal{H}_{j-1}$. Thus, $\sum_y |\mathcal{H}_j(y) - \mathcal{H}_{j-1}(y)| \leq 24\epsilon(1 - 6\epsilon) < 24\epsilon$ and the Lemma follows. $\square$

With Lemma 4.3, we can now bound the statistical distance between the responses observed when the input function is drawn from $\mathcal{D}_{yes}$ or $\mathcal{D}_{no}$.

Lemma 4.4. *Let Q be a sequence of q queries containing t pairs of twins. Let $\mathcal{R}_{yes}$ and $\mathcal{R}_{no}$ be the distributions of the responses to the queries in Q when the input function is drawn from $\mathcal{D}_{yes}$ or $\mathcal{D}_{no}$, respectively. Then*

$$\sum_{y \in \{0,1\}^q} |\mathcal{R}_{yes}(y) - \mathcal{R}_{no}(y)| \leq \frac{24t\epsilon}{k+1} .$$

Proof. Since $\mathcal{R}_{yes}$ is a mixture distribution over $\mathcal{R}_{yes}^{(1)}, \ldots, \mathcal{R}_{yes}^{(k+1)}$, then

$$\sum_y |\mathcal{R}_{yes}^{(i)}(y) - \mathcal{R}_{no}(y)| = \sum_y |\mathcal{H}_{t_i}(y) - \mathcal{H}_0(y)| \leq \sum_{j=1}^{t_i} \sum_y |\mathcal{H}_j(y) - \mathcal{H}_{j-1}(y)|.$$

By Lemma 4.3, the above equation is upper bounded by $\frac{1}{k+1} \sum_{i=1}^{k+1} 24t_i\epsilon$, where t_i represents the number of i-twins in Q. Lemma 4.4 then follows from the fact that $t = \sum_{i=1}^{k+1} t_i$. $\square$

The previous Lemma bounds the statistical distance between the responses observed from a function drawn from $\mathcal{D}_{yes}$ or $\mathcal{D}_{no}$ when we have a bound on the number of twins in the queries. The following Lemma shows that the number of pairs of twins in a sequence of q queries can not be larger than $q \log q$.

Lemma 4.5. *Let $\{x_1, \ldots, x_q\} \subseteq \{0,1\}^n$ be a set of q distinct queries to a function $f : \{0,1\}^n \to \{0,1\}$. Then there are at most $q \log q$ pairs (x_i, x_j) such that x_i and x_j are twins.*

Proof. A natural combinatorial representation for a query $x \in \{0,1\}^n$ is as a vertex on the n-dimensional boolean hypercube. In this representation, a pair of twins corresponds to a pair of vertices connected by an edge on the hypercube. So the number of pairs of twins in a set of queries is equal to the number of edges contained in the corresponding subset of vertices on the hypercube. The Lemma then follows from the Edge-Isoperimetric Inequality of Harper [13], Bernstein [3], and Hart [14] (see also [7, §16]), which states that any subset S of q vertices in the boolean hypercube contains at most $q \log q$ edges.[5] $\square$

We can now combine the above Lemmas to prove Theorem 1.2.

---

[5] The result of Harper, Bernstein, and Hart is slightly tighter, giving a bound of $\sum_{i=1}^{q} h(i)$, where $h(i)$ is the number of ones in the binary representation of i.

Theorem 1.2. *Any non-adaptive algorithm for ϵ-testing k-juntas must make at least $\min\left(\Omega\left(\frac{k/\epsilon}{\log k/\epsilon}\right), \Omega\left(\frac{2^k}{k}\right)\right)$ queries.*

Proof. Let us first consider the case where $\epsilon \geq k/2^k$. Let $\mathcal{A}$ be any non-adaptive deterministic algorithm for testing k-juntas with $q = \frac{k/600\epsilon}{\log k/600\epsilon}$ queries. By Lemma 4.5, there can be at most $q \log q = \frac{k}{600\epsilon}$ pairs of twins in the q queries. By Lemma 4.4, this means that the statistical distance between the response distributions $\mathcal{R}_{yes}$ and $\mathcal{R}_{no}$ is at most $\frac{k}{600\epsilon} \cdot \frac{24\epsilon}{k+1} < \frac{1}{25}$. So the algorithm $\mathcal{A}$ can not predict which distribution generated a given input with accuracy greater than $\frac{1}{2} + \frac{1}{2} \cdot \frac{1}{25} = \frac{26}{50}$. By Lemma 4.2, a function drawn from $\mathcal{D}_{no}$ fails to be ϵ-far from being a k-junta with probability at most $\frac{1}{12}$. So the success rate of $\mathcal{A}$ is at most $\frac{26}{50} + \frac{1}{12} < \frac{2}{3}$. Therefore, by Yao's Minimax Principle, any algorithm for ϵ-testing k-juntas requires $\Omega\left(\frac{k/\epsilon}{\log k/\epsilon}\right)$ queries.

When $\epsilon < k/2^k$, we can repeat the above argument with $\epsilon' = k/2^k$ instead of ϵ. This yields a lower bound of $\Omega\left(\frac{k/\epsilon'}{\log k/\epsilon'}\right) = \Omega\left(\frac{2^k}{k}\right)$ queries. $\square$

5   Conclusion

Our results have improved the upper bound for the query complexity for testing juntas and the lower bound for testing juntas with non-adaptive algorithms. The results stated in this article are all presented in the context of testing functions with boolean domains, but we note that the results also generalize to the context of testing of functions $f : X^n \to \{0, 1\}$ for any finite domain X.

The results also suggest some interesting problems for future work.

Open Problem 5.1 *What is the query complexity of the junta testing problem? In particular, can we ϵ-tests k-juntas non-adaptively with $\tilde{O}(k/\epsilon)$ queries?*

Open Problem 5.1 has some relevance to the study of quantum algorithms in property testing: while Theorem 1.1 improves on all known upper bounds for the query complexity of classical algorithms for testing juntas, it still does not match the query complexity of $O(k/\epsilon)$ obtained by Atıcı and Servedio [1] for a non-adaptive algorithm with access to quantum examples.

Open Problem 5.2 *Is there a gap between the query complexity of adaptive and non-adaptive algorithms for testing juntas?*

Gonen and Ron [11] showed that such a gap exists for some property testing problems in the dense graph model. A positive answer to Open Problem 5.2 would provide an interesting example of a similar gap in the context of testing function properties.

Open Problem 5.3 *Can improved query bounds for testing juntas yield better bounds for testing other properties of boolean functions?*

The work of Diakonikolas et al. [9] strongly suggests a positive answer to Open Problem 5.3, since the junta test plays a central role in their generic algorithm for testing many properties of boolean functions.

Acknowledgments. The author wishes to thank Ryan O'Donnell for many valuable discussions and suggestions during the course of this research. The author also thanks Anupam Gupta, Yi Wu, and the anonymous referees for many helpful suggestions on earlier drafts of this article.

References

1. Atıcı, A., Servedio, R.A.: Quantum algorithms for learning and testing juntas. Quantum Information Processing 6(5), 323–348 (2007)
2. Bellare, M., Goldreich, O., Sudan, M.: Free bits, PCPs and non-approximability – towards tight results. SIAM J. Comput. 27(3), 804–915 (1998)
3. Bernstein, A.J.: Maximally connected arrays on the n-cube. SIAM J. Appl. Math. 15(6), 1485–1489 (1967)
4. Blum, A.: Relevant examples and relevant features: thoughts from computational learning theory. In: AAAI Fall Symposium on 'Relevance' (1994)
5. Blum, A.: Learning a function of r relevant variables. In: Proc. 16th Conference on Computational Learning Theory, pp. 731–733 (2003)
6. Blum, A., Langley, P.: Selection of relevant features and examples in machine learning. Artificial Intelligence 97(2), 245–271 (1997)
7. Bollobás, B.: Combinatorics, Cambridge (1986)
8. Chockler, H., Gutfreund, D.: A lower bound for testing juntas. Information Processing Letters 90(6), 301–305 (2004)
9. Diakonikolas, I., Lee, H.K., Matulef, K., Onak, K., Rubinfeld, R., Servedio, R.A., Wan, A.: Testing for concise representations. In: Proc. 48th Symposium on Foundations of Computer Science, pp. 549–558 (2007)
10. Fischer, E., Kindler, G., Ron, D., Safra, S., Samorodnitsky, A.: Testing juntas. J. Comput. Syst. Sci. 68(4), 753–787 (2004)
11. Gonen, M., Ron, D.: On the benefits of adaptivity in property testing of dense graphs. In: Proc. 11th Workshop RANDOM, pp. 525–539 (2007)
12. Guijarro, D., Tarui, J., Tsukiji, T.: Finding relevant variables in PAC model with membership queries. In: Proc. 10th Conference on Algorithmic Learning Theory, pp. 313–322 (1999)
13. Harper, L.H.: Optimal assignments of numbers to vertices. SIAM J. Appl. Math. 12(1), 131–135 (1964)
14. Hart, S.: A note on the edges of the n-cube. Disc. Math. 14, 157–163 (1976)
15. Kahn, J., Kalai, G., Linial, N.: The influence of variables on boolean functions. In: Proc. 29th Sym. on Foundations of Computer Science, pp. 68–80 (1988)
16. Lipton, R.J., Markakis, E., Mehta, A., Vishnoi, N.K.: On the Fourier spectrum of symmetric boolean functions with applications to learning symmetric juntas. In: Proc. 20th Conference on Computational Complexity, pp. 112–119 (2005)
17. Mossel, E., O'Donnell, R., Servedio, R.A.: Learning functions of k relevant variables. J. Comput. Syst. Sci. 69(3), 421–434 (2004)
18. Parnas, M., Ron, D., Samorodnitsky, A.: Testing basic boolean formulae. SIAM J. Discret. Math. 16(1), 20–46 (2003)
19. Rubinfeld, R., Sudan, M.: Robust characterizations of polynomials with applications to program testing. SIAM J. Comput. 25(2), 252–271 (1996)
20. Yao, A.C.: Probabilistic computations: towards a unified measure of complexity. In: Proc. 18th Sym. on Foundations of Comput. Sci., pp. 222–227 (1977)

The Complexity of Distinguishing Markov Random Fields

Andrej Bogdanov[1,*], Elchanan Mossel[2,**], and Salil Vadhan[3,***]

[1] Institute for Theoretical Computer Science, Tsinghua University
andrejb@tsinghua.edu.cn
[2] Dept. of Statistics and Dept. of Computer Sciences, U.C. Berkeley
mossel@stat.berkeley.edu
[3] School of Engineering and Applied Sciences, Harvard University
salil@eecs.harvard.edu

Abstract. Markov random fields are often used to model high dimensional distributions in a number of applied areas. A number of recent papers have studied the problem of reconstructing a dependency graph of bounded degree from independent samples from the Markov random field. These results require observing samples of the distribution at all nodes of the graph. It was heuristically recognized that the problem of reconstructing the model where there are hidden variables (some of the variables are not observed) is much harder.

Here we prove that the problem of reconstructing bounded-degree models with hidden nodes is hard. Specifically, we show that unless NP = RP,

- It is impossible to decide in randomized polynomial time if two models generate distributions whose statistical distance is at most 1/3 or at least 2/3.
- Given two generating models whose statistical distance is promised to be at least 1/3, and oracle access to independent samples from one of the models, it is impossible to decide in randomized polynomial time which of the two samples is consistent with the model.

The second problem remains hard even if the samples are generated efficiently, albeit under a stronger assumption.

1 Introduction

We study the computational complexity of reconstructing a Markov random field of bounded degree from independent and identically distributed samples at a subset of the nodes.

* This work was supported in part by the National Natural Science Foundation of China Grant 60553001, and the National Basic Research Program of China Grants 2007CB807900 and 2007CB807901.
** Supported by a Sloan fellowship in Mathematics, by NSF Career award DMS-0548249, NSF grant DMS-0528488 and ONR grant N0014-07-1-05-06.
*** Work done while visiting U.C. Berkeley, supported by the Miller Institute for Basic Research in Science, a Guggenheim Fellowship, US-Israel BSF grant 2002246, and ONR grant N00014-04-1-0478.

A. Goel et al. (Eds.): APPROX and RANDOM 2008, LNCS 5171, pp. 331–342, 2008.

The problem of reconstructing Markov random fields (MRF) has been recently considered as Markov random fields provide a a very general framework for defining high dimensional distributions. Much of the interest emanates from the use of such models in biology, see e.g. [1] and a list of related references [2].

Reconstructing Markov random fields where the generating model is a bounded-degree *tree* is one of the major computational problems in evolutionary biology, see e.g. [3,4]. For tree models the problem of sampling from a given model or calculating the probability of observing a specific sample for a given model are well known to be computationally feasible using simple recursions (also termed "dynamic programming" and "peeling"). Moreover, in the last decade it was shown that the problem of reconstructing a tree model given samples at a subset of the nodes is computationally feasible under mild non-degeneracy conditions, see e.g. [5,6,7] for some of the best results of this type. (These results often assume that the samples are observed at the leaves of the tree, but they easily extend to the case where some of the observables are internal nodes.)

Following extensive experimental work, Abbeel *et al.* [8] considered the problem of reconstructing bounded-degree (non-tree) graphical models based on factor graphs, and proposed an algorithm with polynomial time and sample complexity. The goal of their algorithm was not to reconstruct the true structure, but rather to produce a distribution that is close in Kullback-Leibler divergence to the true distribution.

In a more recent work [9], it was shown that the generating graph of maximal degree d on n nodes can be efficiently reconstructed in time $n^{O(d)}$ under mild non-degeneracy conditions. Other results on reconstructing the graph have appeared in [10].

Note that all of the results for non-tree models assume that there are no hidden variables. This is consistent with our results described next which show that the problem of reconstructing models with hidden variables is computationally hard.

1.1   Definitions and Main Results

Fix an alphabet Σ. An *undirected model* M over Σ^n consists of an undirected graph G with n vertices and a collection of weight functions $w_e : \Sigma^2 \to \mathbb{R}^{\geq 0}$, one for each edge $e \in E(G)$. The degree of the model is the degree of the underlying graph. To each undirected model M we associate the probability distribution μ_M on Σ^n given by

$$\Pr_{X \sim \mu_M}[X = a] = \frac{\prod_{(u,v) \in E(G)} w_{(u,v)}(a_u, a_v)}{Z_M} \tag{1}$$

where Z_M is the *partition function*

$$Z_M = \sum_{a \in \Sigma^n} \prod_{(u,v) \in E(G)} w_{(u,v)}(a_u, a_v).$$

This probability distribution μ_M is called the *Markov Random Field* of M. (Throughout, we will only work with models where $Z_M \neq 0$ so that μ_M is well-defined.)

As an example, consider the special case that $\Sigma = \{0, 1\}$ and all the weight functions are the NAND function. Then an assignment a has nonzero weight iff it is the characteristic vector of an independent set in the graph, Z_M counts the number of independent sets in the graph, and μ_M is the uniform distribution on the independent sets in the graph. For even this special case, it is NP-hard to compute Z_M given M is NP-hard, even approximately [11] and in bounded-degree graphs [12]. Due to the close connection between approximate counting and sampling [13], it follows that given a bounded-degree model M, it is infeasible to sample from the distribution μ_M (unless NP = RP). Here, we are interested in computational problems of the reverse type: given samples, determine M. Nevertheless, our techniques are partly inspired by the line of work on the complexity of counting and sampling.

We note that in standard definitions of Markov Random Fields, there is a weight function w_C for every *clique* C in the graph (not just edges), and the probability given to an assignment a is proportional to the product of the weights of all cliques in the graph. Our definition corresponds to the special case where all cliques of size greater than 2 have weight functions that are identically one. This restriction only makes our hardness results stronger. (Note that in bounded-degree graphs, there are only polynomially many cliques and they are all of bounded size, so our restriction has only a polynomial effect on the representation size.)

Markov Random fields model many stochastic processes. In several applications of interest one is given samples from the distribution μ_M and is interested in "reconstructing" the underlying model M. Often the observer does not have access to all the vertices of M, but only to a subset $V \subseteq \{1, \ldots, n\}$ of "revealed" vertices. We call this a model with *hidden nodes* $M \mid V$ and denote the corresponding distribution by $\mu_{M|V}$.

We are interested in the computational complexity of reconstructing the model M given samples from $\mu_{M|V}$. Of course, the model M may not be uniquely specified by $\mu_{M|V}$ (e.g. M may have a connected component that is disjoint from V), so one needs to formalize the question more carefully. Since we are interested in proving hardness results, we take a minimalist view of reconstruction: Any algorithm that claims to reconstruct M given samples from $\mu_{M|V}$ should in particular be able to distinguish two models M and M' when their corresponding distributions $\mu_{M|V}$ and $\mu_{M'|V}$ are statistically far apart.

As a first step towards understanding this question, we consider the following computational problem:

Problem. dDIST

INPUT: Two models M_0 and M_1 over Σ^n of degree d, a set $V \subseteq \{1, \ldots, n\}$.

PROMISE: Z_{M_0} and Z_{M_1} are nonzero.

YES INSTANCES: The statistical distance between $\mu_{M_0|V}$ and $\mu_{M_1|V}$ is at most $1/3$.

NO INSTANCES: The statistical distance between $\mu_{M_0|V}$ and $\mu_{M_1|V}$ is at least $2/3$.

Here, the *statistical distance* (a.k.a. total variation distance) between two distributions μ and ν on a set Ω is the quantity

$$\mathrm{sd}(\mu, \nu) = \max_{T \subseteq \Omega} |\mathrm{Pr}_{X \sim \mu}[X \in T] - \mathrm{Pr}_{X \sim \nu}[X \in T]|.$$

The computational problem dDIST, and all others we consider in this paper, are *promise problems*, which are decision problems where the set of inputs are restricted in some way, and we do not care what answer is given on inputs that are neither yes or no instances or violate the promise. Languages are special cases where all strings are either yes or no instances. For more about promise problems, see the survey by Goldreich [14].

Next, we consider a problem that seems much more closely related to (and easier than) reconstructing a model from samples. Here, the distinguisher is given two candidate models for some probabilistic process, as well as access to samples coming from this process. The goal of the distinguisher then is to say which is the correct model for this process.

Problem. dSAMP

INPUT: Two models M_0 and M_1 over Σ^n of degree d, a set $V \subseteq \{1, \ldots, n\}$.

PROMISE: Z_{M_0} and Z_{M_1} are nonzero, and the statistical distance between $\mu_0 = \mu_{M_0|V}$ and $\mu_1 = \mu_{M_1|V}$ is at least $1/3$.

PROBLEM: Given oracle access to a sampler S that outputs independent samples from either μ_0 or μ_1, determine which is the case.

More precisely, the distinguishing algorithm D is required to satisfy the condition

$$\Pr[D^{S_b}(M_0, M_1, V) = b] > 2/3 \qquad \text{for } b \in \{0, 1\} \tag{2}$$

where S_b denotes the sampler for μ_b and the probability is taken both over the randomness of the sampler and the randomness of D.

Our main results are that both of these problems are hard:

Theorem 1. *If there is a deterministic (resp., randomized) polynomial-time algorithm for 3DIST, then* NP = P *(resp.,* NP = RP*). This holds even if we restrict to models over the alphabet* $\Sigma = \{0, 1\}$.

Theorem 2. *If there is a randomized polynomial-time algorithm for 3SAMP, then* NP = RP. *This holds even if we restrict to models over the alphabet* $\Sigma = \{0, 1\}$.

These characterizations are the best possible: If NP = RP, both DIST and SAMP have efficient algorithms. See Appendix A.

The proofs of the two theorems are based on the fact that the Markov Random Field of a suitably chosen model can approximate the uniform distribution over satisfying assignments of an arbitrary boolean circuit. By revealing one node, we can then use an algorithm for either dDIST or dSAMP to distinguish the case that the first variable is 1 in all satisfying assignments from the case that the first variable is 0 in all satisfying assignments, which is an NP-hard problem.

2   Sampling Satisfying Assignments with a Markov Random Field

In this section, we establish the key lemma that is used in all of our hardness results — given a boolean circuit C, we can construct a model whose Markov Random Field corresponds to the uniform distribution on satisfying assignments of C.

Lemma 1. *There is a polynomial-time algorithm R that on input a circuit C : $\{0,1\}^n \to \{0,1\}$ produces an undirected model M of degree 3 over alphabet $\{0,1\}$ with a collection of special vertices $v_1, \ldots, v_n$ such that $Z_M \neq 0$ and if C is satisfiable, then the statistical distance between a random satisfying assignment of C and the Markov Random Field of M restricted to $v_1, \ldots, v_n$ is at most 2^{-n}.*

This proof is an extension of the standard reduction from circuit satisfiability to independent set: For each gate in the circuit and every possible assignment to the wires at this gate we have a vertex in the graph, and we put an edge between vertices corresponding to inconsistent assignments. (For the output gate, we remove those vertices corresponding to non-satisfying assignments.) Then the uniform distribution on *maximum* independent sets in the graph corresponds exactly to the uniform distribution on satisfying assignments in the circuit. However, the independent set model also gives weight to independent sets that are not maximum.

The weight corresponding to maximum independent sets can be magnified using the "blow-up" technique of [13,11], where we clone every vertex polynomially many times and replace each edge with complete bipartite graph between the clones of the endpoints. However, this results in a graph of polynomially large degree. In order to obtain a degree 3 model, we use the more general weight functions allowed in a Markov Random Field to achieve the same blow-up effect with many fewer edges. Specifically, we can force all clones of a vertex to have the same value by connecting them in a cycle with appropriate weight functions, and can also use the weights to magnify the weight of large sets. Then we can spread out the edges of the original graph among the clones in a way that the degree increases only by 1.

Proof. Consider the following polynomial-time algorithm that, on input a circuit C of size s, produces an undirected model M over alphabet $\{0,1\}$. We assume without loss of generality that each gate has fanin two and that all NOT gates are at the input level. For each gate g of C, including the input gates, and each consistent assignment α of values to the wires incident to this gate, the model M has $r = 8s$ vertices $v_{g,\alpha,1}, \ldots, v_{g,\alpha,r}$. (Note that for each gate g, there are at most $2^3 = 8$ possible assignments α.) For the output gate, we only consider assignments consistent with the circuit accepting. For every i, connect the vertices $v = v_{g,\alpha,i}$ and $u = v_{g,\alpha,i+1}$ by an edge with the following weighted "inner constraint":

$$
w_{in}(a_u, a_v) = \begin{cases} 1 & \text{if } a_u = a_v = 0 \\ 2 & \text{if } a_u = a_v = 1 \\ 0 & \text{otherwise.} \end{cases}
$$

For any pair of gates g, h where either $g = h$ or g and h are connected, and any pair of assignments α for g and β for h that are inconsistent, add the following "outer constraint' between $v = v_{g,\alpha,i}$ and $u = v_{h,\beta,j}$, where i (resp. j) is the first index that has not been used in any outer constraint for g (resp. h):

$$w_{out}(a_u, a_v) = \begin{cases} 0 & \text{if } a_u = a_v = 1 \\ 1 & \text{otherwise.} \end{cases}$$

The first type of constraint ensures that all representatives of the same gate-assignment pair are given the same value, and favors values that choose the assignment. The second type of constraint ensures that the assignments to the vertices of the model are consistent with circuit evaluation.

Assume that C is satisfiable, and look at the distribution induced by the Markov Random Field of M on the vertices $v_1, \ldots, v_n$, where $v_i = v_{x_i,1,1}$ represent the inputs of C. For every satisfying assignment α of C, consider the corresponding assignment α' of M that assigns value 1 to all vertices representing gate-assignment pairs consistent with the evaluation of C on input α, and 0 to all others. This gives α' relative weight 2^{sr} in the Markov Random Field.

We now argue that the combined weight of all other assignments of M cannot exceed $2^{-s} \cdot 2^{sr}$, and the claim follows easily from here. By construction, every assignment of M with nonzero weight assigns 1 to at most one group of vertices $v_{g,\alpha,1}, \ldots, v_{g,\alpha,r}$ for every gate g, and if the assignment does not represent a satisfying assignment of C then at least one gate must have no group assigned 1. For each group assigned 1, there are at most 8 ways to choose the assignment from each group, and each such assignment contributes a factor of 2^r to the weight, so the total weight of non-satisfying assignments is at most

$$\sum_{k=0}^{s-1} \binom{s}{k} \cdot (8 \cdot 2^r)^k \leq 2^s \cdot 8^s \cdot 2^{(s-1)r} \leq 2^{-s} \cdot 2^{sr}$$

by our choice of r.   □

3   Hardness of 3DIST and 3SAMP

In this section we prove Theorems 1 and 2. For both, we will reduce from the following NP-hard problem.

Problem. CKTDIST

INPUT: A circuit C (with AND, OR, NOT gates) over $\{0,1\}^n$.

PROMISE: C is satisfiable.

YES INSTANCES: All satisfying assignments of C assign the first variable 1.

NO INSTANCES: All satisfying assignments of C assign the first variable 0.

Lemma 2. *If* CKTDIST *has a polynomial-time (resp., randomized polynomial-time) algorithm, then* NP $=$ P *(resp.,* NP $=$ RP*).*

Proof. This follows from a result of Even, Selman, and Yacobi [15], who showed that given two circuits (C_0, C_1) where it is promised that exactly one is satisfiable, it is NP-hard to distinguish the case that C_0 is satisfiable from the case that C_1 is satisfiable. This problem is easily seen to be equivalent to CKTDIST by setting $C(b, x) = C_b(x)$. (The interest of [15] in this problem was the fact that it is in the promise-problem analogue NP ∩ coNP, whereas there cannot be NP-hard languages in NP ∩ coNP unless NP = coNP.)   □

Now we use Lemma 1 to reduce CKTDIST to 3DIST and 3SAMP.

Proof (of Theorem 1). To prove Theorem 1, let's assume for sake of contradiction that there is an efficient algorithm D for 3DIST. For simplicity, we assume that D is deterministic; the extension to randomized algorithms is straightforward.

Given a satisfiable circuit C, we will to use the distinguishing algorithm D to distinguish the case that all satisfying assignments assign the first variable 1 from the case that all satisfying assignments assign the first variable 0. First, using Lemma 1, we turn the circuit C into an undirected model M and let v be the variable corresponding to the first variable of C. Then $\mu_{M|\{v\}}$ is a Bernoulli random variable that outputs 1 with probability approximately equal (within $\pm 2^{-n}$) to the fraction of satisfying assignments that assign the first variable 1.

Next, let M' be any model where the node v is always assigned 1 in $\mu_{M'}$. (For example, we can have a single edge (u, v) with weight function $w_{(u,v)}$ $(a_u, a_v) = a_u a_v$.)

Then $\mu_{M|\{v\}}$ and $\mu_{M'|\{v\}}$ have statistical distance at most $2^{-n} \leq 1/3$ if C is a NO instance of CKTDIST, and have statistical distance at least $1 - 2^{-n} \geq 2/3$ if C is a YES instance. Thus, $D(M, M', \{v\})$ correctly decides CKTDIST, and NP = P.   □

Proof (of Theorem 2). Similarly to the previous proof, we reduce CKTDIST to 3SAMP: Given a circuit C, define the circuits $C_0(x_1, x_2, \ldots, x_n) = C(x_1, \ldots, x_n)$ and $C_1(x_1, x_2, \ldots, x_n) = C(\neg x_1, x_2, \ldots, x_n)$. Note that if all satisfying assignments of C assign the first variable value b, then all satisfying assignments to C_b assign the $x_1 = 0$ and all satisfying assignments to $C_{\neg b}$ assign $x_1 = 1$. Now, we apply Lemma 1 to construct models M_0 and M_1 corresponding to C_0 and C_1, and we reveal only the vertex $V = \{v_1\}$ corresponding to the variable x_1. (Note that $\mu_{M_0|V}$ and $\mu_{M_1|V}$ have statistical distance at least $1 - 2 \cdot 2^{-n}$.) Given a randomized algorithm A for 3SAMP, we run $A^S(M_0, M_1)$ where S is the sampler that always outputs 0. If all satisfying assignments of C assign $x_1 = b$ then S is 2^{-n}-close in statistical distance to $S_b \sim \mu_{M_b|V}$. Thus

$$\Pr[A^S(M_0, M_1) = b] \geq \Pr[A^{S_b}(M_0, M_1) = b] - \text{poly}(n) \cdot 2^{-n} \geq 2/3 - o(1)$$

and the construction gives a randomized algorithm for CKTDIST.   □

4   On the Samplability of the Models

One possible objection to the previous results is that the Markov Random Fields in question are not required to be samplable. In some of the applications we have in mind, the model represents a natural (physical, biological, sociological,...) process. If we believe that nature itself is a computationally efficient entity, then it makes sense to assume that the models we are trying to reconstruct will be efficiently samplable. It is natural to ask if the problem of distinguishing Markov Random Fields remains hard in this setting too.

Problem. EFFSAMP

INPUT: Two models M_0 and M_1 over Σ^n of degree d, a set $V \subseteq \{1, \ldots, n\}$, and a parameter s in unary.

PROMISE: Z_{M_0} and Z_{M_1} are nonzero, the statistical distance between $\mu_0 = \mu_{M_0|V}$ and $\mu_1 = \mu_{M_1|V}$ is at least $1/3$, and both μ_{M_0} and μ_{M_1} are 2^{-n}-close in statistical distance to distributions samplable by circuits of size at most s.

PROBLEM: Given oracle access to a sampler S that outputs independent samples from either μ_0 or μ_1, determine which is the case.

We have the following hardness result for EFFSAMP. Here CZK is the class of decision problems that have "computational zero-knowledge proofs". (See [16] for a definition.)

Theorem 3. *If* EFFSAMP *has a polynomial-time randomized algorithm, then* CZK = BPP.

A slightly weaker version of this theorem says that if EFFSAMP has a polynomial-time randomized algorithm, then one-way functions, or equivalently pseudorandom generators [17], do not exist. (See [16] for definitions of both one-way functions and pseudorandom generators.) To prove this, we observe that an algorithm for EFFSAMP can be used to break any candidate pseudorandom generator G: Convert G into an undirected model $M_0 \mid V$ using Lemma 1, and let $M_1 \mid V$ be a model whose Markov Random Field is uniform. Then the algorithm for EFFSAMP can be used to tell if a sample came from the pseudorandom generator or from the uniform distribution, thereby breaking the generator. Theorem 3 is stronger because it is known that if one-way functions exist, then CZK = PSPACE $\neq$ BPP [18,19,20].

To prove the actual theorem, we use a result of Ostrovsky and Wigderson [21], which says that if CZK $\neq$ BPP then there must exist an "auxiliary-input pseudorandom generator", which can also be broken by the same argument.

Proof. Suppose that CZK $\neq$ BPP. Then by Ostrovsky and Wigderson [21], there exists an *auxiliary-input one-way function*: This is a polynomial-time computable function $f : \{0,1\}^n \times \{0,1\}^n \to \{0,1\}^n$ such that for every polynomial p and polynomial-size circuit C, there exist infinitely many a such that

$$\Pr_{x \sim \{0,1\}^n}[f(a, C(a, f(a, x))) = f(a, x)] < 1/p(n)$$

where n is the length of a. By Håstad et al. [17], it follows that there is also an *auxiliary-input pseudorandom generator*: This is a polynomial-time computable function $G : \{0,1\}^n \times \{0,1\}^n \to \{0,1\}^{n+1}$ such that for every polynomial-size circuit family D and every polynomial p, there exist infinitely many a such that

$$\left| \Pr_{y \sim \{0,1\}^{n+1}}[D(a,y)] - \Pr_{x \sim \{0,1\}^n}[D(a,G(a,x))] \right| < 1/p(n).$$

It follows by a standard hybrid argument that for every polynomial-size oracle circuit D whose oracle provides independent samples from a given distribution we have that

$$\left| \Pr[D^U(a)] - \Pr[D^{G_a}(a)] \right| < 1/p(n).$$

for infinitely many a, where U is (a sampler for) the uniform distribution on $\{0,1\}^{n+1}$ and G_a is the output distribution of $G(a,x)$ when x is chosen uniformly from $\{0,1\}^n$. We show that if EFFSAMP has a polynomial-time randomized algorithm A, then for every polynomial-time computable G there is a circuit D such that

$$\left| \Pr[D^U(a)] - \Pr[D^{G_a}(a)] \right| > 1/4.$$

for every a. Fix an a of length n, let $C_a(x,y)$ be the circuit

$$C_a(x,y) = \begin{cases} 1 & \text{if } y = G(a,x) \\ 0 & \text{otherwise} \end{cases}$$

Apply Lemma 1 to circuit C_a to obtain a model M_a, and let V be the set of nodes of M_a corresponding to the input y of C_a. Then the Markov Random Field of M_a is 2^{-n} close to the distribution G_a. Let $M' \mid V$ be a model whose Markov Random Field is the uniform distribution over $\{0,1\}^{n+1}$. Then $D^?(a) = A^?(M_a, M')$ is the desired circuit. $\qquad\square$

Acknowledgments

We thank the anonymous referees for helpful comments on the presentation.

References

1. Friedman, N.: Infering cellular networks using probalistic graphical models. Science (2004)
2. Kasif, S.: Bayes networks and graphical models in computational molecular biology and bioinformatics, survey of recent research (2007),
   http://genomics10.bu.edu/bioinformatics/kasif/bayes-net.html
3. Felsenstein, J.: Inferring Phylogenies. Sinauer, New York (2004)
4. Semple, C., Steel, M.: Phylogenetics. Mathematics and its Applications series, vol. 22. Oxford University Press, Oxford (2003)
5. Erdös, P.L., Steel, M.A., Székely, L.A., Warnow, T.A.: A few logs suffice to build (almost) all trees (part 1). Random Structures Algorithms 14(2), 153–184 (1999)
6. Mossel, E.: Distorted metrics on trees and phylogenetic forests. IEEE Computational Biology and Bioinformatics 4, 108–116 (2007)

7. Daskalakis, C., Mossel, E., Roch, S.: Optimal phylogenetic reconstruction. In: Proceedings of the thirty-eighth annual ACM symposium on Theory of computing (STOC 2006), pp. 159–168 (2006)
8. Abbeel, P., Koller, D., Ng, A.Y.: Learning factor graphs in polynomial time and sampling complexity. Journal of Machine Learning Research 7, 1743–1788 (2006)
9. Bresler, G., Mossel, E., Sly, A.: Reconstruction of Markov random fields from samples: Some easy observations and algorithms. These proceedings (2008), http://front.math.ucdavis.edu/0712.1402
10. Wainwright, M.J., Ravikumar, P., Lafferty, J.D.: High dimensional graphical model selection using ℓ_1-regularized logistic regression. In: Proceedings of the NIPS (2006)
11. Sinclair, A.: Algorithms for Random Generation and Counting: A Markov chain Approach. In: Progress in Theoretical Computer Science. Birkhäuser, Basel (1993)
12. Luby, M., Vigoda, E.: Fast convergence of the Glauber dynamics for sampling independent sets. Random Struct. Algorithms 15(3–4), 229–241 (1999)
13. Jerrum, M., Valiant, L.G., Vazirani, V.V.: Random generation of combinatorial structures from a uniform distribution. Theor. Comput. Sci. 43, 169–188 (1986)
14. Goldreich, O.: On promise problems: a survey. In: Goldreich, O., Rosenberg, A.L., Selman, A.L. (eds.) Theoretical Computer Science. LNCS, vol. 3895, pp. 254–290. Springer, Heidelberg (2006)
15. Even, S., Selman, A.L., Yacobi, Y.: The complexity of promise problems with applications to public-key cryptography. Information and Control 61, 159–173 (1984)
16. Goldreich, O.: Foundations of cryptography (Basic tools). Cambridge University Press, Cambridge (2001)
17. Håstad, J., Impagliazzo, R., Levin, L.A., Luby, M.: A pseudorandom generator from any one-way function. SIAM Journal on Computing 28(4), 1364–1396 (1999)
18. Goldreich, O., Micali, S., Wigderson, A.: Proofs that yield nothing but their validity, or All languages in NP have zero-knowledge proof systems. Journal of the Association for Computing Machinery 38(3), 691–729 (1991)
19. Impagliazzo, R., Yung, M.: Direct minimum-knowledge computations (extended abstract). In: Pomerance, C. (ed.) CRYPTO 1987. LNCS, vol. 293, pp. 40–51. Springer, Heidelberg (1988)
20. Ben-Or, M., Goldreich, O., Goldwasser, S., Håstad, J., Kilian, J., Micali, S., Rogaway, P.: Everything provable is provable in zero-knowledge. In: Goldwasser, S. (ed.) CRYPTO 1988. LNCS, vol. 403, pp. 37–56. Springer, Heidelberg (1990)
21. Ostrovsky, R., Wigderson, A.: One-way functions are essential for non-trivial zero knowledge. In: Proc. 2nd Israel Symp. on Theory of Computing and Systems, pp. 3–17. IEEE Computer Society Press, Los Alamitos (1993)
22. Sahai, A., Vadhan, S.: A complete problem for statistical zero knowledge. Journal of the ACM 50(2), 196–249 (2003)
23. Babai, L., Moran, S.: Arthur-Merlin games: A randomized proof system and a hierarchy of complexity classes. Journal of Computer and System Sciences 36, 254–276 (1988)

A   Converse Theorem

Theorem 4. *If* NP $=$ RP, *then for every d there are randomized polynomial-time algorithms for dDIST and dSAMP.*

To prove Theorem 4, we use the following results of Jerrum, Valiant, Vazirani [13].

Theorem 5. *Assume* NP = RP. *Then there exists*

1. *A randomized polynomial-time sampling algorithm* Sample *that on input a satisfiable circuit* $C : \{0,1\}^n \to \{0,1\}$ *and* $\varepsilon > 0$ *(represented in unary), has an output distribution that is* ε-*close in statistical distance to the uniform distribution on the satisfying assignments of* C.
2. *A randomized polynomial-time sampling algorithm* Count *that on input a circuit* $C : \{0,1\}^n \to \{0,1\}$ *and* $\varepsilon > 0$ *(represented in unary) such that with high probability*

$$|C^{-1}(1)| \leq \text{Count}(C, \varepsilon) \leq (1 + \varepsilon)|C^{-1}(1)|$$

Now we assume NP = RP and describe algorithms for dDIST and dSAMP.

Algorithm for DIST: Using part (1) of Theorem 5, we can sample from a distribution close to the Markov Random Field $M \mid V$. To see this, consider the circuit C that takes as inputs an assignment $x \in \Sigma^n$, and numbers $t_e \in \mathbb{N}$, one for each edge e of M and outputs

$$C(x, e, w) = \begin{cases} 1, & \text{if } t_e \leq w_e(x_e) \text{ for all } e \\ 0, & \text{otherwise.} \end{cases}$$

Conditioned on $C(x, e, w) = 1$, for a uniformly chosen triple (x, e, w) the input $x \sim \Sigma^n$ follows exactly the distribution μ_M. Using the above theorem, there is then an algorithm which on input (M, V) outputs a sample from a distribution that is $1/9$-close (in statistical distance) to $\mu_{M|V}$. Let us use $C_{M,V}$ as the sampling circuit obtained by hardwiring M and V as inputs to the algorithm A.

Now given an input M_0, M_1, V for dDIST, we produce the circuits $C_0 = C_{M_0,V}$ and $C_1 = C_{M_1,V}$. Note that if $\text{sd}(\mu_0, \mu_1) > 2/3$ then the statistical distance between the output distributions of these two circuits is $> 2/3 - 1/9 = 5/9$, and if $\text{sd}(\mu_0, \mu_1) < 1/3$ then the distance is $< 1/3 + 1/9 = 4/9$. The problem of distinguishing circuits with large statistical distance from those with small statistical distance is known to be in the complexity class AM [22], which collapses to BPP under the assumption that NP = RP [23].

Algorithm for SAMP: First, we may assume that the statistical distance between the distributions μ_0 and μ_1 is as large as $9/10$: Instead of working with the original models, take 40 independent copies of each model; now each sample of this new model will correspond to 40 independent samples of the original model. The statistical distance increases from $1/3$ to $9/10$ by the following inequality:

Claim. Let μ, ν be arbitrary distributions, and μ^k, ν^k consist of k independent copies of μ, ν, respectively. Then

$$1 - \exp(k \cdot \text{sd}(\mu, \nu)^2 / 2) \leq \text{sd}(\mu^k, \nu^k) \leq k \cdot \text{sd}(\mu, \nu).$$

Using part (2) of Theorem 5, for every partial configuration $a \in \Sigma^V$, we can efficiently compute approximations $p_0(a), p_1(a)$ such that

$$p_0(a) \leq \mu_0(a) \leq 2p_0(a) \qquad \text{and} \qquad p_1(a) \leq \mu_1(a) \leq 2p_1(a),$$

where $\mu_i(a) = \Pr_{X \sim \mu_i}[X = a]$. Now consider the following algorithm D: On input M_0, M_1, V, generate a sample a from S, output 0 if $p_0(a) > p_1(a)$ and 1 otherwise. Then, assuming the counting algorithm of Theorem 5 returns the correct answer, we have:

$$\Pr[D^{S_0}(M_0, M_1, V) = 0] \geq \sum_{a: \mu_0(a) > 2\mu_1(a)} \mu_0(a)$$

$$\geq \sum_{a: \mu_0(a) > \mu_1(a)} \mu_0(a) - \sum_{a: 2\mu_1(a) \geq \mu_0(a) > \mu_1(a)} \mu_0(a).$$

The first term is at least as large as $\mathrm{sd}(\mu_0, \mu_1) \geq 9/10$. For the second term, we have

$$\sum_{a: 2\mu_1(a) \geq \mu_0(a) > \mu_1(a)} \mu_0(a) \leq \sum_{a: 2\mu_1(a) \geq \mu_0(a) > \mu_1(a)} 2\mu_1(a)$$

$$\leq 2 \cdot \sum_{a: \mu_0(a) > \mu_1(a)} \mu_1(a)$$

$$\leq 2 \cdot (1 - \mathrm{sd}(\mu_0, \mu_1)) = 1/5.$$

It follows that $\Pr[D^{S_0}(M_0, M_1, V) = 0] > 2/3$, and by the same argument $\Pr[D^{S_1}(M_0, M_1, V) = 1] > 2/3$.

Reconstruction of Markov Random Fields from Samples:
Some Observations and Algorithms

Guy Bresler[1,*], Elchanan Mossel[2,**], and Allan Sly[3,***]

[1] Dept. of Electrical Engineering and Computer Sciences, U.C. Berkeley
gbresler@eecs.berkeley.edu
[2] Dept. of Statistics and Dept. of Electrical Engineering and Computer Sciences,
U.C. Berkeley
mossel@stat.berkeley.edu
[3] Dept. of Statistics, U.C. Berkeley
sly@stat.berkeley.edu

Abstract. Markov random fields are used to model high dimensional
distributions in a number of applied areas. Much recent interest has
been devoted to the reconstruction of the dependency structure from
independent samples from the Markov random fields. We analyze a simple algorithm for reconstructing the underlying graph defining a Markov
random field on n nodes and maximum degree d given observations. We
show that under mild non-degeneracy conditions it reconstructs the generating graph with high probability using $\Theta(d \log n)$ samples which is
optimal up to a multiplicative constant. Our results seem to be the first
results for general models that guarantee that *the* generating model is
reconstructed. Furthermore, we provide an explicit $O(dn^{d+2} \log n)$ running time bound. In cases where the measure on the graph has correlation
decay, the running time is $O(n^2 \log n)$ for all fixed d. In the full-length
version we also discuss the effect of observing noisy samples. There we
show that as long as the noise level is low, our algorithm is effective. On
the other hand, we construct an example where large noise implies non-
identifiability even for generic noise and interactions. Finally, we briefly
show that in some cases, models with hidden nodes can also be recovered.

1 Introduction

In this paper we consider the problem of reconstructing the graph structure
of a Markov random field from independent and identically distributed samples.
Markov random fields (MRF) provide a very general framework for defining high
dimensional distributions and the reconstruction of the MRF from observations

---

* Supported by a Vodafone US-Foundation fellowship and NSF Graduate Research
Fellowship.
** Supported by a Sloan fellowship in Mathematics, by NSF Career award DMS-
0548249, NSF grant DMS-0528488 and ONR grant N0014-07-1-05-06.
*** Supported by NSF grants DMS-0528488 and DMS-0548249.

A. Goel et al. (Eds.): APPROX and RANDOM 2008, LNCS 5171, pp. 343–356, 2008.
© Springer-Verlag Berlin Heidelberg 2008

has attracted much recent interest, in particular in biology, see e.g. [9] and a list of related references [10].

1.1 Our Results

We give sharp, up to a multiplicative constant, estimates for the number of independent samples needed to infer the underlying graph of a Markov random field. In Theorem 2 we use a simple information-theoretic argument to show that $\Omega(d \log n)$ samples are required to reconstruct a randomly selected graph on n vertices with maximum degree at most d. Then in Theorems 4 and 5 we propose two algorithms for reconstruction that use only $O(d \log n)$ samples assuming mild non-degeneracy conditions on the probability distribution. The two theorems differ in their running time and the required non-degeneracy conditions. It is clear that non-degeneracy conditions are needed to insure that there is a unique graph associated with the observed probability distribution.

Chickering [2] showed that maximum-likelihood estimation of the underlying graph of a Markov random field is NP-complete. This does not contradict our results which assume that the data is generated from a model (or a model with a small amount of noise). Although the algorithm we propose runs in time polynomial in the size of the graph, the dependence on degree (the run-time is $O(dn^{d+2} \log n)$) may impose too high a computational cost for some applications. Indeed, for some Markov random fields exhibiting a decay of correlation a vast improvement can be realized: a modified version of the algorithm runs in time $O(dn^2 \log n)$. This is proven in Theorem 8.

In addition to the fully-observed setting in which samples of all variables are available, we extend our algorithm in several directions. These sections are omitted due to space constraints; we refer the reader to the full version [14] for the discussion on these topics. In Section 5 of [14] we consider the problem of noisy observations. We first show by way of an example that if some of the random variables are perturbed by noise then it is in general impossible to reconstruct the graph structure with probability approaching 1. Conversely, when the noise is relatively weak as compared to the coupling strengths between random variables, we show that the algorithms used in Theorems 4 and 5 reconstruct the graph with high probability. Furthermore, we study the problem of reconstruction with partial observations, i.e. samples from only a subset of the nodes are available, and provide sufficient conditions on the probability distribution for correct reconstruction.

1.2 Related Work

Chow and Liu [1] considered the problem of estimating Markov random fields whose underlying graphs are trees, and provided an efficient (polynomial-time) algorithm based on the fact that in the tree case maximum-likelihood estimation amounts to the computation of a maximum-weight spanning tree with edge weights equal to pairwise empirical mutual information. Unfortunately, their approach does not generalize to the estimation of Markov random fields whose graphs have cycles or hidden nodes. Much work in mathematical biology is devoted to reconstructing tree Markov fields when there are hidden nodes. For trees, given data that is generated from the model, the tree can be reconstructed efficiently from samples at a subset of the nodes given mild non-degeneracy conditions. See [12,13,11] for some of the most recent and tightest results in this setup.

Abbeel, et al [3] considered the problem of reconstructing graphical models based on factor graphs, and proposed a polynomial time and sample complexity algorithm. However, the goal of their algorithm was not to reconstruct the true structure, but rather to produce a distribution that is close in Kullback-Leibler divergence to the true distribution. In applications it is often of interest to reconstruct the true structure which gives some insight into the underlying structure of the inferred model.

Note furthermore that two networks that differ only in the neighborhood of one node will have $O(1)$ KL distance. Therefore, even in cases where it is promised that the KL distance between the generating distribution and any other distribution defined by another graph is as large as possible, the lower bounds on the KL distance is $\Omega(1)$. Plugging this into the bounds in [3] yields a polynomial sampling complexity in order to find the generating network compared to our logarithmic sampling complexity. For other work based on minimizing the KL divergence see the references in [3].

Essentially the same problem as in the present work (but restricted to the Ising model) was studied by Wainwright, et al [5], where an algorithm based on ℓ_1-regularization was introduced. In that work, sufficient conditions—different than ours—for correct reconstruction were given. They require a condition (called A2) where the neighborhood of every vertex is only weakly affected by their neighbors. Verifying when the condition holds seems hard and no example is given in the paper where the condition holds. The simulation studies in the paper are conducted for graphs consisting of small disconnected components. In this setting the running time of their algorithm is $O(n^5)$. The result [5] is best compared to our result showing that under standard decay of correlation (e.g., for models satisfying the Dobrushin condition, which is satisfied for the models simulated in their work), the running time of our algorithm is $O(n^2 \log n)$ as given in Theorem 8. The algorithm of [5] has suboptimal sample complexity, requiring $\Theta(d^5 \log n)$ samples for reconstruction.

Subsequent to our work being posted on the Arxiv, Santhanam and Wainwright [4] again considered essentially the same problem for the Ising model, producing nearly matching lower and upper bounds on the asymptotic sampling complexity. A key difference from our work is that they restrict attention to the Ising model, i.e. Markov random fields with pairwise potentials and where each variable takes two values. Also, they consider models with a fixed number of total edges, and arbitrary node degree, in contrast to our study of models with bounded node degrees and an arbitrary number of edges. We note that their results are limited to determining the information theoretic sampling complexity for reconstruction, and provide no efficient algorithm.

2   Preliminaries

We begin with the definition of Markov random field.

Definition 1. *On a graph $G = (V, E)$, a Markov random field is a distribution X taking values in $\mathcal{A}^V$, for some finite set $\mathcal{A}$ with $\mid \mathcal{A} \mid = A$, which satisfies the Markov property*

$$P(X(W), X(U) \mid X(S)) = P(X(W) \mid X(S))P(X(U) \mid X(S)) \qquad (1)$$

when W, U, and S are disjoint subsets of V such that every path in G from W to U passes through S and where $X(U)$ denotes the restriction of X from $\mathcal{A}^V$ to $\mathcal{A}^U$ for $U \subset V$.

Famously, by the Hammersley-Clifford Theorem, such distributions can be written in a factorized form as

$$P(\sigma) = \frac{1}{Z} \exp \left[\sum_a \Psi_a(\sigma_a) \right] , \qquad (2)$$

where Z is a normalizing constant, a ranges over the cliques in G, and $\Psi_a \colon \mathcal{A}^{|a|} \to \mathbb{R} \cup \{-\infty\}$ are functions called *potentials*.

The problem we consider is that of reconstructing the graph G, given k independent samples $\underline{X} = \{X^1, \ldots, X^k\}$ from the model. Denote by $\mathcal{G}_d$ the set of labeled graphs with maximum degree at most d. We assume that the graph $G \in \mathcal{G}_d$ is from this class. A structure estimator (or reconstruction algorithm) $\widehat{G} \colon \mathcal{A}^{kn} \to \mathcal{G}_d$ is a map from the space of possible sample sequences to the set of graphs under consideration. We are interested in the asymptotic relationship between the number of nodes in the graph, n, the maximum degree d, and the number of samples k that are required. An algorithm using number of samples $k(n)$ is deemed successful if in the limit of large n the probability of reconstruction error approaches zero.

3   Lower Bound on Sample Complexity

Suppose G is selected uniformly at random from $\mathcal{G}_d$. The following theorem gives a lower bound of $\Omega(d \log n)$ on the number of samples necessary to reconstruct the graph G. The argument is information theoretic, and follows by comparing the number of possible graphs with the amount of information available from the samples.

Theorem 2. *Let the graph G be drawn according to the uniform distribution on $\mathcal{G}_d$. Then there exists a constant $c = c(A) > 0$ such that if $k \leq cd \log n$ then for any estimator $\widehat{G} \colon \underline{X} \to \mathcal{G}_d$, the probability of correct reconstruction is $P(\widehat{G} = G) = o(1)$.*

Remark 1. Note that the theorem above doesn't need to assume anything about the potentials. The theorem applies for any potentials that are consistent with the generating graph. In particular, it is valid both in cases where the graph is "identifiable" given many samples and in cases where it isn't.

Proof. To begin, we note that the probability of error is minimized by letting $\widehat{G}$ be the maximum a posteriori (MAP) decision rule,

$$\widehat{G}_{\mathrm{MAP}}(\underline{X}) = \mathrm{argmax}_{g \in \mathcal{G}} P(G = g \mid \underline{X}).$$

By the optimality of the MAP rule, this bounds the probability of error using any estimator. Now, the MAP estimator $\widehat{G}_{\mathrm{MAP}}(\underline{X})$ is a deterministic function of $\underline{X}$. Clearly, if a graph g is not in the range of $\widehat{G}$ then the algorithm always makes an error when $G = g$. Let S be the set of graphs in the range of $\widehat{G}_{\mathrm{MAP}}$, so $P(\mathrm{error} \mid g \in S^c) = 1$. We have

$$
\begin{aligned}
P(\mathrm{error}) &= \sum_{g \in \mathcal{G}} P(\mathrm{error} \mid G = g) P(G = g) \\
&= \sum_{g \in S} P(\mathrm{error} \mid G = g) P(G = g) + \sum_{g \in S^c} P(\mathrm{error} \mid G = g) P(G = g) \\
&\geq \sum_{g \in S^c} P(G = g) = 1 - \sum_{g \in S} |\mathcal{G}|^{-1} \\
&\geq 1 - \frac{A^{nk}}{|\mathcal{G}|},
\end{aligned}
$$

(3)

where the last step follows from the fact that $|S| \leq |\underline{X}| \leq A^{nk}$. It remains only to express the number of graphs with max degree at most d, $|\mathcal{G}_d|$, in terms of the parameters n, d. The following lemma gives an adequate bound.

Lemma 3. *Suppose $d \leq n^\alpha$ with $\alpha < 1$. Then the number of graphs with max degree at most d, $|\mathcal{G}_d|$, satisfies*

$$\log |\mathcal{G}_d| = \Omega(nd \log n).$$

(4)

Proof. To make the dependence on n explicit, let $U_{n,d}$ be the number of graphs with n vertices with maximum degree at most d. We first bound $U_{n+2,d}$ in terms of $U_{n,d}$. Given a graph G with n vertices and degree at most d, add two vertices a and b. Select d distinct neighbors $v_1, \ldots, v_d$ for vertex a, with d labeled edges; there are $\binom{n}{d} d!$ ways to do this. If v_i already has degree d in G, then v_i has at least one neighbor u that is not a neighbor of a, since there are only $d - 1$ other neighbors of a. Remove the edge (v_i, u) and place an edge labeled i from vertex b to u. This is done for each vertex $v_1, \ldots, v_d$, so b has degree at most d. The graph G can be reconstructed from the resulting labeled graph on $n + 2$ vertices as follows: remove vertex a, and return the neighbors of b to their correct original neighbors (this is possible because the edges are labeled).

Removing the labels on the edges from a and b sends at most $d!^2$ edge-labeled graphs of this type on $n + 2$ vertices to the same unlabeled graph. Hence, the number of graphs with max degree d on $n + 2$ vertices is lower bounded as

$$U_{n+2,d} \geq U_{n,d} \binom{n}{d} d! \frac{1}{d!^2} = U_{n,d} \binom{n}{d} \frac{1}{d!}.$$

It follows that for n even (and greater than $2d + 4$)

$$U_{n,d} \geq \times_{i=1}^{n/2} \binom{n-2i}{d} \frac{1}{d!} \geq \left(\binom{n/2}{d} \frac{1}{d!} \right)^{n/4}. \tag{5}$$

If n is odd, it suffices to note that $U_{n+1,d} \geq U_{n,d}$. Taking the logarithm of equation (5) yields

$$\log U_{n,d} = \Omega(nd(\log n - \log d)) = \Omega(nd \log n), \tag{6}$$

assuming that $d \leq n^\alpha$ with $\alpha < 1$.   □

Together with equation (3), Lemma 3 implies that for small enough c, if the number of samples $k \leq cd \log n$, then

$$P(\text{error}) \geq 1 - \frac{A^{nk}}{|\mathcal{G}|} = 1 - o(1).$$

This completes the proof of Theorem 2.   □

4   Reconstruction

We now turn to the problem of reconstructing the graph structure of a Markov random field from samples. For a vertex v we let $N(v) = \{u \in V \setminus \{v\} : (u, v) \in E\}$ denote the set of neighbors of v. Determining the neighbors of v for every vertex in the graph is sufficient to determine all the edges of the graph and hence reconstruct the graph. Our algorithms reconstruct the graph by testing each candidate neighborhood of size at most d by using the Markov property, which states that for each $w \in V \setminus (N(v) \cup \{v\})$

$$P(X(v) \mid X(N(v)), X(w)) = P(X(v) \mid X(N(v))). \tag{7}$$

We give two algorithms for reconstructing networks; they differ in their non-degeneracy conditions and their running time. The first one, immediately below, has more stringent non-degeneracy conditions and faster running time.

4.1   Conditional Two Point Correlation Reconstruction

The first algorithm requires the following non-degeneracy condition:

Condition N1: There exist $\epsilon, \delta > 0$ such that for all $v \in V$, if $U \subset V \setminus \{v\}$ with $\mid U \mid \leq d$ and $N(v) \not\subset U$ then there exist values $x_v, x_w, x_w', x_{u_1}, \ldots, x_{u_l}$ such that for some $w \in V \setminus (U \cup \{v\})$

$$\begin{aligned} \big| P(X(v) = x_v \mid X(U) = x_U, X(w) = x_w) \\ - P(X(v) = x_v \mid X(U) = x_U, X(w) = x_w') \big| > \epsilon \end{aligned} \tag{8}$$

and

$$|P(X(U) = x_U, X(w) = x_w)| > \delta,$$
$$|P(X(U) = x_U, X(w) = x'_w)| > \delta. \qquad (9)$$

Remark 2. Condition (8) captures the notion that each edge should have sufficient strength. Condition (9) is required so that we can accurately calculate the empirical conditional probabilities.

We now describe the reconstruction algorithm, with the proof of correctness given by Theorem 4. In the following, $\widehat{P}$ denotes the empirical probability measure from the k samples.

Algorithm SIMPLERECON(*Input:* k i.i.d. samples from MRF; *Output:* estimated graph G)

- Initialize $E = \varnothing$.
- For each vertex v do
  - For each $U \subseteq V \setminus \{v\}$ with $|U| \leq d$, $w \in V \setminus (U \cup \{v\})$, and $x_1, \ldots, x_l, x_w, x'_w, x_v \in \mathcal{A}$
    - If
    $$|\widehat{P}(X(U) = x_U, X(w) = x_w)| > \delta/2$$
    and
    $$|\widehat{P}(X(U) = x_U, X(w) = x'_w)| > \delta/2 ,$$
    then compute
    $$r(U, x_U, w, x_w, x'_w) = |\widehat{P}(X(v) = x_v | X(U) = x_U, X(w) = x_w)$$
    $$- \widehat{P}(X(v) = x_v | X(U) = x_U, X(w) = x'_w)| .$$
  - Let $N(v)$ be the minimum cardinality U such that $\max_{x_U, w, x_w, x'_w} r(U, x_U, w, x_w, x'_w) < \epsilon/2$.
  - Add the edges incident to v: $E = E \cup \{(v, u) : u \in N(v)\}$.
- Return the graph $G = (V, E)$.

Run-time analysis. The analysis of the running time is straightforward. There are n nodes, and for each node we consider $O(n^d)$ neighborhoods U. For each candidate neighborhood, we check $O(n)$ nodes x_w and perform a correlation test of complexity $O(d \log n)$. The run-time of SIMPLERECON is thus $O(dn^{d+2} \log n)$ operations.

We now give the main theorem.

Theorem 4 (Correctness of SimpleRecon). *Suppose the MRF satisfies condition* **N1**. *Then with the constant* $C = \left(\frac{81(d+2)}{\epsilon^2 \delta^4 2d} + C_1 \right)$, *when* $k > Cd \log n$, *the estimator* SIMPLERECON *correctly reconstructs with probability at least* $1 - O(n^{-C_1})$.

Proof. Azuma's inequality gives that if $Y \sim \text{Bin}(k, p)$ then

$$P(\mid Y - kp \mid > \gamma k) \le 2 \exp(-2\gamma^2 k)$$

and so for any collection $U = \{u_1, \dots, u_l\} \subseteq V$ and $x_1, \dots, x_l \in \mathcal{A}$ we have

$$P\left(\left| \widehat{P}(X(U) = x_U) - P(X(U) = x_U) \right| \le \gamma \right) \le 2 \exp(-2\gamma^2 k). \tag{10}$$

There are $A^l \binom{n}{l} \le A^l n^l$ such choices of $u_1, \dots, u_l$ and $x_1, \dots, x_l$. An application of the union bound implies that with probability at least $1 - A^l n^l 2 \exp(-2\gamma^2 k)$ it holds that

$$\left| \widehat{P}(X(U) = x_U) - P(X(U) = x_U) \right| \le \gamma \tag{11}$$

for all $\{u_i\}_{i=1}^l$ and $\{x_i\}_{i=1}^l$. If we additionally have $l \le d+2$ and $k \ge C(\gamma)d \log n$, then equation (11) holds with probability at least $1 - A^{d+2} n^{d+2} 2/n^{2\gamma^2 C(\gamma)d}$. Choosing $C(\gamma) = \frac{d+2}{\gamma^2 2d} + C_1$, equation (11) holds with probability at least $1 - 2A^{d+2}/n^{C_1}$.

For the remainder of the proof assume (11) holds. Taking

$$\gamma(\epsilon, \delta) = \epsilon\delta^2/9 , \tag{12}$$

we can bound the error in conditional probabilities as

$$\mid \widehat{P}(X(v) = x_v \mid X(U) = x_U) - P(X(v) = x_v \mid X(U) = x_U) \mid$$

$$= \left| \frac{\widehat{P}(X(v) = x_v, X(U) = x_U)}{\widehat{P}(X(U) = x_U)} - \frac{P(X(v) = x_v, X(U) = x_U)}{P(X(U) = x_U)} \right|$$

$$\le \left| \frac{\widehat{P}(X(v) = x_v, X(U) = x_U)}{P(X(U) = x_U)} - \frac{P(X(v) = x_v, X(U) = x_U)}{P(X(U) = x_U)} \right|$$

$$+ \left| \frac{1}{\widehat{P}(X(U) = x_U)} - \frac{1}{P(X(U) = x_U)} \right|$$

$$\le \frac{\gamma}{\delta} + \frac{\gamma}{(\delta - \gamma)\delta} \le \frac{\epsilon\delta^2}{9\delta} + \frac{\epsilon\delta^2}{9(\delta - \frac{\epsilon\delta^2}{9})\delta} = \frac{\epsilon\delta}{9} + \frac{\epsilon}{(9 - \epsilon\delta)} < \frac{\epsilon}{4} . \tag{13}$$

For each vertex $v \in V$ we consider all candidate neighborhoods for v, subsets $U \subset V \setminus \{v\}$ with $\mid U \mid \le d$. The estimate (13) and the triangle inequality imply that if $N(v) \subseteq U$, then by the Markov property,

$$\left| \widehat{P}(X(v) = x_v \mid X(U) = x_U, X(w) = x_w) \right.$$
$$\left. - \widehat{P}(X(v) = x_v \mid X(U) = x_U, X(w) = x'_w) \right| < \epsilon/2 \tag{14}$$

for all $w \in V$ and $x_1, \dots, x_l, x_w, x'_w, x_v \in \mathcal{A}$ such that

$$\left| \widehat{P}(X(U) = x_U, X(w) = x_w) \right| > \delta/2,$$

$$\left| \widehat{P}(X(U) = x_U, X(w) = x'_w) \right| > \delta/2. \tag{15}$$

Conversely by condition **N1** and (9) and the estimate (13), we have that for any U with $N(v) \not\subseteq U$ there exists some $w \in V$ and $x_{u_1}, \ldots, x_{u_l}, x_w, x'_w, x_v \in \mathcal{A}$ such that equation (15) holds but equation (14) does not hold. Thus, choosing the smallest set U such that (14) holds gives the correct neighborhood.

To summarize, with number of samples

$$k = \left(\frac{81(d+2)}{\epsilon^2 \delta^4 2d} + C_1 \right) d \log n$$

the algorithm correctly determines the graph G with probability

$$P(\text{SIMPLERECON}(X) = G) \geq 1 - 2A^{d+2}/n^{C_1} .$$

$\square$

4.2   General Reconstruction

While the algorithm SIMPLERECON applies to a wide range of models, condition **N1** may occasionally be too restrictive. One setting in which condition **N1** does not apply is if the marginal spin at some vertex v is independent of the marginal spin at each of the other vertices, (i.e for all $u \in V \setminus \{v\}$ and all $x, y \in \mathcal{A}$ we have $P(X(v) = x, X(u) = y) = P(X(v) = x)P(X(u) = y)$. In this case the algorithm would incorrectly return the empty set for the neighborhood of v. The weaker condition for GENERALRECON holds on essentially all Markov random fields. In particular, (16) says that the potentials are non-degenerate, which is clearly a necessary condition in order to recover the graph. Equation (17) holds for many models, for example all models with soft constraints. This additional generality comes at a computational cost, with SIMPLERECON having a faster running time, $O(dn^{d+2} \log n)$, versus $O(dn^{2d+1} \log n)$ for GENERALRECON.

We use the following notation in describing the non-degeneracy conditions. For an assignment $x_U = (x_{u_1}, \ldots, x_{u_l})$ and $x'_{u_i} \in \mathcal{A}$, define

$$x_U^i(x'_{u_i}) = (x_{u_1}, \ldots, x'_{u_i}, \ldots, x_{u_l})$$

to be the assignment obtained from x_U by replacing the ith element by x'_{u_i}.

Condition N2: There exist $\epsilon, \delta > 0$ such that the following holds: for all $v \in V$, if $N(v) = u_1, \ldots, u_l$, then for each $i, 1 \leq i \leq l$ and for any set $W \subset V \setminus (v \cup N(v))$ with $\mid W \mid \leq d$ there exist values $x_v, x_{u_1}, \ldots, x_{u_i}, \ldots, x_{u_l}, x'_{u_i} \in \mathcal{A}$ and $x_W \in \mathcal{A}^{\mid W \mid}$ such that

$$\begin{aligned} \big| P(X(v) = x_v \mid X(N(v)) = x_{N(v)}) \\ - P(X(v) = x_v \mid X(N(v)) = x_{N(v)}^i(x'_{u_i})) \big| > \epsilon \end{aligned} \quad (16)$$

and

$$\begin{aligned} \mid P(X(N(v)) = x_{N(v)}, X(W) = x_W) \mid > \delta, \\ \mid P(X(N(v)) = x_{N(v)}^i(x'_{u_i}), X(W) = x_W) \mid > \delta. \end{aligned} \quad (17)$$

We now give the algorithm GENERALRECON.

Algorithm GENERALRECON(*Input:* k i.i.d. samples from MRF; *Output:* estimated graph G)

- Initialize $E = \varnothing$.
- For each vertex v do
  - Initialize $N(v) = \varnothing$.
  - For each $U \subseteq V \setminus \{v\}$ with $l = |U| \leq d$, $W \in V \setminus (U \cup \{v\})$ with $|W| \leq d$, each i, $1 \leq i \leq l$, and $x_v, x_W, x_U, x'_{u_i} \in \mathcal{A}$
    * If

$$\widehat{P}(X(W) = x_W, X(U) = x_U) > \delta/2$$
$$\widehat{P}(X(W) = x_W, X(U) = x^i_U(x'_{u_i})) > \delta/2$$

    then compute

$$r(U, W, i, x_v, x_W, x_U, x'_{u_i})$$
$$= \big| \widehat{P}(X(v) = x_v | X(W) = x_W, X(U) = x_U)$$
$$- \widehat{P}(X(v) = x_v | X(W) = x_W, X(U) = x^i_U(x'_{u_i})) \big|.$$

  - Let $N(v)$ be the maximum cardinality set U such that $\min_{W,i} \max_{x_v, x_W, x_U, x'_{u_i}} r(U, W, i, x_v, x_W, x_U, x'_{u_i}) > \epsilon/2$.
  - Add the edges incident to v: $E = E \cup \{(v, u) : u \in N(v)\}$.
- Return the graph $G = (V, E)$.

Run-time analysis. The analysis of the running time is similar to the previous algorithm. The run-time of GENERALRECON is $O(dn^{2d+1} \log n)$.

Theorem 5 (Correctness of GeneralRecon). *Suppose condition* **N2** *holds with ϵ and δ. Then for the constant $C = \frac{81(2d+1)}{\epsilon^2 \delta^4 2d} + C_1$, if $k > Cd \log n$ then the estimator* GENERALRECON *correctly reconstructs with probability at least $1 - O(n^{-C_1})$.*

Proof. As in Theorem 4 we have that with high probability

$$\left| \widehat{P}(X(U) = x_U) - P(X(U) = x_U) \right| \leq \gamma \tag{18}$$

for all $\{u_i\}_{i=1}^l$ and $\{x_i\}_{i=1}^l$ when $l \leq 2d+1$ and $k \geq C(\gamma)d \log n$; we henceforth assume that (18) holds. For each vertex $v \in V$ we consider all candidate neighborhoods for v, subsets $U = \{u_1, \ldots, u_l\} \subset V \setminus \{v\}$ with $0 \leq l \leq d$. For each candidate neighborhood U, the algorithm computes a score

$$f(v; \ U) =$$
$$min_{W,i} \ max_{x_v, x_W, x_U, x'_{u_i}} \big| \widehat{P}(X(v) = x_v \mid X(W) = x_W, X(U) = x_U)$$
$$- \widehat{P}(X(v) = x_v \mid X(W) = x_W, X(U) = x^i_U(x'_{u_i})) \big|,$$

where for each W, i, the maximum is taken over all x_v, x_W, x_U, x'_{u_i}, such that

$$\widehat{P}(X(W) = x_W, X(U) = x_U) > \delta/2 \tag{19}$$

$$\widehat{P}(X(W) = x_W, X(U) = x_U^i(x'_{u_i})) > \delta/2$$

and $W \subset V \setminus (U \cup \{v\})$ is an arbitrary set of nodes of size d, $x_W \in \mathcal{A}^d$ is an arbitrary assignment of values to the nodes in W, and $1 \leq i \leq l$.

The algorithm selects as the neighborhood of v the largest set $U \subset V \setminus \{v\}$ with $f(v; \ U) > \epsilon/2$. It is necessary to check that if U is the true neighborhood of v, then the algorithm accepts U, and otherwise the algorithm rejects U.

Taking $\gamma(\epsilon, \delta) = \epsilon \delta^2/9$, it follows exactly as in Theorem 4 that the error in each of the relevant empirical conditional probabilities satisfies

$$| \ \widehat{P}(X(v) = x_v \mid X(W) = x_W, X(U) = x_U)$$

$$- P(X(v) = x_v \mid X(W) = x_W, X(U) = x_U) \ | < \frac{\epsilon}{4} \ . \tag{20}$$

If $U \not\subset N(v)$, choosing $u_i \in U - N(v)$, we have when $N(v) \subset W \cup U$ that

$$\left| P(X(v) = x_v \mid X(W) = x_W, X(U) = x_U) \right.$$

$$\left. - P(X(v) = x_v \mid X(W) = x_W, X(U) = x_U^i(x'_{u_i})) \right|$$

$$= \left| P(X(v) = x_v \mid X(N(v)) = x_{N(v)}) - P(X(v) = x_v \mid X(N(v)) = x_{N(v)}) \right|$$

$$= 0 \ ,$$

by the Markov property (7). Assuming that equation (18) holds with γ chosen as in (12), the estimation error in $f(v; \ U)$ is at most $\epsilon/2$ by equation (20) and the triangle inequality, and it holds that $f(v; \ U) < \epsilon/2$ for each $U \not\subset N(v)$. Thus all $U \not\subset N(v)$ are rejected. If $U = N(v)$, then by the Markov property (7) and the conditions (16) and (17), for any i and $W \subset V$,

$$\left| P(X(v) = x_v \mid X(W) = x_W, X(U) = x_U) \right.$$

$$\left. - P(X(v) = x_v \mid X(W) = x_W, X(U) = x_U^i(x'_{u_i})) \right|$$

$$= \left| P(X(v) = x_v \mid X(N(v)) = x_{N(v)}) - P(X(v) = x_v \mid X(N(v)) = x_{N(v)}^i(x'_{u_i})) \right|$$

$$> \epsilon$$

for some x_v, x_W, x_U, x'_{u_i}. The error in $f(v; \ U)$ is less than $\epsilon/2$ as before, hence $f(v; \ U) > \epsilon/2$ for $U = N(v)$. Since $U = N(v)$ is the largest set that is not rejected, the algorithm correctly determines the neighborhood of v for every $v \in V$ when (18) holds.

To summarize, with number of samples

$$k = \left(\frac{81(2d + 1)}{\epsilon^2 \delta^4 2d} + C_1 \right) d \log n$$

the algorithm correctly determines the graph G with probability

$$P(\text{GENERALRECON}(X) = G) \geq 1 - 2A^{2d+1}/n^{C_1} \ .$$

$\square$

4.3   Non-degeneracy of Models

We can expect condition **N2** to hold in essentially all models of interest. The following proposition shows that the condition holds for any model with soft constraints.

Proposition 6 (Models with soft constraints). *In a graphical model with maximum degree d given by equation (2) suppose that all the potentials Ψ_{uv} satisfy $\|\Psi_{uv}\|_\infty \leq K$ and*

$$max_{x_1,x_2,x_3,x_4 \in \mathcal{A}} |\Psi_{uv}(x_1, x_2) - \Psi_{uv}(x_3, x_2) - \Psi_{uv}(x_1, x_4) + \Psi_{uv}(x_3, x_4)| > \gamma, \tag{21}$$

for some $\gamma > 0$. Then there exist $\epsilon, \delta > 0$ depending only on d, K and γ such that condition N2 holds.

Proof. It is clear that for some sufficiently small $\delta = \delta(d, m, K) > 0$ we have that for all $u_1, \ldots, u_{2d+1} \in V$ and $x_{u_1}, \ldots, x_{u_{2d+1}} \in \mathcal{A}$ that

$$P(X(u_1) = x_{u_1}, \ldots, X(u_{2d+1}) = x_{u_{2d+1}}) > \delta. \tag{22}$$

Now suppose that $u_1, \ldots, u_l$ is the neighborhood of v. Then for any $1 \leq i \leq l$ it follows from equation (21) that there exists $x_v, x_v', x_{u_i}, x_{u_i}' \in \mathcal{A}$ such that for any $x_{u_1} \ldots, x_{u_{i-1}}, x_{u_{i+1}}, \ldots, x_{u_l} \in \mathcal{A}$,

$$\frac{P(X(v) = x_v \mid X(u_1) = x_{u_1}, \ldots, X(u_i) = x_{u_i}', \ldots, X(u_l) = x_{u_l})}{P(X(v) = x_v' \mid X(u_1) = x_{u_1}, \ldots, X(u_i) = x_{u_i}', \ldots, X(u_l) = x_{u_l})}$$
$$\geq e^\gamma \frac{P(X(v) = x_v \mid X(u_1) = x_{u_1}, \ldots, X(u_i) = x_{u_i}, \ldots, X(u_l) = x_{u_l})}{P(X(v) = x_v' \mid X(u_1) = x_{u_1}, \ldots, X(u_i) = x_{u_i}, \ldots, X(u_l) = x_{u_l})}.$$

Combining with equation (22), equation (16) follows, showing that condition **N2** holds.   □

Although the results to follow hold more generally, for ease of exposition we will keep in mind the example of the Ising model with no external magnetic field,

$$P(x) = \frac{1}{Z} \exp\left(\sum_{(u,v) \in E} \beta_{uv} x_u x_v \right), \tag{23}$$

where $\beta_{uv} \in \mathbb{R}$ are coupling constants and Z is a normalizing constant.

The following lemma gives explicit bounds on ϵ and δ in terms of bounds on the coupling constants in the Ising model, showing that condition **N2** can be expected to hold quite generally.

Proposition 7. *Consider the Ising model with all parameters satisfying*

$$0 < c < |\beta_{ij}| < C$$

on a graph G with max degree at most d. Then condition N2 is satisfied with

$$\epsilon \geq \frac{\tanh(2c)}{2C^2 + 2C^{-2}} \quad and \quad \delta \geq \frac{e^{-4dC}}{2^{2d}}.$$

Proof. We refer the reader to the full version [14] for the proof.   □

4.4 $O(n^2 \log n)$ Algorithm for Models with Correlation Decay

The reconstruction algorithms SIMPLERECON and GENERALRECON run in polynomial time $O(dn^{d+2} \log n)$ and $O(dn^{2d+1} \log n)$, respectively. It would be desirable for the degree of the polynomial to be independent of d, and this can be achieved for Markov random fields with exponential decay of correlations. For two vertices $u, v \in V$, let $d(u, v)$ denote the graph distance and let $d_C(u, v)$ denote the correlation between the spins at u and v defined as

$$d_C(u, v) = \sum_{x_u, x_v \in \mathcal{A}} |P(X(u) = x_u, X(v) = x_v) - P(X(u) = x_u)P(X(v) = x_v)|.$$

If the interactions are sufficiently weak, the graph will satisfy the Dobrushin-Shlosman condition (see e.g. [8]) and there will be exponential decay of correlations between vertices, i.e. $d_C(u, v) \leq \exp(-\alpha d(u, v))$ for some $\alpha > 0$.

The following theorem shows that by restricting the candidate neighborhoods of the GENERALRECON algorithm to those nodes with sufficiently high correlation, one can achieve a run-time of $O(dn^2 \log n)$.

Theorem 8 (Reconstruction with correlation decay). *Suppose that G and X satisfy the hypothesis of Theorem 5 and that for all $u, v \in V$, $d_C(u, v) \leq \exp(-\alpha d(u, v))$ and there exists some $\kappa > 0$ such that for all $(u, v) \in E$, $d_C(u, v) > \kappa$. Then for some constant $C = C(\alpha, \kappa, \epsilon, \delta) > 0$, if $k > Cd \log n$ then there exists an estimator $\widehat{G}(\underline{X})$ such that the probability of correct reconstruction is $P(G = \widehat{G}(\underline{X})) = 1 - o(1)$ and the algorithm runtime is $O(nd^{\frac{d \ln(4/\kappa)}{\alpha}} + dn^2 \ln n)$ with high probability.*

Proof. Denote the correlation neighborhood of a vertex v as $N_C(v) = \{u \in V : \widehat{d_C}(u, v) > \kappa/2\}$ where $\widehat{d_C}(u, v)$ is the empirical correlation of u and v. For large enough C, with high probability for all $v \in V$, we have that $N(v) \subseteq N_C(v) \subseteq \{u \in V : d(u, v) \leq \frac{\ln(4/\kappa)}{\alpha}\}$. Now, we have the estimate $|\{u \in V : d(u, v) \leq \frac{\ln(4/\kappa)}{\alpha}\}| \leq d^{\frac{\ln(4/\kappa)}{\alpha}}$, which is independent of n.

When reconstructing the neighborhood of a vertex v we modify GENERALRECON to only test candidate neighborhoods U and sets W which are subsets of $N_C(v)$. The algorithm restricted to the smaller range of possible neighborhoods correctly reconstructs the graph since the true neighborhood of a vertex is always in its correlation neighborhood. For each vertex v the total number of choices of candidate neighborhoods U and sets W the algorithm has to check is $O(d^{\frac{d \ln(4/\kappa)}{\alpha}})$, so the reconstruction algorithm takes $O(nd^{\frac{d \ln(4/\kappa)}{\alpha}})$ operations. It takes $O(dn^2 \ln n)$ operations to calculate all the correlations, which for large n dominates the run time.   □

Acknowledgment. E.M. thanks Marek Biskup for helpful discussions on models with hidden variables.

References

1. Chow, C.K., Liu, C.N.: Approximating discrete probability distributions with dependence trees. IEEE Trans. Info. Theory IT-14, 462–467 (1968)
2. Chickering, D.: Learning Bayesian networks is NP-complete. In: Proceedings of AI and Statistics (1995)
3. Abbeel, P., Koller, D., Ng, A.: Learning factor graphs in polynomial time and sample complexity. Journal of Machine Learning Research 7, 1743–1788 (2006)
4. Santhanam, N., Wainwright, M.J.: Information-theoretic limits of graphical model selection in high dimensions (submitted, January 2008)
5. Wainwright, M.J., Ravikumar, P., Lafferty, J.D.: High-dimensional graphical model selection using ℓ_1-regularized logistic regression. In: NIPS 2006, Vancouver, BC, Canada (2006)
6. Baldassi, C., Braunstein, A., Brunel, N., Zecchina, R.: Efficient supervised learning in networks with binary synapses; arXiv:0707.1295v1
7. Mahmoudi, H., Pagnani, A., Weigt, M., Zecchina, R.: Propagation of external and asynchronous dynamics in random Boolean networks; arXiv:0704.3406v1
8. Dobrushin, R.L., Shlosman, S.B.: Completely analytical Gibbs fields. In: Fritz, J., Jaffe, A., Szasz, D. (eds.) Statistical mechanics and dynamical systems, pp. 371–403. Birkhauser, Boston (1985)
9. Friedman, N.: Infering cellular networks using probalistic graphical models. In: Science (February 2004)
10. Kasif, S.: Bayes networks and graphical models in computational molecular biology and bioinformatics, survey of recent research (2007),
    http://genomics10.bu.edu/bioinformatics/kasif/bayes-net.html
11. Daskalakis, C., Mossel, E., Roch, S.: Optimal phylogenetic reconstruction. In: STOC 2006: Proceedings of the 38th Annual ACM Symposium on Theory of Computing, pp. 159–168. ACM, New York (2006)
12. Erdős, P.L., Steel, M.A., Székely, L.A., Warnow, T.A.: A few logs suffice to build (almost) all trees (part 1). Random Struct. Algor. 14(2), 153–184 (1999)
13. Mossel, E.: Distorted metrics on trees and phylogenetic forests. IEEE/ACM Trans. Comput. Bio. Bioinform. 4(1), 108–116 (2007)
14. Bresler, G., Mossel, E., Sly, A.: Reconstruction of Markov Random Fields from Samples: Some Observations and Algorithms; arXiv:0712.1402v1

Tight Bounds for Hashing Block Sources*

Kai-Min Chung** and Salil Vadhan***

School of Engineering & Applied Sciences
Harvard University
Cambridge, MA
{kmchung,salil}@eecs.harvard.edu

Abstract. It is known that if a 2-universal hash function H is applied to elements of a *block source* $(X_1, \ldots, X_T)$, where each item X_i has enough min-entropy conditioned on the previous items, then the output distribution $(H, H(X_1), \ldots, H(X_T))$ will be "close" to the uniform distribution. We provide improved bounds on how much min-entropy per item is required for this to hold, both when we ask that the output be close to uniform in statistical distance and when we only ask that it be statistically close to a distribution with small collision probability. In both cases, we reduce the dependence of the min-entropy on the number T of items from $2 \log T$ in previous work to $\log T$, which we show to be optimal. This leads to corresponding improvements to the recent results of Mitzenmacher and Vadhan (SODA '08) on the analysis of hashing-based algorithms and data structures when the data items come from a block source.

1 Introduction

A *block source* is a sequence of items $\mathbf{X} = (X_1, \ldots, X_T)$ in which each item has at least some k bits of "entropy" conditioned on the previous ones [CG88]. Previous works [CG88, Zuc96, MV08] have analyzed what happens when one applies a 2-universal hash function to each item in such a sequence, establishing results of the following form:

Block-Source Hashing Theorems (informal): *If $(X_1, \ldots, X_T)$ is a block source with k bits of "entropy" per item and H is a random hash function from a 2-universal family mapping to $m \ll k$ bits, then $(H(X_1), \ldots, H(X_T))$ is "close" to the uniform distribution.*

In this paper, we prove new results of this form, achieving improved (in some cases, optimal) bounds on how much entropy k per item is needed to ensure that

---

* A full version of this paper can be found on [CV08].
** Work done when visiting U.C. Berkeley, supported by US-Israel BSF grant 2006060 and NSF grant CNS-0430336.
*** Work done when visiting U.C. Berkeley, supported by the Miller Institute for Basic Research in Science, a Guggenheim Fellowship, US-Israel BSF grant 2006060, and ONR grant N00014-04-1-0478.

A. Goel et al. (Eds.): APPROX and RANDOM 2008, LNCS 5171, pp. 357–370, 2008.
© Springer-Verlag Berlin Heidelberg 2008

the output is close to uniform, as a function of the other parameters (the output length m of the hash functions, the number T of items, and the "distance" from the uniform distribution). But first we discuss the two applications that have motivated the study of Block-Source Hashing Theorems.

1.1   Applications of Block-Source Hashing

Randomness Extractors. A *randomness extractor* is an algorithm that extracts almost-uniform bits from a source of biased and correlated bits, using a short *seed* of truly random bits as a catalyst [NZ96]. Extractors have many applications in theoretical computer science and have played a central role in the theory of pseudorandomness. (See the surveys [NT99, Sha04, Vad07].) Block-source Hashing Theorems immediately yield methods for extracting randomness from block sources, where the seed is used to specify a universal hash function. The gain over hashing the entire T-tuple at once is that the blocks may be much shorter than the entire sequence, and thus a much shorter seed is required to specify the universal hash function. Moreover, many subsequent constructions of extractors for general sources (without the block structure) work by first converting the source into a block source and performing block-source hashing.

Analysis of Hashing-Based Algorithms. The idea of hashing has been widely applied in designing algorithms and data structures, including hash tables [Knu98], Bloom filters [BM03], summary algorithms for data streams [Mut03], etc. Given a stream of data items $(x_1, \ldots, x_T)$, we first hash the items into $(H(x_1), \ldots, H(x_T))$, and carry out a computation using the hashed values. In the literature, the analysis of a hashing algorithm is typically a worst-case analysis on the input data items, and the best results are often obtained by unrealistically modelling the hash function as a truly random function mapping the items to uniform and independent m-bit strings. On the other hand, for realistic, efficiently computable hash functions (eg., 2-universal or $O(1)$-wise independent hash functions), the provable performance is sometimes significantly worse. However, such gaps seem to not show up in practice, and even standard 2-universal hash functions empirically seem to match the performance of truly random hash functions. To explain this phenomenon, Mitzenmacher and Vadhan [MV08] have suggested that the discrepancy is due to worst-case analysis, and propose to instead model the input items as coming from a block source. Then Block-Source Hashing Theorems imply that the performance of universal hash functions is close to that of truly random hash functions, provided that each item has enough bits of entropy.

1.2   How Much Entropy Is Required?

A natural question about Block-Source Hashing Theorems is: how large does the "entropy" k per item need to be to ensure a certain amount of "closeness" to uniform (where both the entropy and closeness can be measured in various ways). This also has practical significance for the latter motivation regarding hashing-based algorithms, as it corresponds to the amount of entropy we need

Table 1. Our Results: Each entry denotes the min-entropy (actually, Renyi entropy) required per item when hashing a block source of T items to m-bit strings to ensure that the output has statistical distance at most ε from uniform (or from having collision probability within a constant factor of uniform). Additive constants are omitted for readability.

Setting	Previous Results	Our Results
2-universal hashing ε-close to uniform	$m + 2\log T + 2\log(1/\varepsilon)$ [CG88, ILL89, Zuc96]	$m + \log T + 2\log(1/\varepsilon)$
2-universal hashing ε-close to small cp.	$m + 2\log T + \log(1/\varepsilon)$ [MV08]	$m + \log T + \log(1/\varepsilon)$
4-wise indep. hashing ε-close to small cp.	$\max\{m + \log T,$ $1/2(m + 3\log T + \log 1/\varepsilon)\}$ [MV08]	$\max\{m + \log T,$ $1/2(m + 2\log T + \log(1/\varepsilon))\}$

to assume in data items. In [MV08], they provide bounds on the entropy required for two measures of closeness, and use these as basic tools to bound the required entropy in various applications. The requirement is usually some small constant multiple of $\log T$, where T is the number of items in the source, which can be on the borderline between a reasonable and unreasonable assumption about real-life data. Therefore, it is interesting to pin down the optimal answers to these questions. In what follows, we first summarize the previous results, and then discuss our improved analysis and corresponding lower bounds.

A standard way to measure the distance of the output from the uniform distribution is by *statistical distance*.[1] In the randomness extractor literature, classic results [CG88, ILL89, Zuc96] show that using 2-universal hash functions, $k = m + 2\log(T/\varepsilon) + O(1)$ bits of min-entropy (or even Renyi entropy)[2] per item is sufficient for the output distribution to be ε-close to uniform in statistical distance. Sometimes a less stringent closeness requirement is sufficient, where we only require that the output distribution is ε-close to a distribution having "small" *collision probability*[3]. A result of [MV08] shows that $k = m + 2\log T + \log(1/\varepsilon) + O(1)$ suffices to achieve this requirement. Using 4-wise independent hash functions, [MV08] further reduce the required entropy to $k = \max\{m + \log T, 1/2(m + 3\log T + \log(1/\varepsilon))\} + O(1)$.

Our Results. We reduce the entropy required in the previous results, as summarized in Table 1. Roughly speaking, we save an additive $\log T$ bits of min-entropy (or Renyi entropy) for all cases. We show that using universal hash functions, $k = m + \log T + 2\log 1/\varepsilon + O(1)$ bits per item is sufficient for the output to be

---

[1] The *statistical distance* of two random variables X and Y is $\Delta(X, Y) = \max_T |\Pr[X \in T] - \Pr[Y \in T]|$, where T ranges over all possible events.

[2] The *min-entropy* of a random variable X is $H_\infty(X) = \min_x \log(1/\Pr[X = x])$. All of the results mentioned actually hold for the less stringent measure of *Renyi entropy* $H_2(X) = \log(1/E_{x \leftarrow X}[\Pr[X = x]])$.

[3] The *collision probability* of a random variable X is $\sum_x \Pr[X = x]^2$. By "small collision probability," we mean that the collision probability is within a constant factor of the collision probability of uniform distribution.

ε-close to uniform, and $k = m + \log(T/\varepsilon) + O(1)$ is enough for the output to be ε-close to having small collision probability. Using 4-wise independent hash functions, the entropy k further reduces to $\max\{m + \log T, 1/2(m + 2\log T + \log 1/\varepsilon)\} + O(1)$. The results hold even if we consider the joint distribution $(H, H(X_1), \ldots, H(X_T))$ (corresponding to "strong extractors" in the literature on randomness extractors). Substituting our improved bounds in the analysis of hashing-based algorithms from [MV08], we obtain similar reductions in the min-entropy required for every application with 2-universal hashing. With 4-wise independent hashing, we obtain a slight improvement for Linear Probing, and for the other applications, we show that the previous bounds can already be achieved with 2-universal hashing. The results are summarized in Table 2.

Although the $\log T$ improvement seems small, we remark that it could be significant for practical settings of parameter. For example, suppose we want to hash 64 thousand internet traffic flows, so $\log T \approx 16$. Each flow is specified by the 32-bit IP addresses and 16-bit port numbers for the source and destination plus the 8-bit transport protocol, for a total of 104 bits. There is a noticeable difference between assuming that each flow contains $3 \log T \approx 48$ vs. $4 \log T \approx 64$ bits of entropy as they are only 104 bits long, and are very structured.

We also prove corresponding lower bounds showing that our upper bounds are almost tight. Specifically, we show that when the data items have not enough entropy, then the joint distribution $(H, H(X_1), \ldots, H(X_T))$ can be "far" from uniform. More precisely, we show that if $k = m + \log T + 2\log 1/\varepsilon - O(1)$, then there exists a block source $(X_1, \ldots, X_T)$ with k bits of min-entropy per item such that the distribution $(H, H(X_1), \ldots, H(X_T))$ is ε-far from uniform in statistical distance (for H coming from any hash family). This matches our upper bound up to an additive constant. Similarly, we show that if $k = m + \log T - O(1)$, then there exists a block source $(X_1, \ldots, X_T)$ with k bits of min-entropy per item such that the distribution $(H, H(X_1), \ldots, H(X_T))$ is 0.99-far from having small collision probability (for H coming from any hash family). This matches our upper bound up to an additive constant in case the statistical distance parameter ε is constant; we also exhibit a specific 2-universal family for which the $\log(1/\varepsilon)$ in our upper bound is nearly tight — it cannot be reduced below $\log(1/\varepsilon) - \log\log(1/\varepsilon)$. Finally, we also extend all of our lower bounds to the case that we only consider distribution of hashed values $(H(X_1), \ldots, H(X_T))$, rather than their joint distribution with Y. For this case, the lower bounds are necessarily reduced by a term that depends on the size of the hash family. (For standard constructions of universal hash functions, this amounts to $\log n$ bits of entropy, where n is the bit-length of an individual item.)

Techniques. At a high level, all of the previous analyses for hashing block sources were loose due to summing error probabilities over the T blocks. Our improvements come from avoiding this linear blow-up by choosing more refined measures of error. For example, when we want the output to have small statistical distance from uniform, the classic Leftover Hash Lemma [ILL89] says that min-entropy $k = m + 2\log(1/\varepsilon_0)$ suffices for a single hashed block to be ε_0-close to uniform, and then a "hybrid argument" implies that the joint distribution of T hashed

Table 2. Applications: Each entry denotes the min-entropy (actually, Renyi entropy) required per item to ensure that the performance of the given application is "close" to the performance when using truly random hash functions. In all cases, the bounds omit additive terms that depend on how close a performance is desired, and we restrict to the (standard) case that the size of the hash table is linear in the number of items being hashed. That is, $m = \log T + O(1)$.

Type of Hash Family	Previous Results [MV08]	Our Results
Linear Probing		
2-universal hashing	$4 \log T$	$3 \log T$
4-wise independence	$2.5 \log T$	$2 \log T$
Balanced Allocations with d Choices		
2-universal hashing	$(d+2) \log T$	$(d+1) \log T$
4-wise independence	$(d+1) \log T$	—
Bloom Filters		
2-universal hashing	$4 \log T$	$3 \log T$
4-wise independence	$3 \log T$	—

blocks is $T\varepsilon_0$-close to uniform [Zuc96]. Setting $\varepsilon_0 = \varepsilon/T$, this leads to a min-entropy requirement of $k = m + 2\log(1/\varepsilon) + 2\log T$ per block. We obtain a better bound, reducing $2\log T$ to $\log T$, by using *Hellinger distance* to analyze the error accumulation over blocks, and only passing to statistical distance at the end.

For the case where we only want the output to be close to having small collision probability, the previous analysis of [MV08] worked by first showing that the expected collision probability of each hashed block $h(X_i)$ is "small" even conditioned on previous blocks, then using Markov's Inequality to deduce that each hashed block has small collision probability except with some probability ε_0, and finally doing a union bound to deduce that all hashed blocks have small collision probability except with probability $T\varepsilon_0$. We avoid the union bound by working with more refined notions of "conditional collision probability," which enable us to apply Markov's Inequality on the entire sequence rather than on each block individually.

The starting point for our negative results is the tight lower bound for randomness extractors due to Radhakrishnan and Ta-Shma [RT00]. Their methods show that if the min-entropy parameter k is not large enough, then for any hash family, there exists a (single-block) source X such that $h(X)$ is "far" from uniform (in statistical distance) for "many" hash functions h. We then take our block source $(X_1, \ldots, X_T)$ to consist of T iid copies of X, and argue that the statistical distance from uniform grows sufficiently fast with the number T of copies taken. For example, we show that if two distributions have statistical distance ε, then their T-fold products have statistical distance $\Omega(\min\{1, \sqrt{T} \cdot \varepsilon\})$, strengthening a previous bound of Reyzin [Rey04], who proved a bound of $\Omega(\min\{\varepsilon^{1/3}, \sqrt{T} \cdot \varepsilon\})$. Due to space constraints, we skip the precise statements and proofs of our negative results. Please refer to the full version of this paper [CV08] for details.

2   Preliminaries

Notations. All logs are based 2. We use the convention that $N = 2^n$, $K = 2^k$, and $M = 2^m$. We think of a data item X as a random variable over $[N] = \{1, \ldots, N\}$, which can be viewed as the set of n-bit strings. A hash function $h : [N] \to [M]$ hashes an item to a m-bit string. A *hash function family* $\mathcal{H}$ is a multiset of hash functions, and H will usually denote a uniformly random hash function drawn from $\mathcal{H}$. $U_{[M]}$ denotes the uniform distribution over $[M]$. Let $\mathbf{X} = (X_1, \ldots, X_T)$ be a sequence of data items. We use $X_{<i}$ to denote the first $i - 1$ items $(X_1, \ldots, X_{i-1})$. We refer to X_i as an item or a block interchangeably. Our goal is to study the distribution of hashed sequence $(H, \mathbf{Y}) = (H, Y_1, \ldots, Y_T) \overset{\text{def}}{=} (H, H(X_1), \ldots, H(X_T))$.

Hash Families. The *truly random hash family* $\mathcal{H}$ is the set of all functions from $[N]$ to $[M]$. A hash family $\mathcal{H}$ is *s-universal* if for every sequence of distinct elements $x_1, \ldots, x_s \in [N]$, $\Pr_H[H(x_1) = \cdots = H(x_s)] \leq 1/M^s$. $\mathcal{H}$ is *s-wise* independent if for every sequence of distinct elements $x_1, \ldots, x_s \in [N]$, $H(x_1), \ldots, H(x_s)$ are independent and uniform random variables over $[M]$.

Block Sources and Collision Probability. For a random variable X, the collision probability of X is $\mathrm{cp}(X) = \Pr[X = X'] = \sum_x \Pr[X = x]^2$, where X' is an independent copy of X. The *Renyi entropy* $H_2(X) = \log(1/\mathrm{cp}(X))$ can be viewed as a measure of the amount of randomness in X (In the randomness extractor literature, the entropy is measured by *min-entropy* $H_\infty(X) = \min_{x \in \mathrm{supp}(X)} \log(1/\Pr[X = x])$, but using the less stringent measure Renyi entropy makes our results stronger since $H_2(X) \geq H_\infty(X)$.) For an event E, $(X|_E)$ is the random variable defined by conditioning X on E.

Definition 2.1 (Block Sources). *A sequence of random variables $(X_1, \ldots, X_T)$ over $[N]^T$ is a block K-source if for every $i \in [T]$, and every $x_{<i}$ in the support of $X_{<i}$, we have $\mathrm{cp}(X_i|X_{<i} = x_{<i}) \leq 1/K$. That is, each item X_i has at least $k = \log K$ bits of Renyi entropy even after conditioning on the previous items.*

Let $\mathbf{X} = (X_1, \ldots, X_T)$ be a sequence of random variables over $[M]^T$. We are interested in bounding the overall collision probability $\mathrm{cp}(\mathbf{X})$ by the collision probability of each blocks. Suppose all X_i's are independent, then $\mathrm{cp}(\mathbf{X}) = \prod_{i=1}^T \mathrm{cp}(X_i)$. The following lemma generalizes Lemma 4.2 in [MV08], which says that if for every $\mathbf{x} \in \mathbf{X}$, the average collision probability of every block X_i conditioning on $X_{<i} = x_{<i}$ is small, then the overall collision probability $\mathrm{cp}(\mathbf{X})$ is also small. In particular, if $\mathbf{X}$ is a block K-source, then $\mathrm{cp}(\mathbf{X}) \leq 1/K^T$.

Lemma 2.2. *Let $\mathbf{X} = (X_1, \ldots, X_T)$ be a sequence of random variables such that for every $\mathbf{x} \in \mathrm{supp}(\mathbf{X})$,*

$$\frac{1}{T} \sum_{i=1}^T \mathrm{cp}(X_i|X_{<i} = x_{<i}) \leq \alpha.$$

Then the overall collision probability satisfies $\mathrm{cp}(\mathbf{X}) \leq \alpha^T$.

Proof. By Arithmetic Mean-Geometric Mean Inequality, the inequality in the premise implies

$$\prod_{i=1}^{T} \text{cp}(X_i | X_{<i} = x_{<i}) \leq \alpha^T.$$

Therefore, it suffices to prove

$$\text{cp}(\mathbf{X}) \leq \max_{\mathbf{x} \in \text{supp}(\mathbf{X})} \prod_{i=1}^{T} \text{cp}(X_i | X_{<i} = x_{<i}).$$

We prove it by induction on T. The base case $T = 1$ is trivial. Suppose the lemma is true for $T - 1$. We have

$$\text{cp}(\mathbf{X}) = \sum_{x_1} \Pr[X_1 = x_1]^2 \cdot \text{cp}(X_2, \ldots, X_T | X_1 = x_1)$$

$$\leq \left(\sum_{x_1} \Pr[X_1 = x_1]^2 \right) \cdot \max_{x_1} \text{cp}(X_2, \ldots, X_T | X_1 = x_1)$$

$$\leq \text{cp}(X_1) \cdot \max_{x_1} \left(\max_{x_2, \ldots, x_T} \prod_{i=2}^{T} \text{cp}(X_i | X_{<i} = x_{<i}) \right)$$

$$= \max_{\mathbf{x}} \prod_{i=1}^{T} \text{cp}(X_i | X_{<i} = x_{<i}),$$

as desired.

Statistical Distance. The statistical distance is a standard way to measure the distance of two distributions. Let X and Y be two random variables. The *statistical distance* of X and Y is $\Delta(X, Y) = \max_T |\Pr[X \in T] - \Pr[Y \in T]| = (1/2) \cdot \sum_x |\Pr[X = x] - \Pr[Y = x]|$, where T ranges over all possible events. When $\Delta(X, Y) \leq \varepsilon$, we say that X is ε-*close* to Y. Similarly, if $\Delta(X, Y) \geq \varepsilon$, then X is ε-*far* from Y. The following standard lemma says that if X has small collision probability, then X is close to uniform in statistical distance.

Lemma 2.3. *Let X be a random variable over $[M]$ such that $\text{cp}(X) \leq (1+\varepsilon)/M$. Then $\Delta(X, U_{[M]}) \leq \sqrt{\varepsilon}$.*

Conditional Collision Probability. Let (X, Y) be jointly distributed random variables. We can define the conditional Renyi entropy of X conditioning on Y as follows.

Definition 2.4. *The* conditional collision probability *of X conditioning on Y is $\text{cp}(X|Y) = \mathbb{E}_{y \leftarrow Y}[\text{cp}(X|_{Y=y})]$. The* conditional Renyi entropy *is $\text{H}_2(X|Y) = \log 1/\text{cp}(X|Y)$.*

The following lemma says that as in the case of Shannon entropy, conditioning can only decrease the entropy.

Lemma 2.5. *Let (X, Y, Z) be jointly distributed random variables. We have $\text{cp}(X) \leq \text{cp}(X|Y) \leq \text{cp}(X|Y, Z)$.*

Proof. For the first inequality, we have

$$\mathrm{cp}(X) = \sum_x \Pr[X = x]^2$$

$$= \sum_{y,y'} \Pr[Y = y] \cdot \Pr[Y = y'] \cdot \left(\sum_x \Pr[X = x|Y = y] \cdot \Pr[X = x|Y = y'] \right)$$

$$\leq \sum_{y,y'} \Pr[Y = y] \cdot \Pr[Y = y'] \cdot$$

$$\left(\sum_x \Pr[X = x|Y = y]^2 \right)^{1/2} \cdot \left(\sum_x \Pr[X = x|Y = y']^2 \right)^{1/2}$$

$$= \mathop{\mathrm{E}}_{y \leftarrow Y} \left[\mathrm{cp}(X|Y = y)^{1/2} \right]^2$$

$$\leq \mathrm{cp}(X|Y)$$

For the second inequality, observe that for every y in the support of Y, we have $\mathrm{cp}(X|_{Y=y}) \leq \mathrm{cp}((X|_{Y=y})|(Z|_{Y=y}))$ from the first inequality. It follows that

$$\mathrm{cp}(X|Y) = \mathop{\mathrm{E}}_{y \leftarrow Y}[\mathrm{cp}(X|_{Y=y})]$$

$$\leq \mathop{\mathrm{E}}_{y \leftarrow Y}[\mathrm{cp}((X|_{Y=y})|(Z|_{Y=y}))]$$

$$= \mathop{\mathrm{E}}_{y \leftarrow Y}[\mathop{\mathrm{E}}_{z \leftarrow (Z|Y=y)}[\mathrm{cp}(X|_{Y=y,Z=z})]]$$

$$= \mathrm{cp}(X|Y, Z)$$

3   Positive Results: How Much Entropy Is Sufficient?

In this section, we present our positive results, showing that the distribution of hashed sequence $(H, \mathbf{Y}) = (H, H(X_1), \ldots, H(X_T))$ is close to uniform when H is a random hash function from a 2-universal hash family, and $\mathbf{X} = (X_1, \ldots, X_T)$ has sufficient entropy per block. The new contribution is that we will not need $K = 2^k$ to be as large as in previous works, and so save the required randomness in the block source $\mathbf{X} = (X_1, \ldots, X_T)$.

3.1   Small Collision Probability Using 2-Universal Hash Functions

Let $H : [N] \to [M]$ be a random hash function from a 2-universal family $\mathcal{H}$. We first study the conditions under which $(H, \mathbf{Y}) = (H, H(X_1), \ldots, H(X_T))$ is ε-close to having collision probability $O(1/(|\mathcal{H}| \cdot M^T))$. This requirement is less stringent than $(H, \mathbf{Y})$ being ε-close to uniform in statistical distance, and so requires less bits of entropy. Mitzenmacher and Vadhan [MV08] show that this guarantee suffices for some hashing applications. They show that $K \geq MT^2/\varepsilon$ is enough to satisfy the requirement. We save a factor of T, and show that in fact, $K \geq MT/\varepsilon$, is sufficient. (Taking logs yields the first entry in Table 1, i.e. it suffices to have Renyi entropy $k = m + \log T + \log(1/\varepsilon)$ per block.) Formally, we prove the following theorem.

Theorem 3.1. *Let* $H : [N] \rightarrow [M]$ *be a random hash function from a 2-universal family* $\mathcal{H}$. *Let* $\mathbf{X} = (X_1, \ldots, X_T)$ *be a block K-source over* $[N]^T$. *For every* $\varepsilon > 0$, *the hashed sequence* $(H, \mathbf{Y}) = (H, H(X_1), \ldots, H(X_T))$ *is ε-close to a distribution* $(H, \mathbf{Z}) = (H, Z_1, \ldots, Z_T)$ *such that*

$$\mathrm{cp}(H, \mathbf{Z}) \le \frac{1}{|\mathcal{H}| \cdot M^T} \left(1 + \frac{M}{K\varepsilon}\right)^T.$$

In particular, if $K \ge MT/\varepsilon$, *then* $(H, \mathbf{Z})$ *has collision probability at most* $(1 + 2MT/K\varepsilon)/(|\mathcal{H}| \cdot M^T)$.

To analyze the distribution of the hashed sequence $(H, \mathbf{Y})$, the starting point is the following version of the Leftover Hash Lemma [BBR85, ILL89], which says that when we hash a random variable X with enough entropy using a 2-universal hash function H, the conditional collision probability of $H(X)$ conditioning on H is small.

Lemma 3.2 (The Leftover Hash Lemma). *Let* $H : [N] \rightarrow [M]$ *be a random hash function from a 2-universal family* $\mathcal{H}$. *Let X be a random variable over $[N]$ with* $\mathrm{cp}(X) \le 1/K$. *We have* $\mathrm{cp}(H(X)|H) \le 1/M + 1/K$.

We now sketch how the hashed block source $\mathbf{Y} = (Y_1, \ldots, Y_T) = (H(X_1), \ldots, H(X_T))$ is analyzed in [MV08], and how we improve the analysis. The following natural approach is taken in [MV08]. Since the data $\mathbf{X}$ is a block K-source, the Leftover Hash Lemma tells us that for every block $i \in [T]$, if we condition on the previous blocks $X_{<i} = x_{<i}$, then the hashed value $(Y_i|_{X_{<i}=x_{<i}})$ has small conditional collision probability, i.e. $\mathrm{cp}((Y_i|_{X_{<i}=x_{<i}})|H) \le 1/M + 1/K$. This is equivalent to saying that the average collision probability of $(Y_i|_{X_{<i}=x_{<i}})$ over the choice of the hash function H is small, i.e.,

$$\underset{h \leftarrow H}{\mathrm{E}}[\mathrm{cp}(h(X_i)|_{X_{<i}=x_{<i}})] = \mathrm{cp}((Y_i|_{X_{<i}=x_{<i}})|H) \le \frac{1}{M} + \frac{1}{K}.$$

We can then use a Markov argument to say that for every block, with probability at least $1 - \varepsilon/T$ over $h \leftarrow H$, the collision probability is at most $1/M + T/(K\varepsilon)$. We can then take a union bound to say that for every $\mathbf{x} \in \mathrm{supp}(\mathbf{X})$, at least $(1-\varepsilon)$-fraction of hash functions h are good in the sense that $\mathrm{cp}(h(X_i)|_{X_{<i}=x_{<i}})$ is small for all blocks $i = 1, \ldots, T$. [MV08] shows that if this condition is true for every $(h, \mathbf{x}) \in \mathrm{supp}(H, \mathbf{X})$, then $\mathbf{Y}$ is a block $(1/M + T/(K\varepsilon))$-source, and thus the overall collision probability is at most $(1 + MT/K\varepsilon)^T/M^T$. [MV08] also shows how to modify an ε-fraction of the distribution to fix the bad hash functions, and thus complete the analysis.

The problem of the above analysis is that taking a Markov argument for each block, and then taking a union bound incurs a loss of factor T. To avoid this, we want to apply Markov argument only once to the whole sequence. For example, a natural thing to try is to sum over blocks to get

$$\underset{h \leftarrow H}{\mathrm{E}}\left[\frac{1}{T}\sum_{i=1}^{T}\mathrm{cp}(h(X_i)|_{X_{<i}=x_{<i}})\right] = \frac{1}{T}\sum_{i=1}^{T}\mathrm{cp}((Y_i|_{X_{<i}=x_{<i}})|H) \le \frac{1}{M} + \frac{1}{K},$$

and use a Markov argument to deduce that for every $\mathbf{x} \in \mathrm{supp}(\mathbf{X})$, with probability $1 - \varepsilon$ over $h \leftarrow H$, the average collision probability per block satisfies

$$\frac{1}{T} \cdot \sum_{i=1}^{T} \mathrm{cp}(h(X_i)|_{X_{<i}=x_{<i}}) \leq \frac{1}{M} + \frac{1}{K\varepsilon}.$$

We need to bound the collision probability of $\mathbf{Y}$ using this information. We may try to apply Lemma 2.2, but it requires the information on $(1/T) \sum_i \mathrm{cp}(Y_i|_{Y_{<i}=y_{<i}})$ instead of $(1/T) \sum_i \mathrm{cp}(h(X_i)|_{X_{<i}=x_{<i}})$. That is, Lemma 2.2 requires us to condition on previous *hashed values* $Y_{<i}$, whereas the above argument refers to conditioning on the un-hashed values $X_{<i}$. The difficulty with directly reasoning about the former is that conditioned on the hashed values $Y_{<i}$, the hash function H may no longer be uniform (as it is correlated with $Y_{<i}$) and thus the Leftover Hash Lemma no longer applies.

To get around with the issues, we work with the averaged form of conditional collision probability $\mathrm{cp}(Y_i|H, Y_{<i})$, as from Definition 2.4. Our key observation is that now we can apply Lemma 2.5 to deduce that for every block $i \in [T]$, the conditional collision probability satisfies $\mathrm{cp}(Y_i|H, Y_{<i}) \leq \mathrm{cp}(Y_i|H, X_{<i}) \leq 1/M + 1/K$. Then, by a Markov argument, it follows that with probability $1 - \varepsilon$ over $(h, \mathbf{y}) \leftarrow (H, \mathbf{Y})$, the average collision probability satisfies

$$\frac{1}{T} \sum_{i=1}^{T} \mathrm{cp}(Y_i|_{(H,Y_{<i})=(h,y_{<i})}) \leq \frac{1}{M} + \frac{1}{K\varepsilon}.$$

We can then modify an ε-fraction of distribution, and apply Lemma 2.2 to complete the analysis.

The following lemma formalizes our claim about that the conditional collision probability of every block of $(H, \mathbf{Y})$ is small.

Lemma 3.3. *Let $H : [N] \to [M]$ be a random hash function from a 2-universal family $\mathcal{H}$. Let $\mathbf{X} = (X_1, \ldots, X_T)$ be a block K-source over $[N]^T$. Let $(H, \mathbf{Y}) = (H, H(X_1), \ldots, H(X_T))$. Then $\mathrm{cp}(H) = 1/|\mathcal{H}|$ and for every $i \in [T]$, $\mathrm{cp}(Y_i|H, Y_{<i}) \leq 1/M + 1/K$.*

Proof. $\mathrm{cp}(H) = 1/|\mathcal{H}|$ is trivial since H is the uniform distribution. Fix $i \in [T]$. By the definition of block K-source, for every $x_{<i}$ in the support of $X_{<i}$, $\mathrm{cp}(X_i|_{X_{<i}=x_{<i}}) \leq 1/K$. By the Leftover Hash Lemma, we have $\mathrm{cp}((Y_i|_{X_{<i}=x_{<i}})|(H|_{X_{<i}=x_{<i}})) \leq 1/M + 1/K$ for every $x_{<i}$. It follows that $\mathrm{cp}(Y_i|H, X_{<i}) \leq 1/M+1/K$. Now, we can think of $(Y_i|H, X_{<i})$ as Y_i first conditioning on $(H, Y_{<i})$, and then further conditioning on $X_{<i}$. By Lemma 2.5, we have

$$\mathrm{cp}(Y_i|H, Y_{<i}) \leq \mathrm{cp}(Y_i|H, Y_{<i}, X_{<i}) = \mathrm{cp}(Y_i|H, X_{<i}) \leq 1/M + 1/K,$$

as desired.

The remaining part of the proof follows the above sketch closely. Details can be found in the full version of this paper[CV08].

3.2 Small Collision Probability Using 4-Wise Independent Hash Functions

As discussed in [MV08], using 4-wise independent hash functions $H : [N] \to [M]$ from $\mathcal{H}$, we can further reduce the required randomness in the data $\mathbf{X} = (X_1, \ldots, X_T)$. [MV08] shows that in this case, $K \geq MT + \sqrt{2MT^3/\varepsilon}$ is enough for the hashed sequence $(H, \mathbf{Y})$ to be ε-close to having collision probability $O(1/|\mathcal{H}| \cdot M^T)$. As discussed in the previous subsection, by avoiding using union bounds, we show that $K \geq MT + \sqrt{2MT^2/\varepsilon}$ suffices. (Taking logs yields the second entry in Table 1, i.e. it suffices to have Renyi entropy $k = \max\{m + \log T, (1/2) \cdot (m + 2 \log T + \log(1/\varepsilon))\} + O(1)$ per block.) Formally, we prove the following theorem.

Theorem 3.4. *Let $H : [N] \to [M]$ be a random hash function from a 4-wise independent family $\mathcal{H}$. Let $\mathbf{X} = (X_1, \ldots, X_T)$ be a block K-source over $[N]^T$. For every $\varepsilon > 0$, the hashed sequence $(H, \mathbf{Y}) = (H, H(X_1), \ldots, H(X_T))$ is ε-close to a distribution $(H, \mathbf{Z}) = (H, Z_1, \ldots, Z_T)$ such that*

$$\mathrm{cp}(H, \mathbf{Z}) \leq \frac{1}{|\mathcal{H}| \cdot M^T} \left(1 + \frac{M}{K} + \sqrt{\frac{2M}{K^2 \varepsilon}} \right)^T .$$

In particular, if $K \geq MT + \sqrt{2MT^2/\varepsilon}$, then $(H, \mathbf{Z})$ has collision probability at most $(1 + \gamma)/(|\mathcal{H}| \cdot M^T)$ for $\gamma = 2 \cdot (MT + \sqrt{2MT^2/\varepsilon})/K$.

The improvement of Theorem 3.4 over Theorem 3.1 comes from that when we use 4-wise independent hash families, we have a concentration result on the conditional collision probability for each block . For the proof of the theorem, please refer to [CV08].

3.3 Statistical Distance to Uniform Distribution

Let $H : [N] \to [M]$ be a random hash function form a 2-universal family $\mathcal{H}$. Let $\mathbf{X} = (X_1, \ldots, X_T)$ be a block K-source over $[N]^T$. In this subsection, we study the statistical distance between the distribution of hashed sequence $(H, \mathbf{Y}) = (H, H(X_1), \ldots, H(X_T))$ and the uniform distribution $(H, U_{[M]^T})$. Classic results of [CG88, ILL89, Zuc96] show that if $K \geq MT^2/\varepsilon^2$, then $(H, \mathbf{Y})$ is ε-close to uniform. The proof idea is as follows. The Leftover Hash Lemma together with Lemma 2.3 tells us that the joint distribution of hash function and a hashed value $(H, Y_i) = (H, H(X_i))$ is $\sqrt{M/K}$-close to uniform $U_{[M]}$ even conditioning on the previous blocks $X_{<i}$. One can then use a hybrid argument to show that the distance grows linearly with the number of blocks, so $(H, \mathbf{Y})$ is $T \cdot \sqrt{M/K}$-close to uniform. Taking $K \geq MT^2/\varepsilon^2$ completes the analysis.

   We save a factor of T, and show that in fact, $K = MT/\varepsilon^2$ is sufficient. (Taking logs yields the third entry in Table 1, i.e. it suffices to have Renyi entropy $k = m + \log T + 2 \log(1/\varepsilon)$ per block.) Formally, we prove the following theorem.

Theorem 3.5. *Let $H : [N] \to [M]$ be a random hash function from a 2-universal family $\mathcal{H}$. Let $\mathbf{X} = (X_1, \ldots, X_T)$ be a block K-source over $[N]^T$. For every $\varepsilon > 0$ such that $K > MT/\varepsilon^2$, the hashed sequence $(H, \mathbf{Y}) = (H, H(X_1), \ldots, H(X_T))$ is ε-close to uniform $(H, U_{[M]^T})$.*

Recall that the previous analysis goes by passing to statistical distance first, and then measuring the growth of distance using statistical distance. This incurs a quadratic dependency of K on T. Since without further information, the hybrid argument is tight, to save a factor of T, we have to measure the increase of distance over blocks in another way, and pass to statistical distance only in the end. It turns out that the *Hellinger distance* (cf., [GS02]) is a good measure for our purposes:

Definition 3.6 (Hellinger distance). *Let X and Y be two random variables over $[M]$. The Hellinger distance between X and Y is*

$$d(X,Y) \overset{\text{def}}{=} \left(\frac{1}{2} \sum_i (\sqrt{\Pr[X=i]} - \sqrt{\Pr[Y=i]})^2 \right)^{1/2} = \sqrt{1 - \sum_i \sqrt{\Pr[X=i] \cdot \Pr[Y=i]}}.$$

Like statistical distance, Hellinger distance is a distance measure for distributions, and it takes value in $[0, 1]$. The following standard lemma says that the two distance measures are closely related. We remark that the lemma is tight in both directions even if Y is the uniform distribution.

Lemma 3.7 (cf., [GS02]). *Let X and Y be two random variables over $[M]$. We have*

$$d(X,Y)^2 \leq \Delta(X,Y) \leq \sqrt{2} \cdot d(X,Y).$$

In particular, the lemma allows us to upper-bound the statistical distance by upper-bounding the Hellinger distance. Since our goal is to bound the distance to uniform, it is convenient to introduce the following definition.

Definition 3.8 (Hellinger Closeness to Uniform). *Let X be a random variable over $[M]$. The Hellinger closeness of X to uniform $U_{[M]}$ is*

$$C(X) \overset{\text{def}}{=} \frac{1}{M} \sum_i \sqrt{M \cdot \Pr[X=i]} = 1 - d(X, U_{[M]})^2.$$

Note that $C(X, Y) = C(X) \cdot C(Y)$ when X and Y are independent random variables, so the Hellinger closeness is well-behaved with respect to products (unlike statistical distance). By Lemma 3.7, if the Hellinger closeness $C(X)$ is close to 1, then X is close to uniform in statistical distance. Recall that collision probability behaves similarly. If the collision probability $\mathrm{cp}(X)$ is close to $1/M$, then X is close to uniform. In fact, by the following normalization, we can view the collision probability as the 2-norm of X, and the Hellinger closeness as the $1/2$-norm of X.

Let $f(i) = M \cdot \Pr[X=i]$ for $i \in [M]$. In terms of $f(\cdot)$, the collision probability is $\mathrm{cp}(X) = (1/M^2) \cdot \sum_i f(i)^2$, and Lemma 2.3 says that if the "2-norm" $M \cdot \mathrm{cp}(X) = \mathrm{E}_i[f(i)^2] \leq 1 + \varepsilon$ where the expectation is over uniform $i \in [M]$, then

$\Delta(X, U) \leq \sqrt{\varepsilon}$,. Similarly, Lemma 3.7 says that if the "1/2-norm" $C(X) = \mathrm{E}_i[\sqrt{f(i)}] \geq 1 - \varepsilon$, then $\Delta(X, U) \leq \sqrt{\varepsilon}$.

We now discuss our approach to prove Theorem 3.5. We want to show that $(H, \mathbf{Y})$ is close to uniform. All we know is that the conditional collision probability $\mathrm{cp}(Y_i | H, Y_{<i})$ is close to $1/M$ for every block. If all blocks are independent, then the overall collision probability $\mathrm{cp}(H, \mathbf{Y})$ is small, and so $(H, \mathbf{Y})$ is close to uniform. However, this is not true without independence, since 2-norm tends to over-weight heavy elements. In contrast, the 1/2-norm does not suffer this problem. Therefore, our approach is to show that small conditional collision probability implies large Hellinger closeness. Formally, we have the following lemma. The main idea is to use Hölder's inequality to relate two different norms.

Lemma 3.9. *Let* $\mathbf{X} = (X_1, \ldots, X_T)$ *be jointly distributed random variables over* $[M_1] \times \cdots \times [M_T]$ *such that* $\mathrm{cp}(X_i | X_{<i}) \leq \alpha_i / M_i$ *for every* $i \in [T]$. *Then the Hellinger closeness satisfies*

$$C(\mathbf{X}) \geq \sqrt{\frac{1}{\alpha_1 \ldots \alpha_T}}.$$

The proof of this lemma can be found in the full version of this paper[CV08]. With this lemma, the proof of Theorem 3.5 is immediate.

Proof of Theorem 3.5. By Lemma 3.3, $\mathrm{cp}(H) = 1/|\mathcal{H}|$, and $\mathrm{cp}(Y_i | H, Y_{<i}) \leq (1 + M/K)/M$ for every $i \in [T]$. By Lemma 3.9, the Hellinger closeness satisfies $C(H, \mathbf{Y}) \geq (1 + M/K)^{-T/2} \geq 1 - MT/2K$ (recall that $K \geq MT/\varepsilon^2$). It follows by Lemma 3.7 that

$$\Delta((H, \mathbf{Y}), (H, U_{[M]^T})) \leq \sqrt{2} \cdot d((H, \mathbf{Y}), (H, U_{[M]^T}))$$
$$= \sqrt{2} \cdot \sqrt{1 - C(H, \mathbf{Y})} \leq \sqrt{MT/K} \leq \varepsilon.$$

■

Acknowledgments

We thank Wei-Chun Kao for helpful discussions in the early stages of this work, David Zuckerman for telling us about Hellinger distance, and Michael Mitzenmacher for suggesting parameter settings useful in practice.

References

[BBR85] Bennett, C.H., Brassard, G., Robert, J.-M.: How to reduce your enemy's information (extended abstract). In: Williams, H.C. (ed.) CRYPTO 1985. LNCS, vol. 218, pp. 468–476. Springer, Heidelberg (1986)

[BM03]   Broder, A.Z., Mitzenmacher, M.: Survey: Network applications of bloom filters: A survey. Internet Mathematics 1(4) (2003)

[CG88]   Chor, B., Goldreich, O.: Unbiased bits from sources of weak randomness and probabilistic communication complexity. SIAM J. Comput. 17(2), 230–261 (1988)

[CV08]   Chung, K.-M., Vadhan, S.: Tight bounds for hashing block sources (2008), http://www.citebase.org/abstract?id=oai:arXiv.org:0806.1948

[GS02]   Gibbs, A.L., Su, F.E.: On choosing and bounding probability metrics. International Statistical Review 70, 419 (2002)

[ILL89]   Impagliazzo, R., Levin, L.A., Luby, M.: Pseudo-random generation from one-way functions (extended abstracts). In: Proceedings of the Twenty First Annual ACM Symposium on Theory of Computing, Seattle, Washington, May 15–17, 1989, pp. 12–24 (1989)

[Knu98]   Knuth, D.E.: The art of computer programming. Sorting and Searching, vol. 3. Addison-Wesley Longman Publishing Co., Inc, Boston (1998)

[MV08]   Mitzenmacher, M., Vadhan, S.: Why simple hash functions work: Exploiting the entropy in a data stream. In: Proceedings of the 19th Annual ACM-SIAM Symposium on Discrete Algorithms (SODA 2008), January 20–22, 2008, pp. 746–755 (2008)

[Mut03]   Muthukrishnan, S.: Data streams: algorithms and applications. In: SODA, p. 413 (2003)

[NT99]   Nisan, N., Ta-Shma, A.: Extracting randomness: A survey and new constructions. J. Comput. Syst. Sci. 58(1), 148–173 (1999)

[NZ96]   Nisan, N., Zuckerman, D.: Randomness is linear in space. Journal of Computer and System Sciences 52(1), 43–52 (1996)

[RT00]   Radhakrishnan, J., Ta-Shma, A.: Bounds for dispersers, extractors, and depth-two superconcentrators. SIAM Journal on Discrete Mathematics 13(1), 2–24 (2000) (electronic)

[Rey04]   Reyzin, L.: A note on the statistical difference of small direct products. Technical Report BUCS-TR-2004-032, Boston University Computer Science Department (2004)

[Sha04]   Shaltiel, R.: Recent developments in extractors. In: Paun, G., Rozenberg, G., Salomaa, A. (eds.) Current Trends in Theoretical Computer Science. Algorithms and Complexity, vol. 1. World Scientific, Singapore (2004)

[Vad07]   Vadhan, S.: The unified theory of pseudorandomness. SIGACT News 38(3) (September 2007)

[Zuc96]   Zuckerman, D.: Simulating BPP using a general weak random source. Algorithmica 16(4/5), 367–391 (1996)

Improved Separations between Nondeterministic and Randomized Multiparty Communication

Matei David[1,*], Toniann Pitassi[1,**], and Emanuele Viola[2,***]

[1] Department of Computer Science, University of Toronto
[2] Computer Science Department, Columbia University

Abstract. We exhibit an explicit function $f : \{0,1\}^n \to \{0,1\}$ that can be computed by a nondeterministic number-on-forehead protocol communicating $O(\log n)$ bits, but that requires $n^{\Omega(1)}$ bits of communication for randomized number-on-forehead protocols with $k = \delta \cdot \log n$ players, for any fixed $\delta < 1$. Recent breakthrough results for the Set-Disjointness function (Sherstov, STOC '08; Lee Shraibman, CCC '08; Chattopadhyay Ada, ECCC '08) imply such a separation but only when the number of players is $k < \log \log n$.

We also show that for any $k = A \log \log n$ the above function f is computable by a small circuit whose depth is constant whenever A is a (possibly large) constant. Recent results again give such functions but only when the number of players is $k < \log \log n$.

1 Introduction

Number-on-forehead communication protocols are a fascinating model of computation where k collaborating players are trying to evaluate a function $f : (\{0,1\}^n)^k \to \{0,1\}$. The players are all-powerful, but the input to f is partitioned into k pieces of n bits each, $x_1, \ldots, x_k \in \{0,1\}^n$, and x_i is placed, metaphorically, on the forehead of player i. Thus, each player only sees $(k-1)n$ of the $k \cdot n$ input bits. In order to compute f, the players communicate by writing bits on a shared blackboard, and the complexity of the protocol is the number of bits that are communicated (i.e., written on the board). This model was introduced in [CFL83] and has found applications in a surprising variety of areas, including circuit complexity [HG91, NW93], pseudorandomness [BNS92], and proof complexity [BPS07].

In this model, a protocol is said to be *efficient* if it has complexity $\log^{O(1)} n$. Correspondingly, P_k^{cc}, RP_k^{cc}, BPP_k^{cc} and NP_k^{cc} are the number-on-forehead communication complexity analogs of the standard complexity classes [BFS86], see also [KN97]. For example, RP_k^{cc} is the class of functions having efficient one-sided-error randomized communication protocols. One of the most fundamental questions in NOF communication complexity, and the main question addressed in this paper, is to separate these classes. In [BDPW07], Beame et al. give an exponential separation between randomized and deterministic protocols for $k \leq n^{O(1)}$ players (in particular, $\mathsf{RP}_k^{cc} \neq \mathsf{P}_k^{cc}$ for

* Research supported by NSERC.
** Research supported by NSERC.
*** Research supported by grants NSF award CCF-0347282 and NSF award CCF-0523664.

A. Goel et al. (Eds.): APPROX and RANDOM 2008, LNCS 5171, pp. 371–384, 2008.
© Springer-Verlag Berlin Heidelberg 2008

$k \leq n^{O(1)}$). The breakthrough work by Sherstov [She07, She08a] sparked a flurry of exciting results in communication complexity [Cha07, LS08, CA08] which gave an exponential separation between nondeterministic and randomized protocols for $k < \log\log n$ players (in particular, $\text{NP}_k^{cc} \not\subseteq \text{BPP}_k^{cc}$ for $k < \log\log n$). Our main result is to improve the latter separation to larger values of k.

Theorem 1 (Main Theorem; $\text{NP}_{\delta\log n}^{cc} \not\subseteq \text{BPP}_{\delta\log n}^{cc}$). *For every fixed $\delta < 1$, sufficiently large n and $k = \delta \cdot \log n$, there is an explicit function $f : (\{0,1\}^n)^k \to \{0,1\}$ such that: f can be computed by k-player nondeterministic protocols communicating $O(\log n)$ bits, but f cannot be computed by k-player randomized protocols communicating $n^{o(1)}$ bits.*

We note that the number of players $k = \delta \cdot \log n$ in the above Theorem 1 is state-of-the-art: it is a major open problem in number-on-forehead communication complexity to determine if every explicit function on n bits can be computed by $k = \log_2 n$ players communicating $O(\log n)$ bits. We also note that Theorem 1 in particular implies an exponential separation between nondeterministic and deterministic protocols (hence, $\text{NP}_k^{cc} \not\subseteq \text{P}_k^{cc}$ for $k = \delta\log n$ players). Similar separations follow from [BDPW07], but only for non-explicit functions.

We also address the challenge of exhibiting functions computable by small (unbounded fan-in) constant-depth circuits that require high communication for k-player protocols, which is relevant to separating various circuit classes (see, [HG91, RW93]). Previous results [Cha07, LS08, CA08] give such functions for $k < \log\log n$. We offer a slight improvement and achieve $k = A\log\log n$ for any (possibly large) constant A, where the depth of the circuit computing the function depends on A.

Theorem 2 (Some constant-depth circuits require high communication). *For every constant $A > 1$ there is a constant B such that for sufficiently large n and $k := A\log\log n$ there is a function $f : (\{0,1\}^n)^k \to \{0,1\}$ which satisfies the following: f can be computed by circuits of size n^B and depth B, but f cannot be computed by k-player randomized protocols communicating $n^{o(1)}$ bits.*

1.1   Techniques

In this section we discuss the technical challenges presented by our theorems and how we have overcome them, building on previous work. An exposition of previous works and of some of the ideas in this paper also appears in the survey by Sherstov [She08b]. For concreteness, in our discussion we focus on separating nondeterministic from deterministic (as opposed to randomized) protocols, a goal which involves all the main difficulties.

Until very recently, it was far from clear how to obtain communication lower bounds in the number-on-forehead model for any explicit function f with efficient nondeterministic protocols. The difficulty can be described as follows. The standard method for obtaining number-on-forehead lower bounds is what can be called the "correlation method" [BNS92, CT93, Raz00, VW07].[1] This method goes by showing that f has

---

[1] This method is sometimes called the "discrepancy method." We believe that lower bound proofs are easier to understand when presented in terms of correlation rather than discrepancy, cf. [VW07].

exponentially small $(2^{-n^{\Omega(1)}})$ correlation with efficient (deterministic) protocols, and this immediately implies that f does not have efficient protocols (the correlation is w.r.t. some probability distribution which in general is not uniform). The drawback of this method is that, although for the conclusion that f does not have efficient protocols it is clearly enough to show that the correlation of f with such protocols is strictly less than one, the method actually proves the stronger exponentially small correlation bound. This is problematic in our setting because it is not hard to see that every function that has an efficient nondeterministic protocol also has *noticeable* $(\geq 2^{-\log^{O(1)}n})$ correlation with an efficient (deterministic) protocol, and thus this method does not seem useful for separating nondeterministic from deterministic protocols.

In recent work, these difficulties were overcome to obtain a surprising lower bound for a function with an efficient nondeterministic protocol: the Set-Disjointness function [LS08, CA08]. The starting point is the work by Sherstov [She08a] who applies the correlation method in a more general way for the 2-player model in order to overcome the above difficulties. This *generalized* correlation method is then adapted to handle more players $(k \gg 2)$ in [LS08, CA08]. The high-level idea of the method is as follows. Suppose that we want to prove that some specific function f does not have efficient protocols. The idea is to come up with another function f' and a distribution λ such that: (1) f and f' have constant correlation, say f and f' disagree on at most $1/10$ mass of the inputs with respect to λ, and (2) f' has exponentially small $(2^{-n^{\Omega(1)}})$ correlation with efficient protocols with respect to λ. The combination of (1) and (2) easily implies that f also has correlation at most $1/10 + 2^{-n^{\Omega(1)}} < 1$ with efficient protocols, which gives the desired lower bound for f. This method is useful because for f' we can use the correlation method, and on the other hand the correlation of f with efficient protocols is *not* shown to be exponentially small, only bounded away from 1 by a constant. Thus it is conceivable that f has efficient nondeterministic protocols, and in fact this is the case in [LS08, CA08] and in this work.

Although a framework similar to the above is already proposed in previous papers, e.g. [Raz87, Raz03], it is the work by Sherstov [She08a] that finds a way to successfully apply it to functions f with efficient nondeterministic protocols. For this, [She08a] uses two main ideas, generalized to apply to the number-on-forehead setting in [Cha07, LS08, CA08]. The first is to consider a special class of functions $f := \mathrm{Lift}(\mathrm{OR}, \phi)$ with efficient nondeterministic protocols. These are obtained by combining the "base" function OR on m bits with a "selection" function ϕ as described next. It is convenient to think of $f = \mathrm{Lift}(\mathrm{OR}, \phi)$ as a function on $(k+1)n$ bits distributed among $k+1$ players as follows: Player 0 receives an n-bit vector x, while Player i, for $1 \leq i \leq k$, gets an n-bit vector y_i. The selection function ϕ takes as input $y_1, \dots, y_k$ and outputs an m-bit subset of $\{1, \dots, n\}$. We view ϕ as selecting m bits of Player 0's input x, denoted $x|\phi(y_1, \dots, y_k)$. $\mathrm{Lift}(\mathrm{OR}, \phi)$ outputs the value of OR on those m bits of x:

$$\mathrm{Lift}(\mathrm{OR}, \phi)(x, y_1, \dots, y_k) := \mathrm{OR}(x|\phi(y_1, \dots, y_k)).$$

The second idea is to apply to such a function $f := \mathrm{Lift}(\mathrm{OR}, \phi)$ a certain orthogonality principle to produce a function f' that satisfies the points (1) and (2) above. The structure of $f = \text{Lift}(\text{OR}, \phi)(x, y_1, \ldots, y_k)$ is crucially exploited to argue that f' satisfies (2), and it is here that previous works require $k < \log\log n$ [Cha07, LS08, CA08].

So far we have rephrased previous arguments. We now discuss the main new ideas in this paper.

Ideas for the proof of Theorem 1. To prove Theorem 1 we start by noting that regardless of what function ϕ is chosen, $\text{Lift}(\text{OR}, \phi)$ has an efficient nondeterministic protocol: Player 0 simply guesses an index j that is one of the indices chosen by ϕ (she can do so because she knows the input to ϕ) and then any of the other players can easily verify whether or not x_j is 1 in that position. In previous work [LS08, CA08], ϕ is the bitwise AND function, and this makes $\text{Lift}(\text{OR}, \phi)$ the Set-Disjointness function. By contrast, *in this work we choose the function ϕ uniformly at random* and we argue that, for almost all ϕ, $\text{Lift}(\text{OR}, \phi)$ does not have efficient randomized protocols, whenever k is at most $\delta \log n$ for a fixed $\delta < 1$.

The above argument gives a *non-explicit* separation, due to the random choice of ϕ. To make it explicit, we derandomize the choice of ϕ. Specifically, we note that the above argument goes through as long as ϕ is 2^k-wise independent, i.e. as long as ϕ comes from a distribution such that for every 2^k fixed inputs $\bar{y}^1, \ldots, \bar{y}^{2^k} \in (\{0,1\}^n)^k$ the values $\phi(\bar{y}^1), \ldots, \phi(\bar{y}^{2^k})$ are uniform and independent (over the choice of ϕ). Known constructions of such distributions [ABI86, CG89] only require about $n \cdot 2^k = n^{O(1)}$ random bits, which can be given as part of the input. Two things should perhaps be stressed. The first is that giving a description of ϕ as part of the input does not affect the lower bound in the previous paragraph which turns out to hold even against protocols that depend on ϕ. The second is that, actually, using 2^k-wise independence seems to add the constraint $k < 1/2(\log n)$; to achieve $k = \delta \log n$ for every $\delta < 1$ we use a distribution on ϕ that is *almost 2^k-wise independent* [NN93].

Ideas for the proof of Theorem 2. To prove Theorem 2 we show how to implement the function given by Theorem 1 by small constant-depth circuits when k is $A \log\log n$ for a fixed, possibly large, constant A. In light of the above discussion, this only requires computing a 2^k-wise independent function by small constant-depth circuits, a problem which is studied in [GV04, HV06]. Specifically, dividing up ϕ in blocks it turns out that it is enough to compute 2^k-wise independent functions $g : \{0,1\}^t \to \{0,1\}^t$ where t is also about 2^k. When $k = A\log\log n$, g is a $(2^k = \log^A n)$-wise independent function on $\log^A n$ bits, and [HV06] shows how to compute it with circuits of size n^B and depth B where B depends on A only – and this dependence of B on A is tight even for almost 2-wise independence. This gives Theorem 2. Finally, we note that [HV06] gives explicit (a.k.a. uniform) circuits, and that we are not aware of an alternative to [HV06] even for non-explicit circuits.

Organization. The organization of the paper is as follows. In Section 2 we give necessary definitions and background. We present the proof of our main result Theorem 1 in two stages. First, in Section 3 we present a non-explicit separation obtained by selecting ϕ at random. Then, in Section 4 we derandomize the choice of ϕ in order to give an explicit separation and prove Theorem 1. Finally, in Section 5 we prove our results about constant-depth circuits, Theorem 2.

2 Preliminaries

Correlation. Let $f, g : X \to \mathbb{R}$ be two functions, and let μ be a distribution on X. We define the *correlation between* f *and* g *under* μ to be $\mathrm{Cor}_\mu(f, g) := \mathbb{E}_{x \sim \mu}[f(x)g(x)]$. Let $\mathcal{G}$ be a class of functions $g : X \to \mathbb{R}$ (e.g. efficient communication protocols). We define the *correlation between* f *and* $\mathcal{G}$ *under* μ to be $\mathrm{Cor}_\mu(f, \mathcal{G}) := \max_{g \in \mathcal{G}} \mathrm{Cor}_\mu(f, g)$. Note that, whenever $\mathcal{G}$ is closed under complements, which will always be the case in this paper, this correlation is non-negative. Whenever we omit to mention a specific distribution when computing the correlation, an expected value or a probability, it is to be assumed that we are referring to the uniform distribution, which we denote by $\mathcal{U}$.

Communication Complexity. In the number-on-forehead (NOF) multiparty communication complexity model [CFL83], k players are trying to collaborate to compute a function $f : X_1 \times \ldots \times X_k \to \{-1, 1\}$. For each i, player i knows the values of all of the inputs $(x_1, \ldots, x_k) \in X_1 \times \ldots \times X_k$ except for x_i (which conceptually is thought of as being placed on Player i's forehead). The players exchange bits according to an agreed-upon protocol, by writing them on a public blackboard. A *protocol* specifies what each player writes as a function of the blackboard content and the inputs seen by that player, and whether the protocol is over, in which case the last bit written is taken as the output of the protocol. The *cost* of a protocol is the maximum number of bits written on the blackboard.

In a *deterministic protocol*, the blackboard is initially empty. A *randomized protocol* is a distribution on deterministic protocols such that for every input a protocol selected at random from the distribution errs with probability at most $1/3$. In a *nondeterministic protocol*, an initial guess string is written on the blackboard at the beginning of the protocol (and counted towards communication) and the players are trying to verify that the output of the function is -1 (representing *true*) in the usual sense: There exists a guess string where the output of the protocol is -1 if and only if the output of the function is -1. The *communication complexity* of a function f under one of the above types of protocols is the minimum cost of a protocol of that type computing f. In line with [BFS86], a k-player protocol computing $f : (\{0, 1\}^n)^k \to \{-1, 1\}$ is considered to be *efficient* if its cost is at most poly-logarithmic, $\log^{O(1)} n$. Equipped with the notion of efficiency, one has the NOF communication complexity classes BPP_k^{cc} and NP_k^{cc} that are analogues of the corresponding complexity classes.

Definition 1. *We denote by* $\Pi^{k,c}$ *the class of all deterministic k-player NOF communication protocols of cost at most c.*

The following immediate fact allows us to derive lower bounds on the randomized communication complexity of f from upper bounds on the correlation between f and the class $\Pi^{k,c}$ [KN97, Theorem 3.20].

Fact 3. *If there exists a distribution* μ *such that* $\mathrm{Cor}_\mu(f, \Pi^{k,c}) \leq 1/3$ *then every randomized protocol (with error* $1/3$*) for f must communicate at least c bits.*

In order to obtain upper bounds on the correlation between f and the class $\Pi^{k,c}$, we use the following result, which is also standard. Historically, it was first proved by Babai, Nisan and Szegedy [BNS92] using the notion of *discrepancy* of a function. It has since been rewritten in many ways [CT93, Raz00, FG05, VW07]. The formulation we use appears in [VW07], except that in [VW07] one also takes two copies of x; it is easy to modify the proof in [VW07] to obtain the following lemma.

Lemma 1 (The standard BNS argument). *Let* $f : X \times Y_1 \times \cdots \times Y_k \to \mathbb{R}$. *Then,*

$$\text{Cor}_U(f, \Pi^{k+1,c})^{2^k} \leq 2^{c \cdot 2^k} \cdot \mathbb{E}_{\substack{(y_1^0,\ldots,y_k^0) \in Y_1 \times \cdots \times Y_k \\ (y_1^1,\ldots,y_k^1) \in Y_1 \times \cdots \times Y_k}} \left[\left| \mathbb{E}_{x \in X} \left[\prod_{u \in \{0,1\}^k} f(x, y_1^{u_1}, \ldots, y_k^{u_k}) \right] \right| \right].$$

We later write $\bar{y}$ for $(y_1, \ldots, y_k)$.

Degree. The ε-*approximate degree of* f is the smallest d for which there exists a multivariate real-valued polynomial g of degree d such that $\max_x |f(x) - g(x)| \leq \varepsilon$. We will use the following result of Nisan and Szegedy; see [Pat92] for a result that applies to more functions.

Lemma 2 ([NS94]). *There exists a constant* $\gamma > 0$ *such that the* $(5/6)$-*approximate degree of the* OR *function on* m *bits is at least* $\gamma \cdot \sqrt{m}$.

The following key result shows that if a function f has ε-approximate degree d then there is another function g and a distribution μ such that g is orthogonal to degree-d polynomials and g has correlation ε with f. Sherstov [She08a] gives references in the mathematics literature and points out a short proof by duality.

Lemma 3 (Orthogonality Lemma). *If* $f : \{0,1\}^m \to \{-1,1\}$ *is a function with* ε-*approximate degree* d, *there exist a function* $g : \{0,1\}^m \to \{-1,1\}$ *and a distribution* μ *on* $\{0,1\}^m$ *such that:*

(i) $\text{Cor}_\mu(g, f) \geq \varepsilon$; *and*
(ii) $\forall T \subseteq [m]$ *with* $|T| \leq d$ *and* $\forall h : \{0,1\}^{|T|} \to \mathbb{R}$, $\mathbb{E}_{x \sim \mu}[g(x) \cdot h(x|T)] = 0$,

where $x|T$ *denotes the* m *bits of* x *indexed by* T.

3 Non-explicit Separation

In this section we prove a *non-explicit* separation between nondeterministic and randomized protocols. As mentioned in the introduction, we restrict our attention to analyzing the communication complexity of certain functions constructed from a *base* function $f : \{0,1\}^m \to \{-1,1\}$, and a *selection* function ϕ. The base function we will work with is the OR function, which takes on the value -1 if and only if any of its input bits is 1.

We now give the definition of the function we prove the lower bound for, and then the statement of the lower bound.

Definition 2 (Lift). *Let* ϕ *be a function that takes as input* k *strings* $y_1, \ldots, y_k$ *and outputs an* m-*element subset of* $[n]$. *Let* f *be a function on* m *bits. We construct a* lifted *function* $\text{Lift}(f, \phi)$ *as follows. On input* $(x \in \{0,1\}^n, y_1, \ldots, y_k)$, $\text{Lift}(f, \phi)$ *evaluates* ϕ *on the latter* k *inputs to select a set of* m *bits in* x *and returns the value of* f *on those* m *bits. Formally,*

$$\text{Lift}(f, \phi)(x, y_1, \ldots, y_k) := f(x|\phi(y_1, \ldots, y_k)),$$

where for a set $S = \{i_1,\ldots,i_m\} \subseteq [n]$, $x|S$ denotes the substring $x_{i_1} \cdots x_{i_m}$ of x indexed by the elements in S, where $i_1 < i_2 < \ldots < i_m$.

The inputs to $\mathrm{Lift}(f,\phi)$ are partitioned among $k+1$ players as follows: Player 0 is given x and, for all $1 \le i \le k$, Player i is given y_i.

The following is the main theorem proved in this section.

Theorem 4. *For every $\delta < 1$ there are constants $\varepsilon, \alpha > 0$ such that for sufficiently large n, for $k = \delta \cdot \log n$, and for $m = n^\varepsilon$, the following holds. There is a distribution λ such that if we choose a random selection function $\phi : (\{0,1\}^n)^k \to \binom{[n]}{m}$, we have:*

$$\mathbb{E}_\phi[\mathrm{Cor}_\lambda(\mathrm{Lift}(\mathrm{OR},\phi), \Pi^{k+1,n^\alpha})] \le 1/3.$$

3.1 Overview of the Proof

We obtain our lower bound on the randomized communication complexity of the function $\mathrm{Lift}(\mathrm{OR},\phi)$ using an analysis that follows [CA08]. In their paper, Chattopadhyay and Ada analyze the Set-Disjointness function, and for that reason, their selection function ϕ must be the AND function. In our case, we allow ϕ to be a random function. While our results no longer apply to Set-Disjointness, we still obtain a separation between randomized and nondeterministic communication (BPP_k^{cc} and NP_k^{cc}) because, no matter what selection function is used, $\mathrm{Lift}(\mathrm{OR},\phi)$ always has an efficient nondeterministic protocol.

At a more technical level, the results of [CA08] require $k < \log\log n$ because of the relationship between n (the size of player 0's input) and m (the number of bits the base function OR gets applied to.) For their analysis to go through, they need $n > 2^{2^k} \cdot m^{O(1)}$. In our case, $n = 2^k \cdot m^{O(1)}$ is sufficient, and this allows our results to be non-trivial for $k \le \delta \log n$ for any $\delta < 1$.

As mentioned earlier, we will start with the base function OR on m input bits, $m = n^\varepsilon \ll n$. We lift the base function OR in order to obtain the lifted function $\mathrm{Lift}(\mathrm{OR},\phi)$. Recall that $\mathrm{Lift}(\mathrm{OR},\phi)$ is a function on $(k+1)n$ inputs with small nondeterministic complexity, and is obtained by applying the base function (in this case the OR function) to the selected bits of Player 0's input, x. We want to prove that for a random ϕ, $\mathrm{Lift}(\mathrm{OR},\phi)$ has high randomized communication complexity.

We start with a result of Nisan and Szegedy [NS94] who prove a lower bound on the approximate degree of the OR function. By Lemma 3 this implies that there exists a function g (also on m bits) and a distribution μ such that the functions g and OR are highly correlated over μ and, furthermore, g is orthogonal to low-degree polynomials. Now we lift the function g in order to get the function $\mathrm{Lift}(g,\phi)$, and we define λ to be a distribution over all $(k+1)n$-bit inputs that chooses the y_i's uniformly at random and x also uniformly at random except on the bits indexed by $\phi(y_1,\ldots,y_k)$ which are selected according to μ. Since g and OR are highly correlated with respect to μ, it is not hard to see that the lifted functions $\mathrm{Lift}(f,\phi)$ and $\mathrm{Lift}(g,\phi)$ are also highly correlated with respect to λ. Therefore, to prove that $\mathrm{Lift}(f,\phi)$ has low correlation with c-bit protocols it suffices to prove that $\mathrm{Lift}(g,\phi)$ has. To prove this, we use the correlation method. This involves bounding the average value of $\mathrm{Lift}(g,\phi)$ on certain k-dimensional cubes (cf. Lemma 1). For this, we need to analyze the distribution of the 2^k sets that arise from evaluating ϕ on the 2^k points of the cube. Specifically, we are interested in how much these 2^k sets are "spread out," as measured by the size of their union. If the sets are not spread out, we use in Lemma 4 the fact that g is orthogonal to low-degree polynomials to bound the average value of $\mathrm{Lift}(g,\phi)$ on the cubes. This step is similar to [She07, Cha07, LS08, CA08]. The main novelty in our analysis is that since we choose ϕ at random, we can prove good upper bounds (Lemma 6) on the probability that the sets are spread out.

3.2 Proof of Theorem 4

Let $m := n^\varepsilon$ for a small $\varepsilon > 0$ to be determined later. Combining Lemma 2 and 3, we see that there exists a function g and a distribution μ such that:

(i) $\mathrm{Cor}_\mu(g, \mathrm{OR}) \geq 5/6$; and
(ii) $\forall T \subseteq [m], |T| \leq \gamma\sqrt{m}$ and $\forall h : \{0,1\}^{|T|} \to \mathbb{R}, \mathbb{E}_{x \sim \mu}[g(x)h(x|T)] = 0$.

Define the distribution λ on $\{0,1\}^{(k+1)n}$ as follows. For $x, y_1, \ldots, y_k \in \{0,1\}^n$, let

$$\lambda(x, y_1, \ldots, y_k) := \frac{\mu(x|\phi(y_1, \ldots, y_k))}{2^{(k+1)n-m}},$$

in words we select $y_1, \ldots, y_k$ uniformly at random and then we select the bits of x indexed by $\phi(y_1, \ldots, y_k)$ according to μ and the others uniformly.

It can be easily verified that $\mathrm{Cor}_\lambda(\mathrm{Lift}(g,\phi), \mathrm{Lift}(\mathrm{OR},\phi)) = \mathrm{Cor}_\mu(g, \mathrm{OR}) \geq 5/6$. Consequently, for every ϕ and c,

$$\mathrm{Cor}_\lambda(\mathrm{Lift}(\mathrm{OR},\phi), \Pi^c) \leq \mathrm{Cor}_\lambda(\mathrm{Lift}(g,\phi), \Pi^c) + 2 \cdot \Pr_\lambda[\mathrm{Lift}(\mathrm{OR},\phi) \neq \mathrm{Lift}(g,\phi)]$$

$$\leq \mathrm{Cor}_\lambda(\mathrm{Lift}(g,\phi), \Pi^c) + 1/6, \quad (1)$$

where in the last inequality we use that $\mathrm{Cor}_\lambda(\mathrm{Lift}(\mathrm{OR},\phi), \mathrm{Lift}(g,\phi)) = E_\lambda[\mathrm{Lift}(\mathrm{OR},\phi) \cdot \mathrm{Lift}(g,\phi)] \geq 5/6$. Therefore, we only have to upper bound $\mathrm{Cor}_\lambda(\mathrm{Lift}(g,\phi), \Pi^c)$, and this is addressed next. We have, by the definition of λ and then Lemma 1:

$$\mathrm{Cor}_\lambda(\mathrm{Lift}(g,\phi), \Pi^c)^{2^k} = 2^{m \cdot 2^k} \mathrm{Cor}_{\mathcal{U}}(\mu(x|\phi(y_1,\ldots,y_k))g(x|\phi(y_1,\ldots,y_k)), \Pi^c)^{2^k}$$

$$\leq 2^{(c+m)2^k} \mathbb{E}_{\bar{y}^0, \bar{y}^1} \left[\left| \mathbb{E}_x \left[\prod_{u \in \{0,1\}^k} \mu(x|\phi(y_1^{u_1},\ldots,y_k^{u_k}))g(x|\phi(y_1^{u_1},\ldots,y_k^{u_k})) \right] \right| \right], \quad (2)$$

for every ϕ.

Our analysis makes extensive use of the following notation.

Definition 3. *Let $\mathcal{S} = (S_1, \ldots, S_z)$ be a multiset of m-element subsets of $[n]$. Let the range of $\mathcal{S}$, denoted by $\bigcup\mathcal{S}$, be the set of indices from $[n]$ that appear in at least one set in $\mathcal{S}$. Let the boundary of $\mathcal{S}$, denoted by $\partial\mathcal{S}$, be the set of indices from $[n]$ that appear in exactly one set in the collection $\mathcal{S}$.*

For $u \in \{0,1\}^k$, define $S_u = S_u(\bar{y}^0, \bar{y}^1, \phi) = \phi(y_1^{u_1}, \ldots, y_k^{u_k})$. Let $\mathcal{S} = \mathcal{S}(\bar{y}^0, \bar{y}^1, \phi)$ be the multiset $(S_u : u \in \{0,1\}^k)$. We define the number of conflicts in $\mathcal{S}$ to be $q(\mathcal{S}) := m \cdot 2^k - |\bigcup\mathcal{S}|$.

Intuitively, $|\bigcup S|$ measures the range of S, while $m2^k$ is the maximum possible value for this range. We use the following three lemmas to complete our proof. The first Lemma 4 deals with the case where the multiset S has few conflicts. In this case, we argue that one of the sets $S_u \in S$ has a very small intersection with the rest of the other sets, which allows us to apply Property (ii) of g and μ to obtain the stated bound. A variant of Lemma 4 appears in [CA08].

Lemma 4. *For every* $\bar{y}^0, \bar{y}^1$ *and* ϕ, *if* $q(S(\bar{y}^0, \bar{y}^1, \phi)) < \gamma \cdot \sqrt{m} \cdot 2^k / 2$, *then*

$$\mathbb{E}_x \left[\prod_{u \in \{0,1\}^k} \mu(x | S_u(\bar{y}^0, \bar{y}^1, \phi)) g(x | S_u(\bar{y}^0, \bar{y}^1, \phi)) \right] = 0.$$

Lemma 5 gives a bound in terms of the number of conflicts in S which only uses the fact that μ is a probability distribution. A slightly weaker version of this lemma appeared originally in [CA08]. Independently of our work, Chattopadhyay and Ada have subsequently also derived the stronger statement we give below.

Lemma 5. *For every* $\bar{y}^0, \bar{y}^1$ *and* ϕ:

$$\mathbb{E}_x \left[\prod_{u \in \{0,1\}^k} \mu(x | S_u(\bar{y}^0, \bar{y}^1, \phi)) \right] \leq \frac{2^{q(S(\bar{y}^0, \bar{y}^1, \phi))}}{2^{m \cdot 2^k}}.$$

Lemma 6 is the key place where we exploit the fact that ϕ is chosen at random to obtain an upper bound on the probability of having a given number of conflicts in S.

Lemma 6. *For every* $q > 0$ *and uniformly chosen* $\bar{y}^0, \bar{y}^1, \phi$:

$$\Pr_{\bar{y}^0, \bar{y}^1, \phi} [q(S(\bar{y}^0, \bar{y}^1, \phi)) = q] \leq \left(\frac{m^3 \cdot 2^{2k}}{q \cdot n} \right)^q.$$

We defer the proofs of Lemmas 4, 5 and 6 to the Appendix. We now complete the proof of our Theorem 4. Using Equation 2, Lemmas 4, 5 and 6, along with standard derivations that we defer to the Appendix, we obtain that, for a uniformly chosen ϕ,

$$\mathbb{E}_\phi \left[\mathrm{Cor}_\lambda (\mathrm{Lift}(g, \phi), \Pi^c) \right]^{2^k} \leq 2^{c \cdot 2^k} \cdot \sum_{q \geq \gamma \sqrt{m} 2^k / 2} \left(\frac{2 \cdot m^3 \cdot 2^{2k}}{q \cdot n} \right)^q.$$

Furthermore, using $q \geq \gamma \sqrt{m} 2^k / 2$, $k = \delta \log n$ where $\delta < 1$, and taking $m = n^\varepsilon$ for a sufficiently small ε, we get,

$$\mathbb{E}_\phi \left[\mathrm{Cor}_\lambda (\mathrm{Lift}(g, \phi), \Pi^c) \right]^{2^k} \leq 2^{c \cdot 2^k} \cdot \sum_{q \geq \gamma \sqrt{m} 2^k / 2} \left(\frac{1}{2} \right)^q \leq 2^{c \cdot 2^k + 1 - \gamma \sqrt{m} 2^k / 2} \leq 2^{2^k (c - n^{\Omega(1)})}.$$

Therefore, when c is a sufficiently small power of n we have $\mathbb{E}_\phi [\mathrm{Cor}_\lambda (\mathrm{Lift}(g, \phi), \Pi^c)] \leq 1/6$. Combining this with Equation (1), completes the proof of Theorem 4.

4 Explicit Separation

In this section we prove our main Theorem 1. We proceed as follows. First, we prove a derandomized version of Theorem 4 from the previous section. This derandomized version is such that the distribution on ϕ can be generated using only n random bits r. Then, we observe how including the random bits r as part of the input gives an explicit function for the separation, thus proving Theorem 1. As we mentioned in the introduction, the idea is that the only property of the distribution over ϕ that the previous construction was using is that such a distribution is 2^k-wise independent. That is, the evaluations of ϕ at any 2^k points, fixed and distinct, are jointly uniformly distributed, over the choice of ϕ (cf. the proof of Lemma 6). The most straightforward way to obtain explicit constructions from our previous results is thus to replace a random ϕ with a 2^k-wise independent distribution, and then include a description of ϕ as part of the input. However, this raises some technicalities, one being that the range of our ϕ was a size-m subset of $[n]$, and it is not immediate how to give constructions with such a range. We find it more simple to use a slightly different block-wise approach as we describe next.

We think of our universe of n bits as divided in $m := n^\varepsilon$ blocks of $b := n^{1-\varepsilon}$ bits each, where as before ε is a sufficiently small constant. We consider functions $\phi(y_1, \ldots, y_k)$ whose output is a subset of $[n]$ that contains exactly one bit per block. That is, $\phi(y_1, \ldots, y_k) \in [b]^m$. The building block of our distribution is the following result about almost t-wise independent functions. We defer its proof to the Appendix. We say that two distributions X and Y on the same support are ε-close in statistical distance if for every event E we have $|\Pr[E(X)] - \Pr[E(Y)]| \le \varepsilon$.

Lemma 7 (almost t-wise independence; [NN93]). *There is a universal constant $a > 0$ such that for every t, b (where b is a power of 2) there is a polynomial-time computable map*

$$h : \{0,1\}^t \times \{0,1\}^{a \cdot t \cdot \log b} \to [b]$$

such that for every t distinct $x_1, \ldots, x_t \in \{0,1\}^t$, the distribution $(h(x_1; r), \ldots, h(x_t; r)) \in [b]^t$, over the choice of $r \in \{0,1\}^{a \cdot t \cdot \log b}$, is $(1/b)^t$-close in statistical distance to the uniform distribution over $[b]^t$.

We now define our derandomized distribution on ϕ. This is the concatenation of m of the above functions using independent random bits, a function per block. Specifically, for each of the m blocks of b bits, we are going to use the above function h where $t := k \cdot 2^k \cdot (1 + \log b)$. Jumping ahead, the large input length t is also chosen so that the probability (over the choice of the y's) that we do not obtain 2^k distinct inputs drops down exponentially with 2^k, which is needed in the analysis. On input $y_1, \ldots, y_k$ and randomness r, we break up each y_i in m blocks and also r in m blocks. The value of ϕ in the j-th block depends only on the j-th blocks of the y_i's and on the j-th block of r.

Definition 4 (Derandomized distribution on ϕ, given parameters n, $m = n^\varepsilon$, $b = n^{1-\varepsilon}$, $k = \delta \cdot \log n$; and a universal constant from Lemma 7). *Let $l := 2^k \cdot (1 + \log b)$, $t := l \cdot k$. We define*

$$\phi : \{0,1\}^{m \cdot t} \times \{0,1\}^{m \cdot a \cdot t \cdot \log b} \to [b]^m$$

as follows. On input $(y_1, \ldots, y_k) \in \{0,1\}^{m \cdot t}$ and randomness $r \in \{0,1\}^{m \cdot a \cdot t \cdot \log b}$, think of each $y_i \in \{0,1\}^{m \cdot l}$ as divided in m blocks of l bits each, i.e. $(y_i = (y_i)_1 \circ \cdots \circ (y_i)_m)$, and *r as divided in m blocks of a · t · log b bits each, i.e. ($r = r_1 \circ \cdots \circ r_m$). The j-th output of φ in [b] is then*

$$\phi(y_1,\ldots,y_k;r)_j := h(\underbrace{(y_1)_j,\ldots,(y_k)_j;}_{l\cdot k=t\ bits}\ \underbrace{r_j}_{a\cdot t\cdot\log b\ bits}) \in [b].$$

The distribution on φ is obtained by selecting a uniform $r \in \{0,1\}^{m\cdot a\cdot t\cdot\log b}$ and then considering the map $(y_1,\ldots,y_k) \to \phi(y_1,\ldots,y_k;r) \in [b]^m$.

Note that, in the above definition, the input length of each y_i is $m \cdot l$ which up to poly-logarithmic factors is $n^\varepsilon \cdot 2^k = n^{1-\Omega(1)}$, for a sufficiently small ε depending on δ.

Theorem 5. *For every $\delta < 1$ there are constants $\varepsilon, \alpha > 0$ such that for sufficiently large n, $k := \delta \cdot \log n$, and $m = n^\varepsilon$, the following holds.*

There is a distribution λ such that if $\phi : \{0,1\}^{m\cdot t} \to [b]^m$ is distributed according to Definition 4 we have:

$$\mathbb{E}_\phi[\mathrm{Cor}_\lambda(\mathrm{Lift}(\mathrm{OR},\phi),\Pi^{k+1,n^\alpha})] \le 1/3.$$

Proof. The proof follows very closely that of Theorem 4. A minor difference is that now the y_i's are over $m \cdot l$ bits as opposed to n in Theorem 4, but the definition of the distribution λ in Theorem 4 immediately translates to the new setting – λ just selects the y_i's at random. The only other place where the proofs differ is in Lemma 6, which is where the properties of φ are used. Thus we only need to verify the following Lemma.

Lemma 8. *For every $q > 0$ and φ distributed as in Definition (4):*

$$\Pr_{\bar{y}^0,\bar{y}^1,\phi}[q(S(\bar{y}^0,\bar{y}^1,\phi)) = q] \le \left(\frac{m^2 \cdot 2^{2k}}{q \cdot b}\right)^q = \left(\frac{m^3 \cdot 2^{2k}}{q \cdot n}\right)^q.$$

We defer its proof to the Appendix, and we prove the main Theorem of this work.

Theorem 1 (Main Theorem; $\mathrm{NP}^{cc}_{\delta\log n} \not\subset \mathrm{BPP}^{cc}_{\delta\log n}$). (Restated.) *For every fixed $\delta < 1$, sufficiently large n and $k = \delta \cdot \log n$, there is an explicit function $f : (\{0,1\}^n)^k \to \{0,1\}$ such that: f can be computed by k-player nondeterministic protocols communicating $O(\log n)$ bits, but f cannot be computed by k-player randomized protocols communicating $n^{o(1)}$ bits.*

Proof. Let $f(x,(y_1,r),y_2,\ldots,y_k) := \mathrm{OR}(x|\phi(y_1,\ldots,y_k;r))$, where φ is as in Definition 4. We partition an input $(x,(y_1,r),y_2,\ldots,y_k)$ as follows: Player 0 gets x, Player 1 gets the pair (y_1,r), where r is to be thought of as selecting which φ to use, and player $i > 1$ gets y_i. Let p be the distribution obtained by choosing r uniformly at random, and independently $(x,y_1,\ldots,y_k)$ according to the distribution λ in Theorem 5.

It is not hard to see that f has a nondeterministic protocol communicating $O(\log n)$ bits: We guess a bit position i and then the player that sees $(y_1,r),y_2,\ldots,y_k$ verifies that the position i belongs to $\phi(y_1,\ldots,y_k;r)$, and another player verifies that $x_i = 1$.

To see the second item observe that:

$$\mathrm{Cor}_p(f, \Pi^{k+1,n^\alpha}) = \max_{\pi \in \Pi^{k+1,n^\alpha}} \mathbb{E}_r[\mathbb{E}_{(x,\bar{y}) \sim \lambda}[\mathrm{OR}(x|\phi(\bar{y};r)) \cdot \pi(x,\bar{y},r)]]$$

$$\leq \mathbb{E}_r[\max_{\pi \in \Pi^{k+1,n^\alpha}} \mathbb{E}_{(x,\bar{y}) \sim \lambda}[\mathrm{OR}(x|\phi(\bar{y};r)) \cdot \pi(x,\bar{y},r)]] \leq 1/3,$$

where the last inequality follows by Theorem 5. Again, the claim about randomized communication follows by standard techniques, cf. Fact 3.

To conclude, we need to verify that we can afford to give r as part of the input without affecting the bounds. Specifically, we need to verify that $|(y_1, r)| \leq n$. Indeed, $|(y_1, r)| \leq m \cdot l + O(m \cdot t \cdot \log b) = m \cdot 2^k(1 + \log b) + O(m \cdot 2^k(1 + \log b)k \cdot \log b)$ which is less than n when $k = \delta \log n$ for a fixed $\delta < 1$, $m = n^\varepsilon$ for a sufficiently small ε, and n is sufficiently large (recall $b \cdot m = n$, and in particular $b \leq n$.)

As is apparent from the proofs, and similarly to previous works [She08b], our lower bound Theorems 4 and 5 hold more generally for any function of the form $\mathrm{Lift}(f, \phi)$ for an arbitrary base function f. The communication bound is then expressed in terms of the approximate degree of f. In our paper, we focused on $f = \mathrm{OR}$ for concreteness. However, also note that the choice of $f = \mathrm{OR}$ is essential in Theorem 1 in order for $\mathrm{Lift}(f, \phi)$ to have a cheap nondeterministic protocol.

5   Communication Bounds for Constant-Depth Circuits

In this section we point out how Theorem 5 from the previous section gives us some new communication bounds for functions computable by constant-depth circuits. Specifically, the next theorem, which was also stated in the introduction, gives communication bounds for up to $k = A \cdot \log \log n$ players for functions computable by constant-depth circuits (whose parameters depend on A), whereas previous results [Cha07, LS08, CA08] require $k < \log \log n$.

Theorem 2 (Some constant-depth circuits require high communication). *(Restated.)* *For every constant $A > 1$ there is a constant B such that for sufficiently large n and $k := A \log \log n$ there is a function $f : (\{0,1\}^n)^k \to \{0,1\}$ which satisfies the following: f can be computed by circuits of size n^B and depth B, but f cannot be computed by k-player randomized protocols communicating $n^{o(1)}$ bits.*

Proof. Use the function from the proof of Theorem 1. This only requires computing $(2^k = \log^A n)$-wise independent functions on $\log^{O(A)} n$ bits. (As mentioned before, although Theorem 5 uses the notion of *almost* t-wise independence, for small values of k, such as those of interest in the current proof, we can afford to use *exact* t-wise independence, i.e. set the distance from uniform distribution to 0). Such functions can be computed by circuits of size n^B and depth B, for a constant B that depends on A only. To see this, one can use the standard constructions based on arithmetic over finite fields [CG89, ABI86] and then the results from [HV06, Corollary 6]. Equivalently, "scale down" [HV06, Theorem 14] as described in [HV06, Section 3].

It is not clear to us how to prove a similar result for $k = \omega(\log\log n)$. This is because our approach would require computing almost $(2^k = \log^{\omega(1)} n)$-wise independent functions on $\log^{\omega(1)} n$ bits by $n^{O(1)}$-size circuits of constant depth, which cannot be done (even for almost 2-wise independence). The fact that this cannot be done follows from the results in [MNT90] or known results on the noise sensitivity of constant-depth circuits [LMN93, Bop97].

We point out that Theorem 2 can be strengthened to give a function that has correlation $2^{-n^{\Omega(1)}}$ with protocols communicating $n^{o(1)}$ bits. This can be achieved using the Minsky-Papert function instead of OR (cf. [She07, Cha07]).

Finally, Troy Lee (personal communication, May 2008) has pointed out to us that the analogous of our Theorem 2 for *deterministic* protocols can be easily obtained from the known lower bound for generalized inner product (GIP) [BNS92]. This is because it is not hard to see that for every constant c there is a circuit of depth $B = B(c)$ and size n^B that has correlation at least $\exp(-n/\log^c n)$ with GIP – just compute the parity in GIP by brute-force on blocks of size $\log^c n$ – but on the other hand low-communication k-party protocols have correlation at most $\exp(-\Omega(n/4^k))$ with GIP [BNS92]. However, this idea does not seem to give a bound for randomized protocols or a correlation bound, whereas our results do.

Acknowledgements. We thank Sasha Sherstov and Troy Lee for helpful comments on the write-up. Matei David and Toniann Pitassi gratefully acknowledge Arkadev Chattopadhyay and Anil Ada for several very insightful conversations. Emanuele Viola is especially grateful to Troy Lee for many stimulating conversations on communication complexity.

References

[ABI86]   Alon, N., Babai, L., Itai, A.: A fast and simple randomized algorithm for the maximal independent set problem. Journal of Algorithms 7, 567–583 (1986)

[BDPW07] Beame, P., David, M., Pitassi, T., Woelfel, P.: Separating deterministic from nondet, nof multiparty communication complexity. In: Arge, L., Cachin, C., Jurdziński, T., Tarlecki, A. (eds.) ICALP 2007. LNCS, vol. 4596, pp. 134–145. Springer, Heidelberg (2007)

[BFS86]   Babai, L., Frankl, P., Simon, J.: Complexity classes in communication complexity theory (preliminary version). In: FOCS, pp. 337–347. IEEE, Los Alamitos (1986)

[BNS92]   Babai, L., Nisan, N., Szegedy, M.: Multiparty protocols, pseudorandom generators for logspace, and time-space trade-offs. J. Comput. System Sci. 45(2), 204–232 (1992)

[Bop97]   Boppana, R.B.: The average sensitivity of bounded-depth circuits. Inform. Process. Lett. 63(5), 257–261 (1997)

[BPS07]   Beame, P., Pitassi, T., Segerlind, N.: Lower bounds for lovász–schrijver systems and beyond follow from multiparty communication complexity. SIAM J. Comput. 37(3), 845–869 (2007)

[CA08]   Chattopadhyay, A., Ada, A.: Multiparty communication complexity of disjointness. ECCC, Technical Report TR08-002 (2008)

[CFL83]   Chandra, A.K., Furst, M.L., Lipton, R.J.: Multi-party protocols. In: STOC (1983)

[CG89]   Chor, B., Goldreich, O.: On the power of two-point based sampling. Journal of Complexity 5(1), 96–106 (1989)

[Cha07]   Chattopadhyay, A.: Discrepancy and the power of bottom fan-in in depth-three circuits. In: FOCS, October 2007, pp. 449–458. IEEE, Los Alamitos (2007)

[CT93]   Chung, F.R.K., Tetali, P.: Communication complexity and quasi randomness. SIAM J. Discrete Math. 6(1), 110–123 (1993)

[FG05]   Ford, J., Gál, A.: Hadamard tensors and lower bounds on multiparty communication complexity. In: Caires, L., Italiano, G.F., Monteiro, L., Palamidessi, C., Yung, M. (eds.) ICALP 2005. LNCS, vol. 3580, pp. 1163–1175. Springer, Heidelberg (2005)

[GV04]   Gutfreund, D., Viola, E.: Fooling parity tests with parity gates. In: Jansen, K., Khanna, S., Rolim, J., Ron, D. (eds.) RANDOM 2004. LNCS, vol. 3122, pp. 381–392. Springer, Heidelberg (2004)

[HG91]   Håstad, J., Goldmann, M.: On the power of small-depth threshold circuits. Comput. Complexity 1(2), 113–129 (1991)

[HV06]   Healy, A., Viola, E.: Constant-depth circuits for arithmetic in finite fields of characteristic two. In: Durand, B., Thomas, W. (eds.) STACS 2006. LNCS, vol. 3884, pp. 672–683. Springer, Heidelberg (2006)

[KN97]   Kushilevitz, E., Nisan, N.: Communication complexity. Cambridge University Press, Cambridge (1997)

[LMN93]   Linial, N., Mansour, Y., Nisan, N.: Constant depth circuits, Fourier transform, and learnability. J. Assoc. Comput. Mach. 40(3), 607–620 (1993)

[LS08]   Lee, T., Shraibman, A.: Disjointness is hard in the multi-party number on the forehead model. In: CCC. IEEE, Los Alamitos (2008)

[MNT90]   Mansour, Y., Nisan, N., Tiwari, P.: The computational complexity of universal hashing. In: STOC, pp. 235–243. ACM Press, New York (1990)

[NN93]   Naor, J., Naor, M.: Small-bias probability spaces: efficient constructions and applications. SIAM J. Comput. 22(4), 838–856 (1993)

[NS94]   Nisan, N., Szegedy, M.: On the degree of boolean functions as real polynomials. Computational Complexity 4, 301–313 (1994)

[NW93]   Nisan, N., Wigderson, A.: Rounds in communication complexity revisited. SIAM J. Comput. 22(1), 211–219 (1993)

[Pat92]   Paturi, R.: On the degree of polynomials that approximate symmetric boolean functions (preliminary version). In: STOC, pp. 468–474. ACM, New York (1992)

[Raz87]   Razborov, A.A.: Lower bounds on the dimension of schemes of bounded depth in a complete basis containing the logical addition function. Mat. Zametki 41(4), 598–607, 623 (1987)

[Raz00]   Raz, R.: The BNS-Chung criterion for multi-party communication complexity. Comput. Complexity 9(2), 113–122 (2000)

[Raz03]   Razborov, A.: Quantum communication complexity of symmetric predicates. Izvestiya: Mathematics 67(1), 145–159 (2003)

[RW93]   Razborov, A., Wigderson, A.: $n^{\Omega(\log n)}$ lower bounds on the size of depth-3 threshold circuits with AND gates at the bottom. Inform. Process. Lett. 45(6), 303–307 (1993)

[She07]   Sherstov, A.: Separating AC^0 from depth-2 majority circuits. In: STOC 2007: Proceedings of the 39th Annual ACM Symposium on Theory of Computing (2007)

[She08a]   Sherstov, A.: The pattern matrix method for lower bounds on quantum communication. In: STOC (2008)

[She08b]   Sherstov, A.A.: Communication lower bounds using dual polynomials. Electronic Colloquium on Computational Complexity, Technical Report TR08-057 (2008)

[VW07]   Viola, E., Wigderson, A.: Norms, xor lemmas, and lower bounds for GF(2) polynomials and multiparty protocols. In: CCC. IEEE, Los Alamitos (2007); Theory of Computing (to appear)

Quantum and Randomized Lower Bounds for Local Search on Vertex-Transitive Graphs

Hang Dinh and Alexander Russell

Department of Computer Science & Engineering
University of Connecticut
Storrs, CT 06269, USA
{hangdt,acr}@engr.uconn.edu

Abstract. We study the problem of *local search* on a graph. Given a real-valued black-box function f on the graph's vertices, this is the problem of determining a local minimum of f—a vertex v for which $f(v)$ is no more than f evaluated at any of v's neighbors. In 1983, Aldous gave the first strong lower bounds for the problem, showing that any randomized algorithm requires $\Omega(2^{n/2-o(1)})$ queries to determine a local minima on the n-dimensional hypercube. The next major step forward was not until 2004 when Aaronson, introducing a new method for query complexity bounds, both strengthened this lower bound to $\Omega(2^{n/2}/n^2)$ and gave an analogous lower bound on the quantum query complexity. While these bounds are very strong, they are known only for narrow families of graphs (hypercubes and grids). We show how to generalize Aaronson's techniques in order to give randomized (and quantum) lower bounds on the query complexity of local search for the family of vertex-transitive graphs. In particular, we show that for any vertex-transitive graph G of N vertices and diameter d, the randomized and quantum query complexities for local search on G are $\Omega\left(\frac{\sqrt{N}}{d \log N}\right)$ and $\Omega\left(\frac{\sqrt[4]{N}}{\sqrt{d \log N}}\right)$, respectively.

1 Introduction

The *local search* problem is that of determining a local minimum of a function defined on the vertices of a graph. Specifically, given a real-valued black-box function f on the vertices of a graph, this is the problem of determining a vertex v at which $f(v)$ is no more than f evaluated at any of v's neighbors. The problem provides an abstract framework for studying local search heuristics that have been widely applied in combinatorial optimization, heuristics that typically combine random selection with steepest descent. The performance of these heuristic algorithms, as recognized in [1], *"was generally considered to be satisfactory, partly based on experience, partly based on a belief in some physical or biological analogy, ..."* Ideally, of course, we would evaluate practical results in the context of crisp theoretical bounds on the complexity of these problems! Moreover, as pointed out in [2], the complexity of the local search problem is also central for understanding a series of complexity classes which are subclasses of

A. Goel et al. (Eds.): APPROX and RANDOM 2008, LNCS 5171, pp. 385–401, 2008.
© Springer-Verlag Berlin Heidelberg 2008

the *total function* class TFNP, including PPP (Polynomial Pigeohole Principle), PODN (Polynomial Odd-Degree Node), and PLS (Polynomial Local Search).

Local search has been the subject of a sizable body of theoretical work, in which complexity is typically measured by *query complexity*: the total number of queries made to the black-box function f in order to find a local minimum. The first strong lower bounds were established in 1983 by Aldous [3], who showed that $2^{n/2-o(n)}$ queries are necessary, in general, in order for a randomized algorithm to find a local minimum of a function on the hypercube $\{0,1\}^n$. His proof constructs a rich collection of unimodal functions (that is, functions with a unique minimum) using hitting times of random walks. Llewellyn et al. [6] improved the bound for deterministic query complexity to $\Omega(2^n/\sqrt{n})$ using an adversary argument characterized by vertex cuts.

With the advent of quantum computing, these black-box problems received renewed interest [2, 11, 8, 9, 10]. Most notably, Aaronson [2] introduced a query lower bound method tuned for such problems, the *relational adversary method*. Though his principal motivation was, no doubt, to provide *quantum* lower bounds for local search, his techniques felicitously demonstrated improved bounds on randomized query complexity. In particular, he established a $\Omega(2^{n/2}n^{-2})$ lower bound for randomized local search on the Boolean hypercube $\{0,1\}^n$ and the first nontrivial lower bound of $\Omega\left(n^{d/2-1}/(d\log n)\right)$ for randomized local search on a d-dimensional grid $[n]^d$ with $d \geqslant 3$.

These two lower bounds of Aaronson's have been recently improved by Zhang [11]: refining Aaronson's framework, he established randomized query complexity lower bounds of $\Theta(2^{n/2}n^{1/2})$ on the hypercube and $\Theta(n^{d/2})$ on the grid $[n]^d$, $d \geqslant 4$. Additionally, Zhang's method can be applied to certain classes of product graphs, though it provides a rather complicated relationship between the lower bound and the product decomposition.[1] A remaining hurdle in this direct line of research was to establish strong bounds for grids of small dimension. Sun and Yao [9] have addressed this problem, proving that the quantum query complexity is $\Omega\left(n^{1/2-c}\right)$ for $[n]^2$ and $\Omega(n^{1-c})$ for $[n]^3$, for any fixed constant $c > 0$. Focusing on general graphs, Santha and Szegedy [8] established quantum lower bounds of $\Omega(\log N)$ and $\Omega\left(\sqrt[8]{\frac{s}{\delta}}/\log N\right)$ for local search on connected N-vertex graphs with maximal degree δ and *separation number* s.[2] We remark that $s/\delta \leq N$ and

---

[1] Zhang [11]'s general lower bounds for a product graph $G_w \times G_c$ involve the length L of the longest self-avoiding path in the "clock" graph G_c, and parameters $p(u,v,t)$'s of a regular random walk W on G_w, where $p(u,v,t)$ is the probability that the random walk W starting at u ends up at v after exactly t steps. In particular, he showed that $\mathsf{RLS}(G_w \times G_c) = \Omega\left(\dfrac{L}{\sum_{t=1}^{L/2} \max_{u,v} p(u,v,t)}\right)$ and $\mathsf{QLS}(G_w \times G_c) = \Omega\left(\dfrac{L}{\sum_{t=1}^{L/2} \sqrt{\max_{u,v} p(u,v,t)}}\right)$.

[2] Santha and Szegedy define the *separation number* $s(G)$ of a graph $G = (V,E)$ to be: $s(G) = \max\limits_{H \subseteq V} \min\limits_{S \subseteq H, |H|/4 \leq |S| \leq 3|H|/4} |\partial_H S|$, where $\partial_H S = \{v \in H \setminus S : \exists u \in S, (v,u) \in E\}$ is the boundary of S in the subgraph of G restricted to H.

these bounds for general graphs are, naturally, much weaker than those obtained for the highly structured families of graphs above.

In this article, we show off the flexibility of Aaronson's framework by extending it to arbitrary vertex-transitive graphs. Recall that a graph $G = (V, E)$ is *vertex-transitive* if the automorphism group of G acts transitively on the vertices: for any pair of vertices $x, y \in V$, there is a graph automorphism $\phi : V \rightarrow V$ for which $\phi(x) = y$. In particular, all Cayley graphs are vertex-transitive, so this class of graphs contains the hypercubes of previous interest and the looped grids (tori).

Our lower bounds depend only on the size and diameter of the graph:

Theorem 1. *Let G be a connected, vertex-transitive graph with N vertices and diameter d. Then*

$$\mathsf{RLS}(G) = \Omega\left(\frac{\sqrt{N}}{d \log N}\right), \qquad and \qquad \mathsf{QLS}(G) = \Omega\left(\frac{\sqrt[4]{N}}{\sqrt{d \log N}}\right)$$

where $\mathsf{RLS}(G)$ and $\mathsf{QLS}(G)$ are the randomized and quantum query complexities of local search on G, respectively.

Thus the vertex transitive graphs, compromising between the specific families of graphs addressed by [2, 11] and the general results of Santha and Szegedy, still provide enough structure to support strong lower bounds.

2   Definitions and Notation

As in [2, 11], we focus on the local search problem stated precisely as follows: given a graph $G = (V, E)$ and a black-box function $f : V \rightarrow \mathbb{R}$, find a local minimum of f on G, i.e. find a vertex $v \in V$ such that $f(v) \leq f(w)$ for all neighbors w of v. While the graph is known to the algorithm, the values of f may only be accessed through an oracle. For an algorithm $\mathcal{A}$ that solves the local search problem on G, let $T(\mathcal{A}, G)$ be the maximum number of queries made to the black-box function by $\mathcal{A}$ before it returns a local minimum, this maximum taken over all functions f on G. Given a graph G, the randomized query complexity for Local Search on G is defined as $\min_{\mathcal{A}} T(\mathcal{A}, G)$, where the minimum ranges over all randomized algorithms $\mathcal{A}$ that output a local minimum with probability at least $2/3$. The quantum query complexity is defined similarly, except that in the quantum case, $T(\mathcal{A}, G)$ is the maximum number of unitary query transformations of the error-bounded quantum algorithm $\mathcal{A}$. The randomized (resp. quantum) query complexity for local search on G will be denoted by $\mathsf{RLS}(G)$ (resp. $\mathsf{QLS}(G)$).

As mentioned in the introduction, we focus on the vertex-transitive graphs, those whose automorphism groups act transitively on their vertex sets. Perhaps the most important subclass of the vertex-transitive graphs are the Cayley graphs. Let G be a group (finite, in this article, and written multiplicatively) and Γ a set of generators for G. The Cayley graph $C(G, \Gamma)$ is the graph with vertex set G and edges $E = \{(g, g\gamma) \mid g \in G, \gamma \in \Gamma \cup \Gamma^{-1}\}$. Note that with this definition for the edges, $(a, b) \in E \Leftrightarrow (b, a) \in E$ even when G is nonabelian, and we may consider the graph to be undirected.

If X is a sequence of vertices in a graph, we write $X_{i \to j}$ to denote the subsequence of X from position i to position j ($i \le j$). If $X = (x_1, \ldots, x_t)$ is a sequence of vertices in a Cayley graph and g is a group element, then we use gX to denote the sequence $(gx_1, \ldots, gx_t)$. More generally, for any automorphism σ of a vertex-transitive graph G and any sequence $X = (x_1, \ldots, x_t)$ of vertices in G, we let σX denote the sequence $(\sigma(x_1), \ldots, \sigma(x_t))$.

The distance between two vertices u, v of a graph G shall be denoted by $\Delta_G(u, v)$; when G is understood from context we abbreviate to $\Delta(u, v)$. The statistical distance between two distributions D_1 and D_2 on the same set Ω is defined as the distance in total variation:

$$\|D_1 - D_2\|_{\text{t.v.}} = \max_{E \subset \Omega} |D_1(E) - D_2(E)| = \frac{1}{2} \sum_{\omega \in \Omega} |D_1(\omega) - D_2(\omega)|.$$

We say that the distribution D_1 is δ-*close* to distribution D_2 if $\|D_1 - D_2\|_{\text{t.v.}} \le \delta$.

3   Generalizing Aaronson's Snakes

Aaronson's [2] application of the quantum and relational adversary methods to local search problems involved certain families of walks on a graph he called "snakes." We begin by presenting Aaronson's snake method, adjusted to suit our generalization. Throughout this section, let G be a graph. A *snake* X of length L is a sequence $(x_0, \ldots, x_L)$ of vertices in G such that each x_{i+1} is either equal to x_i or a neighbor of x_i. The subsequence $X_{0 \to j}$ shall be referred to as the j-length "head" of the snake X. Suppose $\mathcal{D}_{x_0, L}$ is a distribution over snakes of length L starting at x_0, and X is a snake drawn from $\mathcal{D}_{x_0, L}$. In Aaronson's parlance, the snake X "flicks" its tail by choosing a position j uniformly at random from the set $\{0, \ldots, L-1\}$, and then drawing a new snake Y from $\mathcal{D}_{x_0, L}$ conditioned on the event that $Y_{0 \to j} = X_{0 \to j}$, that is, that Y has the same j-length head as X. In order to simplify the proof for vertex-transitive graphs below, we consider a generalization in which a snake flicks its tail according to a distribution $\mathbf{D}_L$, which may be nonuniform, on the set $\{0, \ldots, L-1\}$. We shall relax, also, Aaronson's original condition that, aside from adjacent repetition of a vertex v, snakes be non-self-intersecting.

Let $X = (x_0, \ldots, x_L)$ be a snake. Define the function f_X on G as follows: for each vertex v of G,

$$f_X(v) = \begin{cases} L - \max\{i : x_i = v\} & \text{if } v \in X, \\ L + \Delta(x_0, v) & \text{if } v \notin X. \end{cases}$$

In other words, $f_X(x_L) = 0$, and for any $i < L$, $f_X(x_i) = L - i$ if $x_i \notin \{x_{i+1}, \ldots, x_L\}$. Clearly f_X has a unique local minimum at x_L.

Let X and Y be snakes of length L starting at x_0. A vertex v is called a *disagreement* between X and Y if $v \in X \cap Y$ and $f_X(v) \neq f_Y(v)$. We say X and Y are *consistent* if there is no disagreement between X and Y. Observe that so long as X and Y are consistent, $f_X(v) \neq f_Y(v) \iff set_X(v) \neq set_Y(v)$ for all vertices v, where set_X is the function on G defined as $set_X(v) = 1$ if $v \in X$ and 0 otherwise.

Fix a distribution $\mathcal{D}_{x_0,L}$ for snakes of length L starting at x_0 and a distribution $\mathbf{D}_L$ on the set $\{0, \ldots, L-1\}$. With these in place, we let $\Pr_{j,X}[\cdot]$ denote the probability of an event over the distribution determined by independently selecting j according to $\mathbf{D}_L$ and X from $\mathcal{D}_{x_0,L}$.

We record Aaronson's definition of *good* snakes, replacing the uniform distribution on the set $\{0, \ldots, L-1\}$ with the distribution $\mathbf{D}_L$, and requiring a good snake's endpoint to be different from those of most other snakes.

Definition 1. *A snake $X \in \mathcal{D}_{x_0,L}$ is ϵ-good w.r.t. distribution $\mathbf{D}_L$ if it satisfies the following:*

1. *X is 0.9-consistent:* $\Pr_{j,Y}[X \ and \ Y \ are \ consistent, \ and \ x_L \neq y_L | Y_{0\to j} = X_{0\to j}] \geq 0.9$.
2. *X is ϵ-hitting: For all $v \in G$, $\Pr_{j,Y}[v \in Y_{j+1\to L} \mid Y_{0\to j} = X_{0\to j}] \leq \epsilon$.*

Our lower bounds will depend on the following adaptation of Aaronson's theorem of [2]:

Theorem 2. *Assume a snake X drawn from $\mathcal{D}_{x_0,L}$ is ϵ-good w.r.t. $\mathbf{D}_L$ with probability at least 0.9. Then*

$$\mathsf{RLS}(G) = \Omega(1/\epsilon) \ and \ \mathsf{QLS}(G) = \Omega(\sqrt{1/\epsilon}).$$

Proof. To begin, we reduce the local search problem to a decision problem. For each snake $X \in \mathcal{D}_{x_0,L}$ and a bit $b \in \{0,1\}$, define the function $g_{X,b}$ on G as follows: $g_{X,b}(v) = (f_X(v), -1)$ for all vertices $v \neq x_L$, and $g_{X,b}(x_L) = (0,b)$. Then, an input of the decision problem for local search on G is an ordered pair $(X, g_{X,b})$, where $X \in \mathcal{D}_{x_0,L}$ and $b \in \{0,1\}$ is an answer bit. However, the "snake part" X in the input cannot be queried—it appears in the input as a bookkeeping tool. Given such an input $(X, g_{X,b})$, the decision problem is to output the answer bit b. Observe that the randomized (resp. quantum) query complexity of the decision problem is a lower bound for that of the original local search problem. This incorporation of X into the input of the decision problem induces a natural one-to-one correspondence between an input set of the same answer bit and the set of snakes appearing in the input set. (Thanks to Scott Aaronson for suggesting this convention to us!) In Aaronson's original version, since the input part X is omitted, the snakes must be non-self-intersecting in order to obtain such a one-to-one correspondence. Santha and Szegedy [8] have presented an alternate approach for eliminating self-intersecting snakes while following Aaronson's proof scheme, though their technique only applies to the quantum case.

The remaining part of the proof, which establishes lower bounds for the decision problem using the rational and quantum adversary methods, is similar to Aaronson's proof with the exception of some technical details due to the adjustments in the definition of good snakes. We have relegated the full proof to the appendix.

4   Lower Bounds for Vertex-Transitive Graphs

For simplicity, we first apply the snake framework for Cayley graphs, and then extend the approach for vertex-transitive graphs.

4.1   Lower Bounds for Cayley Graphs

Consider a Cayley graph $C(G, \Gamma)$ of group G determined by a generating set Γ. Our goal is to design a good snake distribution for $C(G, \Gamma)$. Our snakes will consist of a series of "chunks" so that the endpoint of each chunk looks almost random given the preceding chunks. The locations at which a snake flicks its tail will be chosen randomly from the locations of the chunks' endpoints. Each chunk is an "extended" shortest path connecting its endpoint with the endpoint of the previous one. The relevant properties of these snakes depends on the length of each chunk as well as the number of chunks in each snake. To determine these parameters, we begin with the following definitions.

Let $B(s)$ be the ball of radius s centered at the group identity, i.e., $B(s)$ is the set of vertices v for which $\Delta(1, v) \leq s$. We say that Cayley graph $C(G, \Gamma)$ is s-mixing if there is a distribution over the ball $B(s)$ that is $O\left(s/|G|^{3/2}\right)$-close to the uniform distribution over G. Clearly, every Cayley graph of diameter d is d-mixing.

Now we assume $C(G, \Gamma)$ is s-mixing, and let D_s be a distribution over $B(s)$ so that the extension of D_s to be over G is δ-close to the uniform distribution over G, where $s \leq \sqrt{|G|}$ and $\delta = 0.1s/|G|^{3/2}$. For each group element $g \in B(s)$, we fix a shortest path $(1, g_1, \ldots, g_r)$ in $C(G, \Gamma)$ from the group identity to g (here $r = \Delta(1, g)$). Then let $S(g)$ denote the sequence $(g_1, g_2, \ldots, g_s)$, where $g_i = g$ for $i \geq r$.

Fix $\ell = \sqrt{|G|}/(200s)$ and let $L = (\ell+1)s$. We formally define our snake distribution $\mathcal{D}_{x_0, L}$ for snakes $X = (x_0, \ldots, x_L)$ as follows. For any $k \in \{0, \ldots, \ell\}$, choose g_k independently according to the distribution D_s, and let the kth "chunk" $(x_{sk+1}, \ldots, x_{sk+s})$ be identical to the sequence $x_{sk} S(g_k)$.

Proposition 1. A snake X drawn from $\mathcal{D}_{x_0, L}$ δ-mixes by s steps in the sense that for any k and any $t \geq s$, x_{sk+t} is δ-close to uniform over G given x_{sk}.

We define distribution $\mathbf{D}_L$ on $\{0, \ldots, L-1\}$ as the uniform distribution on the set $\{s, 2s, \ldots, \ell s\}$. So, unlike Aaronson's snakes whose tails may be flicked at any location, our snakes can not "break" in the middle of any chunk and only flick their tails at the chunk endpoints.

To show that most of our snakes are good, we start by showing that most snakes X and Y are consistent and have different endpoints.

Proposition 2. *Let j be chosen according to $\mathbf{D}_L$. Let X, Y be drawn from $\mathcal{D}_{x_0,L}$ conditioned on $Y_{0\to j} = X_{0\to j}$. Then*

$$\Pr_{X,j,Y}[X \text{ and } Y \text{ are consistent, and } x_L \neq y_L \mid Y_{0\to j} = X_{0\to j}] \geq 0.9999 - \frac{2}{|G|}.$$

Proof. Fix $j \in \{s, 2s, \dots, \ell s\}$. Suppose v is a disagreement between X and Y, letting $t = \max\{i : x_i = v\}$ and $t' = \max\{i : y_i = v\}$, then $t \neq t'$ and $t', t \geq j$. We can't have both $t < j + s$ and $t' < j + s$, because otherwise we would have $v \neq x_{j+s}$ and $v \neq y_{j+s}$ which implies that both $t - j$ and $t' - j$ equal the distance from x_j to v.

If there is a disagreement, there must exist t and t' such that $x_t = y_{t'}$ and either $t \geq j + s$ or $t' \geq j + s$. In the case $t \geq j + s$, we have x_t is δ-close to uniform given $y_{t'}$, which implies

$$\Pr_{X,Y_{j\to L}}[x_t = y_{t'}] \leq \delta + \frac{1}{|G|} \leq \frac{2}{|G|}.$$

Similarly, in the case $t' \geq j + s$, we also have $\Pr_{X,Y_{j\to L}}[x_t = y_{t'}] \leq \frac{2}{|G|}$. Summing up for all possible pairs of t and t' yields

$$\Pr_{X,Y_{j\to L}}[\text{there is a disagreement between } X \text{ and } Y] \leq \frac{2(L-s)^2}{|G|} \leq 0.0001.$$

Averaging over j produces

$$\Pr_{X,j,Y}[X \text{ and } Y \text{ are not consistent} \mid Y_{0\to j} = X_{0\to j}] \leq 0.0001.$$

To complete the proof, observe that

$$\Pr_{X,j,Y}[x_L = y_L \mid Y_{0\to j} = X_{0\to j}] \leq \delta + \frac{1}{|G|} \leq \frac{2}{|G|}.$$

since y_L is δ-close to uniform given x_L.

By Markov's inequality, we obtain:

Corollary 1. *Let X be drawn from $\mathcal{D}_{x_0,L}$. Then*

$$\Pr_X[X \text{ is } 0.9\text{-consistent}] \geq 1 - \frac{0.0001 + 2/|G|}{0.1} = 0.999 - \frac{20}{|G|}.$$

We now turn our attention to bounding the hitting probability when a snake flicks its tail. Following Aaronson, we introduce a notion of ϵ-*sparseness* for snakes and show that *(i)* if a snake is ϵ-sparse then it is $O(\epsilon)$-hitting, and that *(ii)* most snakes are ϵ-sparse.

Formally, we define:

Definition 2. *For each $x \in G$, let $P(x) = \Pr_{g \in D_s}[x \in S(g)]$. A snake X drawn from $\mathcal{D}_{x_0,L}$ is called ϵ-sparse if for all vertex $v \in G$,*

$$\sum_{k=1}^{\ell} P(x_{sk}^{-1}v) \leq \epsilon\ell.$$

Intuitively, the sparseness of a snake means that if the snake flicks a random chunk, it is unlikely to hit any fixed vertex.

Proposition 3. *For $\epsilon \geq \frac{2(L-s)}{|G|}$, if snake X is ϵ-sparse then X is 2ϵ-hitting.*

Proof. Fix a snake X, and fix $j \in \{s, 2s \ldots, \ell s\}$. Let Y be drawn from $\mathcal{D}_{x_0,L}$ conditioned on the event that $Y_{0 \to j} = X_{0 \to j}$. Since y_t is δ-close to uniform for all $t \geq j + s$,

$$\Pr_Y[v \in Y_{j+s \to L} \mid Y_{0 \to j} = X_{0 \to j}] \leq (L-s)(\delta + \frac{1}{|G|}) \leq \frac{2(L-s)}{|G|}.$$

On the other hand,

$$\Pr_Y[v \in Y_{j+1 \to j+s} \mid Y_{0 \to j} = X_{0 \to j}] = \Pr_{g \in D_s}[v \in x_j S(g)] = P(x_j^{-1}v).$$

Hence,

$$\Pr_{j,Y}[v \in Y_{j+1 \to L} \mid Y_{0 \to j} = X_{0 \to j}] \leq \frac{1}{\ell}\sum_{k=1}^{\ell} P(x_{sk}^{-1}v) + \frac{2(L-s)}{|G|} \leq 2\epsilon.$$

It remains to show that a snake drawn from $\mathcal{D}_{x_0,L}$ is ϵ-sparse with high probability. Firstly, we consider for the "ideal" case in which the endpoints of the chunks in a snake are independently uniform.

Lemma 1. *Let $u_1, \ldots, u_\ell$ be independently and uniformly random vertices in G. If $\frac{s}{|G|} \leq \epsilon^2/6$ then*

$$\Pr_{u_1,\ldots,u_\ell}\left[\sum_{i=1}^{\ell} P(u_i) > 2\ell\epsilon\right] \leq 2^{-\ell\epsilon}.$$

Proof. We will use a Chernoff bound to show that there are very few u_i's for which $P(u_i)$ is large. To do this, we first need an upper bound on the expectation of $P(u_i)$. Let u be a uniformly random vertex in G. For any given $g \in G$, we have $\Pr_u[u \in S(g)] = \frac{\Delta(1,g)}{|G|} \leq \frac{s}{|G|}$. Averaging over $g \in D_s$ yields $\Pr_{g,u}[u \in S(g)] \leq \frac{s}{|G|}$, where g is chosen from D_s independently to u. Since $\mathbb{E}_u[P(u)] = \Pr_{u,g}[u \in S(g)]$, we have $\mathbb{E}_u[P(u)] \leq \frac{s}{|G|}$.

Let $Z = |\{i : P(u_i) \geq \epsilon\}|$. By Markov's inequality,

$$\mathbb{E}[Z] = \ell\Pr_u[P(u) \geq \epsilon] \leq \frac{\ell\,\mathbb{E}_u[P(u)]}{\epsilon} \leq \frac{\ell s}{|G|\epsilon} = \mu.$$

By a Chernoff bound, for any $\lambda \geq 2e$

$$\Pr_u[Z \geq \lambda\mu] \leq \left(\frac{e^{\lambda-1}}{\lambda^\lambda}\right)^\mu = \left(\frac{e}{\lambda}\right)^{\lambda\mu} e^{-\mu} \leq 2^{-\lambda\mu-\mu}.$$

Note that if $Z < \lambda\mu$ then

$$\sum_{i=1}^{\ell} P(u_i) \leq (\ell - Z)\epsilon + Z \leq \ell\epsilon + \lambda\mu.$$

Setting $\lambda\mu = \ell\epsilon$, which satisfies $\lambda \geq 2e$ due to the assumption that $\frac{s}{|G|} \leq \epsilon^2/6$, we have

$$\Pr_{u_1,\ldots,u_\ell}\left[\sum_{i=1}^{\ell} P(u_i) > 2\ell\epsilon\right] \leq \Pr_{u_1,\ldots,u_\ell}[Z \geq \ell\epsilon] \leq 2^{-\ell\epsilon}.$$

In order to apply this to our scenario without strict independence, we record the following fact about distance in total variation.

Proposition 4. *Let $X_1,\ldots,X_n$ and $Y_1,\ldots,Y_n$ be discrete random variables so that X_i and Y_i have the same value range. Let $(X_i \mid A_1,\ldots,A_{i-1})$ denote the distribution of X_i given that $X_1 \in A_1,\ldots,X_{i-1} \in A_{i-1}$; similarly let $(Y_i \mid A_1,\ldots,A_{i-1})$ denote the distribution of Y_i given that $Y_1 \in A_1,\ldots,Y_{i-1} \in A_{i-1}$. Then*

$$\|(X_1,\ldots,X_n) - (Y_1,\ldots,Y_n)\|_{t.v.} \leq \|X_1 - Y_1\|_{t.v.} + \sum_{i=2}^{n} \Delta_i$$

where

$$\Delta_i = \max_{A_1,\ldots,A_{i-1}} \|(X_i \mid A_1,\ldots,A_{i-1}) - (Y_i \mid A_1,\ldots,A_{i-1})\|_{t.v.}.$$

A detailed proof of Proposition 4 can be found in the appendix.

Lemma 2. *Suppose $\frac{s}{|G|} \leq \epsilon^2/6$. Then a snake X drawn from $\mathcal{D}_{x_0,L}$ is 2ϵ-sparse with probability at least $1 - |G|2^{-\ell\epsilon} - 1/2000$.*

Proof. The proof for the lemma follows immediately by observing that for any vertex v, the variables $x_s^{-1}v,\ldots,x_{s\ell}^{-1}v$ satisfy that $x_{s(k+1)}^{-1}v$ is δ-close to uniform given $x_{sk}^{-1}v$. By Proposition 4,

$$\left|\Pr_X\left[\sum_{k=1}^{\ell} P(x_{sk}^{-1}v) > 2\ell\epsilon\right] - \Pr_{u_1,\ldots,u_\ell}\left[\sum_{i=1}^{\ell} P(u_i) > 2\ell\epsilon\right]\right| \leq \ell\delta \leq \frac{1}{2000|G|}.$$

From Lemma 1,

$$\Pr_X\left[\sum_{k=1}^{\ell} P(x_{sk}^{-1}v) > 2\ell\epsilon\right] \leq 2^{-\ell\epsilon} + \frac{1}{2000|G|}.$$

Summing up over $v \in G$ gives $\Pr_X[X$ is *not* 2ϵ-sparse$] \leq |G|2^{-\ell\epsilon} + 1/2000$.

We need to choose ϵ such that $|G|2^{-\ell\epsilon} \le 1/2000$, or $\epsilon \ge \frac{\log|G|+O(1)}{\ell}$.

Corollary 2. *A snake X drawn from $\mathcal{D}_{x_0,L}$ is $O\left(\frac{s\log|G|}{\sqrt{|G|}}\right)$-hitting with probability at least 0.999.*

Putting all the pieces together and applying Theorem 2, we have

Theorem 3. *For $s = O(\sqrt{|G|})$, if Cayley graph $C(G,\Gamma)$ is s-mixing, then*

$$\mathrm{RLS}(C(G,\Gamma)) = \Omega\left(\frac{\sqrt{|G|}}{s\log|G|}\right), \qquad \mathrm{QLS}(C(G,\Gamma)) = \Omega\left(\frac{\sqrt[4]{|G|}}{\sqrt{s\log|G|}}\right).$$

In particular, any Cayley graph $C(G,\Gamma)$ of diameter d has

$$\mathrm{RLS}(C(G,\Gamma)) = \Omega\left(\frac{\sqrt{|G|}}{d\log|G|}\right), \qquad \mathrm{QLS}(C(G,\Gamma)) = \Omega\left(\frac{\sqrt[4]{|G|}}{\sqrt{d\log|G|}}\right).$$

For comparison, applying Aldous's randomized upper bound [3] and Aaronson's quantum upper bound [2] for arbitrary Cayley graph $C(G,\Gamma)$, we have

$$\mathrm{RLS}(C(G,\Gamma)) = O\left(\sqrt{|G||\Gamma|}\right) \text{ and } \mathrm{QLS}(C(G,\Gamma)) = O\left(\sqrt[3]{|G|}\sqrt[6]{|\Gamma|}\right).$$

For example, for constant degree expanding Cayley graphs, this randomized lower bound is tight to within $O(\log^2|G|)$ of Aldous's upper bound.

Random Cayley graphs. In fact, it can be showed that most Cayley graphs are s-mixing for $s = \Omega(\log|G|)$. Let $g_1, \ldots, g_s$ be a sequence of group elements. Following [5], we call an element of the form $g_1^{a_1} \cdots g_s^{a_s}$, where $a_i \in \{0,1\}$, a *subproduct* of the sequence $g_1, \ldots, g_s$. A *random subproduct* of this sequence is a subproduct obtained by independently choosing a_i as a fair coin flip. A sequence $g_1, \ldots, g_s$ is called *a sequence of δ-uniform Erdős-Rényi (E-R) generators* if its random subproductors are δ-uniformly distributed over G in the sense that

$$(1-\delta)\frac{1}{|G|} \le \Pr_{a_1,\ldots,a_s}[g_1^{a_1} \cdots g_s^{a_s} = g] \le (1+\delta)\frac{1}{|G|} \qquad \text{for all } g \in G.$$

Theorem 4. *(Erdős and Rényi, See also [5])* For $s \ge 2\log|G| + 2\log(1/\delta) + \lambda$, a sequence of s random elements of G is a sequence of δ-uniform E-R generators with probability at least $1 - 2^{-\lambda}$.

Clearly, any Cayley graph determined by an s-length sequence of δ-uniform E-R generators is s-mixing. So applying our lower bounds for arbitrary Cayley graphs and the E-R theorem, we have

Proposition 5. *Let $s \ge 5\log|G| - 2\log s + \lambda$. With probability at least $1 - 2^{-\lambda}$, a random Cayley graph $C(G,\Gamma)$ determined by a sequence of s random group elements has*

$$O(\sqrt{|G|s}) \ge \mathrm{RLS}(C(G,\Gamma)) \ge \Omega\left(\frac{\sqrt{|G|}}{s\log|G|}\right) \text{ and } O(\sqrt[3]{|G|}\sqrt[6]{s}) \ge \mathrm{QLS}(C(G,\Gamma)) \ge \Omega\left(\frac{\sqrt[4]{|G|}}{\sqrt{s\log|G|}}\right).$$

4.2   Extending to Vertex-Transitive Graphs

Our approach above for Cayley graphs can be easily extended to vertex-transitive graphs. We shall describe here how to define a snake distribution $\mathcal{D}_{x_0,L}$ similar to that for a Cayley graph. Consider a vertex-transitive graph $G = (V, E)$ with $N = |V|$, and let d be the diameter of G. We fix an arbitrary vertex $v_0 \in V$. For each vertex $v \in V$, we also fix an extended shortest path $S(v) = (v_1, \dots, v_d)$ of length d from v_0 to v. (v_0 is omitted in $S(v)$ for technical reasons.) That is, $(v_0, \dots, v_r)$ is the actual shortest path from v_0 to v, where $r = \Delta(v_0, v)$, and $v_i = v$ for all $i \geq r$.

Since the automorphism group of G acts transitively on V, we can fix an automorphism σ_x, for each $x \in V$, so that $\sigma_x(v_0) = x$. Hence, for any $x, v \in V$, the sequence $\sigma_x S(v)$ is the extended shortest path from x to $\sigma_x(v)$. So now we can determine the kth chunk of a snake as the sequence $\sigma_{x_{dk}} S(u_k)$, where x_{dk} is the endpoint of the $(k-1)$th chunk of the snake, and u_k is an independently and uniformly random vertex. Let $P(x) = \Pr_u[x \in S(u)]$, where u is chosen from V uniformly at random. The condition for a snake $X = (x_0, \dots, x_{(\ell+1)d})$ to be ϵ-sparse is now redefined as

$$\sum_{k=1}^{\ell} P\left(\sigma_{x_{dk}}^{-1}(v)\right) \leq \ell\epsilon \quad \text{for all } v \in V.$$

Observe that, given x_{dk}, the endpoint $x_{dk+k} = \sigma_{x_{dk}}(u_k)$ of the kth chunk is a uniformly random vertex, since $\sigma_{x_{dk}}$ is a bijective. Also clearly, $\sigma_x \neq \sigma_y$ for any $x \neq y$ because σ_x and σ_y send v_0 to different places. This means there is a one-to-one correspondence $x \leftrightarrow \sigma_x$ between V and the set of automorphisms $\{\sigma_x : x \in V\}$. Therefore, if x is uniformly distributed over V, then so is the vertex at any given position in $\sigma_x S$, for any sequence S of vertices. It follows that in our snake $X = (x_0, \dots, x_L)$, for all $t \geq k$, x_{dk+t} is uniformly distributed over V given x_{dk}. With this snake distribution, we can similarly follow the proof for Cayley graphs to prove the lower bounds for vertex-transitive graphs as given in Theorem 1.

Acknowledgements

We gratefully acknowledge Scott Aaronson for discussing his previous work with us and showing us the trick for removing the requirement of snake non-self-intersection. We would like to thank anonymous referees for many helpful comments.

References

[1] Aardal, K., van Hoesel, S., Lenstra, J.K., Stougie, L.: A decade of combinatorial optimization. In: CWI Tracts, vol. 122, pp. 5–14 (1997)
[2] Aaronson, S.: Lower bounds for local search by quantum arguments. In: STOC 2004: Proceedings of the 36th Annual ACM Symposium on Theory of Computing (2004)

[3] Aldous, D.: Minimization algorithms and random walk on the d-cube. Annals of Probability 11(2), 403–413 (1983)

[4] Ambainis, A.: Quantum lower bounds by quantum arguments. In: STOC 2000: Proceedings of the thirty-second annual ACM symposium on Theory of computing (2000)

[5] Babai, L.: Local expansion of vertex-transitive graphs and random generation in finite groups. In: STOC 1991: Proceedings of the twenty-third annual ACM symposium on Theory of computing, pp. 164–174. ACM, New York (1991)

[6] Llewellyn, D.C., Tovey, C., Trick, M.: Local optimization on graphs. Discrete Appl. Math. 23(2), 157–178 (1989)

[7] Mohar, B., Thomassen, C.: Graphs on Surfaces. The Johns Hopkins University Press (2001)

[8] Santha, M., Szegedy, M.: Quantum and classical query complexities of local search are polynomially related. In: STOC 2004: Proceedings of the thirty-sixth annual ACM symposium on Theory of computing, pp. 494–501. ACM, New York (2004)

[9] Sun, X., Yao, A.C.: On the quantum query complexity of local search in two and three dimensions. In: FOCS 2006: Proceedings of the 47th Annual IEEE Symposium on Foundations of Computer Science, Washington, DC, USA, pp. 429–438. IEEE Computer Society, Los Alamitos (2006)

[10] Verhoeven, Y.F.: Enhanced algorithms for local search. Inf. Process. Lett. 97(5), 171–176 (2006)

[11] Zhang, S.: New upper and lower bounds for randomized and quantum local search. In: STOC 2006: Proceedings of the thirty-eighth annual ACM symposium on Theory of computing, pp. 634–643. ACM, New York (2006)

A   Appendix

A.1   Quantum and Relational Adversary Methods

The quantum adversary method [4] is a powerful tool underlying many proofs of quantum lower bounds. The classical counterpart applied above is the relational adversary method [2]. The central intuition of these adversary methods is to make it hard to distinguish "related" input sets. Technically, consider two input sets $\mathcal{A}$ and $\mathcal{B}$ for a function $F : I^n \to [m]$ so that $F(A) \neq F(B)$ for all $A \in \mathcal{A}$ and $B \in \mathcal{B}$. Here, an input to function F is a black-box function $A : [n] \to I$. The oracle for an input A answers queries of the form $A(x) =?$. If A and B are the two inputs that have the same value at every queryable location, then we must have $F(A) = F(B)$. Define a "relation" function $R(A, B) \geq 0$ on $\mathcal{A} \times \mathcal{B}$. Two inputs A and B are said to be related if $R(A, B) > 0$. Then, for $A \in \mathcal{A}$, $B \in \mathcal{B}$, and a queryable location $x \in [n]$, let

$$M(A) = \sum_{B' \in \mathcal{B}} R(A, B'), \qquad M(B) = \sum_{A' \in \mathcal{A}} R(A', B)$$
$$M(A, x) = \sum_{B' \in \mathcal{B}: A(x) \neq B'(x)} R(A, B'), \quad M(B, x) = \sum_{A' \in \mathcal{A}: A'(x) \neq B(x)} R(A', B).$$

Intuitively, the fraction $M(A, x)/M(A)$ (resp. $M(B, x)/M(B)$) measures how hard it is to distinguish input A (resp. B) with related inputs in $\mathcal{B}$ (resp. $\mathcal{A}$) by queying at location x. Formally, if there are such input sets $\mathcal{A}, \mathcal{B}$ and relation function $R(A, B)$, then

Theorem 5. *(Ambainis) The number of quantum queries needed to evaluate F with probability at least 0.9 is $\Omega(M_{\text{geom}})$, where*

$$M_{\text{geom}} = \min_{\substack{A \in \mathcal{A}, B \in \mathcal{B}, x \\ R(A,B)>0, A(x) \neq B(x)}} \sqrt{\frac{M(A)}{M(A,x)} \frac{M(B)}{M(B,x)}} \, .$$

Theorem 6. *(Aaronson) The number of randomized queries needed to evaluate F with probability at least 0.9 is $\Omega(M_{\text{max}})$, where*

$$M_{\text{max}} = \min_{\substack{A \in \mathcal{A}, B \in \mathcal{B}, x \\ R(A,B)>0, A(x) \neq B(x)}} \max \left\{ \frac{M(A)}{M(A,x)}, \frac{M(B)}{M(B,x)} \right\} \, .$$

A.2 Proofs

Continued proof for Aaronson's theorem (Theorem 2)

Proof. To apply the quantum and relational adversary method for the decision problem, define the input sets $\mathcal{A} = \{(X, g_{X,0}) : X \in \mathcal{D}^*\}$ and $\mathcal{B} = \{(Y, g_{Y,1}) : Y \in \mathcal{D}^*\}$, where $\mathcal{D}^*$ denotes the set of ϵ-good snakes drawn from $\mathcal{D}_{x_0,L}$. For simplicity, we write A_X as $(X, g_{X,0})$, and B_Y as $(Y, g_{Y,1})$. For $A_X \in \mathcal{A}$ and $B_Y \in \mathcal{B}$, define relation function $R(A_X, B_Y) = w(X, Y)$ if X and Y are consistent and $x_L \neq y_L$, and $R(A_X, B_Y) = 0$ otherwise, where $w(X, Y)$ is determined as follows. Let $p(X)$ be the probability of drawing snake X from $\mathcal{D}_{x_0,L}$, and let

$$w(X, Y) = p(X) \Pr_{j,Z}[Z = Y \mid Z_{0 \to j} = X_{0 \to j}] \, .$$

Claim. For any snakes $X, Y \in \mathcal{D}_{x_0,L}$, we have $w(X, Y) = w(Y, X)$.

Proof. (of the claim) Fix $j \in \{0, \ldots, L-1\}$ and let $q_j(X, Y) = \Pr_Z[Z = Y | Z_{0 \to j} = X_{0 \to j}]$. We want to show

$$p(X) q_j(X, Y) = p(Y) q_j(Y, X) \, .$$

Assume $X_{0 \to j} = Y_{0 \to j}$, otherwise $q_j(X, Y) = q_j(Y, X) = 0$. Then letting Z be drawn from $\mathcal{D}_{x_0,L}$ and let E be the event $Z_{0 \to j} = X_{0 \to j} = Y_{0 \to j}$, we have

$$p(X) q_j(X, Y) = \Pr_Z[E] \cdot \Pr_Z[Z_{j+1 \to L} = X_{j+1 \to L} | E] \cdot \Pr_Z[Z_{j+1 \to L} = Y_{j+1 \to L} | E]$$

$$= \Pr_Z[E] \cdot \Pr_Z[Z_{j+1 \to L} = Y_{j+1 \to L} | E] \cdot \Pr_Z[Z_{j+1 \to L} = X_{j+1 \to L} | E]$$

$$= p(Y) q_j(Y, X) \, .$$

completing the proof for the claim.

As in Aaronson's original proof, we won't be able to take the whole input sets $\mathcal{A}$ and $\mathcal{B}$ defined above because of the fact that not all snakes are good. Instead, we will take only a subset of each of these input sets that would be hard enough to distinguish. This is done by applying Lemma 8 in [2], which states as follows.

Lemma 3. *Let* $p(1), \ldots, p(m)$ *be positive reals such that* $\sum_i p(i) \leq 1$. *Let* $R(i, j)$, *for* $i, j \in \{1, \ldots, m\}$, *be nonnegative reals satisfying* $R(i, j) = R(j, i)$ *and* $\sum_{i,j} R(i, j) \geq r$. *Then there exists a nonempty subset* $U \in \{1, \ldots, m\}$ *such that* $\sum_{j \in U} R(i, j) \geq rp(i)/2$ *for all* $i \in U$.

To apply this lemma, we need a lower bound for the sum $\sum_{X,Y \in \mathcal{D}^*} R(A_X, B_Y)$. Let $E(X, Y)$ denote the event that snakes X and Y are consistent and $x_L \neq y_L$. For any $X \in \mathcal{D}^*$, we have

$$\sum_{Y:E(X,Y)} w(X, Y) = p(X) \Pr_{j,Y}[E(X, Y) \mid Y_{0 \to j} = X_{0 \to j}] \geq 0.9p(X).$$

Hence, since a snake drawn from $\mathcal{D}_{x_0, L}$ is good with probability at least 0.9,

$$\sum_{X,Y:E(X,Y)} w(X, Y) \geq 0.9 \sum_{X \in \mathcal{D}^*} p(X) \geq 0.9 \times 0.9 \geq 0.8.$$

By the union bound,

$$\sum_{X,Y \in \mathcal{D}^*} R(A_X, B_Y) \geq \sum_{X,Y:E(X,Y)} w(X, Y) - \sum_{X \notin \mathcal{D}^*} p(X) - \sum_{Y \notin \mathcal{D}^*} p(Y) \geq 0.8 - 0.1 - 0.1 = 0.6.$$

So, by Lemma 3, there exists a nonempty subset $\tilde{\mathcal{D}} \subset \mathcal{D}^*$ so that for all $X, Y \in \tilde{\mathcal{D}}$,

$$\sum_{Y' \in \tilde{\mathcal{D}}} R(A_X, B_{Y'}) \geq 0.3p(X),$$

$$\sum_{X' \in \tilde{\mathcal{D}}} R(A_{X'}, B_Y) \geq 0.3p(Y).$$

So now we take the input sets $\tilde{\mathcal{A}} = \left\{ A_X : X \in \tilde{\mathcal{D}} \right\}$ and $\tilde{\mathcal{B}} = \left\{ B_Y : Y \in \tilde{\mathcal{D}} \right\}$. We have shown that $M(A_X) \geq 0.3p(X)$ and $M(B_Y) \geq 0.3p(Y)$ for any $A_X \in \tilde{\mathcal{A}}$ and $B_Y \in \tilde{\mathcal{B}}$. Since the snake part in the inputs can not be queried, we only care about the measure for distinguishing A_X, B_Y with their related inputs by querying the function part (i.e. $g_{X,0}$ or $g_{Y,1}$) in the inputs. Formally, we focus on lower-bounding $M(A_X, v)$ and $M(B_Y, v)$ for inputs $A_X \in \tilde{\mathcal{A}}, B_Y \in \tilde{\mathcal{B}}$ for which $R(A_X, B_Y) > 0$ and $g_{X,0}(v) \neq g_{Y,1}(v)$. We remark that since $R(A_X, B_Y) > 0$, the event $E(X, Y)$ must hold, which implies that for all vertex v,

$$g_{X,0}(v) \neq g_{Y,1}(v) \iff f_X(v) \neq f_Y(v) \iff set_X(v) \neq set_Y(v).$$

Applying the quantum and randomized adversary method, we will have $\mathsf{RLS}(G) \geq \Omega(M_{\max})$ and $\mathsf{QLS}(G) \geq \Omega(M_{\text{geom}})$, where

$$M_{\max} = \min_{\substack{A_X \in \tilde{\mathcal{A}}, B_Y \in \tilde{\mathcal{B}}, v \\ R(A_X, B_Y) > 0, set_X(v) \neq set_Y(v)}} \max \left\{ \frac{M(A_X)}{M(A_X, v)}, \frac{M(B_Y)}{M(B_Y, v)} \right\}$$

$$M_{\text{geom}} = \min_{\substack{A_X \in \widetilde{\mathcal{A}}, B_Y \in \widetilde{\mathcal{B}}, v \\ R(A_X, B_Y) > 0, set_X(v) \neq set_Y(v)}} \sqrt{\frac{M(A_X)}{M(A_X, v)} \frac{M(B_Y)}{M(B_Y, v)}} \, .$$

Let $A_X \in \widetilde{\mathcal{A}}, B_Y \in \widetilde{\mathcal{B}}$ be inputs for which $set_X(v) \neq set_Y(v)$. Then $v \notin X$ or $v \notin Y$. Assuming the case $v \notin X$, we will show $M(A_X, v)$ is small. We have

$$
\begin{aligned}
M(A_X, v) &\leq \sum_{Y' \in \widetilde{\mathcal{D}}: set_X(v) \neq set_{Y'}(v)} w(X, Y') \\
&\leq \sum_{Y': v \in Y'} p(X) \Pr_{j, Z}[Z = Y' \mid Z_{0 \to j} = X_{0 \to j}] \\
&= p(X) \Pr_{j, Z}[v \in Z \mid Z_{0 \to j} = X_{0 \to j}] \\
&= p(X) \Pr_{j, Z}[v \in Z_{j+1 \to L} \mid Z_{0 \to j} = X_{0 \to j}] \quad (\text{since } v \notin X) \\
&\leq p(X)\epsilon \quad (\text{since } X \text{ is } \epsilon\text{-hitting}).
\end{aligned}
$$

In the case $v \notin Y$, we can also obtain $M(B_Y, v) \leq p(Y)\epsilon$ due to symmetry. Hence,

$$\max\left\{ \frac{M(A_X)}{M(A_X, v)}, \frac{M(B_Y)}{M(B_Y, v)} \right\} \geq 0.3/\epsilon$$

$$\sqrt{\frac{M(A_X)}{M(A_X, v)} \frac{M(B_Y)}{M(B_Y, v)}} \geq \sqrt{0.3/\epsilon}.$$

The latter inequality is obtained due to the fact that $M(A_X, v) \leq M(A_X)$ and $M(B_Y, v) \leq M(B_Y)$. Consequently, $M_{\max} = \Omega(1/\epsilon)$ and $M_{\text{geom}} = \Omega(\sqrt{1/\epsilon})$, completing the proof for Theorem 2.

Proof of Proposition 4

Proof. We prove by induction on n. The case $n = 2$ can be easily obtained by applying the following simple fact:

Fact 1. *Let* x_1, x_2, y_1, y_2 *be any real numbers in* $[0, 1]$. *Then*

$$|x_1 x_2 - y_1 y_2| = |(x_1 - y_1)x_2 + (x_2 - y_2)y_1| \leq |x_1 - y_1|x_2 + |x_2 - y_2|y_1 \leq |x_1 - y_1| + |x_2 - y_2|.$$

In particular, applying the above fact, we have for any pair of events (A, B),

$$
\left| \Pr[X_1 \in A, X_2 \in B] - \Pr[Y_1 \in A, Y_2 \in B] \right| \leq \left| \Pr[X_1 \in A] - \Pr[Y_1 \in A] \right| +
$$
$$
\left| \Pr[X_2 \in B | X_1 \in A] - \Pr[Y_2 \in B | Y_1 \in A] \right|.
$$

Recall that by definition of total variation,

$$\|(X_1, X_2) - (Y_1, Y_2)\|_{t.v.} = \max_{A,B} \left| \Pr[X_1 \in A, X_2 \in B] - \Pr[Y_1 \in A, Y_2 \in B] \right| \quad \text{and}$$

$$\Delta_2 = \max_{A,B} \left| \Pr[X_2 \in B | X_1 \in A] - \Pr[Y_2 \in B | Y_1 \in A] \right|.$$

Hence,

$$\|(X_1, X_2) - (Y_1, Y_2)\|_{t.v.} \leq \|X_1 - Y_1\|_{t.v.} + \Delta_2.$$

Now we can apply this result and get

$$\|(X_1, \ldots, X_n) - (Y_1, \ldots, Y_n)\|_{t.v.} \leq \|(X_1, \ldots, X_{n-1}) - (Y_1, \ldots, Y_{n-1})\|_{t.v.} + \Delta_n$$

which establishes the proposition by induction.

A.3   Upper Bounds for Local Search

Various upper bounds for both quantum and classical query complexities have been given for general graphs. For any graph G of N vertices and maximal degree δ, it has been showed that $\mathsf{RLS}(G) = O(\sqrt{N\delta})$ [3] and $\mathsf{QLS}(G) = O(N^{1/3}\delta^{1/6})$ [2]. The idea for designing local search algorithms in [3, 2] is random sampling followed by steepest descent. More specifically, these algorithms start off by sampling a subset of vertices, find the best vertex v (i.e., the one with the minimum f value) in the sampled set, and finally performing steepest descent beginning at the chosen vertex v.

Zhang [11] later introduced new quantum and randomized algorithms for local search on general graphs, providing upper bounds that depend on the graph diameter and the expansion speed. While Zhang's upper bounds can only work well for graphs with slow expansion speed, such as hypecubes, many vertex-transitive graphs, unfortunately, do not possess this property. Also, Zhang's randomized upper bound is no better than $O\left(\frac{N}{d} \log \log d\right)$, and his quantum upper bound is no better than $O\left(\sqrt{\frac{N}{d}} (\log \log d)^{1.5}\right)$, except for the line or cycle graphs, where d is the diameter of the graph. This means Zhang's upper bounds do not seem to beat Aaronson and Aldous's bounds, especially for graphs with small degrees and small diameters. Note that there are Cayley graphs of non-abelian simple groups which have constant degrees and have diameters no larger than $O(\log N)$. While Zhang's upper bounds fail for graphs of small diameters, Aldous and Aaronson's upper bounds fail for graphs of large degrees. So, a question to ask is whether there is a better upper bound for graphs with large degrees and small diameters?

Recently, Verhoeven [10] has proposed another deterministic algorithm and enhanced Zhang's quantum algorithm, improving upper bounds on deterministic and quantum query complexities of Local Search that depend on the graph's degrees and genus. Precisely, he showed that for any N-vertex graph G of genus g and maximal degree δ, the deterministic (thus, randomized) and query complexities of Local Search on G are $\delta + O(\sqrt{g})\sqrt{N}$ and $O(\sqrt{\delta}) + O(\sqrt[4]{g})\sqrt[4]{N}\log\log N$, respectively. However, these bounds fail for the class of graphs we are caring about: vertex-transitive graphs, since every vertex-transitive graph is regular and it has been shown that the genus of an N-vertex m-egde connected graph is at least $\lceil \frac{m}{6} - \frac{N}{2} + 1 \rceil$ (see [7, p114]).

On the Query Complexity of Testing Orientations for Being Eulerian[*]

Eldar Fischer[1,**], Oded Lachish[2], Ilan Newman[3,**], Arie Matsliah[1],
and Orly Yahalom[1]

[1] Computer Science Department, Technion - Israel Institute of Technology, Haifa 32000, Israel
{eldar,ariem,oyahalom}@cs.technion.ac.il
[2] Centre for Discrete Mathematics and its Applications (DIMAP), Warwick, UK
o.lachish@warwick.ac.uk
[3] Computer Science Department, Haifa University, Haifa 31905 Israel
ilan@cs.haifa.ac.il

Abstract. We consider testing directed graphs for being Eulerian in the orientation model introduced in [15]. Despite the local nature of the property of being Eulerian, it turns out to be significantly harder for testing than other properties studied in the orientation model. We show a non-constant lower bound on the query complexity of 2-sided tests and a linear lower bound on the query complexity of 1-sided tests for this property. On the positive side, we give several 1-sided and 2-sided tests, including a sub-linear query complexity 2-sided test for general graphs. For special classes of graphs, including bounded-degree graphs and expander graphs, we provide improved results. In particular, we give a 2-sided test with constant query complexity for dense graphs, as well as for expander graphs with a constant expansion parameter.

1 Introduction

Property testing deals with the following relaxation of decision problems: Given a property $\mathcal{P}$, an input structure S and $\epsilon > 0$, distinguish between the case where S satisfies $\mathcal{P}$ and the case where S is ϵ-far from satisfying $\mathcal{P}$. Roughly speaking, an input S is said to be ϵ-*far* from satisfying a property $\mathcal{P}$ if more than an ϵ-fraction of its values must be modified in order to make it satisfy the property. Algorithms which distinguish with high probability between the two cases are called *property testers* or simply *testers* for $\mathcal{P}$. Furthermore, a tester for $\mathcal{P}$ is said to be *1-sided* if it never rejects an input that satisfies $\mathcal{P}$. Otherwise, the tester is called *2-sided*. We say that a tester is *adaptive* if some of the choices of the locations for which the input is queried may depend on the returned values (answers) of previous queries. Otherwise, the tester is called *non-adaptive*. Property testing normally deals with problems involving a very large input or a costly retrieval procedure. Thus, the number of queries of input values, rather than the computation time, is considered to be the most expensive resource.

[*] A full version is available at http://www.cs.technion.ac.il/˜oyahalom/EulerianOrientations.pdf.
[**] Research supported in part by an ISF grant number 1101/06.

A. Goel et al. (Eds.): APPROX and RANDOM 2008, LNCS 5171, pp. 402–415, 2008.

Property testing has been a very active field of research since it was initiated by Blum, Luby and Rubinfeld [5]. The general definition of property testing was formulated by Rubinfeld and Sudan [25], who were interested mainly in testing algebraic properties. The study of property testing for combinatorial objects, and mainly for labelled graphs, began in the seminal paper of Goldreich, Goldwasser and Ron [12]. They introduced the *dense graph model*, where the graph is represented by an adjacency matrix, and the distance function is computed accordingly. For comprehensive surveys on property testing see [24,8].

The dense graph model is in a sense too lenient, since for n-vertex graphs, the distance function allows adding and removing $o(n^2)$ edges, regardless of the number of actual edges in the graph. Thus, many interesting properties, such as connectivity in undirected or directed graphs, are trivially testable in this model, as all the graphs are close to satisfying the property. In recent years, researchers have studied several alternative models for graph testing, including the *bounded-degree* graph model of [13], in which a sparse representation of sparse graphs is considered, and the *general density* model (also called the *mixed* model) of [21] and [17]. In these models, the distance function allows edge insertions and deletions whose number is at most a fraction of the number of the edges in the original graph.

Property testing of directed graphs has also been studied in the context of the above models [1,3]. Here we continue the study of testing properties of directed graphs in the orientation model, which started in [15] and followed in [14] and [7]. In this model, an underlying undirected graph $G = (V, E)$ is given in advance, and the actual input is an orientation $\overrightarrow{G}$ of G, in which every edge in E has a direction. Our testers may access the input using edge queries. That is, every query concerns an edge $e \in E$, and the answer to the query is the direction of e in $\overrightarrow{G}$. An orientation $\overrightarrow{G}$ of G is called ϵ-*close* to a property $\mathcal{P}$ if it can be made to satisfy $\mathcal{P}$ by inverting at most an ϵ-fraction of the edges of G, and otherwise $\overrightarrow{G}$ is said to be ϵ-*far* from $\mathcal{P}$.

Note that the distance function in the orientation model naturally depends on the size of the underlying graph and is independent of representation details. Moreover, the testing algorithm may strongly depend on the structure of the underlying graph. The model is strict in that the distance function allows only edge inversions, but no edge insertions or deletions. On the other hand, we assume that our algorithms have a full knowledge of the underlying graph, whose size is roughly the same as the input size. Viewing the underlying graph as a parameter that the testing algorithm receives in advance, we say that the orientation model is an example of a *massively parameterized* model. Other examples of massively parameterized models appear in [20], where the property is represented by a known bounded-width branching program, in [9], where the input is a vertex-coloring of a known graph, and in other works.

In this paper we consider the property of being Eulerian, which was presented in [14] as one of the natural orientation properties whose query complexity was still unknown. A directed graph $\overrightarrow{G}$ is called *Eulerian* if for every vertex v in the graph, the in-degree of v is equal to its out-degree. An undirected graph G has an Eulerian orientation $\overrightarrow{G}$ if and only if all the degrees of G are even. Such an undirected graph is called Eulerian also. Throughout the paper we assume that our underlying undirected graph G is Eulerian. We note that it is common to require an Eulerian graph to be connected. However, we may ignore this requirement, as all our algorithms and proofs work equally well whether G is connected or not. Moreover, as G is given as a parameter, its connectivity can be tested in a preprocessing stage.

Eulerian graphs and Eulerian orientations have attracted researchers since the dawn of graph theory in 1736, when Leonard Euler published his solution for the famous "Königsberg bridge problem". Throughout the years, Eulerian graphs have been the subject of extensive research (e.g. [23,18,26,19,6,2]; see [10,11] for an extensive survey). Aside from their appealing theoretic characteristics, Eulerian graphs have been studied in the context of networking [16] and genetics [22].

Testing for being Eulerian in the orientation model is equivalent to the following problem. We have a known network (e.g. a communication network or a transportation system) where every edge can transport a unit of "flow" in both directions. Our goal is to know whether the network is "balanced", or far from being balanced, where being balanced means that the number of flows entering every node in the network is equal to the number of flows exiting it. To examine the network, we detect the flow direction in selected individual edges, and this is deemed to be the expensive operation.

The main difficulty in testing orientations for being Eulerian arises from the fact that an orientation might have a small number of unbalanced vertices, and each of them with a small imbalance, and yet be far from being Eulerian. This is since trying to balance an unbalanced vertex by inverting some of its incident edges may violate the balance of its balanced neighbors. Thus, we must continue to invert edges along a directed path between a vertex with a positive imbalance and a vertex with a negative imbalance. We call such a path a *correction path*. A main component of our work is giving upper bounds for the length of the correction paths. We note that Babai [2] showed that the ratio between the diameter of digraphs and the diameter of their underlying undirected graphs is $\Omega(n^{1/3})$ for an infinite family of Eulerian graphs.

Our upper bounds are based on three "generic" tests, one 1-sided test and two 2-sided tests. Instead of receiving ϵ as a parameter, the generic tests receive a parameter p, which stands for the number of required correction paths in an orientation that is far from being Eulerian. We hence call these tests p-*tests*. We later derive ϵ-tests from the p-tests by proving two lower bounds for p. The first one gives an efficient test for dense graphs and the second one gives an efficient test for expander graphs. Finally, we show how to use variations of the expander tests for obtaining a 1-sided test and a 2-sided test for general graphs, using a decomposition ("chopping") procedure into subgraphs that are roughly expanders. The 2-sided test that we obtain this way has a sub-linear query complexity for every graph. Unfortunately, our chopping procedure is adaptive and has an exponential computational time in $|E|$. All of our other algorithms are non-adaptive and their computational complexity is of the same order as their query complexity.

On the negative side, we provide several lower bounds. We show that any 1-sided test for being Eulerian must use $\Omega(m)$ queries for some graphs. For bounded-degree graphs, we use the toroidal grid to prove non-constant 1-sided and 2-sided lower bounds. These bounds are noteworthy, as bounded-degree graphs have a constant size witness for not being Eulerian, namely the edges incident with one unbalanced vertex. In contrast, the st-connectivity property, whose witness must include a cut in the graph, is testable with a constant number of queries in the orientation model [7]. In other testing models there

Table 1. Upper bounds

Result	1-sided tests	2-sided tests	
Graphs with large d	$O\left(\frac{\Delta m}{\epsilon^2 d^2}\right)$	$\min\left\{\tilde{O}\left(\frac{m^3}{\epsilon^6 d^6}\right), \tilde{O}\left(\frac{\sqrt{\Delta}m}{\epsilon^2 d^2}\right)\right\}$	
α-expanders (Section 4)	$O\left(\frac{\Delta \log(1/\epsilon)}{\alpha\epsilon}\right)$	$\min\left\{\tilde{O}\left(\left(\frac{\log(1/\epsilon)}{\alpha\epsilon}\right)^3\right), \tilde{O}\left(\frac{\sqrt{\Delta}\log(1/\epsilon)}{\alpha\epsilon}\right)\right\}$	
General graphs (Section 6)	$O\left(\frac{(\Delta m \log m)^{2/3}}{\epsilon^{4/3}}\right)$	$\min\left\{\tilde{O}\left(\frac{\Delta^{1/3}m^{2/3}}{\epsilon^{4/3}}\right), \tilde{O}\left(\frac{\Delta^{3/16}m^{3/4}}{\epsilon^{5/4}}\right)\right\}$	

Table 2. Lower bounds

Result	1-sided tests	2-sided tests
General graphs (Section 2)	$\Omega(m)$	—
Bounded-degree graphs, non-adaptive tests	$\Omega(m^{1/4})$	$\Omega\left(\sqrt{\frac{\log m}{\log\log m}}\right)$
Bounded-degree graphs, adaptive tests	$\Omega(\log m)$	$\Omega(\log\log m)$

are known super-constant lower bounds also for properties which have constant-size witness, e.g., [4] prove a linear lower bound for testing whether a truth assignment satisfies a known 3CNF formula. However, most of these bounds are for properties that have stronger expressive power than that of being Eulerian.

Tables 1 and 2 summarize our upper and lower bounds, respectively. Here and throughout the paper, we set $n = |V|$ and $m = |E|$, let Δ be the maximum vertex-degree in G, and set $d \overset{\text{def}}{=} m/n$. The tilde notation hides polylogarithmic factors. Due to space limitations, our upper bounds for dense graphs and lower bounds for bounded-degree graphs are omitted from this version, and most of the proofs are given as sketches.

2   Preliminaries and the 1-Sided Lower Bound

In this section we introduce basic definitions, notations and lemmas to be used in the sequel. Throughout the paper, we assume a fixed and known underlying graph $G = (V, E)$ which is Eulerian, that is, for every $v \in V$, the degree $\deg(v)$ of v is even. Given an orientation $\overrightarrow{G} = (V, \overrightarrow{E})$ and a vertex $v \in V$, let $\text{indeg}_{\overrightarrow{G}}(v)$ denote the in-degree of v with respect to $\overrightarrow{G}$ and let $\text{outdeg}_{\overrightarrow{G}}(v)$ denote the out-degree of v with respect to $\overrightarrow{G}$. We define the *imbalance* of v in $\overrightarrow{G}$ as $\text{ib}_{\overrightarrow{G}}(v) \overset{\text{def}}{=} \text{outdeg}_{\overrightarrow{G}}(v) - \text{indeg}_{\overrightarrow{G}}(v)$. In the following, we sometimes omit the subscript $\overrightarrow{G}$ whenever it is obvious from the context. We say that a vertex $v \in V$ is a *spring* in $\overrightarrow{G}$ if $\text{ib}_{\overrightarrow{G}}(v) > 0$. We say that v is a *drain* in $\overrightarrow{G}$ if $\text{ib}_{\overrightarrow{G}}(v) < 0$. If $\text{ib}_{\overrightarrow{G}}(v) = 0$ then we say that v is *balanced* in $\overrightarrow{G}$. We say that $\overrightarrow{G}$ is *Eulerian* if all its vertices are balanced. Since all the vertices of G are of even degree, there always exists some Eulerian orientation $\overrightarrow{G}$ of G.

Given a set $U \subseteq V$, let:

$$E(U) \overset{\text{def}}{=} \{\{u, v\} \in E \mid u, v \in U\} \quad \text{and} \quad \overrightarrow{E}(U) \overset{\text{def}}{=} \{(u, v) \in \overrightarrow{E} \mid u, v \in U\},$$

$$\partial U \overset{\text{def}}{=} \{\{u, v\} \in E \mid u \in U, v \notin U\} \quad \text{and} \quad \overrightarrow{\partial} U \overset{\text{def}}{=} \{(u, v) \in \overrightarrow{E} \mid u \in U, v \notin U\}.$$

Given two disjoint sets $U, W \subseteq V$, let $E(U, W) \stackrel{\text{def}}{=} \{\{u, w\} \in E \mid u \in U, w \in W\}$ and $\overrightarrow{E}(U, W) \stackrel{\text{def}}{=} \{(u, w) \in \overrightarrow{E} \mid u \in U, w \in W\}$.

Lemma 1. *Suppose that $\overrightarrow{H}$ is a knowledge graph that does not contain invalid cuts. Then $\overrightarrow{H}$ is extensible to an Eulerian orientation $\overrightarrow{G} = (V, \overrightarrow{E_G})$ of G. That is, $\overrightarrow{E_H} \subseteq \overrightarrow{E_G}$. Consequently, a witness that an orientation $\overrightarrow{G}$ is not Eulerian must contain at least half of the edges of some invalid cut with respect to $\overrightarrow{G}$.*

Proof sketch. We extend the knowledge graph to an orientation of the entire graph by orienting the edges one by one. In each step we prove using counting arguments that if a certain orientation of an edge would invalidate one of the cuts, then orienting it in the other direction would not invalidate any of the other cuts. $\qquad\square$

Theorem 2. *There exists an infinite family of graphs for which every 1-sided test for being Eulerian must use $\Omega(m)$ queries.*

Proof. For every even n, let $G_n \stackrel{\text{def}}{=} K_{2,n-2}$, namely, the graph with a set of vertices $V = \{v_1, \ldots, v_n\}$ and a set of edges $E = \{\{v_i, v_j\} \mid i \in \{1, 2\}, j \in \{3, \ldots, n\}\}$. Clearly, G_n is Eulerian and $n = \Omega(m)$. Consider the orientation $\overrightarrow{G}_n$ of G_n in which all the edges incident with v_1 are outgoing and all the edges incident with v_2 are incoming. Clearly, $\overrightarrow{G}_n$ is $\frac{1}{2}$-far from being Eulerian. According to Lemma 1, every 1-sided test must query at least half of the edges in some unbalanced cut (because otherwise it would clearly not obtain an invalid cut in the knowledge graph). However, one can easily see that every cut which does not separate v_1 and v_2 is balanced, while every cut which separates v_1 and v_2 is of size $n - 2 = \Omega(m)$. $\qquad\square$

Let $\overrightarrow{G}$ be an orientation of G. Given a subgraph $\overrightarrow{H} = (V_H, \overrightarrow{E_H})$ of $\overrightarrow{G}$ (that is, a directed graph where $V_H \subseteq V$ and $\overrightarrow{E_H} \subseteq \overrightarrow{E}$) we define $\overrightarrow{G}_{\overleftarrow{H}} \stackrel{\text{def}}{=} (V, \overrightarrow{E}_{\overleftarrow{H}})$ to be the orientation of G derived from $\overrightarrow{G}$ by inverting all the edges of $\overrightarrow{H}$. Namely, $\overrightarrow{E}_{\overleftarrow{H}} = \overrightarrow{E} \setminus \overrightarrow{E_H} \cup \{(v, u) \in (V_H)^2 \mid (u, v) \in \overrightarrow{E_H}\}$. We say that $\overrightarrow{H}$ is a *correction subgraph* of $\overrightarrow{G}$ if $\overrightarrow{G}_{\overleftarrow{H}}$ is Eulerian. Note that in such a case, $\overrightarrow{G}$ is $|\overrightarrow{E_H}|/m$-close to being Eulerian. Since we assume that G is Eulerian, there exists some correction subgraph $\overrightarrow{H}$ for any $\overrightarrow{G}$. Furthermore, it is not difficult to show that any correction subgraph $\overrightarrow{H}$ of $\overrightarrow{G}$ has an acyclic subgraph which is also a correction subgraph of $\overrightarrow{G}$. Let S be the set of springs in $\overrightarrow{G}$ and let T be the set of drains in $\overrightarrow{G}$. We say that a directed path $\overrightarrow{P} = \langle u_0, \ldots, u_k \rangle$ in $\overrightarrow{G}$ is a *spring-drain path* if $u_0 \in S$ and $u_k \in T$. It is easy to show that for any correction subgraph $\overrightarrow{H}$ of $\overrightarrow{G}$, u_0 is a spring in $\overrightarrow{H}$ and u_k is a drain in $\overrightarrow{H}$.

Lemma 3. *If $\overrightarrow{G}$ is not Eulerian then any acyclic correction subgraph $\overrightarrow{H}$ of $\overrightarrow{G}$ is a union of $p = \frac{1}{4} \sum_{u \in V} |\text{ib}(u)|$ edge-disjoint spring-drain paths.*

Proof sketch. Suppose that $\overrightarrow{G}$ is not Eulerian and let $\overrightarrow{H}$ be an acyclic correction subgraph of $\overrightarrow{G}$. By definition, if we invert all the edges of $\overrightarrow{H}$ in $\overrightarrow{G}$ then we obtain an Eulerian orientation of G. It can easily be seen that, since $\overrightarrow{G}$ is not Eulerian, $\overrightarrow{H}$ contains a spring-drain path. We thus invert the edges of $\overrightarrow{H}$ along one spring-drain path at a time, until we obtain an Eulerian orientation. One can see that $\overrightarrow{H}$ is thus decomposed to the inverted paths. The value of p is computed by noting that by inverting a spring-drain path, we reduce the sum $\sum_{u \in V} |\text{ib}(u)|$ by exactly four.   □

Let p be some positive number. If every correction subgraph of an orientation $\overrightarrow{G}$ is a union of at least p disjoint spring-drain paths, we say that $\overrightarrow{G}$ is p-far from being Eulerian. An algorithm is called a p-test for being Eulerian if it accepts an Eulerian orientation with probability at least $2/3$ and rejects a p-far orientation with probability at least $2/3$. Similarly to ϵ-tests, if a p-test accepts every Eulerian orientation with probability 1 then it is called 1-sided, and otherwise it is called 2-sided.

Given $\beta > 0$, we say that a vertex v is β-small if $\deg(v) \leq \beta$ and β-big if $\deg(v) > \beta$. An orientation $\overrightarrow{G}$ is called β-Eulerian if all the β-small vertices in V are balanced in $\overrightarrow{G}$. Note that for $\beta \geq \Delta$, $\overrightarrow{G}$ is β-Eulerian if and only if $\overrightarrow{G}$ is Eulerian. All our lemmas and observations for Eulerian orientations may be adapted to β-Eulerian orientations. In particular, we can show that modifying an orientation $\overrightarrow{G}$ to become β-Eulerian requires inverting edges along at least $\frac{1}{4} \sum_{u \in V, \deg(u) \leq \beta} |\text{ib}(u)|$ spring-drain paths in which at least one of the spring and the drain is β-small. We call such paths β-spring-drain paths.

An algorithm is called a (p, β)-test for being Eulerian for some positive number p if it accepts a β-Eulerian orientation with probability at least $2/3$ and rejects an orientation that is p-far from being β-Eulerian with probability at least $2/3$. As usual, a (p, β)-test is said to be 1-sided if it accepts every β-Eulerian orientation with probability 1. Otherwise, the test is said to be 2-sided.

3   Generic Tests

We present a p-test and two (p, β)-tests for being Eulerian. In later sections we devise several lower bounds on p for every orientation $\overrightarrow{G}$ that is ϵ-far from being Eulerian, thus obtaining corresponding upper bounds on the tests below.

We begin with a simple 2-sided p-test whose query complexity is independent of the maximum degree Δ. The algorithm uses probabilistic methods, as well as the characterization of p given given in Lemma 3, in order to detect an unbalanced vertex with high probability. To simplify notation, we denote $\delta \stackrel{\text{def}}{=} \frac{p}{4m}$.

Algorithm 4. *SIMPLE-2($\overrightarrow{G}, p$):*

- *Repeat $\frac{4}{\delta}$ times independently:*
  - *Select an edge $e \in E(G)$ uniformly and query it. Denote the start vertex of e in $\overrightarrow{E}$ by u and the end vertex of e in $\overrightarrow{E}$ by v.*
  - *Query $\frac{16 \ln(12/\delta)}{\delta^2}$ edges incident with u uniformly and independently and reject if the sample contains at least $(1 + \delta) \frac{8 \ln(12/\delta)}{\delta^2}$ outgoing edges.*
- *Accept if the input was not rejected earlier.*

Lemma 5. *SIMPLE-2 is a 2-sided p-test for being Eulerian with query complexity $\widetilde{O}\left(\frac{1}{\delta^3}\right) = \widetilde{O}\left(\frac{m^3}{p^3}\right)$.*

We next give a simple 1-sided (p, β)-test, which has a better query complexity than SIMPLE-2 for $\Delta \ll \frac{m^2}{p^2} \ln(\frac{m}{p})$. Note that the test checks only β-small vertices for being unbalanced.

Algorithm 6. *GENERIC-1($\overrightarrow{G}, p, \beta$):*

1. *Repeat $\frac{\ln 3\, m}{p}$ times independently:*
   - *Select an edge $e \in E(G)$ uniformly and query it. Denote the start vertex of e in $\overrightarrow{E}$ by u and the end vertex of e in $\overrightarrow{E}$ by v.*
   - *If $\deg(u) \leq \beta$ then query all the edges $\{u, w\} \in E$ and reject if u is unbalanced.*
2. *Repeat $\frac{\ln 3\, m}{p}$ times independently:*
   - *Select an edge $e \in E(G)$ uniformly and query it. Denote the start vertex of e in $\overrightarrow{E}$ by u and the end vertex of e in $\overrightarrow{E}$ by v.*
   - *If $\deg(v) \leq \beta$ then query all the edges $\{w, v\} \in E$ and reject if v is unbalanced.*
3. *Accept if the input was not rejected by the above.*

Lemma 7. *GENERIC-1 is a 1-sided (p, β)-test for being Eulerian with query complexity $O\left(\frac{\beta m}{p}\right)$. In particular, for $\beta = \Delta$, GENERIC-1 is a 1-sided p-test with query complexity $O\left(\frac{\Delta m}{p}\right)$.*

We conclude this section with a 2-sided (p, β)-test, which gives better query complexity than GENERIC-1 for $\beta \gg \log^2 m$ and better query complexity than SIMPLE-2 for $p \ll \frac{m}{\sqrt{\beta}}$. The main idea of the algorithm is to perform roughly $O((\log \beta)^2)$ testing stages, each designed to detect unbalanced β-small vertices whose degree and imbalance lie in a certain interval. In the following, log denotes the logarithm with base 2.

Algorithm 8. *MULTISTAGE-2($\overrightarrow{G}, p, \beta$):*

For $i = 1, \ldots, \lceil \log \beta \rceil - 1$, do:

1. *Let $V_i \stackrel{\text{def}}{=} \{u \in V \mid \deg(u) \in [2^i, 2^{i+1})\}$ and $n_i \stackrel{\text{def}}{=} |V_i|$.*
2. *Let $j = \lceil i/2 \rceil$. If $2^j \cdot n_i > \frac{2p}{(\log \beta)^2}$ then:*
   - *Sample $x_{ij} = \frac{\ln 12 \,(\log \beta)^2 \, 2^{j+1} \, n_i}{2p}$ vertices in V_i uniformly and independently.*
   - *For every sampled vertex u, query all the edges incident with u, and reject if u is unbalanced.*
3. *For every $j \in \{\lceil i/2 \rceil + 1, \ldots, i - 1\}$ such that $2^j \cdot n_i > \frac{2p}{(\log \beta)^2}$ do:*
   - *Sample $x_{ij} = \frac{\ln 12 \,(\log \beta)^2 \, 2^{j+1} \, n_i}{2p}$ vertices in V_i uniformly and independently.*
   - *For every sampled vertex u, query $q_{ij} = 256 \cdot \ln(\, 6(\log \beta)^2 \, x_{ij}) \cdot 2^{2(i-j)}$ edges adjacent to u, uniformly and independently, and reject if the absolute difference between the number of incoming and outgoing edges in the sample is at least $\frac{q_{ij}}{4 \cdot 2^{i-j}}$.*

Accept if the input was not rejected earlier.

Lemma 9. *MULTISTAGE-2 is a 2-sided (p, β)-test for being Eulerian with query complexity $\widetilde{O}\left(\frac{\sqrt{\beta}\,m}{p}\right)$. In particular, for $\beta = \Delta$, it is a 2-sided p-test for being Eulerian with query complexity $\widetilde{O}\left(\frac{\sqrt{\Delta}\,m}{p}\right)$.*

4   Testing Orientations of Expander Graphs

In this section we show how to apply our generic tests for expander graphs. A graph $G = (V, E)$ is called an α-*expander* for some $\alpha > 0$, if it is connected and for every $U \subseteq V$ such that $0 < |E(U)| \leq m/2$ we have $|\partial U| \geq \alpha |E(U)|$. Note that while the diameter of G is $O(\log_{(1+\alpha)} m)$, the "oriented-diameter" of $\overrightarrow{G}$ is not necessarily low, even if we assume that the orientation is Eulerian, as was shown by [2].

In the following, $\log_b^{(k)}(x)$ denotes the k-nested logarithm with base b of x, i.e., $\log_b^{(1)}(x) \overset{\text{def}}{=} \log_b(x)$ and $\log_b^{(k+1)}(x) \overset{\text{def}}{=} \log_b(\log_b^{(k)}(x))$ for any natural $k \geq 1$.

Lemma 10. *Let G be an Eulerian α-expander and let $k \geq 1$ be a natural number such that $\log_{(1+\alpha/2)}^{(k-1)} m \geq \log_{(1+\alpha/2)}\left(\frac{4}{\epsilon}\right)$. Then: (1) Every non-Eulerian orientation $\overrightarrow{G}$ of G contains a spring-drain path of length at most $\ell_k \overset{\text{def}}{=} 2 \cdot \log_{(1+\alpha/2)}^{(k)} m + 2 \cdot \log_{(1+\alpha/2)}\left(\frac{4}{\epsilon}\right)$; (2) Every orientation $\overrightarrow{G}$ of G that is ϵ-far from being Eulerian is p_k-far from being Eulerian for $p_k \overset{\text{def}}{=} \frac{\epsilon m}{\ell_k}$.*

Proof sketch. We prove the lemma by induction on k. In each inductive step, we use the known bounds of ℓ_k and p_k to bound ℓ_{k+1} and p_{k+1} in an iterative manner. To prove Item 1 of the lemma for $k = 1$, let $\overrightarrow{G}$ be a non-Eulerian orientation of $\overrightarrow{G}$. Consider a BFS traversal of $\overrightarrow{G}$ starting from the set S of springs. For every $i \geq 0$, let L_i be the ith level of the traversal, where $L_0 = S$, and let $U_{<i} \overset{\text{def}}{=} \bigcup_{0 \leq j < i} L_j$ and $U_{\geq i} \overset{\text{def}}{=} \bigcup_{j \geq i} L_j$. For every $i > 0$, let f_i be the number of directed edges going from L_{i-1} to L_i. Let L_ℓ be the first level that contains a drain. By the expander property of G, for every $i > 0$ while $|E(U_{<i})| \leq m/2$ we have $|\partial(U_{<i})| \geq \alpha |E(U_{<i})|$. Note that for every $i \leq \ell$, the set $U_{<i}$ contains no drains, and all the directed edges that exit it are from L_{i-1} to L_i. Hence, for every $0 < i \leq \ell$ while $|E(U_{<i})| \leq m/2$, we have $f_i > \frac{1}{2}|\partial(U_{<i})| > \frac{\alpha}{2}|E(U_{<i})|$ and therefore $|E(U_{<i+1})| > \left(1 + \frac{\alpha}{2}\right)|E(U_{<i})|$. By induction, we have $|E(U_{<i})| > \left(1 + \frac{\alpha}{2}\right)^{i-1} f_1 \geq \left(1 + \frac{\alpha}{2}\right)^{i-1}$ for every $0 < i \leq \ell$ for which $|E(U_{<i})| \leq m/2$.

Now, if for every $0 < i \leq \ell$ we have $|E(U_{<i})| \leq m/2$, then clearly, $|E(U_{<\ell})| > \left(1 + \frac{\alpha}{2}\right)^{\ell-1}$, and hence $\ell = \ell_1 < \log_{(1+\alpha/2)} m$. Otherwise, let $r > 0$ be the minimal index for which $|E(U_{<r})| > m/2$. Using similar arguments to the above, we show that $|E(U_{\geq i-1})| > \left(1 + \frac{\alpha}{2}\right)^{\ell-i+1} |E(U_{\geq \ell})| \geq \left(1 + \frac{\alpha}{2}\right)^{\ell-i+1}$ for every $r \leq i \leq \ell$, which yields $\ell_1 < 2 \cdot \log_{(1+\alpha/2)} m$.

To prove Item 2 of the lemma for $k = 1$, let $\overrightarrow{G}$ be an orientation of G that is ϵ-far from being Eulerian. While $\overrightarrow{G}$ is not Eulerian, choose a shortest spring-drain path in $\overrightarrow{G}$ and invert all its edges. By Item 1, every chosen spring-drain path is of length at most ℓ_1. Let $\overrightarrow{H}$ be the union of the paths inverted. Clearly, $\overrightarrow{H}$ is a correction subgraph of $\overrightarrow{G}$. As $\overrightarrow{G}$ is ϵ-far from being Eulerian, $\overrightarrow{H}$ contains at least ϵm edges, and thus it is necessarily a union of at least $p_1 = \frac{\epsilon m}{\ell_1}$ disjoint spring-drain paths. By Lemma 3, every correction subgraph of $\overrightarrow{G}$ contains the same number of disjoint spring-drain paths, which completes the base case.

Assuming that the lemma holds for some natural $k \geq 1$, the proof of the lemma for $k+1$ is very similar to that of the base case. However, we now know that $f_1 \geq p_k$ and $|E(U_{\geq \ell})| \geq p_k$, and so we use our known lower bound for p_k (instead of 1 in the base case). Item 1 is now proved using standard arithmetics, as well as the condition $\log_{(1+\alpha/2)}^{(k)} m \geq \log_{(1+\alpha/2)}\left(\frac{4}{\epsilon}\right)$. The proof of Item 2 is the same as for the base case. $\square$

Lemma 11. *Let G be an Eulerian α-expander. Let $\overrightarrow{G}$ be an orientation of G that is ϵ-far from being Eulerian. Then $\overrightarrow{G}$ is p-far from being Eulerian for $p = \Omega\left(\frac{\alpha \epsilon m}{\log(\frac{1}{\epsilon})}\right)$.*

Proof sketch. The proof considers the first natural number k such that the condition of Lemma 10 does not apply, namely $\log_{(1+\alpha/2)}^{(k)} m < \log_{(1+\alpha/2)}\left(\frac{4}{\epsilon}\right)$. The proof is similar to that of Lemma 10 for smaller k's. However, since $\log_{(1+\alpha/2)}^{(k)} m$ is sufficiently small, we are able to give the stated upper bound, which is independent of k. $\square$

Substituting the lower bound for p of Lemma 11 in Lemmas 5, 7, and 9, we obtain the following theorem. Note that for a constant α, the query complexity of SIMPLE-2 depends only on ϵ.

Theorem 12. *Let G be an α-expander (for some $\alpha > 0$) with m edges and maximum degree Δ. Then:*

1. *SIMPLE-2$\left(\overrightarrow{G}, \Omega\left(\frac{\alpha \epsilon m}{\log(1/\epsilon)}\right)\right)$ is a 2-sided ϵ-test for being Eulerian with query complexity $\tilde{O}\left(\left(\frac{\log(1/\epsilon)}{\alpha \epsilon}\right)^3\right)$.*

2. *GENERIC-1$\left(\overrightarrow{G}, \Omega\left(\frac{\alpha \epsilon m}{\log(1/\epsilon)}\right), \Delta\right)$ is a 1-sided ϵ-test for being Eulerian with query complexity $O\left(\frac{\Delta \log(1/\epsilon)}{\alpha \epsilon}\right)$.*

3. *MULTISTAGE-2$\left(\overrightarrow{G}, \Omega\left(\frac{\alpha \epsilon m}{\log(1/\epsilon)}\right), \Delta\right)$ is a 2-sided ϵ-test for being Eulerian with query complexity $\tilde{O}\left(\frac{\sqrt{\Delta} \log(1/\epsilon)}{\alpha \epsilon}\right)$.*

5   Testing Orientations of "Lame" Directed Expanders

In this section we discuss a variation of the expander test, which will serve us in Section 6 for devising tests for general graphs. Given an orientation $\overrightarrow{G}$ of G, we now test a subgraph $\overrightarrow{G}[U]$ of $\overrightarrow{G}$, induced by a subset $U \subseteq V$. We refer to the edges in $E(U)$ as the *internal edges* of $\overrightarrow{G}[U]$, and denote $m_U \stackrel{\text{def}}{=} |E(U)|$. We say that $\overrightarrow{G}[U]$ is *Eulerian* if and only if all the vertices in U are balanced in $\overrightarrow{G}$. We say that $\overrightarrow{G}[U]$ is β-*Eulerian* if and only if all the β-small vertices in U are balanced in $\overrightarrow{G}$. Note that these definitions rely also on the edges in ∂U, which we will henceforth call *external edges*. We assume that the orientations of all the external edges are known, and furthermore, we use a distance function that does not allow inverting external edges. Namely, we will say that $\overrightarrow{G}[U]$ is ϵ-*close* to being Eulerian if and only if it has a correction subgraph of size at most ϵm_U which includes only internal edges. Otherwise, we say that $\overrightarrow{G}[U]$ is ϵ-*far* from being Eulerian. Note that we can view the external edges as comprising a knowledge graph (see Section 2). We always assume that all the cuts in $\overrightarrow{G}$ are valid with respect to the orientation $\overrightarrow{\partial}U$ of the external edges. This condition ensures that $\overrightarrow{G}[U]$ can be made Eulerian (or β-Eulerian) by inverting internal edges only.

We will be interested in induced subgraphs $\overrightarrow{G}[U]$ that are "lame directed expanders". Formally, given a subset $U \subseteq V$ and a parameter $\beta > 0$, we say that a cut (A, B) of U is a β-*cut of* U if $|E(B)| \geq |E(A)| \geq \beta$. Given $\alpha, \beta > 0$, we say that the subgraph $\overrightarrow{G}[U]$ of G is an (α, β)-*expander* if for every β-cut (A, B) of U:

$$|E(A, B)| - \left||\overrightarrow{E}(V \setminus U, A)| - |\overrightarrow{E}(A, V \setminus U)|\right| \geq 2\alpha|E(A)|. \tag{1}$$

Lemma 13. *Let* $\alpha, \beta, \epsilon > 0$ *be parameters and let* $U \subseteq V$ *be such that* $\overrightarrow{G}[U]$ *is an* (α, β)-*expander. Denote* $m_U \stackrel{def}{=} |E(U)|$ *and* $\Delta_U \stackrel{def}{=} \max\{\deg(u) \mid u \in U\}$. *Assume that the external edges of* U *are known and do not induce an invalid cut. Then:*

1. *There exists a 1-sided ϵ-test for whether* $\overrightarrow{G}[U]$ *is Eulerian, GEN-1($\overrightarrow{G}[U], \alpha, \beta, \epsilon$), whose query complexity is* $O\left(\frac{\Delta_U \log m_U}{\epsilon\alpha} + \frac{\beta \cdot \min\{\beta, \Delta_U\}}{\epsilon}\right)$.

2. *There exists a 2-sided ϵ-test for whether* $\overrightarrow{G}[U]$ *is Eulerian, MULTI-2($\overrightarrow{G}[U], \alpha, \beta, \epsilon$), whose query complexity is* $\tilde{O}\left(\frac{\sqrt{\Delta_U} \log m_U}{\epsilon\alpha} + \frac{\beta \cdot \sqrt{\min\{\beta, \Delta_U\}}}{\epsilon}\right)$.

Proof sketch. GEN-1 is based on at most two calls to GENERIC-1 (Algorithm 6) and MULTI-2 is based on at most two calls to MULTISTAGE-2 (Algorithm 8). The parameters in these calls are computed by analyzing two possible cases in which $\overrightarrow{G}[U]$ is ϵ-far from being Eulerian.

In the first case, $\overrightarrow{G}[U]$ is $\frac{\epsilon}{2}$-far from being 2β-Eulerian, which means that we need to invert many 2β-spring-drain paths in $\overrightarrow{G}[U]$ in order to make it 2β-Eulerian. Using an analysis similar to that used in the proof of Lemma 10 (with our condition for lame expansion instead of the condition for undirected expansion), we obtain a lower bound $p' = \Omega\left(\frac{\epsilon m_U}{\log m_U/\alpha + \beta}\right)$ for the number of these 2β-spring-drain paths. Thus, to take care of this case we call GENERIC-1 or MULTISTAGE-2 to test whether $\overrightarrow{G}[U]$ is $(p', 2\beta)$-Eulerian. Note that p' differs from our bound for expander graphs in the addition of β to the denominator, which indicates an addition of β to the upper bound on the length of a correction path. This arises from the fact that the lame expansion condition applies only for β-cuts, and thus, it might not apply in the first and last β BFS layers.

As for the second case, if $\overrightarrow{G}[U]$ is ϵ-far from being Eulerian, but $\frac{\epsilon}{2}$-close to being 2β-Eulerian, we consider a 2β-Eulerian orientation $\overrightarrow{G}'[U]$ that is $\frac{\epsilon}{2}$-close to $\overrightarrow{G}[U]$. Clearly, $\overrightarrow{G}'[U]$ is $\frac{\epsilon}{2}$-far from being Eulerian. However, since it is 2β-Eulerian, we can show that it can be made Eulerian by inverting edges along paths between β-big springs and β-big drains. We next use a similar analysis as for Lemma 10. However, since the spring and drain in each of our paths are 2β-big, it can be seen that all the cuts between our BFS layers are β-cuts, and thus, we obtain a lower bound $p'' = \Omega\left(\frac{\epsilon m_U}{\log m_U/\alpha}\right)$ for the number of spring-drain paths. Hence, to take care of the second case, we call GENERIC-1 or MULTISTAGE-2 to test whether $\overrightarrow{G}[U]$ is p''-Eulerian (namely, we use $\beta = \Delta_U$).

The correctness of our algorithms now follows from Lemmas 7 and 9. The query complexity bounds are obtained from these lemmas, noting also that the second case discussed above is only possible for $\beta < \frac{\Delta_U}{2}$.   □

6   General Tests Based on Chopping

We provide a 1-sided test and a 2-sided test as follows. Given an orientation $\overrightarrow{G}$ of G, we show how to decompose $\overrightarrow{G}$ into a collection of (α, β)-expanders with a relatively small number of edges that are outside the (α, β)-expanders, called henceforth *external edges*. We will find this "chopping" adaptively while querying external edges only. If we do not find a witness showing that $\overrightarrow{G}$ is not Eulerian during the chopping procedure, then we sample a few (α, β)-expanders and test them using GEN-1 or MULTI-2 (see Lemma 13), obtaining a 1-sided test or a 2-sided test respectively.

Lemma 14 (The chopping lemma). *Given an orientation $\overrightarrow{G}$ as input and parameters $\alpha, \beta > 0$, we can either find a witness showing that $\overrightarrow{G}$ is not Eulerian, or find non-empty induced subgraphs $\overrightarrow{G}_i = (V_i, \overrightarrow{E}_i = \overrightarrow{E}(V_i))$ of $\overrightarrow{G}$ (where $i = 1, \ldots, k$ for some k), which we call (α, β)-components (or simply components), that satisfy the following:*

1. *The vertex sets $V_1, \ldots, V_k$ of the components are mutually disjoint.*
2. *$|\overrightarrow{E}_i| \geq \beta$ for $i = 1, \ldots k$.*
3. *All the components $\overrightarrow{G}_i$ are (α, β)-expanders.*
4. *The total number of external edges satisfies $|\overrightarrow{E} \setminus \bigcup_{i=1,\ldots,k} \overrightarrow{E}(V_i)| = O(\alpha m^2 \log m/\beta)$.*

During the chopping procedure, we query only external edges, i.e., edges that are not in any component G_i. The query complexity is in the same order also if we find a witness that $\overrightarrow{G}$ is not Eulerian.

Proof sketch. The chopping procedure proceeds as follows. At first, we define $\overrightarrow{G} = \overrightarrow{G}[V]$ as our single component. Then, at each step, we decompose a component $\overrightarrow{G}[U]$ into two separate components $\overrightarrow{G}[A]$ and $\overrightarrow{G}[B]$, if (A, B) is a β-cut of U and Inequality (1) above does *not* apply. When decomposing, we query the edges of the cut (A, B) and mark them as external edges. Note that we need not query any additional edges to decide on cutting a component, as all the required information is given by the underlying graph G and by the orientation of the external edges that were queried in previous steps. After each stage, we check whether the orientations of the edges queried so far invalidate any of the cuts in the graph (see Section 2), in which case we conclude that $\overrightarrow{G}$ is not Eulerian and return the invalid cut.

The procedure terminates once there is no cut of any component that satisfies the chopping conditions. The components are clearly disjoint throughout the procedure. Since we only chopped components across β-cuts, every final component contains at least β edges. Moreover, note that a component is always chopped by the procedure unless all its β-cuts satisfy Inequality (1). Hence, if the algorithm terminates without finding a witness that $\overrightarrow{G}$ is not Eulerian, then every G_i is an (α, β)-expander. It remains to prove the upper bound for the number of external edges and the query complexity of the chopping procedure.

Suppose that the chopping procedure has not found a witness that $\overrightarrow{G}$ is not Eulerian. Consider a component U and a β-cut (A, B) of U whose edges were queried in some step of the lemma. Using the chopping criterion and the fact that all the cuts in the knowledge graph are valid, we obtain

$$\min\left\{|\overrightarrow{E}(A, B)|, |\overrightarrow{E}(B, A)|\right\} < \alpha|E(A)|. \tag{2}$$

We refer to the edges in the minimal cut among $\overrightarrow{E}(A, B)$ and $\overrightarrow{E}(B, A)$ as *rare edges*, and to the edges in the other direction as *common edges*. We then prove that the total number of rare external edges is $O(\alpha m \log m)$, by "charging" a cost of α on every edge $e \in E(A)$. The proof uses Inequality (2) and the fact that, by definition, $|E(A)| \leq |E(B)|$. To complete the proof of the upper bound, we show that the ratio between the number of common edges and the number of rare edges is $O(m/\beta)$. This is done by observing that the multigraph defined by the components $\overrightarrow{G}_i$ is Eulerian, and so decomposable into edge-disjoint directed cycles. Every cycle contains at least one rare edge because the subgraph of common edges is acyclic. The proof follows since the number of components is $O(m/\beta)$. Finally, it is easy to see that the query complexity is not larger in the case where the procedure terminates after finding an invalid cut. □

Algorithm 15. *CHOP-1$(\overrightarrow{G}, \epsilon, \alpha, \beta)$:*

1. *Use Lemma 14 (the chopping lemma) for finding (α, β)-components $\overrightarrow{G}_1, \ldots, \overrightarrow{G}_k$ and querying their external edges, or reject and terminate if an invalid cut is found in the process.*
2. *Sample $3 \ln 3/\epsilon$ (α, β)-components $\overrightarrow{G}_i$ randomly and independently, where the probability of selecting a component $\overrightarrow{G}_i$ in a sample is proportional to $m_i \stackrel{\text{def}}{=} |E(V_i)|$.*
3. *Test every selected component $\overrightarrow{G}_i$ using GEN-1$(\overrightarrow{G}_i, \alpha, \beta, \epsilon/2)$ (see Lemma 13). Reject if the test rejects for at least one of the components selected.*
4. *Accept if the input was not rejected by any of the above steps.*

Theorem 16. *CHOP-1 is a 1-sided test for being Eulerian with query complexity $O\left(\frac{\alpha m^2 \log m}{\beta} + \frac{\Delta \log m}{\epsilon^2 \alpha} + \frac{\beta \cdot \min\{\beta, \Delta\}}{\epsilon^2}\right)$. In particular, for $\alpha = \frac{(\Delta \log m)^{1/3}}{(\epsilon m)^{2/3}}$ and $\beta = \frac{(\epsilon m \log m)^{2/3}}{\Delta^{1/3}}$, the query complexity is $O\left(\frac{(\Delta m \log m)^{2/3}}{\epsilon^{4/3}}\right)$.*

Finally, we obtain a similar 2-sided test, CHOP-2, by replacing the calls to GEN-1 in Step 3 of CHOP-1 with calls to MULTI-2 and using slightly different constants.

Theorem 17. *CHOP-2 is a 2-sided test for being Eulerian with query complexity*
$O\left(\frac{\alpha m^2 \log m}{\beta}\right) + \widetilde{O}\left(\frac{\sqrt{\Delta}\log m}{\epsilon^2 \alpha} + \frac{\beta \cdot \sqrt{\min\{\beta,\Delta\}}}{\epsilon^2}\right)$. *In particular, if* $\Delta \leq (\epsilon m)^{4/7}$, *then,*
for $\alpha = \frac{\Delta^{1/6}}{(\epsilon m)^{2/3}}$ *and* $\beta = \frac{(\epsilon m)^{2/3}}{\Delta^{1/6}}$, *the query complexity is* $\widetilde{O}\left(\frac{\Delta^{1/3} m^{2/3}}{\epsilon^{4/3}}\right) = \widetilde{O}\left(\frac{m^{6/7}}{\epsilon^{8/7}}\right)$.
If $(\epsilon m)^{4/7} < \Delta \leq m$, *then, for* $\alpha = \frac{\Delta^{5/16}}{(\epsilon m)^{3/4}}$ *and* $\beta = \Delta^{1/8}\sqrt{\epsilon m}$, *the query complexity*
is $\widetilde{O}\left(\frac{\Delta^{3/16} m^{3/4}}{\epsilon^{5/4}}\right) = \widetilde{O}\left(\frac{m^{15/16}}{\epsilon^{5/4}}\right)$.

7 Concluding Comments and Open Problems

We have shown a test with a sub-linear number of queries for all graphs. However, excepting the special cases of dense graphs and expander graphs, this should be only considered as a first step for this problem.

The procedure of our general test is surprisingly involved considering the problem statement. The question arises as to whether we can reduce the computational complexity from exponential to polynomial in m. Also, to make the test truly attractive, most of the calculations should be performed in a preprocessing stage, where the amount of calculations done while making the queries should ideally be also sub-linear in m.

Related to the preprocessing question is the unresolved question of adaptivity. We would like to think that a sub-linear query complexity *non-adaptive* test also exists for all graphs. Other adaptive versus non-adaptive gaps, such as the one concerning the 2-sided lower bounds, need also be addressed.

References

1. Alon, N., Shapira, A.: Testing subgraphs in directed graphs. J. Comput. Syst. Sci. 69(3), 354–382 (2004) (a preliminary version appeared in Proc. of STOC 2003, pp. 700-709)
2. Babai, L.: On the diameter of Eulerian orientations of graphs. In: Proceedings of the 17th SODA, pp. 822–831 (2006)
3. Bender, M., Ron, D.: Testing properties of directed graphs: Acyclicity and connectivity. Random Structures and Algorithms, 184–205 (2002)
4. Ben-Sasson, E., Harsha, P., Raskhodnikova, S.: Some 3CNF properties are hard to test. SIAM J. Computing 35(1), 1–21 (2005) (a preliminary version appeared in Proc.35th STOC, 2003)
5. Blum, M., Luby, M., Rubinfeld, R.: Self-testing/correcting with applications to numerical problems. Journal of Computer and System Sciences 47, 549–595 (1993) (a preliminary version appeared in Proc. 22nd STOC, 1990)
6. Brightwell, G.R., Winkler, P.: Counting Eulerian circuits is #P-complete. In: Demetrescu, C., Sedgewick, R., Tamassia, R. (eds.) Proc. 7th ALENEX and 2nd ANALCO 2005 (Vancouver BC), pp. 259–262. SIAM Press, Demetrescu (2005)
7. Chakraborty, S., Fischer, E., Lachish, O., Matsliah, A., Newman, I.: Testing st-Connectivity. In: Proceedings of the 11th RANDOM and the 10th APPROX, pp. 380–394 (2007)

8. Fischer, E.: The art of uninformed decisions: A primer to property testing. In: Paun, G., Rozenberg, G., Salomaa, A. (eds.) Current Trends in Theoretical Computer Science: The Challenge of the New Century, vol. I, pp. 229–264. World Scientific Publishing, Singapore (2004)

9. Fischer, E., Yahalom, O.: Testing convexity properties of tree colorings. In: Thomas, W., Weil, P. (eds.) STACS 2007. LNCS, vol. 4393, pp. 109–120. Springer, Heidelberg (2007)

10. Fleishcner, H.: Eulerian graphs and related topics, Part 1. Vol. 1. Annals of Discrete Mathematics 45 (1990)

11. Fleishcner, H.: Eulerian graphs and related topics, Part 1. Vol. 2. Annals of Discrete Mathematics 50 (1991)

12. Goldreich, O., Goldwasser, S., Ron, D.: Property testing and its connection to learning and approximation. JACM 45(4), 653–750 (1998)

13. Goldreich, O., Ron, D.: Property testing in bounded degree graphs. Algorithmica 32(2), 302–343 (2002)

14. Halevy, S., Lachish, O., Newman, I., Tsur, D.: Testing properties of constraint-graphs. In: Proceedings of the 22nd IEEE Annual Conference on Computational Complexity (CCC 2007), pp. 264–277 (2007)

15. Halevy, S., Lachish, O., Newman, I., Tsur, D.: Testing orientation properties, technical report, Electronic Colloquium on Computational Complexity (ECCC), 153 (2005)

16. Ibaraki, T., Karzanov, A.V., Nagamochi, H.: A fast algorithm for finding a maximum free multiflow in an inner Eulerian network and some generalizations. Combinatorica 18(1), 61–83 (1988)

17. Kaufman, T., Krivelevich, M., Ron, D.: Tight bounds for testing bipartiteness in general graphs. SICOMP 33(6), 1441–1483 (2004)

18. Lovász, L.: On some connectivity properties of Eulerian graphs. Acta Math. Hung. 28, 129–138 (1976)

19. Mihail, M., Winkler, P.: On the number of Eulerian orientations of a graph. Algorithmica 16(4/5), 402–414 (1996)

20. Newman, I.: Testing of Functions that have small width Branching Programs. SIAM J. Computing 31(5), 1557–1570 (2002) (a preliminary version appeared in Proc. 41st FOCS, 2000)

21. Parnas, M., Ron, D.: Testing the diameter of graphs. Random Struct. and Algorithms 20(2), 165–183 (2002)

22. Pevzner, P.A., Tang, H., Waterman, M.S.: An Eulerian path approach to DNA fragment assembly. Proc. Natl. Acad. Sci. USA 98, 9748–9753 (2001)

23. Robinson, R.W.: Enumeration of Euler graphs. In: Harary, F. (ed.) Proof Techniques in Graph Theory, pp. 147–153. Academic Press, New York (1969)

24. Ron, D.: Property testing (a tutorial). In: Rajasekaran, S., Pardalos, P.M., Reif, J.H., Rolim, J.D.P. (eds.) Handbook of Randomized Computing, vol. II, ch.15. Kluwer Press, Dordrecht (2001)

25. Rubinfeld, R., Sudan, M.: Robust characterization of polynomials with applications to program testing. SIAM J. Computing 25, 252–271 (1996) (first appeared as a technical report, Cornell University, 1993)

26. Tutte, W.T.: Graph theory. Addison-Wesley, New York (1984)

Approximately Counting Embeddings into Random Graphs

Martin Fürer* and Shiva Prasad Kasiviswanathan

Computer Science and Engineering, Pennsylvania State University
{furer,kasivisw}@cse.psu.edu

Abstract. Let H be a graph, and let $C_H(G)$ be the number of (subgraph isomorphic) copies of H contained in a graph G. We investigate the fundamental problem of estimating $C_H(G)$. Previous results cover only a few specific instances of this general problem, for example, the case when H has degree at most one (monomer-dimer problem). In this paper, we present the first general subcase of the subgraph isomorphism counting problem which is almost always efficiently approximable. The results rely on a new graph decomposition technique. Informally, the new decomposition is a labeling of the vertices generating a sequence of bipartite graphs. The decomposition permits us to break the problem of counting embeddings of large subgraphs into that of counting embeddings of small subgraphs. Using this, we present a simple randomized algorithm for the counting problem. For all decomposable graphs H and all graphs G, the algorithm is an unbiased estimator. Furthermore, for all graphs H having a decomposition where each of the bipartite graphs generated is small and almost all graphs G, the algorithm is a fully polynomial randomized approximation scheme.

We show that the graph classes of H for which we obtain a fully polynomial randomized approximation scheme for almost all G includes graphs of degree at most two, bounded-degree forests, bounded-width grid graphs, subdivision of bounded-degree graphs, and major subclasses of outerplanar graphs, series-parallel graphs and planar graphs, whereas unbounded-width grid graphs are excluded.

1 Introduction

Given a *template* graph H and a *base* graph G, we call an injection φ between vertices of H and vertices of G an *embedding* of H into G if φ maps every edge of H into an edge of G. In other words, φ is an isomorphism between H and a subgraph (not necessarily induced) of G. Deciding whether such an injection exists is known as the subgraph isomorphism problem. Subgraph isomorphism is an important and general form of pattern matching. It generalizes many interesting graph problems, including Clique, Hamiltonian Path, Maximum Matching, and Shortest Path. This problem arises in application areas ranging from text processing to physics and chemistry [1,2,3,4]. The general subgraph isomorphism problem is NP-complete, but there are various special cases which are known to be fixed-parameter tractable in the size of H [5].

---

* Supported in part by NSF award CCF-0728921.

In this work, we consider the related fundamental problem of counting the number of copies of a template graph in another graph. By a *copy* of H in G we mean any, not necessarily induced subgraph of G, isomorphic to H. In general the problem is #P-complete (introduced by Valiant [6]). The class #P is defined as $\{f : \exists$ a non-deterministic polynomial time Turing machine M such that on input x, the computation tree of M has exactly $f(x)$ accepting leaves$\}$. Problems complete for this class are presumably very difficult, especially since Toda's result [7] implies that a call to a #P-oracle suffices to solve any problem in the polynomial hierarchy in polynomial time.

Fixed-parameter tractability of this counting problem has been well-studied with negative results for exact counting [8] and positive results for some special cases of approximate counting [9]. In this paper, we are interested in the more general problem of counting copies of large subgraphs. Exact counting is possible for very few classes of non-trivial large subgraphs. Two key examples are spanning trees in a graph, and perfect matchings in a planar graph [10]. A few more problems such as counting perfect matchings in a bipartite graph (a.k.a. (0-1) permanent) [11], counting all matchings in a graph [12], counting labeled subgraphs of a given degree sequence in a bipartite graph [13], counting combinatorial quantities encoded by the Tutte polynomial in a dense graph [14], and counting Hamilton cycles in dense graphs [15], can be done approximately. But problems like counting perfect matchings in general graphs are still open.

Since most of the other interesting counting problems are hopelessly hard to solve (in many cases even approximately) [16], we investigate whether there exists a *fully polynomial randomized approximation scheme* (henceforth, abbreviated as fpras) that works well for *almost all graphs*. The statement can be made precise as: Let G_n be a graph chosen uniformly at random from the set of all n-vertex graphs. We say that a predicate $\mathcal{P}$ holds for almost all graphs if $\Pr[\mathcal{P}(G_n) = true] \rightarrow 1$ as $n \rightarrow \infty$ (probability over the choice of a random graph). By fpras we mean a randomized algorithm that produces a result that is correct to within a relative error of $1 \pm \epsilon$ with high probability (i.e., probability tending to 1). The algorithm must run in time $\text{poly}(n, \epsilon^{-1})$, where n is the input size. We call a problem *almost always efficiently approximable* if there is a randomized polynomial time algorithm producing a result within a relative error of $1 \pm \epsilon$ with high probability for almost all instances.

Previous attempts at solving these kinds of problems have not been very fruitful. For example, even seemingly simple problems like counting cycles in a random graph have remained open for a long time (also stated as an open problem in the survey by Frieze and McDiarmid [17]). In this paper we present new techniques that can not only handle simple graphs like cycles, but also major subclasses of more complicated graph classes like outerplanar, series-parallel, planar etc.

The theory of random graphs was initiated by Erdős and Rényi [18]. The most commonly used models of random graphs are $\mathcal{G}(n, p)$ and $\mathcal{G}(n, m)$. Both models specify a distribution on n-vertex graphs. In $\mathcal{G}(n, p)$ each of the $\binom{n}{2}$ edges is added to the graph independently with probability p and $\mathcal{G}(n, m)$ assigns equal probability to all graphs with exactly m edges. Unless explicitly stated otherwise, the default model addressed in this paper is $\mathcal{G}(n, p)$.

There has been a lot of interest in using random graph models for analyzing typical cases (beating the pessimism of worst-case analysis). Here, we mention some of these results relevant to our counting problem (see the survey of Frieze and McDiarmid [17] for more). One of the most well-studied problem is that of counting perfect matchings in graphs. For this problem, Jerrum and Sinclair [19] have presented a simulation of a Markov chain that almost always is an fpras (extended to all bipartite graphs in [11]). Similar results using other approaches were obtained later in [20,21,22,23]. Another well-studied problem is that of counting Hamiltonian cycles in random digraphs. For this problem, Frieze and Suen [24] have obtained an fpras, and later Rasmussen [21] has presented a simpler fpras. Afterwards, Frieze *et al.* [25] have obtained similar results in random regular graphs. Randomized approximation schemes are also available for counting the number of cliques in a random graph [26]. However, there are no general results for counting copies of an arbitrary given graph.

1.1   Our Results and Techniques

In this paper, we remedy this situation by presenting the first general subcase of the subgraph isomorphism counting problem which is almost always efficiently approximable. For achieving this result we introduce a new graph decomposition that we call an *ordered bipartite decomposition*. Informally, an ordered bipartite decomposition is a labeling of vertices such that every edge is between vertices with different labels and for every vertex all neighbors with a higher label have identical labels. The labeling implicitly generates a sequence of bipartite graphs and the crucial part is to ensure that each of the bipartite graphs is of small size. The size of the largest bipartite graph defines the *width* of the decomposition. The decomposition allows us to obtain general results for the counting problem which could not be achieved using the previous methods. It also leads to a relatively simple and elegant analysis. We will show that many graph classes have such decomposition, while at the same time many simple small graphs (like a triangle) may not possess a decomposition.

The actual algorithm itself is based on the following simple sampling idea (known as importance sampling in statistics): let $\mathcal{S} = \{x_1, \ldots, x_z\}$ be a large set whose cardinality we want to estimate. Assume that we have a randomized algorithm $\mathcal{A}$ that picks each element x_i with non-zero known probability p_i. Then, the function Count (Fig. 1) produces an estimate for the cardinality of $\mathcal{S}$. The following proposition shows that the estimate is unbiased, i.e., $\mathbb{E}[Z] = |\mathcal{S}|$.

Proposition 1. *The Function Count is an unbiased estimator for the cardinality of* $\mathcal{S}$.

Proof. It suffices to show that each element x_i has an expected contribution of 1 towards $|\mathcal{S}|$. This holds because on picking x_i (an event that happens with probability p_i), we set Z to the inverse probability of this event happening. Therefore, $\mathbb{E}[Z] = \sum_i p_i \cdot \frac{1}{p_i} = |\mathcal{S}|$.

Function Count: Assume $p_i > 0$ for all i and $\sum_{i=1}^{z} p_i \leq 1$
If some element x_i is picked by $\mathcal{A}$ then output $Z = \frac{1}{p_i}$
Else output $Z = 0$

Fig. 1. Unbiased estimator for the cardinality of $\mathcal{S}$

Similar schemes of counting have previously been used by Hammersley [27] and Knuth [28] in other settings. Recently, this scheme has been used by Rasmussen for approximating the permanent of a (0-1) matrix [21], and later for approximately counting cliques in a graph [26]. A variant of this scheme has also been used by the authors to provide a near linear-time algorithm for counting perfect matchings in random graphs [29,23]. This is however the first generalization of this simple idea to the general problem of counting graph embeddings. Another nice feature of such schemes is that they also seem to work well in practice [30].

Our randomized algorithm will try to embed H into G. If the algorithm succeeds in finding an embedding of H in G, it outputs the inverse probability of finding this embedding. The interesting question here is not only to ensure that each embedding of H in G has a positive probability of being found but also to pick each embedding with approximately equal probability to obtain a low variance. For this purpose, the algorithm considers an increasing subsequence of subgraphs $\bar{H}_1 \subset \bar{H}_2 \subset \cdots \subset \bar{H}_\ell = H$ of H. The algorithm starts by randomly picking an embedding of $\bar{H}_1$ in G, then randomly an embedding of $\bar{H}_2$ in G containing the embedding of $\bar{H}_1$ and so on. It is for defining the increasing sequence of subgraphs that our decomposition is useful.

The algorithm is always an unbiased estimator for $C_H(G)$. The decomposition provides a natural sufficient condition for the class of algorithms based on the principle of the function Count to be an unbiased estimator. Additionally, if the base graph is a random graph from $\mathcal{G}(n, p)$ with constant p and if the template graph has an ordered bipartite decomposition of bounded width, we show that the algorithm is an fpras. The interesting case of the result is when $p = 1/2$. Since the $\mathcal{G}(n, 1/2)$ model assigns a uniform distribution over all graphs of n given vertices, an fpras (when the base graph is from $\mathcal{G}(n, 1/2)$) can be interpreted as an fpras for almost all base graphs. This result is quite powerful because now to prove that the number of copies of a template graph can be well-approximated for most graphs G, one just needs to show that the template graph has an ordered bipartite decomposition of bounded width.

The later half of the paper is devoted to showing that a lot of interesting graph classes naturally have an ordered bipartite decomposition of bounded width. Let $\mathcal{C}_k$ denote a cycle of length k. In this extended abstract, we show that graphs of degree at most two, bounded-degree forests, bounded-width grid (lattice) graphs, subdivision of bounded-degree graphs, bounded-degree outerplanar graphs of girth at least four, and bounded-degree $[\mathcal{C}_3, \mathcal{C}_5]$-free series-parallel graphs, planar graphs of girth at least 16 have an ordered bipartite decomposition of bounded width. Using this we obtain the following result (proved in Theorems 3 and 4).

Theorem 1 (Main Result). *Let H be a simple graph where each connected component is one of the following: graph of degree at most two, bounded-degree tree, bounded-width grid, subdivision of a bounded-degree graph, bounded-degree C_3-free outerplanar graph, bounded-degree $[C_3, C_5]$-free series-parallel graph, or planar graphs of girth at least 16. Then, for almost all graphs G, there exists an fpras for estimating the number of copies of H in G.*

Even when restricted to graphs of degree at most two, this theorem recovers most of the older results. It also provides simpler, unified proofs for (some of) the results in [20,21,22,24]. For example, to count matchings of cardinality k one could use a template consisting of k disjoint edges. Similarly, to count all cycles of length k the template is a cycle of that length. By varying k and boosting the success probability, the algorithm can easily be extended to count all matchings or all cycles. This provides the first fpras for counting all cycles in a random graph (solving an open problem of Frieze and McDiarmid [17]). We omit further discussion of this problem.

For template graphs coming from the other classes, our result supplies the first efficient randomized approximation scheme for counting copies of them in almost all base graphs. For example, it was not known earlier how to even obtain an fpras for counting the number of copies of a given bounded-degree tree in a random graph. For the simpler graph classes the decomposition follows quite straightforwardly, but for graph classes such as subdivision, outerplanar, series-parallel, and planar, constructing the decomposition requires several new ideas. Even though our techniques can be extended to other interesting graph classes, we conclude by showing that our techniques can't be used to count the copies of an unbounded-width grid graph in a random graph.

2   Definitions and Notation

Let Q be some function from the set of input strings Σ^* to natural numbers. A fully polynomial randomized approximation scheme for Q is a randomized algorithm that takes input $x \in \Sigma^*$ and an accuracy parameter $\epsilon \in (0, 1)$ and outputs a number Z (a random variable depending on the coin tosses of the algorithm) such that,

$$\Pr[(1 - \epsilon)Q(x) \leq Z \leq (1 + \epsilon)Q(x)] \geq 3/4,$$

and runs in time polynomial in $|x|$, ϵ^{-1}. The success probability can be boosted to $1 - \delta$ by running the algorithm $O(\log \delta^{-1})$ times and taking the median [31].

Automorphisms are edge respecting permutations on the set of vertices, and the set of automorphisms form a group under composition. For a graph H, we use $aut(H)$ to denote the size of its automorphism group. For a bounded-degree graph H, $aut(H)$ can be evaluated in polynomial time [32]. Most of the other graph-theoretic concepts which we use, such as planarity are covered in standard text books (see, e.g., [33]).

Throughout this paper, we use G to denote a base random graph on n vertices. The graph H is the template whose copies we want to count in G. We can assume without loss of generality that the graph H also contains n vertices, otherwise we just add isolated vertices to H. The number of isomorphic images remains unaffected. Let $\triangle = \triangle(H)$ denote the maximum degree of H.

For a graph F, we use V_F and E_F to denote its vertex set and edge set, respectively. Furthermore, we use $v_F = |V_F|$ and $e_F = |E_F|$ for the number of vertices and edges. For a subset S of vertices of F, $N_F(S) = \{v \in V_F - S : \exists u \in S \text{ such that } (u, v) \in E_F\}$ denotes the neighborhood of S in F. $F[S]$ denotes the subgraph of F induced by S. We use $C_H(G)$ to denote the number of copies of H in G. Let $L_H(G) = C_H(G) \cdot aut(H)$ denote the number of embeddings (or labeled copies) of H in G. For a random graph G, we will be interested in quantities $\mathbb{E}[C_H(G)^2]$ and $\mathbb{E}[C_H(G)]^2$.

Our algorithm is randomized. The output of the algorithm is denoted by Z, which is an unbiased estimator of $C_H(G)$, i.e., $C_H(G) = \mathbb{E}_{\mathcal{A}}[Z]$ (expectation over the coin tosses of the algorithm). As the output of our algorithm depends on both the input graph, and the coin tosses of the algorithm, we use terms such as $\mathbb{E}_{\mathcal{G}}[\mathbb{E}_{\mathcal{A}}[Z]]$. Here, the inner expectation is over the coin-tosses of the algorithm, and the outer expectation is over the graphs of $\mathcal{G}(n, p)$. Note that $\mathbb{E}_{\mathcal{A}}[Z]$ is a random variable defined on the set of graphs.

Because of space constraints proofs are omitted from the following presentation; all omitted proofs can be found in [34].

3   Approximation Scheme for Counting Copies

We define a new graph decomposition technique which is used for embedding the template graph into the base graph. As stated earlier our algorithm for embedding works in stages and our notion of decomposition captures this idea.

Ordered Bipartite Decomposition. An ordered bipartite decomposition of a graph $H = (V_H, E_H)$ is a sequence $V_1, \ldots, V_\ell$ of subsets of V_H such that:

① $V_1, \ldots, V_\ell$ form a partition of V_H.
② Each of the V_i (for $i \in [\ell] = \{1, \ldots, \ell\}$) is an independent set in H.
③ $\forall i \, \exists j$ such that $v \in V_i$ implies $N_H(v) \subseteq \left(\bigcup_{k<i} V_k\right) \cup V_j$.

Property ③ just states that if a neighbor of a vertex $v \in V_i$ is in some V_j ($j > i$), then all other neighbors of v which are not in $V_1 \cup \cdots \cup V_{i-1}$, are in V_j. Property ③ will be used in the analysis for random graphs to guarantee that in every stage, the base graph used for embedding is still random with the original edge probability. Let $V^i = \bigcup_{j \le i} V_j$. Define $U_i = N_H(V_i) \cap V^{i-1}$. U_i is the set of neighbors of V_i in $V_1 \cup \cdots \cup V_{i-1}$. Define H_i to be the subgraph of H induced by $U_i \cup V_i$. Let E_{H_i}

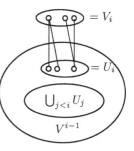

denote the edge set of graph H_i. The *width* of an ordered bipartite decomposition is the size (number of edges) of the largest H_i.

The U_i's will play an important role in our analysis. Note that given a U_j, its corresponding V_j has the property that $V_j \supseteq N_H(U_j) - V^{j-1}$. Hereafter, when the context is clear, we just use *decomposition* to denote an ordered bipartite decomposition. In general, the decomposition of a graph needn't be unique. The following lemma describes some important consequences of the decomposition.

Lemma 1. *Let $V_1, \ldots, V_\ell$ be a decomposition of a graph $H = (V_H, E_H)$. Then, the following assertions are true. (i) Each of the U_i is an independent set in H (H_i is a bipartite graph). (ii) The edge set E_H is partitioned into $E_{H_1}, \ldots, E_{H_\ell}$.*

Every graph has a trivial decomposition satisfying properties ① and ②, but the situation changes if we add property ③ (C_3 is the simplest graph which has no decomposition). Every bipartite graph though has a simple decomposition, but not necessarily of bounded width. Note that the bipartiteness of H is a sufficient condition for it to have an ordered bipartite decomposition, but not a necessary one.

We will primarily be interested in cases where the decomposition is of bounded width. This can only happen if Δ is a constant. In general, if Δ grows as a function of n, no decomposition could possibly have a bounded width ($\Delta/2$ is always a trivial lower-bound for the width). For us the parameter ℓ plays no role.

ALGORITHM EMBEDDINGS(G,H)

Initialize $X \leftarrow 1$, $Mark(0) \leftarrow \varnothing$, $\varphi(\varnothing) \leftarrow \varnothing$
let $V_1, \ldots, V_\ell$ denote an ordered bipartite decomposition of H
for $i \leftarrow 1$ to ℓ do
    let $G_f \leftarrow G[V_G - Mark(i-1) \cup \varphi(U_i)]$
    compute X_i (the number of embeddings of H_i in G_f with U_i mapped by φ)
    pick an embedding u.a.r. (if one exists) and use it to update φ
    if no embedding exists, then set Z to 0 and terminate
    $X \leftarrow X \cdot X_i$
    $Mark(i) \leftarrow Mark(i-1) \cup \varphi(V_i)$
$Z \leftarrow X/aut(H)$
output Z

The input to the algorithm Embeddings is the template graph H together with its decomposition and the base graph G. The algorithm tries to construct a bijection φ between the vertices of H and G. V_i represents the set of vertices of H which get embedded into G during the ith-stage, and the already constructed mapping of U_i is used to achieve this. For a subset of vertices $S \subseteq V_H$, $\varphi(S)$ denotes the image of S under φ. If $X > 0$, then the function φ represents an embedding of H in G (consequence of properties ① and ②), and the output X represents the inverse probability of this event happening. Since every embedding has a positive probability of being found, X is an unbiased estimator for the number of embeddings of H in G (Proposition 1), and Z is an unbiased estimator for the number of copies of H in G.

The actual procedure for computing the X_i's is not very relevant for our results, but note that the X_i's can be computed in polynomial time if H has a decomposition of bounded width. In this case the algorithm Embeddings runs in polynomial time.

3.1   FPRAS for Counting in Random Graphs

Since the algorithm Embeddings is an unbiased estimator, use of Chebyshev's inequality implies that repeating the algorithm $O(\epsilon^{-2}\mathbb{E}_{\mathcal{A}}[Z^2]/\mathbb{E}_{\mathcal{A}}[Z]^2)$ times and taking the mean of the outputs results in a randomized approximation scheme for estimating $C_H(G)$. From here on, we abbreviate $C_H(G)$ as C. The ratio $\mathbb{E}_{\mathcal{A}}[Z^2]/\mathbb{E}_{\mathcal{A}}[Z]^2$ is commonly referred to as the *critical ratio*.

We now concentrate on showing that for random graphs the algorithm is an fpras. A few of the technical details of our proof are somewhat similar to previous applications of this sampling idea, such as that for counting perfect matchings [21,23]. The simpler techniques in these previous results, however, are limited to handling one edge per stage (therefore, working only when H is a matching). Our algorithm embeds a small sized subgraph at every stage. The key for obtaining an fpras is to guarantee that the factor contributed to the critical ratio at every stage is very small (which is now involved because it is no longer a simple ratio of binomial moments as in [21,23]). Adding this to the fact that we can do a stage-by-stage analysis of the critical ratio (thanks to the decomposition property which ensures the graph stays essentially random), provides the ingredients for the fpras.

The analysis will be done for a worst-case graph H under the assumption that the sizes of the bipartite graphs H_i's are bounded by a universal constant w, and a random graph G. Here, instead of investigating the critical ratio, we investigate the much simpler ratio $\mathbb{E}_{\mathcal{G}}[\mathbb{E}_{\mathcal{A}}[Z^2]]/\mathbb{E}_{\mathcal{G}}[\mathbb{E}_{\mathcal{A}}[Z]]^2$, which we call the *critical ratio of averages*. We use the second moment method to show that these two ratios are closely related. For this purpose, we take a detour through the $\mathcal{G}(n,m)$ model. The ratio $\mathbb{E}[C^2]/\mathbb{E}[C]^2$ plays an important role here and for bounding it we use a recent result of Riordan [35]. The result (stated below) studies the related question of when a random graph G is likely to have a spanning subgraph isomorphic to H.

In the following, N is used to denote $\binom{n}{2}$. We say an event holds with high probability (w.h.p.), if it holds with probability tending to 1 as $n \to \infty$.

Theorem 2. *(Riordan [35]) Let H be a graph on n vertices. Let $e_H = \alpha N = \alpha(n)N$, and let $p = p(n) \in (0,1)$ with pN an integer. Suppose that the following conditions hold: $\alpha N \geq n$, and $pN, (1-p)\sqrt{n}, np^\gamma/\triangle^4 \to \infty$, where*

$$\gamma = \gamma(H) = \max_{3 \leq s \leq n} \{\max\{e_F : F \subseteq H, v_F = s\}/(s-2)\}.$$

Then, w.h.p. a random graph $G \in \mathcal{G}(n, pN)$ has a spanning subgraph isomorphic to H.

The quantity γ is closely related to twice the maximum average degree of a subgraph of H. The idea behind the proof is to use Markov's inequality to bound $\Pr[C = 0]$ in terms of $\mathbb{E}[C]$ and $Var[C]$. The main thrust lies in proving that $\mathbb{E}[C^2]/\mathbb{E}[C]^2 = 1 + o(1)$. Now by just following Riordan's proof, we obtain the following result.

Proposition 2. *Let H be a graph on n vertices. Let $e_H = \alpha N = \alpha(n)N$, and let $p = p(n) \in (0,1)$ with pN an integer. Let $\nu = \max\{2, \gamma\}$. Suppose that the following conditions hold: $pN, np^\nu/\triangle^4 \to \infty$ and $\alpha^3 N p^{-2} \to 0$. Then, w.h.p. a random graph $G \in \mathcal{G}(n, pN)$ satisfies $\mathbb{E}[C^2]/\mathbb{E}[C]^2 = 1 + o(1)$. In particular, if H is a bounded-degree graph on n vertices. Then, w.h.p. a random graph $G \in \mathcal{G}(n, \Omega(n^2))$ satisfies $\mathbb{E}[C^2]/\mathbb{E}[C]^2 = 1 + o(1)$.*

Note that some of the conditions in Proposition 2 are rephrased from Theorem 2. These are the conditions in the proof of Theorem 2 that are needed for bounding $\mathbb{E}[C^2]/\mathbb{E}[C]^2$. We will be interested in bounded-degree graphs H. For a bounded-degree graph H, both $\triangle$ and γ are constants. Additionally, we will be interested in dense random graphs (where the conditions of Proposition 2 are satisfied). Interpreting Proposition 2 in the $\mathcal{G}(n, p)$ model by using known results for asymptotic equivalence between $\mathcal{G}(n, m)$ and $\mathcal{G}(n, p)$ models (e.g., see Proposition 1.12 of [36]) yields

Lemma 2. *Let H be a bounded-degree graph on n vertices. Let $\omega = \omega(n)$ be any function tending to ∞ as $n \to \infty$, and let p be a constant. Then, w.h.p. a random graph $G \in \mathcal{G}(n, p)$ satisfies $C \geq \mathbb{E}[C]/\omega$.*

Remark: Since C is fairly tightly concentrated around its mean, a rudimentary approximation for C is just $\mathbb{E}[C] = \frac{n! p^{e_H}}{aut(H)}$ (as $v_H = n$). However, this naive approach doesn't produce for *any* $\epsilon > 0$, an $(1 \pm \epsilon)$-approximation for C (see, e.g., [21,20,24,26]).

Using the above result we investigate the performance of algorithm Embeddings when G is a random graph. In this extended abstract, we don't try to optimize the order of the polynomial arising in the running time analysis. Even though for simple template instances such as matchings or cycles, one could easily determine the exact order. The proof idea is to break the critical ratio analysis of the large subgraph into a more manageable critical ratio analysis of small subgraphs.

Theorem 3 (Main Theorem). *Let H be a n-vertex graph with a decomposition of width w (a constant). Let Z be the output of algorithm Embeddings, and let p be a constant. Then, w.h.p. for a random graph $G \in \mathcal{G}(n, p)$ the critical ratio $\mathbb{E}_A[Z^2]/\mathbb{E}_A[Z]^2$ is polynomially bounded in n.*

Summarizing: if H has a decomposition of bounded width w, then for almost all graphs G, running the algorithm Embeddings poly$(n)\epsilon^{-2}$ times and taking the mean, results in an $(1 \pm \epsilon)$-approximation for C. Here, poly(n) is a polynomial in n depending on w and p. Since each run of the algorithm also takes polynomial time (as H has bounded width), we get an fpras.

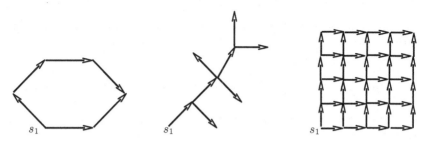

Fig. 2. Decomposition of a cycle, tree, and grid. The graphs are actually undirected. The arrows just connect the vertices of U_i to their neighbors in V_i. All out-neighbors of a vertex are in the same V_i, and all in-neighbors of a vertex are in the same U_i.

4   Graphs with Ordered Bipartite Decomposition

We divide this section into subsections based on the increasing complexity of the graph classes. Some of the later graph classes include the ones that will be covered earlier. We will prove the following result in this section.

Theorem 4. *Let H be a graph where each connected component is one of the following: graph of degree at most two, tree, bounded-width grid, subdivision graph, C_3-free outerplanar graph, $[C_3, C_5]$-free series-parallel graph, or planar graph of girth at least 16. Then, there exists an ordered bipartite decomposition of H. Furthermore, if H has bounded degree, then the decomposition has bounded width.*

Decompositions of subdivision graphs, $[C_3, C_5]$-free series-parallel graphs are omitted in this extended abstract (see [34]). From now onwards, we concentrate on connected components of the graph H. If H is disconnected a decomposition is obtained by combining the decomposition of all the connected components (in any order). We will abuse notation and let H stand for both the graph and a connected component in it. $\triangle$ is the maximum degree in H. For constructing the decomposition, the following definitions are useful, $U^i = \bigcup_{j \leq i} U_j, V^i = \bigcup_{j \leq i} V_j$, and $D^i = V^i - U^i$.

4.1   Some Simple Graph Classes

We start off by considering simple graph classes such as graphs of degree at most two (paths and cycles), trees, and grid graphs. Fig. 2 illustrates some examples.

- **Paths:** Let H represent a path $(s_1, \ldots, s_{k+1})$ of length $k = k(n)$. Then the decomposition is, $V_i = \{s_i\}$ for $1 \leq i \leq k+1$.
- **Cycles:** First consider the cycles of length four or greater. Let $s_1, \ldots, s_k$ be the vertices of a cycle H of length $k = k(n)$ enumerated in cyclic order. In the decomposition, $V_1 = \{s_1\}$, $V_2 = \{s_2, s_k\}$, and $V_i = \{s_i\}$ for $3 \leq i \leq k-1$. Cycles of length three (triangles) don't have a decomposition, but can easily be handled separately (see [34]). Actually, if $H = H_1 \cup H_2$, where graphs

H_1 and H_2 are disjoint, H_1 has a decomposition of bounded width, and H_2 consists of a vertex disjoint union of triangles, then again, there exists an fpras for estimating C. This also completes the claim for graphs of degree at most two in Theorem 1.

- **Trees:** For a tree H, $V_1 = \{s_1\}$, where s_1 is any vertex in H. For $i \geq 2$, let U_i be any vertex from D^{i-1}, then V_i is the set of neighbors of this vertex which are not in V^{i-1}. Intuitively, V_i is the set of children of the vertex in U_i, if one thinks of H as a tree rooted at s_1. The width of the decomposition is at most $\triangle$.
- **Grids:** Let w_0 be the width of the grid graph H. Set $V_1 = \{s_1\}$, where s_1 is any corner vertex in H. Later on, V_i is the set of all vertices which are at a lattice (Manhattan) distance i from s_1. Since for each i, there are at most w_0 vertices at distance i from from s_1, the sizes of the V_i's are bounded if w_0 is bounded. Consequently, the width of the decomposition is bounded if w_0 is bounded. This construction also extends to higher dimensional grid graphs.

4.2   Outerplanar Graphs

A graph is outerplanar if it has a planar embedding such that all vertices are on the same face. Let H be a C_3-free outerplanar graph. The idea behind the decomposition is that vertices in U_i partitions the outer face into smaller intervals, each of which can then be handled separately.

Before we formally describe the decomposition, we need some terminology. Let $s_1, \ldots, s_k$ be the vertices around the outer face with $k = k(n)$ (ordering defined by the outerplanar embedding). For symmetry, we add two dummy vertices s_0, s_{k+1} without neighbors and define $U_1 = \{s_0, s_{k+1}\}$, and $V_1 = \{s_1\}$ (the dummy vertices play no role and can be removed before running the algorithm Embeddings). For $i \geq 1$, two vertices s_{j_0}, s_{j_1} with $j_0 < j_1$, define a stage i interval if $s_{j_0}, s_{j_1} \in U^i$, but for $j_0 < l < j_1, s_l \notin U^i$. If the interval is defined it is the sequence of vertices between s_{j_0}, s_{j_1} (including the endpoints). Let a_i be a median vertex of $I \cap V^i$ (median based on the ordering), where I is a stage i interval. Define U_{i+1} as the smallest subset of V^i containing $\{a_i\}$ and also $N_H(N_H(U_{i+1}) - V^i) \cap V^i$. Define V_{i+1} as $N_H(U_{i+1}) - V^i$. We now argue that this is indeed a decomposition. Consider a stage i interval I, with s_{j_0}, s_{j_1} as the defining end points, and a_i as the median of $I \cap V^i$.

Lemma 3. *For every $i \geq 1$, there is a stage i interval I with $U_{i+1} \subseteq I$ and* $|U_{i+1}| \leq |I \cap V^i| \leq 2\triangle$.

The properties ① and ③ of the decomposition are guaranteed by the construction. Lemma 3 implies that the width of the decomposition is most $2\triangle^2$. Property ② holds because there are no triangles in H.

4.3   Planar Graphs

Define a *thread* as an induced path in H whose vertices are all of degree 2 in H. A k-thread is a thread with k vertices. Let H be a planar graph of girth at least 16. We first prove a structural result on planar graphs.

Lemma 4. *Let H be a planar graph of minimum degree 2 and girth at least 16, then H always contains a 3-thread.*

In order to define a decomposition, we define a 3-thread partition $X_1, \ldots, X_c$ of a planar graph H as a partition of V_H such that each X_i satisfies

$$X_i = \begin{cases} \{a_i\}, \text{where } a_i \text{ is a degree 0 or 1 vertex in } H[V_H - \bigcup_{j<i} X_j], \text{or} \\ \{a_i, b_i, c_i\}, \text{where } a_i, b_i, c_i \text{ form a 3-thread in } H[V_H - \bigcup_{j<i} X_j]. \end{cases}$$

Remember that for a subset of vertices S of H, $H[S]$ denotes the subgraph of H induced by S. By Lemma 4, every planar graph with girth at least 16 has a 3-thread partition. As earlier, we say, a vertex is selected if we add it to some V_k. Set $i = 1$. Using the 3-thread partition (which can be constructed using Lemma 4), a decomposition of a planar graph of girth at least 16 can be constructed by repeating the following procedure until all vertices are selected.

   i. Find the largest index l such that X_l contains a vertex z_l which has not yet been selected, but is adjacent to an already selected vertex.
   ii. Define $U_i = N_H(z_l) \cap D^{i-1}$ and $V_i = N_H(U_i) - V^{i-1}$.
   iii. Increment i.

Lemma 5. *Let H be a planar graph of girth at least 16. Then the above procedure finds a decomposition of H of width at most $2\triangle$.*

5   Negative Result for Ordered Bipartite Decomposition

As mentioned earlier only graphs of bounded degree have a chance of having a decomposition of bounded width. So a natural question to ask is whether all bounded-degree graphs with a decomposition have one of bounded width. In this section, we answer this question negatively by showing that every unbounded-width grid graph fails to satisfy this condition. For simplicity, we will only consider $\sqrt{n} \times \sqrt{n}$ grid graphs, but our proof extend to other cases as well.

Let $H = (V_H, E_H)$ be a $\sqrt{n} \times \sqrt{n}$ grid graph with $V_H = \{(i,j) : 0 \le i,j \le \sqrt{n}-1\}$ and $E_H = \{((i,j),(i',j')) : i = i' \text{ and } |j-j'| = 1 \text{ or } |i-i'| = 1 \text{ and } j = j'\}$. We now show that any decomposition of H has a width of at least $\sqrt{n}$. Let $V_1, \ldots, V_\ell$ be any decomposition of H. Consider any 2×2 square of H defined by vertices a, b, c, d. The two neighbors a, b of the vertex c with the smallest label l always have the same label $l' > l$. The fourth vertex d has any label l'' with $l'' \ge l$ and $l'' \ne l'$. We define a new graph $H' = (V_H, E_{H'})$ on the same set of vertices by putting the edge (a, b) into $E_{H'}$. Note that all vertices in a connected component have the same label thus are chosen together.

Let $\mathcal{H}_D$ be a class of graphs on vertex set V_H with exactly one diagonal in every 2×2 square (and no other edges). That is any graph $H_D = (V_H, E_D)$ from $\mathcal{H}_D$ has for every (i,j) with $0 \le i,j \le \sqrt{n} - 2$ exactly one of the edges $((i,j),(i+1,j+1)),((i,j+1),(i+1,j))$ in E_D and no other edges are in E_D. Note that $H' \in \mathcal{H}_D$. The following theorem shows that any graph $H_D \in \mathcal{H}_D$ has the property that there is a connected component touching top and bottom or left and right, which in turn implies the desired result.

Theorem 5. *There exists a connected component of H_D that contains at least one vertex from every row or at least one vertex from every column. Therefore, there exists no decomposition of a $\sqrt{n} \times \sqrt{n}$-grid graph H of width $\sqrt{n} - 1$.*

Acknowledgments

We thank Andrzej Ruciński for pointing us to [35] and Piotr Berman for simplifying the proofs in Section 5. The authors would also like to thank Sofya Raskhodnikova, Adam Smith, and Martin Tancer for helpful discussions.

References

1. Dong, H., Wu, Y., Ding, X.: An ARG representation for chinese characters and a radical extraction based on the representation. In: International Conference on Pattern Recognition, pp. 920–922 (1988)
2. Artymiuk, P.J., Bath, P.A., Grindley, H.M., Pepperrell, C.A., Poirrette, A.R., Rice, D.W., Thorner, D.A., Wild, D.J., Willett, P., Allen, F.H., Taylor, R.: Similarity searching in databases of three-dimensional molecules and macromolecules. Journal of Chemical Information and Computer Sciences 32, 617–630 (1992)
3. Stahs, T., Wahl, F.M.: Recognition of polyhedral objects under perspective views. Computers and Artificial Intelligence 11, 155–172 (1992)
4. Levinson, R.: Pattern associativity and the retrieval of semantic networks. Computers & Mathematics with Applications 23(6–9), 573–600 (1992)
5. Alon, N., Yuster, R., Zwick, U.: Color-coding. Journal of the ACM 42(4), 844–856 (1995)
6. Valiant, L.G.: The complexity of computing the permanent. Theoretical Computer Science 8, 189–201 (1979)
7. Toda, S.: On the computational power of PP and $\oplus P$. In: FOCS, pp. 514–519. IEEE, Los Alamitos (1989)
8. Flum, J., Grohe, M.: The parameterized complexity of counting problems. SIAM Journal of Computing 33(4), 892–922 (2004)
9. Arvind, V., Raman, V.: Approximation algorithms for some parameterized counting problems. In: Bose, P., Morin, P. (eds.) ISAAC 2002. LNCS, vol. 2518, pp. 453–464. Springer, Heidelberg (2002)
10. Kasteleyn, P.W.: Graph theory and crystal physics. In: Harary, F. (ed.). Academic Press, London (1967)
11. Jerrum, M., Sinclair, A., Vigoda, E.: A polynomial-time approximation algorithm for the permanent of a matrix with nonnegative entries. Journal of ACM 51(4), 671–697 (2004)
12. Jerrum, M., Sinclair, A.: Polynomial-time approximation algorithms for the Ising model. SIAM Journal on Computing 22(5), 1087–1116 (1993)
13. Bezáková, I., Bhatnagar, N., Vigoda, E.: Sampling binary contingency tables with a greedy start. In: SODA, pp. 414–423. SIAM, Philadelphia (2006)
14. Alon, N., Frieze, A.M., Welsh, D.: Polynomial time randomized approximation schemes for Tutte-Gröthendieck invariants: The dense case. Random Structures & Algorithms 6(4), 459–478 (1995)
15. Dyer, M., Frieze, A.M., Jerrum, M.: Approximately counting Hamilton paths and cycles in dense graphs. SIAM Journal on Computing 27(5), 1262–1272 (1998)

16. Jerrum, M.: Counting, sampling and integrating: algorithms and complexity. Birkhäuser, Basel (2003)
17. Frieze, A.M., McDiarmid, C.: Algorithmic theory of random graphs. Random Structures & Algorithms 10(1-2), 5–42 (1997)
18. Erdős, P., Rényi, A.: On the evolution of random graphs. Publ. Math. Inst. Hung. Acad. Sci. 5, 17–61 (1960)
19. Jerrum, M., Sinclair, A.: Approximating the permanent. SIAM Journal on Computing 18(6), 1149–1178 (1989)
20. Frieze, A.M., Jerrum, M.: An analysis of a Monte Carlo algorithm for estimating the permanent. Combinatorica 15(1), 67–83 (1995)
21. Rasmussen, L.E.: Approximating the permanent: A simple approach. Random Structures & Algorithms 5(2), 349–362 (1994)
22. Chien, S.: A determinant-based algorithm for counting perfect matchings in a general graph. In: SODA, pp. 728–735. SIAM, Philadelphia (2004)
23. Fürer, M., Kasiviswanathan, S.P.: Approximately counting perfect matchings in general graphs. In: ALENEX/ANALCO, pp. 263–272. SIAM, Philadelphia (2005)
24. Frieze, A.M., Suen, S.: Counting the number of Hamilton cycles in random digraphs. Random Structures & Algorithms 3(3), 235–242 (1992)
25. Frieze, A.M., Jerrum, M., Molloy, M.K., Robinson, R., Wormald, N.C.: Generating and counting Hamilton cycles in random regular graphs. Journal of Algorithms 21(1), 176–198 (1996)
26. Rasmussen, L.E.: Approximately counting cliques. Random Structures & Algorithms 11(4), 395–411 (1997)
27. Hammersley, J.M.: Existence theorems and Monte Carlo methods for the monomer-dimer problem. Research Papers in Statistics, 125–146 (1966)
28. Knuth, D.E.: Estimating the efficiency of backtrack programs. Mathematics of Computation 29(129), 121–136 (1975)
29. Fürer, M., Kasiviswanathan, S.P.: An almost linear time approximation algorithm for the permanent of a random (0-1) matrix. In: Lodaya, K., Mahajan, M. (eds.) FSTTCS 2004. LNCS, vol. 3328, pp. 263–274. Springer, Heidelberg (2004)
30. Sankowski, P.: Alternative algorithms for counting all matchings in graphs. In: Alt, H., Habib, M. (eds.) STACS 2003. LNCS, vol. 2607, pp. 427–438. Springer, Heidelberg (2003)
31. Jerrum, M., Valiant, L., Vazirani, V.: Random generation of combinatorial structures from a uniform distribution. Theoretical Computer Science 43, 169–188 (1986)
32. Luks, E.M.: Isomorphism of graphs of bounded valence can be tested in polynomial time. Journal of Computer and System Sciences 25, 42–65 (1982)
33. Diestel, R.: Graph theory, 2nd edn. Springer, Heidelberg (2000)
34. Fürer, M., Kasiviswanathan, S.P.: Approximately counting embeddings into random graphs. CoRR, arXiv:0806.2287 [cs.DS] (2008)
35. Riordan, O.: Spanning subgraphs of random graphs. Combinatorics, Probability & Computing 9(2), 125–148 (2000)
36. Janson, S., Łuczak, T., Ruciński, A.: Random graphs. Wiley-Interscience, Chichester (2000)

Increasing the Output Length of Zero-Error Dispersers

Ariel Gabizon[1] and Ronen Shaltiel[2]

[1] Department of Computer Science, Weizmann institute of science, Rehovot, Israel[*]
[2] Department of Computer Science, Haifa University, Haifa, Israel[**]

Abstract. Let $\mathcal{C}$ be a class of probability distributions over a finite set Ω. A function $D : \Omega \mapsto \{0,1\}^m$ is a *disperser* for $\mathcal{C}$ with *entropy threshold* k and *error* ϵ if for any distribution X in $\mathcal{C}$ such that X gives positive probability to at least 2^k elements we have that the distribution $D(X)$ gives positive probability to at least $(1-\epsilon)2^m$ elements. A long line of research is devoted to giving explicit (that is polynomial time computable) dispersers (and related objects called "extractors") for various classes of distributions while trying to maximize m as a function of k.

In this paper we are interested in explicitly constructing *zero-error dispersers* (that is dispersers with error $\epsilon = 0$). For several interesting classes of distributions there are explicit constructions in the literature of zero-error dispersers with "small" output length m and we give improved constructions that achieve "large" output length, namely $m = \Omega(k)$.

We achieve this by developing a general technique to improve the output length of zero-error dispersers (namely, to transform a disperser with short output length into one with large output length). This strategy works for several classes of sources and is inspired by a transformation that improves the output length of extractors (which was given in [29] building on earlier work by [15]). Nevertheless, we stress that our techniques are different than those of [29] and in particular give non-trivial results in the errorless case.

Using our approach we construct improved zero-error dispersers for the class of *2-sources*. More precisely, we show that for any constant $\delta > 0$ there is a constant $\eta > 0$ such that for sufficiently large n there is a poly-time computable function $D : \{0,1\}^n \times \{0,1\}^n \mapsto \{0,1\}^{\eta n}$ such that for any two independent distributions X_1, X_2 over $\{0,1\}^n$ such that both of them support at least $2^{\delta n}$ elements we get that the output distribution $D(X_1, X_2)$ has full support. This improves the output length of previous constructions by [2] and has applications in Ramsey Theory and in constructing certain data structures [13].

We also use our techniques to give explicit constructions of zero-error dispersers for bit-fixing sources and affine sources over polynomially large fields. These constructions improve the best known explicit constructions due to [26,14] and achieve $m = \Omega(k)$ for bit-fixing sources and $m = k - o(k)$ for affine sources.

[*] Research supported by Binational Science Foundation (BSF) grant and by Minerva Foundation grant.
[**] Research supported by Binational US-Israel Science Foundation (BSF) grant 2004329 and Israel Science Foundation (ISF) grant 686/07.

A. Goel et al. (Eds.): APPROX and RANDOM 2008, LNCS 5171, pp. 430–443, 2008.

1   Introduction

1.1   Background

Randomness extractors and dispersers are functions that refine the randomness in "weak sources of randomness" that "contain sufficient entropy". Various variants of extractors and dispersers are closely related to expander graphs, error correcting codes and objects from Ramsey theory. A long line of research is concerned with explicit constructions of these objects and these constructions have many applications in many areas of computer science and mathematics (e.g. network design, cryptography, pseudorandomness, coding theory, hardness of approximation, algorithm design and Ramsey theory).

Randomness extractors and dispersers. We start with formal definitions of extractors and dispersers. (We remark that in this paper we consider the "seedless version" of extractors and dispersers).

Definition 1 (min-entropy and statistical distance). *Let Ω be a finite set. The* min-entropy *of a distribution X on Ω is defined by $H_\infty(X) = min_{x \in \Omega} \log_2 \frac{1}{\Pr[X=x]}$. For a class C of distributions on Ω we use C_k to denote the class of all distributions $X \in C$ such that $H_\infty(X) \geq k$. We say that two distributions X, Y on Ω are ϵ-close if $\frac{1}{2} \sum_{w \in \Omega} |\Pr[X = w] - \Pr[Y = w]| \leq \epsilon$.*

When given a class C of distributions (which we call "sources") the goal is to design one function that refines the randomness of any distribution X in C. An *extractor* produces a distribution that is (close to) uniform whereas a *disperser* produces a distribution with (almost) full support. A precise definition follows:

Definition 2 (Extractors and Dispersers). *Let C be a class of distributions on a finite set Ω.*

- *A function $E : \Omega \mapsto \{0,1\}^m$ is an* extractor *for C with* entropy threshold k and error $\epsilon > 0$ *if for every $X \in C_k$, $E(X)$ is ϵ-close to the uniform distribution on $\{0,1\}^m$.*
- *A function $D : \Omega \mapsto \{0,1\}^m$ is a* disperser *for C with* entropy threshold k and error $\epsilon > 0$ *if for every $X \in C_k$, $|\text{Supp}(D(X))| \geq (1 - \epsilon)2^m$ (where $\text{Supp}(Z)$ denotes the support of the random variable Z).*

We remark that every extractor is in particular a disperser and that the notion of dispersers only depends on the support of the distributions in C. A long line of research is concerned with designing extractors and dispersers for various classes of sources. For a given class C we are interested in designing extractors and dispersers with as small as possible entropy threshold k, as large as possible output length m and as small as possible error ϵ. (We remark that it easily follows that $m \leq k$ whenever $\epsilon < 1/2$).

It is often the case that the probabilistic method gives that a randomly chosen function E is an excellent extractor. (This is in particular true whenever the class C contains "not too many" sources). However, most applications of extractors and dispersers require *explicit constructions*, namely functions that can be computed in time polynomial in their input length. Much of the work done in this area can be described as an attempt of matching the parameters obtained by existential results using explicit constructions.

Some related work

Classes of sources. Various classes C of distributions were studied in the literature: The first construction of deterministic extractors can be traced back to von Neumann [33] who showed how to use many independent tosses of a biassed coin (with unknown bias) to obtain an unbiased coin. Blum [5] considered sources that are generated by a finite Markov-chain. Santha and Vazirani [28], Vazirani [28,32], Chor and Goldreich [8], Dodis et al. [11], Barak, Impagliazzo and Wigderson [1], Barak et al. [2], Raz [27], Rao [25], Bourgain [6], Barak et al. [3], and Shaltiel [29] studied sources that are composed of several independent samples from "high entropy" distributions. Chor et al. [9], Ben-Or and Linial [4], Cohen and Wigderson [10], Mossel and Umans [22], Kamp and Zuckerman [20], Gabizon, Raz and Shaltiel [15], and Rao [26] studied bit-fixing sources which are sources in which a subset of the bits are uniformly distributed. Trevisan and Vadhan [31] and Kamp et al. [19] studied sources which are "samplable" by "efficient" procedures. Barak et al. [2], Bourgain [7], Gabizon and Raz [14], and Rao [26] studied sources which are uniform over an affine subspace. Dvir, Gabizon and Wigderson [12] studied a generalization of affine sources to sources which are sampled by low degree multivariate polynomials.

Seeded extractors and dispersers. A different variant of extractors and dispersers are *seeded* extractors and dispersers (defined by Nisan and Zuckerman [23]). Here the class C is the class of all distributions on $\Omega = \{0,1\}^n$. It is easy to verify that there do not exist extractors or dispersers for C (even when $k = n - 1$, $m = 1$ and $\epsilon < 1/2$). However, if one allows the extractor (or disperser) to receive an additional independent uniformly distributed input (which is called "a seed") then extraction is possible as long as the seed is of length $\Theta(\log(n/\epsilon))$. More precisely, a seeded extractor (or disperser) with entropy threshold k and error ϵ is a function $F : \{0,1\}^n \times \{0,1\}^t \mapsto \{0,1\}^m$ such that for any distribution X on $\{0,1\}^n$ with $H_\infty(X) \geq k$ the distribution $F(X,Y)$ (where Y is an independent uniformly distributed variable) satisfies the guarantees of Definition 2. A long line of research is concerned with explicit constructions of seeded extractors and dispersers (the reader is referred to [30] for a survey article and to [21,18] for the current milestones in explicit constructions of extractors).

Zero-error dispersers. In this paper we are interested in *zero-error dispersers*. These are dispersers where the output distribution has full support. That is for every source X in the class C:

$$\{D(x) : x \in \text{Supp}(X)\} = \{0,1\}^m$$

We also consider a stronger variant which we call *strongly-hitting disperser* in which every output element $z \in \{0,1\}^m$ is obtained with "not too small" probability. A precise definition follows:

Definition 3 (Zero-error dispersers and strongly hitting dispersers).
Let C be a class of distributions on a finite set Ω.

- *A function D is a* zero-error disperser *for C with* entropy threshold k *if it is a disperser for C with entropy threshold k and error $\epsilon = 0$.*
- *A function $D : \Omega \mapsto \{0,1\}^m$ is a μ-strongly hitting disperser for C with entropy threshold k if for every $X \in C_k$ and for every $z \in \{0,1\}^m$, $\Pr[D(X) = z] \geq \mu$.*

Note that a μ-strongly hitting disperser with $\mu > 0$ is in particular a zero-error disperser and that any μ-strongly hitting disperser has $\mu \leq 2^{-m}$. The following facts immediately follow:

Fact 1. *Let $f : \Omega \mapsto \{0,1\}^m$ be a function and let $\epsilon \leq 2^{-(m+1)}$.*

- *If f is a disperser with error ϵ then f is a zero-error disperser (for the same class C and entropy threshold k).*
- *If f is an extractor with error ϵ then f is a $2^{-(m+1)}$-strongly hitting disperser (for the same class C and entropy threshold k).*

It follows that extractors and dispersers with small ϵ immediately translate into zero-error dispersers (as one can truncate the output length to $m' = \log(1/\epsilon) - 1$ bits and such a truncation preserves the output guarantees of extractors and dispersers).

1.2   Increasing the Output Length of Zero-Error Dispersers

For several interesting classes of sources there are explicit constructions of dispersers with "large" error (which by Fact 1 give zero-error dispersers with "short" output length). In this paper we develop techniques to construct zero-error dispersers with large output length.

The composition approach. The following methodology for increasing the output length of extractors was suggested in [15,29]: When given an extractor E' with "small" output length t (for some class C) consider the function $E(x) = F(x, E'(x))$ where F is a seeded extractor. Shaltiel [29] (building on earlier work by Gabizon et al. [15]) shows that if E' and F fulfill certain requirements then this construction yields an extractor for C with large output length. The high level idea is that if certain conditions are fulfilled then the distribution $F(X, E(X))$ (in which the two inputs of F are *dependent*) is close to the distribution $F(X, Y)$ (where Y is an independent uniformly distributed variable) and note that the latter distribution is close to uniform by the definition of seeded extractors. This technique proved useful for several interesting classes of sources.

We would like to apply an analogous idea to obtain zero-error dispersers. However, by the lower bounds of [23,24] if F is a seeded extractor (or seeded disperser) then its seed length is at least $\log(1/\epsilon)$. This means that if we want $F(X, Y)$ to output m bits with error $\epsilon < 1/2^m$ we need seed length larger than

m. This in turn means that we want E' to have output length $t > m$ which makes the transformation useless.

There are also additional problems. The argument in [29] requires the "original function" E' to be an extractor (and it does not go through if E' is a disperser) and furthermore the error of the "target function" E is at least as large as that of the "original function" E' (and once again we don't gain when shooting for zero-error dispersers).

Summing up we note that if we want to improve the output length of a zero-error disperser D' by a composition of the form $D(x) = F(x, D'(x))$ we need to use a function F with different properties (a seeded extractor or disperser will not do) and we need to use a different kind of analysis.

Composing zero-error dispersers. In this paper we imitate the method of [29] and give a general method to increase the output length of zero-error dispersers. That is when given:

- A zero-error disperser $D' : \Omega \mapsto \{0,1\}^t$ for a class $\mathcal{C}$ and "small" output length t.
- A function $F : \Omega \times \{0,1\}^t \mapsto \{0,1\}^m$ for "large" output length m.

We identify properties of F that are sufficient so that the construction

$$D(x) = F(x, D'(x))$$

gives a zero-error disperser. (The argument is more general and transforms $2^{-(t+O(1))}$-strongly hitting dispersers into $2^{-(m+O(1))}$-strongly hitting dispersers). We then use this technique to give new constructions of zero-error dispersers and strongly-hitting dispersers.

Subsource hitters. As explained earlier we cannot choose F to be a seeded extractor. Instead, we introduce a new object which we call a *subsource hitter*. The definition of subsource hitters is somewhat technical and is tailored so that the construction $D(x) = F(x, D'(x))$ indeed produces a disperser.

Definition 4 (subsource hitter). *A distribution X' on Ω is a subsource of a distribution X on Ω if there exist $\alpha > 0$ and a distribution X'' on Ω such that X can be expressed as a convex combination $X = \alpha X' + (1 - \alpha) X''$.*

Let $\mathcal{C}$ be a class of distributions on Ω. A function $F : \Omega \times \{0,1\}^t \mapsto \{0,1\}^m$ is a subsource-hitter for $\mathcal{C}$ with entropy threshold k and subsource entropy $k - v$ if for any $X \in \mathcal{C}_k$ and $z \in \{0,1\}^m$ there exists a $y \in \{0,1\}^t$ and a distribution $X' \in \mathcal{C}_{k-v}$ that is a subsource of X such that for every $x \in \text{Supp}(X')$ we have that $F(x, y) = z$.

A subsource hitter has the property that for any $z \in \{0,1\}^m$ there exist $y \in \{0,1\}^t$ and $x \in \text{Supp}(X)$ such that $F(x, y) = z$ and in particular

$$\{F(x,y) : x \in \text{Supp}(X), y \in \{0,1\}^t\} = \{0,1\}^m$$

In addition a subsource hitter has the stronger property that there exists a subsource X' of X (which is itself a source in $\mathcal{C}$) such that for any $z \in \{0,1\}^m$ there exists $y \in \{0,1\}^t$ such that for *any* $x \in \text{Supp}(X') \subseteq \text{Supp}(X)$, $F(x,y) = z$.

This property allows us to show that $D(x) = F(x, D'(x))$ is a zero-error disperser with entropy threshold k whenever D' is a zero-error disperser with entropy threshold $k - v$. This is because when given a source $X \in \mathcal{C}_k$ and $z \in \{0,1\}^m$ we can consider the seed $y \in \{0,1\}^t$ and subsource X' guaranteed in the definition. We have that D' is a zero-error disperser and that X' meets the entropy threshold of D'. It follows that there exist $x \in \text{Supp}(X') \subseteq \text{Supp}(X)$ such that $D'(x) = y$. It follows that:

$$D(x) = F(x, D'(x)) = F(x,y) = z$$

and this means that D indeed outputs z. (We remark that a more complicated version of this argument shows that the composition applies to strongly-hitting dispersers). The exact details are given in the full version. It is interesting to note that this argument is significantly simpler than that of [29]. Indeed, the definition of subsource hitters is specifically tailored to make the composition argument go through and the more complicated task is to design subsource hitters. This is in contrast to [29] in which the function F is in most cases an "off the shelf" seeded extractor and the difficulty is to show that the composition succeeds.

1.3 Applications

We use the new composition technique to construct zero-error dispersers with large output length for various classes of sources. We discuss these constructions and some applications below.

Zero-error 2-source dispersers. The class of *2-sources* is the class of distributions $X = (X_1, X_2)$ on $\Omega = \{0,1\}^n \times \{0,1\}^n$ such that X_1, X_2 are independent. It is common to consider the case where each of the two distributions X_1, X_2 has min-entropy at least some threshold k.

Definition 5 (2-source extractors and dispersers). *A function $f : \{0,1\}^n \times \{0,1\}^n \mapsto \{0,1\}^m$ is a 2-source extractor (resp. disperser) with entropy threshold $2 \cdot k$ and error $\epsilon \geq 0$ if for every two independent distributions X_1, X_2 on $\{0,1\}^n$ both having min-entropy at least k, $f(X_1, X_2)$ is ϵ-close to the uniform distribution on $\{0,1\}^m$ (resp. $|\text{Supp}(f(X_1, X_2))| \geq (1-\epsilon)2^m$). We say that f is a zero-error disperser if it is a disperser with error $\epsilon = 0$. We say that f is a μ-strongly hitting disperser if for every X_1, X_2 as above and every $z \in \{0,1\}^m$, $\Pr[f(X_1, X_2) = z] \geq \mu$.*

Background. The probabilistic method gives 2-source extractors with $m = 2 \cdot k - O(\log(1/\epsilon))$ for any $k \geq \Omega(\log n)$. However, until 2005 the best explicit constructions [8,32] only achieved $k > n/2$. The current best extractor construction [6] achieves entropy threshold $k = (1/2 - \alpha)n$ for some constant $\alpha > 0$. Improved constructions of dispersers for entropy threshold $k = \delta n$ (for an arbitrary constant $\delta > 0$) were given in [2]. These dispersers can output any constant number of bits with zero-error (and are μ-strongly hitting for some constant $\mu > 0$).[1] Subsequent work by [3] achieved entropy threshold to $k = n^{o(1)}$ and gives zero-error dispersers that output one bit.

Our results. We use our composition techniques to improve the output length in the construction of [2]. We show that:

Theorem 2 (2-source zero-error disperser). *For every $\delta > 0$ there exists a $\nu > 0$ and $\eta > 0$ such that for sufficiently large n there is a $poly(n)$-time computable $(\nu 2^{-m})$-strongly hitting 2-source disperser $D : (\{0,1\}^n)^2 \mapsto \{0,1\}^m$ with entropy threshold $2 \cdot \delta n$ and $m = \eta n$.*

Note that our construction achieves an output length that is optimal up to constant factors for this entropy threshold. For lower entropy threshold our techniques gives that any explicit construction of a zero-error 2-source disperser D' with entropy threshold k and output length $t = \Omega(\log n)$ can be transformed into an explicit construction of a zero-error 2-source disperser D with entropy threshold $2 \cdot k$ and output length $m = \Omega(k)$. (See the full version for a precise formulation that also considers strongly hitting dispersers). This cannot be applied on the construction of [3] that achieves entropy threshold $k = n^{o(1)}$ as this construction only outputs one bit. Nevertheless, this means that it suffices to extend the construction of [3] so that it outputs $\Theta(\log n)$ bits in order to obtain an output length of $m = \Omega(k)$ for low entropy threshold k.

We prove Theorem 2 by designing a subsource hitter for 2-sources and using our composition technique. The details are given in the full version and a high level outline appears next.

Outline of the argument. We want to design a function $F : \{0,1\}^n \times \{0,1\}^n \times \{0,1\}^t \mapsto \{0,1\}^m$ such that for any 2-source $X = (X_1, X_2)$ with sufficient min-entropy and for any $z \in \{0,1\}^m$ there exists a "seed" $y \in \{0,1\}^t$ and a subsource X' of X such that $X' = (X_1', X_2')$ is a 2-source with roughly the same min-entropy as X and $\Pr[F(X_1', X_2', y) = z] = 1$. We will be shooting for $m = \Omega(n)$ and $t = O(\log n)$.

We construct the seed obtainer F using ideas from [2,3]. Let E be a seeded extractor with seed length $t = O(\log n)$, output length $v = \Omega(k)$ and error $\epsilon_E = 1/100$ (such extractors were constructed in [21,18]). When given inputs x_1, x_2, y we consider $r_1 = E(x_1, y)$ and $r_2 = E(x_2, y)$. By using a stronger variant of seeded extractors called "strong extractors" it follows that there exists a "good seed" $y \in \{0,1\}^t$ such that $R_1 = E(X_1, y)$ and $R_2 = E(X_2, y)$ are ϵ_E-close to uniform. We then use a 2-source extractor $H : \{0,1\}^v \times \{0,1\}^v \mapsto \{0,1\}^m$ for *very high* entropy threshold (say entropy threshold $2 \cdot 0.9v$) and very low error

---

[1] In [25] it is pointed out that by enhancing the technique of [2] using ideas from [3] and replacing some of the components used in the construction with improved components that are constructed in [25] it is possible to increase the output length and achieve a zero-error disperser with output length $m = k^{\Omega(1)}$ for the same entropy threshold k.

(say error $2^{-(m+1)}$ for output length $m = \Omega(v) = \Omega(k)$). Such extractors were constructed in [32]. Our final output is given by:

$$F(x_1, x_2, y) = H(E(x_1, y), E(x_2, y))$$

This seems strange at first sight as it is not clear why running H on inputs R_1, R_2 that are already close to uniform helps. Furthermore, the straightforward analysis only gives that $H(R_1, R_2)$ is ϵ-close to uniform for *large* error $\epsilon \geq \epsilon_E = 1/100$ and this means that the output of F may miss a large fraction of strings in $\{0,1\}^m$.

The point to notice is that both R_1, R_2 are close to uniform and therefore have large support $(1 - \epsilon_E)2^v \geq 2^{0.9v}$. Using Fact 1 we can think of H as a zero-error disperser. Recall that for dispersers are oblivious to the precise probability distribution of R_1, R_2 and it is sufficient that R_1, R_2 have large support. It follows that indeed every string $z \in \{0,1\}^m$ is hit by $H(R_1, R_2)$.

This does not suffice for our purposes as we need that any string z is hit with probability one on a subsource $X' = (X_1', X_2')$ of X in which the two distributions X_1' and X_2' are independent. For any output string $z \in \{0,1\}^m$ we consider a pair of values (r_1, r_2) for R_1, R_2 on which $H(r_1, r_2) = z$ (we have just seen that such a pair exists) and set $X_1' = (X_1 | E(X_1, y) = r_1)$ and $X_2' = (X_2 | E(X_2, y) = r_2)$. Note that these two distributions are indeed independent (as each depends only on one of the original distributions X_1, X_2) and that on every $x_1' \in \text{Supp}(X_1')$ and $x_2' \in \text{Supp}(X_2')$ we have that:

$$F(x_1', x_2', y) = H(E(x_1', y), E(x_2', y)) = H(r_1, r_2) = z$$

Furthermore, for a typical choice of (r_1, r_2) we can show that both X_1', X_2' have min-entropy roughly $k - v$. Thus, setting v appropriately, X' is a subsource of X with the required properties. (A more careful version of this argument can be used to preserve the "strongly hitting" property).

Interpretation in Ramsey Theory. A famous theorem in Ramsey Theory (see [17]) states that for sufficiently large N and any 2-coloring of the edges of the complete graph on N vertices there is an induced subgraph on $K = \Theta(\log N)$ vertices which is "monochromatic" (that is all edges are of the same color).

Zero-error 2-source dispersers (with output length $m = 1$) can be seen as providing counterexamples to this statement for larger values of K in the following way: When given a zero-error 2-source disperser $D : \{0,1\}^n \times \{0,1\}^n \mapsto \{0,1\}^m$ with entropy threshold $2 \cdot k$ we can consider coloring the edges of the full graph on $N = 2^n$ vertices with 2^m colors by coloring an edge (v_1, v_2) by $D(v_1, v_2)$. (A technicality is that $D(v_1, v_2)$ may be different than $D(v_2, v_1)$ and to avoid this problem the coloring is defined by ordering the vertices according to some order and coloring the edge (v_1, v_2) where $v_1 \leq v_2$ by $D(v_1, v_2)$). The disperser guarantee can be used to show that any induced subgraph with $K = 2^{k+1}$ vertices contains edges of *all* 2^m colors.[2]

---

[2] In fact, Dispersers translate into a significantly stronger guarantee that discusses colorings of the edges of the complete N by N bipartite graph such that any induced K by K subgraph has all colors.

Note that dispersers with $m > 1$ translate into colorings with more colors and that in this context of Ramsey Theory the notion of a zero-error disperser seems more natural than one that allows error. Our constructions achieve $m = \Omega(k)$ and thus the number of colors in the coloring approaches the size of the induced subgraph.

Generalizing this relation between dispersers and Ramsey theory we can view any zero-error disperser for a class $\mathcal{C}$ as a coloring of all $x \in \Omega$ such that any set S that is obtained as the support of a distribution in $\mathcal{C}$ is colored by all possible 2^m colors.

Rainbows and implicit $O(1)$-probe search. As we now explain, explicit constructions of zero-error 2-source dispersers can be used to construct certain data structures (this connection is due to [13]).

Consider the following problem: We are given a set $S \subseteq \{0, 1\}^n$ of size 2^k. We want to store the elements of S in a table T of the same size where every entry in the table contains a single element of S (and so the only freedom is in ordering the elements of S in the table T). We say that T supports q-queries if given $x \in \{0, 1\}^n$ we can determine whether $x \in S$ using q queries to T (note for example that ordered tables and binary search support $q = k$ queries). Yao [34] and Fiat and Naor [13] showed that it is impossible to achieve $q = O(1)$ when n is large enough relative to k. (This result can be seen as a kind of Ramsey Theorem).

Fiat and Naor [13] gave explicit constructions of tables that support $q = O(1)$ queries when $k = \delta \cdot n$ for any constant $\delta > 0$. This was achieved by reducing the implicit probe search problem to the task of explicitly constructing a certain combinatorial object that they call a "rainbow".

Loosely speaking a rainbow is a zero-error disperser for the class of distributions X that are composed of q independent copies of a high min-entropy distribution. We stress that for this application one needs (strongly-hitting) dispersers with large output length. More precisely, in order to support $q = O(1)$ queries one requires such dispersers that have output length m that is a *constant fraction* of the entropy threshold.

Our techniques can be used to explicitly construct rainbows which in turn allow implicit probe schemes that support $q = O(1)$ queries for smaller values of k than previously known. More precisely for any constant $\delta > 0$ and $k = n^\delta$ there is a constant q and a scheme that supports q queries. The precise details are given in the full version. (We remark that one can also achieve the same results by using the technique of [13] and plugging in recent constructions of seeded dispersers).

Zero-error dispersers for bit-fixing sources. The class of *bit-fixing sources* is the class of distributions X on $\Omega = \{0, 1\}^n$ such that there exists a set $S \subseteq [n]$ such that X_S (that is X restricted to the indices in S) is uniformly distributed and $X_{[n]\setminus S}$ is constant. Note that for such a source X, $\mathrm{H}_\infty(X) = |S|$. (We remark that these sources are sometimes called "oblivious bit-fixing sources" to differentiate them from "non-oblivious bit-fixing sources" in which $X_{[n]\setminus S}$ is allowed to be a function of X_S).

Background. The function $Parity(x)$ (that is the exclusive-or of the bits of x) is obviously an extractor for bit-fixing sources with entropy threshold $k = 1$, error $\epsilon = 0$ and output length $m = 1$. It turns out that there are no errorless extractors for $m = 2$. More precisely, [9] showed that for $k < n/3$ there are no extractors for bit-fixing sources with $\epsilon = 0$ and $m = 2$. For larger values of k, [9] give constructions with $m > 1$ and $\epsilon = 0$. For general entropy threshold k the current best explicit construction of extractors for bit-fixing sources is due to [26] (in fact, this extractor works for a more general class of "low weight affine sources"). These extractors work for any entropy threshold $k \geq (\log n)^c$ for some constant c, and achieve output length $m = (1 - o(1))k$ for error $\epsilon = 2^{-k^{\Omega(1)}}$. Using Fact 1 this gives a zero-error disperser with output length $m = k^{\Omega(1)}$.

Our results. We use our composition techniques to construct zero-error dispersers for bit-fixing sources with output length $m = \Omega(k)$. We show that:

Theorem 3 (Zero-error disperser for bit-fixing sources). *There exist $c > 1$ and $\eta > 0$ such that for sufficiently large n and $k \geq (\log n)^c$ there is a poly(n)-time computable zero-error disperser $D : \{0,1\}^n \mapsto \{0,1\}^m$ for bit-fixing sources with entropy threshold k and output length $m = \eta k$.*

Note that our construction achieves an output length that is optimal up to constant factors. We prove Theorem 3 by designing a subsource hitter for bit-fixing sources and using our composition technique. The details are given in the full version and a high level outline appears next.

Outline of the argument. Our goal is to design a subsource hitter $G : \{0,1\}^n \times \{0,1\}^t \mapsto \{0,1\}^m$ for bit-fixing sources with entropy threshold k, output length $m = \Omega(k)$ and "seed length" $t = O(\log n)$. We make use of the subsource hitter for 2-sources $F : \{0,1\}^n \times \{0,1\}^n \times \{0,1\}^{O(\log n)} \mapsto \{0,1\}^m$ that we designed earlier. We apply it for entropy threshold $k' = k/8$ and recall that it has output length $m = \Omega(k') = \Omega(k)$.

When given a seed $y \in \{0,1\}^t$ for G we think about it as a pair of strings (y', y'') where y' is a seed for F and y'' is a seed for an explicit construction of pairwise independent variables $Z_1, \ldots, Z_n$ where for each i, Z_i takes values in $\{1, 2, 3\}$ (indeed there are such constructions with seed length $O(\log n)$). When given such a seed y'' we can use the values $Z_1, \ldots, Z_n$ to partition the set $[n]$ into three disjoint sets T_1, T_2, T_3 by having each index $i \in [n]$ belong to T_{Z_i}. We construct G as follows:

$$G(x, (y', y'')) = F(x_{T_1}, x_{T_2}, y')$$

In words, we use y'' to partition the given n bit string into three strings and we run F on the first two strings (padding each of them to length n) using the seed y'.

We need to show that for any bit-fixing source X of min-entropy k and for any $z \in \{0,1\}^m$ there exist a seed $y = (y', y'')$ and a subsource X' of X such that X' is a bit-fixing source with roughly the same min-entropy as X and $\Pr[G(X', (y', y'')) = z] = 1$.

We have that X is a bit-fixing source and let $S \subseteq [n]$ be the set of its "good indices". Note that $|S| \geq k$. By the "sampling properties" of pairwise independent distributions (see e.g. [16] for a survey on "averaging samplers") it follows that there exists a y'' such that for every $i \in [3]$, $|S \cap T_i| \geq k/8$. It follows that $X_{T_1}, X_{T_2}, X_{T_3}$ are bit-fixing sources with min-entropy at least $k/8$ (and note that these three distributions are independent). Thus, by the properties of the subsource hitter F there exist x_1, x_2, y' such that $F(x_1, x_2, y') = z$ (note that here we're only using the property that F "hits z" and do not use the stronger property that F "hits z on a subsource"). Consider the distribution

$$X' = (X|X_{T_1} = x_1 \wedge X_{T_2} = x_2)$$

This is a subsource of X which is a bit-fixing source with min-entropy at least $k/8$ (as we have not fixed the $k/8$ good bits in T_3). It follows that for every $x \in \mathrm{Supp}(X')$

$$G(x, (y', y'')) = F(x_1, x_2, y') = z$$

and G is indeed a subsource hitter for bit-fixing sources.

Affine sources. The class of *affine sources* is the class of distributions X on $\Omega = \mathbb{F}_q^n$ (where $\mathbb{F}_q$ is the finite field of q elements) such that X is uniformly distributed over an affine subspace V in $\mathbb{F}_q^n$. Note that such a source X has min-entropy $\log q \cdot dim(V)$. Furthermore, any bit-fixing source is an affine source over $\mathbb{F}_2$.

Background. For $\mathbb{F}_2$ the best explicit construction of extractors for affine sources was given in [7]. This construction works for entropy threshold $k = \delta n$ (for any fixed $\delta > 0$) and achieves output length $m = \Omega(k)$ with error $\epsilon < 2^{-m}$.

Extractors for lower entropy thresholds were given by [14] in the case that $q = n^{\Theta(1)}$. For any entropy threshold $k > \log q$ these extractors can output $m = (1 - o(1))k$ bits with error $\epsilon = n^{-\Theta(1)}$. Using Fact 1 this gives zero-error dispersers with output length $m = \Theta(\log n)$.

Our results. Our composition techniques can be applied on affine sources. We focus on the case of large fields (as in that case we can improve the results of [14]). We remark that our techniques also apply when q is small (however, at the moment we do not gain by applying them on the existing explicit constructions). We prove the following theorem:

Theorem 4. *Fix any prime power q and integers n, k such that $q \geq n^{18}$ and $2 \leq k < n$. There is a poly$(n, \log q)$-time computable zero-error disperser $D : \mathbb{F}_q^n \mapsto \{0, 1\}^m$ for affine sources with entropy threshold $k \cdot \log q$ and $m = (k - 1) \cdot \log q$.*

Outline of the argument. We use our composition techniques to give a different analysis of the construction of [14] which shows that this construction also gives a zero-error disperser. The construction of [14] works by first constructing an affine source extractor D' with small output length $m = \Theta(\log n)$ and then composing it with some function F to obtain an extractor $D(x) = F(x, D'(x))$ that extracts many bits (with rather large error). We observe that the function

F designed in [14] is in fact a subsource hitter for affine sources and therefore our composition technique gives that the final construction is a zero-error disperser.

2   Open Problems

2-sources. One of the most important open problems in this area is to give constructions of extractors for entropy threshold $k = o(n)$. Such constructions are not known even for $m = 1$ and large error ϵ.

There are explicit constructions of zero-error dispersers with $k = n^{o(1)}$ [3]. However, these dispersers only output one bit. Improving the output length in these constructions to $\Theta(\log n)$ bits will allow our composition techniques to achieve output length $m = \Omega(k)$.

Another intriguing problem is that for the case of zero-error (or strongly hitting) dispersers we do not know whether the existential results proven via the probabilistic method achieve the best possible parameters. More precisely, a straightforward application of the probabilistic method gives zero-error 2-source dispersers which on entropy threshold $2 \cdot k$ output $m = k - \log(n - k) - O(1)$ bits. On the other hand the lower bounds of [23,24] can be used to show that any zero-error 2-source disperser with entropy threshold $2 \cdot k$ has $m \leq k + O(1)$.[3]

O(1)-sources, rainbows and implicit probe search. When allowing ℓ-sources for $\ell = O(1)$ we give constructions of zero-error dispersers which on entropy threshold $k = n^{\Omega(1)}$ achieve output length $m = \Omega(k)$. An interesting open problem is to try and improve the entropy threshold. As explained in the full version this immediately implies improved implicit probe search schemes.

Bit-fixing sources. We give constructions of zero-error dispersers which on entropy threshold k achieve $m = \Omega(k)$. A straightforward application of the probabilistic method gives zero-error dispersers with $m = k - \log n - o(\log n)$. We do not know how to match these parameters with explicit constructions.

Affine sources. We constructed a subsource hitter for affine sources over relatively large fields (that is $q = n^{\Theta(1)}$). It is interesting to try and construct subsource hitters for smaller fields.

Finally, it is also natural to ask whether our composition approach applies to other classes of sources.

Acknowledgements

We are grateful to Ran Raz for his support.

---

[3] Radhakrishnan and Ta-Shma [24] show that any seeded disperser $D : \{0,1\}^n \times \{0,1\}^t \to \{0,1\}^m$ that is nontrivial in the sense that $m \geq t+1$ has $t \geq \log(1/\epsilon) - O(1)$. A zero-error 2-source disperser D' with entropy threshold k can be easily transformed into a seeded disperser with seed length $t = k$ by setting $D(x,y) = D'(x,y')$ where y' is obtained by padding the k bit long "seed" y with $n - k$ zeroes. The bound follows as D' has error smaller than 2^{-m}.

References

1. Barak, B., Impagliazzo, R., Wigderson, A.: Extracting randomness using few independent sources. SIAM J. Comput 36(4), 1095–1118 (2006)
2. Barak, B., Kindler, G., Shaltiel, R., Sudakov, B., Wigderson, A.: Simulating independence: New consturctions of condenesers, ramsey graphs, dispersers, and extractors. In: STOC 2005, pp. 1–10 (2005)
3. Barak, B., Rao, A., Shaltiel, R., Wigderson, A.: 2-source dispersers for subpolynomial entropy and Ramsey graphs beating the Frankl–Wilson construction. In: STOC 2006, pp. 671–680 (2006)
4. Ben-Or, M., Linial, N.: Collective coin flipping. ADVCR: Advances in Computing Research 5, 91–115 (1989)
5. Blum, M.: Independent unbiased coin flips from a correlated biased source-a finite stae markov chain. Combinatorica 6(2), 97–108 (1986)
6. Bourgain, J.: More on the sum-product phenomenon in prime fields and its applications. International Journal of Number Theory 1, 1–32
7. Bourgain, J.: On the construction of affine extractors. Geometric And Functional Analysis 17(1), 33–57 (2007)
8. Chor, B., Goldreich, O.: Unbiased bits from sources of weak randomness and probabilistic communication complexity. SIAM Journal on Computing 17(2), 230–261 (1988); Special issue on cryptography
9. Chor, B., Goldreich, O., Hastad, J., Friedman, J., Rudich, S., Smolensky, R.: The bit extraction problem or t-resilient functions. In: FOCS 1985, pp. 396–407 (1985)
10. Cohen, A., Wigderson, A.: Dispersers, deterministic amplification and weak random sources. In: FOCS 1989, pp. 14–25 (1989)
11. Dodis, Y., Elbaz, A., Oliveira, R., Raz, R.: Improved randomness extraction from two independent sources. In: Jansen, K., Khanna, S., Rolim, J., Ron, D. (eds.) RANDOM 2004. LNCS, vol. 3122, pp. 334–344. Springer, Heidelberg (2004)
12. Dvir, Z., Gabizon, A., Wigderson, A.: Extractors and rank extractors for polynomial sources. In: FOCS 2007, pp. 52–62 (2007)
13. Fiat, A., Naor, M.: Implicit O(1) probe search. SICOMP: SIAM Journal on Computing 22 (1993)
14. Gabizon, A., Raz, R.: Deterministic extractors for affine sources over large fields. In: FOCS 2005, pp. 407–418 (2005)
15. Gabizon, A., Raz, R., Shaltiel, R.: Deterministic extractors for bit-fixing sources by obtaining an independent seed. SICOMP: SIAM Journal on Computing 36(4), 1072–1094 (2006)
16. Goldreich, O.: A sample of samplers – A computational perspective on sampling (survey). In: ECCCTR: Electronic Colloquium on Computational Complexity, technical reports (1997)
17. Graham, R.L., Rothschild, B.L., Spencer, J.H.: Ramsey Theory. Wiley, Chichester (1980)
18. Guruswami, V., Umans, C., Vadhan, S.P.: Unbalanced expanders and randomness extractors from parvaresh-vardy codes. In: CCC 2007, pp. 96–108 (2007)
19. Kamp, J., Rao, A., Vadhan, S., Zuckerman, D.: Deterministic extractors for small-space sources. In: STOC 2006, pp. 691–700 (2006)
20. Kamp, J., Zuckerman, D.: Deterministic extractors for bit-fixing sources and exposure-resilient cryptography. SIAM J. Comput. 36(5), 1231–1247 (2007)
21. Lu, C., Reingold, O., Vadhan, S., Wigderson, A.: Extractors: Optimal up to constant factors. In: STOC 2003, pp. 602–611 (2003)

22. Mossel, E., Umans, C.: On the complexity of approximating the vc dimension. In: CCC 2001, pp. 220–225 (2001)
23. Nisan, N., Zuckerman, D.: Randomness is linear in space. Journal of Computer and System Sciences 52(1), 43–52 (1996)
24. Radhakrishnan, J., Ta-Shma, A.: Bounds for dispersers, extractors, and depth-two superconcentrators. SIAM Journal on Discrete Mathematics 13(1), 2–24 (2000)
25. Rao, A.: Extractors for a constant number of polynomially small min-entropy independent sources. In: STOC 2006, pp. 497–506 (2006)
26. Rao, A.: Extractors for low weight affine sources (unpublished manuscript) (2008)
27. Raz, R.: Extractors with weak random seeds. In: STOC 2005, pp. 11–20 (2005)
28. Santha, M., Vazirani, U.V.: Generating quasi-random sequences from semi-random sources. Journal of Computer and System Sciences 33, 75–87 (1986)
29. Shaltiel, R.: How to get more mileage from randomness extractors. In: CCC 2006, pp. 46–60 (2006)
30. Shaltiel, R.: Recent developments in explicit constructions of extractors. Bulletin of the EATCS 77, 67–95 (2002)
31. Trevisan, L., Vadhan, S.: Extracting randomness from samplable distributions. In: FOCS 2000, pp. 32–42 (2000)
32. Vazirani, U.: Strong communication complexity or generating quasi-random sequences from two communicating semi-random sources. Combinatorica 7, 375–392 (1987)
33. von Neumann, J.: Various techniques used in connection with random digits. Applied Math Series 12, 36–38 (1951)
34. Yao, A.C.-C.: Should tables be sorted? J. ACM 28(3), 615–628 (1981)

Euclidean Sections of ℓ_1^N with Sublinear Randomness and Error-Correction over the Reals

Venkatesan Guruswami[1,*], James R. Lee[2,**], and Avi Wigderson[3]

[1] Department of Comp. Sci. & Eng., University of Washington, and (on leave at) School of Mathematics, Institute for Advanced Study, Princeton
[2] Department of Comp. Sci. & Eng., University of Washington
[3] School of Mathematics, Institute for Advanced Study, Princeton

Abstract. It is well-known that $\mathbb{R}^N$ has subspaces of dimension proportional to N on which the ℓ_1 and ℓ_2 norms are uniformly equivalent, but it is unknown how to construct them explicitly. We show that, for any $\delta > 0$, such a subspace can be generated using only N^δ random bits. This improves over previous constructions of Artstein-Avidan and Milman, and of Lovett and Sodin, which require $O(N \log N)$, and $O(N)$ random bits, respectively.

Such subspaces are known to also yield error-correcting codes over the reals and compressed sensing matrices. Our subspaces are defined by the kernel of a relatively sparse matrix (with at most N^δ non-zero entries per row), and thus enable compressed sensing in near-linear $O(N^{1+\delta})$ time.

As in the work of Guruswami, Lee, and Razborov, our construction is the continuous analog of a Tanner code, and makes use of expander graphs to impose a collection of local linear constraints on vectors in the subspace. Our analysis is able to achieve *uniform* equivalence of the ℓ_1 and ℓ_2 norms (independent of the dimension). It has parallels to iterative decoding of Tanner codes, and leads to an analogous near-linear time algorithm for error-correction over reals.

1 Introduction

Given $x \in \mathbb{R}^N$, one has the straightforward inequality $\|x\|_2 \leq \|x\|_1 \leq \sqrt{N}\|x\|_2$. Classical results of Figiel, Lindenstrauss, and Milman [FLM77] and Kasin [Kas77] show that for every $\eta > 0$, there exists a constant $C(\eta)$ and a subspace $X \subseteq \mathbb{R}^N$ with $\dim(X) \geq (1 - \eta)N$ such that for every $x \in X$,

$$C(\eta)\sqrt{N}\|x\|_2 \leq \|x\|_1 \leq \sqrt{N}\|x\|_2.$$

We say that such a subspace has distortion at most $C(\eta)$, where for a subspace $X \subseteq \mathbb{R}^N$, we define the *distortion* of X as the quantity

$$\Delta(X) = \sup_{\substack{x \in X \\ x \neq 0}} \frac{\sqrt{N}\|x\|_2}{\|x\|_1}.$$

* Supported by a Packard Fellowship and NSF grant CCR-0324906 to the IAS.
** Research supported by NSF CAREER award CCF-0644037.

A. Goel et al. (Eds.): APPROX and RANDOM 2008, LNCS 5171, pp. 444–454, 2008.

The distortion always lies in the range $1 \leqslant \Delta(X) \leqslant \sqrt{N}$, and describes the extent to which the ℓ_2 mass of vectors in X is spread among the coordinates.

It is known that subspaces with good distortion give rise to error-correcting codes over the reals and compressed sensing matrices [KT07, Don06]. When viewed as embeddings of ℓ_2^n into ℓ_1^N (hence the terminology "distortion"), they are useful for problems like high-dimensional nearest-neighbor search [Ind06]. We discuss connections of our work to coding over reals in Section 3.

The existence proofs of [FLM77, Kas77] proceed by showing that a random subspace (for various notions of "random") satisfies the above conditions with positive probability. The problem of *explicit* constructions of these subspaces has been raised by a number of authors; see, e.g. [Sza06, Sec. 4], [Mil00, Prob. 8], [JS01, Sec. 2.2].

Although no explicit construction is known, there has been progress on reducing the amount of randomness needed to construct such subspaces. Kasin's proof [Kas77] is particularly amenable to such analysis because the subspaces he constructs are kernels of uniformly random sign matrices. This immediately gives rise to an algorithm which produces such a matrix using $O(N^2)$ random bits. Previous approaches to partial derandomization construct sign matrices whose entries are non-independent random signs; indeed, Artstein-Avidan and Milman reduce the randomness requirement to $O(N \log N)$ using random walks on expander graphs [AAM06], and Lovett and Sodin [LS07] improve this to $O(N)$ random bits by employing, in addition, families of $\Theta(\log N)$-wise independent random variables. We remark that the pseudorandom generator approach of Indyk [Ind06] can be used to efficiently construct such subspaces using $O(N \log^2 N)$ random bits.

As pointed out in [LS07], since these direct approaches require taking a union bound over an (exponentially large) ε-net, it is unlikely that they can be pushed beyond a linear dependence on the number of random bits. In contrast, the approaches of Indyk [Ind07] and Guruswami, Lee, and Razborov [GLR08] are inspired by work in randomness extraction and the theory of error-correcting codes. The latter paper uses a continuous variant of *expander codes* to deterministically construct a subspace $X \subseteq \mathbb{R}^N$ satisfying $\mathsf{dim}(X) \geq (1 - o(1))N$ and,

$$\Delta(X) \leqslant (\log N)^{O(\log \log \log N)} \ .$$

Even using sub-linear randomness, they are only able to achieve a distortion of $\mathsf{poly}(\log N)$. In the present paper, we continue with the expander codes approach of [GLR08], with a different construction and analysis. As a result, we are able to produce almost-Euclidean sections of ℓ_1^N with constant distortion and proportional dimension, while using only N^δ random bits for any $\delta > 0$. In Section 2 and Remark 1, we discuss how our analysis overcomes some difficulties from [GLR08].

In the next section, we show that given a subspace $X \subseteq \mathbb{R}^n$ with $\mathsf{dim}(X) \geq (1-\eta)n$, for every $N \geq n$ there exists a simple, explicit construction of a subspace $X' \subseteq \mathbb{R}^N$ satisfying $\mathsf{dim}(X') \geq (1 - 2\eta)N$ and $\Delta(X') \leq N^{O(\frac{1}{\log n})}\Delta(X)$. If X is the kernel of a sign matrix, then so is X'. By choosing $n = N^{\delta/2}$ and generating X as the kernel of a random sign matrix, we achieve a construction with $\Delta(X') = 2^{O(1/\delta)}$ distortion using at most N^δ random bits.

1.1 Preliminaries

We use $[M]$ to denote the set $\{1, 2, \ldots, M\}$. For $x \in \mathbb{R}^N$ and a subset $I \subseteq [N]$, we denote by $x_I \in \mathbb{R}^{|I|}$ the restriction of x to the coordinates in I.

Definition 1 (Well spread subspaces). *A subspace $L \subseteq \mathbb{R}^m$ is said to be (b, ρ)-spread if for every $y \in L$, and every set $S \subseteq [m]$ size at least $m - b$, $\|y_S\|_2 \geqslant \rho \|y\|_2$.*

As stated below, there is a straightforward relation between spread subspaces and distortion (see, e.g. [GLR08, Lemma 2.11]), but the former notion is a well-suited to our arguments

Lemma 1. *Suppose $X \subseteq \mathbb{R}^N$.*

1. If X is (b, ρ)-spread then

$$\Delta(X) \leq \sqrt{\frac{N}{b}} \cdot \rho^{-2};$$

2. Conversely, X is $\left(\frac{N}{2\Delta(X)^2}, \frac{1}{4\Delta(X)}\right)$-spread.

We will make use of the following (non-constructive) result on the existence of well-spread subspaces; it is due to Kasin [Kas77], with the optimal bound obtained by Garnaev and Gluskin for uniformly random subspaces [GG84]. The proof that sign matrices suffice is now standard, given the covering estimates of Schütt [Sch84]; see e.g. [LS07, Lemma B].

Theorem 2. *For all integers $1 \leqslant k < d$, there exists a subspace $Y \subseteq \mathbb{R}^d$ of dimension at least $d - k$, specified as the kernel of a $k \times d$ sign matrix, such that $\Delta(Y) \leqslant O\left(\sqrt{\frac{d}{k}} \log \frac{d}{k}\right)$, and so by Lemma 1, Y is*

$$\left(\Omega\left(\frac{k}{\log(d/k)}\right), \Omega\left(\sqrt{\frac{k}{d \log(d/k)}}\right)\right) \text{-spread.}$$

In fact, a random such matrix has this property with probability $1 - o_d(1)$.

Definition 3 (Subspaces from regular graphs). *Given an undirected d-regular graph $G = (V, E)$ with N edges and a subspace $L \subseteq \mathbb{R}^d$, we define the subspace $T(G, L) \subseteq \mathbb{R}^N$ by*

$$T(G, L) = \{x \in \mathbb{R}^N \mid x_{\Gamma(v)} \in L \text{ for every } v \in V\}. \tag{1}$$

where $\Gamma(v)$ is the set of d edges incident on v in some fixed order.

The definition of the subspace $T(G, L)$ is inspired by the construction of *expander codes,* following Sipser and Spielman [SS96] and Tanner [Tan81] in the case of finite fields, and its adaptation to the reals by Guruswami, Lee, and Razborov [GLR08].

Definition 4 (Expander). *A simple, undirected graph G is said to be an (n, d, λ)-expander if G has n vertices, is d-regular, and the second largest eigenvalue of the adjacency matrix of G in absolute value is at most λ.*

For a graph $G = (V, E)$ (which will be clear from context) and $W \subseteq V$, we denote by $E(W)$ the set of edges both of whose endpoints lie in W. For two subsets $X, Y \subseteq V$ (which could intersect), we denote by $E(X, Y)$ the (multi)set of edges with one endpoint in X and the other in Y. Recall that for a vertex $v \in V$, $\Gamma(v) \subseteq E$ is the set of edges incident upon v.

2 Derandomized Sections

Following [GLR08], we now show that if L is well-spread and G is an expander graph, then $T(G, L)$ is itself well-spread. This immediately implies the ability to create large dimensional low-distortion subspaces from those with smaller dimension. In Remark 1, we discuss how our analysis is able to overcome the apparent barrier in [GLR08]. Finally, in Section 2.2, we present a construction of Noga Alon which shows that our analysis is tight amongst a certain class of approaches.

2.1 Spread Boosting

The following is the analog of the spread-boosting theorem in [GLR08], except we only care about the mass outside edges in *induced* subgraphs of G (and not an arbitrary collection of edges of certain size).

Lemma 2. *Let $G = (V, E)$ be an (n, d, λ)-expander, and let $L \subseteq \mathbb{R}^d$ be a $(d/B, \rho)$-spread subspace for some parameters $B > 1$ and $\rho < 1$. Then, for all $W \subseteq V$, $|W| \leqslant \frac{n}{2B}$, there exists a subset $Z \subseteq W$, $|Z| \leqslant \left(\frac{2\lambda B}{d}\right)^2 |W|$ such that for every $x \in T(G, L)$ the following holds:*

$$\sum_{e \notin E(W)} x_e^2 \geqslant \rho^2 \sum_{e \notin E(Z)} x_e^2 . \tag{2}$$

Proof. Given W, we define Z as follows:

$$Z = \left\{ w \in W \ : \ |N_G(w) \cap W| \geqslant \frac{d}{B} \right\} .$$

By definition, $|E(Z, W)| \geqslant \frac{d}{B}|Z|$. On the other hand, by the expander mixing lemma (see, e.g. [HLW06, §2.4]),

$$|E(Z, W)| \leqslant d|Z|\frac{|W|}{n} + \lambda\sqrt{|Z||W|} \leqslant \frac{d|Z|}{2B} + \lambda\sqrt{|Z||W|} .$$

Combining the two bounds, $|Z| \leqslant \left(\frac{2\lambda B}{d}\right)^2 |W|$. By definition of Z and the $(d/B, \rho)$-spread property of L, it follows easily that

$$\sum_{e \in E(W, \overline{W})} x_e^2 \geqslant \rho^2 \sum_{v \in W \setminus Z} \|x_{\Gamma(v)}\|_2^2 . \tag{3}$$

Now

$$\sum_{v \in W \setminus Z} \|x_{\Gamma(v)}\|_2^2 \geqslant \|x\|_2^2 - \sum_{e \in E(\overline{W})} x_e^2 - \sum_{e \in E(Z)} x_e^2 = \sum_{e \notin E(Z)} x_e^2 - \sum_{e \in E(\overline{W})} x_e^2 . \tag{4}$$

Combining the previous two bounds, we get the desired conclusion (2).

Remark 1 (Comparison to [GLR08]). A generalization of the $T(G, L)$ construction (and, indeed, the natural setting for expander codes) is to consider a *bipartite* graph $H = (V_L, V_R, E)$ with $N = |V_L|$, in which every node of V_R has degree exactly d. In this case, given $L \subseteq \mathbb{R}^d$, one defines the subspace

$$X(H, L) = \left\{x \in \mathbb{R}^N : x_{\Gamma_H(j)} \in L \text{ for every } j \in V_R\right\},$$

where now $\Gamma_H(j) \subseteq V_L$ denotes the neighbors of a vertex $j \in V_R$. Clearly $T(G, L) = X(H, L)$, where H is defined by $V_L = E(G)$, $V_R = V(G)$, and the edges of H are naturally given by edge-vertex incidence in G.

The paper [GLR08] analyzes the well-spreadness of $X(H, L)$ in terms of the well-spreadness of L and the combinatorial expansion properties of H (i.e. how large are the neighbor sets of subsets $S \subseteq V_L$). There does not exist a bipartite expander graph H with properties strong enough to imply our main result (Corollary 7), if one requires expansion from *every* subset $S \subseteq V_L$, and uses only the iterative spreading analysis of [GLR08]. We overcome this by structuring the iteration so that only certain subsets arise in the analysis, and we only require strong expansion properties from these subsets. Lemma 2 represents this idea, applied to the subspace $T(G, L)$. Here, the special subsets are precisely those edge sets which arise from induced subgraphs of G (as opposed to arbitrary subsets of the edges).

Iterating Lemma 2 yields the following.

Corollary 5. *Let $G = (V, E)$ be an (n, d, λ)-expander, and $L \subseteq \mathbb{R}^d$ be a $(d/B, \rho)$-spread subspace. Let ℓ be an integer such that*

$$\left(\frac{d}{2\lambda B}\right)^{2\ell} \geqslant \frac{n}{2B} .$$

Then for all $x \in T(G, L)$ with $\|x\|_2 = 1$ and every $W \subseteq V$ with $|W| < \frac{n}{2B}$, we have

$$\sum_{e \notin E(W)} x_e^2 \geqslant \rho^{2\ell} .$$

We now come to the proof of our main theorem.

Theorem 6 (Main). *Let $G = (V, E)$ be an (n, d, λ)-expander with $N = nd/2$ edges. Let $L \subseteq \mathbb{R}^d$ be a $(d/B, \rho)$-spread subspace of co-dimension at most ηd for some parameters $\rho, \eta < 1$ and $B > 1$. Then the subspace $T(G, L) \subseteq \mathbb{R}^N$ has dimension at least $N(1 - 2\eta)$ and it is*

$$\left(\frac{N}{2B^2}, \frac{\rho}{\sqrt{2}} n^{\frac{-\log(1/\rho)}{\log(d/(2\lambda B))}} \right) \text{-spread} .$$

In particular, by Lemma 1, this implies

$$\Delta(T(G, L)) \leqslant \frac{2\sqrt{2}B}{\rho^2} \cdot n^{\frac{2\log(1/\rho)}{\log(d/(2\lambda B))}} .$$

Proof. The claim about the dimension of $T(G, L)$ is obvious. Fix an arbitrary $x \in T(G, L)$ with $\|x\|_2 = 1$. Let $F \subseteq E$ be an arbitrary set with $|F| \leqslant \frac{N}{2B^2} = \frac{nd}{4B^2}$. We need to prove that

$$\sum_{e \notin F} x_e^2 \geqslant \frac{\rho^2}{2} n^{\frac{-2\log(1/\rho)}{\log(d/(2\lambda B))}} . \tag{5}$$

Define

$$W = \left\{ v \in V : |F \cap \Gamma(v)| > d/B \right\} . \tag{6}$$

Since $2|F| > |W|d/B$, we have $|W| < \frac{2|F|B}{d} \leqslant \frac{n}{2B}$. We can now apply Corollary 5 with a choice of ℓ that satisfies $\left(\frac{d}{2\lambda B}\right)^{2\ell} \leqslant nd \leqslant n^2$, and conclude that

$$\sum_{e \notin E(W)} x_e^2 \geqslant \rho^{2\ell} \geqslant n^{-\frac{2\log(1/\rho)}{\log(d/(2\lambda B))}} \tag{7}$$

We have the chain of inequalities

$$2 \sum_{e \notin F} x_e^2 = \sum_{v \in V} \|x_{\Gamma(v) \setminus F}\|_2^2 \geqslant \sum_{v \notin W} \|x_{\Gamma(v) \setminus F}\|_2^2 \geqslant \rho^2 \sum_{v \notin W} \|x_{\Gamma(v)}\|_2^2 \geqslant \rho^2 \sum_{e \notin E(W)} x_e^2 , \tag{8}$$

where the last but one step follows since L is $(d/B, \rho)$-spread, and $|\Gamma(v) \cap F| \leqslant d/B$ when $v \notin W$. Combining (7) and (8) gives our desired goal (5).

The main application of the above theorem is the following result.

Corollary 7 (Constant distortion with sub-linear randomness). *For every $\delta, \eta > 0$ and every $N \geq 1$, there is a randomized (Monte Carlo) construction of a subspace $X \subseteq \mathbb{R}^N$ using N^δ random bits that has dimension at least $(1 - \eta)N$ and distortion $\Delta(X) \leqslant \left(\frac{1}{\eta}\right)^{O(1/\delta)}$.*

Proof. For every $N' \geq 1$, it is known how to construct an explicit (n, d, λ)-expander, with $\lambda \leq d^{0.9}$, $n^{\delta/4} \leq d \leq n^{\delta/2}$ such that $N' \leq N \leq 10N'$, where $N = nd/2$ [LPS88] (also see [HLW06, §2.6.3] for a discussion of explicit constructions of expander graphs).

Let $L \subset \mathbb{R}^d$ be the kernel of a random $(\eta/2)d \times d$ sign matrix. Constructing L requires at most n^δ random bits, and by Theorem 2, L is $(\eta^{O(1)}d, \eta^{-O(1)})$-spread with high probability. When this happens, the subspace $T(G, L) \subseteq \mathbb{R}^N$ has distortion at most $(\frac{1}{\eta})^{O(1/\delta)}$ by Theorem 6.

The above description only works for (infinitely many) values N of a certain form. With some combinatorial manipulations, it can be made to work for all N; see [GLR08, §2.2.2].

2.2   Optimality of Myopic Analysis

The analysis of the previous section is myopic in the sense that it only cares about the expansion properties of G, and the spreading properties of the local subspace L. The following construction, suggested to us by Noga Alon, shows that if we only use the fact that G is an expander, and that every vector induced on the neighbors of a vertex is well-spread, then asymptotically our analysis is tight. The point is that, unlike in the boolean setting, real numbers come in scales, allowing them to decay slowly while still satisfying the local constraints.

Theorem 8 ([Alo08]). *For every $d \geq 4$, and infinitely many $n \in \mathbb{N}$, there exists an $(n, d, O(\sqrt{d}))$-expander $G = (V, E)$ with $N = |E|$, and a point $x \in \mathbb{R}^N$, $x \neq 0$, such that*

$$\|x_{\Gamma(v)}\|_1 \geq \frac{\sqrt{d}}{2}\|x_{\Gamma(v)}\|_2$$

for every $v \in V$, but $\|x\|_1 \leq N^{\frac{1}{2} - \Omega(\frac{1}{\log d})}\|x\|_2$.

Proof. Let $n = 2(d-1)^k$ for some $k \in \mathbb{N}$, and let H be an $(n, d-1, O(\sqrt{d}))$-expander. Let T' and T'' be two disjoint, complete $(d-1)$-ary rooted trees of depth k. Let T be the tree that results from adding an edge e_0 between the roots of T' and T''. Finally, define $G = (V, E)$ as the d-regular graph that results from identifying the n leaves of T in an arbitrary way with the nodes of H. It is easy to check that G is an $(n', d, O(\sqrt{d}))$-expander with $n' = \Theta(d^k)$ and $N = |E| = \Theta(d^{k+1})$.

We may think of $x \in \mathbb{R}^N$ as indexed by edges $e \in E$. For $e \in E$, let $h(e)$ be the distance from e to e_0, and put $x_e = (2\sqrt{d})^{-h(e)}$. It is straightforward to verify that, for every $v \in V$, one has

$$\frac{\|x_{\Gamma(v)}\|_1}{\|x_{\Gamma(v)}\|_2} = \frac{2\sqrt{d} + d - 1}{\sqrt{5d - 1}} \geq \frac{\sqrt{d}}{2}.$$

Clearly we have $\|x\|_2 \geq 1$, whereas

$$\|x\|_1 = 1 + 2\left[\sum_{h=1}^{k}\left(\frac{d-1}{2\sqrt{d}}\right)^h\right] + \left(\frac{d-1}{2\sqrt{d}}\right)^{k+1}$$

$$= O(2^{-k}d^{(k+1)/2}) = O(2^{-k}\sqrt{N}) \leq N^{\frac{1}{2} - \Omega(\frac{1}{\log d})}.$$

We remark that the vector $x \in \mathbb{R}^N$ exhibited in the preceding theorem lies in $T(G, L)$, where $L = \mathrm{span}(1, \frac{1}{2\sqrt{d}}, \ldots, \frac{1}{2\sqrt{d}})$, as long as we choose the ordering of the neighborhoods $\Gamma(v)$ appropriately (recall that one has to fix such an ordering in defining $T(G, L)$). In light of this obstruction to our analysis, the following question seems fundamental.

Question: Is there a number $K \in \mathbb{N}$ such that for infinitely many N, there exists an $\frac{N}{2} \times N$ $\{0, 1\}$-matrix A, with at most K ones per row, and such that $\Delta(\ker(A)) = O(1)$?

Theorem 6 shows that one can take $K = N^\delta$ for any $\delta > 0$, but this is the best current bound.

3   Error-Correction over Reals

In this section, we discuss connections of our work to compressed sensing, or equivalently to error-correcting codes over reals. We will use the coding terminology to describe these connections.

3.1   Background

A w-error-correcting code of dimension m and block length N over the reals is given by a linear map $C : \mathbb{R}^m \to \mathbb{R}^N$, such that for each $f \in \mathbb{R}^m$, $f \neq 0$, $\|Cf\|_0 > 2w$. The rate of the code is the ratio m/N. Given a received word $y = Cf + e$ with $\|e\|_0 \leqslant w$, one can recover the message f as the solution x to optimization problem:

$$\min_{x \in \mathbb{R}^m} \|y - Cx\|_0 .$$

The above non-convex optimization problem is NP-hard to solve in general. Quite remarkably, if the code C meets certain conditions, one can recover f by solving the linear program (LP)

$$\min_{x \in \mathbb{R}^m} \|y - Cx\|_1 .$$

(The above ℓ_1-minimization task, which is easily written as a linear program, is often called *basis pursuit* in the literature.) Note that we are not restricting the magnitude of erroneous entries in e, only that their number is at most w.

Candes and Tao [CT05] studied the above error-correction problem and proved that ℓ_1-minimization works if the code has a certain restricted isometry property. A sufficient condition for ℓ_1-minimization to work is also implied by the distortion property of the image of C, as formalized in the following lemma by Kashin and Temlyakov [KT07].

Lemma 3 ([KT07]). *Let $X = \{Cx \mid x \in \mathbb{R}^m\} \subseteq \mathbb{R}^N$ be the image of C. Then C is a w-error-correcting code provided $w < \frac{N}{4\Delta(X)^2}$, and moreover, given $y \in \mathbb{R}^N$ such that $\|y - Cf\|_0 \leqslant w$ for some $f \in \mathbb{R}^m$, the signal f can be recovered efficiently by solving the (LP) $\min_{x \in \mathbb{R}^m} \|y - Cx\|_1$.*

By picking X to be the kernel of a random $\gamma N \times N$ sign matrix, and plugging in the distortion bound of Theorem 2, gives codes of rate at least $(1 - \gamma)$ that are, w.h.p, w-error-correcting with an efficient algorithm for $w = \Theta(\frac{\gamma N}{\log(1/\gamma)})$. This is not far from the best possible bound, achieved non-constructively, of $w = \Theta(\gamma N)$.

The ℓ_1-minimization decoding algorithm, while polynomial time, requires solving a dense linear program of size N. It is of interest to develop faster methods for decoding, ideally with near-linear complexity.

3.2   Near-linear Time Decoding

Before we begin, let us remark that in the compressed sensing setting, since $T(G,L)$ is the kernel of a relatively sparse matrix A (with at most N^δ non-zero entries per row), the sensing, i.e., the computation of Ax for $x \in \mathbb{R}^N$, can be done in $O(N^{1+\delta})$ time. This sparsity also makes interior point methods for basis pursuit more efficient by a similar factor. Note that in the language of codes, "sensing" corresponds to syndrome computation, while signal recovery (recovering x from a noisy version of Ax) corresponds to syndrome decoding.

We now turn to algorithms for our construction $X = T(G,L)$. Efficient encoding follows immediately from sparsity of the encoding matrix (in fact, we achieve linear-time encoding by choosing $d = O(1)$ in what follows). Near-linear time decoding will be achieved using a natural iterative algorithm for Tanner codes.

For technical reasons that help with the decoding, we will take $G = (V_\ell, V_r, E)$ to be a d-regular $n \times n$ bipartite graph. Specifically, we will take G to be the double cover of an (n, d, λ)-expander, and $L \subseteq \mathbb{R}^d$ to be the kernel of a random $\gamma d/2 \times d$ matrix (for a constant $\gamma > 0$). With this choice, $X \subseteq \mathbb{R}^N$ satisfies $\dim(X) \geqslant (1 - \gamma)N$ where $N = nd$, and our code has rate at least $(1 - \gamma)$. With high probability, L will be ζd-error-correcting via ℓ_1-minimization for $\zeta = \Theta(\frac{\gamma}{\log(1/\gamma)})$, and we will assume this is the case.

There is a natural iterative algorithm for Tanner codes [SS96], specifically for the version when the underlying graph is bipartite [Z01] (see also [GI05, Sec. 2.2]), which can be adapted in a straightforward manner to the coding over reals setting. An adaptation of the related sequential "bit-flipping" algorithm for expander based low-density parity-check codes appears in [XH07]; using lossless expanders, their approach also leads to a near-linear time decoding algorithm.

Our algorithm for decoding $T(G,L)$ proceeds in several rounds, alternating between "left" and "right" rounds. In a left round, on input a string $y \in \mathbb{R}^N$ from the previous decoding round (or the noisy codeword at the start of decoding), we locally decode $y_{\Gamma(u)}$ for each $u \in V_\ell$ to its closest vector in L in ℓ_1-sense. In the subsequent "right" round, we do the same for each $v \in V_r$, and then switch back to a left round. The key point is that if the number of errors in the local neighborhood $y_{\Gamma(u)}$ of a vertex is less than ζd, then the local ℓ_1-minimization will correct those errors and thus fix this local neighborhood. The decoder terminates when either all the local projections $y_{\Gamma(u)}$, $u \in V_\ell \cup V_r$, belong to L (so that globally we have decoded to a codeword of $T(G,L)$), or more than $\Omega(\log N)$ iterations have passed without convergence to a global codeword (in this case, there must have been too many errors in the original input).

Only $O(\log N)$ iterations suffice because the number of local neighborhoods which are not yet fully decoded decays geometrically in each round, as long as the initial number of errors is small enough. Arguing as in [ZÓ1], one can show the following.

Theorem 9. *If G is the double cover of an (n, d, λ)-expander with $\lambda \leq \gamma^2 d$, and $N = nd$, then the above algorithm decodes $T(G, L)$ and corrects up to $w = O\left(\frac{\gamma^2}{\log^2(1/\gamma)} N\right)$ errors. Further, the algorithm runs in $O(Nt(d) \log_d N)$ time (or in $O(t(d) \log_d N)$ parallel time with $O(N)$ processors), where $t(d)$ is the time to perform ℓ_1-minimization for the subspace $L \subseteq \mathbb{R}^d$ (and is thus a constant if d is a constant).*

Thus if we settle for a slightly worse fraction of errors, namely $w/N = \Theta\left(\frac{\gamma^2}{\log^2(1/\gamma)}\right)$ instead of $\Theta\left(\frac{\gamma}{\log(1/\gamma)}\right)$, then the decoding can be performed in near-linear time.

An argument similar to the one in Section 2.1 can be used to show that $\Delta(X) \leqslant N^{O\left(\frac{\log(1/\gamma)}{\log d}\right)}$. In fact, the sequence of sets that arise in the repeated application of Lemma 2, starting from the set W defined in (6), would correspond to the subsets of vertices, alternating between the left and right sides, that arise in decoding a vector supported on $F \subseteq E$ to the all-zeroes vector. Note that by the connection mentioned in Lemma 3, this would only enable correcting w errors for $w \leqslant N^{1 - \Omega(1/\log d)}$ via global ℓ_1-minimization, compared to the $\Omega(N)$ errors handled by the iterative decoder.

However, there is a substantial shortcoming of the error model used in Theorem 9. In practice, it is not reasonable to assume that the error is only supported on w positions. It is more reasonable to allow small non-zero noise even in the remaining positions (recall that we assume no bound on the magnitude of noise in the w erroneous positions); this error model is used in previous works like [CT05]. The ℓ_1-minimization works also in this setting; specifically, if $w \leqslant O(N/\Delta(X)^2)$ and the error vector e satisfies $\|e - \sigma_w(e)\|_1 \leqslant \varepsilon$ where $\sigma_w(e)$ is the vector with the w largest components of e and rest equal to zero, then ℓ_1-minimization recovers a string z such that $\|z - Cf\|_2 \leqslant \frac{\Delta(X)}{\sqrt{N}} \varepsilon$ (see [KT07]).

Extending iterative decoding to the above setting is an interesting challenge that we hope to study in future work.

Acknowledgments

We are grateful to Noga Alon for his help with the proof of Theorem 8 and his permission to include it here.

References

[AAM06] Artstein-Avidan, S., Milman, V.D.: Logarithmic reduction of the level of randomness in some probabilistic geometric constructions. J. Funct. Anal. 235(1), 297–329 (2006)

[Alo08] Alon, N.: Personal communication (2008)

[CT05]   Candes, E.J., Tao, T.: Decoding by linear programming. IEEE Trans. Information. Theory 51(12), 4203–4215 (2005)

[Don06]   Donoho, D.L.: Compressed sensing. IEEE Transactions on Information Theory 52, 1289–1306 (2006)

[FLM77]   Figiel, T., Lindenstrauss, J., Milman, V.D.: The dimension of almost spherical sections of convex bodies. Acta Math. 139(1-2), 53–94 (1977)

[GG84]   Garnaev, A., Gluskin, E.D.: The widths of Euclidean balls. Doklady An. SSSR. 277, 1048–1052 (1984)

[GI05]   Guruswami, V., Indyk, P.: Linear-time encodable/decodable codes with near-optimal rate. IEEE Trans. Inform. Theory 51(10), 3393–3400 (2005)

[GLR08]   Guruswami, V., Lee, J.R., Razborov, A.: Almost Euclidean subspaces of ℓ_1^N via expander codes. In: SODA 2008: Proceedings of the nineteenth annual ACM-SIAM symposium on Discrete algorithms, Philadelphia, PA, USA. Society for Industrial and Applied Mathematics, pp. 353–362 (2008)

[HLW06]   Hoory, S., Linial, N., Wigderson, A.: Expander graphs and their applications. Bull. Amer. Math. Soc. (N.S.) 43(4), 439–561 (2006)

[Ind06]   Indyk, P.: Stable distributions, pseudorandom generators, embeddings, and data stream computation. Journal of the ACM 53(3), 307–323 (2006)

[Ind07]   Indyk, P.: Uncertainty principles, extractors, and explicit embeddings of L_2 into L_1. In: Proceedings of the 39th Annual ACM Symposium on the Theory of Computing, pp. 615–620 (2007)

[JS01]   Johnson, W.B., Schechtman, G.: Finite dimensional subspaces of L_p. In: Handbook of the geometry of Banach spaces, vol. I, pp. 837–870. North-Holland, Amsterdam (2001)

[Kas77]   Kashin, B.S.: The widths of certain finite-dimensional sets and classes of smooth functions. Izv. Akad. Nauk SSSR Ser. Mat. 41(2), 334–351, 478 (1977)

[KT07]   Kashin, B.S., Temlyakov, V.N.: A remark on compressed sensing (2007), http://www.dsp.ece.rice.edu/cs/KT2007.pdf

[LPS88]   Lubotzky, A., Phillips, R., Sarnak, P.: Ramanujan graphs. Combinatorica 8(3), 261–277 (1988)

[LS07]   Lovett, S., Sodin, S.: Almost Euclidean sections of the N-dimensional cross-polytope using O(N) random bits. Electronic Colloquium on Computational Complexity, Report TR07-012 (2007)

[Mil00]   Milman, V.: Topics in asymptotic geometric analysis. Geom. Funct. Anal., 792–815 (2000) (Special Volume, Part II); GAFA 2000 (Tel Aviv, 1999)

[Sch84]   Schütt, C.: Entropy numbers of diagonal operators between symmetric Banach spaces. J. Approx. Theory 40(2), 121–128 (1984)

[SS96]   Sipser, M., Spielman, D.A.: Expander codes. IEEE Trans. Inform. Theory 42(6, part 1), 1710–1722 (1996) (Codes and complexity)

[Sza06]   Szarek, S.: Convexity, complexity, and high dimensions. In: International Congress of Mathematicians, vol. II, pp. 1599–1621. Eur. Math. Soc., Zürich (2006)

[Tan81]   Tanner, R.M.: A recursive approach to low complexity codes. IEEE Transactions on Information Theory 27(5), 533–547 (1981)

[XH07]   Xu, W., Hassibi, B.: Efficient compressive sensing with determinstic guarantees using expander graphs. In: IEEE Information Theory Workshop (September 2007)

[Zó1]   Zémor, G.: On expander codes. IEEE Transactions on Information Theory 47(2), 835–837 (2001)

The Complexity of Local List Decoding

Dan Gutfreund[1],[*] and Guy N. Rothblum[2],[**]

[1] Department of Mathematics and CSAIL, MIT
danny@math.mit.edu
[2] CSAIL, MIT
rothblum@csail.mit.edu

Abstract. We study the complexity of locally list-decoding binary error correcting codes with good parameters (that are polynomially related to information theoretic bounds). We show that computing majority over $\Theta(1/\epsilon)$ bits is essentially equivalent to locally list-decoding binary codes from relative distance $1/2 - \epsilon$ with list size at most $\mathrm{poly}(1/\epsilon)$. That is, a local-decoder for such a code can be used to construct a circuit of roughly the same size and depth that computes majority on $\Theta(1/\epsilon)$ bits. On the other hand, there is an explicit locally list-decodable code with these parameters that has a very efficient (in terms of circuit size and depth) local-decoder that uses majority gates of fan-in $\Theta(1/\epsilon)$.

Using known lower bounds for computing majority by constant depth circuits, our results imply that every constant-depth decoder for such a code must have size almost exponential in $1/\epsilon$ (this extends even to sub-exponential list sizes). This shows that the list-decoding radius of the constant-depth local-list-decoders of Goldwasser *et al.* [STOC07] is essentially optimal.

Keywords: locally-decodable codes, list-decodable codes, constant-depth circuits.

1 Introduction

Error correcting codes are highly useful combinatorial objects that have found numerous applications both in practical settings as well as in many areas of theoretical computer science and mathematics. In the most common setting of error-correcting codes we have a message space that contains strings over some finite alphabet Σ (for simplicity we assume that all strings in the message space are of the same length). The goal is to design a function, which we call the *encoding function*, that encodes every message in the message space into a *codeword* such that even if a fairly large fraction of symbols in the codeword are

---

[*] Research was partially supported by NSF grant CCF-0514167. Part of this research was done while the author was at Harvard University and supported by ONR grant N00014-04-1-0478 and NSF grant CNS-0430336.

[**] Supported by NSF Grants CCF-0635297, NSF-0729011, CNS-0430336 and by a Symantec Graduate Fellowship.

A. Goel et al. (Eds.): APPROX and RANDOM 2008, LNCS 5171, pp. 455–468, 2008.
© Springer-Verlag Berlin Heidelberg 2008

corrupted it is still possible to recover from it the original message. The procedure that recovers the message from a possibly corrupted codeword is called *decoding*.

It is well known that beyond a certain fraction of errors, it is impossible to recover the original message, simply because the relatively few symbols that are not corrupted do not carry enough information to specify (uniquely) the original message. Still, one may hope to recover a list of candidate messages, one of which is the original message. Such a procedure is called *list-decoding*.

Typically, the goal of the decoder is to recover the entire message (or list of candidate messages) by reading the entire (possibly corrupted) codeword. There are settings, however, in which the codeword is too long to be read as a whole. Still, one may hope to recover any given individual symbol of the message, by reading only a small number of symbols from the corrupted codeword. This setting is called *local-decoding*, and both the unique and list decoding variants (as discussed above) can be considered.

Locally decodable codes, both in the unique and list decoding settings, have found many applications in theoretical computer science, most notably in private information retrieval [3,12], and worst-case to average-case hardness reductions [17] (we elaborate on this application below). Furthermore, they have the potential of being used for practical applications, such as reliably storing a large static data file, only small portions of which need to be read at a time.

1.1 This Work

In this work we study the complexity of locally list decoding binary codes (i.e. where the alphabet is $\{0,1\}$). Let us proceed more formally. Let $C : \{0,1\}^M \to \{0,1\}^N$ be the encoding function of an error-correcting code.[1] A local list-decoder D for a code C gets oracle access to a corrupted codeword, and outputs a "list" of ℓ local-decoding circuits $D_1, \ldots, D_\ell$. Each D_a is itself a probabilistic circuit with oracle access to the corrupted codeword. On input an index $j \in [M]$, a circuit D_a from the list tries to output the j-th bit of the message. We say that the decoder is a $(1/2 - \epsilon, \ell)$-local-list-decoder, if for every $y \in \{0,1\}^N$ and $m \in \{0,1\}^M$, such that the fractional Hamming distance between $C(m)$ and y is at most $1/2 - \epsilon$, with high probability at least one of the D_a's successfully decodes *every* bit of the message y. Note that here $1/2 - \epsilon$ refers to the "noise rate" (or the list-decoding radius) from which the decoder recovers, and ℓ is the "list size": the number of decoding circuits, one of which makes the decoder recover every index correctly (with high probability). The quantity N/M which measures the amount of redundancy in the code is called *the rate* of the code.

Throughout this paper we think of a local list-decoder as receiving an "advice" index $a \in [\ell]$, running D to output D_a, and then running D_a to retrieve the j-th message bit. Note that by giving both D and the D_a's oracle access to the received word, and requiring them to decode individual symbols, we can hope

---

[1] Formally, we consider a family of codes one for each message length M. The parameters listed above and below, e.g. N, ϵ, ℓ, should all be thought of as functions of M. For the exact definition of locally list-decodable codes see Definition 3.

for decoders whose size is much smaller than N (in particular we can hope for size that is poly-logarithmic in N). See Definition 3 for a formal definition of locally list decodable codes.[2]

It is well known that for every (non-trivial) $(1/2 - \epsilon, \ell)$-locally-list-decodable code, it must hold that $\ell = \Omega(1/\epsilon^2)$ [1,9] (in fact this bound holds even for standard, non-local, list decoding). Thus, aiming to stay within polynomial factors of the best possible information theoretic parameters, our primary goal is to understand the complexity of decoding $(1/2-\epsilon, \text{poly}(1/\epsilon))$-locally-list-decodable binary codes that have polynomial rate (i.e. where $N(M) = \text{poly}(M)$). We consider such codes to have "good" parameters (we elaborate on this choice below).

An explicit code with good parameters was given by Sudan, Trevisan and Vadhan [17]. The local-decoder for this code (namely the algorithm D as well as the circuits D_a) is in the complexity class NC^2 (i.e. its depth is poly-logarithmic in its input length). Explicit codes with local-decoders in the (strictly lower) class AC^0 (i.e. constant depth unbounded fan-in decoders) are also known [7,5]. However, these codes do not have good parameters.[3] Specifically the Hadamard code has such a decoder [7], but its rate is exponential in M. This was improved by Goldwasser et al. [5] who showed codes with AC^0 local-list-decoders in which the rate is exponential in $1/\epsilon$ (but not in M). Furthermore, the circuit size of the AC^0 decoders for both these codes is exponential in $1/\epsilon$. For comparison, the size of the NC^2 decoder of [17] is $\text{poly}(\log M, 1/\epsilon)$. In other words, the constant-depth decoder of [5] only matches the parameters of [17] (in terms of circuit size and information theoretic parameters) when $\epsilon \geq 1/\text{poly}\log\log M$ (while in general ϵ can be as small as $1/\text{poly}(M)$).

Our results. Our goal in this work is to understand the complexity of local-list-decoders of binary codes with good parameters. Specifically, we ask whether the exponential dependency on $1/\epsilon$ in both the rate and decoder size in [5] can be improved, and whether the parameters of [17] can be achieved by (small) constant-depth decoders. We show that while the rate of codes with constant-depth local-list-decoders can be improved, their circuit size cannot.

These conclusions follow from our main technical result, which shows that computing the majority function on $\Theta(1/\epsilon)$ bits is essentially equivalent to $(1/2-\epsilon, \text{poly}(1/\epsilon))$-local-list-decoding binary codes: Any circuit for a local-decoder of such a code can be used to construct a circuit of roughly the same size and depth that computes majority on $\Theta(1/\epsilon)$ bits. In the other direction, there is an explicit $(1/2 - \epsilon, \text{poly}(1/\epsilon))$-locally-list-decodable code with a very efficient (in terms of size and depth) local-decoder that uses majority gates of fan-in $\Theta(1/\epsilon)$. This is stated (informally) in the following theorem.

---

[2] We would like to point out that we use $(1/2-\epsilon, \ell)$ to denote the relative distance and list size, whereas previous work (e.g. [17]) used (ϵ, ℓ) to denote the same quantities (for binary codes). We find this notation more useful, especially when we work with non-binary codes (which come up in our construction).

[3] We note that for non-binary codes, i.e. codes with large alphabets, one can construct codes with constant-depth local list-decoders and "good" parameters, see [5].

Theorem 1 (Informal). *If there exists a binary code with a $(1/2-\epsilon, poly(1/\epsilon))$-local-list-decoder of size s and depth d, then there exists a circuit of size $poly(s)$ and depth $O(d)$ that computes majority on $\Theta(1/\epsilon)$ bits.*

In the other direction, there exist (for every $\epsilon \geq 1/2^{\sqrt{\log(M)}}$) explicit binary codes of polynomial rate, with a $(1/2-\epsilon, poly(1/\epsilon))$-local-list-decoder. The decoder is a constant depth circuit of size $poly(\log M, 1/\epsilon)$ with majority gates of fan-in $\Theta(1/\epsilon)$.

The upper bound follows by replacing one of the ingredients in the construction of Goldwasser et al. [5], with a modification of the recent de-randomized direct-product construction of Impagliazzo *et al.* [10], thus improving the code's rate. Our main technical contribution is in the lower bound, where we show a reduction from computing majority over inputs of size $\Omega(1/\epsilon)$ to local-list-decoding binary codes with good parameters. In fact, our lower bound holds for any $(1/2 - \epsilon, poly(1/\epsilon))$-local-list-decodable binary code, regardless of its rate. By known lower bounds on the size of constant-depth circuits that compute majority [15,16], we obtain the following corollary.

Corollary 2 (Informal). *Any constant-depth $(1/2 - \epsilon, poly(1/\epsilon))$-local list decoder for a binary code, must have size almost exponential in $1/\epsilon$. This holds even if the decoder is allowed   mod q gates, where q is an arbitrary prime number.*

In particular, this result shows that the noise rate from which the constant-depth local-list-decoders of [5] recover is essentially optimal. And thus we get an exact characterization of what is possible with constant-depth decoders: up to radius $1/2 - 1/poly \log \log M$ locally-list-decodable codes with constant-depth decoders and good parameters exist, and beyond this radius they do not. We note that in fact we prove a stronger result in terms of the list size. We show that $(1/2-\epsilon, \ell)$-local-list-decoding with a decoder of size s and depth d, implies a circuit of size $poly(s, \ell)$ and depth d that computes majority on $O(1/\epsilon)$ bits. This means that even if the list size is *sub-exponential* in $1/\epsilon$, the size of the decoder still must be nearly exponential in $1/\epsilon$ (even if the decoder is allowed   mod q gates).

Hardness amplification. Hardness amplification is the task of obtaining from a Boolean function f that is somewhat hard on the average, a Boolean function f' that is very hard on the average. By a beautiful sequence of works [17,20,19,21], it is well known that there is a tight connection between binary locally (list) decodable codes and hardness amplification. Using this connection, we obtain limits (in the spirit of Corollary 2) on (black-box) hardness amplification procedures. We defer the statement of these results and a discussion to Section 5.

1.2   Related Work

The question of lower bounding the complexity of local-list-decoders was raised by Viola [22]. He conjectured that $(1/2-\epsilon, \ell)$-locally-list-decodable codes require computing majority over $O(1/\epsilon)$ bits,[4] even when the list size ℓ is exponential

---

[4] By "require" we mean that the decoding circuit can be used to construct a circuit of comparable size and depth that computes the majority function on $O(1/\epsilon)$ bits.

in $1/\epsilon$. Note that while exponential lists are not commonly considered in the coding setting (the focus instead is on polynomial or even optimal list sizes), they do remain interesting for applications to (non-uniform) worst-case to average-case hardness reductions. In particular, lower bounds for local-list-decoding with exponential lists, imply impossibility results for *non-uniform* black-box worst-case to average-case hardness reductions (see Section 5). In this paper we prove the conjecture for the case of sub-exponential size lists. While a proof of the full-blown conjecture remains elusive, there are results for other (incomparable) special cases:

Known Results for Non-Local *Decoders.* Viola [22] gave a proof (which he attributed to Madhu Sudan) of the conjecture for the special case of the standard *non-local* list-decoding setting. It is shown that a list-decoder from distance $1/2 - \epsilon$ can be used to compute majority on $\Theta(1/\epsilon)$ bits, with only a small blow-up in the size and depth of the decoder. This result rules out, for example, constant-depth list-decoders whose size is $\text{poly}(1/\epsilon)$. Note, however, that in the non-local list decoding setting the size of the decoder is at least N (the codeword length) because it takes as input the entire (corrupted) codeword. This means that the bound on the size of constant-depth decoders does not have consequences for fairly large values of ϵ. For example, when $\epsilon \geq 1/\log N$, the only implication that we get from [22], is that there is a constant-depth circuit of size at least $N \geq 2^{1/\epsilon}$ that computes majority on instances of size $1/\epsilon$. But this is trivially true, and thus we do not get any contradiction. In the local-decoding setting the decoders' circuits are much smaller and thus we can obtain limitations for much larger ϵ's. Indeed in this paper we rule out constant-depth decoders for $(1/2 - \epsilon, \text{poly}(1/\epsilon))$-local-list-decoders for any ϵ smaller than $1/\text{poly}\log\log N$ (and recall that this matches the construction of [5]).

Known Results for Specific *Codes.* Viola [22] also proved that there are no constant-depth decoders (with polynomial-size lists) for *specific* codes, such as the Hadamard and Reed-Muller codes. We, on the other hand, show that there are no such decoders for *any* code (regardless of the code's rate, and even with sub-exponential list size).

Known Results for Non-Adaptive *Decoders.* Recently (simultaneously and independently of our work), Shaltiel and Viola [18] gave a beautiful proof of the conjecture for the local-decoding setting, with ℓ exponential in $1/\epsilon$, but for the special case that the decoder is restricted to have *non-adaptive* access to the received word. (I.e., they give a lower bound for decoders that make all their queries to the received word simultanuously.) Our result is incomparable to [18]: we prove Viola's conjecture only for the case that ℓ is sub-exponential in $1/\epsilon$, but do so for *any* decoder, even an adaptive one. We emphasize that for important ranges of parameters the best codes known to be decodable in constant depth use *adaptive* decoders. In particular, the constant depth decoder of [5], as well as its improvement in this work, are adaptive. In light of this, it is even more important to show lower bounds for adaptive decoders.

1.3   On the Choice of Parameters

In this work codes with polynomial-rate are considered to have "good" parameters. Usually in the standard coding-theory literature, "good" codes are required to have *constant* rate.[5] We note that, as far as we know, there are no known locally-decodable codes (both in the unique and list decoding settings) with constant rate (let alone codes that have both constant rate *and* have decoders that are in the low-level complexity classes that we consider here). The best binary locally decodable codes known have polynomial rate [17]. It is an interesting open question to find explicit codes with constant or even polylogarithmic rate.

Finally, we note that in this work we do not (explicitly) consider the query complexity of the decoder. The only bound on the number of queries the decoder makes to the received word comes from the bound on the size of the decoding circuit. The reason is that known codes with much smaller query complexity than the decoder size (in particular constant query complexity) have a very poor rate (see e.g. [25]). Furthermore, there are negative results that suggest that local-decoding with small query complexity may *require* large rate [12,4,14,11,23,6].

2   Preliminaries

For a string $m \in \{0,1\}^*$ we denote by $m[i]$ the i'th bit of m. $[n]$ denotes the set $\{1, \ldots, n\}$. For a finite set S we denote by $x \in_R S$ that x is a sample uniformly chosen from S. For a finite alphabet Γ we denote by Δ_Γ the relative (or fractional) Hamming distance between strings over Γ. That is, let $x, y \in \Gamma^n$ then $\Delta_\Gamma(x, y) = \Pr_{i \in_R [n]}[x[i] \neq y[i]]$, where $x[i], y[i] \in \Gamma$. Typically, Γ will be clear from the context, we will then drop it from the subscript.

2.1   Circuit Complexity Classes

Boolean circuits in this work always have NOT gates at the bottom and unbounded AND and OR gates. Such circuits may output more than one bit. Whenever we use circuits with gates that compute other functions, we explicitly state so. For a positive integer $i \geq 0$, AC^i circuits are Boolean circuits of size $\text{poly}(n)$, depth $O(\log^i n)$, and unbounded fan-in AND and OR gates (where n is the length of the input). $AC^i[q]$ (for a prime q) are similar to AC^i circuits, but augmented with mod q gates. Throughout, we extensively use oracle circuits: circuits that have (unit cost) access to an oracle computing some function. We sometimes interpret this function as a string, in which case the circuit queries and index and receives from the oracle the symbol in that location in the string.

2.2   Locally List-Decodable Codes

Definition 3 (Locally list-decodable codes). *Let Γ be a finite alphabet. An ensemble of functions $\{C_M : \{0,1\}^M \to \Gamma^{N(M)}\}_{M \in \mathbb{N}}$ is a $(d(M), \ell(M))$-locally-list-decodable code, if there is an oracle Turing machine $D[\cdot, \cdot, \cdot, \cdot]$ that takes as*

---

[5] We do remark that for applications such as worst-case to average-case reductions, polynomial or even quasi-polynomial rates suffice.

input an index $i \in [M]$, an "advice" string $a \in [\ell(M)]$ and two random strings r_1, r_2,[6] and the following holds: for every $y \in \Gamma^{N(M)}$ and $x \in \{0,1\}^M$ such that $\Delta_\Gamma(C_M(x), y) \leq d(M)$,

$$\Pr_{r_1} \left[\exists a \in [\ell] \text{ s.t. } \forall i \in [M] \; \Pr_{r_2}[D^y(a, i, r_1, r_2) = x[i]] > 9/10 \right] > 99/100 \quad (1)$$

If $|\Gamma| = 2$ we say that the code is binary. If $\ell = 1$ we say that the code is uniquely decodable. We say that the code is explicit if C_M can be computed in time $poly(N(M))$.

Remark 4. One should think of the decoder's procedure as having two stages: first it tosses coins r_1 and generates a sequence of ℓ circuits $\{C_a(\cdot, \cdot)\}_{a \in [\ell]}$, where $C_a(i, r_2) = D(a, i, r_1, r_2)$. In the second stage, it uses the advice a to pick the probabilistic circuit C_a and use it (with randomness r_1) to decode the message symbol at index i. In [17] the two-stage process is part of the definition, for us it is useful to encapsulate it in one machine (D).

In the sequel it will be convenient to simplify things by ignoring the first stage, and consider D as a probabilistic circuit (taking randomness r_2) with two inputs: the advice a and the index to decode i, with the property that (always) for at least one $a \in [\ell]$, $D(a, \cdot)$ decodes correctly every bit of the message (with high probability over r_2). Indeed if we hardwire any "good" r_1 (chosen in the first stage) into D then we are in this situation. This happens with probability at least $99/100$. Thus in our proofs we will assume that this is the case, while (implicitly) adding $1/100$ to the bound on the overall probability that the decoder errs. This simplification makes our proofs much clearer (since we do not have to deal with the extra randomness r_1).

2.3   Majority and Related Functions

We use the promise problem Π, defined in [22] as follows:
$\Pi_{Yes} = \{x : x \in \{0,1\}^{2k} \text{ for some } k \in \mathbb{N} \text{ and } weight(x) \leq k - 1\}$
$\Pi_{No} = \{x : x \in \{0,1\}^{2k} \text{ for some } k \in \mathbb{N} \text{ and } weight(x) = k\}$
where $weight(x)$ is the number of bits in x which are 1.

We will extensively use the fact, proven in [22], that computing the *promise* problem Π on $2k$ bit inputs is (informally) "as hard" (in terms of circuit depth) as computing majority of $2k$ bits. This is stated formally in the claim below:

Lemma 5 ([22]). *Let $\{C\}_{M \in \mathbb{N}}$ be a circuit family of size $S(M)$ and depth $d(M)$ that solves the promise problem Π on inputs of size M. Then, for every $M \in \mathbb{N}$, there exists a circuit B_M of size $poly(S(M))$ and depth $O(d(M))$ that computes majority on M bits. The types of gates used by the B_M circuit are identical to those used by C_M. E.g., if C_M is an $AC^0[q]$ circuit, then so is B_M.*

---

[6] The length of these random strings lower-bounds D's running time. Later in this work, when we consider D's with bounded running time, the length of these random strings will also be bounded.

3 Local-List-Decoding Requires Computing Majority

Theorem 6. *Let $\{C_M : \{0,1\}^M \to \{0,1\}^{N(M)}\}_{M\in\mathbb{N}}$ be a $(1/2 - \epsilon(M), \ell(M))$-locally-list-decodable code, such that $\ell(M) \leq 2^{\kappa \cdot M}$, and $1/N^{\delta_1} \leq \epsilon(M) \leq \delta_2$ for universal constants $\kappa, \delta_1, \delta_2$. Let D be the local decoding machine, of size $S(M)$ and depth $d(M)$.*

Then, for every $M \in \mathbb{N}$, there exists a circuit A_M of size $poly(S(M), \ell(M))$ and depth $O(d(M))$, that computes majority on $\Theta(1/\epsilon(M))$ bits. The types of gates used by the circuit A_M are identical to those used by D. E.g., if D is an $AC^0[q]$ circuit, then so is A_M.

Proof Intuition for Theorem 6. Fix a message length M and $\epsilon = \epsilon(M)$. We will describe a circuit B with the stated parameters that decides the promise problem Π on inputs of length roughly $1/\epsilon$. By Lemma 5 this will also give a circuit for computing majority.

We start with a simple case: assume that the (local) decoder D makes only non-adaptive queries to the received word. In this case we proceed using ideas from the proof of Theorem 6.4 in [22]. Take m to be a message that cannot be even approximately decoded[7] from random noise with error rate $1/2$. Such a word exists by a counting argument. Let $C(m)$ be the encoding of m. Let $x \in \Pi_{Yes} \cup \Pi_{No}$ be a Π-instance of size $1/2\epsilon$ (we assume w.l.o.g. throughout that $1/\epsilon$ is an integer). B uses x to generate a noisy version of $C(m)$, by XORing each one of its bits with some bit of x that is chosen at random. It then uses D to decode this noisy version of $C(m)$. If $x \in \Pi_{No}$, this adds random noise (error rate $1/2$), and the decoding algorithm cannot recover most of m's bits. If $x \in \Pi_{Yes}$, then each bit is noisy with probability less than $1/2 - 2\epsilon$, which means that w.h.p. the fraction of errors is at most $1/2 - \epsilon$, and the decoding algorithm successfully recovers every bit of m.

By comparing the answers of the decoding algorithm (or more precisely, every decoding algorithm in the list, by trying every possible advice) and the real bits of m in a small number of random locations, the algorithm B distinguishes w.h.p. whether $x \in \Pi_{Yes}$ or $x \in \Pi_{No}$.

Note, however, that B as described above is *not* a standard algorithm for Π. This is because we gave B access to the message m as well as its encoding. Both of these are strings that are much larger than we want B itself to be. So our next goal is to remove (or at least minimize) B's access to m and $C(m)$, making B a standard circuit for Π. Observe that B as described above distinguishes whether x is in Π_{Yes} or in Π_{No} with high probability over the choices of D's random coins, the random locations in which we compare D's answers against m, and the random noise generated by sampling bits from x. In particular, there exists a fixing of D's random string as well as the (small number of) testing locations of m that maintains the advantage in distinguishing whether x comes from Π_{Yes} or Π_{No}, where now the probability is only over the randomness used to sample bits from x. So now we can hardwire the bits of m used to test whether

---

[7] By this we mean that no decoder can recover (w.h.p.) a string that is, say, 1/3-close to m.

D decodes the noisy version of $C(m)$ correctly (i.e. we got rid of the need to store the whole string m). Furthermore, after we fix D's randomness, *by the fact that it is non-adaptive*, we get that the positions in which B queries the noisy $C(m)$ are now also fixed, and *independent of x*. So we also hardwire the values of $C(m)$ in these positions (and only these positions) into B. For any x, we now have all the information to run B and conclude whether x is in Π_{Yes} or Π_{No}.

Next we want to deal with adaptive decoders. If we proceed with the ideas described above, we run into the following problem: suppose the circuit has two (or more) levels of adaptivity. The queries in the second level do not only depend on the randomness of the decoder, but also on the values read from the received word at the first level, and in particular they also depend on *the noise*. The noise in our implementation depends on the specific Π-instance x. This means that we cannot hardwire the values of $C(m)$ that are queried at the second level because they depend on x!

To solve this problem, we analyze the behavior of the decoder when its error rate changes in the middle of its execution. Specifically, suppose that the decoder D queries the received word in d levels of adaptivity. For every $0 \leq k \leq d$, we consider the behavior of the decoder when up to level k we give it access to the encoded message corrupted with error-rate $1/2 - 2\epsilon$, and above the k'th level we give it access to the encoded message corrupted with error-rate $1/2$. By a hybrid argument, there exists some level k, in which the decoder has a significant advantage in decoding correctly when up to the k'th level it sees error rate $1/2 - 2\epsilon$ (and error-rate $1/2$ above it), over the case that up to the $(k-1)$'th level the error-rate is $1/2 - 2\epsilon$ (and $1/2$ from k and up). We now fix and hardwire randomness for the decoder, as well as noise for the first $k - 1$ levels (chosen according to error-rate $1/2 - 2\epsilon$), such that this advantage is preserved. Once the randomness of D and the noise for the first $k - 1$ levels are fixed, the queries at the k-th level (but not their answers) are also fixed. For this k-th level we can proceed as in the non-adaptive case (i.e. choose noise according to x and hardwire the fixed positions in $C(m)$). We now have to deal with queries above the k'th level. At first glance it is not clear that we have gained anything, because we still have to provide answers for these queries, and as argued above, these may now depend on the input x and therefore the query locations as well as the restriction of $C(m)$ to these locations cannot be hard-wired. The key point is that for these "top" layers the error rate has changed to $1/2$. So while we have no control on the query locations (as they depend on x) we do know their answers: they are completely random bits that have nothing to do with m or $C(m)$! Thus, B can continue to run the decoder, answering its queries (in the levels above the k'th) with random values. We thus obtain a circuit that decides membership in Π correctly with a small advantage. Since the number of adaptivity levels is only d (the circuit depth of the decoder), the distinguishing advantage of the k-th hybrid is at least $O(1/d)$, and in particular this advantage can now be amplified by using only additional depth of $O(\log(d))$. This gives a circuit that computes Π and concludes the proof. Due to space constraints we defer the formal proof of Theorem 6 to the full version of this paper [8].

By using known lower bounds for computing the majority function by $AC^0[q]$ circuits (for a prime q) [15,16], we obtain the following corollary.

Corollary 7. *Let $\{C_M : \{0,1\}^M \to \{0,1\}^{N(M)}\}_{M \in \mathbb{N}}$ be a $(1/2-\epsilon, \ell)$-locally-list-decodable code (where ϵ is in the range specified in Theorem 6) with a decoder that can be implemented by a family of $AC^0[q]$ circuits of size $s = s(M)$ and depth $d = d(M)$. Then $s = 2^{(1/\epsilon)^{\Omega(1/d)}}/poly(\ell)$.*

4 Majority Suffices for Local-List-Decoding

Theorem 8. *For every $2^{-\Theta(\sqrt{\log M})} \le \epsilon = \epsilon(M) < 1/2$, there exists a $(1/2 - \epsilon, poly(1/\epsilon))$-locally-list-decodable code $\{C_M : \{0,1\}^M \to \{0,1\}^{poly(M)}\}_{M \in \mathbb{N}}$ with a local-decoder that can be implemented by a family of constant depth circuits of size $poly(\log M, 1/\epsilon)$ using majority gates of fan-in $\Theta(1/\epsilon)$ (and AND/OR gates of unbounded fan-in).*

Remark 9. The construction above only applies for $\epsilon \ge 2^{-\Theta(\sqrt{\log M})}$. Thus we fall slightly short of covering the whole possible range (since one can hope to get such codes for $\epsilon = 1/M^\delta$ for a small constant δ). We note, however, that the range of ϵ which is most interesting for us is between $1/poly \log M$ and $1/poly \log \log M$ (see the discussion in the introduction) which we do cover. We also mention that if one insists on codes with $\epsilon = 1/M^\delta$, then we can construct such codes with quasi-polynomial rate (in the full version [8] we state without proof the exact parameters of these codes).

The proof of Theorem 8 is omitted due to lack of space. In a nutshell, to prove the theorem we combine three codes. The first, by [5], is a binary locally-decodable code that can be uniquely decoded from a constant relative distance. The second code that we need is a modification of the de-randomized direct product code of [10]. The main reason that we need to modify the code of [10], is that in their construction the decoder needs to manipulate small dimension affine subspaces over finite fields. We do not know of a concise and unique representation of low-dimensional affine sub-spaces that can be computed and manipulated in AC^0. We instead represent such subspaces using randomly selected basis vectors and a shift vector (a concise, but not unique representation). This changes the code and allows decoding in AC^0. The third code in our construction is the well known Hadamard code with its local list-decoder given by Goldreich and Levin [7].

Informally, our code for Theorem 8 first encodes the message using the first code, it then encodes this encoding using the second code. Finally, it concatenates this code (i.e. encodes every symbol of it) using the third and final code. For details of the construction and the parameters it achieves see the full version of this work [8].

5 Hardness Amplification

In this section we describe our results regarding hardness amplification of Boolean functions. A more thorough discussion and extensions of our results appear in the full version [8].

Functions that are hard to compute *on the average* (by a given class of algorithms or circuits) have many applications, for example in cryptography or for de-randomization via the construction of pseudo-random generators (the "hardness vs. randomness" paradigm [2,24,13]). Typically, for these important applications, one needs a function that no algorithm (or circuit) in the class can compute on random inputs much better than a random guess. Unfortunately, however, it is often the case that one does not have or cannot assume access to such a "hard on the average" function, but rather only to a function that is "somewhat hard": every algorithm in the class fails to compute it and errs, but only on relatively few inputs (e.g. a small constant fraction, or sometimes even just a single input for every input length). A fundamental challenge is to obtain functions that are very hard on the average from functions that are somewhat hard on the average (or even just hard on the worst-case).

Let us be more precise. We say that a Boolean function $f : \{0,1\}^* \to \{0,1\}$ is δ-hard on the average for a circuit class $\mathcal{C} = \{\mathcal{C}_n\}_{n \in \mathbb{N}}$ (where circuits in the set $\mathcal{C}_n$ have input length n), if for every large enough n, for every circuit $C_n \in \mathcal{C}_n$;

$$\Pr_{x \in_R U_n} [C_n(x) = f(x)] \leq 1 - \delta$$

The task of obtaining from a function f that is δ-hard for a class $\mathcal{C}$, a function f' that is δ'-hard for the class $\mathcal{C}$, where $\delta' > \delta$ is called hardness amplification from δ-hardness to δ'-hardness (against the class $\mathcal{C}$). Typical values for δ are small constants (close to 0), and sometimes even 2^{-n}, in which case the hardness amplification is from worst-case hardness. Typical values for δ' (e.g. for cryptographic applications) are $1/2 - n^{-\omega(1)}$.

The most commonly used approach to prove hardness amplification results is via reductions, showing that if there is a sequence of circuits in $\mathcal{C}$ that computes f' on more than a $1 - \delta'$ fraction of the inputs, then there is a sequence of circuits in $\mathcal{C}$ that computes f on more than $1 - \delta$ fraction of the inputs. An important family of such reductions are so-called fully-black-box reductions which we define next.

Definition 10. *A (δ, δ')-fully-black-box hardness amplification from input length k to input length $n = n(k, \delta, \delta')$, is defined by an oracle Turing machine Amp that computes a Boolean function on n bits, and an oracle Turing machine Dec that takes non-uniform advice of length $a = a(k, \delta, \delta')$. It holds that for every $f : \{0,1\}^k \to \{0,1\}$, for every $A : \{0,1\}^n \to \{0,1\}$ for which*

$$\Pr_{x \in_R U_n} [A(x) = Amp^f(x)] > 1 - \delta'$$

there is an advice string $\alpha \in \{0,1\}^a$ such that

$$\Pr_{x \in_R U_k} [Dec^A(\alpha, x) = f(x)] > 1 - \delta$$

where $Dec^A(\alpha, x)$ denotes running Dec with oracle access to A on input x and with advice α.

If Dec does not take non-uniform advice ($a = |\alpha| = 0$), then we say that the hardness amplification is uniform. *If Dec can ask all its queries to A in parallel (i.e. no query depends on answers to previous queries) then we say that the hardness amplification is* non-adaptive.

The complexity of Dec determines against which classes (of Boolean circuits or algorithms) we measure hardness (when we translate the reduction to a hardness amplification result). In particular if one wants to obtain hardness amplification against AC^0 or $AC^0[q]$ circuits, Dec must be implemented by such circuits.

It is well known [17,20,19,21] that there is a tight connection between $(2^{-k}, \delta')$-fully-black-box hardness amplification (or in other words worst-case to average-case reductions) and binary locally (list) decodable codes. In particular a lower bound on the complexity of local-list-decoders implies a lower bound on the complexity of Dec in Definition 10. Using Theorem 6 we can show that worst-case to average-case hardness amplification with small non-uniform advice requires computing majority. This is stated formally in the theorem below:

Theorem 11. *If there is a $(2^{-k}, 1/2 - \epsilon(k))$-fully-black-box hardness amplification from length k to length $n(k)$ where Dec takes $a(k)$ bits of advice and can be implemented by a circuit of size $s(k)$ and depth $d(k)$, then for every $k \in \mathbb{N}$ there exists a circuit of size $poly(s(k), 2^{a(k)})$ and depth $O(d(k))$, that computes majority on $O(1/\epsilon(k))$ bits.*

It is known [15,16] that low complexity classes cannot compute majority. Thus, Theorem 11 shows limits on the amount of hardness amplification that can be achieved by fully-black-box worst-case to average-case reductions (that do not use too many bits of advice), in which Dec can be implemented in low-level complexity classes.

Finally, we note that the worst-case lower bounds (which are actually mildly average-case lower bounds) of [15,16] hold against *non-uniform* $AC^0[q]$. This means that it may be possible to get strong average-case hardness (e.g. as required for pseudo-randomness) by using a lot of non-uniformity in a fully-black-box reduction (i.e. a reduction in which Dec takes $poly(k)$ bits of advice). Shaltiel and Viola [18] rule out such non-uniform fully-black-box reductions in the special case that Dec has only non-adaptive access to A.

As we mentioned, in the full version [8] we give a more thorough discussion of hardness amplification, together with extensions of Theorem 11 to hardness amplification from functions that are mildly hard on the average (rather than worst-case hard), as well as to reductions that are black-box but not necessarily fully-black-box.

Acknowledgements

We thank Shafi Goldwasser and Salil Vadhan for their assistance and insightful comments. Thanks to Russell Impagliazzo for helpful discussions and for supplying us with a manuscript of [10]. We also thank Ronen Shaltiel and Emanuele

Viola for helpful discussions on the topics addressed in this work, and for supplying us with a manuscript of [18]. Finally, we thank the anonymous referees for their many helpful and insightful comments.

References

1. Blinkovsky, V.M.: Bounds for codes in the case of list decoding of finite volume. Problems of Information Transmission 22(1), 7–19 (1986)
2. Blum, M., Micali, S.: How to generate cryptographically strong sequences of pseudo-random bits. SIAM Journal on Computing 13(4), 850–864 (1984)
3. Chor, B., Kushilevitz, E., Goldreich, O., Sudan, M.: Private information retrieval. Journal of the ACM 45(6), 965–981 (1998)
4. Deshpande, A., Jain, R., Kavitha, T., Radhakrishnan, J., Lokam, S.V.: Better Lower Bounds for Locally Decodable Codes. In: Proceedings of the IEEE Conference on Computational Complexity, pp. 184–193 (2002)
5. Goldwasser, S., Gutfreund, D., Healy, A., Kaufman, T., Rothblum, G.N.: Verifying and decoding in constant depth. In: Proceedings of the 39th Annual ACM Symposium on Theory of Computing, pp. 440–449 (2007)
6. Goldreich, O., Karloff, H.J., Schulman, L.J., Trevisan, L.: Lower bounds for linear locally decodable codes and private information retrieval. Computational Complexity 15(3), 263–296 (2006)
7. Goldreich, O., Levin, L.A.: A hard-core predicate for all one-way functions. In: Proceedings of the 21st Annual ACM Symposium on Theory of Computing, pp. 25–32 (1989)
8. Gutfreund, D., Rothblum, G.N.: The complexity of local list decoding. Technical Report TR08-034, Electronic Colloquium on Computational Complexity (2008)
9. Guruswami, V., Vadhan, S.: A lower bound on list size for list decoding. In: Chekuri, C., Jansen, K., Rolim, J., Trevisan, L. (eds.) RANDOM 2005. LNCS, vol. 3624, pp. 318–329. Springer, Heidelberg (2005)
10. Impagliazzo, R., Jaiswal, R., Kabanets, V., Wigderson, A.: Uniform direct-product theorems: Simplified, optimized, and derandomized. In: Proceedings of the 40th Annual ACM Symposium on Theory of Computing, pp. 579–588 (2008)
11. Kerenidis, I., de Wolf, R.: Exponential lower bound for 2-query locally decodable codes via a quantum argument. Journal of Computer and System Sciences 69(3), 395–420 (2004)
12. Katz, J., Trevisan, L.: On the efficiency of local decoding procedures for error-correcting codes. In: Proceedings of the 32nd Annual ACM Symposium on Theory of Computing, pp. 80–86 (2000)
13. Nisan, N., Wigderson, A.: Hardness vs. randomness. Journal of Computer and System Sciences 49, 149–167 (1994)
14. Obata, K.: Optimal Lower Bounds for 2-Query Locally Decodable Linear Codes. In: Proceedings of the 5th International Workshop on Randomization and Computation (RANDOM), pp. 39–50 (2002)
15. Razborov, A.A.: Lower bounds on the dimension of schemes of bounded depth in a complete basis containing the logical addition function. Akademiya Nauk SSSR. Matematicheskie Zametki 41(4), 598–607, 623 (1987)
16. Smolensky, R.: Algebraic methods in the theory of lower bounds for boolean circuit complexity. In: Proceedings of the 19th Annual ACM Symposium on Theory of Computing, pp. 77–82 (1987)

17. Sudan, M., Trevisan, L., Vadhan, S.: Pseudorandom generators without the XOR Lemma. Journal of Computer and System Sciences 62(2), 236–266 (2001)
18. Shaltiel, R., Viola, E.: Hardness amplification proofs require majority. In: Proceedings of the 40th Annual ACM Symposium on Theory of Computing, pp. 589–598 (2008)
19. Trevisan, L.: List-decoding using the XOR lemma. In: Proceedings of the 44th Annual IEEE Symposium on Foundations of Computer Science, pp. 126–135 (2003)
20. Trevisan, L., Vadhan, S.: Pseudorandomness and average-case complexity via uniform reductions. Computational Complexity 16(4), 361–364 (2007)
21. Viola, E.: The complexity of constructing pseudorandom generators from hard functions. Computational Complexity 13(3-4), 147–188 (2005)
22. Viola, E.: The complexity of hardness amplification and derandomization. PhD thesis, Harvard University (2006)
23. Wehner, S., de Wolf, R.: Improved Lower Bounds for Locally Decodable Codes and Private Information Retrieval. In: Caires, L., Italiano, G.F., Monteiro, L., Palamidessi, C., Yung, M. (eds.) ICALP 2005. LNCS, vol. 3580, pp. 1424–1436. Springer, Heidelberg (2005)
24. Yao, A.C.: Theory and applications of trapdoor functions. In: Proceedings of the 23rd Annual IEEE Symposium on Foundations of Computer Science, pp. 80–91 (1982)
25. Yekhanin, S.: Towards 3-query locally decodable codes of subexponential length. In: Proceedings of the 39th Annual ACM Symposium on Theory of Computing, pp. 266–274 (2007)

Limitations of Hardness vs. Randomness under Uniform Reductions

Dan Gutfreund[1,*] and Salil Vadhan[2,**]

[1] Department of Mathematics and CSAIL
MIT
danny@math.mit.edu
[2] School of Engineering and Applied Sciences
Harvard University
salil@eecs.harvard.edu

Abstract. We consider (uniform) reductions from computing a function f to the task of distinguishing the output of some pseudorandom generator G from uniform. Impagliazzo and Wigderson [10] and Trevisan and Vadhan [24] exhibited such reductions for every function f in PSPACE. Moreover, their reductions are "black box," showing how to use *any* distinguisher T, given as oracle, in order to compute f (regardless of the complexity of T). The reductions are also adaptive, but with the restriction that queries of the same length do not occur in different levels of adaptivity. Impagliazzo and Wigderson [10] also exhibited such reductions for every function f in EXP, but those reductions are not black-box, because they only work when the oracle T is computable by small circuits.

Our main results are that:

- *Nonadaptive* black-box reductions as above can only exist for functions f in BPP^{NP} (and thus are unlikely to exist for all of PSPACE).
- *Adaptive* black-box reductions, with the same restriction on the adaptivity as above, can only exist for functions f in PSPACE (and thus are unlikely to exist for all of EXP).

Beyond shedding light on proof techniques in the area of hardness vs. randomness, our results (together with [10,24]) can be viewed in a more general context as identifying techniques that overcome limitations of black-box reductions, which may be useful elsewhere in complexity theory (and the foundations of cryptography).

Keywords: pseudorandom generators, derandomization, black-box reductions.

* Research was partially supported by NSF grant CCF-0514167. Part of this research was done while the author was at Harvard University and supported by ONR grant N00014-04-1-0478 and NSF grant CNS-0430336.
** Work done in part while visiting UC Berkeley, supported by the Miller Institute for Basic Research in Science and the Guggenheim Foundation. Also supported by ONR grant N00014-04-1-0478 and US-Israel BSF grant 2002246.

A. Goel et al. (Eds.): APPROX and RANDOM 2008, LNCS 5171, pp. 469–482, 2008.
© Springer-Verlag Berlin Heidelberg 2008

1 Introduction

A central goal in the theory of computation is to identify relations between the complexities of different computational tasks. Indeed, some of the greatest discoveries in computational complexity (and the foundations of cryptography) are surprising connections between the complexities of tasks that seem very different in nature. The traditional way of showing such relations is by various forms of "black-box" reductions. That is, the proofs exhibit an efficient (oracle) algorithm R that when given oracle access to any function f that solves task T_1, R^f solves task T_2. Such an algorithm R proves that if T_1 has an efficient solution then so does task T_2. In complexity theory, such algorithms R are often referred to as just "reductions" (or "Cook reductions"), but here we use "black box" to emphasize that R only has oracle access to the function f solving task T_1 and is required to work regardless of the complexity of this oracle.

There are also a number of negative results about black-box reductions, showing that for certain pairs of tasks T_1, T_2, it is unlikely that such an algorithm R exists. The natural interpretation of such negative results is that proving the desired relation between T_1 and T_2 may be a difficult task that is beyond the reach of "current techniques." However, black-box reductions are not the only way computational relations can be obtained, and by now there are several proofs of relations that are unlikely to be provable using black-box reductions. A classic example is the result that an efficient algorithm for SAT implies an efficient algorithm for every language in the polynomial-time hierarchy; this is proven via a non-black-box argument and indeed a black-box reduction showing this relation seems unlikely (as it would imply the collapse of the polynomial-time hierarchy). It is useful to isolate and identify techniques like this, which overcome limitations of black-box reductions, because they may enable overcoming barriers elsewhere in complexity theory.

In this paper, we study black-box reductions in the area of hardness vs. randomness (i.e. constructing pseudorandom generators from hard functions). Specifically, we take T_1 to be the task of distinguishing the output distribution of a pseudorandom generator G from the uniform distribution, and T_2 to be the task of computing some supposedly 'hard' function g, and we are interested in reductions R such that R^f computes g if f is any function distinguishing G from uniform. We show limitations of such black-box reductions, and together with [10,24] (which do show that such connections exist), point to two ways in which black-box limitations can be overcome:

- Allowing the reduction R to make *adaptive* queries to its oracle f. We show that a relation established in [10,24] using an adaptive reduction is unlikely to be provable using a nonadaptive reduction. This may be interpreted as a hope to overcome other barriers known for nonadaptive reductions, such as worst-case/average-case connections for NP [4] and strong hardness amplification for constant-depth circuits [22]. (We mention though that for the latter connection, even adaptive black-box reductions are ruled out, unless they are highly non-uniform [5].)

– Using the efficiency of the oracle f in the *analysis* of the reduction R. Namely, we consider redcutions that use f as an oracle, but their analysis relies on the fact that f can be computed efficiently.[1] We observe that a reduction of this type is implicit in [10], and we show that it is unlikely to have an analogous reduction that works regardless of the complexity of the oracle. (A similar separation was given in the context of worst-case/average-case reductions for NP [6,7].)

We hope that these realizations will prove useful in obtaining stronger results in hardness vs. randomness and in overcoming limitations elsewhere in complexity theory.

Hardness vs. Randomness. We start with some background on hardness vs. randomness and the use of reductions in this area. The hardness versus randomness paradigm, first developed by Blum, Micali, Yao, Nisan, and Wigderson [3,26,16], is one of the most exciting achievements of the field of computational complexity. It shows how to use the hardness of a function f (computable in exponential time) to construct a pseudorandom generator G, which can then be used to derandomize probabilistic algorithms. By now there are many varieties of such results, trading off different assumptions on the function f, different types of probabilistic algorithms (e.g. BPP algorithms or AM proof systems), and different levels of derandomization.

For many years, all of the results of this type (based on the hardness of an arbitrary exponential-time computable function) required the function f to be hard for even nonuniform algorithms, e.g. $f \notin \mathrm{P}/\mathrm{poly}$. Nearly a decade ago, Impagliazzo and Wigderson [10] overcame this barrier, showing how to construct pseudorandom generators assuming only the existence of an exponential-time computable function f that is hard for uniform probabilistic algorithms, i.e. assuming $\mathrm{EXP} \neq \mathrm{BPP}$.[2] This result and some work that followed it have raised the hope that we may be able to prove an equivalence between uniform and nonuniform hardness assumptions (since in some cases derandomization implies non-uniform lower bounds [9,12,18]), or even obtain unconditional derandomization and new lower bounds.

The work of Impagliazzo and Wigderson [10], as well as the subsequent ones on derandomization from uniform assumptions, have used a number of ingredients that were not present in earlier works on hardness vs. randomness. In this paper, following [24], we explore the extent to which these new ingredients are really necessary. The hope is that such an understanding will help point the way to

---

[1] Here the reduction does not need to use the *code* of the algorithm for f, but just the fact that an efficient algorithm exists. This is in contrast to the example of SAT vs. PH mentioned above.

[2] The generator of [10], as well as other generators that are based on uniform hardness assumptions, are weaker than those that are based on nonuniform assumptions, in the sense that they only fool (uniform) Turing machines and hence only imply an average-case derandomization of probabilistic classes.

even stronger results,[3] and also, as we mentioned above, highlight techniques that might be used to overcome barriers in other parts of complexity theory. We now describe the new ingredients introduced by Impagliazzo and Wigderson [10].

Black-box reductions. Classic results on hardness vs. randomness can be formulated as "black box" constructions. That is, they are obtained by providing two efficient oracle algorithms G and R. The *construction* G uses oracle access to a (supposedly hard) function f to compute a *generator* G^f, which stretches a short seed to a long sequence of bits. The *reduction* R is meant to show that the output of G^f is pseudorandom if the function f is hard. Specifically, we require that for every statistical test T that distinguishes the output of G^f from uniform, there exists an "advice string" z such that $R^T(z, \cdot)$ computes the function f. Note that if T is efficient, then by hardwiring z, we obtain a small circuit computing f. Put in the contrapositive, this says that if f cannot be computed by small circuits, then there cannot exist an efficient test T distinguishing the output of G^f from uniform.

Note that the above notion require both the construction G and the reduction R to be black box, and requires that they work for every function f and statistical test T, regardless of the complexity of f and T. In the taxonomy of [17], these are referred to as *fully black-box* constructions. The advice string z that we provide to the reduction R is what makes the reduction *nonuniform*, and thereby require a nonuniform hardness assumption on the function f to deduce that G^f is pseudorandom. If the advice string could be eliminated, then we would get results based on uniform assumptions, like those of [10]. Unfortunately, as shown in [24], it is impossible to have a fully black-box construction of a pseudorandom generator without a significant amount of advice. Thus the Impagliazzo–Wigderson construction necessarily deviates from the fully black-box framework.

The most obvious way in which the Impagliazzo–Wigderson [10] construction is not fully black box is that it is not proven to work for every function f, and rather the construction (and its proof of correctness) makes use of the fact that f is in EXP or some other complexity class such as $P^{\#P}$ or PSPACE [24]. For example, in the case of $P^{\#P}$ or PSPACE, it uses the fact that f can be reduced to a function f' that is both downward self-reducible and self-correctible (e.g. f' is the PERMANENT), which is then used to construct the pseudorandom generator. That is, the construction algorithm G is not black box. Whether the Impagliazzo–Wigderson reduction algorithm R is or is not black box (i.e. works for every test T given as oracle) depends on which class f is taken from. For functions in $P^{\#P}$ or PSPACE, R is black box. But if we are only given a function in EXP, then the reduction relies on the fact that the test T is efficiently computable. Another interesting aspect of the reduction R is that it makes *adaptive* queries to the statistical test T, whereas earlier reductions this

---

[3] A seemingly modest but still elusive goal is a "high-end" version of [10], whereby one can construct a pseudorandom generator with exponential stretch from the assumption that EXP does not have subexponential-time probabilistic algorithms.

area were nonadaptive. (There are subsequent reductions, due to Shaltiel and Umans [21,25], that are also adaptive.)

Our results. Our main results provide evidence that some of these new ingredients are necessary. Specifically, we consider arbitrary (non-black-box) constructions of a pseudorandom generator G from a function f, and uniform reductions R (i.e. with no advice) from computing f to distinguishing the output of G from uniform. For simplicity, we also assume that the generator G is computable in time exponential in its seed length and that it stretches by a factor of at least 4. More general statements are given in the body of the paper (See Theorems 7 and 8).

Our first result shows that adaptivity is likely to be necessary unless we assume the function is in PH (rather than PSPACE or EXP).

Theorem 1 (informal). *If there is a nonadaptive, uniform, black-box reduction R from distinguishing a generator G to computing a function f, then f is in* $\mathrm{BPP}^{\mathrm{NP}}$.

Next, we consider reductions R that are *adaptive*, but with the restriction that all the queries of a particular length must be made simultaneously (they may depend on answers of the statistical test on queries of other lengths). (We call this *1-adaptive* later in the paper, as a special case of a more general notion (see Definition 5).) The Impagliazzo–Wigderson reduction for functions f in $\mathrm{P}^{\#\mathrm{P}}$ or PSPACE is 1-adaptive. We show that this property is unlikely to extend to EXP.

Theorem 2 (informal). *If there is a 1-adaptive, uniform, black-box reduction R from distinguishing a generator G to computing a function f, then f is in* PSPACE.

Thus, to obtain a result for arbitrary functions f in EXP, the reduction must either be non-black-box or "more adaptive." Impagliazzo and Wigderson exploit the former possibility, giving a non-black-box reduction, and their method for doing so turns out to have a substantial price — a statistical test running in time $t(n)$ yields an algorithm computing f that runs in time roughly $t(t(n))$, rather than something polynomially related to t, which is what is needed for a "high end" result (See [24]). Theorem 2 suggests that their result might be improved by employing reductions with greater adaptivity, such as [21,25]. Alternatively, it would be interesting to rule out such an improvement by strengthening Theorem 2 to hold for arbitrary adaptive reductions.

Finally, we consider "how non-black-box" the Impagliazzo–Wigderson reduction is for EXP. Specifically, we observe that even though the analysis of the reduction R relies on the fact that T is efficient (i.e. computable by small size circuits), the reduction itself only needs oracle access to T (i.e., it does not need the description of the circuits). We call such reductions *size-restricted black-box reductions*. Reductions of this type were recently studied by Gutfreund and Ta-Shma [7].[4]

---

[4] There are subtle differences between the reductions that we consider and the ones in [7], see the remark following Definition 6.

They exhibited such a reduction (based on [6]) for a worst-case/average-case connection that cannot be established via standard black-box reductions. Theorem 2, together with Theorem 3 below (which is implicit in [10]), provides another example of a size-restricted black-box reduction that bypasses black-box limitations. For technical reasons, we state the [10] result in terms of *hitting-set generators*, which are a natural weakening of pseudorandom generators that suffice for derandomizing probabilistic algorithms with 1-sided error (i.e. RP rather than BPP). Theorems 1 and 2 above can be strengthened to apply also to hitting-set generators.

Theorem 3 (implicit in [10], informal). *For every function f in EXP, there is a generator G and a 1-adaptive, uniform, size-restricted black-box reduction from distinguishing G as a hitting set to computing f.*

A final result of ours is an "infinitely-often" version of the Impagliazzo–Wigderson reduction [10]. The original versions of their reductions are guaranteed to compute f correctly on *all* input lengths assuming that the statistical test T successfully distinguishes the generator on *all* input lengths. Unlike most other results in the area, it is not known how to obtain reductions that compute f correctly on infinitely many input lengths when the test T is only guaranteed to succeed on infinitely many input lengths. We observe that such a result can be obtained for constructing *hitting-set generators* (and derandomizing RP) from hard problems in PSPACE rather than constructing pseudorandom generators (and derandomizing BPP) from hard problems in EXP as done in [10]. Due to space limitations, the statement and proof of this result is deferred to the full version of this paper [8].

Perspective. As discussed above, one motivation for studying the limitations of black-box reductions is to help identify potential approaches to overcoming apparent barriers. Another motivation is that black-box reductions sometimes have advantages over non-black-box reductions, and thus it is informative to know when these advantages cannot be achieved. For example, Trevisan's realization that fully black-box constructions of pseudorandom generators yield *randomness extractors* [23] yielded substantial benefits for both the study of pseudorandom generators and extractors. Similarly, Klivans and van Melkebeek [14] observed that black-box constructions of pseudorandom generators extend naturally to derandomize classes other than BPP, such as AM.

Unfortunately, as we have mentioned, results showing the limitations of black-box reductions are often interpreted as saying that proving certain results are outside the reach of "current techniques". We strongly disagree with these kinds of interpretations, and indeed hope that our results together with [10] will serve as another reminder that such limitations can be overcome.

2   Preliminaries

We assume that the reader is familiar with standard complexity classes such as EXP, BPP, the polynomial-time hierarchy etc., as well as standard models of computation such as probabilistic Turing Machines and Boolean circuits.

For a class C of *algorithms,* we denote by $io - C$ the class of languages L such that an algorithm from C correctly decides L for infinitely many input lengths.

For $n \in \mathbb{N}$, we denote by U_n the uniform distribution over $\{0,1\}^n$. For a distribution D, we denote by $x \leftarrow D$ that x is a sample drawn from D.

2.1   Pseudorandom Generators and Hardness vs. Randomness

Definition 4. *Let $b : \mathbb{N} \rightarrow \mathbb{N}$ be such that for every a, $b(a) > a$. Let $\mathcal{G} = \{G_a : \{0,1\}^a \rightarrow \{0,1\}^{b(a)}\}_{a \in \mathbb{N}}$ be a sequence of functions, and let $\mathcal{T} = \{T : \{0,1\}^* \rightarrow \{0,1\}\}$ be a family of Boolean functions (which we call* statistical tests*). For $\delta > 0$ we say that,*

1. *$\mathcal{G}$ is a sequence of* pseudorandom generators *(PRGs for short) that δ-fools $\mathcal{T}$ i.o. (infinitely often), if for every $T \in \mathcal{T}$, there are infinitely many $a \in \mathbb{N}$ such that*

$$| \Pr_{y \leftarrow U_a} [T(G_a(y)) = 1] - \Pr_{x \leftarrow U_{b(a)}} [T(x) = 1]| < \delta \qquad (1)$$

2. *$\mathcal{G}$ is a sequence of* hitting-set generators *(HSGs for short) that δ-hits $\mathcal{T}$ i.o., if for every $T \in \mathcal{T}$, there are infinitely many $a \in \mathbb{N}$ such that*

$$\Pr_{x \leftarrow U_{b(a)}} [T(x) = 0] \geq \delta \Rightarrow \Pr_{y \leftarrow U_a} [T(G_a(y)) = 0] > 0 \qquad (2)$$

If a function $T : \{0,1\}^* \rightarrow \{0,1\}$ violates (1) (respectively (2)), we say that it δ-distinguishes $\mathcal{G}$ from uniform a.e. (almost everywhere).

δ-fooling (respectively δ-hitting) a.e. and δ-distinguishing i.o. are defined analogously with the appropriate changes in the quantification over input lengths.

Note that if G is a PRG that δ-fools $\mathcal{T}$ i.o. (respectively a.e.) then it is also a HSG that δ-hits $\mathcal{T}$ i.o. (respectively a.e.).

Definition 5. *A (uniform)* black-box reduction *from deciding a language L to δ-distinguishing a.e. a family of (either pseudorandom or hitting-set) generators $\mathcal{G} = \{G_a : \{0,1\}^a \rightarrow \{0,1\}^{b(a)}\}_{a \in \mathbb{N}}$, is a probabilistic polynomial-time oracle Turing Machine (TM) R, such that for every statistical test T that δ-distinguishes $\mathcal{G}$ a.e., for every large enough $n \in \mathbb{N}$ and for every $x \in \{0,1\}^n$,*

$$\Pr[R^T(x) = L(x)] > 2/3$$

where the probability is over the random coins of R, and $R^T(x)$ denotes the execution of R on input x and with oracle access to T.

We say that such a reduction asks single-length *queries if for every n, there exist $a = a(n)$ such that on every execution of R on instances of length n, all the queries that R makes are of length exactly $b(a)$.*

We say that the reduction has $k = k(n)$ levels of adaptivity *if on every execution of R on inputs of length n and every statistical test T, the queries to T can* be partitioned to $k+1$ subsets (which are called the levels of adaptivity), such that each query in the i'th set is a function of the input x, the randomness of R, the index of the query within the i'th set (as well as i itself), and the answers that T gives on queries in the sets $1, \ldots, i-1$. We say that a reduction is nonadaptive if it has zero levels of adaptivity.

Finally, we say that the reduction is $k(a, n)$-adaptive if for every statistical test T, every instance of length n and every a, there are at most $k(a, n)$ levels of adaptivity in which queries of length $b(a)$ appear with positive probability (over the randomness of R when it is given oracle access to T).

We now define a different notion of reductions that still only have oracle access to the distinguishers, however the correctness of the reduction is only required to hold when the distinguisher is restricted to be a function that is computable by polynomial-size circuits.

Definition 6. A (uniform) size-restricted black-box reduction *from deciding a language L to δ-distinguishing a.e. a family of (pseudorandom or hitting-set) generators $\mathcal{G} = \{G_a : \{0,1\}^a \to \{0,1\}^{b(a)}\}_{a \in \mathbb{N}}$, is a probabilistic polynomial-time oracle TM R, such that for every statistical test T that δ-distinguishes $\mathcal{G}$ a.e.*, and is computable by a sequence of quadratic-size circuits[5], *for every large enough $n \in \mathbb{N}$ and for every $x \in \{0,1\}^n$,*

$$\Pr[R^T(x) = L(x)] > 2/3$$

where the probability is over the random coins of R.

Quantifiers over query length and adaptivity are defined as in the black-box case.

A remark about the quadratic size bound. The quadratic bound on the circuit size of the distinguishers is arbitrary and can be any (fixed) polynomial. The reason for our quadratic choice is that restricting the attention to distinguishers of this size is enough for derandomization.

A comparison to the definition of [7]. The restricted black-box reductions that we consider here run in any arbitrary polynomial time bound, which in particular can be larger than the fixed (quadratic) polynomial bound on the size of the distinguishers. In contrast, the notion of *class-specific black-box reductions* defined in [7], considers reductions that run in a fixed polynomial-time that is independent of the running time (or the circuit size) of the oracle (i.e. the oracle function can be computed by algorithms that run in arbitrary polynomial time).

3   Nonadaptive Reductions

In this section we show that any black-box nonadaptive reduction from deciding a language L to distinguishing a generator implies that L is in the polynomial-time hierarchy.

---

[5] Recall that the distinguishers' circuit size is measured with respect to their input length, which is the output length of the generator.

Theorem 7. *Let $L \subseteq \{0,1\}^*$ be a language, and let $\mathcal{G} = \{G_a : \{0,1\}^a \to \{0,1\}^{b(a)}\}_{a \in \mathbb{N}}$ be a family of hitting-set generators such that G_a is computable in time $2^{O(a)}$, and $b(a) > 4a$. If there is a nonadaptive black-box reduction R from L to $\frac{1}{2}$-distinguishing $\mathcal{G}$ a.e., then L is in $\mathrm{BPP}^{\mathrm{NP}}$. If we remove the time bound condition on computing G_a then L is in $\mathrm{P}^{\mathrm{NP}}/\mathrm{poly}$.*

Proof outline. We give here the main ideas in the proof. For the formal details refer to the full version of this paper [8]. Let us concentrate on the single-length case. We describe a $\mathrm{BPP}^{\mathrm{NP}}$ algorithm that decides L. Fix an input $x \in \{0,1\}^n$, and let $a \in \mathbb{N}$ be such that R queries its oracle on instances of length $b = b(a)$ when given inputs of length n.

The basic idea is to define, based on x, a statistical test T (that may not be efficiently computable) with the following properties:

1. T $\frac{1}{2}$-distinguishes G_a. This means that R^T decides L correctly on every instance of length n.
2. There is a function T' that can be computed in $\mathrm{BPP}^{\mathrm{NP}}$, such that R^T and $R^{T'}$ behave almost the same on the input x. This means that $R^{T'}$ decides correctly the membership of x in L (since so does R^T), but now the procedure *together* with the oracle computations can be implemented in $\mathrm{BPP}^{\mathrm{NP}}$.

Before we explain how to construct T and T', we want to stress that these functions depend on the specific input x, and the fact that R^T and $R^{T'}$ behave almost the same is only guaranteed when we run them on that x. I.e. every instance determines different functions T and T' (we avoid using the notation T_x and T'_x because in the proof there are other parameters involved and the notations become cumbersome). The point is that given any instance x, the answers of the oracle T', that is determined by x, can be computed (from scratch) in $\mathrm{BPP}^{\mathrm{NP}}$.

Now, if G_a were computable in time $\mathrm{poly}(b(a))$, we could simply take $T = T' = \mathrm{Im}(G_a)$. Indeed, $\mathrm{Im}(G_a)$ is the optimal distinguisher for G_a, and membership in $\mathrm{Im}(G_a)$ can be decided in nondeterministic polynomial time if G_a is efficiently computable (by guessing a corresponding seed). However, as in [16,10], we allow the generator to run in time $2^{O(a)} \gg b(a)$, since this suffices when pseudorandom generators are used for derandomization. In such a case, deciding membership in $\mathrm{Im}(G_a)$ may not be feasible in the polynomial hierarchy. So instead we will take $T = \mathrm{Im}(G_a) \cup H$ and $T' = H$ where H is a "small" set defined so that R^T and $R^{T'}$ behave almost the same.

To construct such a set H, we classify queries that R makes on input x, according to the probability that they come up in the reduction (where the probability is over R's randomness). (A similar idea appears in [4].) We call a query *heavy* if the probability it comes up is at least 2^{-t} and *light* otherwise, where t is the average of a and $b = b(a)$. Note that the classification to heavy/light is well defined and is *independent* of any oracle that R may query, because the reduction is nonadaptive. We define H to be the set of heavy queries.

First, we argue that $T = H \cup \mathrm{Im}(G_a)$ $\frac{1}{2}$-distinguishes G_a. This is because clearly it is always 1 on a sample taken by G_a. On the other hand, the number of elements for which T is 1 is small relative to the universe $\{0,1\}^b$. This is because there are only 2^a elements in the image of G_a, and at most 2^t heavy elements. Recall that both a and t are smaller than b.

Next, we argue that the behavior of $R^T(x)$ is roughly the same as $R^{T'}(x)$, where $T' = H$. Note that the only difference between T and T' is on light elements in the image set of G_a (T gives them the value 1, while T' gives them the value 0). When we run R on input x, the probability that such elements appear is small because their number is small (at most 2^a) and each one appears with small probability (because it is light). So R, on input x, behaves roughly the same when it has oracle access to either T or T'. We therefore conclude that $R^{T'}$ decides correctly the membership of x in L.

Finally, to show that $T' = H$ is computable in $\mathrm{BPP}^{\mathrm{NP}}$, we use the fact that approximate counting can be done in $\mathrm{BPP}^{\mathrm{NP}}$ [19,20,11], which allows us to approximate the weight of queries made by R and thus simulate its run with the oracle T'. Since for every query we only get an approximation of its weight, we cannot handle a sharp threshold between heavy and light queries. To that end, instead of defining the threshold t to be the average of a and b, we define two thresholds (both of which are a weighted average of a and b), such that those queries with weight below the low threshold are considered light, those with weight above the high threshold are considered heavy, and those in between can be classified arbitrarily. We now need more subtle definitions of T and T', but still the outline described above works.

4   Adaptive Reductions

In this section we show that any black-box reduction, from a language L to distinguishing a generator, that is adaptive with the restriction that queries of the same length do not appear in too many different levels, implies that L is in PSPACE.

Theorem 8. *Let $L \subseteq \{0,1\}^*$ be a language, and let $\mathcal{G} = \{G_a : \{0,1\}^a \to \{0,1\}^{b(a)}\}_{a \in \mathbb{N}}$ be a family of hitting-set generators such that G_a is computable in time $2^{O(a)}$, and $b(a) > 4a$. If there is a $\ell(a,n)$-adaptive black-box reduction R from L to $\frac{1}{2}$-distinguishing $\mathcal{G}$ a.e., where $\ell(a,n) \leq \frac{b(a)-a}{40 \log n}$ for $a \geq 15 \log n$, then L is in PSPACE. If we remove the time bound condition on computing G_a then L is in PSPACE/poly.*

Proof outline. We give here the main ideas in the proof. For the formal details refer to the full version of this paper [8]. Our starting point is the proof of Theorem 7 (see proof outline in Section 3). Our aim is to construct, based on an input x, the functions T and T' as before. The problem that we face when trying to implement the same ideas is that now, because the reduction is adaptive, the property of a query being light or heavy depends on the oracle that R queries (this is because queries above the first level depend on answers of the oracle). We therefore cannot define T in the same manner (such a definition would be circular). Instead, we classify queries to light and heavy separately for each level of adaptivity (i.e. a query can be light for one level and heavy for another). We do that inductively as follows. For the first level we set a threshold 2^{-t_1} (where t_1 is a weighted average of a and $b = b(a)$). We then define light and heavy with respect to this threshold. The distribution over queries at the first level is independent of any oracle, so the classification is well defined. We then define a function T_1 to be 1 on queries that are heavy for the first level and 0 otherwise. We can now proceed to define light and heavy for the second level when considering the distribution over queries at the second level when running $R(x)$ with oracle access to T_1 at the first level. We continue with this process inductively to define light/heavy at level i, with respect to the distribution obtained by running $R(x)$ with oracles $T_1, \ldots, T_{i-1}$ (each at the corresponding level). Here T_j is defined to be 1 on queries that are heavy for at least one of levels from the j'th down (and 0 otherwise). For each level i we define a different threshold 2^{-t_i}, with the property that the thresholds gradually increase with the levels (the reason for this will soon be clear).

We now define the statistical test T to be 1 on elements that are heavy for at least one of the levels as well as on elements in the image set of G_a (and 0 otherwise). The argument showing that T $\frac{1}{2}$-distinguishes G_a, is similar to the one in the proof of Theorem 7.

In the next step, instead of defining a T' as in the proof of Theorem 7, we directly compare the outcomes of running $R(x)$ with T as an oracle and running $R(x)$ with oracles $T_1, \ldots, T_\ell$ (where ℓ is the number of adaptivity levels), each at the corresponding level. We argue that the two runs should be roughly the same (in the sense that the distributions over the outputs will be close). To do that, we observe that at each level i, the answer of T on a query q differs from the answer of T_i on this query if one of the following occurs:

1. q is in the image set of G_a and it is light for levels $1, \ldots, i$.
2. q is light for all levels $1, \ldots, i$ but heavy for at least one of the levels $i + 1, \ldots, \ell$.

In both cases T will give q the value 1, while T_i the value 0. We bound the probability that queries as above are generated by $R(x)$ when it is given the oracles $T_1, \ldots, T_\ell$. The argument that bounds the probability that queries of the first type are generated is similar to the argument in the proof of Theorem 7. The probability that queries of the second type are generated at the i'th level is bounded as follows: the total number of heavy elements for levels above the i'th is small (it is at most the reciprocal of their weight, which is high). Of these elements, those that are light at level i have small probability to be generated at level i by virtue of them being light for that level. When we take the union bound over all such queries we still get a small probability of at least one of them being generated. The point is that the number of elements in the union bound is computed according to thresholds of levels above the i'th, while their probability is taken according to the threshold of the i'th level. By the fact that thresholds increase with the levels, we get that the number of elements in the union bound is much smaller than the reciprocal of their probabilities, and therefore the overall probability of such an event is small.

We conclude that the output distributions of running R with oracle access to T and running $R(x)$ with oracle access to $T_1, \ldots, T_\ell$ are very close, and therefore the latter decides correctly the membership of x in L.

Finally we show that $T_1, \ldots, T_\ell$ can be implemented in PSPACE and thus the whole procedure of running $R(x)$ and computing these oracles is in PSPACE. To compute the answers of the oracle T_i (at level i) we compute the exact weight of the query. We do that by a recursive procedure that computes the exact weights of all the queries (at levels below the i'th) that appear along the way. The fact that T_i only depends on T_j for $1 \leq j < i$ allows this procedure to run in polynomial-space.

5   Comparison to Known Reductions

In this section we contrast our negative results regarding black-box reductions to known relations between deciding languages and distinguishing pseudorandom (and hitting-set) generators. Impagliazzo and Wigderson [10] showed such a reduction from every language in the class $P^{\#P}$. This was extended by Trevisan and Vadhan [24] to languages in PSPACE. These reductions are black-box and adaptive (1-adaptive to be precise, see Definition 5).

Theorem 9. *[10,24] For every language L in PSPACE there exists a polynomial function $k(\cdot)$ such that for every polynomial function $b(\cdot)$, there is a uniform black-box reduction from deciding L to distinguishing a certain family of pseudorandom generators $\mathcal{G} = \{G_a : \{0,1\}^a \rightarrow \{0,1\}^{b(a)}\}_{a \in \mathbb{N}}$ a.e., where G_a is computable in time $2^{O(a)}$. The reduction is 1-adaptive and has $k(n)$ levels of adaptivity.*

We conclude by Theorem 7 that the black-box reduction from the theorem above is *inherently* adaptive, unless $PSPACE = BPP^{NP}$.

Next we turn our attention to reductions from languages in the class EXP. Such a reduction was given by Impagliazzo and Wigderson [10]. Their reduction is not black-box but rather *size-restricted* black-box (see Definition 6). We refer the reader to the full version of this paper [8] where we explain how the fact that the distinguisher can be computed by small-size circuits plays a role in this reduction.

Theorem 10. *(implicit in [10]) For every language L in EXP and polynomial function $b(\cdot)$, there is a polynomial function $k(\cdot)$ and a uniform size-restricted black-box reduction from deciding L to distinguishing a certain family of hitting-set generators $\mathcal{G} = \{G_a : \{0,1\}^a \rightarrow \{0,1\}^{b(a)}\}_{a \in \mathbb{N}}$ a.e., where G_a is computable in time $2^{O(a)}$. The reduction is 1-adaptive and has $k(n)$ levels of adaptivity.*

Theorem 10 should be contrasted with Theorem 8, which says that any reduction that is 1-adaptive cannot be black box (unless $EXP = PSPACE$). That is, the 'size-restricted' aspect of Theorem 10 cannot be removed.

A remark about reductions from computing a function on the average. Typically, constructions of pseudorandom (resp. hitting-set) generators from hard functions (both against uniform and non-uniform classes) combine two reductions: the first reduces the task of computing f (the supposedly hard function) on every instance to computing some related function $\hat{f}$ on the average. The second reduces computing $\hat{f}$ on the average to distinguishing the generator. In particular, the proof of [10] takes this form. We mention that our negative results about black-box reductions can be strengthened to show the same limitations for reductions from computing a function on the average to distinguishing a generator from the uniform distribution. In other words, it is really the fact that we reduce to distinguishing a generator that makes it impossible to do with black-box reductions, and not the fact that we start from a worst-case hardness assumption. In fact, *nonadaptive* (and uniform, black-box) worst-case to average-case reductions for PSPACE-complete and EXP-complete functions are known [1,2,24].

Acknowledgements

We thank Ronen Shaltiel and the anonymous reviewers for helpful comments on the write-up.

References

1. Babai, L., Fortnow, L., Lund, C.: Non-deterministic exponential time has two-prover interactive protocols. In: Proceedings of the 31st Annual IEEE Symposium on Foundations of Computer Science, pp. 16–25 (1990)
2. Babai, L., Fortnow, L., Nisan, N., Wigderson, A.: BPP has subexponential simulation unless EXPTIME has publishable proofs. Computational Complexity 3, 307–318 (1993)
3. Blum, M., Micali, S.: How to generate cryptographically strong sequences of pseudo-random bits. SIAM Journal on Computing 13(4), 850–864 (1984)
4. Bogdanov, A., Trevisan, L.: On worst-case to average-case reductions for NP problems. In: Proceedings of the 44th Annual IEEE Symposium on Foundations of Computer Science, pp. 308–317 (2003)
5. Gutfreund, D., Rothblum, G.: The complexity of local list decoding. Technical Report TR08-034, Electronic Colloquium on Computational Complexity (2008)
6. Gutfreund, D., Shaltiel, R., Ta-Shma, A.: If NP languages are hard in the worst-case then it is easy to find their hard instances. Computational Complexity 16(4), 412–441 (2007)
7. Gutfreund, D., Ta-Shma, A.: Worst-case to average-case reductions revisited. In: Charikar, M., Jansen, K., Reingold, O., Rolim, J. (eds.) RANDOM 2007. LNCS, vol. 4627, pp. 569–583. Springer, Heidelberg (2007)
8. Gutfreund, D., Vadhan, S.: Limitations of hardness vs. randomness under uniform reductions. Technical Report TR08-007, Electronic Colloquium on Computational Complexity (2008)
9. Impagliazzo, R., Kabanets, V., Wigderson, A.: In search of an easy witness: Exponential time vs. probabilistic polynomial time. Journal of Computer and System Sciences 65(4), 672–694 (2002)

10. Impagliazzo, R., Wigderson, A.: Randomness vs. time: de-randomization under a uniform assumption. Journal of Computer and System Sciences 63(4), 672–688 (2001)
11. Jerrum, M., Valiant, L., Vazirani, V.: Random generation of combinatorial structures from a uniform distribution. Theor. Comput. Sci. 43, 169–188 (1986)
12. Kabanets, V., Impagliazzo, R.: Derandomizing polynomial identity tests means proving circuit lower bounds. Computational Complexity 13(1-2), 1–46 (2004)
13. Karp, R.M., Lipton, R.J.: Some connections between nonuniform and uniform complexity classes. In: Proceedings of the 12th Annual ACM Symposium on Theory of Computing, pp. 302–309 (1980)
14. Klivans, A.R., van Melkebeek, D.: Graph nonisomorphism has subexponential size proofs unless the polynomial-time hierarchy collapses. SIAM Journal on Computing 31(5), 1501–1526 (2002)
15. Lipton, R.: New directions in testing. In: Proceedings of DIMACS workshop on distributed computing and cryptography, vol. 2, pp. 191–202 (1991)
16. Nisan, N., Wigderson, A.: Hardness vs. randomness. Journal of Computer and System Sciences 49, 149–167 (1994)
17. Reingold, O., Trevisan, L., Vadhan, S.: Notions of reducibility between cryptographic primitives. In: Naor, M. (ed.) TCC 2004. LNCS, vol. 2951, pp. 1–20. Springer, Heidelberg (2004)
18. Santhanam, R.: Circuit lower bounds for arthur–merlin classes. In: Proceedings of the 39th Annual ACM Symposium on Theory of Computing, pp. 275–283 (2007)
19. Sipser, M.: A complexity theoretic approach to randomness. In: Proceedings of the 15th Annual ACM Symposium on Theory of Computing, pp. 330–335 (1983)
20. Stockmeyer, L.: On approximation algorithms for $\sharp P$. SIAM Journal on Computing 14(4), 849–861 (1985)
21. Shaltiel, R., Umans, C.: Simple extractors for all min-entropies and a new pseudorandom generator. Journal of the ACM 52(2), 172–216 (2005)
22. Shaltiel, R., Viola, E.: Hardness amplification proofs require majority. In: Proceedings of the 40th Annual ACM Symposium on Theory of Computing, pp. 589–598 (2008)
23. Trevisan, L.: Construction of extractors using pseudo-random generators. Journal of the ACM 48(4), 860–879 (2001)
24. Trevisan, L., Vadhan, S.: Pseudorandomness and average-case complexity via uniform reductions. Computational Complexity 16(4), 331–364 (2007)
25. Umans, C.: Pseudo-random generators for all hardnesses. Journal of Computer and System Sciences 67(2), 419–440 (2003)
26. Yao, A.C.: Theory and applications of trapdoor functions. In: Proceedings of the 23rd Annual IEEE Symposium on Foundations of Computer Science, pp. 80–91 (1982)

Learning Random Monotone DNF

Jeffrey C. Jackson[1,*], Homin K. Lee[2], Rocco A. Servedio[2,**],
and Andrew Wan[2]

[1] Duquesne University, Pittsburgh, PA 15282
jacksonj@duq.edu
[2] Columbia University, New York, NY 10027
homin@cs.columbia.edu, rocco@cs.columbia.edu, atw12@cs.columbia.edu

Abstract. We give an algorithm that with high probability properly learns random monotone DNF with $t(n)$ terms of length $\approx \log t(n)$ under the uniform distribution on the Boolean cube $\{0,1\}^n$. For any function $t(n) \leq \text{poly}(n)$ the algorithm runs in time $\text{poly}(n, 1/\epsilon)$ and with high probability outputs an ϵ-accurate monotone DNF hypothesis. This is the first algorithm that can learn monotone DNF of arbitrary polynomial size in a reasonable average-case model of learning from random examples only.

1 Introduction

Motivation and background. Any Boolean function $f : \{0,1\}^n \to \{0,1\}$ can be expressed as a disjunction of conjunctions of Boolean literals, i.e. as an OR of ANDs. Such a logical formula is said to be a *disjunctive normal formula*, or DNF. Learning polynomial-size DNF formulas (disjunctions of $\text{poly}(n)$ many conjunctions) from random examples is an outstanding open question in computational learning theory, dating back more than 20 years to Valiant's introduction of the PAC (Probably Approximately Correct) learning model [Val84].

The most intensively studied variant of the DNF learning problem is PAC learning DNF under the uniform distribution. In this problem the learner must generate a high-accuracy hypothesis with high probability when given uniform random examples labeled according to the unknown target DNF. Despite much effort, no polynomial-time algorithms are known for this problem.

A tantalizing question that has been posed as a goal by many authors (see e.g. [Jac97, JT97, BBL98, Blu03b, Ser04]) is to learn *monotone* DNF, which only contain unnegated Boolean variables, under the uniform distribution. Besides being a natural restriction of the uniform distribution DNF learning problem, this problem is interesting because several impediments to learning general DNF under uniform – known lower bounds for Statistical Query based algorithms [BFJ+94], the apparent hardness of learning the subclass of $\log(n)$-juntas [Blu03a] – do not apply in the monotone case. This paper solves a natural average-case version of this problem.

---

* Supported in part by NSF award CCF-0209064.
** Supported in part by NSF award CCF-0347282, by NSF award CCF-0523664, and by a Sloan Foundation Fellowship.

A. Goel et al. (Eds.): APPROX and RANDOM 2008, LNCS 5171, pp. 483–497, 2008.
© Springer-Verlag Berlin Heidelberg 2008

Previous work. Many partial results have been obtained on learning monotone DNF under the uniform distribution. Verbeurgt [Ver90] gave an $n^{O(\log n)}$-time uniform distribution algorithm for learning any poly(n)-term DNF, monotone or not. Several authors [KMSP94, SM00, BT96] have given results on learning monotone t-term DNF for larger and larger values of t; most recently, [Ser04] gave a uniform distribution algorithm that learns any $2^{O(\sqrt{\log n})}$-term monotone DNF to any constant accuracy $\epsilon = \Theta(1)$ in poly(n) time. O'Donnell and Servedio [OS06] have recently shown that poly(n)-leaf *decision trees* that compute monotone functions (a subclass of poly(n)-term monotone DNF) can be learned to any constant accuracy under uniform in poly(n) time. Various other problems related to learning different types of monotone functions under uniform have also been studied, see e.g. [KLV94, BBL98, Ver98, HM91, AM02].

Aizenstein and Pitt [AP95] first proposed a model of random DNF formulas and gave an exact learning algorithm that learns random DNFs generated in this way. As noted in [AP95] and [JS06], this model admits a trivial learning algorithm in the uniform distribution PAC setting. Jackson and Servedio [JS05] gave a uniform distribution algorithm that learns log-depth decision trees on average in a natural random model. Previous work on average-case uniform PAC DNF learning, also by Jackson and Servedio, is described below.

Our results. The main result of this paper is a polynomial-time algorithm that can learn random poly(n)-term monotone DNF with high probability. (We give a full description of the exact probability distribution defining our random DNFs in Section 4; briefly, the reader should think of our random t-term monotone DNFs as being obtained by independently drawing t monotone conjunctions uniformly from the set of all conjunctions of length $\log_2 t$ over variables $x_1, \ldots, x_n$. Although many other distributions could be considered, this seems a natural starting point. Some justification for the choice of term length is given in Sections 4 and 6.)

Theorem 1. [Informally] *Let $t(n) \leq poly(n)$, and let $c > 0$ be any fixed constant. Then random monotone $t(n)$-term DNFs are PAC learnable (with failure probability $\delta = n^{-c}$) to accuracy ϵ in poly($n, 1/\epsilon$) time under the uniform distribution. The algorithm outputs a monotone DNF as its hypothesis.*

In independent and concurrent work, Sellie [Sel08] has given an alternate proof of this theorem using different techniques.

Our technique. Jackson and Servedio [JS06] showed that for any $\gamma > 0$, a result similar to Theorem 1 holds for random t-term monotone DNF with $t \leq n^{2-\gamma}$. The main open problem stated in [JS06] was to prove Theorem 1. Our work solves this problem by using the previous algorithm to handle $t \leq n^{3/2}$, developing new Fourier lemmas for monotone DNF, and using these lemmas together with more general versions of techniques from [JS06] to handle $t \geq n^{3/2}$.

The crux of our strategy is to establish a connection between the term structure of certain monotone DNFs and their low-order Fourier coefficients. There is an extensive body of research on Fourier properties of monotone Boolean functions [BT96, MO03, BBL98], polynomial-size DNF [Jac97, Man95], and related classes. These results typically establish that *every* function in the class has a Fourier spectrum with certain properties; unfortunately, the Fourier properties that have been obtainable to date for general statements of this sort have not been sufficient to yield polynomial-time learning algorithms.

We take a different approach by carefully defining a set of conditions, and showing that *if a monotone DNF f satisfies these conditions then the structure of the terms of f will be reflected in the low-order Fourier coefficients of f*. In [JS06], the degree two Fourier coefficients were shown to reveal the structure of the terms for certain (including random) monotone DNFs having at most $n^{2-\gamma}$ terms. In this work we develop new lemmas about the Fourier coefficients of more general monotone DNF, and use these new lemmas to establish a connection between term structure and constant degree Fourier coefficients for monotone DNFs with any polynomial number of terms. Roughly speaking, this connection holds for monotone DNF that satisfy the following conditions:

- each term has a reasonably large fraction of assignments which satisfy it and no other term;
- for each small tuple of distinct terms, only a small fraction of assignments simultaneously satisfy all terms in the tuple; and
- for each small tuple of variables, only a few terms contains the entire tuple.

The "small" tuples referred to above should be thought of as tuples of constant size. The constant degree coefficients capture the structure of the terms in the following sense: tuples of variables that all co-occur in some term will have a large magnitude Fourier coefficient, and tuples of variables that do not all co-occur in some term will have a small magnitude Fourier coefficient (even if subsets of the tuple do co-occur in some terms). We show this in Section 2.

Next we show a reconstruction procedure for obtaining the monotone DNF from tuple-wise co-occurrence information. Given a hypergraph with a vertex for each variable, the procedure turns each co-occurrence into a hyperedge, and then searches for all hypercliques of size corresponding to the term length. The hypercliques that are found correspond to the terms of the monotone DNF hypothesis that the algorithm constructs. This procedure is described in Section 3; we show that it succeeds in constructing a high-accuracy hypothesis if the monotone DNF f satisfies a few additional conditions. This generalizes a reconstruction procedure from [JS06] that was based on finding cliques in a graph (in the $n^{2-\gamma}$-term DNF setting, the algorithm deals only with co-occurrences of pairs of variables so it is sufficient to consider only ordinary graphs rather than hypergraphs).

The ingredients described so far thus give us an efficient algorithm to learn any monotone DNF that satisfies all of the required conditions. Finally, we show that random monotone DNF satisfy all the required conditions with high probability. We do this in Section 4 via a fairly delicate probabilistic argument. Section 5 combines the above ingredients to prove Theorem 1. We close the paper by showing that our technique lets us easily recapture the result of [HM91] that read-k monotone DNF are uniform-distribution learnable in polynomial time.

Preliminaries. We write $[n]$ to denote the set $\{1, \ldots, n\}$ and use capital letters for subsets of $[n]$. We will use calligraphic letters such as $\mathcal{C}$ to denote sets of sets and script letters such as $\mathscr{X}$ to denote sets of sets of sets. We write log to denote $\log_2$ and ln to denote the natural log. We write U_n to denote the uniform distribution over the Boolean cube $\{0, 1\}^n$.

A Boolean function $f : \{0, 1\}^n \to \{0, 1\}$ is *monotone* if changing the value of an input bit from 0 to 1 never causes the value of f to change from 1 to 0. We denote the input variables to f as $x_1, \ldots, x_n$. A *t-term monotone DNF* is a t-way OR of ANDs of Boolean variables (no negations). Recall that every monotone Boolean function has a unique representation as a reduced monotone DNF. We say that a term T of such a monotone DNF is *uniquely satisfied* by input x if x satisfies T and no other term of f.

Our learning model is an "average-case" variant of the well-studied uniform distribution PAC learning model. Let D_C be a probability distribution over some fixed class C of Boolean functions over $\{0, 1\}^n$, and let f (drawn from D_C) be an unknown target function. A learning algorithm A for D_C takes as input an accuracy parameter $0 < \epsilon < 1$ and a confidence parameter $0 < \delta < 1$. During its execution, algorithm A has access to a *random example oracle* $EX(f, U_n)$, which, when queried generates a random labeled example $(x, f(x))$, where x is drawn from U_n. The learning algorithm outputs a hypothesis h, which is a Boolean function over $\{0, 1\}^n$. The error of this hypothesis is defined to be $\mathrm{Pr}_{U_n}[h(x) \neq f(x)]$. We say that A *learns* D_C *under* U_n if for every $0 < \epsilon, \delta < 1$, with probability at least $1 - \delta$ (over both the random examples used for learning and the random draw of f from D_C) algorithm A outputs a hypothesis h which has error at most ϵ.

2 Fourier Coefficients and Monotone DNF Term Structure

Throughout this section let $f(x_1, \ldots, x_n)$ be a monotone DNF and let $S \subseteq \{1, \ldots, n\}$ be a fixed subset of variables. We write s to denote $|S|$ throughout this section. The Fourier coefficient, written $\hat{f}(S)$, measures the correlation between f and the parity of the variables in S.

The main result of this section is Lemma 3, which shows that under suitable conditions on f, the value $|\hat{f}(S)|$ is "large" if and only if f has a term containing all the variables of S. To prove this, we observe that the inputs which uniquely satisfy such a term will make a certain contribution to $\hat{f}(S)$. (In Section 2.1 we explain this in more detail and show how to view $\hat{f}(S)$ as a sum of contributions from inputs to f.) It remains then to show that the contribution from other inputs is small. The main technical novelty comes in Sections 2.2 and 2.3, where we show that all other inputs which make a contribution to $\hat{f}(S)$ must satisfy the terms of f in a special way, and use this property to show that under suitable conditions on f, the fraction of such inputs must be small.

2.1 Rewriting $\hat{f}(S)$

We observe that $\hat{f}(S)$ can be expressed in terms of 2^s conditional probabilities, each of which is the probability that f is satisfied conditioned on a particular setting of the variables in S. That is:

$$\hat{f}(S) \stackrel{\text{def}}{=} \mathbf{E}_{x \in U^n}\left[(-1)^{\sum_{i \in S} x_i} \cdot f(x)\right] = \frac{1}{2^n} \sum_{x \in \{0,1\}^n} (-1)^{\sum_{i \in S} x_i} \cdot f(x)$$

$$= \frac{1}{2^n} \sum_{U \subseteq S} (-1)^{|U|} \sum_{x \in Z_S(U)} f(x) = \frac{1}{2^s} \sum_{U \subseteq S} (-1)^{|U|} \Pr_x[f(x) = 1 \mid x \in Z_S(U)],$$

where $Z_S(U)$ denotes the set of those $x \in \{0,1\}^n$ such that $x_i = 1$ for all $i \in U$ and $x_i = 0$ for all $i \in S \setminus U$. If f has some term T containing all the variables in S, then $\Pr_x[f(x) = 1 \mid x \in Z_S(S)]$ is at least as large as $\Pr_x[T$ is uniquely satisfied in $f \mid x \in Z_S(S)]$. On the other hand, if f has no such term, then $\Pr_x[f(x) = 1 \mid x \in Z_S(S)]$ does not receive this contribution. We will show that this contribution is the chief determinant of the magnitude of $\hat{f}(S)$.

It is helpful to rewrite $\hat{f}(S)$ as a sum of contributions from each input $x \in \{0,1\}^n$. To this end, we decompose f according to the variables of S. Given a subset $U \subseteq S$, we will write g_U to denote the disjunction of terms in f that contain every variable indexed by $U \subseteq S$ and no variable indexed by $S \setminus U$, but with the variables indexed by U removed from each term. (So for example if $f = x_1 x_2 x_4 x_6 \vee x_1 x_2 x_5 \vee x_1 x_2 x_3 \vee x_3 x_5 \vee x_1 x_5 x_6$ and $S = \{1, 2, 3\}$ and $U = \{1, 2\}$, then $g_U = x_4 x_6 \vee x_5$.) Thus we can split f into disjoint sets of terms: $f = \bigvee_{U \subseteq S}(t_U \wedge g_U)$, where t_U is the term consisting of exactly the variables indexed by U.

Suppose we are given $U \subseteq S$ and an x that belongs to $Z_S(U)$. We have that $f(x) = 1$ if and only if $g_{U'}(x)$ is true for some $U' \subseteq U$. (Note that $t_{U'}(x)$ is true for every $U' \subseteq U$ since x belongs to $Z_S(U)$.) Thus we can rewrite the Fourier coefficients $\hat{f}(S)$ as follows: (Below we write $I(P)$ to denote the indicator function that takes value 1 if predicate P is true and value 0 if P is false.)

$$\hat{f}(S) = \frac{1}{2^n} \sum_{U \subseteq S} (-1)^{|U|} \sum_{x \in Z_S(U)} f(x) = \sum_{U \subseteq S} (-1)^{|U|} \frac{1}{2^n} \sum_{x \in Z_S(U)} I\left(\bigvee_{U' \subseteq U} g_{U'}(x)\right)$$

$$= \sum_{x \in \{0,1\}^n} \frac{1}{2^s} \frac{1}{2^n} \sum_{U \subseteq S} (-1)^{|U|} I\left(\bigvee_{U' \subseteq U} g_{U'}(x)\right).$$

We can rewrite this as $\hat{f}(S) = \sum_{x \in \{0,1\}^n} \text{Con}_S(x)$, where

$$\text{Con}_S(x) \stackrel{\text{def}}{=} \frac{1}{2^s} \frac{1}{2^n} \sum_{U \subseteq S} (-1)^{|U|} I\left(\bigvee_{U' \subseteq U} g_{U'}(x)\right). \tag{1}$$

The value $\text{Con}_S(x)$ may be viewed as the "contribution" that x makes to $\hat{f}(S)$. Recall that when f has a term T which contains all the variables in S, those $x \in Z_S(S)$ which uniquely satisfy T will contribute to $\hat{f}(S)$. We will show that under suitable conditions on f, the other x's make little or no contribution.

2.2 Bounding the Contribution to $\hat{f}(S)$ from Various Inputs

The variable $\mathcal{C}$ will denote a subset of $\mathcal{P}(S)$, the power set of S; i.e. $\mathcal{C}$ denotes a collection of subsets of S. We may view $\mathcal{C}$ as defining a set of g_U's (those g_U's for which U belongs to $\mathcal{C}$).

We may partition the set of inputs $\{0,1\}^n$ into $2^{|\mathcal{P}(S)|} = 2^{2^s}$ parts according to what subset of the 2^s functions $\{g_U\}_{U \subseteq S}$ each $x \in \{0,1\}^n$ satisfies. For $\mathcal{C}$ a subset of $\mathcal{P}(S)$ we denote the corresponding piece of the partition by $P_{\mathcal{C}}$; so $P_{\mathcal{C}}$ consists of precisely those $x \in \{0,1\}^n$ that satisfy $\left(\bigwedge_{U \in \mathcal{C}} g_U\right) \wedge \left(\bigwedge_{U \notin \mathcal{C}} \bar{g}_U\right)$. Note that for any given fixed $\mathcal{C}$, each x in $P_{\mathcal{C}}$ has exactly the same contribution $\text{Con}_S(x)$ to the Fourier coefficient $\hat{f}(S)$ as every other x' in $P_{\mathcal{C}}$; this is simply because x and x' will satisfy exactly the same set of $g_{U'}$'s in (1). More generally, we have the following (proved in the full version):

Lemma 1. *Let $\mathcal{C}$ be any subset of $\mathcal{P}(S)$. Suppose that there exist $U_1, U_2 \in \mathcal{C}$ such that $U_1 \subsetneq U_2$. Then for any y, z where $y \in P_{\mathcal{C}}$ and $z \in P_{\mathcal{C} \setminus U_2}$, we have that:* $\text{Con}_S(y) = \text{Con}_S(z)$.

Given a collection $\mathcal{C}$ of subsets of S, let $\text{Con}_S(\mathcal{C})$ denote $\sum_{x \in P_{\mathcal{C}}} \text{Con}_S(x)$, and we refer to this quantity as the contribution that $\mathcal{C}$ makes to the Fourier coefficient $\hat{f}(S)$. It is clear that we have $\hat{f}(S) = \sum_{\mathcal{C} \subseteq \mathcal{P}(S)} \text{Con}_S(\mathcal{C})$.

The following lemma, proved in the full version establishes a broad class of $\mathcal{C}$'s for which $\text{Con}_S(\mathcal{C})$ is zero:

Lemma 2. *Let $\mathcal{C}$ be any collection of subsets of S. If $\bigcup_{U \in \mathcal{C}} U \neq S$ then $\text{Con}_S(x) = 0$ for each $x \in P_{\mathcal{C}}$ and hence $\text{Con}_S(\mathcal{C}) = 0$.*

It remains to analyze those $\mathcal{C}$'s for which $\bigcup_{U \in \mathcal{C}} U = S$; for such a $\mathcal{C}$ we say that $\mathcal{C}$ *covers* S.

Recall from the previous discussion that $\text{Con}_S(\mathcal{C}) = |P_{\mathcal{C}}| \cdot \text{Con}_S(x)$ where x is any element of $P_{\mathcal{C}}$. Since $|\text{Con}_S(x)| \leq \frac{1}{2^n}$ for all $x \in \{0,1\}^n$, for any collection $\mathcal{C}$, we have that

$$|\text{Con}_S(\mathcal{C})| \leq \Pr_{x \in U_n}[x \in P_{\mathcal{C}}] = \Pr_{x \in U_n}[(\bigwedge_{U \in \mathcal{C}} g_U) \wedge (\bigwedge_{U \notin \mathcal{C}} \bar{g}_s)] \leq \Pr_{x \in U_n}[(\bigwedge_{U \in \mathcal{C}} g_U)].$$

We are interested in bounding this probability for $\mathcal{C} \neq \{S\}$ (we will deal with the special case $\mathcal{C} = \{S\}$ separately later). Recall that each g_U is a disjunction of terms; the expression $\bigwedge_{U \in \mathcal{C}} g_U$ is satisfied by precisely those x that satisfy at least one term from each g_U as U ranges over all elements of $\mathcal{C}$. For $j \geq 1$ let us define a quantity B_j as follows

$$B_j \stackrel{\text{def}}{=} \max_{i_1,\dots,i_j} \Pr_{x \in U_n}[x \text{ simultaneously satisfies terms } T_{i_1},\dots,T_{i_j} \text{ in } \vee_{U \subseteq S}(g_U)]$$

where the max is taken over all j-tuples of distinct terms in $\vee_{U \subseteq S}(g_U)$. Then it is not hard to see that by a union bound, we have

$$|\mathrm{Cons}_S(\mathcal{C})| \leq B_{|\mathcal{C}|} \prod_{U \in \mathcal{C}} (\#g_U), \tag{2}$$

where $\#g_U$ denotes the number of terms in the monotone DNF g_U.

The idea of why (2) is a useful bound is as follows. Intuitively, one would expect that the value of B_j decreases as j (the number of terms that must be satisfied) increases. One would also expect the value of $\#g_U$ to decrease as the size of U increases (if U contains more variables then fewer terms in f will contain all of those variables). This means that there is a trade-off which helps us bound (2): if $|\mathcal{C}|$ is large then $B_{|\mathcal{C}|}$ is small, but if $|\mathcal{C}|$ is small then (since we know that $\bigcup_{U \in \mathcal{C}} U = S$) some U is large and so $\prod_{U \in \mathcal{C}} \#g_U$ will be smaller.

2.3   Bounding $\hat{f}(S)$ Based on Whether S Co-occurs in a Term of f

We are now ready to state formally the conditions on $\hat{f}$ that allow us to detect a co-occurrence of variables in the value of the corresponding Fourier coefficient.

Lemma 3. *Let $f : \{0,1\}^n \to \{-1,1\}$ be a monotone DNF. Fix a set $S \subset [n]$ of size $|S| = s$ and let*

$$\mathscr{Y} = \{\mathcal{C} \subseteq \mathcal{P}(S) : \mathcal{C} \text{ covers } S \text{ and } S \notin \mathcal{C}\}.$$

Suppose that we define $\alpha, \beta_1, \ldots, \beta_{2^s}$ and $\Phi : \mathscr{Y} \to \mathbb{R}$ so that:

C1. *Each term in f is uniquely satisfied with probability at least α;*
C2. *For $1 \leq j \leq 2^s$, each j-tuple of terms in f is simultaneously satisfied with probability at most β_j; and*
C3. *For every $\mathcal{C}_{\mathscr{Y}} \in \mathscr{Y}$ we have $\prod_{U \in \mathcal{C}_{\mathscr{Y}}} (\#g_U) \leq \Phi(\mathcal{C}_{\mathscr{Y}})$.*

Then

1. *If the variables in S do not simultaneously co-occur in any term of f, then*

$$|\hat{f}(S)| \leq \Upsilon \quad \text{where} \quad \Upsilon := \sum_{\mathcal{C}_{\mathscr{Y}} \in \mathscr{Y}} \left(2^s \beta_{|\mathcal{C}_{\mathscr{Y}}|} \Phi(\mathcal{C}_{\mathscr{Y}})\right);$$

2. *If the variables in S do simultaneously co-occur in some term of f, then $|\hat{f}(S)| \geq \frac{\alpha}{2^s} - 2 \cdot \Upsilon$.*

Using Lemma 3, if f satisfies conditions **C1** through **C3** with values of β_j and $\Phi(\cdot)$ so that there is a "gap" between $\alpha/2^s$ and 3Υ, then we can determine whether all the variables in S simultaneously co-occur in a term by estimating the magnitude of $\hat{f}(S)$.

Proof. Let $\mathcal{C}^\star$ denote the 'special' element of $\mathcal{P}(S)$ that consists solely of the subset S, i.e. $\mathcal{C}^\star = \{S\}$, and let $\mathscr{X} = \{\mathcal{C} \subseteq \mathcal{P}(S) : \mathcal{C} \text{ covers } S \text{ and } S \in \mathcal{C} \text{ and } \mathcal{C} \neq \mathcal{C}^\star\}$. Using Lemma 2, we have

$$\hat{f}(S) = \mathrm{Cons}_S(\mathcal{C}^\star) + \sum_{\mathcal{C}_\mathcal{Y} \in \mathcal{Y}} \mathrm{Cons}_S(\mathcal{C}_\mathcal{Y}) + \sum_{\mathcal{C}_\mathcal{X} \in \mathcal{X}} \mathrm{Cons}_S(\mathcal{C}_\mathcal{X}). \qquad (3)$$

We first prove point 1. Suppose that the variables of S do not simultaneously co-occur in any term of f. Then g_S is the empty disjunction and $\#g_S = 0$, so $\mathrm{Cons}_S(\mathcal{C}) = 0$ for any $\mathcal{C}$ containing S. Thus in this case we have $\hat{f}(S) = \sum_{\mathcal{C}_\mathcal{Y} \in \mathcal{Y}} \mathrm{Cons}_S(\mathcal{C}_\mathcal{Y})$; using (2) and condition **C3**, it follows that $|\hat{f}(S)|$ is at most $\sum_{\mathcal{C}_\mathcal{Y} \in \mathcal{Y}} B_{|\mathcal{C}_\mathcal{Y}|} \Phi(\mathcal{C}_\mathcal{Y})$. It is not hard to see that $B_{|\mathcal{C}_\mathcal{Y}|} \le 2^s \beta_{|\mathcal{C}_\mathcal{Y}|}$ (we give a proof in the full version). So in this case we have

$$|\hat{f}(S)| \le \sum_{\mathcal{C}_\mathcal{Y} \in \mathcal{Y}} |\mathrm{Cons}_S(\mathcal{C}_\mathcal{Y})| \le \sum_{\mathcal{C}_\mathcal{Y} \in \mathcal{Y}} B_{|\mathcal{C}_\mathcal{Y}|} \Phi(\mathcal{C}_\mathcal{Y}) \le \sum_{\mathcal{C}_\mathcal{Y} \in \mathcal{Y}} \left(2^s \beta_{|\mathcal{C}_\mathcal{Y}|} \Phi(\mathcal{C}_\mathcal{Y}) \right) = \varUpsilon.$$

Now we turn to point 2. Suppose that the variables of S do co-occur in some term of f. Let x be any element of $P_{\mathcal{C}^\star}$, so x satisfies g_U if and only if $U = S$. It is easy to see from (1) that for such an x we have $\mathrm{Cons}_S(x) = (-1)^{|S|}/(2^n 2^s)$. We thus have that

$$\mathrm{Cons}_S(\mathcal{C}^\star) = \frac{(-1)^{|S|}}{2^s} \cdot \Pr[x \in P_{\mathcal{C}^\star}] = \frac{(-1)^{|S|}}{2^s} \Pr[g_S \wedge (\bigwedge_{U \subsetneq S} \bar{g}_U)]. \qquad (4)$$

Since S co-occurs in some term of f, we have that g_S contains at least one term T. By condition **C1**, the corresponding term $(T \wedge (\wedge_{i \in S} x_i))$ of f is uniquely satisfied with probability at least α. Since each assignment that uniquely satisfies $(T \wedge (\wedge_{i \in S} x_i))$ (among all the terms of f) must satisfy $g_S \wedge (\bigwedge_{U \subsetneq S} \bar{g}_U)$, we have that the magnitude of (4) is at least $\alpha/2^s$.

We now show that $|\sum_{\mathcal{C}_\mathcal{X} \in \mathcal{X}} \mathrm{Cons}_S(\mathcal{C}_\mathcal{X})| \le \varUpsilon$, which completes the proof, since we already have that $|\sum_{\mathcal{C}_\mathcal{Y} \in \mathcal{Y}} \mathrm{Cons}_S(\mathcal{C}_\mathcal{Y})| \le \sum_{\mathcal{C}_\mathcal{Y} \in \mathcal{Y}} |\mathrm{Cons}_S(\mathcal{C}_\mathcal{Y})| \le \varUpsilon$. First note that if the set $\mathcal{C}_\mathcal{X} \setminus \{S\}$ does not cover S, then by Lemmas 1 and 2 we have that $\mathrm{Cons}_S(x) = 0$ for each $x \in P_{\mathcal{C}_\mathcal{X}}$ and thus $\mathrm{Cons}_S(\mathcal{C}_\mathcal{X}) = 0$. So we may restrict our attention to those $\mathcal{C}_\mathcal{X}$ such that $\mathcal{C}_\mathcal{X} \setminus \{S\}$ covers S. Now since such a $\mathcal{C}_\mathcal{X} \setminus \{S\}$ is simply some $\mathcal{C}_\mathcal{Y} \in \mathcal{Y}$, and each $\mathcal{C}_\mathcal{Y} \in \mathcal{Y}$ is obtained as $\mathcal{C}_\mathcal{X} \setminus \{S\}$ for at most one $\mathcal{C}_\mathcal{X} \in \mathcal{X}$, we have

$$\left| \sum_{\mathcal{C}_\mathcal{X} \in \mathcal{X}} \mathrm{Cons}_S(\mathcal{C}_\mathcal{X}) \right| \le \sum_{\mathcal{C}_\mathcal{Y} \in \mathcal{Y}} |\mathrm{Cons}_S(\mathcal{C}_\mathcal{Y})| \le \varUpsilon.$$

3 Hypothesis Formation

In this section, we show that if a target monotone DNF f satisfies the conditions of Lemma 3 and two other simple conditions stated below (see Theorem 2), then it is possible to learn f from uniform random examples.

Theorem 2. *Let f be a t-term monotone DNF. Fix $s \in [n]$. Suppose that*

- *For all sets $S \subset [n], |S| = s$, conditions **C1** through **C3** of Lemma 3 hold for certain values α, β_j, and $\Phi(\cdot)$ satisfying $\Delta > 0$, where $\Delta := \alpha/2^s - 3 \cdot \Upsilon$. (Recall that $\Upsilon := \sum_{C_{\mathcal{Y}} \in \mathcal{Y}} (2^s \beta_{|C_{\mathcal{Y}}|} \Phi(C_{\mathcal{Y}}))$, where $\mathcal{Y} = \{C \subseteq \mathcal{P}(S) : C \text{ covers } S \text{ and } S \notin C\}$.)*

C4. *Every set S of s co-occurring variables in f appears in at most γ terms (here $\gamma \geq 2$); and*

C5. *Every term of f contains at most κ variables (note that $s \leq \kappa \leq n$).*

Then algorithm $\mathcal{A}$ (described formally in the full version) PAC learns f to accuracy ϵ with confidence $1 - \delta$ given access to $EX(f, U_n)$, and runs in time $poly(n^{s+\gamma}, t, 1/\Delta, \gamma^\kappa, 1/\epsilon, \log(1/\delta))$.

Proof. Lemma 3 implies that for each set $S \subset [n], |S| = s$,

- if the variables in S all co-occur in some term of f, then $|\hat{f}(S)|$ is at least $\Delta/2$ larger than $\Upsilon + \Delta/2$;
- if the variables in S do not all co-occur in some term of f, then $|\hat{f}(S)|$ is at least $\Delta/2$ smaller than $\Upsilon + \Delta/2$.

A straightforward application of Hoeffding bounds (to estimate the Fourier coefficients using a random sample of uniformly distributed examples) shows that Step 1 of Algorithm $\mathcal{A}$ can be executed in $poly(n^s, 1/\Delta, \log(1/\delta))$ time, and that with probability $1 - \delta/3$ the S's that are marked as "good" will be precisely the s-tuples of variables that co-occur in some term of f.

Conceptually, the algorithm next constructs the hypergraph G_f that has one vertex per variable in f and that includes an s-vertex hyperedge if and only if the corresponding s variables co-occur in some term of f. Clearly there is a k-hyperclique in G_f for each term of k variables in f. So if we could find all of the k-hypercliques in G_f (where again k ranges between s and κ), then we could create a set HC_f of monotone conjunctions of variables such that f could be represented as an OR of t of these conjunctions. Treating each of the conjunctions in HC_f as a variable in the standard elimination algorithm for learning disjunctions (see e.g. Chapter 1 of [KV94]) would then enable us to properly PAC learn f to accuracy ϵ with probability at least $1 - \delta/3$ in time polynomial in n, t, $|HC_f|$, $1/\epsilon$, and $\log(1/\delta)$. Thus, $\mathcal{A}$ will use a subalgorithm $\mathcal{A}'$ to find all the k-hypercliques in G_f and will then apply the elimination algorithm over the corresponding conjunctions to learn the final approximator h.

We now explain the subalgorithm $\mathcal{A}'$ for locating the set HC_f of k-hypercliques. For each set S of s co-occurring variables, let $N_S \subseteq ([n] \setminus S)$ be defined as follows: a variable x_i is in N_S if and only if x_i is present in some term that contains all of the variables in S. Since by assumption there are at most γ terms containing such variables and each term contains at most κ variables, this means that $|N_S| < \kappa\gamma$. The subalgorithm will use this bound as follows. For each set S of s co-occurring variables, $\mathcal{A}'$ will determine the set N_S using a procedure $\mathcal{A}''$ described shortly. Then, for each $s \leq k \leq \kappa$ and each $(k - s)$-element subset N' of N_S, $\mathcal{A}'$ will test whether or not $N' \cup S$ is a k-hyperclique in G_f. The set of all k-hypercliques found in this way is HC_f. For each S, the number of sets tested in this process is at most

$$\sum_{i=0}^{\kappa} \binom{|N_S|}{i} \leq \sum_{i=0}^{\kappa} \binom{\kappa\gamma}{i} \leq \left(\frac{e\kappa\gamma}{\kappa}\right)^{\kappa} = (e\gamma)^{\kappa}.$$

Thus, $|HC_f| = O(n^s (e\gamma)^{\kappa})$, and this is an upper bound on the time required to execute Step 2 of subalgorithm $\mathcal{A}'$.

Finally, we need to define the procedure $\mathcal{A}''$ for finding N_S for a given set S of s co-occurring variables. Fix such an S and let N_γ be a set of at most γ variables in $([n] \setminus S)$ having the following properties:

P1. In the projection $f_{N_\gamma \leftarrow 0}$ of f in which all of the variables of N_γ are fixed to 0, the variables in S do not co-occur in any term; and

P2. For every set $N'_\gamma \subset N_\gamma$ such that $|N'_\gamma| = |N_\gamma| - 1$, the variables in S do co-occur in at least one term of $f_{N'_\gamma \leftarrow 0}$.

We will use the following claim (proved in the full version):

Claim. N_S is the union of all sets N_γ of cardinality at most γ that satisfy **P1** and **P2**.

There are only $O(n^\gamma)$ possible candidate sets N_γ to consider, so our problem now reduces to the following: given a set N of at most γ variables, determine whether the variables in S co-occur in $f_{N \leftarrow 0}$.

Recall that since f satisfies the three conditions **C1**, **C2** and **C3**, Lemma 3 implies that $|\hat{f}(S)|$ is either at most Υ (if the variables in S do not co-occur in any term of f) or at least $\frac{\alpha}{2^s} - 2 \cdot \Upsilon$ (if the variables in S do co-occur in some term). We now claim that the function $f_{N \leftarrow 0}$ has this property as well: i.e., $|\widehat{f_{N \leftarrow 0}}(S)|$ is either at most the same value Υ (if the variables in S do not co-occur in any term of $f_{N \leftarrow 0}$) or at least the same value $\frac{\alpha}{2^s} - 2 \cdot \Upsilon$ (if the variables in S do co-occur in some term of $f_{N \leftarrow 0}$). To see this, observe that the function $f_{N \leftarrow 0}$ is just f with some terms removed. Since each term in f is uniquely satisfied with probability at least α (this is condition **C1**), the same must be true of $f_{N \leftarrow 0}$ since removing terms from f can only increase the probability of being uniquely satisfied for the remaining terms. Since each j-tuple of terms in f is simultaneously satisfied with probability at most β_j (this is condition **C2**), the same must be true for j-tuples of terms in $f_{N \leftarrow 0}$. Finally, for condition **C3**, the value of $\#g_U$ can only decrease in passing from f to $f_{N \leftarrow 0}$. Thus, the upper bound of Υ that follows from applying Lemma 3 to f is also a legitimate upper bound when the lemma is applied to $|\widehat{f_{N \leftarrow 0}}(S)|$, and similarly the lower bound of $\frac{\alpha}{2^s} - 2 \cdot \Upsilon$ is also a legitimate lower bound when the lemma is applied to $f_{N \leftarrow 0}$. Therefore, for every $|N| \leq \gamma$, a sufficiently accurate (within $\Delta/2$) estimate of $\widehat{f_{N \leftarrow 0}}(S)$ (as obtained in Step 1 of subalgorithm $\mathcal{A}''$) can be used to determine whether or not the variables in S co-occur in any term of $f_{N \leftarrow 0}$.

To obtain the required estimate for $\widehat{f_{N \leftarrow 0}}$, observe that for a given set N, we can simulate a uniform example oracle for $f_{N \leftarrow 0}$ by filtering the examples from the uniform oracle for f so that only examples setting the variables in N to 0 are accepted. Since $|N| \leq \gamma$, the filter accepts with probability at least $1/2^\gamma$. A Hoeffding bound argument then shows that the Fourier coefficients $\widehat{f_{N \leftarrow 0}}(S)$ can be estimated (with probability of failure no more than a small fraction of δ) from an example oracle for f in time polynomial in n, 2^γ, $1/\Delta$, and $\log(1/\delta)$.

Algorithm $\mathcal{A}''$, then, estimates Fourier coefficients of restricted versions of f, using a sample size sufficient to ensure that all of these coefficients are sufficiently accurate over all calls to $\mathcal{A}''$ with probability at least $1 - \delta/3$. These estimated coefficients are then used by $\mathcal{A}''$ to locate the set N_S as just described. The overall algorithm $\mathcal{A}$ therefore succeeds with probability at least $1 - \delta$, and it is not hard to see that it runs in the time bound claimed.

Required parameters. In the above description of Algorithm $\mathcal{A}$, we assumed that it is given the values of $s, \alpha, \Upsilon, \gamma$, and κ. In fact it is not necessary to assume this; a standard argument gives a variant of the algorithm which succeeds without being given the values of these parameters.

The idea is simply to have the algorithm "guess" the values of each of these parameters, either exactly or to an adequate accuracy. The parameters s, γ and κ take positive integer values bounded by $\text{poly}(n)$. The other parameters α, Υ take values between 0 and 1; a standard argument shows that if approximate values α' and Υ' (that differ from the true values by at most $1/\text{poly}(n)$) are used instead of the true values, the algorithm will still succeed. Thus there are at most $\text{poly}(n)$ total possible settings for $(s, \gamma, \kappa, \alpha, \Upsilon)$ that need to be tried. We can run Algorithm $\mathcal{A}$ for each of these candidate parameter settings, and test the resulting hypothesis; when we find the "right" parameter setting, we will obtain a high-accuracy hypothesis (and when this occurs, it is easy to recognize that it has occurred, simply by testing each hypothesis on a new sample of random labeled examples). This parameter guessing incurs an additional polynomial factor overhead. Thus Theorem 2 holds true for the extended version of Algorithm $\mathcal{A}$ that takes only ϵ, δ as input parameters.

4   Random Monotone DNF

The random monotone DNF model. Let $\mathcal{M}_n^{t,k}$ be the probability distribution over monotone t-term DNF induced by the following process: each term is independently and uniformly chosen at random from all $\binom{n}{k}$ monotone ANDs of size exactly k over $x_1, \ldots, x_n$.

Given a value of t, throughout this section we consider the $\mathcal{M}_n^{t,k}$ distribution where $k = \lfloor \log t \rfloor$ (we will relax this and consider a broader range of values for k in Section 6). To motivate this choice, consider a random draw of f from $\mathcal{M}_n^{t,k}$. If k is too large relative to t then a random $f \in \mathcal{M}_n^{t,k}$ will likely have $\Pr_{x \in U_n}[f(x) = 1] \approx 0$, and if k is too small relative to t then a random $f \in \mathcal{M}_n^{t,k}$ will likely have $\Pr_{x \in U_n}[f(x) = 1] \approx 1$; such functions are trivial to learn to high accuracy using either the constant-0 or constant-1 hypothesis. A straightforward analysis (see e.g. [JS06]) shows that for $k = \lfloor \log t \rfloor$ we have that $\mathbf{E}_{f \in \mathcal{M}_n^{t,k}}[\Pr_{x \in U_n}[f(x) = 1]]$

is bounded away from both 0 and 1, and thus we feel that this is an appealing and natural choice.

Probabilistic analysis. In this section we will establish various useful probabilistic lemmas regarding random monotone DNF of polynomially bounded size. **Assumptions.** Throughout the rest of Section 4 we assume that $t(n)$ is any function such that $n^{3/2} \leq t(n) \leq \text{poly}(n)$. To handle the case when $t(n) \leq n^{3/2}$, we will use the results from [JS06]. Let $a(n)$ be such that $t(n) = n^{a(n)}$. For brevity we write t for $t(n)$ and a for $a(n)$ below, but the reader should keep in mind that a actually denotes a function $\frac{3}{2} \leq a = a(n) \leq O(1)$. Because of space limitations all proofs are given in the full version.

The first lemma provides a bound of the sort needed by condition **C3** of Lemma 3:

Lemma 4. *Let $|S| = s = \lfloor a \rfloor + 2$. Fix any $C_{\mathscr{Y}} \in \mathscr{Y}$. Let $\delta_{\text{terms}} = n^{-\Omega(\log n)}$. With probability at least $1 - \delta_{\text{terms}}$ over the random draw of f from $\mathcal{M}_n^{t,k}$, we have that for some absolute constant c and all sufficiently large n,*

$$\prod_{U \in C_{\mathscr{Y}}} (\#g_U) \leq c \cdot \frac{t^{|C_{\mathscr{Y}}|-1}k^{2^s}}{\sqrt{n}}. \tag{5}$$

The following lemma shows that for f drawn from $\mathcal{M}_n^{t,k}$, with high probability each term is "uniquely satisfied" by a noticeable fraction of assignments as required by condition **C1**. (Note that since $k = O(\log n)$ and $t > n^{3/2}$, we have $\delta_{\text{usat}} = n^{-\Omega(\log\log n)}$ in the following.)

Lemma 5. *Let $\delta_{\text{usat}} := \exp(\frac{-tk}{3n}) + t^2(\frac{k}{n})^{\log\log t}$. For n sufficiently large, with probability at least $1 - \delta_{\text{usat}}$ over the random draw of $f = T_1 \vee \cdots \vee T_t$ from $\mathcal{M}_n^{t,k}$, f is such that for all $i = 1,\ldots,t$ we have $\Pr_x[T_i$ is satisfied by x but no other T_j is satisfied by $x] \geq \frac{\Theta(1)}{2^k}$.*

We now upper bound the probability that any j distinct terms of a random DNF $f \in \mathcal{M}_n^{t,k}$ will be satisfied simultaneously (condition **C2**). (In the following lemma, note that for $j = \Theta(1)$, since $t = n^{\Theta(1)}$ and $k = \Theta(\log n)$ we have that the quantity δ_{simult} is $n^{-\Theta(\log\log n)}$.)

Lemma 6. *Let $1 \leq j \leq 2^s$, and let $\delta_{\text{simult}} := \frac{t^j e^{jk-\log k}(jk-\log k)^{\log k}}{n^{\log k}}$. With probability at least $1 - \delta_{\text{simult}}$ over the random draw of $f = T_1 \vee \cdots \vee T_t$ from $\mathcal{M}_n^{t,k}$, for all $1 \leq \iota_1 < \cdots < \iota_j \leq t$ we have $\Pr[T_{\iota_1} \wedge \ldots \wedge T_{\iota_j}] \leq \beta_j$, where $\beta_j := \frac{k}{2^{jk}}$.*

Finally, the following lemma shows that for all sufficiently large n, with high probability over the choice of f, every set S of s variables appears in at most γ terms, where γ is independent of n (see condition **C4**).

Lemma 7. *Fix any constant $c > 0$. Let $s = \lfloor a \rfloor + 2$ and let $\gamma = a + c + 1$. Let $\delta_\gamma = n^{-c}$. Then for n sufficiently large, with probability at least $1 - \delta_\gamma$ over the random draw of f from $\mathcal{M}_n^{t,k}$, we have that every s-tuple of variables appears in at most γ terms of f.*

5   Proof of Theorem 1

Theorem 1 [Formally] *Let $t(n)$ be any function such that $t(n) \leq poly(n)$, let $a(n) = O(1)$ be such that $t(n) = n^{a(n)}$, and let $c > 0$ be any fixed constant. Then for any $n^{-c} < \delta < 1$ and $0 < \epsilon < 1$, $\mathcal{M}_n^{t(n), \lfloor \log t(n) \rfloor}$ is PAC learnable under U_n in $poly(n^{2a(n)+c+3}, (a(n) + c + 1)^{\log t(n)}, t(n), 1/\epsilon, \log 1/\delta)$ time.*

Proof. The result is proved for $t(n) \leq n^{3/2}$ already in [JS06], so we henceforth assume that $t(n) \geq n^{3/2}$. We use Theorem 2 and show that for $s = \lfloor a(n) \rfloor + 2$, random monotone $t(n)$-term DNFs, with probability at least $1 - \delta$, satisfy conditions **C1–C5** with values $\alpha, \beta_j, \Phi(\cdot), \Delta, \gamma$, and κ such that $\Delta > 0$ and the quantities $n^{s+\gamma}, 1/\Delta$, and γ^κ are polynomial in n. This will show that the extended version of Algorithm $\mathcal{A}$ defined in Section 3 PAC learns random monotone $t(n)$-term DNFs in time $poly(n, 1/\epsilon)$. Let $t = t(n)$ and $k = \lfloor \log t \rfloor$, and let f be drawn randomly from $\mathcal{M}_n^{t,k}$. By Lemmas 4–7, with probability at least $1 - \delta_{\text{usat}} - \delta_\gamma - 2^{2^s} \delta_{\text{terms}} - \delta_{\text{simult}}$, f will satisfy **C1–C5** with the following values:

$$\textbf{C1}\ \alpha > \frac{\Theta(1)}{2^k}; \quad \textbf{C2}\ \beta_j \leq \frac{k}{2^{jk}} \text{ for } 1 \leq j \leq 2^s;$$
$$\textbf{C3}\ \Phi(\mathcal{C}_\mathcal{Y}) \leq O(1) \frac{t^{|\mathcal{C}_\mathcal{Y}|-1} k^{2^s}}{\sqrt{n}} \text{ for all } \mathcal{C}_\mathcal{Y} \in \mathcal{Y}; \quad \textbf{C4}\ \gamma \leq a(n) + c + 1;$$
$$\textbf{C5}\ \kappa = k = \lfloor \log t \rfloor,$$

which gives us that $n^{s+\gamma} = n^{2a+c+3}$ and $\gamma^\kappa = (a + c + 1)^{\lfloor \log t \rfloor}$. Finally, we show that $\Delta = \Omega(1/t)$ so $1/\Delta$ is polynomial in n:

$$\Delta = \alpha/2^s - 3 \cdot \Upsilon = \frac{\Theta(1)}{t2^s} - 3 \sum_{\mathcal{C}_\mathcal{Y} \in \mathcal{Y}} 2^s \beta_{|\mathcal{C}_\mathcal{Y}|} \Phi(\mathcal{C}_\mathcal{Y})$$
$$\geq \frac{\Theta(1)}{t2^s} - \Theta(1) \sum_{\mathcal{C}_\mathcal{Y} \in \mathcal{Y}} 2^s \frac{k}{t^{|\mathcal{C}_\mathcal{Y}|}} \cdot \frac{t^{|\mathcal{C}_\mathcal{Y}|-1} k^{2^s}}{\sqrt{n}}$$
$$= \frac{\Theta(1)}{t2^s} - \frac{\Theta(1) k^{2^s+1}}{t\sqrt{n}} = \Omega(1/t).$$

6   Discussion

Robustness of parameter settings. Throughout Sections 4 and 5 we have assumed for simplicity that the term length k in our random t-term monotone DNF is exactly $\lfloor \log t \rfloor$. In fact, the results extend to a broader range of k's; one can straightforwardly verify that by very minor modifications of the given proofs, Theorem 1 holds for $\mathcal{M}_n^{t,k}$ for any $(\log t) - O(1) \leq k \leq O(\log t)$.

Relation to previous results. Our results are powerful enough to subsume some known "worst-case" results on learning restricted classes of monotone DNF formulas. Hancock and Mansour [HM91] have shown that read-k monotone DNF (in which each Boolean variable x_i occurs in at most k terms) are learnable under the uniform distribution in $poly(n)$ time for constant k. Their result extends an earlier result of Kearns *et al.* [KLV94] showing that read-once DNF (which can be assumed monotone without loss of generality) are polynomial-time learnable under the uniform distribution. It is not hard to see that (a very restricted special case of) our algorithm can be used to learn read-k monotone DNF in polynomial time; we give some details in the full version.

References

[AM02]   Amano, K., Maruoka, A.: On learning monotone boolean functions under the uniform distribution. In: Proc. 13th ALT, pp. 57–68 (2002)

[AP95]   Aizenstein, H., Pitt, L.: On the learnability of disjunctive normal form formulas. Machine Learning 19, 183–208 (1995)

[BBL98]   Blum, A., Burch, C., Langford, J.: On learning monotone boolean functions. In: Proc. 39th FOCS, pp. 408–415 (1998)

[BFJ+94]   Blum, A., Furst, M., Jackson, J., Kearns, M., Mansour, Y., Rudich, S.: Weakly learning DNF and characterizing statistical query learning using Fourier analysis. In: Proc. 26th STOC, pp. 253–262 (1994)

[Blu03a]   Blum, A.: Learning a function of r relevant variables (open problem). In: Proc. 16th COLT, pp. 731–733 (2003)

[Blu03b]   Blum, A.: Machine learning: a tour through some favorite results, directions, and open problems. In: FOCS 2003 tutorial slides (2003)

[BT96]   Bshouty, N., Tamon, C.: On the Fourier spectrum of monotone functions. Journal of the ACM 43(4), 747–770 (1996)

[HM91]   Hancock, T., Mansour, Y.: Learning monotone k-μ DNF formulas on product distributions. In: Proc. 4th COLT, pp. 179–193 (1991)

[Jac97]   Jackson, J.: An efficient membership-query algorithm for learning DNF with respect to the uniform distribution. JCSS 55, 414–440 (1997)

[JS05]   Jackson, J., Servedio, R.: Learning random log-depth decision trees under the uniform distribution. SICOMP 34(5), 1107–1128 (2005)

[JS06]   Jackson, J., Servedio, R.: On learning random DNF formulas under the uniform distribution. Theory of Computing 2(8), 147–172 (2006)

[JT97]   Jackson, J., Tamon, C.: Fourier analysis in machine learning. In: ICML/COLT 1997 tutorial slides (1997)

[KLV94]   Kearns, M., Li, M., Valiant, L.: Learning Boolean formulas. Journal of the ACM 41(6), 1298–1328 (1994)

[KMSP94]   Kučera, L., Marchetti-Spaccamela, A., Protassi, M.: On learning monotone DNF formulae under uniform distributions. Information and Computation 110, 84–95 (1994)

[KV94]   Kearns, M., Vazirani, U.: An introduction to computational learning theory. MIT Press, Cambridge (1994)

[Man95]   Mansour, Y.: An $O(n^{\log \log n})$ learning algorithm for DNF under the uniform distribution. JCSS 50, 543–550 (1995)

[MO03]   Mossel, E., O'Donnell, R.: On the noise sensitivity of monotone functions. Random Structures and Algorithms 23(3), 333–350 (2003)

[OS06]   O'Donnell, R., Servedio, R.: Learning monotone decision trees in polynomial time. In: Proc. 21st CCC, pp. 213–225 (2006)

[Sel08]   Sellie, L.: Learning Random Monotone DNF Under the Uniform Distribution. In: Proc. 21st COLT (to appear, 2008)

[Ser04]   Servedio, R.: On learning monotone DNF under product distributions. Information and Computation 193(1), 57–74 (2004)

[SM00]   Sakai, Y., Maruoka, A.: Learning monotone log-term DNF formulas under the uniform distribution. Theory of Computing Systems 33, 17–33 (2000)

[Val84]   Valiant, L.: A theory of the learnable. CACM 27(11), 1134–1142 (1984)

[Ver90]   Verbeurgt, K.: Learning DNF under the uniform distribution in quasi-polynomial time. In: Proc. 3rd COLT, pp. 314–326 (1990)

[Ver98]   Verbeurgt, K.: Learning sub-classes of monotone DNF on the uniform distribution. In: Proc. 9th ALT, pp. 385–399 (1998)

Breaking the ε-Soundness Bound of the Linearity Test over GF(2)

Tali Kaufman[1,*], Simon Litsyn[2,**], and Ning Xie[3,***]

[1] IAS, Princeton, NJ 08540, USA
kaufmant@mit.edu
[2] Department of Electrical Engineering-Systems,
Tel Aviv University, Tel Aviv 69978, Israel
litsyn@eng.tau.ac.il
[3] CSAIL, MIT, Cambridge, MA 02139, USA
ningxie@csail.mit.edu

Abstract. For Boolean functions that are ε-away from the set of linear functions, we study the lower bound on the rejection probability (denoted by $\text{REJ}(\epsilon)$) of the linearity test suggested by Blum, Luby and Rubinfeld. This problem is one of the most extensively studied problems in property testing of Boolean functions.

The previously best bounds for $\text{REJ}(\epsilon)$ were obtained by Bellare, Coppersmith, Håstad, Kiwi and Sudan. They used Fourier analysis to show that $\text{REJ}(\epsilon) \geq \epsilon$ for every $0 \leq \epsilon \leq \frac{1}{2}$. They also conjectured that this bound might not be tight for ε's that are close to 1/2. In this paper we show that this indeed is the case. Specifically, we improve the lower bound of $\text{REJ}(\epsilon) \geq \epsilon$ by an additive term that depends only on ε: $\text{REJ}(\epsilon) \geq \epsilon + \min\{1376\epsilon^3(1-2\epsilon)^{12}, \frac{1}{4}\epsilon(1-2\epsilon)^4\}$, for every $0 \leq \epsilon \leq \frac{1}{2}$. Our analysis is based on a relationship between $\text{REJ}(\epsilon)$ and the weight distribution of a coset of the Hadamard code. We use both Fourier analysis and coding theory tools to estimate this weight distribution.

1 Introduction

Property testing [22,12] studies the robust characterizations of various algebraic and combinatorial objects. It often leads to a new understanding of some well-studied problems and yields insight to other areas of computer science (see survey articles [11,21,23] for more on property testing). The first problem that was studied under the framework of property testing, as well as being one of the most extensively investigated property testing problems, is linearity testing. A function $f : \{0,1\}^m \to \{0,1\}$ is called *linear* if for all $x, y \in \{0,1\}^m$, $f(x) + f(y) = f(x+y)$. A function f is said to be ε-away from linear functions if one needs to change f's value on an ε-fraction of its domain to make f linear.

* Research supported in part by the NSF Awards CCF-0514167 and NSF-0729011.
** Research supported by ISF Grant 1177/06.
*** Research done while the author was at State Univ. of New York at Buffalo and visiting CSAIL, MIT. Research supported in part by NSF grant 0514771.

A. Goel et al. (Eds.): APPROX and RANDOM 2008, LNCS 5171, pp. 498–511, 2008.

Blum, Luby and Rubinfeld [9] considered the following randomized algorithm (henceforth referred to as the "BLR test") to test if a function is linear : Given a function $f : \{0,1\}^m \rightarrow \{0,1\}$, choose uniformly at random $x, y \in \{0,1\}^m$ and reject if $f(x)+f(y) \neq f(x+y)$. We call the probability of the test accepting linear functions the *completeness* of the test while the probability of rejecting nonlinear functions *soundness*. Note that in general, among other things, soundness depends on the distance parameter ϵ.

In retrospect, it is quite surprising that the analysis of such a natural test turned out to be far from simple. Much effort has been devoted to understanding the rejection probability behavior of the BLR test [9,3,6,4] due to its relation to the hardness of approximating some NP-hard problems [10,6,7,5]. Other line of works considered the optimal tradeoff between query complexity and soundness of some variants of the BLR test [29,27,25,14,26] and randomness needed for linearity tests over various groups [8,28]. Many generalizations and extension of the BLR test were also studied; for example, testing linear consistency among multiple functions [2], testing polynomials of higher degree or polynomials over larger fields (generalizing the linear case in the BLR test) [22,1,19,15,24], and testing Long Codes [5,13].

It is clear that the completeness of the BLR test is one, i.e., if f is linear, then the BLR test always accepts. The most important quantity for the BLR test (and for many other tests as well) is the soundness, since this parameter indicates how *robust* the test characterizes the objects being tested. The soundness analysis of the BLR test was found to be pretty involved. Indeed, various papers studied the following question: For every integer $m > 0$, real number $\epsilon \in [0, 1/2]$ and all Boolean functions $f : \{0,1\}^m \rightarrow \{0,1\}$ that are ϵ-away from linear functions, what is the minimum rejection probability of the BLR linearity test. We denote this lower bound by REJ(ϵ). That is, if we denote the probability that the BLR test rejects f by Rej(f) and denote the set of linear functions by LIN, then

$$\text{REJ}(\epsilon) \stackrel{\text{def}}{=} \min_{\text{dist}(f,\text{LIN})=\epsilon} \text{Rej}(f).$$

Understanding REJ(ϵ) is important not only because its relation to the hardness of approximating some NP-hard problems but also due to the fact that it is a natural and fundamental combinatorial problem. The hardest cases are those where $\frac{1}{4} \leq \epsilon < \frac{1}{2}$.

In this paper, by combining Fourier analysis and coding theoretic tools, we improve the previously best known bound of REJ(ϵ) by an additive term depending only on ϵ for all $\epsilon \in [1/3, 1/2)$. Our result shows that the celebrated Fourier analysis based soundness bound [4], REJ$(\epsilon) \geq \epsilon$, is suboptimal by an additive term that depends only on ϵ for *all* $\epsilon \in (0, \frac{1}{2})$. That is, for every constant $\epsilon \in [\frac{1}{4}, \frac{1}{2})$, there exists a constant $\delta(\epsilon) > 0$ that is independent of m such that REJ$(\epsilon) \geq \epsilon + \delta$.

A key ingredient of our proof is viewing the Fourier coefficients in terms of the weight distributions of codewords and applying coding bounds to them. It is hoped that techniques developed in coding theory may find other places to improve results on Boolean functions obtained by Fourier analysis.

1.1   Related Research

Blum, Luby and Rubinfeld [9] first suggested the BLR linearity test and showed that for every ϵ, $\text{REJ}(\epsilon) \geq \frac{2}{9}\epsilon$ based on a self-correction approach. Using a combinatorial argument, Bellare et al. [6] proved that $\text{REJ}(\epsilon) \geq 3\epsilon - 6\epsilon^2$. This bound is optimal for small ϵ but is very weak for ϵ's that are close to $\frac{1}{2}$. Bellare and Sudan [7] further showed that $\text{REJ}(\epsilon) \geq \frac{2}{3}\epsilon$ when $\epsilon \leq \frac{1}{3}$ and $\text{REJ}(\epsilon) \geq \frac{2}{9}$ when $\epsilon > \frac{1}{3}$. All these mentioned results hold over general fields. This series of works culminated in [4], where Fourier transform techniques found their first use in PCP-related analysis. The results obtained by [4] hold for binary field and they are the following.

$$
\text{REJ}(\epsilon) \geq \begin{cases} 3\epsilon - 6\epsilon^2 & 0 \leq \epsilon \leq \frac{5}{16}; \\ \frac{45}{128} & \frac{5}{16} \leq \epsilon \leq \frac{45}{128}; \\ \epsilon & \frac{45}{128} \leq \epsilon < \frac{1}{2} \end{cases}
$$

The results of [4] show that the bounds are tight for $\epsilon \leq \frac{5}{16}$. Numerical simulation results of [4] suggested that the lower bound $\text{REJ}(\epsilon) \geq \epsilon$ for $\epsilon > \frac{5}{16}$ may be improved, but not by too much. Kiwi [16] and Kaufman and Litsyn [17] gave alternative proofs for the fact that $\text{REJ}(\epsilon) \geq \epsilon$ for every ϵ (up to an additive term of $O(\frac{1}{2^m})$). Their proofs are more coding theory oriented. Specifically, the proofs are based on studying the weight distribution of the Hadamard code and its ϵ-away coset as well as various properties of Krawtchouk polynomials.

1.2   The Main Result

In the following, we present our main result showing an improved bound for $\text{REJ}(\epsilon)$. Specifically, we prove

Theorem 1. *Let* $\Delta(\gamma) = \frac{5\gamma}{8} - \frac{\gamma^2}{32}$. *For all* ϵ, $1/4 \leq \epsilon \leq 1/2$ *and for all* γ, $0 < \gamma \leq 1$,

$$
\text{REJ}(\epsilon) \geq \epsilon + \min\{4096(1 - \Delta(\gamma))^3 \epsilon^3 (1 - 2\epsilon)^{12}, \frac{\gamma}{2}\epsilon(1 - 2\epsilon)^4\}.
$$

As a simple corollary by plugging in $\gamma = 1/2$ and combining our new result with known bounds for $0 \leq \epsilon \leq \frac{1}{4}$ (i.e., $\text{REJ}(\epsilon) \geq 3\epsilon - 6\epsilon^2$), we get

Corollary 2. *For all* ϵ, $0 \leq \epsilon \leq 1/2$,

$$
\text{REJ}(\epsilon) \geq \epsilon + \min\{1376\epsilon^3(1 - 2\epsilon)^{12}, \frac{1}{4}\epsilon(1 - 2\epsilon)^4\}.
$$

Note that for every constant $\epsilon \in [\frac{1}{4}, \frac{1}{2})$, Theorem 1 improves upon $\text{REJ}(\epsilon) \geq \epsilon$ by an additive *constant*. Our result improves over all previously known bounds for every $\epsilon \in [\frac{45}{128}, \frac{1}{2})$, but only by a very small quantity. For example, for $\epsilon = 0.4$, our improvement of REJ(ϵ) is about 1.024×10^{-7}. We believe our bound can be further improved systematically (we remark that our current approach already gives bounds better than that stated in the Main Theorem for ϵ's such that $1/(1-2\epsilon)^2$ are far from integers). However, as the numerical results shown in [4], one can not expect to see too much improvement over REJ(ϵ) $\geq \epsilon$. Our improvement over REJ(ϵ) $\geq \epsilon$ vanishes at $\epsilon = \frac{1}{2}$. This is indeed as expected since we know that REJ($\frac{1}{2}$) $= \frac{1}{2}$.

1.3   Proof Overview

The proof has three key ingredients. We will use C to denote the Hadamard code of block length $n = 2^m$ whose codewords are exactly the set of all linear functions.

The coset code $C + v$. There are two equivalent ways of viewing the BLR test: one is to think f as a Boolean function mapping $\{0,1\}^m$ to $\{0,1\}$ and the BLR test simply picks x and y uniformly at random and check if $f(x) + f(y) = f(x + y)$. This functional viewpoint leads naturally to the beautiful Fourier analysis approach of [4], which shows that $1 - 2$REJ(ϵ) can be exactly expressed as a cubic sum of Fourier coefficients of the function $(-1)^f$. Another way to study the BLR test, first suggested in [16] and followed by [17], is to treat f as a vector v of length n with $n = 2^m$. (Due to this fact, from now on, we will use vector v and function f interchangeably.) Since the set of linear functions may be viewed as the set of codewords of the Hadamard code C, the BLR test can be viewed as picking a random weight-3 codeword from $C^\perp$ (which denotes the dual code of C) and check if it is orthogonal to v. We combine these two viewpoints together by reinterpreting the Fourier analytic result in the coding theoretic setting. Our simple but important observation is that the Fourier coefficients of f are equivalent to the weights of the codewords in a *coset* of C. Therefore $1 - 2$REJ(ϵ) can be expressed as a simple function of the weight distribution of the code $C + v$. That is, $1 - 2$REJ(ϵ) can be written as a normalized sum of cubes $\sum_i x_i^3$, each x_i is the weight of a codeword in $C + v$, where $C + v$ is an ϵ-away coset[1] of the Hadamard code C.

Maximization Problem. In order to obtain a lower bound on REJ(ϵ), we need to obtain an upper bound on function that involves the weight distribution of $C + v$. To this end, we reformulate our problem as a *Maximal Sum of Cubes Problem*, in which we look for an upper bound on the sum of cubes of a set of integers under certain constraints. The bound REJ(ϵ) $= \epsilon$ obtained by [4] corresponds

---

[1] To make this clear, we remind the reader that the weight distribution of a code C is a set of integers that represent the numbers of codewords in C of different weights, where the weight of a codeword is the number of coordinates at which the codeword is non-zero. A vector v is ϵ-away from a code C if one needs to change an ϵ-fraction of v's bits to make it belong to C. An ϵ-away coset of C is obtained by adding a vector v to every codeword in C, where v is ϵ-away from C.

to the simple optimal configuration in which all the codewords of $C + v$ are of weight $\frac{1}{2}n$ except a constant number $(\frac{1}{(1-2\epsilon)^2})$ of them are of weight ϵn (one can use coding theory argument to show that there can't be more than $\frac{1}{(1-2\epsilon)^2}$ codewords of weight ϵn). Moreover, this is the unique configuration that meets the bound $\text{REJ}(\epsilon) = \epsilon$. Any deviation from the optimal configuration implies an improved lower bound on $\text{REJ}(\epsilon)$. Our strategy thus is to show that this optimal weight distribution is not achievable for $C + v$ due to some special properties of the code $C + v$. In particular, we will focus on the following two ways in which the optimal configuration may break down:

1. There exists a codeword of weight larger than $\frac{n}{2}$ in $C + v$.
2. The number of codewords in $C + v$ of weight at most $(\epsilon + \eta)n$ is less than $\frac{1}{(1-2\epsilon)^2}$, for some positive number η.

A natural tool to show that one of the above properties holds is the well-known Johnson Bound. Roughly speaking, the Johnson bound offers a bound on the maximum number of codewords of a specific weight in a code with some specific minimum distance. However, it turns out that Johnson bound does *allow* the optimal configuration for the code $C + v$ (which yields $\text{REJ}(\epsilon) = \epsilon$ as discussed above), so we fail to get any improvement by applying it directly to $C + v$. The way we overcome this is by considering a new code $C|_\mathcal{V}$ of *shorter block length* and applying to it a slightly stronger variant of the commonly used Johnson bound (a variant which enables us to bound the number of codewords of *at least* (or *at most*) a specific weight). The possible switch from the code $C + v$ to the code $C|_\mathcal{V}$ turns out to be crucial in our analysis.

From the code $C + v$ to the code $C|_\mathcal{V}$. We consider the code $C|_\mathcal{V}$ of block length $n' = \epsilon n$, obtained from C by restricting it to the ϵn non-zero coordinates of v. This code is a linear code. It has the same number of codewords as the original code $C + v$. More precisely, we show that if it contains fewer codewords then an improved lower bound on $\text{REJ}(\epsilon)$ is immediate. A nice property of this new code is that there is a one-to-one correspondence between the weight of a codeword in $C|_\mathcal{V}$ and the weight of the corresponding codeword in $C + v$. Since $C|_\mathcal{V}$ is a linear code, its minimum distance equals the minimum weight of its codewords. If this minimum weight is small, then by the one-to-one relation between the weights of $C + v$ and that of $C|_\mathcal{V}$, the heaviest codeword in $C + v$ will have a large weight, which yields an improved lower bound for $\text{REJ}(\epsilon)$ according to Condition 1 from above. However, if the maximum weight of $C + v$ is small, or equivalently, the minimum distance of $C|_\mathcal{V}$ is large, then by applying the Johnson bound to $C|_\mathcal{V}$, we get that the number of codewords lying between weight ϵn and $(\epsilon + \eta)n$ in $C + v$ is less than the optimal bound $(\frac{1}{(1-2\epsilon)^2})$, which also yields an improved lower bound for $\text{REJ}(\epsilon)$ by Condition 2 mentioned before.

The intuitive reason that we benefit from applying the Johnson bound to $C|_\mathcal{V}$ rather than to $C + v$ is straightforward: The block length of $C|_\mathcal{V}$ is much smaller than the block length of $C + v$, but the number of codewords in $C|_{\mathcal{V}}$ is the same as $C + v$.[2]

The relations between the three codes in consideration, namely C, $C + v$, and $C|_{\mathcal{V}}$ (for a code C and a vector v that is ϵ-away from C), as well as the idea of looking at a restricted code of smaller block length in order to get better coding bounds, might have other applications.

1.4   Organization

Section 2 introduces necessary notation and definitions. In section 3 we show that, for every f that is ϵ-away from linear, $\mathrm{Rej}(f)$ can be expressed as a function of the weight distribution of a coset of the Hadamard code. Then we reformulate the problem of lower bounding $\mathrm{REJ}(\epsilon)$ as a maximization problem in section 4. In section 5 we study the weight distribution of a restricted code of the coset code and then provide a proof outline of the Main Theorem in section 6. All the missing proofs may be found in the full version of the paper [18].

2   Preliminaries

We will use $[n]$ to denote the set $\{1, \ldots, n\}$, where n is a positive integer. Let v be a vector in $\{0, 1\}^n$. We use $v(i)$ to denote the ith bit of v for every $1 \le i \le n$. The weight of v, denoted $\mathrm{wt}(v)$ is the number of non-zero bits in v. A code C of block length n is a subset of $\{0, 1\}^n$. C is called a *linear code* if C is a linear subspace. Let $u, v \in \{0, 1\}^n$. The *distance* between u and v is defined to be the number of bits at which they disagree: $\mathrm{dist}(u, v) = |\{i \in [n] | u(i) \ne v(i)\}| = \mathrm{wt}(u - v)$. The minimum distance of a code C is $\min_{u,v \in C} \mathrm{dist}(u, v)$. If C is a linear code, then the minimum distance of C equals the minimum weight of codewords in C. Let C be a code of block length n. The distance of $v \in \{0, 1\}^n$ from code C is the minimum distances between v and codewords in C, i.e., $\mathrm{dist}(v, C) \overset{\mathrm{def}}{=} \min_{c \in C} \mathrm{dist}(v, c)$. With an abuse of notation, in the following, we will use C to denote the Hadamard code and $C^\perp$ to denote its dual Hamming code.

Recall that a function $\ell : \{0, 1\}^m \to \{0, 1\}$ is *linear* if for all $x, y \in \{0, 1\}^m$, $\ell(x) + \ell(y) = \ell(x + y)$. An equivalent characterization is: ℓ is linear if and only if $\ell(x) = \alpha \cdot x = \sum_i^m \alpha_i x_i$ for some $\alpha \in \{0, 1\}^m$, and we denote such a

---

[2] The reason we are able to improve the bound $\mathrm{REJ}(\epsilon) \ge \epsilon$ by a constant is more subtle: For $\frac{1}{4} \le \epsilon \le \frac{1}{2}$, there is a "reciprocal" relationship between the *relative weights* of codeword in C and corresponding codeword in $C|_{\mathcal{V}}$; that is, the smaller the relative weight in C, the larger the relative weight in $C|_{\mathcal{V}}$, and vice versa. Note that the denominator of the expression in Johnson bound is $\frac{d}{n} - 2\frac{w}{n}(1 - \frac{w}{n})$ after dividing by n^2. Therefore Johnson bound will give better bounds when $\frac{w}{n}(1 - \frac{w}{n})$ gets smaller, or, when w/n is very close to either 0 or 1. By switching from C to $C|_{\mathcal{V}}$, $\frac{w}{n}$ is mapped to $\frac{w'}{n'}$. The advantage of changing to $C|_{\mathcal{V}}$ is that it makes the distance between $\frac{w'}{n'}$ and 1 smaller than the distance between $\frac{w}{n}$ and zero. This advantage disappears at $\epsilon = 1/2$, therefore we get no improvement at that point, as expected.

linear function by ℓ_α and denote the set of all such functions by LIN. Let $f, g : \{0,1\}^m \to \{0,1\}$. The (relative) distance between f and g is defined to be the fraction of points at which they disagree: $\text{dist}(f,g) \overset{\text{def}}{=} \Pr_{x \in \{0,1\}^m}[f(x) \neq g(x)]$. The distance between a function f and linear functions is the minimum distance between f and any linear function: $\text{dist}(f, \text{LIN}) \overset{\text{def}}{=} \min_{g \in \text{LIN}} \text{dist}(f,g)$. A function f is said to be ϵ-*away* from linear functions if its distance from linear functions is ϵ, and is said to ϵ-*far* from linear functions if the distance is at least ϵ.

Next we introduce some basic notions in Fourier analysis. We will focus on functions defined over the Boolean cube. Note that the set of functions $f : \{0,1\}^m \to \mathbb{R}$ forms a vector space of dimension 2^m. A convenient orthonormal basis for this vector space is the following functions called characters: $\psi_\alpha(x) = (-1)^{\alpha \cdot x}$, where $\alpha \in \{0,1\}^m$. Consequently, any $f(x) : \{0,1\}^m \to \mathbb{R}$ can be expanded as

$$f(x) = \sum_\alpha \hat{f}_\alpha \psi_\alpha(x),$$

where $\hat{f}_\alpha = \langle f, \psi_\alpha \rangle \overset{\text{def}}{=} \frac{1}{2^m} \sum_{x \in \{0,1\}^m} f(x) \psi_\alpha(x)$ is called the α-th Fourier coefficient of f. Define $h(x) = (-1)^{f(x)}$. Note that the range of $h(x)$ is $\{-1,1\}$.

One can encode f as an $n = 2^m$ bit codeword $v \in \{0,1\}^n$ by enumerating all its values on the Boolean cube. The same encoding applied to the set of linear functions $\{\ell_\alpha\}$ gives rise to the Hadamard code C, in which we denote the codeword corresponding to ℓ_α by c_α.

We are going to use the following two elementary inequality in our analysis. We omit the proofs of these inequalities here and the interested reader may find them in the full version of this paper [18].

Lemma 3. *For all real y with $0 \leq y \leq 1/2$,*

$$\frac{1}{1-y} - y \geq \frac{1}{\sqrt{1-2y^2}}.$$

Lemma 4. *Let γ be a constant with $0 \leq \gamma \leq 1$. Then for all real y with $0 \leq y \leq 1/2$,*

$$\frac{1}{(1-y)^2} - \frac{1}{1-2y^2} - \gamma \frac{y}{1-y} \geq (8 - 5\gamma)y^2.$$

3   The Coset Code $C + v$

Using Fourier analytic tools, Bellare et al. proved the following result in their seminal paper.

Lemma 5 ([4]). *Let $f : \{0,1\}^m \to \{0,1\}$ and $h(x) = (-1)^{f(x)}$. Then (recall that Rej(f) is the probability that BLR test rejects f)*

$$Rej(f) = \frac{1}{2}\left(1 - \sum_{\alpha \in \{0,1\}^m} \hat{h}_\alpha^3\right).$$

Sometimes reformulating a Boolean function problem as a coding theoretic problem offers new perspectives. To this end, we need to introduce the standard notion of coset codes. Let C be a linear code of block length n and let $v \in \{0,1\}^n$ such that $v \notin C$, the v-*coset* of C is $C+v \overset{\text{def}}{=} \{c+v | c \in C\}$. Note that $|C+v| = |C|$. The *weight distribution* or *spectrum* of C is $B^C = (B_0^C, B_1^C, \cdots, B_n^C)$, where $B_i^C = |\{c \in C | \mathrm{wt}(c) = i\}|$.

Now we switch the viewpoint from Boolean functions to vectors in the Boolean cube. That is, we transform Boolean function $f : \{0,1\}^m \to \{0,1\}$ into a vector $v \in \{0,1\}^n$ by evaluating f on every points in the Boolean cube. Since f and v are equivalent, we use $\mathrm{Rej}(v)$ to denote the BLR rejection probability of the Boolean function that corresponds to v. Using the relation between linear functions and Hadamard code, we have the following coding theoretic formula for $\mathrm{Rej}(v)$ (and therefore $\mathrm{Rej}(f)$):

Lemma 6. *Let $v \in \{0,1\}^n$, then*

$$Rej(v) = \frac{1}{2}\left(1 - \frac{1}{n^3}\sum_{i=0}^{n} B_i^{C+v}(n-2i)^3\right).$$

Proof. By the definition of Fourier coefficient,

$$\hat{h}_\alpha = \langle h, \psi_\alpha\rangle = \langle (-1)^f, (-1)^{\ell_\alpha}\rangle = \langle (-1)^v, (-1)^{c_\alpha}\rangle = \frac{1}{2^m}\sum_{x\in\{0,1\}^m}(-1)^{v(x)+c_\alpha(x)}$$

$$= \Pr_x[v(x)=c_\alpha(x)] - \Pr_x[v(x)\neq c_\alpha(x)] = 1 - \frac{2\mathrm{dist}(v,c_\alpha)}{n} = \frac{n-2\mathrm{wt}(v+c_\alpha)}{n},$$

where in the last step we use the fact that, for binary vectors u and v, $\mathrm{dist}(u,v) = \mathrm{wt}(u-v) = \mathrm{wt}(u+v)$. Lemma 5 now gives

$$\mathrm{Rej}(v) = \frac{1}{2}\left(1 - \sum_{\alpha\in\{0,1\}^m}\hat{h}_\alpha^3\right) = \frac{1}{2}\left(1 - \sum_{\alpha\in\{0,1\}^m}\frac{(n-2\mathrm{wt}(v+c_\alpha))^3}{n^3}\right)$$

$$= \frac{1}{2}\left(1 - \sum_{c\in C}\frac{(n-2\mathrm{wt}(v+c))^3}{n^3}\right)$$

$$= \frac{1}{2}\left(1 - \frac{\sum_{i=0}^{n} B_i^{C+v}(n-2i)^3}{n^3}\right),$$

where in the final step we change summation over codewords in C to summation over weights of the codewords in $C + v$. ☐

This relation between the Fourier coefficients of $(-1)^f$ and the weight distribution of coset code $C + v$ seems to be new and may find applications in other places.

Since $\mathrm{Rej}(v)$ is now expressed as a weight distribution of the coset code $C+v$, our next step is to study how the codewords in $C + v$ are distributed so that to make the rejection probability minimum.

4   Maximization Problem

Note that we can rewrite Lemma 6 as

$$\text{Rej}(v) = \frac{1}{2} - \frac{\sum_{c_i \in C} (n - 2\text{wt}(v + c_i))^3}{2n^3} = \frac{1}{2} - \frac{1}{2n^3} \sum_{i=0}^{n-1} x_i^3,$$

where $x_i = n - 2\text{wt}(c_i + v)$, for $c_i \in C$, $0 \leq i \leq n - 1$. In order to prove that $\text{REJ}(\epsilon) \geq \epsilon + \delta$, all we need to show is, for *every* f that is ϵ-away from linear functions, $\text{Rej}(f) \geq \epsilon + \delta$. Hence our goal of getting a better *lower bound* than ϵ for $\text{REJ}(\epsilon)$ is equivalent to, for every vector v with $\text{dist}(v, C) = \epsilon n$, getting a better *upper bound* than $1 - 2\epsilon$ for $\frac{1}{n^3} \sum_{i=0}^{n-1} x_i^3$. This observation motivates the following measure of improvement (gain) and to reformulate the problem of lower bounding $\text{REJ}(\epsilon)$ as a Maximal Sum of Cubes Problem.

Definition 7. *Let $x_i = n - 2wt(c_i + v)$, for $c_i \in C$, $0 \leq i \leq n - 1$. Define*

$$\text{GAIN}(v) = \frac{1}{n^3} \left((1 - 2\epsilon)n^3 - \sum_{i=0}^{n-1} x_i^3 \right).$$

Consequently, if $\text{dist}(v, C) = \epsilon n$, then $\text{Rej}(v) = \epsilon + \frac{1}{2}\text{GAIN}(v)$.

Since v is ϵ-away from C, it follows that $x_i \leq (1 - 2\epsilon)n$, for all $0 \leq i \leq n - 1$. We further observe another constraint on the set of integers $\{x_0, x_1, \ldots, x_{n-1}\}$ is that their Euclidean norm is n^2.

Claim 8. $\sum_{i=0}^{n-1} x_i^2 = n^2$.

This claim follows directly from the Parseval's equality. An alternative proof, based on the norm-preserving property of the Hadamard matrix, was given in [17].

As we will show in the next lemma, if these two constraints are the only constraints on $\{x_0, x_1, \ldots, x_{n-1}\}$, then the bound $\text{REJ}(\epsilon) \geq \epsilon$ is essentially optimal. However, as we will see in the next section, since the x_i's are related to the weight distribution of $C + v$, the properties of the code $C + v$ impose more constraints on x_i's, thus making this optimal bound unattainable.

Lemma 9. *Consider the following Maximal Sum of Cubes Problem: Let $0 < \alpha \leq 1$ be a constant and n be a large enough integer. For a set of n integers $x_0, x_1, \cdots, x_{n-1}$, find the maximum of $x_0^3 + x_1^3 + \cdots + x_{n-1}^3$ under the constraints:*

$$x_0^2 + x_1^2 + \cdots + x_{n-1}^2 = n^2$$

$$\forall i : x_i \leq \alpha n.$$

The maximum is achieved at the following optimal configuration[3]*: $\frac{1}{\alpha^2}$ of the x_i's are assigned the maximum value αn, and the rest are assigned the value zero. The maximum thus obtained is αn^3.*

---

[3] Another requirement necessary to attain the optimal bound is that $\frac{1}{\alpha^2}$ is an integer. Therefore we already see some improvement upon $\text{REJ}(\epsilon) \geq \epsilon$ without any further calculation for all ϵ such that $\frac{1}{(1-2\epsilon)^2}$ is not an integer.

Note that in our setting $x_i = n - 2\mathrm{wt}(c_i + v)$ so $\alpha = 1 - 2\epsilon$ and consequently $\sum_{i=0}^{n-1} x_i^3 \leq (1 - 2\epsilon)n^3$.

Proof. First note that, without loss of generality, we may assume that each x_i is non-negative[4]. Then we have

$$\frac{(\sum_{j=0}^{n-1} x_j^3)^{1/3}}{(\sum_{j=0}^{n-1} x_j^2)^{1/2}} = (\sum_{j=0}^{n-1} \frac{x_j^3}{(\sum_{i=0}^{n-1} x_i^2)^{3/2}})^{1/3} = (\sum_{j=0}^{n-1} (\frac{x_j^2}{\sum_{i=0}^{n-1} x_i^2})^{3/2})^{1/3}.$$

Since $0 \leq x_j^2 \leq \alpha^2 n^2$ for every j, $(\frac{x_j^2}{\sum_{i=0}^{n-1} x_i^2})^{3/2} \leq \alpha \frac{x_j^2}{\sum_{i=0}^{n-1} x_i^2}$. This gives

$$\frac{(\sum_{j=0}^{n-1} x_j^3)^{1/3}}{(\sum_{j=0}^{n-1} x_j^2)^{1/2}} \leq (\alpha \sum_{j=0}^{n-1} \frac{x_j^2}{\sum_{i=0}^{n-1} x_i^2})^{1/3} = \alpha^{1/3}.$$

Moreover, the equality is attained only if all of the values of x_i are either zero or αn. This is possible only if $\frac{1}{\alpha^2}$ of the x_i's equal αn, and the rest are zeros. In that case $x_0^3 + x_1^3 + \cdots + x_{n-1}^3 = \alpha n^3$. □

We will employ the following two lemmas on $\mathrm{GAIN}(v)$ to obtain improvement.

Lemma 10. *If there exists an x_i such that $x_i = -\delta n$ for some $\delta > 0$, then $\mathrm{GAIN}(v) \geq \min\{2\delta^3, 2\alpha^3\}$.*

Proof. We first consider the case that $\delta \leq \alpha$. Note that $\{x_0, \ldots, x_{i-1}, -x_i, x_{i+1}, \ldots, x_{n-1}\}$ satisfies all the constraints in the Maximal Sum of Cubes Problem, so we have

$$\alpha n^3 \geq \sum_{k=0, k \neq i}^{n-1} x_k^3 + (-x_i)^3 = \sum_{k=0}^{n-1} x_k^3 + 2|x_i|^3 = \sum_{k=0}^{n-1} x_k^3 + 2(\delta n)^3.$$

Now if $\delta > \alpha$, we may assume, without loss of generality, that $x_0 = -\delta n$. Note that $\sum_{i=1}^{n-1} x_i^2 = (1 - \delta^2)n^2$, so if we apply Lemma 9 to $x_1, \ldots, x_{n-1}$, we have $\sum_{i=1}^{n-1} x_i^3 \leq \alpha(1 - \delta^2)n^3$. Therefore,

$$\mathrm{GAIN}(v) = \frac{1}{n^3}(\alpha n^3 - \sum_{i=0}^{n-1} x_i^3) \geq \alpha\delta^2 + \delta^3 \geq 2\alpha^3.$$

□

Lemma 11. *Let $\eta > 0$. If the number of x_i's such that $x_i \geq (\alpha - \eta)n$ is at most $\lfloor \frac{1}{\alpha^2} \rfloor - 1$, then $\mathrm{GAIN}(v) \geq \alpha^2 \eta$.*

---

[4] Otherwise, we can do the following to keep $\sum x_i^2 = n^2$ while increasing $\sum x_i^3$. For $-\alpha n \leq x_i < 0$, replace x_i with $-x_i$; for $x_i < -\alpha n$, replace x_i with several smaller positive integers with the same L_2-norm. See the proof of Lemma 10 below for more details.

Proof. Set $M = \lfloor \frac{1}{\alpha^2} \rfloor$. Let $\{y_1, \ldots, y_n\}$ be a permutation of $\{x_0, \ldots, x_{n-1}\}$ such that $\alpha n \geq y_1 \geq \cdots \geq y_n$. We have $y_1^2 + \cdots + y_n^2 = n^2$ and $y_M \leq (\alpha - \eta)n$. Define T to be: $T = y_1^2 + \cdots + y_{M-1}^2$. Then we have $T \leq (M-1)(\alpha n)^2 \leq (\frac{1}{\alpha^2} - 1)\alpha^2 n^2$, and $y_M^2 + \cdots + y_n^2 = n^2 - T$. Therefore,

$$\sum_{i=0}^{n-1} x_i^3 = \sum_{i=1}^{n} y_i^3 \leq \left(\sum_{i=1}^{M-1} y_i^2\right)\alpha n + \left(\sum_{i=M}^{n} y_i^2\right)(\alpha - \eta)n = n^2(\alpha - \eta)n + \eta n T$$

$$\leq n^2(\alpha - \eta)n + \eta n\left(\frac{1}{\alpha^2} - 1\right)\alpha^2 n^2 = \alpha n^3 - \alpha^2 \eta n^3.$$ □

5 From the Code $C + v$ to the Code $C|_\mathcal{V}$

We denote by $\mathcal{V}$ the set of coordinates at which v is non-zero, i.e., $\mathcal{V} = \{i \in [n] | v(i) = 1\}$. Note that $|\mathcal{V}| = \text{wt}(v)$. In the following we consider a code $C|_\mathcal{V}$ which will enable us to get some insight into the weight distribution of the code $C + v$.

First observe that, since we are only interested in the weight distribution of $C + v$, without loss of generality, we may assume that $\text{wt}(v) = \epsilon n$. To see this, suppose that $c_v \in C$ is the closest codeword to v (if there are more than one such codeword, then we may pick one arbitrarily). Since $\text{dist}(v, C) = \epsilon n$, v can be written as $v = c_v + v_{\epsilon n}$, with $\text{wt}(v_{\epsilon n}) = \epsilon n$. Since C is a linear code, $C + v = \{c + v | c \in C\} = \{c + c_v + v_{\epsilon n} | c \in C\} = \{c' + v_{\epsilon n} | c' \in C\} = C + v_{\epsilon n}$, where $c' \stackrel{\text{def}}{=} c + c_v$.

Definition 12. *Let C be a code of block length n and $v \in \{0,1\}^n$ be a vector of weight ϵn. We define the code $C|_\mathcal{V}$ of block length ϵn to be the code obtained by restricting code C to the non-zero coordinates of v. For convenience of notation, we will use $D = C|_\mathcal{V}$ from now on.*

The following lemma shows that there is a one-to-one correspondence between the weight of $c_i + v$ and the weight of the corresponding codeword in D.

Lemma 13. *For $0 \leq i \leq n-1$, let c_i be the ith codeword in the Hadamard code C and $d_i \in D$ be the restriction of c_i to coordinates in $\mathcal{V}$. Let $x_i = n - 2\text{wt}(c_i + v)$, then*

$$x_i = \begin{cases} (1 - 2\epsilon)n, & \text{if } i = 0, \\ 4wt(d_i) - 2\epsilon n, & \text{otherwise.} \end{cases}$$

Proof. For $i = 0$, $\text{wt}(c_0 + v) = \text{wt}(v) = \epsilon n$, hence $x_0 = (1 - 2\epsilon)n$. Since C is a Hadamard code, for all $i > 0$, $\text{wt}(c_i) = n/2$, i.e., there are $n/2$ ones and $n/2$ zeros in each codeword. For each $c_i \in C$, since there are $\text{wt}(d_i)$ ones in $\mathcal{V}$, there are $n/2 - \text{wt}(d_i)$ ones in $[n] \setminus \mathcal{V}$; this also holds for $c_i + v$, since v does not flip the bits at these coordinates. Since $|v| = \epsilon n$, there are $\epsilon n - \text{wt}(d_i)$ zeros in $\mathcal{V}$ for c_i, therefore there are $\epsilon n - \text{wt}(d_i)$ ones in $\mathcal{V}$ for $c_i + v$. It follows that $\text{wt}(c_i + v) = n/2 - \text{wt}(d_i) + \epsilon n - \text{wt}(d_i) = (1/2 + \epsilon)n - 2\text{wt}(d_i)$ and $x_i = 4\text{wt}(d_i) - 2\epsilon n$. □

Lemma 14. *Either D is a linear code or* GAIN$(v) \geq 2(1 - 2\epsilon)^3$.

Proof. Since D is a restriction of linear code C, D is a linear code if and only if all the codewords d_i in D are distinct. If D is not a linear code, then there exist $i \neq j$ such that $d_i = d_j$. This implies that there is a $k \neq 0$ such that $d_k = \mathbf{0}$. By Lemma 13, $x_k = -2\epsilon n$. Since $2\epsilon \geq 1 - 2\epsilon$, by Lemma 10, GAIN$(v) \geq 2(1 - 2\epsilon)^3$. □

Since $2(1 - 2\epsilon)^3$ is always larger than the gain we are going to prove, from now on, we will focus on the case that D is a linear code. Let $n' = \epsilon n$ be the block length of D, and d be the minimum distance of D. Note that D contains n codewords. The following simple bound is useful.

Theorem 15 (Plotkin bound [20]). *Let C be a binary code of block length n and minimum distance d. If $d \geq n/2$, then C has at most $2n$ codewords.*

Now we have

Claim 16. $d < n'/2$.

Proof. Suppose $d \geq n'/2$, then by the Plotkin bound stated in Theorem 15, D has at most $2n' = 2\epsilon n < n$ codewords, a contradiction. □

6   Proof Outline of the Main Theorem

In this section, we give a proof outline of the main theorem. Recall that our main theorem is the following.

Theorem 1 (Main Theorem). *Let $\Delta(\gamma) = \frac{5\gamma}{8} - \frac{\gamma^2}{32}$. For all ϵ, $1/4 \leq \epsilon \leq 1/2$ and for all γ, $0 < \gamma \leq 1$,*

$$\mathrm{REJ}(\epsilon) \geq \epsilon + \min\{4096(1 - \Delta(\gamma))^3 \epsilon^3 (1 - 2\epsilon)^{12}, \frac{\gamma}{2}\epsilon(1 - 2\epsilon)^4\}.$$

Our proof will rely on the following basic coding theorem which bounds the number of codewords of weight at least w. This is a slightly stronger variant of the well-known Johnson bound, for a proof see, e.g., the Appendix in [5].

Theorem 17 (Johnson bound). *Let C be a binary code of block length n and minimum distance d. Let $B'(n, d, w)$ denote the maximum number of codewords in C of weight at least w, then $B'(n, d, w) \leq \frac{nd}{nd - 2w(n-w)}$, provided that $nd > 2w(n - w)$.*

The basic idea of the proof is the following. Since there is a one-to-one correspondence between the weight of codeword in $C + v$ and that of D, we will be working with the spectrum of D. Since D is a linear code, its minimum distance d is equal to the minimum weight of its codewords. If d is small (much smaller than $n'/2$), then there is low weight codeword in D. Consequently, there is an $x_i = -\delta n$ for some positive δ, which implies a large gain by Lemma 10. However, if d is large (very close to $n'/2$), then we can apply the Johnson bound to D to show that the number of x_i such that $x_i \geq (1 - 2\epsilon - \eta)n$ is less than $\frac{1}{(1-2\epsilon)^2}$ for some positive η. This also implies a large gain by Lemma 11. Moreover, one can show that there is a trade-off relation between these two gains: If δ is small then η is large and vice versa. This trade-off enables us to prove that, for every v that is ϵ-away from C, $\text{GAIN}(v) = \Omega(1)$ for every ϵ, $1/4 \leq \epsilon < 1/2$. The complete proof can be found in the full version of the paper [18].

Acknowledgment

N.X. is very grateful to Ronitt Rubinfeld for making his visit to MIT possible. We would like to thank Oded Goldreich, Ronitt Rubinfeld, Madhu Sudan and Luca Trevisan for encouragement, helpful discussions and valuable suggestions. We also wish to thank the anonymous referees for their extensive reports and helpful comments, especially for pointing out a simpler proof of Lemma 10.

References

1. Alon, N., Kaufman, T., Krivelevich, M., Litsyn, S., Ron, D.: Testing low-degree polynomials over GF(2). In: Proceedings of Random 2003, pp. 188–199 (2003)
2. Aumann, Y., Håstad, J., Rabin, M., Sudan, M.: Linear-consistency testing. J. Comp. Sys. Sci. 62(4), 589–607 (2001)
3. Babai, L., Fortnow, L., Lund, C.: Non-deterministic exponential time has twoprover interactive protocols. In: Computational Complexity, vol. 1(1), pp. 3–40 (1991); Earlier version in FOCS 1990
4. Bellare, M., Coppersmith, D., Håstad, J., Kiwi, M., Sudan, M.: Linearity testing over characteristic two. IEEE Transactions on Information Theory 42(6), 1781–1795 (1996); Earlier version in FOCS 1995
5. Bellare, M., Goldreich, O., Sudan, M.: Free bits, PCP and non-approximability - towards tight results. SIAM J. on Comput. 27(3), 804–915 (1998); Earlier version in FOCS 1995
6. Bellare, M., Goldwasser, S., Lund, C., Russell, A.: Efficient probabilistically checkable proofs and applications to approximation. In: Proc. 25th Annual ACM Symposium on the Theory of Computing, pp. 304–294 (1993)
7. Bellare, M., Sudan, M.: Improved non-approximability results. In: Proc. 26th Annual ACM Symposium on the Theory of Computing, pp. 184–193 (1994)
8. Ben-Sasson, E., Sudan, M., Vadhan, S., Wigderson, A.: Randomness-efficient low degree tests and short PCPs via epsilon-biased sets. In: Proc. 35th Annual ACM Symposium on the Theory of Computing, pp. 612–621 (2003)
9. Blum, M., Luby, M., Rubinfeld, R.: Self-testing/correcting with applications to numerical problems. J. Comp. Sys. Sci. 47, 549–595 (1993); Earlier version in STOC 1990
10. Feige, U., Goldwasser, S., Lovász, L., Safra, S., Szegedy, M.: Approximating clique is almost NP-complete. In: Journal of the ACM; Earlier version in FOCS 1991
11. Fischer, E.: The art of uninformed decisions: A primer to property testing. Bulletin of the European Association for Theoretical Computer Science 75, 97–126 (2001)

12. Goldreich, O., Goldwaser, S., Ron, D.: Property testing and its connection to learning and approximation. Journal of the ACM 45, 653–750 (1998); Earlier version in FOCS 1996
13. Håstad, J.: Some optimal inapproximability results. Journal of the ACM 48(4), 798–859 (2001); Earlier version in STOC 1997
14. Håstad, J., Wigderson, A.: Simple analysis of graph tests for linearity and PCP. Random Structures and Algorithms 22, 139–160 (2003)
15. Jutla, C.S., Patthak, A.C., Rudra, A., Zuckerman, D.: Testing low-degree polynomials over prime fields. In: Proc. 45th Annual IEEE Symposium on Foundations of Computer Science, pp. 423–432 (2004)
16. Kiwi, M.: Algebraic testing and weight distributions of codes. Theor. Comp. Sci. 299(1-3), 81–106 (2003); Earlier version appeared as ECCC TR97-010 (1997)
17. Kaufman, T., Litsyn, S.: Almost orthogonal linear codes are locally testable. In: Proc. 46th Annual IEEE Symposium on Foundations of Computer Science, pp. 317–326 (2005)
18. Kaufman, T., Litsyn, S., Xie, N.: Breaking the ϵ-soundness bound of the linearity test over GF(2). Technical Report TR07-098, Electronic Colloquium on Computational Complexity (2007)
19. Kaufman, T., Ron, D.: Testing polynomials over general fields. In: Proc. 45th Annual IEEE Symposium on Foundations of Computer Science, pp. 413–422 (2004)
20. Plotkin, M.: Binary codes with specified minimum distance. IRE Transactions on Information Theory 6, 445–450 (1960)
21. Ron, D.: Property testing (a tutorial). In: Pardalos, P.M., Rajasekaran, S., Reif, J., Rolim, J.D.P. (eds.) Handbook of Randomized Computing, pp. 597–649. Kluwer Academic Publishers, Dordrecht (2001)
22. Rubinfeld, R., Sudan, M.: Robust characterizations of polynomials with applications to program testing. SIAM J. on Comput. 25(2), 252–271 (1996)
23. Rubinfeld, R.: Sublinear time algorithms. In: Proceedings of the International Congress of Mathematicians (2006)
24. Samorodnitsky, A.: Low-degree tests at large distances. In: Proc. 39th Annual ACM Symposium on the Theory of Computing, pp. 506–515 (2007)
25. Samorodnitsky, A., Trevisan, L.: A PCP characterization of NP with optimal amortized query complexity. In: Proc. 32nd Annual ACM Symposium on the Theory of Computing, pp. 191–199 (2000)
26. Samorodnitsky, A., Trevisan, L.: Gower uniformity, influence of variables and PCPs. In: Proc. 38th Annual ACM Symposium on the Theory of Computing, pp. 11–20 (2006)
27. Sudan, M., Trevisan, L.: Probabilistically checkable proofs with low amortized query complexity. In: Proc. 39th Annual IEEE Symposium on Foundations of Computer Science, pp. 18–27 (1998)
28. Shpilka, A., Wigderson, A.: Derandomizing homomorphism testing in general groups. SIAM J. on Comput. 36(4), 1215–1230 (2006); Earlier version in STOC 2004
29. Trevisan, L.: Recycling queries in PCPs and linearity tests. In: Proc. 30th Annual ACM Symposium on the Theory of Computing, pp. 299–308 (1998)

Dense Fast Random Projections and Lean Walsh Transforms

Edo Liberty*,†, Nir Ailon**, and Amit Singer***,†

Abstract. Random projection methods give distributions over $k \times d$ matrices such that if a matrix Ψ (chosen according to the distribution) is applied to a vector $x \in \mathbb{R}^d$ the norm of the resulting vector, $\Psi x \in \mathbb{R}^k$, is up to distortion ϵ equal to the norm of x w.p. at least $1 - \delta$. The Johnson Lindenstrauss lemma shows that such distributions exist over *dense* matrices for k (the target dimension) in $O(\log(1/\delta)/\varepsilon^2)$. Ailon and Chazelle and later Matousek showed that there exist entry-wise i.i.d. distributions over *sparse* matrices Ψ which give the same guaranties for vectors whose ℓ_∞ is bounded away from their ℓ_2 norm. This allows to accelerate the mapping $x \mapsto \Psi x$. We claim that setting Ψ as any column normalized *deterministic dense* matrix composed with random ± 1 diagonal matrix also exhibits this property for vectors whose ℓ_p (for any $p > 2$) is bounded away from their ℓ_2 norm. We also describe a specific tensor product matrix which we term *lean Walsh*. It is applicable to any vector in $\mathbb{R}^d$ in $O(d)$ operations and requires a weaker ℓ_∞ bound on x then the best current result, under comparable running times, using sparse matrices due to Matousek.

Keywords: Random Projections, Lean Walsh Transforms, Johnson Lindenstrauss, Dimension reduction.

1 Introduction

The application of various random matrices has become a common method for accelerating algorithms both in theory and in practice. These procedures are commonly referred to as *random projections*. The critical property of a $k \times d$ random projection matrix, Ψ, is that for any vector x the mapping $x \mapsto \Psi x$ is such that $(1 - \varepsilon)\|x\|_2 \le \|\Psi x\|_2 \le (1 + \varepsilon)\|x\|_2$ with probability at least $1 - \delta$ for specified constants $0 < \varepsilon < 1/2$ and $0 < \delta < 1$. The name *random projections* was coined after the first construction by Johnson and Lindenstrauss in [1] who showed that such mappings exist for $k \in O(\log(1/\delta)/\varepsilon^2)$. Since Johnson and Lindenstrauss other distributions for random projection matrices have been discovered [2,3,4,5,6]. Their properties make random projections a key player in

---

* Yale University, Department of Computer Science, Supported by NGA and AFOSR.

** Google Research.

*** Yale University, Department of Mathematics, Program in Applied Mathematics.

† Edo Liberty and Amit Singer thank the Institute for Pure and Applied Mathematics (IPAM) and its director Mark Green for their warm hospitality during the fall semester of 2007.

A. Goel et al. (Eds.): APPROX and RANDOM 2008, LNCS 5171, pp. 512–522, 2008.

rank-k approximation algorithms [7,8,9,10,11,12,13], other algorithms in numerical linear algebra [14,15,16], compressed sensing [17,18,19], and various other applications, e.g, [20,21].

As a remark, random projections are usually used as an approximate isometric mapping from $\mathbb{R}^d$ to $\mathbb{R}^k$ for n vectors $x_1, \ldots, x_n$. By preserving the length of all $\binom{n}{2}$ distance vectors $x = x_i - x_j$ the entire metric is preserved. Taking $\delta = \frac{1}{2}\binom{n}{2}^{-1}$ yields this w.p. at least $1/2$ due to the union bound. The resulting target dimension is $k = O(\log(n)/\varepsilon^2)$.

Considering the usefulness of random projections it is natural to ask the following question: what should be the structure of a random projection matrix, Ψ, such that mapping $x \mapsto \Psi x$ would require the least amount of computational resources? A naïve construction of a $k \times d$ unstructured matrix Ψ would result in an $O(kd)$ application cost.

In [22], Ailon and Chazelle propose the first asymptotically Fast Johnson Lindenstrauss Transform (FJLT). They give a two stage projection process. First, all input vectors are rotated, using a Fourier transform, such that their ℓ_∞ norm is bounded by $O(\sqrt{k/d})$. Then, a sparse random matrix containing only $O(k^3)$ nonzeros[1] is used to project them into $\mathbb{R}^k$. Thus, reducing the running time of dimensionality reduction from $O(kd)$ to $O(d\log(d)+k^3)$. Matousek in [6] generalized the sparse projection process and showed that if the ℓ_∞ norm of all the input vectors is bounded from above by η, they can be projected by a sparse matrix, Ψ, whose entries are nonzero with probability $\max(ck\eta^2, 1)$ for some constant c. The number of nonzeros in Ψ is therefore $O(k^2d\eta^2)$, with high probability. The concentration analysis is done for i.i.d. entries drawn from distributions satisfying mild assumptions.

Recently, Ailon and Liberty [23] improved the running time to $O(d\log(k))$ for $k \leq d^{1/2-\zeta}$ for any arbitrarily small ζ. They replaced the sparse i.i.d. projection matrix, Ψ, with a deterministic dense code matrix, A, composed with a random ± 1 diagonal matrix[2], D_s. They showed that a careful choice of A results in AD_s being a good random projection for the set of vectors such that $\|x\|_4 \in O(d^{-1/4})$. Here, we analyze this result for general $k \times d$ deterministic matrices. Our concentration result is very much in the spirit of [23]. We claim that any column normalized matrix A can be identified with a set $\chi \subset \mathbb{R}^d$ such that for x chosen from χ, AD_s constitutes a random projection w.h.p. The set χ can be thought of as the "good" set for AD_s. We study a natural tradeoff between the possible computational efficiency of applying A and the size of χ: the smaller χ is, the faster A can be applied[3]. We examine the connection between A and χ in Section 2. The set χ should be thought of as a prior assumption on our data, which may come, for example, from a statistical model generating the data.

---

[1] Each entry is drawn from a distribution which is gaussian with probability proportional to k^2/d, and so, for any constant probability, arbitrarily close to 1, the number of nonzeros is smaller than ck^3 for some constant c.

[2] The random isometric preprocessing is also different than that of the FJLT algorithm.

[3] This, however, might require a time costly preprocessing application of Φ.

Table 1. Types of $k \times d$ matrices and the subsets χ of $\mathbb{R}^d$ for which they constitute a random projection. The meaning of the norm $\| \cdot \|_A$ is given in Definition 2. The top two rows give random dense matrices, below are random i.i.d. sparse matrices, and the last three are *deterministic* matrices composed with random ± 1 diagonals.

	The rectangular $k \times d$ matrix A	Application time	$x \in \chi$ if
Johnson, Lindenstrauss [1]	Random k dimensional subspace	$O(kd)$	$x \in \mathbb{R}^d$
Various Authors [2,4,5,6]	Dense i.i.d. entries Gaussian or ± 1	$O(kd)$	$x \in \mathbb{R}^d$
Ailon, Chazelle [22]	Sparse Gaussian distributed entries	$O(k^3)$	$\frac{\|x\|_\infty}{\|x\|_2} = O((d/k)^{-1/2})$
Matousek [6]	Sparse sub-Gaussian symmetric i.i.d. entries	$O(k^2 d\eta^2)$	$\frac{\|x\|_\infty}{\|x\|_2} \leq \eta$
General rule (This work)	**Any deterministic matrix A**		$\frac{\|x\|_A}{\|x\|_2} = O(k^{-1/2})$
Ailon, Liberty [23]	Four-wise independent	$O(d \log k)$	$\frac{\|x\|_4}{\|x\|_2} = O(d^{-1/4})$
This work	Lean Walsh Transform	$O(d)$	$\frac{\|x\|_\infty}{\|x\|_2} = O(k^{-1/2}d^{-\varsigma})$

We propose in Section 3 a new type of fast applicable matrices and in Section 4 explore their corresponding χ. These matrices are constructed using tensor products and can be applied to any vector in $\mathbb{R}^d$ in linear time, i.e., in $O(d)$. Due to the similarity in their construction to Walsh-Hadamard matrices and their rectangular shape we term them *lean Walsh Matrices*[4]. Lean Walsh matrices are of size $\tilde{d} \times d$ where $\tilde{d} = d^\alpha$ for some $0 < \alpha < 1$. In order to reduce the dimension to $k \leq \tilde{d}$, $k = O(\log(1/\delta)/\varepsilon^2))$, we can compose the lean Walsh matrix, A, with a known Johnson Lindenstrauss matrix construction R. Applying R in $O(d)$ requires some relation between d, k and α as explained in subsection 4.1.

2   Norm Concentration and $\chi(A, \varepsilon, \delta)$

We compose an arbitrary deterministic $\tilde{d} \times d$ matrix A with a random sign diagonal matrix D_s and study the behavior of such matrices as random projections. In order for AD_s to exhibit the property of a random projection it is enough for

---

[4] The terms *lean Walsh Transform* or simply *lean Walsh* are also used interchangeably.

it to approximately preserve the length of any single *unit* vector $x \in \mathbb{R}^d$ with high probability:

$$\Pr\left[\mid \|AD_s x\|_2 - 1 \mid \geq \varepsilon\right] < \delta \tag{1}$$

Here D_s is a diagonal matrix such that $D_s(i,i)$ are random signs (i.i.d. ± 1 w.p. $1/2$ each), $0 < \delta < 1$ is a constant acceptable failure probability, and the constant $0 < \varepsilon < 1/2$ is the prescribed precision.

Note that we can replace the term $AD_s x$ with $AD_x s$ where D_x is a diagonal matrix holding on the diagonal the values of x, i.e. $D_x(i,i) = x(i)$ and similarly $s(i) = D_s(i,i)$. Denoting $M = AD_x$, we view the term $\|Ms\|_2$ as a scalar function over the hypercube $\{1,-1\}^d$, from which the variable s is uniformly chosen. This function is convex over $[-1,1]^d$ and Lipschitz bounded. Talagrand [24] proves a strong concentration result for such functions. We give a slightly restated form of his result for our case.

Lemma 1 (Talagrand [24]). *Given a matrix M and a random vector s ($s(i)$ are i.i.d. ± 1 w.p. $1/2$) define the random variable $Y = \|Ms\|_2$. Denote by μ a median of Y, and by $\sigma = \|M\|_{2 \to 2}$ the spectral norm of M. Then*

$$\Pr\left[\mid Y - \mu \mid > t\right] \leq 4e^{-t^2/8\sigma^2} \tag{2}$$

Definition 1. $\|M\|_{p \to q}$ *denoted the norm of M as an operator from ℓ_p to ℓ_q, i.e., $\|M\|_{p \to q} = \sup_{x, \|x\|_p = 1} \|Mx\|_q$. The ordinary spectral norm of M is thus $\|M\|_{2 \to 2}$.*

Lemma 1 asserts that $\|AD_x s\|$ is distributed like a (sub) Gaussian around its median, with standard deviation 2σ.

First, in order to have $E[Y^2] = 1$ it is necessary and sufficient for the columns of A to be normalized to 1 (or normalized in expectancy). To estimate a median, μ, we substitute $t^2 \to t'$ and compute:

$$E[(Y-\mu)^2] = \int_0^\infty \Pr[(Y-\mu)^2] > t']dt'$$

$$\leq \int_0^\infty 4e^{-t'/(8\sigma^2)}dt' = 32\sigma^2$$

Furthermore, $(E[Y])^2 \leq E[Y^2] = 1$, and so $E[(Y-\mu)^2] = E[Y^2] - 2\mu E[Y] + \mu^2 \geq 1 - 2\mu + \mu^2 = (1-\mu)^2$. Combining, $|1-\mu| \leq \sqrt{32}\sigma$. We set $\varepsilon = t + |1-\mu|$:

$$\Pr[|Y-1| > \varepsilon] \leq 4e^{-\varepsilon^2/32\sigma^2} \quad ,\text{for } \varepsilon > 2|1-\mu| \tag{3}$$

If we set $k = 33\log(1/\delta)/\varepsilon^2$ (for $\log(1/\delta)$ larger than a sufficient constant) and set $\sigma \leq k^{-1/2}$, (1) follows from (3). Moreover μ depends on ε such that the condition $\varepsilon > 2|1-\mu|$ is met for any constant ε (given $\log(1/\delta) > 4$). This can be seen by $|1-\mu| \leq \sqrt{32}\sigma < \varepsilon/\sqrt{\log(1/\delta)}$. We see that $\sigma = \|AD_x\|_{2 \to 2} \leq k^{-1/2}$ is sufficient for the projection to succeed w.h.p. This naturally defines χ.

Definition 2. *For a given matrix $A \in \mathbb{R}^{k \times d}$ we define the vector pseudonorm of $x \in \mathbb{R}^d$ with respect to A as $\|x\|_A \equiv \|AD_x\|_{2 \to 2}$ where D_x is a diagonal matrix such that $D_x(i,i) = x(i)$. Remark: If no column of A has norm zero $\| \cdot \|_A$ induces a proper norm on $\mathbb{R}^d$.*

Definition 3. *We define $\chi(A, \varepsilon, \delta)$ as the intersection of the Euclidian unit sphere and a ball of radius $k^{-1/2}$ in the norm $\| \cdot \|_A$*

$$\chi(A, \varepsilon, \delta) = \left\{ x \in \mathbb{S}^{d-1} \mid \|x\|_A \leq k^{-1/2} \right\} \tag{4}$$

for $k = 33 \log(1/\delta)/\varepsilon^2$.

Lemma 2. *For any column normalized matrix, A, and an i.i.d. random ± 1 diagonal matrix, D_s, the following holds:*

$$\forall x \in \chi(A, \varepsilon, \delta) \quad \Pr\left[\left| \|AD_s x\|_2 - 1 \right| \geq \varepsilon \right] \leq \delta \tag{5}$$

Proof. For any $x \in \chi$, by Definition 3, $\|x\|_A = \|AD_x\|_{2 \to 2} = \sigma \leq k^{-1/2}$. The lemma follows from substituting the value of σ into Equation (3).

It is convenient to think about χ as the "good" set of vectors for which AD_s is length preserving with high probability. En route to explore $\chi(A, \varepsilon, \delta)$ for lean Walsh matrices we first turn to formally defining them.

3   Lean Walsh Transforms

The *lean* Walsh Transform, similar to the Walsh Transform, is a recursive tensor product matrix. It is initialized by a constant seed matrix, A_1, and constructed recursively by using Kronecker products $A_{\ell'} = A_1 \otimes A_{\ell'-1}$. The main difference is that the lean Walsh seeds have fewer rows than columns. We formally define them as follows:

Definition 4. *A_1 is a lean Walsh seed (or simply 'seed') if: i) A_1 is a rectangular matrix $A_1 \in \mathbb{C}^{r \times c}$, such that $r < c$; ii) A_1 is absolute valued $1/\sqrt{r}$ entry-wise, i.e., $|A_1(i,j)| = r^{-1/2}$; iii) the rows of A_1 are orthogonal.*

Definition 5. *A_ℓ is a lean Walsh transform, of order ℓ, if for all $\ell' \leq \ell$ we have $A_\ell = A_1 \otimes A_{\ell'-1}$, where $\otimes$ stands for the Kronecker product and A_1 is a seed according to Definition 4.*

The following are examples of seed matrices:

$$A'_1 = \frac{1}{\sqrt{3}} \begin{pmatrix} 1 & 1 & -1 & -1 \\ 1 & -1 & 1 & -1 \\ 1 & -1 & -1 & 1 \end{pmatrix} \quad A''_1 = \frac{1}{\sqrt{2}} \begin{pmatrix} 1 & 1 & 1 \\ 1 & e^{2\pi i/3} & e^{4\pi i/3} \end{pmatrix} \tag{6}$$

These examples are a part of a large family of possible seeds. This family includes, amongst other constructions, sub-Hadamard matrices (like A'_1) or sub-Fourier matrices (like A''_1). A simple construction is given for possible larger seeds.

Fact 1. *Let F be the $c \times c$ Discrete Fourier matrix such that $F(i,j) = e^{2\pi\sqrt{-1}ij/c}$. Define A_1 to be the matrix consisting of the first $r = c - 1$ rows of F normalized by $1/\sqrt{r}$. A_1 is a lean Walsh seed.*

We use elementary properties of Kronecker products to characterize A_ℓ in terms of the number of rows, r, and the number of columns, c, of its seed. The following facts hold true for A_ℓ:

Fact 2. *i) The size of A_ℓ is $d^\alpha \times d$, where $\alpha = \log(r)/\log(c) < 1$ is the skewness of A_1;[5] ii) for all i and j, $A_\ell(i,j) \in \pm\widetilde{d}^{-1/2}$ which means that A_ℓ is column normalized; and iii) the rows of A_ℓ are orthogonal.*

Fact 3. *The time complexity of applying A_ℓ to any vector $z \in \mathbb{R}^d$ is $O(d)$.*

Proof. Let $z = [z_1; \ldots; z_c]$ where z_i are blocks of length d/c of the vector z. Using the recursive decomposition for A_ℓ we compute $A_\ell z$ by first summing over the different z_i according to the values of A_1 and applying to each sum the matrix $A_{\ell-1}$. Denoting by $T(d)$ the time to apply A_ℓ to $z \in \mathbb{R}^d$ we get that $T(d) = rT(d/c) + rd$. A simple calculation yields $T(d) \leq dcr/(c-r)$ and thus $T(d) = O(d)$ for a constant sized seed.

For clarity, we demonstrate Fact 3 for A_1' (Equation (6)):

$$A_\ell' z = A_\ell' \begin{pmatrix} z_1 \\ z_2 \\ z_3 \\ z_4 \end{pmatrix} = \frac{1}{\sqrt{3}} \begin{pmatrix} A_{\ell-1}'(z_1 + z_2 - z_3 - z_4) \\ A_{\ell-1}'(z_1 - z_2 + z_3 - z_4) \\ A_{\ell-1}'(z_1 - z_2 - z_3 + z_4) \end{pmatrix} \tag{7}$$

Remark 1. For the purpose of compressed sensing, an important parameter of the projection matrix is its Coherence. The Coherence of a column normalized matrix is simply the maximal inner product between two different columns. The Coherence of a lean Walsh matrix is equal to the coherence of its seed and the seed coherence can be reduced by increasing its size. For example, the seeds described in Fact 1, of size r by $c = r + 1$, exhibit coherence of $1/r$.

In what follows we characterize $\chi(A, \varepsilon, \delta)$ for a general lean Walsh transform by the parameters of its seed. The abbreviated notation, A, stands for A_ℓ of the right size to be applied to x, i.e., $\ell = \log(d)/\log(c)$. Moreover, we freely use α to denote the skewness $\log(r)/\log(c)$ of the seed at hand.

4   An ℓ_p Bound on $\| \cdot \|_A$

After describing the lean Walsh transforms we turn our attention to exploring their "good" sets χ. We remind the reader that $\|x\|_A \leq k^{-1/2}$ implies $x \in \chi$:

$$\|x\|_A^2 = \|AD_x\|_{2\to2}^2 = \max_{y, \|y\|_2=1} \|y^T AD_x\|_2^2 \tag{8}$$

---

[5] The size of A_ℓ is $r^\ell \times c^\ell$. Since the running time is linear, we can always pad vectors to be of length c^ℓ without effecting the asymptotic running time. From this point on we assume w.l.o.g $d = c^\ell$ for some integer ℓ.

$$= \max_{y, \|y\|_2=1} \sum_{i=1}^{d} x^2(i)(y^T A^{(i)})^2 \tag{9}$$

$$\leq \left(\sum_{i=1}^{d} x^{2p}(i) \right)^{1/p} \left(\max_{y, \|y\|_2=1} \sum_{i=1}^{d} (y^T A^{(i)})^{2q} \right)^{1/q} \tag{10}$$

$$= \|x\|_{2p}^2 \|A^T\|_{2\to 2q}^2 \tag{11}$$

The transition from the second to the third line follows from Hölder's inequality for dual norms p and q, satisfying $1/p + 1/q = 1$. We now compute $\|A^T\|_{2\to 2q}$.

Theorem 1. [Riesz-Thorin] *For an arbitrary matrix B, assume $\|B\|_{p_1\to r_1} \leq C_1$ and $\|B\|_{p_2\to r_2} \leq C_2$ for some norm indices p_1, r_1, p_2, r_2 such that $p_1 \leq r_1$ and $p_2 \leq r_2$. Let λ be a real number in the interval $[0,1]$, and let p, r be such that $1/p = \lambda(1/p_1) + (1-\lambda)(1/p_2)$ and $1/r = \lambda(1/r_1) + (1-\lambda)(1/r_2)$. Then $\|B\|_{p\to r} \leq C_1^{\lambda} C_2^{1-\lambda}$.*

In order to use the theorem, let us compute $\|A^T\|_{2\to 2}$ and $\|A^T\|_{2\to\infty}$. From $\|A^T\|_{2\to 2} = \|A\|_{2\to 2}$ and the orthogonality of the rows of A we get that $\|A^T\|_{2\to 2} = \sqrt{d/\tilde{d}} = d^{(1-\alpha)/2}$. From the normalization of the columns of A we get that $\|A^T\|_{2\to\infty} = 1$. Using the theorem for $\lambda = 1/q$, for any $q \geq 1$, we obtain $\|A^T\|_{2\to 2q} \leq d^{(1-\alpha)/2q}$. It is worth noting that $\|A^T\|_{2\to 2q}$ might actually be significantly lower than the given bound. For a specific seed, A_1, one should calculate $\|A_1^T\|_{2\to 2q}$ and use $\|A_\ell^T\|_{2\to 2q} = \|A_1^T\|_{2\to 2q}^{\ell}$ to achieve a possibly lower value for $\|A^T\|_{2\to 2q}$.

Lemma 3. *For a lean Walsh transform, A, we have that for any $p > 1$ the following holds:*

$$\{x \in \mathbb{S}^{d-1} \mid \|x\|_{2p} \leq k^{-1/2} d^{-\frac{1-\alpha}{2}(1-\frac{1}{p})}\} \subset \chi(A, \varepsilon, \delta) \tag{12}$$

where $k = O(\log(1/\delta)/\varepsilon^2)$ and α is the skewness of A, $\alpha = \log(r)/\log(c)$ (r is the number of rows, and c is the number of columns in the seed of A).

Proof. We combine the above and use the duality of p and q:

$$\|x\|_A \leq \|x\|_{2p} \|A^T\|_{2\to 2q} \tag{13}$$

$$\leq \|x\|_{2p} d^{\frac{1-\alpha}{2q}} \tag{14}$$

$$\leq \|x\|_{2p} d^{\frac{1-\alpha}{2}(1-\frac{1}{p})} \tag{15}$$

The desired property, $\|x\|_A \leq k^{-1/2}$, is achieved if $\|x\|_{2p} \leq k^{-1/2} d^{-\frac{1-\alpha}{2}(1-\frac{1}{p})}$ for any $p > 1$.

Remark 2. Consider a different family of matrices containing $d/\tilde{d}$ copies of a $\tilde{d} \times \tilde{d}$ identity matrices concatenated horizontally. Their spectral norm is the same as that of lean Walsh matrices and they are clearly row orthogonal and column normalized. Considering $p \to \infty$ they require the same ℓ_∞ constraint on x as lean Walsh matrices do. However, their norm as operators from ℓ_2 to ℓ_{2q} ,for q larger than 1 $(p < \infty)$, is large and fixed, whereas that of lean Walsh matrices is still arbitrarily small and controlled by the size of the their seed.

4.1 Controlling α and Choosing R

We see that increasing the skewness of the seed of A, α, is beneficial from the theoretical stand point since it weakens the constraint on $\|x\|_{2p}$. However, the application oriented reader should keep in mind that this requires the use of a larger seed, which subsequently increases the constant hiding in the big O notation of the running time.

Consider the seed constructions described in Fact 1 for which $r = c - 1$. Their skewness $\alpha = \log(r)/\log(c)$ approaches 1 as their size increases. Namely, for any positive constant ζ there exists a constant size seed such that $1 - 2\zeta \leq \alpha \leq 1$.

Lemma 4. *For any positive constant $\zeta > 0$ there exists a lean Walsh matrix, A, such that:*

$$\{x \in \mathbb{S}^{d-1} \mid \|x\|_\infty \leq k^{-1/2}d^{-\zeta}\} \subset \chi(A, \varepsilon, \delta) \tag{16}$$

Proof. Generate A from a seed such that its skewness $\alpha = \log(r)/\log(c) \geq 1 - 2\zeta$ and substitute $p = \infty$ into the statement of Lemma 3.

The skewness α also determines the minimal dimension d (relative to k) for which the projection can be completed in $O(d)$ operations. The reason being that the vectors $z = AD_sx$ must be mapped from dimension $\widetilde{d}$ $(\widetilde{d} = d^\alpha)$ to dimension k in $O(d)$ operations. This can be done using Ailon and Liberty's construction [23] serving as the random projection matrix R. R is a $k \times \widetilde{d}$ Johnson Lindenstrauss projection matrix which can be applied in $\widetilde{d}\log(k)$ operations if $\widetilde{d} = d^\alpha \geq k^{2+\zeta''}$ for arbitrary small ζ''. For the same choice of a seed as in Lemma 4, the condition becomes $d \geq k^{2+\zeta''+2\zeta}$ which can be achieved by $d \geq k^{2+\zeta'}$ for arbitrary small ζ' depending on ζ and ζ''. Therefore for such values of d the matrix R exists and requires $O(d^\alpha \log(k)) = O(d)$ operations to apply.

5 Comparison to Sparse Projections

Sparse random ± 1 projection matrices were analyzed by Matousek in [6]. For completeness we restate his result. Theorem 4.1 in [6] (slightly rephrased to fit our notation) claims the following:

Theorem 2 (Matousek 2006 [6]). *let $\varepsilon \in (0, 1/2)$ and $\eta \in [1/\sqrt{d}, 1]$ be constant parameters. Set $q = C_0\eta^2 \log(1/\delta)$ for a sufficiently large constant C_0. Let S be a random variable such that*

$$S = \begin{cases} +\frac{1}{\sqrt{qk}} & \text{with probability } q/2 \\ -\frac{1}{\sqrt{qk}} & \text{with probability } q/2 \\ 0 & \text{with probability } 1 - q \ . \end{cases} \tag{17}$$

Let k be $C_1 \log(1/\delta)/\varepsilon^2$ for a sufficiently large C_1. Draw the matrix elements of Ψ i.i.d. from S. Then:

$$\Pr[|\|\Psi x\|_2^2 - 1| > \varepsilon] \leq \delta \tag{18}$$

For any $x \in \mathbb{S}^{d-1}$ such that $\|x\|_\infty \leq \eta$.

With constant probability, the number of nonzeros in Ψ is $O(kdq) = O(k^2 d \eta^2)$ (since ε is a constant $\log(1/\delta) = O(k)$). In the terminology of this paper we say that for a sparse Ψ containing $O(k^2 d \eta^2)$ nonzeros on average (as above) $\{x \in \mathbb{S}^{d-1} \mid \|x\|_\infty \leq \eta\} \subset \chi(A, \varepsilon, \delta)$.

A lower bound on the running time of general dimensionality reduction is at least $\Omega(d)$. Our analysis shows that the problem of satisfying the condition $\Phi x \in \chi$ (via a Euclidean isometry Φ) is at least as hard. Indeed, a design of any such fast transformation, applicable in time $T(d)$, would imply a similar upper bound for general dimensionality reduction. We claim that lean Walsh matrices admit a strictly larger χ than that of sparse matrices which could be applied in the same asymptotic complexity. For $q = k^{-1}$ a sparse matrix Ψ as above contains $O(d)$ nonzeros, w.h.p., and thus can be applied in that amount of time. Due to Theorem 2 this value of q requires $\|x\|_\infty \leq O(k^{-1})$ for the length of x to be preserved w.h.p. For d polynomial in k, this is a stronger constraint on the ℓ_∞ norm of x than $\|x\|_\infty \leq O(k^{-1/2}d^{-\varsigma})$ which is obtained by our analysis for lean Walsh transforms.

6   Conclusion and Work in Progress

We have shown that any $k \times d$ (column normalized) matrix, A, can be composed with a random diagonal matrix to constitute a random projection matrix for some part of the Euclidean space, χ. Moreover, we have given sufficient conditions, on $x \in \mathbb{R}^d$, for belonging to χ depending on different $\ell_2 \to \ell_p$ operator norms of A^T and ℓ_p norms of x. We have also seen that lean Walsh matrices exhibit both a "large" χ and a linear time computation scheme which outperforms sparse projective matrices. These properties make them good building blocks for the purpose of random projections.

However, as explained in the introduction, in order for the projection to be complete, one must design a linear time preprocessing matrix Φ which maps all vectors in $\mathbb{R}^d$ into χ (w.h.p.). Achieving such distributions for Φ would be extremely interesting from both the theoretical and practical stand point. Possible choices for Φ may include random permutations, various wavelet/wavelet-like transforms, or any other sparse orthogonal transformation.

In this framework χ was characterized by a bound over ℓ_p ($p > 2$) norms of $x \in \chi$. Understanding distributions over ℓ_2 isometries which reduce other ℓ_p norms with high probability and efficiency is an interesting problem in its own right. However, partial results hint that for lean Walsh transforms if Φ is taken to be a random permutation (which is an ℓ_p isometry for any p) then the ℓ_∞ requirement reduces to $\|x\|_\infty \leq k^{-1/2}$. Showing this however requires a different technique.

Acknowledgments

The authors would like to thank Steven Zucker, Daniel Spielman, and Yair Bartal for their insightful ideas and suggestions.

References

1. Johnson, W.B., Lindenstrauss, J.: Extensions of Lipschitz mappings into a Hilbert space. Contemporary Mathematics 26, 189–206 (1984)
2. Frankl, P., Maehara, H.: The Johnson-Lindenstrauss lemma and the sphericity of some graphs. Journal of Combinatorial Theory Series A 44, 355–362 (1987)
3. Indyk, P., Motwani, R.: Approximate nearest neighbors: Towards removing the curse of dimensionality. In: Proceedings of the 30th Annual ACM Symposium on Theory of Computing (STOC), pp. 604–613 (1998)
4. DasGupta, S., Gupta, A.: An elementary proof of the Johnson-Lindenstrauss lemma. Technical Report, UC Berkeley 99-006 (1999)
5. Achlioptas, D.: Database-friendly random projections: Johnson-lindenstrauss with binary coins. J. Comput. Syst. Sci. 66(4), 671–687 (2003)
6. Matousek, J.: On variants of the Johnson-Lindenstrauss lemma. Private communication (2006)
7. Drineas, P., Mahoney, M.W., Muthukrishnan, S.: Sampling algorithms for ℓ_2 regression and applications. In: Proceedings of the 17th Annual ACM-SIAM Symposium on Discrete Algorithms (SODA), Miami, Florida, United States (2006)
8. Sarlós, T.: Improved approximation algorithms for large matrices via random projections. In: Proceedings of the 47th Annual IEEE Symposium on Foundations of Computer Science (FOCS), Berkeley, CA (2006)
9. Frieze, A.M., Kannan, R., Vempala, S.: Fast monte-carlo algorithms for finding low-rank approximations. In: IEEE Symposium on Foundations of Computer Science, pp. 370–378 (1998)
10. Peled, S.H.: A replacement for voronoi diagrams of near linear size. In: Proceedings of the 42nd Annual IEEE Symposium on Foundations of Computer Science (FOCS), Las Vegas, Nevada, USA, pp. 94–103 (2001)
11. Achlioptas, M.: Fast computation of low rank matrix approximations. In: STOC: ACM Symposium on Theory of Computing (STOC) (2001)
12. Drineas, P., Kannan, R.: Fast monte-carlo algorithms for approximate matrix multiplication. In: IEEE Symposium on Foundations of Computer Science, pp. 452–459 (2001)
13. Liberty, E., Woolfe, F., Martinsson, P.G., Rokhlin, V., Tygert, M.: Randomized algorithms for the low-rank approximation of matrices. In: Proceedings of the National Academy of Sciences (2007)
14. Dasgupta, A., Drineas, P., Harb, B., Kumar, R., Mahoney, M.W.: Sampling algorithms and coresets for ℓ_p regression. In: Proc. of the 19th Annual ACM-SIAM Symposium on Discrete Algorithms (SODA) (2008)
15. Drineas, P., Mahoney, M.W., Muthukrishnan, S., Sarlos, T.: Faster least squares approximation. TR arXiv:0710.1435 (submitted for publication) (2007)
16. Drineas, P., Mahoney, M., Muthukrishnan, S.: Relative-error cur matrix decompositions. TR arXiv:0708.3696 (submitted for publication) (2007)
17. Candes, E.J., Tao, T.: Near-optimal signal recovery from random projections: Universal encoding strategies? Information Theory. IEEE Transactions 52(12), 5406–5425 (2006)

18. Donoho, D.L.: Compressed sensing. IEEE Transactions on Information Theory 52(4), 1289–1306 (2006)
19. Elad, M.: Optimized projections for compressed sensing. IEEE Transactions on Signal Processing 55(12), 5695–5702 (2007)
20. Paschou, P., Ziv, E., Burchard, E., Choudhry, S., Rodriguez-Cintron, W., Mahoney, M.W., Drineas, P.: Pca-correlated snps for structure identification in worldwide human populations. PLOS Genetics 3, 1672–1686 (2007)
21. Paschou, P., Mahoney, M.W., Javed, A., Pakstis, A., Gu, S., Kidd, K.K., Drineas, P.: Intra- and inter-population genotype reconstruction from tagging snps. Genome Research 17(1), 96–107 (2007)
22. Ailon, N., Chazelle, B.: Approximate nearest neighbors and the fast johnson-lindenstrauss transform. In: STOC 2006: Proceedings of the thirty-eighth annual ACM symposium on Theory of computing, pp. 557–563. ACM Press, New York (2006)
23. Ailon, N., Liberty, E.: Fast dimension reduction using rademacher series on dual bch codes. In: SODA, pp. 1–9 (2008)
24. Ledoux, M., Talagrand, M.: Probability in Banach Spaces: Isoperimetry and Processes. Springer, Heidelberg (1991)

Near Optimal Dimensionality Reductions That Preserve Volumes

Avner Magen and Anastasios Zouzias

Department of Computer Science,
University of Toronto

Abstract. Let P be a set of n points in Euclidean space and let $0 < \varepsilon < 1$. A well-known result of Johnson and Lindenstrauss states that there is a projection of P onto a subspace of dimension $O(\varepsilon^{-2} \log n)$ such that distances change by at most a factor of $1 + \varepsilon$. We consider an extension of this result. Our goal is to find an analogous dimension reduction where not only pairs but all subsets of at most k points maintain their volume approximately. More precisely, we require that sets of size $s \leq k$ preserve their volumes within a factor of $(1 + \varepsilon)^{s-1}$. We show that this can be achieved using $O(\max\{\frac{k}{\varepsilon}, \varepsilon^{-2} \log n\})$ dimensions. This in particular means that for $k = O(\log n / \varepsilon)$ we require no more dimensions (asymptotically) than the special case $k = 2$, handled by Johnson and Lindenstrauss. Our work improves on a result of Magen (that required as many as $O(k\varepsilon^{-2} \log n)$ dimensions) and is tight up to a factor of $O(1/\varepsilon)$. Another outcome of our work is an alternative and greatly simplified proof of the result of Magen showing that all distances between points and affine subspaces spanned by a small number of points are approximately preserved when projecting onto $O(k\varepsilon^{-2} \log n)$ dimensions.

1 Introduction

A classical result of Johnson and Lindenstrauss [12] shows that a set of n points in the Euclidean space can be projected onto $O(\varepsilon^{-2} \log n)$ dimensions so that all distances are changed by at most a factor of $1 + \varepsilon$. Many important works in areas such as computational geometry, approximation algorithms and discrete geometry build on this result in order to achieve a computation speed-up, reduce space requirements or simply exploit the added simplicity of a low dimensional space.

However, the rich structure of Euclidean spaces gives rise to many many geometric parameters other than distances between points. For example, we could care about the centre of gravity of sets of points, angles and areas of triangles of triplets of points among a fixed set of points P, and more generally, the volume spanned by some subsets of P or the volume of the smallest ellipsoid containing them. The generalization of the Johnson-Lindenstrauss lemma to the geometry of subsets of bounded size was considered in [14] where it was shown that it is possible to embed an n-point set of the Euclidean space onto an $O(k\varepsilon^{-2} \log n)$-dimensional Euclidean space, such that no set of size $s \leq k$ changes its volume by more than a factor of $(1 + \varepsilon)^{s-1}$. The exponent $s - 1$ should be thought as a natural normalization measure. Notice that scaling a set of size s by a factor of $1 + \varepsilon$ will change its volume by precisely the above factor. In the current work we improve this result by showing that $O(\max\{\frac{k}{\varepsilon}, \varepsilon^{-2} \log n\})$ dimensions suffice in order to get the same guarantee.

A. Goel et al. (Eds.): APPROX and RANDOM 2008, LNCS 5171, pp. 523–534, 2008.

Theorem 1 (Main theorem). *Let $0 < \varepsilon \leq 1/2$ and let k, n, d be positive integers, such that $d = O(\max\{\frac{k}{\varepsilon}, \varepsilon^{-2} \log n\})$. Then for any n-point subset P of the Euclidean space $\mathbb{R}^n$, there is a mapping $f : P \to \mathbb{R}^d$, such that for all subsets S of P, $1 < |S| < k$,*

$$1 - \varepsilon \leq \left(\frac{\text{vol}(f(S))}{\text{vol}(S)} \right)^{\frac{1}{s-1}} \leq 1 + \varepsilon.$$

Moreover, the mapping f can be constructed efficiently in randomized polynomial time using a Gaussian random matrix.

The line of work presented here (as well as in [14]) is related to, however quite different from, Feige's work on volume-respecting embeddings [8]. Feige defined a notion of volume for sets in general metric spaces that is very different than ours, and measured the quality of an embedding from such spaces into Euclidean space. For the image of the points of these embeddings, Feige's definition of volume is identical to the one used here and in [14]. Embeddings that do not significantly change volumes of small sets (of size $\leq k$) are presented in [8] and further it is shown how these embeddings lead to important algorithmic applications (see also [19]). The typical size k of sets in [8] is $O(\log n)$ and therefore our work shows that the embedding that are obtained in [8] can be assumed to use no more than $O(\log n)$ dimensions. Compare this to the $O(n)$ as in the original embedding or the $O(\log^2 n)$ bound that can be obtained by [14].

As was shown by Alon [5], the upper bound on the dimensionality of the projection of Johnson Lindenstauss is nearly tight[1], giving a lower bound of $\Omega(\varepsilon^{-2} \log n / \log(\frac{1}{\varepsilon}))$ dimensions. Further, in our setting it is immediate that at least $k - 1$ dimensions are needed (otherwise the image of sets of size k will not be full dimensional and will not therefore have a positive volume). These two facts provide a dimension lower bound of $\Omega(\max\{\varepsilon^{-2} \log n / \log(\frac{1}{\varepsilon}), k\})$ which makes our upper bound tight up to a factor of $1/\varepsilon$ throughout the whole range of the parameter k.

Similarly to other dimension reduction results our embedding uses random projections. Several variants have been used in the past, each defining 'random' projection in a slightly different way. Originally, Johnson and Lindenstrauss considered projecting onto a random d-dimensional subspace, while Frankl and Mehara [9] used projection onto d independent random unit vectors. In most later works the standard approach has been to use projection onto d n-dimensional Gaussian, an approach that we adopt here.

In our analysis it is convenient to view the projection as applying a matrix multiplication with a random matrix (of appropriate dimensions) whose entries are i.i.d. Gaussians. A critical component in our analysis is the following "invariance" claim.

Claim. Let S be a subset of the Euclidean space and let $V^\pi(S)$ be the volume of the projection of S by a Gaussian random matrix. Then the distribution of $V^\pi(S)$ depends linearly on vol(S) but does not depend on other properties of S.

When S is of size 2 the claim is nothing else but an immediate use of the fact that the projections are rotational invariant: indeed, any set of size 2 is the same up to an orthonormal transformation, translation and scaling. For $|S| > 2$, while the claim is still easy to show, it may seem somewhat counterintuitive from a geometric viewpoint. It is

---

[1] With respect to any embedding, not necessarily a projection.

certainly no longer the case that any two sets with the same volume are the same up to an orthonormal transformation. Specifically, it does not seem clear why a very 'flat' (for example a perturbation of a co-linear set of points) set should behave similarly to a 'round' set (like a symmetric simplex) of the same volume, with respect to the volume of their projections. The question of the distortion of subsets readily reduces to a stochastic question about one particular set S. Essentially, one needs to study the probability that the volume of the projection of this set deviates from its expected value. This makes the effectiveness of the above claim clear: it means that the question can be further reduced to the question of concentration of volume with respect to a *particular set S* of our choice! Since there are roughly n^s sets of size s to consider, we need to bound the probability of a bad event with respect to any arbitrary set by roughly $e^{-\Omega(sd)}$. The previous bound of [14] (implicitly) showed a concentration bound of only $e^{-\Omega(d)}$ which is one way to understand the improvement of the current work.

Other works have extended the Johnson Lindenstauss original work. From the computational perspective, emphasis was placed on derandomizing the embedding [7,18] and on speeding its computation. This last challenge has attracted considerable amount of attention. Achlioptas [1] has shown that projection onto a (randomly selected) set of discrete vectors generated the same approximation guarantee using the same dimensionality. Ailon and Chazelle [3] supplied a method that uses *Sparse* Gaussian matrices for the projection to achieve fast computation (which they call "Fast Johnson Lindenstrauss Transform"). See also [15,4,13] for a related treatment and extensions. On another branch of extentions (closer in flavour to our result) are works that require that the embeddings will preserve richer structure of the geometry. For example, in [2] the authors ask about distance between points that are *moving* according to some algebraically-limited curve; in [17] for affine subspaces, in [11] for sets with bounded doubling dimension, and in [2,6,20] for curves, (smooth) surfaces and manifolds.

2 Notation and Preliminaries

We think of n points in $\mathbb{R}^n$ as an $n \times n$ matrix P, where the rows correspond to the points and the columns to the coordinates. We call the set $\{0, e_1, e_2, \ldots, e_n\}$ i.e., the n-dimensional standard vectors of $\mathbb{R}^n$ with the origin, *regular*. We associate with a set of k points a volume which is the $(k-1)$-dimensional volume of its convex-hull. For $k = 2$ notice that $\text{vol}([x, y]) = d(x, y)$, and for $k = 3$ is the area of the triangle with vertices the points of the set, etc. Throughout this paper we denote the volume of a set S in the Euclidean space by $\text{vol}(S)$.

We use $\| \cdot \|$ to denote the Euclidean norm. Let X_i $i = 1, \ldots, k$ be k independent, normally distributed random variables with zero mean and variance one, then the random variable $\chi_k^2 = \sum_{i=1}^{k} X_i^2$ is a Chi-square random variable with k degrees of freedom. If A is an $r \times s$ matrix and B is a $p \times q$ matrix, then the Kronecker product $A \otimes B$ is the $rp \times sq$ block matrix

$$A \otimes B = \begin{bmatrix} a_{11}B & \ldots & a_{1s}B \\ \vdots & \ddots & \vdots \\ a_{r1}B & \ldots & a_{rs}B \end{bmatrix}.$$

By $\text{vec}(A) = [a_{11}, \ldots, a_{r1}, a_{12}, \ldots, a_{r2}, \ldots, a_{1s}, \ldots, a_{rs}]^t$ we denote the vectorization of the matrix A. We will use P_S to denote a subset S of rows of P. Let X, Y be random variables. We say that X is *stochastically larger* than Y ($X \succeq Y$) if $\Pr[X > x] \geq \Pr[Y > x]$ for all $x \in \mathbb{R}$. Also $X \sim \mathcal{N}(\mu, \sigma^2)$ denotes that X follows the normal distribution with mean μ and variance σ^2, also $\mathcal{N}_n(\mu', \Sigma)$ is the multivariate n dimensional normal distribution with mean vector μ' and covariance matrix Σ. Similarly, we can define the matrix variate Gaussian distribution, $\mathcal{N}_{n,d}(M, \Sigma'_{nd \times nd})$ with mean matrix M and covariance matrix Σ' of dimension $nt \times nt$. Note that the latter definition is equivalent with the multivariate case, considering its vectorization. However, if we restrict the structure of the correlation matrix Σ' we can capture the matrix form of the entries (see the following Definition).

Definition 1 (Gaussian Random Matrix). *The random matrix X of dimensions $n \times d$ is said to have a matrix variate normal distribution with mean matrix M of size $n \times d$ and covariance matrix $\Sigma \otimes \Psi$ (denoted by $X \sim \mathcal{N}_{n,d}(M, \Sigma'_{nd \times nd})$), where Σ, Ψ are positive definite matrices of size $n \times n$ and $d \times d$ respectively, if $\text{vec}(X^t) \sim \mathcal{N}_{nd}(\text{vec}(M^t), \Sigma \otimes \Psi)$.*

A brief explanation of the above definition is the following: The use of the tensor product $(\Sigma \otimes \Psi)$ is chosen to indicate that the correlation[2] between its entries has a specific structure. More concretely, every row is correlated with respect to the Σ covariance matrix and every column with respect to Ψ. Hence, the correlation between two entries of X, say X_{ij} and X_{lk}, $E[X_{ij}X_{lk}]$ is equal to $\Sigma_{il} \cdot \Psi_{jk}$.

3   A Regular Set of Points Preserves Its Volume

Assume that the set in Euclidean space we wish to reduce its dimensionality is the regular one. Consider a subset S of the regular set of size $s \leq k$ *with* the origin. Our goal is to show that the volume of the projection of such a set is very concentrated, assuming s is sufficient small. Also denote by $X \sim \mathcal{N}_{n,d}(0, I_{nd})$ the projection matrix[3]. Notice that since our input set is the regular (identity matrix), the image of their projection is simply X (the projection matrix), and recall that the points that correspond to S are represented by X_S. It is well known that the volume of the *projected* points of S is

$$\sqrt{\det(X_S X_S^t)}/s!.$$

Therefore the question of volumes is now reduced to one about the determinant of the Gram matrix of X_S.

We will use the following lemma which gives a simple characterization of this latter random variable.

Lemma 1 ([16]). *Let $X \sim \mathcal{N}_{k,d}(0, I_{kd})$. The k-dimensional volume of the parallelotope determined by $X_{\{i\}}$, $i = 1, \ldots, k$ is the product of two independent random variables one of which has a χ-distribution with $d - k + 1$ degrees of freedom and the other is distributed as the $k - 1$ dimensional volume of the parallelotope spanned by $k - 1$ independent Gaussian random vectors, i.e. $\mathcal{N}_{k-1,d}(0, I_{(k-1)d})$. Furthermore,*

---

[2] Since the entries have zero mean, the correlation between the entries ij and lk is $E[X_{ij}X_{lk}]$.

[3] For ease of presentation, we will not consider the normalization parameter $d^{-1/2}$ at this point.

$$\det(XX^t) \sim \prod_{i=1}^{k} \chi_{d-i+1}^2.$$

The proof is simple and geometric thus we supply it here for completeness.

Proof. Let $\Delta_d^{(k)} = \sqrt{\det(XX^t)}$ denote the volume of the parallelotope of the k random vectors. Then

$$\Delta_d^{(k)} = a_k \Delta_d^{(k-1)},$$

where $\Delta_d^{(k-1)}$ is the k-dimensional volume of the parallelotope determined by the set of vectors $X_1, X_2, \ldots, X_{k-1}$ and a_k is the distance of X_k from the subspace spanned by $X_1, X_2, \ldots, X_{k-1}$.

Now we will show that a_k is distributed as a Chi random variable with $d - k + 1$ degrees of freedom. Using the spherical symmetry of the distribution of the points we can assume w.l.o.g. that the points X_i $i = 1, \ldots, k - 1$ span the subspace $W = \{x \in \mathbb{R}^d | x(k) = x(k+1) = \cdots = x(d) = 0\}$, i.e. the set of points that the $d - k + 1$ last coordinates are equal to zero. Next we will show that $a_k \sim \chi_{d-k+1}$. Notice that the distance of the point X_k from the subspace that the rest $k - 1$ points span is equal to[4] $\text{dist}(X_k, W) = \sqrt{\sum_{i=k}^{d} X_{ki}^2}$, which is a Chi random variable of $d - k + 1$ degrees of freedom, since $X_{i,j} \sim \mathcal{N}(0, 1)$. Also note that a_k is independent of $\Delta_d^{(k-1)}$. Using the above statement recursively, we conclude that $\det(XX^t) \sim \prod_{i=1}^{k} \chi_{d-i+1}^2$ with the Chi-square random variables being independent. $\square$

Due to the normalization (see Theorem 1), it turns out that the random variable we are actually interested in is $(\det(X_S X_S^t))^{\frac{1}{2s}}$ and so is the geometric mean of a sequence of Chi-square independent random variables with similar numbers of degrees of freedom. This falls under the general framework of law-of-large-numbers, and we should typically expect an amplification of the concentration which grows exponentially with s. This statement is made formal by a concentration result of a (single) Chi-square random variable.

Theorem 2 (Theorem 4, [10]). *Let $u_i := \chi_{d-i+1}^2$ be independent Chi-square random variables for $i = 1, 2, \ldots, s$. If u_i are independent, then the following holds for every $s \geq 1$,*

$$\chi_{s(d-s+1)+\frac{(s-1)(s-2)}{2}}^2 \succeq s \left(\prod_{i=1}^{s} u_i \right)^{1/s} \succeq \chi_{s(d-s+1)}^2. \tag{1}$$

We are now ready to prove that the random embedding $f : \mathbb{R}^n \mapsto \mathbb{R}^d$ defined by $p \mapsto \frac{p^t X}{\sqrt{d}}$, $X \sim \mathcal{N}_{a,d}(0, I_{nd})$ for $p \in \mathbb{R}^n$ preserves the volume of regular sets of bounded size with high enough probability.

Lemma 2. *Let $0 < \varepsilon \leq 1/2$ and let f be the random embedding defined as above. Further, let S be a subset of $\mathbb{R}^n$ that contains the origin and s standard vectors, with $s < k < \frac{d\varepsilon}{2}$. Then we have that*

---

[4] The length of the orthogonal projection of X_k to the subspace W.

$$\Pr\left[1-\varepsilon<\left(\frac{\mathrm{vol}(f(S))}{\mathrm{vol}(S)}\right)^{\frac{1}{s}}<1+\varepsilon\right]\geq 1-2\exp\left(-s(d-(s-1))\frac{\varepsilon^2}{24}\right). \qquad (2)$$

Proof. We define the random variable $Z=\left(\det(X_S X_S^t)\right)^{1/s}$, $U=\frac{1}{s}\chi^2_{sd-\frac{s^2+s}{2}+1}$ its upper stochastic bound and $L=\frac{1}{s}\chi^2_{sd-s^2+s}$ its lower stochastic bound i.e.,

$$U\succeq Z\succeq L$$

holds from Theorem 2. Also note that this implies upper and lower bounds for the expectation of Z, $d-\frac{s+1}{2}+1/s\geq E[Z]\geq d-s+1$, with $E[L]-E[U]\geq-\frac{s}{2}$ for $s\geq 1$. Now we will relate the volume of an arbitrary subset of P with the random variable Z. Using that $\mathrm{vol}(f(S))=\frac{\sqrt{\det(X_S X_S^t)}}{d^{s/2}s!}$ and $\mathrm{vol}(S)=\frac{1}{s!}$, we get for the upper tail:

$$\Pr\left[\left(\frac{\mathrm{vol}(X_S)}{d^{s/2}\cdot\mathrm{vol}(I_S)}\right)^{\frac{1}{s}}>1+\varepsilon\right]=\Pr\left[\sqrt{\frac{Z}{d}}>(1+\varepsilon)\right]$$

$$\leq\Pr\left[\frac{Z}{E[Z]}>(1+\varepsilon)^2\right]$$

$$\leq\Pr\left[\frac{Z}{E[Z]}>1+2\varepsilon\right]$$

using that $d\geq E[Z]$. Similarly, the lower tail becomes

$$\Pr\left[\left(\frac{\mathrm{vol}(X_S)}{d^{s/2}\cdot\mathrm{vol}(I_S)}\right)^{\frac{1}{s}}<1-\varepsilon\right]=\Pr\left[\sqrt{\frac{Z}{d}}<(1-\varepsilon)\right]$$

$$=\Pr\left[\frac{Z}{d}<(1-\varepsilon)^2\right]$$

$$=\Pr\left[\frac{Z}{E[Z]}<\frac{d}{E[Z]}(1-\varepsilon)^2\right]$$

$$\leq\Pr\left[\frac{Z}{E[Z]}<(1+\varepsilon)(1-\varepsilon)^2\right]$$

$$\leq\Pr\left[\frac{Z}{E[Z]}<1-\varepsilon\right]$$

using that $\frac{d}{E[Z]}\leq 1+\varepsilon$, which is true since $d\geq 2k/\varepsilon$ and $\varepsilon\leq 1$. Now we bound the right tail of Z.

$$\Pr[Z-E[Z]\geq 2\varepsilon E[Z]]\leq\Pr[U-E[Z]\geq 2\varepsilon E[Z]]$$
$$=\Pr[U-E[U]\geq 2\varepsilon E[U]+(1+2\varepsilon)(E[Z]-E[U])]$$
$$\leq\Pr[U-E[U]\geq 2\varepsilon E[U]+(1+2\varepsilon)(E[L]-E[U])]$$

using $U\succeq Z$ and $E[Z]\geq E[L]$. Now we bound $(1+2\varepsilon)(E[L]-E[U])$ from below. It is not hard to show that $(1+2\varepsilon)(E[L]-E[U])\geq-\frac{3\varepsilon}{4}E[U]$ since $d\geq 2k/\varepsilon$. Therefore

$$\Pr[Z-E[Z]\geq 2\varepsilon E[Z]]\leq\Pr[U-E[U]\geq \varepsilon E[U]].$$

Now applying Lemma 4 on U we get the bound

$$\Pr[Z \geq (1+2\varepsilon)E[Z]] \leq \exp\left(-(sd - s(s-1)/2 + 1)\frac{\varepsilon^2}{6}\right).$$

For the other tail of the random variable Z, we have that

$$\begin{aligned}
\Pr[Z - E[Z] < -\varepsilon E[Z]] &\leq \Pr[L - E[Z] < -\varepsilon E[Z]] \\
&\leq \Pr[L - E[L] < -\varepsilon E[L] + (1-\varepsilon)(E[Z] - E[L])] \\
&\leq \Pr[L - E[L] < -\varepsilon E[U] + (E[U] - E[L])]
\end{aligned}$$

using that $Z \succeq L$ and $E[Z] \leq E[U]$. Again we bound $(E[U] - E[L])$ from above. It is not hard to show that $(E[U] - E[L]) \leq \frac{3}{8}\varepsilon E[L]$ since $d \geq 2k/\varepsilon$ and $\varepsilon \leq 1/2$, so

$$\Pr[Z - E[Z] < -\varepsilon E[Z]] \leq \Pr[L - E[L] < -\varepsilon/2 E[L]]$$

holds. Therefore applying Lemma 4 on L we get

$$\Pr[Z < (1-\varepsilon)E[Z]] \leq \exp\left(-(sd - s(s-1))\frac{\varepsilon^2}{24}\right).$$

Comparing the upper and lower bound the lemma follows. □

Remark: The bound on k ($k = O(d\varepsilon)$) is tight. While the probabilistic arguments show that the volume of a projection of a subset is concentrated around its mean, we really have to show that it is concentrated around the volume of the set (before the projection). In other words, it is a necessary condition that

$$\frac{\mu_s}{1/s!} = 1 \pm O(\varepsilon) \tag{3}$$

where μ_s is the expected normalized volume of a regular set of size s. As long as we deal with sets of fixed cardinality, we can easily scale Equation 3 making the LHS equal to 1. However, it turns out that $\frac{\mu_s}{1/s!}$ is decreasing in s and furthermore for sufficiently large s it may be smaller than $1 - O(\varepsilon)$. Here is why, $\frac{\mu_s}{1/s!} = E[(\prod_{i=1}^s \chi_{d-i+1}^2)^{\frac{1}{2s}}] \leq (\prod_{i=1}^s E[\chi_{d-i+1}^2]))^{\frac{1}{2s}} \leq \sqrt[2s]{d(d-1)\ldots(d-s+1)} \leq \sqrt{d - (s-1)/2}$ using independence, Jensen's inequality and arithmetic-geometric mean inequality. On the other hand[5], $\frac{\mu_1}{1/1!} = E[\chi_d] \geq \sqrt{d-1}$. Therefore, no matter what scaling is used we must have that $\sqrt{d-(s-1)/2}/\sqrt{d-1} \geq 1 - O(\varepsilon)$ for all $s \leq k$, from which it follows that $k \leq O(d\varepsilon)$.

4 Extension to the General Case

In this section we will show that if we randomly project a set of s points that are in general position, the (distribution of the) volume of the projection depends linearly *only* on

---

[5] A simple calculation using $E[\chi_d] = \sqrt{2}\frac{\Gamma(\frac{d+1}{2})}{\Gamma(\frac{d}{2})}$ and $E[\chi_d] \geq \sqrt{E[\chi_d]E[\chi_{d-1}]}$ gives the result.

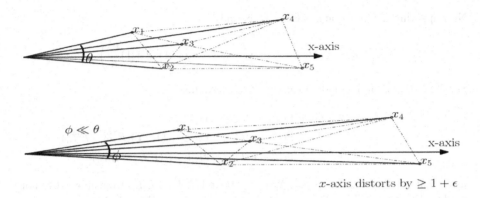

Fig. 1. Example that illustrates the extension of the regular case to the general

the volume of the original set. To gain some intuition, let's consider an example that is essentially as different as possible from the regular case. Consider the one-dimensional set of size s in $\mathbb{R}^n$, $(i,0,\ldots,0)$ with $i = 1,\ldots,s$. By adding a small random perturbation (and changing the location of points by distance at most $\delta \ll \varepsilon$) the points will be in general position, and the perturbed set will have positive volume. Consider a random projection π onto d dimensions, normalized so that in expectation distances do not change. Now, look at the event $E := \{\|\pi(e_1)\| > 1 + \varepsilon\}$ where e_1 is the first standard vector. We know that $\Pr[E] = \exp(-\Theta(d\varepsilon^2))$. But notice that when E occurs then π expands *all* distances in the set by a factor $1 + \varepsilon - O(\delta)$. At this point it may be tempting to conclude that event E implies that the set was roughly scaled by some factor that is at least $1 + \varepsilon$. If that were the case then it would mean that the probability of bad projections for this set would be too big, that is $e^{-\Theta(d\varepsilon^2)}$ instead of $e^{-\Theta(sd\varepsilon^2)}$.

However, this is not the case. The reason is that conditioning on the event E does not provide any information about the expansion or contraction of the perpendicular space of the x-axis. Conditioning on E, we observe that the angles between the x-axis and any two points will decrease, since the x-axis expands (see Figure 1). Therefore the intuition that this set is scaled (conditioned on E) is wrong, since it is "squeezed" in the e_1 direction.

Next we will prove a technical lemma that will allows us to extend the volume concentration from the regular set to a set of points in general position.

Lemma 3. *Let S be a $s \times n$ matrix so that every row corresponds to a point in $\mathbb{R}^n$. Assume Y_S of size $s \times d$ be the projected points of S, $|S| = s \leq d$ then*

$$\frac{\det(Y_S Y_S^t)}{\det(SS^t)} \sim \prod_{i=1}^{s} \chi_{d-i+1}^2. \tag{4}$$

Proof. First, observe that if $X \sim \mathcal{N}_{n,d}(0, I_n \otimes I_d)$ then $Y_S = SX \sim \mathcal{N}_{s,d}(0, (SS^t) \otimes I_d)$. To see this argument, note that any linear (fixed) combination of Gaussian random variables is Gaussian from the stability of Gaussian. Now by the linearity of expectation we can easily show that every entry of SX has expected value zero. Also the correlation between two entries $E[(SX)_{ij}(SX)_{lk}] = E[(\sum_{r=1}^{d} S_{ir}X_{rj})(\sum_{r=i}^{d} S_{lr}X_{rk})]$ is zero if $j \neq k$, and $S_i^t S_l$ otherwise.

We know that $Y_S \sim \mathcal{N}_{s,d}(0, SS^t \otimes I_d)$. Assuming that S has linearly independent rows (otherwise both determinants are zero), there exists an s-by-s matrix R so that $SS^t = RR^t$ (Cholesky Decomposition).

Now we will evaluate $\det(R^{-1}Y_S Y_S^t (R^t)^{-1})$ in two different ways. First note that $R^{-1}, Y_S Y_S^t, (R^t)^{-1}$ are square matrices so

$$\det(R^{-1}Y_S Y_S^t (R^t)^{-1}) = \frac{\det(Y_S Y_S^t)}{(\det(R))^2}. \tag{5}$$

Now note that $R^{-1}Y_S$ is distributed as $\mathcal{N}_{s,d}(0, R^{-1}SS^t(R^t)^{-1} \otimes I_d)$ which is equal to $\mathcal{N}_{s,d}(0, I_s \otimes I_d)$, since $R^{-1}SS^t(R^t)^{-1} = R^{-1}RR^t(R^t)^{-1} = I_s$. Lemma 1 with $R^{-1}Y_S$ implies that

$$\det(R^{-1}Y_S Y_S^t (R^t)^{-1}) \sim \prod_{i=1}^{s} \chi^2_{d-i+1}. \tag{6}$$

Using the fact that $(\det(R))^2 = \det(P_S P_S^t)$ with (5), (6) completes the proof.   □

Remark: A different and simpler proof of the above lemma can be achieved by using the more abstract property of the projections, namely the rotationally invariance property. Consider two sets of s vectors, S and T. Assume for now that $W = \text{span}(S) = \text{span}(T)$. Then for *every* transformation ϕ it holds that $\det^2(A) = \det(\phi(S)\phi(S)^t)/\det(SS^t) = \det(\phi(T)\phi(T)^t)/\det(TT^t)$ where A is the $s \times s$ matrix that describes ϕ using any choice of basis for W and $\phi(W)$. To remove the assumption that $\text{span}(S) = \text{span}(T)$, simply consider a rigid tansformation ψ from $\text{span}(S)$ to $\text{span}(T)$. By rotational invariance of the projection, the distribution of the volume of $\phi(S)$ and that of $\phi(\psi(S))$ is the same, hence we reduce to the case where the span of the sets is the same subspace. Putting it together, this shows that the LHS of (4) distributes the same way for all sets of (linearly independent) vectors of size s, which by Lemma 1, must also be the same as the RHS of (4). We note that we have opted to use the previous proof since Gaussian projections is the tool of choice in our analysis throughout.

To conclude, Lemma 3 implies that the distribution of the volume of any subset of points is *independent* of their geometry up to a multiplicative factor. However, since we are interested in the distortion (fraction) of the volume $\text{vol}(Y_S)/\text{vol}(P_S) = \frac{(\det(Y_S Y_S^t))^{1/2}/s!}{(\det(P_S P_S^t))^{1/2}/s!} = \sqrt{\frac{\det(Y_S Y_S^t)}{\det(P_S P_S^t)}}$ everything boils down to the orthonormal case.

Notice that so far we proved that any subset of the regular set that *contains* the origin gives us a good enough concentration. Combining this fact with the previous Lemma we will show that the general case also holds. Let a subset $P_S = \{p_0, p_1, \ldots, p_{s-1}\}$ of P. We can translate the set P_S (since volume is translation-invariant) so that p_0 is at the origin, and call the resulting set $P_S' = \{0, p_1 - p_0, \ldots, p_{s-1} - p_0\}$. Now it is not hard to see that combining Lemmata 2,3 on the set P_S', we get the following general result.

Theorem 3. *Let $0 < \varepsilon \leq 1/2$ and let $f : \mathbb{R}^n \to \mathbb{R}^d$ be the random embedding defined as above. Further, let S be an arbitrary subset of $\mathbb{R}^n$, with $|S| = s < \frac{d\varepsilon}{2}$. Then we have that*

$$\Pr\left[1 - \varepsilon < \left(\frac{\text{vol}(f(S))}{\text{vol}(S)}\right)^{\frac{1}{s-1}} < 1 + \varepsilon\right] \geq 1 - 2\exp\left(-s(d - (s-1))\frac{\varepsilon^2}{24}\right). \tag{7}$$

A closer look at the proof of Lemma 1 and Lemma 3 implies that the distance between any point and a subset of s points follows a Chi distribution with $d - s + 1$ degrees of freedom. This fact can be used to simplify the proof for the preservation of affine distances as stated in [14], using the same number of dimensions.

5   Proof of the Main Theorem

We now prove the main theorem.

Proof. (of Theorem 1) Let B_S be the event: "The volume of the subset S of P distorts (under the embedding) its volume by more that $(1 + \varepsilon)^{s-1}$". Clearly, the embedding fails if there is any S so that the event B_S occurs. We now bound the failure probability of the embedding from above

$$\Pr\left[\exists\, S : |S| < k,\ B_S\right] \leq \sum_{S:\ |S|<k} \Pr[B_S] \leq$$

$$2\sum_{s=1}^{k-1} \binom{n}{s} \exp\left(-s(d - (s-1))\frac{\varepsilon^2}{24}\right) \leq 2\sum_{s=1}^{k-1} \frac{n^s}{s^s} \exp\left(-s\left[(d - (s-1))\frac{\varepsilon^2}{24} - 1\right]\right)$$

using the union bound, Theorem 3 for any subset of size $s < k$ and bounds on binomial coefficients, i.e. $\binom{n}{s} \leq \left(\frac{ne}{s}\right)^s$. Now if

$$2\sum_{s=1}^{k-1} \frac{n^s}{s^s} \exp\left(-s\left[(d - (s-1))\frac{\varepsilon^2}{24} - 1\right]\right) < 1$$

then the probability that a random projection onto d dimesions doesn't distort the volume of any subset of size at most k by a relative error of ε, is positive.

Since $d > 2k/\varepsilon$, setting $d = 30\varepsilon^{-2}(\log n + 1) + k - 1 = O(\max\{k/\varepsilon, \varepsilon^{-2}\log n\})$ we get that, with positive probability, f has the desired property. $\square$

6   Discussion

We have shown a nearly tight dimension reduction that approximately preserves volumes of sets of size up to k. The main outstanding gap is in the range where $k \geq \log n$ where the dimension required to obtain a k-volume respecting embedding is between k and k/ε. We conjecture that the upper bound we have is tight, and that the lower bound should come from a regular set of points. This conjecture can be phrased as the following linear algebraic statement.

Conjecture. Let A be an $n \times n$ positive semidefinite matrix such that the determinant of every $s \times s$ principal minor ($s \leq k$) is between $(1 - \varepsilon)^{s-1}$ and 1. Then the rank of A is at least $\min\{\Omega(k/\varepsilon), n\}$.

We believe that closing gaps in questions of the type discussed above is particularly important as they will reaffirm a reccuring theme: the oblivious method of random Gaussian projections does as well as any other method. More interesting is to show that this is in fact not the case, and that sophisticated methods can go beyond this standard naive approach.

There is still a gap in our understanding with respect to dimension reduction that preserves all distances to affine subspaces spanned by small sets. Interestingly, this questions seems to be asking whether we can go beyond union bound reasoning when we deal with random projections. An example that captures this issue is a regular set where $\varepsilon < 1/k$. Here, it is implied by the proof in [14] that only $O(\varepsilon^{-2}\log n)$ dimensions are needed. However, the probability of failure for a particular event with this dimensionality is $n^{-O(1)}$, in other words not small enough to supply a proof simply by using the union bound. Does our technique extend to other dimension reduction techniques? Particularly, would projections onto ± 1 vectors provide the same dimension guarantees? Could Ailon and Chazelle's Fast JL transform substitute the original (dense) Gaussian matrix? As was mentioned in [14] the answer is yes when dealing with the weaker result that pays the extra factor of k, simply because the JL lemma is used as a "black box" there. We don't know what are the answers with respect to the stronger result of the current work, and we leave this as an open question.

References

1. Achlioptas, D.: Database-friendly random projections. In: PODS 2001: Proceedings of the twentieth ACM SIGMOD-SIGACT-SIGART symposium on Principles of database systems, pp. 274–281. ACM, New York (2001)
2. Agarwal, P.K., Har-Peled, S., Yu, H.: Embeddings of surfaces, curves, and moving points in euclidean space. In: SCG 2007: Proceedings of the twenty-third annual symposium on Computational geometry, pp. 381–389. ACM, New York (2007)
3. Ailon, N., Chazelle, B.: Approximate nearest neighbors and the fast johnson-lindenstrauss transform. In: STOC 2006: Proceedings of the thirty-eighth annual ACM symposium on Theory of computing, pp. 557–563. ACM, New York (2006)
4. Ailon, N., Liberty, E.: Fast dimension reduction using rademacher series on dual bch codes. In: SODA 2008: Proceedings of the nineteenth annual ACM-SIAM symposium on Discrete algorithms, Philadelphia, PA, USA. Society for Industrial and Applied Mathematics, pp. 1–9 (2008)
5. Alon, N.: Problems and results in extremal combinatorics, i. Discrete Math. (273), 31–53 (2003)
6. Clarkson, K.L.: Tighter bounds for random projections of manifolds. In: SCG 2008: Proceedings of the twenty-fourth annual symposium on Computational geometry, pp. 39–48. ACM, New York (2008)
7. Engebretsen, L., Indyk, P., O'Donnell, R.: Derandomized dimensionality reduction with applications. In: SODA 2002: Proceedings of the thirteenth annual ACM-SIAM symposium on Discrete algorithms, Philadelphia, PA, USA. Society for Industrial and Applied Mathematics, pp. 705–712 (2002)
8. Feige, U.: Approximating the bandwidth via volume respecting embeddings (extended abstract). In: STOC 1998: Proceedings of the thirtieth annual ACM symposium on Theory of computing, pp. 90–99. ACM, New York (1998)
9. Frankl, P., Maehara, H.: The johnson-lindenstrauss lemma and the sphericity of some graphs. J. Comb. Theory Ser. A 44(3), 355–362 (1987)
10. Gordon, L.: Bounds for the distribution of the generalized variance. The Annals of Statistics 17(4), 1684–1692 (1989)

11. Indyk, P., Naor, A.: Nearest-neighbor-preserving embeddings. ACM Trans. Algorithms 3(3), 31 (2007)
12. Johnson, W.B., Lindenstrauss, J.: Extensions of lipschitz mappings into a hilbert space. In: Amer. Math. Soc. (ed.) Conference in modern analysis and probability, pp. 189–206. Providence, RI (1984)
13. Liberty, E., Ailon, N., Singer, A.: Fast random projections using lean walsh transforms. In: RANDOM (to appear, 2008)
14. Magen, A.: Dimensionality reductions in ℓ_2 that preserve volumes and distance to affine spaces. Discrete & Computational Geometry 38(1), 139–153 (2007)
15. Matousek, J.: On the variants of johnson lindenstrauss lemma (manuscript) (2006)
16. Prekopa, A.: On random determinants i. Studia Scientiarum Mathematicarum Hungarica (2), 125–132 (1967)
17. Sarlos, T.: Improved approximation algorithms for large matrices via random projections. In: Proceedings of the 47th Annual IEEE Symposium on Foundations of Computer Science, Washington, DC, USA, pp. 143–152. IEEE Computer Society, Los Alamitos (2006)
18. Sivakumar, D.: Algorithmic derandomization via complexity theory. In: STOC 2002: Proceedings of the thirty-fourth annual ACM symposium on Theory of computing, pp. 619–626. ACM, New York (2002)
19. Vempala, S.: Random projection: A new approach to vlsi layout. In: FOCS 1998: Proceedings of the 39th Annual Symposium on Foundations of Computer Science, Washington, DC, USA, p. 389. IEEE Computer Society, Los Alamitos (1998)
20. Wakin, M.B., Baraniuk, R.G.: Random projections of signal manifolds. In: Proceedings of IEEE International Conference on Acoustics, Speech and Signal Processing. ICASSP 2006, vol. 5, p. V (May 2006)

Appendix

Concentration Bounds for χ^2

Lemma 4 ([1]). *Let* $\chi_t^2 = \sum_{i=1}^{t} X_i^2$, *where* $X_i \sim \mathcal{N}(0,1)$. *Then for every* ε, *with* $0 < \varepsilon \leq 1/2$, *we have that*

$$\Pr\left[\chi_t^2 \leq (1-\varepsilon)E[\chi_t^2]\right] \leq \exp(-t\frac{\varepsilon^2}{6})$$

and

$$\Pr\left[\chi_t^2 \geq (1+\varepsilon)E[\chi_t^2]\right] \leq \exp(-t\frac{\varepsilon^2}{6}).$$

Sampling Hypersurfaces through Diffusion

Hariharan Narayanan and Partha Niyogi

Department of Computer Science, University of Chicago, USA
{hari,niyogi}@cs.uchicago.edu

Abstract. We are interested in efficient algorithms for generating random samples from geometric objects such as Riemannian manifolds. As a step in this direction, we consider the problem of generating random samples from smooth hypersurfaces that may be represented as the boundary ∂A of a domain $A \subset \mathbb{R}^d$ of Euclidean space. A is specified through a membership oracle and we assume access to a blackbox that can generate uniform random samples from A. By simulating a diffusion process with a suitably chosen time constant t, we are able to construct algorithms that can generate points (approximately) on ∂A according to a (approximately) uniform distribution.

We have two classes of related but distinct results. First, we consider A to be a convex body whose boundary is the union of finitely many smooth pieces, and provide an algorithm (Csample) that generates (almost) uniformly random points from the surface of this body, and prove that its complexity is $O^*(\frac{d^4}{\epsilon})$ per sample, where ϵ is the variation distance. Next, we consider A to be a potentially non-convex body whose boundary is a smooth (co-dimension one) manifold with a bound on its absolute curvature and diameter. We provide an algorithm (Msample) that generates almost uniformly random points from ∂A, and prove that its complexity is $O(\frac{R}{\sqrt{\epsilon \tau}})$ where $\frac{1}{\tau}$ is a bound on the curvature of ∂A, and R is the radius of a circumscribed ball.

1 Introduction

Random sampling has numerous applications. They are ingredients in statistical goodness-of-fit tests and Monte-Carlo methods in numerical computation. In computer science, they have been used to obtain approximate solutions to problems that are otherwise intractable. A large fraction of known results in sampling that come with guarantees belong to the discrete setting. A notable exception is the question of sampling convex bodies in $\mathbb{R}^d$. A large body of work has been devoted to this question (in particular [8], [10]) spanning the past 15 years leading to important insights and algorithmic progress.

However, once one leaves the convex domain setting, much less is known. We are interested in the general setting in which we wish to sample a set that may be represented as a submanifold of Euclidean space. While continuous random processes on manifolds have been analyzed in several works, (such as those of P. Matthews [11],[12]), as far as we can see, these do not directly lead to algorithms with complexity guarantees.

A. Goel et al. (Eds.): APPROX and RANDOM 2008, LNCS 5171, pp. 535–548, 2008.

Our interest in sampling a manifold is motivated by several considerations from diverse areas in which such a result would be applicable. In machine learning, the problem of clustering may be posed as finding (on the basis of empirically drawn data points) a partition of the domain (typically $\mathbb{R}^d$) into a finite number of pieces. In the simplest form of this (partition into two pieces) the partition boundary (if smooth) may be regarded as a submanifold of co-dimension one and the best partition is the one with smallest volume (in a certain sense corresponding to a natural generalization of Cheeger's cut of a manifold). More generally, the area of *manifold learning* has drawn considerable attention in recent years within the machine learning community (see [5,18] among others) and many of the questions may be posed as learning geometric and topological properties of a submanifold from randomly drawn samples on it. In scientific computing, one may be interested in numerical methods for integrating functions on a manifold by the Monte Carlo method. Alternatively, in many physical applications, one may be interested in solving partial differential equations where the domain of interest may have the natural structure of a manifold. In contrast to a finite element scheme on a deterministic triangulation (difficult to obtain in high dimensions), one may explore randomized algorithms by constructing a random mesh and solving such PDEs on such a mesh. Finally, in many applications to dynamical systems, one is interested in the topology of the space of attractors which have the natural structure of a manifold (see [13]). In statistics, one is interested in goodness of fit tests for a variety of multivariate random variables. For example, testing for a gamma distribution leads one to consider (positive real valued) random variables $X_1, \ldots, X_n$ such that $\sum_i X_i = a$ and $\prod_j X_j = b$. The set of all $(X_1, \ldots, X_n)$ under these constraints is the boundary of a convex body in the hyperplane defined by $\sum_i X_i = a$. Sampling this is a question that arises naturally in this setting (see [6], [7]).

Thus, we see that building an efficient sampler for a manifold is a problem of fundamental algorithmic significance. Yet, not much is known about this and as a step in this general direction, in the current paper, we address the problem of sampling manifolds that are boundaries of open sets in $\mathbb{R}^d$ from the measure induced by the Lebesgue measure. The particular setting we consider in this paper has direct applications to clustering and goodness of fit tests where co-dimension 1 manifolds naturally arise. In addition, we also provide an algorithm and obtain complexity bounds for sampling the surface of a convex body – a problem to which we have not seen a solution at the present moment.

1.1   Summary of Main Results

We develop algorithms for the following tasks.

Our basic setting is as follows. Consider an open set $A \subset \mathbb{R}^d$ specified through a membership oracle. Assume we have access to an efficient sampler for A and now consider the task of uniformly sampling the (hyper) surface ∂A. We consider two related but distinct problems in this setting.

(i) A is a convex body satisfying the usual constraint of $B_r \subset A \subset B_R$ where B_r and B_R are balls of radius r and R respectively. Then an efficient sampler for

A is known to exist. However, no sampler is known for the surface of the convex body. It is worth noting that a number of intuitively plausible algorithms suggest themselves immediately. One idea may be draw a point x from A, shoot a ray in the direction from 0 to x and find its intersection with the boundary of the object. This will generate non-uniform samples from the surface (and it has been studied under the name Liouville measure.) A second idea may be to consider building a sampler for the set difference of a suitable expansion of the body from itself. This procedure has a complexity of at least $O^*(d^{8.5})$ oracle calls with the present technology because there is no method known to simulate each membership call to the expanded body using less than $O^*(d^{4.5})$ calls (see [4]).

Our main result here (Theorem 1) is to present an algorithm that will generate a sample from an approximately uniform distribution with $O^*(\frac{d^4}{\epsilon})$ calls to the membership oracle where ϵ is the desired variation distance to the target.

Beyond theoretical interest, the surface of the convex body setting has natural applications to many goodness of fit tests in statistics. The example of the gamma distribution discussed earlier requires one to sample from the set $\prod_i X_i = b$ embedded in the simplex (given by $\sum_j X_j = a$). This set corresponds to the boundary of a convex object.

(ii) A is a domain (not necessarily convex) such that its boundary ∂A has the structure of a smooth submanifold of Euclidean space of co-dimension one. A canonical example of such a setting is one in which the submanifold is the zeroset of a smooth function $f : \mathbb{R}^d \to \mathbb{R}$. A is therefore given by $A = \{x | f(x) < 0\}$. In machine learning applications, the function f may often be related to a classification or clustering function. In numerical computation and boundary value problems, one may wish to integrate a function subject to a constraint (given by $f(x) = 0$).

In this setting, we have access to a membership oracle for A (through f) and we assume a sampler for A exists. Alternatively, $A \subset K$ such that it has nontrivial fraction of a convex body K and one can construct a sampler for A sampling from K and using the membership oracle for rejection.

In this non-convex setting, not much is known and our main result (Theorem 2) is an algorithm that generates samples from ∂A that are approximately uniform with complexity $O^*(\frac{R}{\tau\sqrt{\epsilon}})$ where τ is a parameter related to the curvature of the manifold, R is the radius of a circumscribed ball and ϵ is an upper bound on the total variation distance of the output from uniform.

1.2 Notation

Let $\|.\|$ denote the Euclidean norm on $\mathbb{R}^d$. Let λ denote the Lebesgue measure on $\mathbb{R}^d$. The induced measure onto the surface of a manifold $\mathcal{M}$ shall be denoted $\lambda_\mathcal{M}$. Let

$$G^t(x, y) := \frac{1}{(4\pi t)^{\frac{d}{2}}} e^{-\frac{\|x-y\|^2}{4t}}.$$

be the d dimensional gaussian.

Definition 1. *Given two measures μ and ν over $\mathbb{R}^d$, let*

$$\|\mu - \nu\|_{TV} := \sup_{A \subseteq \mathbb{R}^d} |\mu(A) - \nu(A)|$$

denote the total variation distance between μ and ν.

Definition 2. *Given two measures μ and ν on $\mathbb{R}^d$, the transportation distance $d_{TR}(\mu, \nu)$ is defined to be the infimum*

$$\inf_{\gamma} \int \|x - y\| \, d\gamma(x, y).$$

taken over all measures γ on $\mathbb{R}^d \times \mathbb{R}^d$ such that for measurable sets A and B, $\gamma(A \times \mathbb{R}^d) = \mu(A)$, $\gamma(\mathbb{R}^d \times B) = \nu(B)$.

Notation: *We say that $n = O^*(m)$, if $n = O(m \, polylog(m))$. In the complexity analysis, we shall only consider the number of oracle calls made, as is customary in this literature.*

2   Sampling the Surface of a Convex Body

Let B be the unit ball in $\mathbb{R}^d$. Let B_α denote the ball of radius α centred at the origin. Consider a convex body K in $\mathbb{R}^d$ such that

$$B_r \subseteq K \subseteq B_R.$$

Let $\mathcal{B}$ be a source of random samples from K. Our main theorem is

Theorem 1. *Let K be a convex body whose boundary ∂K is a union of finitely many smooth Hypersurfaces.*

1. *The output of Csample has a distribution $\tilde{\mu}$, whose variation distance measured against the uniform distribution $\tilde{\lambda} = \tilde{\lambda}_{\partial K}$ is $O(\epsilon)$,*

$$\|\tilde{\mu} - \nu\|_{TV} \le O(\epsilon).$$

2. *The expected number of oracles calls made by Csample (to $\mathcal{B}$ and the membership oracle of K) for each sample of Csample is $O^*(\frac{d}{\epsilon})$ (, giving a membership query complexity of $O^*(\frac{d^4}{\epsilon})$ for one random sample from ∂K.)*

2.1   Algorithm Csample

Algorithm 1. *Csample*

1. *Estimate (see [15]) with confidence $> 1 - \epsilon$, the smallest eigenvalue κ of the Inertia matrix $A(K) := \mathbb{E}[(x - \overline{x})(x - \overline{x})^T]$ where x is random in uniformly K, to within relative error $1/2$ using $O(d \log^2(d) \log \frac{1}{\epsilon})$ random samples (see Rudelson [16].)*

2. Set

$$\sqrt{t} := \frac{\epsilon\sqrt{\kappa}}{32d}.$$

3. (a) Set $p = $ Ctry (t) .
(b) If $p = \emptyset$, goto (3a). Else output p.

Algorithm 2. *Ctry (t):*

1. *Use $\mathcal{B}$ to generate a random point x from the uniform distribution on K.*
2. *Let $y := Gaussian(x, 2tI)$ be a random vector chosen from a spherical d-dimensional Gaussian distribution with covariance $2tI$ and mean x.*
3. *Let ℓ the segment whose endpoints are x and y.*
4. *If $y \notin K$ output $\ell \cap \partial K$, else output $\emptyset$.*

2.2   Correctness

In our calculations, $z \in \partial K$ will be be a generic point at which ∂K is smooth. In particular for all such z, there is a (unique) tangent hyperplane. Let $\lambda_{\partial K}$ denote the $n - 1$-dimensional surface measure on ∂K. Let S and V denote the surface area and volume, respectively, of K. Let $\mu_{\partial K}$ denote the measure induced by the output of algorithm Csample . Let $|\mu|$ denote the total mass for any measure μ. We shall define a measure $\mu_{\partial K}$ on ∂K related to the "local diffusion" out of small patches. Formally, if Δ a subset of ∂K, the measure assigned to it by $\mu_{\partial K}$ is

$$\mu_{\partial K}(\Delta) := \int_{x \in S} \int_{y \in \mathbb{R}^d \setminus S} G^t(x, y) \mathcal{I}\left[\overline{xy} \cap \Delta \neq \emptyset\right] d\lambda(x) d\lambda(y) \tag{1}$$

where $\mathcal{I}$ is the indicator function and $G^t(x, y)$ is the spherical Gaussian kernel with covariance matrix $2tI$. Note that

$$V\mathbb{P}[\text{Ctry } (t) \in \Delta] = \mu_{\partial K}(\Delta).$$

Theorem 1 (part 1)
The output of Csample has a distribution $\tilde{\mu} = \frac{\mu_{\partial K}}{|\mu_{\partial K}|}$, whose variation distance measured against the uniform distribution $\tilde{\lambda}_{\partial K}$ is $O(\epsilon)$,

$$\|\tilde{\mu} - \tilde{\lambda}_{\partial K}\|_{TV} \leq O(\epsilon).$$

Proof. It follows from lemma 3 to note that at generic points, *locally* the measure generated by one trial of Ctry (t) is always less than the value predicted by its small t asymptotics $\sqrt{\frac{t}{\pi}} \frac{S}{V}$, i.e.

$$\forall \text{ generic } z \in \partial K, \quad \frac{d\mu_{\partial K}}{d\lambda_{\partial k}} < \sqrt{\frac{t}{\pi}} S.$$

Thus we have a local upper bound on $\frac{d\mu_{\partial K}}{d\lambda_{\partial K}} \leq \sqrt{\frac{t}{\pi}}$ uniformly for all generic points $z \in \partial K$. It would now suffice to prove almost matching *global* lower bound on the total measure, of the form

$$|\mu_{\partial K}| > (1 - O(\epsilon))\sqrt{\frac{t}{\pi}} S.$$

This is true by Proposition 4.1 in [3]. This proves that

$$\|\tilde{\mu} - \tilde{\lambda}_{\mathcal{M}}\|_{TV} \leq O(\epsilon.).$$ □

2.3 Complexity

The number of random samples needed to estimate the Inertia matrix is $O^*(d)$ (so that the estimated eigenvalues are all within $(0.5, 1.5)$ of their true values with confidence $1 - \epsilon$) from results of Rudelson ([16]). It is known that a convex body contains a ball of radius $\geq \sqrt{\Lambda_{min}(K)}$. Here $\Lambda_{min}(K)$ is the smallest eigenvalue of $A(K)$. Therefore, K contains a ball of radius r_{in}, where $r_{in}^2 = \frac{9}{10}\kappa$.

Theorem 1 (part 2)
The expected number of oracles calls made by Csample (to $\mathcal{B}$ and the membership oracle of K) for each sample of Csample is $O^*(\frac{d}{\epsilon})$ (, giving a total complexity of $O^*(\frac{d^4}{\epsilon})$ for one random sample from ∂K.)

Proof. The following two results will be used in this proof.

Lemma 1. *Lemma 5.5 in [3]]Suppose x has the distribution of a random vector (point) in K, define $A(K) := \mathbb{E}[(x - \bar{x})(x - \bar{x})^T]$. Let $\frac{5}{2}r_{in}^2$ be greater than the smallest eigenvalue of this (positive definite) matrix, as is the case in our setting. Then, $\frac{V}{S} < 4r_{in}$.*

Define $F_t := \sqrt{\frac{\pi}{t}}|\mu_{\partial K}|$.

Lemma 2 (Lemma 5.4 in [3]). *Suppose K contains a ball of radius r_{in}, (as is the case in our setting) then $S\left(1 - \frac{d\sqrt{\pi t}}{2r_{in}}\right) < F_t$.*

Applying Lemma 2, we see that

$$F_t > (1 - O(\epsilon))S.$$

The probability that Ctry succeeds in one trial is

$$\mathbb{P}[\text{Ctry } (t) \neq \emptyset] = \sqrt{\frac{t}{\pi}\frac{F_t}{V}} \tag{2}$$

$$> \sqrt{\frac{t}{\pi}\frac{S}{V}}(1 - O(\epsilon)) \tag{3}$$

$$> \sqrt{\frac{t}{\pi}\frac{1 - O(\epsilon)}{4r_{in}}} \quad \text{(By Lemma 1)} \tag{4}$$

$$> \Omega(\frac{\epsilon}{d}). \tag{5}$$

Therefore the expected number of calls to $\mathcal{B}$ and the membership oracle is $O^*(\frac{d}{\epsilon})$. By results of Lovász and Vempala ([9]) this number of random samples can be obtained using $O^*(\frac{d^4}{\epsilon})$ calls to the membership oracle. □

2.4 Extensions

S. Vempala [17] has remarked that these results can be extended more generally to sampling certain subsets of the surface ∂K of a convex body such as $\partial K \cap H$ for a halfspace H. In this case $K \cap H$ is convex too, and so Csample can be run on $K \cap H$. In order to obtain complexity guarantees, it is sufficient to bound from below, by a constant, the probability that Csample run on $H \cap K$ outputs a sample from $\partial K \cap H$ rather than $\partial H \cap K$. This follows from the fact that $\partial H \cap K$ is the unique minimal surface spanning $\partial K \cap \partial H$ and so has a surface area that is less than that of $\partial K \cap H$.

3 Sampling Well Conditioned Hypersurfaces

3.1 Preliminaries and Notation

Let $\mathcal{M}$ be a (codimension one) hypersurface.

Definition 3. *Let $\mathcal{M}$ be a codimension 1 hypersurface. The condition number of $\mathcal{M}$ is defined as $\frac{1}{\tau}$ where τ is is the largest number with the following property: No two normals to $\mathcal{M}$ of length less than τ intersect.*

In fact $\frac{1}{\tau}$ is an upper bound on the curvature of $\mathcal{M}$ ([14]). In this paper, we shall restrict attention to a τ-conditioned manifold $\mathcal{M}$ that is also the boundary of a compact subset $U \in \mathbb{R}^d$.

Suppose we have access to a Black-Box $\mathcal{B}$ that produces i.i.d random points $x_1, x_2, \ldots$ from the uniform probability distribution on U. We shall describe a simple procedure to generate almost uniformly distributed points on $\mathcal{M}$.

3.2 Algorithm Msample

The input to Msample is an error parameter ϵ, a guarantee τ on the condition number of $\mathcal{M}$ and a Black-Box $\mathcal{B}$ that generates i.i.d random points from the uniform distribution on U as specified earlier. We are also provided with a membership oracle to U, of which $\mathcal{M}$ is the boundary. We shall assume that U is

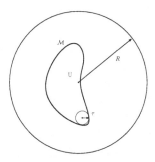

Fig. 1.

contained in a Euclidean ball of radius R, B_R. Msample , like Csample is a Las Vegas algorithm.

Let the probability measure of the output be $\tilde{\mu}_{out}$. The following is the main theorem of this section. Note that given perfectly random samples from U, the output probability distribution is close to the uniform in ℓ_∞, which is *stronger* than a total variation distance bound, and the number of calls to the Black box $\mathcal{B}$ is *independent* of dimension.

Theorem 2. *Let $\mathcal{M}$ be a τ-conditioned hypersurface that is the boundary of an open set contained in a ball of radius R. Let $\tilde{\mu}_{out}$ be the distribution of the output of Msample .*

1. *Let $\tilde{\lambda}_{\mathcal{M}}$ be the uniform probability measure on $\mathcal{M}$. Then, for any subset Δ of $\mathcal{M}$, the probability measure $\tilde{\mu}_{out}$ satisfies*

$$1 - O(\epsilon) < \frac{\tilde{\mu}_{out}(\Delta)}{\tilde{\lambda}_{\mathcal{M}}(\Delta)} < 1 + O(\epsilon).$$

2. *The total expected number of calls to $\mathcal{B}$ and the membership oracle of U is $O(\frac{R(1+\frac{2}{d}\ln\frac{1}{\epsilon})}{\tau\sqrt{\epsilon}})$.*

Algorithm 3. *Msample*

1. Set $\sqrt{t} := \frac{\tau\sqrt{\epsilon}}{4(d+2\ln\frac{1}{\epsilon})}$.
2. Set $p = $ Mtry (t) .
3. If $p = \emptyset$, goto (2). Else output p.

Algorithm 4. *Mtry (t)*

1. Use $\mathcal{B}$ to generate a point x from U.
2. Generate a point $y := Gaussian(x, 2tI)$ from a spherical d-dimensional Gaussian of mean x and covariance matrix $2tI$.
3. If $y \in U$ output $\emptyset$.
   Else output an arbitrary element of $\overline{xy} \cap \mathcal{M}$ using binary search. (Unlike the convex case, $|\overline{xy} \cap \mathcal{M}|$ is no longer only 0 or 1.)

3.3   Correctness

Proof of part (1) of Theorem 2. We shall define a measure $\mu_{\mathcal{M}}$ on $\mathcal{M}$ related to the "local heat flow" out of small patches. Formally, if Δ a subset of $\mathcal{M}$, the measure assigned to it by $\mu_{\mathcal{M}}$ is

$$\mu_{\mathcal{M}}(\Delta) := \int_{x \in U} \int_{y \in \mathbb{R}^d \setminus U} G^t(x, y) \mathcal{I}\left[\overline{xy} \cap \Delta \neq \emptyset\right] d\lambda(x) d\lambda(y) \tag{6}$$

where $\mathcal{I}$ is the indicator function and $G^t(x, y)$ is the spherical Gaussian kernel with covariance matrix $2tI$. For comparison, we shall define μ_{out} by

$$\mu_{out} := V\tilde{\mu}_{out}\mathbb{P}[\text{Mtry } (t) \neq \emptyset].$$

Since Msample outputs at most one point even when $|\overline{xy} \cap \mathcal{M}| > 1$, we see that for all $\Delta \subseteq \mathcal{M}$,

$$\mu_{out}(\Delta) \leq \mu_{\mathcal{M}}(\Delta).$$

The following Lemma provides a uniform *upper* bound on the Radon-Nikodym derivative of $\mu_{\mathcal{M}}$ with respect to the induced Lebesgue measure on $\mathcal{M}$.

Lemma 3. *Let $\lambda_{\mathcal{M}}$ be the measure induced on $\mathcal{M}$ by the Lebesgue measure λ on $\mathbb{R}^d$. Then*

$$\frac{d\mu_{\mathcal{M}}}{d\lambda_{\mathcal{M}}} < \sqrt{\frac{t}{\pi}}.$$

The Lemma below gives a uniform *lower* bound on $\frac{d\mu_{out}}{d\lambda_{\mathcal{M}}}$.

Lemma 4. *Let $\sqrt{t} = \frac{\tau\sqrt{\epsilon}}{4(d+2\ln\frac{1}{\epsilon})}$. Then*

$$\frac{d\mu_{out}}{d\lambda_{\mathcal{M}}} > \sqrt{\frac{t}{\pi}}(1 - O(\epsilon)).$$

Together the above Lemmas prove the first part of the Theorem. Their proofs have been provided below.

3.4   Complexity

Proof of part (2) of Theorem 2. Let S be the surface area of U (or the $d-1$-dimensional volume of $\mathcal{M}$.) Let V be the d-dimensional volume of U. We know that $U \subseteq B_R$. Since of all bodies of equal volume, the sphere minimizes the surface area, and $\frac{S}{V}$ decreases as the body is dilated,

$$\frac{S}{V} \geq \frac{d}{R}.$$

Lemma 4 implies that

$$\mathbb{P}[\text{Mtry }(t) \neq \emptyset] > \frac{S\sqrt{\frac{t}{\pi}}(1 - O(\epsilon))}{V} \tag{7}$$

$$\geq \frac{d}{R}\frac{\tau\sqrt{\epsilon}(1 - O(\epsilon))}{8(d + 2\ln\frac{1}{\epsilon})} \tag{8}$$

$$= \Omega(\frac{\tau\sqrt{\epsilon}}{R(1 + \frac{2}{d}\ln\frac{1}{\epsilon})}). \tag{9}$$

This completes the proof.   □

In our proofs of Lemma 3 and Lemma 4, we shall use the following Theorem of C. Borell.

Theorem 3 (Borell, [2]). *Let $\mu_t = G^t(0, \cdot)$ be the d-dimensional Gaussian measure with mean 0 and covariance matrix $2It$. Let A be any measurable set in $\mathbb{R}^d$ such that $\mu(A) = \frac{1}{2}$. Let A_ϵ be the set of points at a distance $\geq \epsilon$ from A. Then, $\mu_t(A_\epsilon) \geq 1 - e^{\frac{-\epsilon^2}{4t}}$.*

Fact: With μ_t as above, and $B(R)$ the Euclidean ball of radius R centered at 0, $\frac{1}{2} < \mu_t(B(\sqrt{2dt}))$.

Proof of Lemma 3. Let H be a halfspace and ∂H be its hyperplane boundary. Halfspaces are invariant under translations that preserve their boundaries. Therefore for any halfspace H, $\mu_{\partial H}$ is uniform on ∂H. Noting that the image of a Gaussian under a linear transformation is a Gaussian, it is sufficient to consider the 1-dimensional case to compute the $d - 1$-dimensional density $\frac{d\mu_{\partial H}}{d\lambda_{\partial H}}$.

$$\frac{d\mu_{\partial H}}{d\lambda_{\partial H}} = \int_{\mathbb{R}^-} \int_{\mathbb{R}^+} G^t(x, y) d\lambda(x) d\lambda(y), \tag{10}$$

which evaluates to $\sqrt{\frac{t}{\pi}}$ by a direct calculation. For any $z \in \mathcal{M}$, let H_z be the halfspace with the same outer normal as U such that ∂H_z is tangent to $\mathcal{M}$ at z. Let Δ be a small neighborhood of z in $\mathbb{R}^d$, and $|\Delta|$ denote its diameter.

$$
\begin{aligned}
\frac{d\mu_{\mathcal{M}}}{d\lambda_{\mathcal{M}}}(z) &= \lim_{|\Delta| \to 0} \frac{\int_{x \in U} \int_{y \in \mathbb{R}^d \setminus U} G^t(x, y) \mathcal{I}\left[\overline{xy} \cap \Delta \neq \emptyset\right] d\lambda(x) \, d\lambda(y)}{\lambda_{\mathcal{M}}(\Delta)} \\
&= \lim_{|\Delta| \to 0} \frac{\int_{x \in \mathbb{R}^d} \int_{y \in \mathbb{R}^d} G^t(x, y) \mathcal{I}\left[\overline{xy} \cap \Delta \neq \emptyset\right] \mathcal{I}[x \in U \text{ and } y \in \mathbb{R}^d \setminus U] \, d\lambda(x) \, d\lambda(y)}{\lambda_{\mathcal{M}}(\Delta)} \\
&< \lim_{|\Delta| \to 0} \frac{\int_{x \in \mathbb{R}^d} \int_{y \in \mathbb{R}^d} G^t(x, y) \mathcal{I}\left[\overline{xy} \cap \Delta \neq \emptyset\right] d\lambda(x) \, d\lambda(y)}{2\lambda_{\mathcal{M}}(\Delta)} \\
&= \frac{d\mu_{\partial H_z}}{d\lambda_{\partial H_z}}(z) \\
&= \sqrt{\frac{t}{\pi}}.
\end{aligned}
$$

The inequality in the above array of equations is strict because U is bounded. $\qquad\square$

Proof of Lemma 4. Let Δ be a small neighborhood of z in $\mathbb{R}^d$. Since $\mathcal{M}$ is a τ-conditioned manifold, for any $z \in \mathcal{M}$, there exist two balls $B_1 \subseteq U$ and $B_2 \subseteq \mathbb{R}^d \setminus U$ of radius τ that are tangent to $\mathcal{M}$ at z.

$$\frac{d\mu_{out}}{d\lambda_{\mathcal{M}}}(z) > \lim_{|\Delta| \to 0} \frac{\int_{x \in B_1} \int_{y \in B_2} G^t(x, y) \mathcal{I}\left[\overline{xy} \cap \Delta \neq \emptyset\right] d\lambda(x) \, d\lambda(y)}{\lambda_{\mathcal{M}}(\Delta)}.$$

The above is true because $|\overline{xy} \cap \mathcal{M}| = 1$ if $x \in B_1$ and $y \in B_2$. Let us define

$$\mathbb{P}_\tau := \lim_{|\Delta| \to 0} \frac{\int_{x \in B_1} \int_{y \in B_2} G^t(x, y) \mathcal{I}\left[\overline{xy} \cap \Delta \neq \emptyset\right] d\lambda(x) \, d\lambda(y)}{\int_{x \in H_z} \int_{y \in \mathbb{R}^d \setminus H_z} G^t(x, y) \mathcal{I}\left[\overline{xy} \cap \Delta \neq \emptyset\right] d\lambda(x) \, d\lambda(y)}. \tag{11}$$

Then

$$\mathbb{P}_\tau < \sqrt{\frac{\pi}{t} \frac{d\mu_{out}}{d\lambda_{\mathcal{M}}}(z)}.$$

The proof now follows from

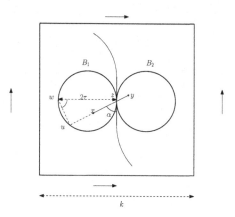

Fig. 2.

Lemma 5. $\mathbb{P}_\tau > 1 - O(\epsilon)$. □

Proof of Lemma 5. In order to obtain bounds on $\mathbb{P}_\tau$, we shall follow the strategy of mapping the picture onto a sufficiently large torus and doing the computations on this torus. This has the advantage that now averaging arguments can be used over the torus by virtue of its being compact (and a symmetric space.) These arguments do not transfer to $\mathbb{R}^d$ in particular because it is not possible to pick a point uniformly at random on $\mathbb{R}^d$.

Consider the natural surjection

$$\phi_k : \mathbb{R}^d \to \mathbb{T}_k \qquad (12)$$

onto a d dimensional torus of side k for $k >> \max(diam(U), \sqrt{t})$. For each point $p \in \mathbb{T}_k$, the fibre $\phi_k^{-1}(p)$ of this map is a translation of $k\mathbb{Z}^d$.

Let x be the origin in $\mathbb{R}^d$, and $e_1, \ldots, e_d$ be the canonical unit vectors. For a fixed k, let

$$\Xi_k := \phi_k(\kappa e_1 + span(e_2, \ldots, e_d)),$$

where κ is a random number distributed uniformly in $[0, k)$, be a random $d-1$-dimensional torus aligned parallel to $\phi_k(span(e_2, \ldots, e_k))$. Let $y := (y_1, \ldots, y_d)$ be chosen from a spherical d-dimensional Gaussian in $\mathbb{R}^d$ centered at 0 having covariance $2tI$.

Define $\mathbb{P}_\tau^{(k)}$ to be

$$\mathbb{P}_\tau^{(k)} := \mathbb{P}[y_2^2 + \cdots + y_d^2 < |y_1|\tau < \tau^2 \,|\, 1 = |\phi_k(\overline{xy}) \cap \Xi_k|] \qquad (13)$$

It makes sense to define B_1 and B_2 on Ξ_k exactly as before i.e. tangent to Ξ_k at $\phi_k(\overline{xy}) \cap \Xi_k$ oriented so that B_1 is nearer to x than B_2 in geodesic distance. For geometric reasons, $\tilde{\mathbb{P}}_\tau^{(k)}$ is a lower bound on the probability that, even when the line segment $\overline{xy}$ in figure 2 is slid along itself to the right until x occupies the position where z is now, y does not leave B_2. Figure 3 illustrates ball B_2 being slid, which is equivalent. In particular, this event would imply that $x \in B_1$ and $y \in B_2$.

$$\limsup_{k \to \infty} \mathbb{P}_\tau^{(k)} \leq \mathbb{P}_\tau.$$

In the light of the above statement, it suffices to prove that for all sufficiently large k,

$$\mathbb{P}_\tau^{(k)} > 1 - O(\epsilon)$$

which will be done in Lemma 6. This completes the proof of this proposition. $\square$

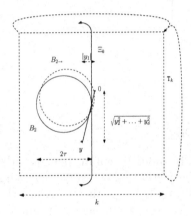

Fig. 3.

Lemma 6. *For all sufficiently large k,*

$$\mathbb{P}_\tau^{(k)} > 1 - O(\epsilon).$$

Proof. Recall that x is the origin and that $y := (y_1, \ldots, y_d)$ is Gaussian$(0, 2tI)$. Denote by E_k the event that

$$|\phi_k(\overline{xy}) \cap \Xi_k| = 1.$$

We note that

$$\mathbb{P}[E_k \mid y_1 = s] = \frac{|s|}{k} \mathcal{I}[|s| < k].$$

By Bayes' rule,

$$\rho[y_1 = s \mid E_k] \, \mathbb{P}[E_k] = \frac{|s|}{k} \left(\frac{e^{-s^2/4t}}{\sqrt{4\pi t}} \right) \mathcal{I}[|s| < k],$$

where $\mathcal{I}$ denotes the indicator function. In other words, there exists a constant $c_k := \frac{\mathbb{P}[E_k]^{-1}}{\sqrt{4\pi t}}$ such that

$$\rho[y_1 = s \,|\, |\Xi_k \cap \phi_k(\overline{xy})| = 1] \;=\; c_k \frac{|s|}{k} e^{-s^2/4t} \mathcal{I}[|s| < k].$$

A calculation tells us that

$$c_k \sim \frac{k}{4t}.$$

Let

$$\mathcal{I}_\tau := \mathcal{I}\left[\tau|y_1| > y_2^2 + \cdots + y_d^2\right] \mathcal{I}[|y_1| < \tau] \mathcal{I}[E_k].$$

By their definitions , $\mathbb{E}[\mathcal{I}_\tau|E_k] = \mathbb{P}_\tau^{(k)}$. Define

$$\mathcal{I}_\shortparallel := \mathcal{I}\left[|y_1| \notin [\sqrt{\epsilon t}, \tau]\right] \mathcal{I}[E_k],$$

and

$$\mathcal{I}_\perp := \mathcal{I}\left[y_2^2 + \ldots y_d^2 > 4t(d + 2\ln\frac{1}{\epsilon})\right] \mathcal{I}[E_k].$$

A direct calculation tells us that $\mathbb{E}[\mathcal{I}_\shortparallel|E_k] = O(\epsilon)$. Similarly $\mathbb{E}[\mathcal{I}_\perp|E_k] = O(\epsilon)$ follows from Theorem 3 and the fact mentioned below it. This Lemma is implied by the following claim. $\qquad\square$

Claim.

$$\mathcal{I}_\tau \geq \mathcal{I}[E_k] - \mathcal{I}_\shortparallel - \mathcal{I}_\perp.$$

Proof.

$$\mathcal{I}_\perp = \mathcal{I}\left[y_2^2 + \cdots + y_d^2 > 4t(d + 2\ln\frac{1}{\epsilon})\right] \mathcal{I}[E_k]$$
$$= \mathcal{I}\left[y_2^2 + \cdots + y_d^2 > \tau\sqrt{\epsilon t}\right] \mathcal{I}[E_k]$$

Therefore

$$\mathcal{I}[E_k] - \mathcal{I}_\shortparallel - \mathcal{I}_\perp \leq \mathcal{I}[E_k]\,[y_2^2 + \cdots + y_d^2 < \tau\sqrt{\epsilon t} < \tau|y_1|]\,\mathcal{I}[|y_1| < \tau] \quad (14)$$
$$\leq \mathcal{I}_\tau \qquad\qquad\qquad\qquad\qquad\qquad\qquad\qquad\qquad\qquad (15)$$

$\qquad\square$

Acknowledgements

We are grateful to David Jerison and Emanuel Milman for numerous very helpful discussions and to Santosh Vempala for pointing out the extension in Section 2.4 and permitting us to include it here. The first author is grateful to AIM for its generous hospitality during the workshop on Algorithmic Convex Geometry and that on Fourier Analytic methods in Convex Geometric Analysis. We would like to thank the anonymous referee for carefully reading an earlier version and pointing out an error in the proof of Lemma 4.

References

1. Ball, K.: An Elementary Introduction to Modern Convex Geometry. Mathematical Sciences Research Institute Publications, vol. 31, pp. 1–58. Cambridge Univ. Press, Cambridge (1997)
2. Borell, C.: The Brunn-Minkowski inequality in Gauss space. Inventiones Math. 30, 205–216 (1975)
3. Belkin, M., Narayanan, H., Niyogi, P.: Heat Flow and a Faster Algorithm to Compute the Surface Area of a Convex Body. In: Proc. of the 44th IEEE Foundations of Computer Science (FOCS 2006) (2006)
4. Bertsimas, D., Vempala, S.: Solving convex programs by random walks. Journal of the ACM (JACM) 51(4), 540–556 (2004); Proc. of the 34th ACM Symposium on the Theory of Computing (STOC 2002), Montreal (2002)
5. Coifman, R.R., Lafon, S.: "Diffusion maps". Applied and Computational Harmonic Analysis: Special issue on Diffusion Maps and Wavelets 21, 5–30 (2006)
6. Diaconis, P.: Generating random points on a Manifold, Berkeley Probability Seminar (Talk based on joint work with S. Holmes and M. Shahshahani)
7. Diaconis, P.: Personal Communication
8. Dyer, M., Frieze, A., Kannan, R.: A random polynomial time algorithm for approximating the volume of convex sets. Journal of the Association for Computing Machinary 38, 1–17 (1991)
9. Lovász, L., Vempala, S.: Hit-and-run from a corner. In: Proc. of the 36th ACM Symposium on the Theory of Computing, Chicago (2004)
10. Lovász, L., Vempala, S.: Simulated annealing in convex bodies and an $O^*(n^4)$ volume algorithm. In: Proc. of the 44th IEEE Foundations of Computer Science (FOCS 2003), Boston (2003)
11. Matthews, P.: Mixing Rates for Brownian Motion in a Convex Polyhedron. Journal of Applied Probability 27(2), 259–268 (1990)
12. Matthews, P.: Covering Problems for Brownian Motion on Spheres. Annals of Probability 16(1), 189–199 (1988)
13. Kaczynski, T., Mischaikov, K., Mrozek, M.: Computational Homology. Springer, New York (2004); (Applied Math. Sci. 157)
14. Niyogi, P., Weinberger, S., Smale, S.: Finding the Homology of Submanifolds with High Confidence from Random Samples. Discrete and Computational Geometry (2004)
15. Pan, V.Y., Chen, Z., Zheng, A.: The Complexity of the Algebraic Eigenproblem. Mathematical Sciences Research Institute, Berkeley (1998) (MSRI Preprint, 1998-71)
16. Rudelson, M.: Random vectors in the isotropic position. J. of Functional Analysis 164(1), 60–72 (1999); Encyclopedia of Mathematics and its Applications. Cambridge University Press (1993)
17. Vempala, S.: Personal Communication
18. Zomorodian, A., Carlsson, G.: Computing persistent homology. Discrete and Computational Geometry 33(2), 247 (2004)

A 2-Source Almost-Extractor for Linear Entropy

Anup Rao[*]

School of Mathematics, Institute for Advanced Study
arao@ias.edu

Abstract. We give an explicit construction of a function that is almost a 2-source extractor for linear entropy, it is a condenser where the output has almost full entropy. Given 2 sources with entropy δn, the output of the condenser is a distribution on m-bit strings that is ϵ-close to having min-entropy $m - \mathrm{poly}(\log(1/\epsilon), 1/\delta)$, where here m is linear in n.

1 Introduction

This paper is about constructing efficiently computable 2-source extractors. These are efficiently computable functions of the type $\mathsf{Ext} : \{0,1\}^n \times \{0,1\}^n \to \{0,1\}^m$ with the property that for any 2 independent distributions X, Y, each with entropy[1] k, the output $\mathsf{Ext}(X, Y)$ is close to uniform. Another way to view this object is as a coloring of the edges of the $N \times N$ complete bipartite graph with M colors that guarantees that in every $K \times K$ complete bipartite subgraph, every set of colors is hit with roughly the right frequency.

This problem was first suggested in the work of Chor and Goldreich [CG88] (see also [SV86]), who gave a simple argument that shows that the inner product function over $GF(2)$ is a good 2 source extractor as long as $k/n > 1/2 + \Omega(1)$. It is easy to generalize this to get many random bits (simply take the inner product over a large enough field). Since then, most work was diverted to the special case of seeded extractors (introduced in [NZ96]), where it is assumed that the second source is much shorter than the first source and is uniformly distributed (a 2-source extractor can be used in this situation just by padding the second source). Here almost optimal results are now known [LRVW03, GUV07].

There was no progress in reducing the entropy requirements for the general case of 2 source extractors until the work of Bourgain [Bou05], almost 20 years after [CG88]. Bourgain used recent results from arithmetic combinatorics [BKT04] to show that if the inputs are viewed as elements of a carefully chosen finite field, and ψ is any non-trivial additive character, the function $\psi(xy + x^2y^2)$ is an extractor even for entropy $0.499n$[2]. Bourgain's result, while seemingly a minor improvement over the previous result, had at least one application that would

---

[*] Supported in part by NSF Grant CCR-0324906.
[1] The definition of entropy we use is *min-entropy*, rather than Shannon entropy.
[2] Note that the inner product function mentioned above can also be viewed as $\psi(xy)$, where x, y are interpreted as elements of $GF(2^n)$ and ψ is a suitably chosen additive character.

A. Goel et al. (Eds.): APPROX and RANDOM 2008, LNCS 5171, pp. 549–556, 2008.

not have been possible using just the ideas of Chor and Goldreich: it led to new constructions of Ramsey graphs with much better parameters than were previously known [BRSW06].

The problem of constructing 2 source extractors for arbitrary linear min-entropy remains open. In this paper we describe some partial progress towards this goal, obtaining an object that seems tantalizingly close to being a 2 source extractor.

1.1   Our Results and Techniques

We prove the following theorem:

Theorem 1. *For every $\delta > 0$ and every ϵ, there exists a polynomial time computable function* $\mathsf{Ext} : \{0,1\}^n \times \{0,1\}^n \to \{0,1\}^m$ *such that if X, Y are independent sources with min-entropy rate δ, $\mathsf{Ext}(X,Y)$ is ϵ close to having min-entropy* $m - \mathrm{poly}(1/\delta, \log(1/\epsilon))$, *with* $m = \Omega(\delta n)$.

The output of this algorithm is close to having such a high min-entropy that we hope that it may still be sufficient for applications where 2-source extractors are required. For instance, if we are willing to make cryptographic assumptions that rely only on secret keys with such high entropy, this extractor may be used in lieu of a 2-source extractor for generating secret keys.

Our result follows by composing several previous explicit constructions. Specifically, we rely on two types of explicit functions from previous work:

2 independent sources $\to$ SR-source. A first observation (already made in [BKS$^+$05]) is that it is possible to use arithmetic combinatoric results to get an explicit function $\mathsf{SExt} : \{0,1\}^n \times \{0,1\}^n \to \{0,1\}^t$, that converts two independent sources into a *somewhere random source*. A distribution on strings is somewhere random if at least one of the strings is distributed uniformly. The above construction combined with some ideas from [Rao06] gives an algorithm that can carry out such a conversion, outputting a somewhere random source with only a constant number of strings, each of length linear in n.

2 independent sources + independent SR-source $\to$ uniform source. It is also easy to use previous work [Raz05, DR05, BKS$^+$05] to get an explicit function $\mathsf{Ext} : \{0,1\}^n \times \{0,1\}^n \times \{0,1\}^t \to \{0,1\}^m$ that can extract randomness from two independent sources with linear entropy and an additional *independent* small somewhere random source.

Our final construction is $\mathsf{Ext}'(X, Y, \mathsf{SExt}(X,Y))$, i.e. we use the somewhere random source generated by the original source to extract random bits from X, Y. At first it may seem like this has very little chance of working, since the somewhere random source is *not independent* of the original sources (in fact it is determined by them). Still, we show that if our goal is just to show that the output has high entropy, something can be salvaged from this approach, giving us our main result. Ideas that superficially seem similar to this one have been used in previous work [GRS04, Sha06]. It is hard to describe the many ideas in those papers succinctly, so in the discussion here we shall be slightly inaccurate in order to convey the gist of the differences between the techniques of those works and the present paper. In the earlier works, the authors first construct a function $\mathsf{DExt} : \{0,1\}^n \rightarrow \{0,1\}^t$ that extracts a few random bits from some class of sources. They then use the extracted random bits to extract many more random bits from the original source. Thus the final algorithm looks like $\mathsf{Ext}'(X, \mathsf{DExt}(X))$ for some carefully chosen function Ext'.

The major difference between the previous works and ours is in the analysis. The previous works carefully controlled the correlations between the extracted bits ($\mathsf{DExt}(X)$) and the original source X. In particular, they carefully chose a random variable in the probability space they were considering and fixed it. Conditioned on this fixing, they were able to argue that the extracted bits (or some subset of the extracted bits) became independent of the original source (or some part of the original source). In this way, after fixing this magic random variable, they were able to obtain two random variables that could be treated as being independent (without paying a too heavy price in terms of lost entropy). In order to make this approach work, they had to carefully exploit the properties of the class of distributions they were building extractors for and the properties of the functions they were constructing.

The ideas in this paper are less delicate and less intricate. In particular, they do not apply just to the case of independent sources. They can be generalized[3] to be used in any situation where we know how to construct an explicit function SExt that can convert a distribution from class $\mathcal{C}_1$ into one from class $\mathcal{C}_2$ with small support size, and an explicit function that can extract random bits (or even high entropy bits) from two independent distributions, one from class $\mathcal{C}_1$ and the other from $\mathcal{C}_2$. In our particular application, $\mathcal{C}_1$ is the class of two independent sources and $\mathcal{C}_2$ is the class of somewhere random sources. In this situation, we simply show how to use the union bound to get a result of the type of Theorem 1.

It is easy to see that if a distribution is far from having high min-entropy, then there must be a small set of the support that has an unusually high probability under it. Fix any subset of the support. In order to show that $\mathsf{Ext}'(X, Y, \mathsf{SExt}(X, Y))$ does not hit this set with such a high probability, we consider the set of *bad* outputs of SExt. Say z is bad if $\mathsf{Ext}'(X, Y, z)$ hits the set with high probability. Then the properties of Ext' guarantee that any somewhere random source has only a small probability of giving such a bad z. On the other hand, since the total number of z's is so small (the output is only a constant number of bits), we can argue that with high probability $\mathsf{Ext}'(X, Y, z)$ does not land in the set for *every good* z. Thus, by the union bound, we can argue that any small enough set is avoided with significant probability.

Since the above argument requires us to use the union bound on as many events as there are elements in the support of SExt, it is crucial that the error

---

[3] Shaltiel [Sha06] also generalized the ideas of [GRS04] to several classes of sources, but there each class he considered required a different construction and a different analysis, though there was a very significant overlap in the various cases.

of the extractor Ext' be significantly small in terms of the number of elements in the support of SExt. Luckily, explicit constructions that we rely on already provide such strong guarantees.

2   Preliminaries

We will be concerned with the treatment of various kinds of distributions that are *nice* in that they contain a lot of usable randomness. Here we discuss some ways to measure this niceness:

Definition 1. *The* min-entropy *of a distribution R is defined to be: $H_\infty(R) = -\log(\max_{x \in R}(R(x))$. The* min-entropy rate *of a distribution R on $\{0,1\}^n$ is $H_\infty(R)/n$.*

Definition 2. *An (n,k)-source denotes some random variable X over $\{0,1\}^n$ with $H_\infty(X) \geq k$.*

Definition 3. *Let D and F be two distributions on a set S. Their* statistical distance *is*

$$|D - F| \stackrel{def}{=} \max_{T \subseteq S}(|D(T) - F(T)|) = \frac{1}{2} \sum_{s \in S} |D(s) - F(s)|$$

If $|D - F| \leq \epsilon$ we shall say that D is ϵ-close to F.

This measure of distance is nice because it is robust in the sense that if two distributions are close in this distance, then applying any functions to them cannot make them go further apart.

Proposition 1. *Let D and F be any two distributions over a set S s.t. $|D-F| \leq \epsilon$. Let g be any function on S. Then $|g(D) - g(F)| \leq \epsilon$.*

A block source is a source broken up into a sequence of blocks, with the property that each block has min-entropy even conditioned on previous blocks.

Definition 4 (Block sources). *A distribution $X = X_1, X_2, \cdots, X_C$ is called a $(k_1, k_2, \ldots, k_C)$-block source if for all $i = 1, \ldots, C$, we have that for all $x_1 \in X_1, \ldots, x_{i-1} \in X_{i-1}$, $H_\infty(X_i|X_1 = x_1, \ldots, X^{i-1} = x_{i-1}) \geq k_i$, i.e., each block has high min-entropy even conditioned on the previous blocks. If $k_1 = k_2 = \cdots = k_C = k$, we say that X is a k-block source.*

We have the following standard lemma:

Lemma 1. *Suppose X is a source with min-entropy k and $f : \{0,1\}^n \rightarrow \{0,1\}^t$ is a function such that $f(X)$ is ϵ close to having min-entropy k'. Then for every ℓ, $(f(X), X)$ is $\epsilon + 2^{-\ell}$ close to being a $k', k - t - \ell$ block source.*

We shall need the concept of a somewhere random distribution.

Definition 5. *A source X is $(t \times r)$* somewhere-random *if it is distribution on $t \times r$ boolean matrices s.t. X is distributed uniformly randomly over one of the rows. Every other row may depend on the random row in arbitrary ways. We say that X has* somewhere min-entropy k *if at least one of the rows has min-entropy k.*

3   Previous Work Needed

Our work relies on several previous constructions. The first object we shall need is the additive number theory based condensers independently constructed by Barak et al. [BKS+05] and Raz [Raz05]:

Lemma 2 ([Raz05, BKS+05]). *For every $\delta > 0$, there exists a polynomial time computable function* $\mathsf{Cond} : \{0,1\}^n \to (\{0,1\}^{n/\mathrm{poly}(1/\delta)})^{\mathrm{poly}(1/\delta)}$, *where the output is interpreted as a* $\mathrm{poly}(1/\delta) \times n/\mathrm{poly}(1/\delta)$ *boolean matrix, such that if X is a source with min-entropy rate δ, $\mathsf{Cond}(X)$ is $2^{-\Omega(\delta^2 n)}$ close to a convex combination of distributions, each of which has some row with min-entropy rate 0.9.*

When this lemma is combined with the merger from Raz's work [Raz05] and the improved analysis of Dvir and Raz (Lemma 3.2 in [DR05]), we get the following lemma:

Lemma 3 ([DR05, Raz05, BKS+05]). *For every $\delta > 0$ and $\epsilon > 2^{-n/10}$, there exists a polynomial-time computable function* $\mathsf{Cond} : \{0,1\}^n \to (\{0,1\}^{n/\mathrm{poly}(1/\delta)})^{2^{\mathrm{poly}(1/\delta)}/\epsilon}$, *where the output is treated as a* $2^{\mathrm{poly}(1/\delta)}/\epsilon \times n/\mathrm{poly}(1/\delta)$ *boolean matrix, such that if X has min-entropy rate δ, $\mathsf{Cond}(X)$ is $2^{-\Omega(\delta^2 n)}$ close to a convex combination of distributions, each of which has at most an ϵ fraction of rows with min-entropy rate less than 0.9.*

We need the following two source extractor of Chor and Goldreich:

Theorem 2 ([CG88]). *For every constant $\delta > 1/2$ there exists a strong two source extractor* $\mathsf{Had} : \{0,1\}^n \times \{0,1\}^n \to \{0,1\}^{\Omega(n)}$ *with error $2^{-\Omega(n)}$ for two independent sources with min-entropy δn.*

We can use X, Y to generate a somewhere random source Z. The following theorem was proved in [BKS+05]:

Theorem 3. *For every δ, there exists $c(\delta) = \mathrm{poly}(1/\delta)$ and a polynomial time computable function* $\mathsf{SExt} : \{0,1\}^n \times \{0,1\}^n \to \{0,1\}^{cn/\mathrm{poly}(1/\delta)}$, *where the output is treated as a $c \times n/\mathrm{poly}(1/\delta)$ boolean matrix, such that if X, Y are independent sources with min-entropy rate δ, SExt is $2^{-\Omega(n/\mathrm{poly}(1/\delta))}$ close to a convex combination of somewhere random sources.*

Proof. Define the $(i,j)'th$ row $\mathsf{SExt}(X,Y)_{i,j} = \mathsf{Had}(\mathsf{Cond}(X)_i, \mathsf{Cond}(Y)_j)$, where Cond is as in Lemma 2 and Had is as in Theorem 2. The theorem follows directly.

Finally, we need the following two source extractor for block sources, that follows from the work of [BKS+05, Rao06]:

Theorem 4 ([BKS+05, Rao06]). *For every $\delta > 0$, there exists a constant $\gamma > 0$ and a polynomial time computable function* $\mathsf{Ext} : (\{0,1\}^n)^4 \to \{0,1\}^m$ *such that if X_1, X_2 is a $\delta n, \delta n$ block source and Y_1, Y_2 is an independent $\delta n, \delta n$ block source,*

$$\Pr_{x_1, x_2} [|\mathsf{Ext}(x_1, x_2, Y_1, Y_2) - U_m| > 2^{-\gamma n}] < 2^{-\gamma n}$$

and

$$\Pr_{y_1, y_2} \left[|\mathsf{Ext}(X_1, X_2, y_1, y_2) - U_m| > 2^{-\gamma n} \right] < 2^{-\gamma n}$$

where here $m = \Omega(n)$ and U_m denotes the uniform distribution on m bit strings.

4   The Condenser

First we show that if we were given a small independent somewhere random source, we can use it to extract random bits from two linear min-entropy independent sources. The idea is that the somewhere random source can be used to turn both of the other sources into block sources, using Lemma 3.

Theorem 5. *For every $1 > \delta, \epsilon_2 > 0$ and $c > 0$, there exists a $t(c, \delta, \epsilon_2) = \mathrm{poly}(c, 1/\delta, \log(1/\epsilon_2))$, a constant $\gamma(\delta)$ and a polynomial time computable function $\mathsf{Ext} : \{0,1\}^n \times \{0,1\}^n \times \{0,1\}^{c \times t} \to \{0,1\}^{\delta n - o(1)}$ such that if X, Y are independent min-entropy rate δ sources and Z is an independent $c \times t$ somewhere random source,*

$$\Pr_z[\Pr_y[\mathsf{Ext}(X, y, z) \text{ is } 2^{-\gamma n} \text{ close to uniform}] > 1 - 2^{-\gamma n}] > 1 - \epsilon_2$$

$$\Pr_z[\Pr_x[\mathsf{Ext}(x, Y, z) \text{ is } 2^{-\gamma n} \text{ close to uniform}] > 1 - 2^{-\gamma n}] > 1 - \epsilon_2$$

Proof. Let $\delta' < \delta$ be a small enough constant so that length of the rows output by Cond in Lemma 3 for error ϵ_2 and min-entropy rate δ' is at most $\delta^2 n/c$. Let 2^t be the number of rows output by Cond for this setting of parameters (so that $t = \mathrm{poly}(1/\delta, c, \log(1/\epsilon_2))$).

Now we treat each row of Z as the name of a row of $\mathsf{Cond}(X)$. Let X_Z denote the string $\mathsf{Cond}(X)_{Z_1}, \ldots, \mathsf{Cond}(X)_{Z_c}$. Similarly let Y_Z denote $\mathsf{Cond}(Y)_{Z_1}, \ldots, \mathsf{Cond}(Y)_{Z_t}$. Then note that X_Z and Y_Z are of length $\delta^2 n$. Further, by the properties of Cond, with high probability over the choice of z, X_z and Y_z are $2^{-\Omega(n)}$ close to having min-entropy rate $0.9/c$. Since X_Z is so short, Lemma 1 implies that (X_z, X) and (Y_z, Y) are $2^{-\Omega(n)}$ close to independent block sources with entropy $0.9\delta^2 n/c, (\delta - \delta^2)n \geq \delta^2 n$. So we can apply the extractor from Theorem 4 to get the result of the lemma.

Now although we don't have access to a somewhere random source Z as above, Theorem 3 tells us that we can generate such a source in polynomial time using the function SExt. So let us define the function $\mathsf{Ext}'(X, Y) \stackrel{def}{=} \mathsf{Ext}(X, Y, \mathsf{SExt}(X, Y))$. It is not at all clear that this function is an extractor, since now X, Y are not independent of the somewhere random source being used (in fact they determine it!). Still, we show that the output of this function must be close to having very high min-entropy.

Before we show this, we need two simple lemmas:

Lemma 4. *Let A be a distribution that is ϵ-far from having min-entropy k. Then, there must be a set H of size at most 2^k such that $\Pr[A \in H] \geq \epsilon$.*

Proof. Set $H = \{h : \Pr[A = h] \geq 2^{-k}\}$. This set clearly has at most 2^k element. The lemma is immediate from the definition of statistical distance.

Lemma 5. *Let $A_1, \ldots, A_l$ be random variables taking values in $\{0,1\}^n$, Z be a random variable taking values in $[l]$ and $G \subset [l]$ be a set such that:*

- *For every $z \in G$, $|A_z - U_n| < \tau$.*
- *$\Pr[Z \in G] > 1 - \epsilon$.*

Then for every integer d, A_Z is $\epsilon + l(\tau + 2^{-d})$ close to having min-entropy $n - d$.

Proof. Suppose not. Then, by Lemma 4, there must be some set of heavy elements $H \subset \{0,1\}^n$ of size at most 2^{n-d} such that $\Pr[A_Z \in H] \geq \epsilon + l(\tau + 2^{-d})$. Now note that $A_Z \in H$ implies that either $Z \notin G$ or one of the good A_i's must have hit H. Thus, by the union bound,

$$\Pr[A_Z \in H] < \Pr[Z \notin G] + \Pr[\exists z \in G \text{ with } A_z \in H]$$
$$\leq \epsilon + |G|(\tau + 2^{-d})$$
$$< \epsilon + l(\tau + 2^{-d})$$

We can now prove the main theorem of this paper.

Proof (Theorem 1). Let A_z denote the random variable $\mathsf{Ext}(X, Y, z)$. Let $\gamma = \Omega_\delta(1)$ be as in Theorem 5, with $\epsilon_2 = \epsilon$. Define

$$G = \{z : \Pr_y[\mathsf{Ext}(X, y, z) \text{ is } 2^{-\gamma n} \text{ close to uniform}] > 1 - 2^{-\gamma n}\}$$

Then we see that $\Pr[Z \in G] > 1 - \epsilon_2$, if Z is somewhere random and independent. Instead we set $Z = \mathsf{SExt}(X, Y)$ (truncating each row to be of length $t = \mathrm{poly}(c, 1/\delta, \log(1/\epsilon))$ as required by Theorem 5).

Thus we have that $\Pr[Z \in G] > 1 - \epsilon_2 - 2^{-\Omega_\delta(n)}$, since $\mathsf{SExt}(X, Y)$ is $2^{-\Omega_\delta(n)}$ close to a convex combination of somewhere random sources. Further, for every $z \in G$, $|A_z - U_n| < 2^{-\Omega_\delta(n)}$. The total number of z's is at most $2^{ct} = 2^{\mathrm{poly}(1/\delta, \log(1/\epsilon_2))}$. Thus, by Lemma 5, setting $d = 100ct/\log(1/\epsilon)$, we have that $\mathsf{Ext}'(X, Y)$ is $2\epsilon_2$ close to having min-entropy $m - \mathrm{poly}(1/\delta, \log(1/\epsilon))$.

Acknowledgements

I would like to thank Boaz Barak, Ronen Shaltiel and Avi Wigderson for useful discussions.

References

[BKS+05] Barak, B., Kindler, G., Shaltiel, R., Sudakov, B., Wigderson, A.: Simulating independence: New constructions of condensers, Ramsey graphs, dispersers, and extractors. In: Proceedings of the 37th Annual ACM Symposium on Theory of Computing, pp. 1–10 (2005)

[BRSW06]   Barak, B., Rao, A., Shaltiel, R., Wigderson, A.: 2 source dispersers for $n^{o(1)}$ entropy and Ramsey graphs beating the Frankl-Wilson construction. In: Proceedings of the 38th Annual ACM Symposium on Theory of Computing (2006)

[Bou05]   Bourgain, J.: More on the sum-product phenomenon in prime fields and its applications. International Journal of Number Theory 1, 1–32 (2005)

[BKT04]   Bourgain, J., Katz, N., Tao, T.: A sum-product estimate in finite fields, and applications. Geometric and Functional Analysis 14, 27–57 (2004)

[CG88]   Chor, B., Goldreich, O.: Unbiased bits from sources of weak randomness and probabilistic communication complexity. SIAM Journal on Computing 17(2), 230–261 (1988)

[DR05]   Dvir, Z., Raz, R.: Analyzing linear mergers. Technical Report TR05-25, ECCC: Electronic Colloquium on Computational Complexity (2005)

[GRS04]   Gabizon, A., Raz, R., Shaltiel, R.: Deterministic extractors for bit-fixing sources by obtaining an independent seed. In: Proceedings of the 45th Annual IEEE Symposium on Foundations of Computer Science (2004)

[GUV07]   Guruswami, V., Umans, C., Vadhan, S.: Unbalanced expanders and randomness extractors from parvaresh-vardy codes. In: Proceedings of the 22nd Annual IEEE Conference on Computational Complexity (2007)

[LRVW03]   Lu, C.J., Reingold, O., Vadhan, S., Wigderson, A.: Extractors: Optimal up to constant factors. In: Proceedings of the 35th Annual ACM Symposium on Theory of Computing, pp. 602–611 (2003)

[NZ96]   Nisan, N., Zuckerman, D.: Randomness is linear in space. Journal of Computer and System Sciences 52(1), 43–52 (1996)

[Rao06]   Rao, A.: Extractors for a constant number of polynomially small min-entropy independent sources. In: Proceedings of the 38th Annual ACM Symposium on Theory of Computing (2006)

[Raz05]   Raz, R.: Extractors with weak random seeds. In: Proceedings of the 37th Annual ACM Symposium on Theory of Computing, pp. 11–20 (2005)

[SV86]   Santha, M., Vazirani, U.V.: Generating quasi-random sequences from semi-random sources. Journal of Computer and System Sciences 33, 75–87 (1986)

[Sha06]   Shaltiel, R.: How to get more mileage from randomness extractors. In: Proceedings of the 21th Annual IEEE Conference on Computational Complexity, pp. 49–60 (2006)

Extractors for Three Uneven-Length Sources

Anup Rao[1],[*] and David Zuckerman[2],[**]

[1] School of Mathematics, Institute for Advanced Study
arao@ias.edu
[2] Department of Computer Science, University of Texas at Austin
diz@cs.utexas.edu

Abstract. We construct an efficient 3-source extractor that requires one of the sources to be significantly shorter than the min-entropy of the other two sources. Our extractors work even when the longer, n-bit sources have min-entropy $n^{\Omega(1)}$ and the shorter source has min-entropy $\log^{10} n$. Previous constructions for independent sources with min-entropy n^{γ} required $\Theta(1/\gamma)$ sources [Rao06]. Our construction relies on lossless condensers [GUV07] based on Parvaresh-Vardy codes [PV05], as well as on a 2-source extractor for a block source and general source [BRSW06].

1 Introduction

Motivated by the widespread use of randomness in computer science, researchers have sought algorithms to extract randomness from a distribution that is only weakly random. A general weak source is one with some min-entropy: a distribution has min-entropy k if all strings have probability at most 2^{-k}. We would like to extract randomness from a weak source knowing only k and not the exact distribution. However, this is impossible, even for more restricted sources [SV86].

Therefore, Santha and Vazirani showed how to extract randomness from two independent restricted weak sources [SV86]. Can we design such efficient *randomness extractors* for general *independent sources*? These are efficiently computable functions $\mathsf{Ext} : (\{0,1\}^n)^{\mathsf{C}} \to \{0,1\}^m$ with the property that for any product distribution $X_1, \ldots, X_{\mathsf{C}}$, the output $\mathsf{Ext}(X_1, \ldots, X_{\mathsf{C}})$ is close to uniformly distributed as long as each X_i has high enough min-entropy. Our primary goals are to minimize the number of sources required and the amount of entropy needed. Secondarily, we'd like to maximize the length of the output and minimize the error, which is the distance of the output from uniform.

Extractors for independent sources have been useful in constructing deterministic extractors for space-bounded sources [KRVZ06] and in new constructions of network extractor protocols [KLRZ08].

1.1 Previous Results

The question of finding such an extractor came up as early as in the works of Santha and Vazirani [SV86] and Chor and Goldreich [CG88]. After that, following

[*] Supported in part by NSF Grants CCF-0634811 and CCR-0324906.
[**] Supported in part by NSF Grant CCF-0634811.

the work of Nisan and Zuckerman [NZ96], most work focused on constructing *seeded* extractors. This is simply a two source extractor where one source is assumed to be very short and uniformly distributed (in this case the problem only makes sense if the extractor outputs more than what's available in the short seed). Extractors for this model have found application in constructions of communication networks and good expander graphs [WZ99, CRVW02], error correcting codes [TZ04, Gur04], cryptographic protocols [Lu04, Vad04], data structures [MNSW98] and samplers [Zuc97]. Seeded extractor constructions are now available that can extract uniform bits from a source with small entropy using a seed of length only $O(\log n)$ [LRVW03, GUV07].

In recent years there have been several works improving the state of the art for independent sources [BIW04, BKS+05, Raz05, Bou05, Rao06, BRSW06]. We now know how to extract from two sources when the entropy in each is at least something like $.4999n$ [Bou05], from three sources if the entropy in each is at least $n^{0.99}$ [Rao06] and from $O(1/\gamma)$ sources if the entropy in each is at least n^{γ} [Rao06, BRSW06]. All of these constructions have exponentially small error, and by a result of Shaltiel [Sha06], the output length can be made almost best possible. Thus, the tradeoff that is most interesting is the one between the number of sources required and the entropy requirements.

1.2   Our Work

In this paper, we construct extractors which require only three sources with polynomial min-entropy. However, there is a key caveat: one of the sources must be significantly shorter than the min-entropy of the other two sources. On the plus side, while the longer, n-bit sources must have min-entropy $n^{\Omega(1)}$, the shorter source need only have min-entropy $\log^{10} n$.

Extractors for uneven-length sources may be more interesting than they appear, for two reasons. First, the extractors with the most applications are seeded extractors, which are extractors for uneven-length sources. Second, in other settings the uneven-length case was more difficult. For example, in the two-source setting with entropy rate bigger than $1/2$, the even length case was known for decades (before extractors were defined), but the uneven-length case was only proved by Raz in 2005 [Raz05].

We now state our three source extractor precisely.

Theorem 1 (3-Source Extractor). *There exists a constant d and a polynomial time computable function* $3\mathsf{Ext} : \{0,1\}^{n_1} \times \{0,1\}^{n_2} \times \{0,1\}^{n_3} \to \{0,1\}^m$ *which is an extractor for three sources with min-entropy requirements $k_1, k_2, k_3 = \sqrt{k_1}$, error $2^{-\Omega(k_2)} + 2^{-k_1^{\Omega(1)}}$ and output length $m = k_1 - o(k_1)$ as long as:*

- $\frac{\log k_1}{\log n_2} > d\frac{\log(n_1+n_3)}{\log k_1}$
- $k_2 > d\log n_1$

One example of how to set parameters is in the following corollary:

Corollary 1. *There exist constants d, h such that for every constant $0 < \gamma < 1$, there is a polynomial time computable function $\mathsf{3Ext} : \{0,1\}^n \times \{0,1\}^n \times \{0,1\}^{n^{\gamma/h}} \to \{0,1\}^{n^\gamma - o(n^\gamma)}$ which is an extractor for three sources with min-entropy requirements $k = n^\gamma, n^\gamma, \log^d n$, and error $2^{-\Omega(\log^{10} n)}$.*

The corollary follows by setting $n_1 = n_3 = n$, $n_2 = n^{\gamma/h}$, $k_1 = n^\gamma, k_2 = \log^d n$ and choosing h to be a large enough constant.

For smaller min-entropy, the first constraint in the theorem forces the length of the shorter source to be much shorter than the other two sources.

It turns out that we don't really need three mutually independent sources to get our results. We obtain an extractor even when we have just two sources, but one of them is a block source with a short block followed by a long block.

Theorem 2 ((Short, Long)-Block-Source Extractor). *There exists a constant d and a polynomial time computable function $\mathsf{3Ext} : \{0,1\}^{n_1} \times \{0,1\}^{n_2} \times \{0,1\}^{n_3} \to \{0,1\}^m$ which is an extractor for a block source with min-entropy requirements k_1, k_2 and an independent source with min-entropy $k_3 = \sqrt{k_2}$, error $2^{-\Omega(k_1)} + 2^{-k_2^{\Omega(1)}}$ and output length $m = k_2 - o(k_2)$ as long as:*

- $\frac{\log k_2}{\log n_1} > d \frac{\log(n_2 + n_3)}{\log k_2}$
- $k_1 > d \log n_2$

Salil Vadhan observed that a slightly different analysis allows us to reverse the order of the short and long sources.

Theorem 3 ((Long, Short)-Block-Source Extractor). *There exists a constant d and a polynomial time computable function $\mathsf{3Ext} : \{0,1\}^{n_1} \times \{0,1\}^{n_2} \times \{0,1\}^{n_3} \to \{0,1\}^m$ which is an extractor for a block source with min-entropy requirements k_1, k_2 and an independent source with min-entropy $k_3 = \sqrt{k_1}$, error $2^{-\Omega(k_2)} + 2^{-k_1^{\Omega(1)}}$ and output length $m = k_1 - o(k_1)$ as long as:*

- $\frac{\log k_1}{\log n_2} > d \frac{\log(n_1 + n_3)}{\log k_1}$
- $k_2 > d \log n_1$

1.3   Techniques

We build on the work of Guruswami et al. [GUV07] and Barak et al. [BRSW06]. Guruswami et al. showed how to build a very good *seeded condenser*. This is a function $\mathsf{Cond} : \{0,1\}^n \times \{0,1\}^d \to \{0,1\}^m$ that guarantees that if X has sufficient min-entropy k and U_d is an independent random variable that's uniform on d bits, $\mathsf{Cond}(X, U_d)$ is close to having very high min-entropy. Inspired by ideas from the list decodable code constructions of [PV05, GR06], they showed that the following function is a condenser: $\mathsf{Cond}(f, y) = f(y), f^h(y), \ldots, f^{h^t}(y)$, where f is a low degree polynomial, f^{h^i} is the powered polynomial modulo a suitably chosen irreducible polynomial and h, t are suitably chosen parameters. Specifically they showed that the output of this function is close to having min-entropy $0.9m$.

We analyze Cond when the second source is *not* uniform, but merely has high entropy (say αd). In this case, we show that the output must have entropy close to $0.9\alpha m$. At first this may not seem useful, since we end up with a distribution where the entropy is even less concentrated than the source that we started with. However, we show that if the output is $m = m_1 + m_2 + \cdots + m_C$ bits, each consecutive block of m_i bits must have entropy close to $0.9\alpha m_i$. Using an idea from [Zuc96], we choose the m_i's to increase geometrically. This implies that the output must be a *block source*: not only does each consecutive block of m_i bits have a reasonable amount of entropy, it has significant entropy conditioned on any fixing of the previous blocks. Intuitively, since each subsequent block is significantly larger than the previous ones, the previous ones cannot contain enough information to significantly reduce the entropy in the current block upon conditioning.

Block sources are a well studied object in extractor constructions. Indeed, earlier works have shown [BKS+05, Rao06, BRSW06] that even two source extractors are easy to obtain under the assumption that both or even one of the sources have some block source structure. In particular, a theorem from [BRSW06] shows that we can extract from one block source and one independent source, if each block and the independent source have entropy n^γ, and the number of blocks in the block source is at least $O(1/\gamma)$.

This completes the construction. We first apply Cond to convert the first two sources into a single block source, and then use the extractor from [BRSW06] and an additional source to get random bits.

2   Preliminaries

For a distribution X, we let $H_\infty(X)$ denote the min-entropy of the distribution. We call a distribution *flat* if it is uniformly distributed on some subset of the universe.

Fact 1. *Every distribution X with min-entropy at least k is a convex combination of flat distributions with min-entropy k.*

Definition 1. *Let D and F be two distributions on a set S. Their statistical distance is*

$$|D - F| \stackrel{def}{=} \max_{T \subseteq S}(|D(T) - F(T)|) = \frac{1}{2}\sum_{s \in S}|D(s) - F(s)|$$

If $|D - F| \le \epsilon$ we shall say that D is ϵ-close to F.

This measure of distance is nice because it is robust in the sense that if two distributions are close in this distance, then applying any functions to them cannot make them go further apart.

Proposition 1. *Let D and F be any two distributions over a set S s.t. $|D-F| \le \epsilon$. Let g be any function on S. Then $|g(D) - g(F)| \le \epsilon$.*

When we manipulate sources which are close to having some min-entropy, it will be convenient to have the following definition. For a distribution D, $D(a)$ denotes the probability that D places on a.

Definition 2. *We call a distribution D ϵ-close to k-light if, when X is chosen according to D, $\Pr[D(X) > 2^{-k}] \leq \epsilon$.*

The following is then immediate:

Lemma 1. *If D is ϵ-close to k-light, then D is ϵ-close to min-entropy k.*

The following lemma gives a sufficient condition to lowerbound the min-entropy of a source.

Lemma 2 ([GUV07]). *Let X be a random variable taking values in a set of size larger than 2^k such that for every set S of size less than $\epsilon 2^k$, $\Pr[X \in S] < \epsilon$. Then X is ϵ-close to k-light.*

Proof. First note that $|\mathsf{supp}(X)| \geq \epsilon 2^k$, or else the hypothesis of the lemma is contradicted by setting $S = \mathsf{supp}(X)$.

Let S be the $\epsilon 2^k$ heaviest elements under X, breaking ties arbitrarily. Then for every $x \notin S$ we must have that $\Pr[X = x] \leq 2^{-k}$, or else every element in S would have weight greater than 2^{-k}, which would contradict the hypothesis. Thus, the set of elements that have weight more than 2^{-k} are hit with probability at most ϵ.

A block source is a source broken up into a sequence of blocks, with the property that each block has min-entropy even conditioned on previous blocks.

Definition 3 (Block sources). *A distribution $X = X_1, X_2, \cdots, X_C$ is called a $(k_1, k_2, \ldots, k_C)$-block source if for all $i = 1, \ldots, C$, we have that for all $x_1 \in X_1, \ldots, x_{i-1} \in X_{i-1}$, $H_\infty(X_i | X_1 = x_1, \ldots, X^{i-1} = x_{i-1}) \geq k_i$, i.e., each block has high min-entropy even conditioned on the previous blocks. If $k_1 = k_2 = \cdots = k_C = k$, we say that X is a k-block source.*

The following lemma is useful to prove that a distribution is close to a block source.

Lemma 3. *Let $X = X_1, \ldots, X_t$ be t dependent random variables. For every $i = 1, 2, \ldots, t$, let X^i denote the concatenation of the first i variables. Suppose each X^i takes values in $\{0, 1\}^{n_i}$ and for every $i = 1, 2, \ldots, t$, X^i is ϵ_i-close to k_i-light, with $\sum_i \epsilon_i < 1/10$. Then for every $\ell > 10 \log t$ we must have that X is $\sum_{i=1}^t \epsilon_i + t2^{-\ell}$-close to a block source, where each block X_i has min-entropy $k_i - n_{i-1} - 1 - \ell$.*

Proof. We will need to define the notion of a *submeasure*. Let $n = n_t$. Say that $M : \{0, 1\}^n \to [0, 1]$ is a submeasure on $\{0, 1\}^n$ if $\sum_{m \in \{0,1\}^n} M(m) \leq 1$. Note that every probability measure is a submeasure. We abuse notation and let $M(x^i)$ denote the marginal measure induced on the first i coordinates.

We say a submeasure on $\{0,1\}^n$ is ϵ-close to k-light if

$$\sum_{m \in \{s : M(s) > 2^{-k}\}} M(m) \leq \epsilon.$$

As usual, for any event $A \subset \{0,1\}^n$, we denote $\Pr[M \in A] = \sum_{m \in A} M(m)$. We now define the submeasures $M_{t+1} = X$, and for $i = t, t-1, t-2, \ldots, 1$,

$$M_i(m) = \begin{cases} 0 & M_{i+1}^i(m^i) > 2^{-k_i} \vee M_{i+1}^i(m^i) < 2^{-n_i - \ell} \\ M_{i+1}(m) & \text{otherwise} \end{cases}$$

Let $M = M_1$. Now note that for every $j < i$, M_i^j is ϵ_j-close to k_j-light, since we only made points lighter in the above process. Further, for all m and $i \leq j$, $M_i^j(m^j) \leq 2^{-k_j}$, since we reduced the weight of all m's that violated this to 0. We also have that for every m, i, $M^i(m^i) = 0$ or $M^i(m^i) \geq 2^{-n_i - \ell}$ by our construction.

Now define the sets $B_i = \{m \in \{0,1\}^n : M_i(m) \neq M_{i+1}(m)\}$. Set $B = \cup_i B_i$. Then note that $\Pr[X \in B] \leq \sum_{i=2}^t \Pr[M_{i+1} \in B_i]$. Each B_i, contains two types of points: points that were removed when moving from M_{i+1} to M_i because they were too heavy, and points that were removed because they were too light. We set $C_i = \{m : M_{i+1}^i(m^i) > 2^{-k_i}\}$, namely the "too heavy" points. We see that $\Pr[M_{i+1} \in C_i] \leq \epsilon_i$, since M_{i+1}^i is ϵ_i-close to k_i-light. Set $D_i = \{m : M_{i+1}^i(m^i) < 2^{-n_i - \ell}\}$, namely the "too light" points. We get $\Pr[M_{i+1} \in D_i] < 2^{-\ell}$ by the union bound. Using both these estimates, we get that $\Pr[X \in B] \leq \sum_{i=1}^t \Pr[M_{i+1} \in B_i] \leq \sum_{i=1}^t \Pr[M_{i+1} \in C_i] + \Pr[M_{i+1} \in D_i] \leq \sum_i \epsilon_i + t2^{-\ell}$.

Now define the distribution $Z = X | X \notin B$. Then Z is $\sum_i \epsilon_i + t2^{-\ell}$-close to X. For every i and $z \in \mathsf{supp}(Z)$, we have that $\Pr[Z^i = z^i | Z^{i-1} = z^{i-1}] = \Pr[Z^i = z^i] / \Pr[Z^{i-1} = z^{i-1}] \leq 2^{-k_i + 1} / 2^{-n_{i-1} - \ell}$ (since every point at most doubles in weight over M), which proves the lemma.

Theorem 4 (Block vs General Source Extractor [BRSW06]). *There exists constants c_1, c_2 such that for every n, k, with $k > \log^{10} n$ there exists a polynomial time computable function $\mathsf{BExt} : \{0,1\}^{\mathsf{C}n} \times \{0,1\}^n \to \{0,1\}^m$ with $\mathsf{C} = O(\frac{\log n}{\log k})$ s.t. , if $X = X^1, \cdots, X^{\mathsf{C}}$ is a k-block source and Y is an independent k-source*

$$\Pr_{x \leftarrow_R X}[|\mathsf{BExt}(x, Y) - U_m| < 2^{-k^{c_1}}] > 1 - 2^{-k^{c_1}},$$

where $m = c_2 k$ and U_m denotes the uniform distribution on m bit strings.

3   The Extractor

In this section we describe our construction. Our extractor uses as a key component a randomness condenser, constructed by Guruswami, Umans and Vadhan [GUV07], which is in turn based on recent constructions of good list decodable codes ([GR06, PV05]), though we give a self contained proof of everything we need in this section.

First let us give a high level description of our algorithm and analysis. Although it seems hard to build extractors for two independent sources, the problem seems considerably easier when one of the sources is a block source. Indeed, our new algorithm will be obtained by reducing to this case. We will give an algorithm that given two independent sources, can turn them into a single block source, with many blocks. Once we have this algorithm, we will simply use one additional source and our extractor from Theorem 4.

3.1 Converting Two Independent Sources into a Block Source

Fix a finite field $\mathbb{F}$. The following algorithm is from [GUV07].

Algorithm 1 ($\mathsf{Cond}(f, y)$)

Input: $f \in \mathbb{F}^{t+1}$, $y \in \mathbb{F}$ and an integer r.
Output: $z \in \mathbb{F}^r$.

Sub-Routines and Parameters:
Let $g \in \mathbb{F}[X]$ be an irreducible polynomial of degree $t + 1$. Set $h = |\mathbb{F}|^{0.8\alpha}$ for some parameter α.

1. We interpret f as a degree t univariate polynomial with coefficients in $\mathbb{F}$.
2. For every $i = 0, 1, \ldots, m - 1$, let $f_i \in \mathbb{F}[x]$ be the polynomial $f^{h^i} \bmod g$.
3. Output $f_0(y), f_1(y), \ldots, f_{r-1}(y)$.

Guruswami et al. were interested in building seeded condensers, so they used the above algorithm with y sampled uniformly at random. Below, we show that the algorithm above is useful even when y is a high min-entropy source. We can prove the following lemma, which is a slight generalization of a lemma in [GUV07]:

Lemma 4. *Suppose F is a distribution on $\mathbb{F}^{t+1}$ with min-entropy k and Y is an independent distribution on $\mathbb{F}$ with min-entropy rate α and*

- $rt < \epsilon |\mathbb{F}|^{0.1\alpha}$
- $k > \log(2/\epsilon) + (0.8\alpha r) \log |\mathbb{F}|$.

Then $\mathsf{Cond}(F, Y)$ is ϵ-close to $.7\alpha r \log |\mathbb{F}|$-light, and hence it is ϵ-close to having min-entropy rate 0.7α.

Remark 1. In order to avoid using too many variables, we have opted to use constants like 0.1 and 0.7 in the proof. We note that we can easily replace the constants $0.7, 0.8$ with constants that are arbitrarily close to 1, at the price of making 0.1 closer to 0.

Proof (Lemma 4). We will repeatedly use the basic fact that any non-zero polynomial of degree d can have at most d roots.

By Fact 1, it suffices to prove the lemma when F and Y are flat sources.

We will prove that the output is close to having high min-entropy via Lemma 2. To do this, we need to show that for every set $S \subset \mathbb{F}^r$ of size $\epsilon|\mathbb{F}|^{0.7\alpha r}$, $\Pr[\mathsf{Cond}(F, Y) \in S] < \epsilon$. Fix a set S.

Let $Q(Z_1, \ldots, Z_r) \in \mathbb{F}[Z_1, \ldots, Z_r]$ be a non-zero r variate polynomial whose degree is at most $h - 1$ in each variable, such that $Q(s) = 0$ for every $s \in S$. Such a polynomial must exist since the parameters have been set up to guarantee $h^r = |\mathbb{F}|^{0.8\alpha r} > |S| = \epsilon|\mathbb{F}|^{0.7r\alpha}$.

Now call $f \in \mathsf{supp}(F)$ bad for S if

$$\Pr_{y \leftarrow_R Y}[\mathsf{Cond}(f, y) \in S] \geq \epsilon/2$$

We will bound the number of bad f's. Fix any such bad f. Then consider the univariate polynomial

$$R(X) = Q(f_0(X), f_1(X), \ldots, f_{r-1}(X)) \in \mathbb{F}[X]$$

This polynomial has degree at most $tr(h - 1)$. But $tr(h - 1) < \epsilon|\mathbb{F}|^{0.1\alpha}|\mathbb{F}|^{0.8\alpha} < \epsilon|\mathbb{F}|^{\alpha}/2 = (\epsilon/2)|\mathsf{supp}(Y)|$, thus this polynomial must be the zero polynomial. In particular, this means that $R(X) = 0 \mod g(X)$. This in turn implies that f must be a root of the polynomial

$$Q'(Z) = Q(Z, Z^h, Z^{h^2}, \ldots, Z^{h^{r-1}}) \in (\mathbb{F}[X]/g(X))[Z]$$

which is a univariate polynomial over the extension field $\mathbb{F}[X]/g(X)$, since $Q'(f(X)) = R(X) \mod g(X)$ by our choice of $f_0, \ldots, f_{r-1}$.

Recall that Q had degree at most $h - 1$ in each variable. This means that Q' has degree at most $h^r - 1$ and is non-zero, since no two monomials can clash when making the substitution Z^i for Z_i in Q. The number of bad f's can be at most $h^r - 1 < |\mathbb{F}|^{0.8\alpha r}$, since every bad f is a root of this low degree non-zero polynomial. This implies that $\Pr[F \text{ is bad}] < |\mathbb{F}|^{0.8\alpha r}/2^k < \epsilon/2$, since the constraint on k implies that $2^k > |\mathbb{F}|^{0.8\alpha r}2/\epsilon$.

Hence

$$\Pr[\mathsf{Cond}(F, Y) \in S] \leq \Pr[F \text{ is bad}] + \Pr[\mathsf{Cond}(F, Y) \in S | F \text{ is not bad}]$$
$$< \epsilon/2 + \epsilon/2 = \epsilon.$$

Note that a seeded condenser corresponds to the special case of $\alpha = 1$ in the above lemma. When α is small, it seems like the lemma doesn't say anything useful, since the min-entropy rate of the output is bounded above by α. But note that the lemma works for a very wide range of r's. The above function is more than a condenser, it *spreads* the entropy out across the output. Specifically, if we look at the first r' symbols in the output, they must also have min-entropy rate close to 0.7α. We can use this to construct a block source with geometrically increasing block lengths, as in the following lemma:

Lemma 5. *Let* Cond, $F, Y, \alpha, r, t, \epsilon$ *be as in Algorithm 1 and Lemma 4. Let* $r_1, r_2, \ldots, r_C = r$ *be positive integers. For $i = 1, 2, \ldots, C$, set Z^i to be the first r_i field elements in the output of* Cond(F, Y). *Then let $Z_1, \ldots, Z_C$ be such that $Z^i = Z_1, \ldots, Z_i$ for every i. Then for every $\ell > 10 \log C$ we have that $Z_1, Z_2, \ldots, Z_C$ is* $C(\epsilon + 2^{-\ell})$-*close to a block source with entropy* $(0.7\alpha r_i - r_{i-1}) \log(|\mathbb{F}|) - 1 - \ell$ *in each block.*

Proof. We will apply Lemma 3.

Note that for each i, Z^i is simply the output of the condenser upto the first r_i elements. Since $r_i \le r$, r_i satisfies the constraints of Lemma 4, so Z^i is ϵ-close to $0.7\alpha |Z^i|$-light. $\qquad \square$

We set parameters to get the following theorem:

Theorem 5. *There exists a polynomial time computable function* BlockConvert : $\{0,1\}^{n_1} \times \{0,1\}^{n_2} \to \{0,1\}^{m_1} \times \{0,1\}^{m_2} \times \cdots \times \{0,1\}^{m_C}$, *such that for every min-entropy k_1 source X over $\{0,1\}^{n_1}$ and every min-entropy k_2 source Y over $\{0,1\}^{n_2}$ satisfying*

- $C(\log \frac{10n_2}{k_2}) + 2 \log(n_1) < 0.095 k_2$
- $\sqrt{k_1} > k_2 (10n_2/k_2)^C$,

BlockConvert(X, Y) *is* $C(2^{-\Omega(k_2)} + 2^{-k_1^{\Omega(1)}})$-*close to a block source with* $\sum_i m_i \le (10n_2/k_2)^C \sqrt{k_1}$ *and min-entropy $2\sqrt{k_1}$ in each block.*

Proof. We show how to set parameters and apply Lemma 5.

Set $\mathbb{F}$ to be the finite field of size 2^{n_2}. Set $t = n_1/n_2, \epsilon = 2^{-0.05 k_2}$ and $k = k_1$. Set $\alpha = k_2/n_2$.

Set $r_i = (10n_2/k_2)^i \sqrt{k_1}$, so $\sum_i m_i = r = k_1^{1/2} (10n_2/k_2)^C$.

Using the first assumption,

$$rt = \sqrt{k_1} \left(\frac{10n_2}{k_2}\right)^C \frac{n_1}{n_2} \le n_1^2 \left(\frac{10n_2}{k_2}\right)^C < 2^{0.095 k_2} = 2^{-0.05 k_2} 2^{0.1 k_2} = \epsilon |\mathbb{F}|^{0.1\alpha}$$

to satisfy the first constraint of Lemma 4.

We have that

$$k_1 = k > 1 + 0.05 k_2 + 0.8 \cdot 10^C k_2 \sqrt{k_1} (n_2/k_2)^C = \log(2/\epsilon) + (0.8 r\alpha) \log(|\mathbb{F}|)$$

to satisfy the second constraint of Lemma 4.

Set $\ell = k_1^{0.1}$. Note that the second constraint implies that $C < \log k_1$.

Then let us use the algorithm Cond as promised by Lemma 5 with the above settings. We get that the final output is $C(\epsilon + 2^{-\ell+1}) \leq C(2^{-\Omega(k_2)} + 2^{-k_1^{\Omega(1)}})$-close to a block source with min-entropy $(0.7\alpha r_i - r_{i-1})\log(|\mathbb{F}|) - 1 - 2\ell$ in each block. We can lower bound this as follows:

$$(0.7\alpha r_i - r_{i-1})\log(|\mathbb{F}|) - 1 - 2\ell$$

$$= \left(0.7\frac{k_2}{n_2}\left(\frac{10n_2}{k_2}\right)^i \sqrt{k_1} - \left(\frac{10n_2}{k_2}\right)^{i-1} \sqrt{k_1}\right) n_2 - 1 - 2k_1^{0.1}$$

$$= (0.7 \cdot 10 - 1)\left(\frac{10n_2}{k_2}\right)^{i-1} n_2\sqrt{k_1} - 1 - 2k_1^{0.1}$$

$$= 6\left(\frac{10n_2}{k_2}\right)^{i-1} n_2\sqrt{k_1} - (1 + 2k_1^{0.1})$$

$$\geq 2\sqrt{k_1}$$

3.2   Putting It All Together

All that remains is to put together the various components to get our extractor.

Algorithm 2 ($\mathsf{IExt}(a, b, c)$)

Input: $a \in \{0,1\}^{n_1}, b \in \{0,1\}^{n_2}, c \in \{0,1\}^{n_3}$.
Output: $z \in \{0,1\}^m$ for a parameter m that we will set.

Sub-Routines and Parameters:
Let BlockConvert be the algorithm promised by Theorem 5, set up to operate on two sources with entropy k_1, k_2 and lengths n_1, n_2 respectively.
Let BExt be the algorithm promised by Theorem 4, set up to extract from a block source with C blocks of length $(10n_2/k_2)^C\sqrt{k_1}$, each with entropy $\sqrt{k_1}$ conditioned on previous blocks, and an independent source with length n_3 and min-entropy k_3.

---

1. Run BlockConvert(a, b) to get the blocks $x = x_1, x_2, \ldots, x_C$.
2. Output BExt(x, c).

We can now prove the main theorem.

Proof (Theorem 1). Let t be a constant so that BExt requires $C = t\log(n_1 + n_3)/\log(k_1)$ blocks to extract bits from an $(n_3, k_3 = \sqrt{k_1})$ source and an independent block source with blocks of length n_1, each with entropy $\sqrt{k_1}$ conditioned on previous blocks. The error of this extractor is promised to be $2^{-k_1^{\Omega(1)}}$.
  We check each of the constraints needed for BlockConvert to succeed.

First we have that

$$\mathsf{C}\left(\log \frac{10n_2}{k_2}\right) + \log n_1$$

$$< \mathsf{C}10\log n_2 + \log n_1$$

$$\leq 10t\frac{\log(n_1 + n_3)}{\log k_1}\log n_2 + \log n_1$$

$$\leq (10t/d)\log k_1 + \log n_1 \qquad\qquad \text{by the first assumption}$$

$$< 0.095d\log n_1 \qquad\qquad\qquad\qquad \text{for } d \text{ large enough}$$

$$< 0.095k_2 \qquad\qquad\qquad\qquad \text{by the second assumption}$$

For the next constraint,

$$\log(k_2(10n_2/k_2)^{\mathsf{C}})$$

$$= \mathsf{C}\log(10n_2/k_2) + \log k_2$$

$$\leq t\frac{\log(n_1 + n_3)}{\log k_1}(\log(n_2) + \log 10) + \log n_2$$

$$< 3(t/d)\log k_1 \qquad\qquad\qquad\qquad \text{by the first assumption}$$

$$< (1/2)\log k_1 \qquad\qquad\qquad\qquad \text{for } d \text{ large enough}$$

We are not yet done, since the algorithm above will only output $m_1 = \sqrt{k_1} - o(\sqrt{k_1})$ bits. However, we do have that:

$$\Pr_{x_1 \leftarrow_R X_1}\left[|\mathsf{IExt}(x_1, Y, Z) - U_{m_1}| > 2^{-\Omega(k_2)} + 2^{-k_1^{\Omega(1)}}\right] < 2^{-\Omega(k_2)} + 2^{-k_1^{\Omega(1)}}$$

since BExt is strong.

Thus we have that $|X, \mathsf{IExt}(X, Y, Z) - X, U_{m_1}| < 2^{-\Omega(k_2)} + 2^{-k_1^{\Omega(1)}}$, which implies that if Ext is any strong seeded extractor set up to extract from a min-entropy k_1 source with seed length m_2, $\mathsf{Ext}(X, U_{m_1})$ is $2^{-\Omega(k_2)} + 2^{-k_1^{\Omega(1)}}$ close to $\mathsf{Ext}(X, \mathsf{IExt}(X, Y, Z))$. This is our final extractor.

3.3   Extension to Block Sources

We now sketch the proofs of Theorems Theorem 2 and Theorem 3.

For Theorem 2, our block source will be (b, c), and our extractor will be $\mathsf{IExt}(a, b, c)$. We will show that $\mathsf{BlockConvert}$ is strong in the sense that with high probability, even conditioned on b, $\mathsf{BlockConvert}(X, b)$ will be a block source. The proof then proceeds as before.

The following lemma shows that any condenser with good parameters is also strong, with slightly weaker parameters.

Lemma 6. *Let $\mathcal{X}$ denote a collection of sources. Suppose the function C is a condenser in that for independent $X \in \mathcal{X}$ and Y with $H_\infty(Y) \geq \ell$, $C(X, Y)$ is ϵ-close to k-light. Then for any such X, Y, when y is chosen from Y,*

$$\Pr_y[C(X, y) \text{ is } \delta\text{-close to } (k - \ell)\text{-light}] \geq 1 - \epsilon/\delta.$$

Proof. Fix any such X and Y. Let $S = \{z \mid \Pr[C(X,Y) = z] > 2^{-k}\}$ denote the set of heavy elements. Note that for any y in the support of Y and $z \notin S$, $\Pr[C(X,y) = z] \leq 2^{\ell-k}$. Now let $p_y = \Pr[C(X,y) \in S]$. Then $E[p_y] \leq \epsilon$, so by Markov $\Pr[p_y \geq \delta] \leq \epsilon/\delta$, which gives the lemma.

We will use this lemma with the condenser Cond and with $\delta = \sqrt{\epsilon}$. We then modify Lemma 5 so that with high probability over the choice of y, $\mathsf{Cond}(F, y)$ is a block source. This immediately yields a strong version of Theorem 5.

Theorem 6. *There exists a polynomial time computable function* BlockConvert $:$ $\{0,1\}^{n_1} \times \{0,1\}^{n_2} \to \{0,1\}^{m_1} \times \{0,1\}^{m_2} \times \cdots \times \{0,1\}^{m_c}$, *such that for every min-entropy k_1 source X over $\{0,1\}^{n_1}$ and every min-entropy k_2 source Y over $\{0,1\}^{n_2}$ satisfying*

- $\mathsf{C}(\log \frac{10n_2}{k_2}) + 2\log(n_1) < 0.095 k_2$
- $\sqrt{k_1} > k_2 (10 n_2 / k_2)^{\mathsf{C}}$,

the following holds. When y is chosen according to Y, with probability $1 - \mathsf{C}(2^{-\Omega(k_2)} + 2^{-k_1^{\Omega(1)}})$, BlockConvert$(X, y)$ *is* $\mathsf{C}(2^{-\Omega(k_2)} + 2^{-k_1^{\Omega(1)}})$*-close to a block source with $\sum_i m_i \leq (10 n_2 / k_2)^{\mathsf{C}} \sqrt{k_1}$ and min-entropy $2\sqrt{k_1}$ in each block.*

Now, when we analyze $\mathsf{IExt}(a, b, c)$ where (b, c) is a block source, we argue that with high probability over the choice of b, we are in the same situation as before, and our proof continues in the same manner.

For Theorem 3, our block source will be (a, b), and our extractor will be $\mathsf{IExt}(a, b, c)$. In the analysis of $\mathsf{Cond}(f, y)$, we analyzed a bad f by counting the number of y that cause $\mathsf{Cond}(f, y) \in S$. The key observation is that this analysis remains unchanged if we choose y from a set Y that depends on f. This is easily verified by looking at the proof of Lemma 4. Hence (f, y) can be from a block source.

Acknowledgements

We would like to thank Salil Vadhan and Chris Umans for useful discussions. In particular, Salil showed us how to get Theorem 3. Thanks to the referees for several useful comments.

References

[BIW04]   Barak, B., Impagliazzo, R., Wigderson, A.: Extracting randomness using few independent sources. In: Proceedings of the 45th Annual IEEE Symposium on Foundations of Computer Science, pp. 384–393 (2004)

[BKS+05]   Barak, B., Kindler, G., Shaltiel, R., Sudakov, B., Wigderson, A.: Simulating independence: New constructions of condensers, Ramsey graphs, dispersers, and extractors. In: Proceedings of the 37th Annual ACM Symposium on Theory of Computing, pp. 1–10 (2005)

[BRSW06]   Barak, B., Rao, A., Shaltiel, R., Wigderson, A.: 2 source dispersers for $n^{o(1)}$ entropy and Ramsey graphs beating the Frankl-Wilson construction. In: Proceedings of the 38th Annual ACM Symposium on Theory of Computing (2006)

[Bou05]   Bourgain, J.: More on the sum-product phenomenon in prime fields and its applications. International Journal of Number Theory 1, 1–32 (2005)

[CRVW02]   Capalbo, M., Reingold, O., Vadhan, S., Wigderson, A.: Randomness conductors and constant-degree lossless expanders. In: Proceedings of the 34th Annual ACM Symposium on Theory of Computing, pp. 659–668 (2002)

[CG88]   Chor, B., Goldreich, O.: Unbiased bits from sources of weak randomness and probabilistic communication complexity. SIAM Journal on Computing 17(2), 230–261 (1988)

[Gur04]   Guruswami, V.: Better extractors for better codes? In: Proceedings of the 36th Annual ACM Symposium on Theory of Computing, pp. 436–444 (2004)

[GR06]   Guruswami, V., Rudra, A.: Explicit capacity-achieving list-decodable codes. In: Proceedings of the 38th Annual ACM Symposium on Theory of Computing (2006)

[GUV07]   Guruswami, V., Umans, C., Vadhan, S.: Unbalanced expanders and randomness extractors from Parvaresh-Vardy codes. In: Proceedings of the 22nd Annual IEEE Conference on Computational Complexity (2007)

[KLRZ08]   Kalai, Y., Li, X., Rao, A., Zuckerman, D.: Network extractor protocols (unpublished manuscript, 2008)

[KRVZ06]   Kamp, J., Rao, A., Vadhan, S., Zuckerman, D.: Deterministic extractors for small space sources. In: Proceedings of the 38th Annual ACM Symposium on Theory of Computing (2006)

[LRVW03]   Lu, C.J., Reingold, O., Vadhan, S., Wigderson, A.: Extractors: Optimal up to constant factors. In: Proceedings of the 35th Annual ACM Symposium on Theory of Computing, pp. 602–611 (2003)

[Lu04]   Lu, C.-J.: Encryption against storage-bounded adversaries from on-line strong extractors. J. Cryptology 17(1), 27–42 (2004)

[MNSW98]   Miltersen, P., Nisan, N., Safra, S., Wigderson, A.: On data structures and asymmetric communication complexity. Journal of Computer and System Sciences 57, 37–49 (1998)

[NZ96]   Nisan, N., Zuckerman, D.: Randomness is linear in space. Journal of Computer and System Sciences 52(1), 43–52 (1996)

[PV05]   Parvaresh, F., Vardy, A.: Correcting errors beyond the guruswami-sudan radius in polynomial time. In: Proceedings of the 46th Annual IEEE Symposium on Foundations of Computer Science, pp. 285–294 (2005)

[Rao06]   Rao, A.: Extractors for a constant number of polynomially small minentropy independent sources. In: Proceedings of the 38th Annual ACM Symposium on Theory of Computing (2006)

[Raz05]   Raz, R.: Extractors with weak random seeds. In: Proceedings of the 37th Annual ACM Symposium on Theory of Computing, pp. 11–20 (2005)

[SV86]   Santha, M., Vazirani, U.V.: Generating quasi-random sequences from semi-random sources. Journal of Computer and System Sciences 33, 75–87 (1986)

[Sha06]   Shaltiel, R.: How to get more mileage from randomness extractors. In: Proceedings of the 21th Annual IEEE Conference on Computational Complexity, pp. 49–60 (2006)

[TZ04]   Ta-Shma, A., Zuckerman, D.: Extractor codes. IEEE Transactions on In-
            formation Theory 50 (2004)
[Vad04]   Vadhan, S.P.: Constructing locally computable extractors and cryptosys-
            tems in the bounded-storage model. J. Cryptology 17(1), 43–77 (2004)
[WZ99]   Wigderson, A., Zuckerman, D.: Expanders that beat the eigenvalue
            bound: Explicit construction and applications. Combinatorica 19(1), 125–
            138 (1999)
[Zuc96]   Zuckerman, D.: Simulating BPP using a general weak random source.
            Algorithmica 16, 367–391 (1996)
[Zuc97]   Zuckerman, D.: Randomness-optimal oblivious sampling. Random Struc-
            tures and Algorithms 11, 345–367 (1997)

The Power of Choice in a Generalized Pólya Urn Model

Gregory B. Sorkin

IBM Watson Research Center
sorkin@watson.ibm.com

Abstract. We establish some basic properties of a "Pólya choice" generalization of the standard Pólya urn process. From a set of k urns, the ith occupied by n_i balls, choose c distinct urns $i_1, \ldots, i_c$ with probability proportional to $n_{i_1}^\gamma \times \cdots \times n_{i_c}^\gamma$, where $\gamma > 0$ is a constant parameter, and increment one with the smallest occupancy (breaking ties arbitrarily). We show that this model has a phase transition. If $0 < \gamma < 1$, the urn occupancies are asymptotically equal with probability 1. For $\gamma > 1$, this still occurs with positive probability, but there is also positive probability that some urns get only finitely many balls while others get infinitely many.

1 Introduction

We introduce an urn model that combines characteristic elements of two well-established models. As a basis of comparison, we first recall the most basic urn model: given a set of urns, each in a sequence of balls is placed in an urn chosen uniformly at random. If there are n urns and n balls, at the end of this process the "maximum load" (occupancy of the fullest urn) whp (with high probability, i.e., with probability tending to 1 as $n \to \infty$), is approximately $\ln n / \ln \ln n$. Of course, if the number of urns k is fixed and the number of balls $n \to \infty$, then whp the loads are asymptotically balanced.

The first model forming the basis of ours is the "power of two choices" urn model, with roots in [KLMadH96, ABKU99]. If at each step a *pair* of urns is selected uniformly, and the ball allocated to the more lightly loaded one, then the maximum load is whp only $\ln \ln n / \ln 2 + \Theta(1)$: having even a small amount of choice is enough to balance the load.

The second is a "rich get richer" Pólya urn model, considered for example in [CHJ03, DEM01]. This model is like the simple urn model except that instead of an urn's being selected uniformly at random, it is selected in proportion to a power γ of its occupancy. (In this model each urn has initial occupancy 1, not 0.) That is, if at step t the ith urn has occupancy $x_i(t)$, it is incremented with probability $x_i(t)^\gamma / \sum_j x_j(t)^\gamma$. (With $\gamma = 1$ this is the usual Pólya process.) With k urns and exponent γ, we denote such a rich-get-richer Pólya model by $P(\gamma, k)$.

Our model synthesizes the two. Considering first a special case, at each step t, a pair i, j of urns, $i \neq j$ ("drawn without replacement"), is chosen with probability proportional to $x_i(t)^\gamma x_j(t)^\gamma$. (Drawing two elements with replacement can

A. Goel et al. (Eds.): APPROX and RANDOM 2008, LNCS 5171, pp. 571–583, 2008.
© Springer-Verlag Berlin Heidelberg 2008

be drastically different: if one urn is much fuller than the others, it is likely that this urn will be drawn twice, spoiling the "power of choice".) The ball is placed into the less full urn, with ties broken arbitrarily.

In general we will choose $c \geq 2$ urns in proportion to the product of the γth powers of their occupancies, and place the next ball in a least occupied urn from this set. We will denote a k-urn γ-exponent Pólya-choice process with c choices by PC(γ, k, c). Note that if we allowed $c = 1$, PC$(\gamma, k, 1)$ would be the same as P(γ, k). We will write just PC and P when the parameters are clear.

We will show that our Pólya-choice (PC) model has a phase transition. If $0 < \gamma < 1$, then with probability equal to 1 the urn occupancies are asymptotically equal. (The Pólya model P(γ, k) also has this property for $\gamma < 1$, so this is not surprising. The $\gamma = 1$ case is treated in [LAS$^+$08], where a much more difficult proof shows that it too is always balanced.) For $\gamma > 1$, asymptotic balance still occurs with positive probability, but there is also positive probability that some urns get only finitely many balls while others get infinitely many. This $\gamma > 1$ case is most interesting. With probability 1, a rich-get-richer Pólya urn process P(γ, k) with $\gamma > 1$ yields a single urn with almost all the balls, while the other urns starve. Adding the power of choice means that while this still happens with positive probability for PC(γ, k, c), there is also positive probability of asymptotic balance.

The PC(γ, k, c) model with $\gamma = 1$ arises in the context of Internet routing as part of a technical analysis of "linked decompositions" of preferential-attachment graphs [LAS$^+$08]. Given the existence of the rich-get-richer Pólya model, generalizing to other values of γ seems natural. Power-law phenomena have received a good deal of attention recently, and it is conceivable that the model could have applicability to, say, the "natural" evolution of a web graph modified by a small amount of engineered network design. The model is sufficiently simple that some variations have already been suggested [McD07].

It must be admitted, however, that our motivation for studying the Pólya-choice model is that both the Pólya model with exponent γ and the "power of two choices" model have resulted in beautiful mathematics, and the combined PC model promises to do the same. In this paper, we describe the two aspects mentioned a moment ago: that the model undergoes a phase transition between $\gamma < 1$ and $\gamma > 1$, and that for $\gamma > 1$ two dramatically different outcomes — balanced or imbalanced — can ensue with positive probability. The possibility of balance for $\gamma > 1$ is our main technical result, given by Theorem 3 in Section 4.

2   Balance for $\gamma < 1$

It is well known that in the Pólya model P(γ, k) with parameter $\gamma < 1$, with probability 1 ("wp 1") the urns are asymptotically balanced. That is, wp 1, for each $i \in [k]$, $\lim_{n \to \infty} n_i/n = 1/k$. A simple and beautiful coupling-based proof is given in [CHJ03]. It is not surprising that the "power of choice" aspect yields even more balanced allocations, and thus leads to the same perfectly balanced long-run outcome.

Theorem 1. *In a* PC(γ, k, c) *process with* $0 < \gamma < 1$, *wp 1, for each* $i \in [k]$, *we have* $\lim_{n \to \infty} n_i/n = 1/k$.

Proof. Assume henceforth that the set of urn occupancies $\{n_i\}$ is sorted in non-decreasing order. Let $\{n_i'\}$ be another set of occupancies, with the same total number of balls, $\sum_{i=1}^{k} n_i = \sum_{i=1}^{k} n_i' = n$. We say that $\{n_i\}$ is "more balanced" than $\{n_i'\}$, if for every i the prefix sum $\sum_{j=1}^{i} n_j$ is at least as large as the corresponding sum $\sum_{j=1}^{i} n_j'$. We will write this as $\{n_i\} \leq \{n_i'\}$, because $\{n_i\}$ is more balanced than $\{n_i'\}$ precisely if $\{n_i\}$ is *majorized* by $\{n_i'\}$.

We exhibit a coupling in which the allocations $\{n_i\}$ from PC(γ, k, c) are always more balanced than the allocations $\{n_i'\}$ from P(γ, k). This suffices, since it is trivial from majorization that if $n_i'/n \to k$ for each i (which holds wp 1), and $\{n_i\} \leq \{n_i'\}$, then also $n_i/n \to k$ for each i.

In both processes, at the initial time $n = k$ each urn is started off with one ball, so it is trivial that PC is more balanced than P: $\{n_i\} \leq \{n_i'\}$, with equality. We prove that if this inequality holds at time n then it also holds at $n + 1$. With reference to the small table immediately below, let variables n_i and m_i indicate occupancies of the process PC at times n and $n + 1$ respectively, n_i' and m_i' the occupancies of the Pólya process P, and m_i'' the occupancies at time $n + 1$ of a Pólya process P$''$ with occupancies n_i (not n_i') at time n.

time	PC	P	P$''$
n	n_i	n_i'	n_i [sic]
$n + 1$	m_i	m_i'	m_i''

We will show that if $\{n_i\} \leq \{n_i'\}$ then $\{m_i''\} \leq \{m_i'\}$ and $\{m_i\} \leq \{m_i''\}$; by transitivity of majorization the desired result $\{m_i\} \leq \{m_i'\}$ follows.

It is shown in [CHJ03] that under the canonical coupling, with the Pólya process P$''$ started in a more balanced state than the Pólya process P, i.e., $\{n_i\} \leq \{n_i'\}$, after one step P$''$ is again in a more balanced state than P, $\{m_i''\} \leq \{m_i'\}$. The coupling is as follows. Generate a real value $\alpha \in [0, 1)$ uniformly at random. For the process P$''$, let $b_i = \sum_{j=1}^{i} n_j^{\gamma}$, with $b_0 = 0$. Place the new ball in urn i'' where i'' is the smallest value such that $\alpha \leq b_i''/b_k$. Similarly, for the process P, let $c_i = \sum_{j=1}^{i} n_i'^{\gamma}$, and use the same value of α to choose the urn i' in this process (that is, the smallest i' such that $\alpha \leq c_i'/c_k$). What [CHJ03] shows is that $i'' \leq i'$ (because the prefix sums b_i for P$''$ are larger than the ones c_i for P), which is to say that the ball is added to an earlier bin in P$''$ than in P, thus keeping the prefix sums of P$''$ larger than those of P, which means that $\{m_i''\} \leq \{m_i'\}$.

It remains only to show a coupling by which $\{m_i\} \leq \{m_i''\}$: that is, starting a Pólya-choice process PC and a Pólya process P$''$ (with the same value of γ) from the same initial allocations $\{n_i\}$, the next-step PC allocation $\{m_i\}$ is more balanced than the next-step P$''$ allocation $\{m_i''\}$. The coupling is given by a coupling between PC tuples $\{i_1, \ldots, i_c\} \in [k]^{(c)}$ (the set of unordered c-tuples of distinct urns) and single urns $\{i\} \in [k]$ such that each has the appropriate marginal probability but, always, $\min\{i_1, \ldots, i_c\} \leq i$. Given such a coupling, in the PC process the ball either goes into the same urn as in P″ or an earlier (less full) one, giving $\{m_i\} \leq \{m_i''\}$ as desired.

For intuitive purposes we remark that if the PC process had urns drawn *with* replacement, the desired coupling would be trivial. If we drew i as the single urn for P″, then for PC we could draw a c-tuple by taking $i_1 = i$ as its first element and drawing the remaining elements independently, guaranteeing that $\min\{i_1, \ldots, i_c\} \leq i_1 = i$. We now show the coupling for draws without replacement.

Let $\mathrm{Pr}_c(i_{\min})$ denote the probability that urn $i_{\min} = \min\{i_1, \ldots, i_c\}$ is selected in the PC process, and $w_i = n_i^\gamma / (\sum_{j=1}^{k} n_j^\gamma)$ the probability that urn i is selected in the Pólya process P″. It suffices to show that for all $i \in [k]$, $\sum_{j=1}^{i} \mathrm{Pr}_c(j) \geq \sum_{j=1}^{i} w_j$: then, coupling by choosing $\alpha \in [0,1]$ uniformly at random and letting $i_{\min} = \min\{i : \alpha \leq \sum_{j=1}^{i} \mathrm{Pr}_c(j)\}$ and $i = \min\{i : \alpha \leq \sum_{j=1}^{i} w_j\}$, we have $i_{\min} \leq i$, implying $\{m_i\} \leq \{m_i''\}$ as claimed.

Note that each feasible c-tuple containing values $i_1 < \cdots < i_c$ has probability proportional to $w_1 \times \cdots \times w_c$, and thus

$$\mathrm{Pr}_c(i_{\min}) = w_{i_{\min}} W_{i_{\min}},$$

where

$$W_{i_{\min}} := \frac{\sum_{i_{\min} < i_2 < \cdots < i_c} w_2 \times \cdots \times w_c}{\sum_{i_1 < i_2 < \cdots < i_c} w_1 \times \cdots \times w_c}.$$

What we wish to show, then, is that the prefix sums of the sequence $w_i W_i$ dominate those of w_i. Since the W_i form a nonincreasing sequence (the numerators decrease while the denominator is fixed), this is intuitively obvious: weighting by the W_i gives greater probability to earlier elements. To confirm this, note that $\sum_{i=1}^{k} w_i W_i = \sum_{i=1}^{k} w_i = 1$ (both $\{w_i W_i\}$ and $\{w_i\}$ are distributions), and thus

$$\sum_{i=1}^{k} w_i(W_i - 1) = 0.$$

We wish to show that, for all $i \in [k]$, we have $\sum_{j=1}^{i} w_j(W_j - 1) \geq 0$. As a function of i, the W_i are nonincreasing, thus the prefix sums are first increasing (as long as $W_i \geq 0$) and then decreasing (when $W_i < 0$), and thus the minimum prefix sum occurs either at the start ($i = 0$) or at the end ($i = k$). At both of these points the prefix sum is 0, therefore it is always nonnegative.

This establishes the existence of the desired coupling, completing the proof. □

As remarked earlier, asymptotic balance for $\gamma = 1$ is established in [LAS$^+$08]. This result cannot be obtained as in our proof of Theorem 1 because the classic Pólya model (two urns, $\gamma = 1$) does not have this property. Rather, for the classic model, wp 1, n_1/n tends to a limit, but the limit itself is uniformly distributed over $[0, 1]$ (one very nice proof is given in [CHJ03]).

3   Imbalance for $\gamma > 1$

We now turn to the more interesting case $\gamma > 1$. Here, there is a positive probability that some urns collect infinitely many balls while the others remain finite, analogous to the P(γ) model where for $\gamma > 1$ one urn always collects an infinite number of balls while the others starve. However, for the PC(γ) model, there is also positive probability that all urns have the same number of balls asymptotically.

The phrase "with positive probability" means with probability bounded away from 0. Specifically, for the PC(γ, k, c) process, we mean that the event in question holds with probability at least $g(\gamma, k, c)$, for some function g with $g(\gamma, k, c) > 0$ for all $\gamma > 0$, $k > c > 1$.

In this section we prove the "imbalance" result.

Theorem 2. *In the* PC(γ, k, c) *Pólya-choice urn process with $\gamma > 1$, with positive probability exactly c urns collect all balls.*

Before giving the proof we present an elegant way of thinking about the process $P(\gamma, k)$. A lovely observation attributed to Herman Rubin is that this discrete process can be replaced by a Poisson arrival process for each urn *independently*. Over continuous time, associate with urn i an arrival process where the waiting time from ball x to $x + 1$ is exponentially distributed with parameter x^γ (mean $x^{-\gamma}$). It is easy to check that, at any given time t, the *next* arrival occurs in urn i with probability proportional to $x_i(t)^\gamma$, so if we look only at moments when a ball arrives, this continuous-time process is an instantiation of the parameter-γ Pólya process. The expected time until an urn acquires infinitely many balls is $\sum_{x=1}^{\infty} x^{-\gamma}$, which for $\gamma > 1$ is finite. Of course there is some variance in the actual time at which this occurs, so that there comes a moment when one urn has infinitely many balls and the others have only finitely many. In the discrete, ball-arrival model, this means that, after a last arrival to some urn, *every* new ball goes into the winning urn: the outcome is as unbalanced as can be.

While our proof of Theorem 2 will not use the Poisson-process machinery, the intuition is helpful. Consider the case $c = 2$, $k = 3$. Since the smaller of the two selected urns is always incremented, the fullest urn must always have a "companion" which is equally full, to within 1 ball. Reversing the sorting of the previous section, without loss of generality imagine the leaders to be urns 1 and 2, and define $x = (n_1 + n_2)/2$. This value x completely describes the leaders: if x is integral then $n_1 = n_2 = x$, while if it is half-integral then $n_1 = x + 1/2$ and $n_2 = x - 1/2$.

With $k = 3$, we are in a situation very close to that of the P(k, γ) Pólya model, except that there is an asymmetry. Speaking extremely informally, the leading (combined) urn is incremented by $\frac{1}{2}$ with Poisson intensity roughly $x^{2\gamma}$ while the follower is incremented by 1 with Poisson intensity $2x^\gamma n_3^\gamma$, which (since only the ratio matters) is roughly equivalent to incrementing the leader with intensity $\frac{1}{2}x^\gamma$ and the follower with intensity $2n_3^\gamma$. This much (the asymmetry between $\frac{1}{2}$ and 2 for the two pseudo-urns) can be accommodated by Rubin's view, but if the two processes are treated independently then the follower can overtake the leader, which is forbidden (if that occurs their roles must be exchanged).

However, modeling the leader and follower independently is valid for a sample path in which they happen never to cross. Per the Poisson model, there is positive probability that the leader acquires infinitely many balls while the follower gets *none*, and in this case the sample paths do not cross.

We will prove precisely this — that there is positive probability that the first 2 urns acquire all the balls — but without explicitly introducing the Poisson-process machinery, and for general k.

Proof. Suppose that at some time the first c urns together have x balls and the remaining $k - c$ urns have one ball each. Since at every step up to this point we put a ball in a least-full selected urn, these c urns have at most two occupancy levels, and $\lfloor x/c \rfloor \leq n_i \leq \lceil x/c \rceil$. For x sufficiently large, then,

$$\frac{1}{2} \left(\frac{x}{c} \right)^{\gamma c} \leq n_1^{\gamma} \cdot n_2^{\gamma} \cdot \ldots \cdot n_c^{\gamma} \leq 2 \left(\frac{x}{c} \right)^{\gamma c}.$$

Let Y_i be the event that the first c urns are chosen in round i. (Round 1 starts with k balls altogether, and ends with $k + 1$.) If event Y_i occurs for all $1 \leq i \leq x - c$, then the first c urns have x balls altogether and the remaining urns have one ball each. Thus by the above, for sufficiently large x,

$$\mathbb{P}(Y_{x-c} \mid \bigcap_{i<x-c} Y_i) \geq \frac{n_1^{\gamma} n_2^{\gamma} \ldots n_c^{\gamma}}{n_1^{\gamma} n_2^{\gamma} \ldots n_c^{\gamma} + \sum_{i=1}^{c} \binom{c}{c-i} \binom{k-c}{i} \lceil x/c \rceil^{\gamma(c-i)}}$$

$$\geq \frac{n_1^{\gamma} n_2^{\gamma} \ldots n_c^{\gamma}}{n_1^{\gamma} n_2^{\gamma} \ldots n_c^{\gamma} + c2^c 2^k (x/c)^{\gamma(c-1)}} \geq 1 - \frac{4c2^k 2^c}{(x/c)^{\gamma}}$$

For any fixed value of x_0, however large, there is positive probability that for all $x < x_0 - c$ the first c urns were chosen, that is, $\mathbb{P}(\bigcap_{i<x_0-c} Y_i) > 0$. The probability that for any $x \geq x_0$, we fail to choose the first c urns is the probability that we fail for x, conditioned upon never having failed before. Setting $m = 4c2^k 2^c c^{\gamma}$ we obtain

$$\mathbb{P}\left(\bigcap_{i=1}^{\infty} Y_i \right) = \mathbb{P}(\bigcap_{i<x_0-c} Y_i) \prod_{i=x_0-c}^{\infty} \mathbb{P}\left(Y_i \mid \bigcap_{j<i} Y_j \right)$$

$$\geq \mathbb{P}(\bigcap_{i<x_0-c} Y_i) \prod_{x=x_0-c}^{\infty} (1 - mx^{-\gamma})$$

$$\geq \mathbb{P}(\bigcap_{i<x_0-c} Y_i) \left(1 - \sum_{x=x_0-c}^{\infty} mx^{-\gamma} \right)$$

$$\geq \mathbb{P}(\bigcap_{i<x_0-c} Y_i) \left(1 - m(x_0 - c - 1)^{1-\gamma} \right)$$

which is bounded away from 0 if we choose x_0 sufficiently large. $\qquad\square$

4   Balance for $\gamma > 1$

While, as just shown, it may happen that c urns swallow all the balls, there is also positive probability that all urns have an asymptotically equal number of balls. Note that the following theorem actually applies to any γ (in equation (3), γ is seen to be irrelevant to our proof) but since Theorem 1 and [LAS+08] show that for $\gamma < 1$ and $\gamma = 1$ there is *always* asymptotic balance, the case $\gamma > 1$ is the one of interest here.

Theorem 3. *In a* $PC(\gamma, k, c)$ *Pólya-choice urn process with any γ, with positive probability, for each $i \in [k]$, we have $\lim_{n \to \infty} n_i/n = 1/k$.*

Proof. To outline the proof, we will define a "potential function" $\Gamma(\{n_i\})$ on the occupancies, with the property that the potential is large if and only if the occupancies are significantly different from one another. If the potential is even modestly large, then it will tend to decrease in the next step of the process. We will show that if the potential reaches some "triggering" threshold (still a small amount), then after some additional steps it is likely to have become smaller (below triggering) without ever having become much larger; the probabilities are sufficiently good that with positive probability this succeeds every time. That the potential is never large and thus the occupancies are always approximately equal simplifies the analysis, for it means that the sampling of c urns is nearly uniform.

We now begin the proof proper. When there are n balls, let $\bar{n} = n/k$ be the average occupancy, and for the individual occupancies write $n_i = (1+\varepsilon_i)\bar{n}$ (some values ε_i will be positive and some negative, with $\sum \varepsilon_i = 0$). Henceforth, assume that the urns are sorted in nondecreasing order, so that ε_1 is the smallest value and ε_k the largest (and one or the other is largest in absolute value), and let $\varepsilon = \max\{|\varepsilon_1|, |\varepsilon_k|\}$. Define the potential

$$\Gamma(\{n_i\}) = \sum_{i=1}^{k} (\ln \bar{n} - \ln n_i).$$

Note that

$$\Gamma(\{n_i\}) = -\sum_{i=1}^{k} \ln(n_i/\bar{n}) = -\sum_{i=1}^{k} \ln(1 + \varepsilon_i) = \sum_{i=1}^{k} \left(-\varepsilon_i + \tfrac{1}{2}\varepsilon_i^2 - \tfrac{1}{3}\varepsilon_i^3 + \cdots\right)$$

$$= -\sum_{i=1}^{k} \varepsilon_i + \sum_{i=1}^{k} \left(\tfrac{1}{2}\varepsilon_i^2 - \tfrac{1}{3}\varepsilon_i^3 + \cdots\right)$$

$$= (1 + O(\varepsilon)) \cdot \frac{1}{2} \sum_{i=1}^{k} \varepsilon_i^2,$$

since $\sum_{i=1}^{k} \varepsilon_i = 0$ and for ε small the Taylor series expansions are dominated by their leading terms.

Then

$$\frac{1}{2}(1 + O(\varepsilon))\varepsilon^2 \leq \Gamma \leq \frac{1}{2}k(1 + O(\varepsilon))\varepsilon^2. \tag{1}$$

To within constant factors, then, Γ summarizes the maximum deviation of any urn from the mean.

From $\Gamma = k \ln \bar{n} - \sum \ln n_i$, for large values of $\bar{n}$, the change in Γ when urn i is incremented (and $\bar{n}$ increases by $1/k$) is

$$
\begin{aligned}
\Delta_i &:= [k \ln(\bar{n} + 1/k) - k \ln \bar{n}] - [\ln(n_i + 1) - \ln n_i] \\
&= k \ln(1 + 1/(k\bar{n})) - \ln(1 + 1/n_i) \\
&= k \left(\frac{1}{k\bar{n}} + O\left(\frac{1}{k^2 \bar{n}^2}\right) \right) - \frac{1}{n_i} + O\left(\frac{1}{n_i^2}\right) \\
&= \frac{n_i - \bar{n}}{n_i \bar{n}} + O\left(\frac{1}{\bar{n}^2}\right) \\
&= \frac{\varepsilon_i}{\bar{n}} + O\left(\frac{\varepsilon_i^2}{\bar{n}}\right) + O\left(\frac{1}{\bar{n}^2}\right).
\end{aligned} \tag{2}
$$

For $i \in \{1, \ldots, k - c + 1\}$, let $K_i = \{i + 1, \ldots, k\}$. By our assumption that $n_1 \leq \cdots \leq n_k$, for any $j \in K_i$, we have $n_j \geq n_i$. To compute the *expected* change in Γ we first give upper and lower bounds on the probability for the event A_i that urn i is the smallest of all chosen urns. To do so we will make use of the fact that with ε small, the sampling of c urns is nearly uniform, because

$$(1 + \varepsilon_i)^{c\gamma} = 1 + O(\varepsilon), \tag{3}$$

with the constant $c\gamma$ subsumed in the $O(\cdot)$. Specifically, for $i \in \{1, \ldots, k - c + 1\}$,

$$
\begin{aligned}
\mathbb{P}(A_i) &= \frac{\sum_{S \subseteq K_i, |S|=c-1} \prod_{s \in S}(1 + \varepsilon_s)^\gamma \bar{n}^\gamma}{\sum_{T \subseteq K_0, |T|=c} \prod_{t \in T}(1 + \varepsilon_t)^\gamma \bar{n}^\gamma} \\
&\geq \frac{(1 - \varepsilon)^{c\gamma} \binom{k-i}{c-1}}{(1 + \varepsilon)^{c\gamma} \binom{k}{c}} \\
&= (1 - O(\varepsilon)) \frac{\binom{k-i}{c-1}}{\binom{k}{c}},
\end{aligned}
$$

and similarly

$$\mathbb{P}(A_i) \leq (1 + O(\varepsilon)) \frac{\binom{k-i}{c-1}}{\binom{k}{c}}.$$

We thus have $\mathbb{P}(A_i) = (1 + O(\varepsilon))\frac{\binom{k-i}{c-1}}{\binom{k}{c}}$ and

$$
\mathbb{E}(\Delta\Gamma) = \sum_{i=1}^{k-c+1} \Delta_i \mathbb{P}(A_i)
$$

$$
= \sum_{i=1}^{k-c+1} \left(\frac{\varepsilon_i}{\bar{n}} + O\left(\frac{\varepsilon_i^2}{\bar{n}}\right) + O\left(\frac{1}{\bar{n}^2}\right)\right) \cdot (1 + O(\varepsilon))\frac{\binom{k-i}{c-1}}{\binom{k}{c}}
$$

$$
= \sum_{i=1}^{k-c+1} \frac{\varepsilon_i}{\bar{n}} \frac{\binom{k-i}{c-1}}{\binom{k}{c}} + O\left(\frac{\varepsilon^2}{\bar{n}}\right) + O\left(\frac{1}{\bar{n}^2}\right)
$$

$$
\leq \sum_{i=1}^{k-1} \frac{\varepsilon_i}{\bar{n}} \frac{k-i}{\binom{k}{2}} + O\left(\frac{\varepsilon^2}{\bar{n}}\right) + O\left(\frac{1}{\bar{n}^2}\right). \tag{4}
$$

To see the last inequality, $\binom{k-i}{c-1}\big/\binom{k}{c}$ is the probability that element i is the smallest of c elements chosen *uniformly* without replacement, so increasing c from 2 to its true value would only decrease the value of i selected, in turn decreasing the expectation (the ε_i are nondecreasing); conversely, by replacing c with 2 we can only increase the expectation.

To estimate the numerator $\sum_{i=1}^{k-1} \varepsilon_i(k-i)$ of the first term in (4), add a kth term $\varepsilon_k \cdot 0$ so that the sum runs from 1 to k, pair up terms i and $k+1-i$ (e.g., the terms $i = 1$ and $i = k$), and observe that

$$
(k-i)\varepsilon_i + (i-1)\varepsilon_{k+1-i} = \frac{k-1}{2}(\varepsilon_i + \varepsilon_{k-i+1}) + \frac{k-2i+1}{2}(\varepsilon_i - \varepsilon_{k-i+1}).
$$

Summing the *first* terms in each of these expressions over $i \leq \lfloor k/2 \rfloor$ (and, if k is odd, a non-paired term for $i = (k+1)/2$, namely $\varepsilon_i(k-i) = \frac{k-1}{2}\varepsilon_{(k+1)/2}$) gives 0, because by definition $\sum \varepsilon_i = 0$. Each *second* term is nonpositive, because the ε_i are nondecreasing. (Even if k is odd, no unpaired second term is called for.) The second term coming from $i = 1$ is a relatively large negative value, $\frac{k-1}{2}(\varepsilon_1 - \varepsilon_k)$. This establishes that $\sum_{i=1}^{k} \varepsilon_i(k-i) \leq \frac{k-1}{2}(\varepsilon_1 - \varepsilon_k) \leq -\frac{k-1}{2}\varepsilon$ and in turn that

$$
\mathbb{E}(\Delta\Gamma) \leq -\frac{\varepsilon}{\bar{n}k} + O\left(\frac{\varepsilon^2}{\bar{n}}\right) + O\left(\frac{1}{\bar{n}^2}\right). \tag{5}
$$

Returning to (1), for ε smaller than some constant ε_0 (equivalently $\Gamma \leq \Gamma_0$), we have $\varepsilon^2 \leq 3\Gamma$ and $\varepsilon \geq \Gamma^{1/2} k^{-1/2}$. Then

$$
\mathbb{E}(\Delta\Gamma) \leq \frac{\Gamma^{1/2}}{\bar{n}k^{3/2}} + O\left(\frac{3\Gamma}{\bar{n}}\right) + O\left(\frac{1}{\bar{n}^2}\right). \tag{6}
$$

We will show that with positive probability, for all n sufficiently large, $\Gamma_n \leq \frac{2}{\bar{n}}$ (defining $\bar{n} = n/k$). (For intuitive purposes we note that the corresponding ε is of order $n^{-1/2}$, while perfect balance to within an additive $O(1)$ for integrality would have ε of order n^{-1}.) Fix an n_0 sufficiently large that that $\frac{2}{\bar{n}_0} \leq \Gamma_0$, and to assure (7) and (11) below. With positive probability, $\Gamma_{n_0} \leq \frac{1}{\bar{n}_0}$. Let n_1 be the first "trigger" time for which $\Gamma_{n_1} > \frac{1}{\bar{n}_1}$. If n_1 is undefined then for all $n \geq n_0$, $\Gamma_n \leq \frac{1}{\bar{n}} \leq \frac{2}{\bar{n}}$ and we are done.

Otherwise, by (2), Γ_{n_1} is very nearly $\frac{1}{\bar{n}_1}$: $\Gamma_{n_1} = \frac{1}{\bar{n}_1} + O\left(\frac{\Gamma_{n_1-1}^{1/2}}{n_1}\right) + O\left(\frac{1}{\bar{n}_1^2}\right)$.
For $n \geq n_1$, as long as Γ_n is between $\frac{1}{2\bar{n}}$ and $\frac{2}{\bar{n}}$, by (6),

$$\mathbb{E}(\Delta\Gamma) \leq -\frac{1}{4}\bar{n}^{-3/2}k^{-3/2} + O\left(\bar{n}^{-2}\right)$$

$$\leq -\frac{1}{5}\bar{n}^{-3/2}k^{-3/2}. \tag{7}$$

Starting at n_1, run for s steps, $s = k^{3/2}\bar{n}_1^{1/2}$, finishing at $n_2 = n_1 + s$. Assuming that (7) holds for all $n \in \{n_1, n_1 + s - 1\}$, the *expected* value of Γ at time n_2 satisfies

$$\mathbb{E}(\Gamma_{n_2}) \leq \Gamma_{n_1} - \sum_{i=0}^{s-1}\frac{1}{5}(\bar{n} + \frac{i}{k})^{-3/2}k^{-3/2}$$

$$\leq \frac{5}{6}(\bar{n}_1 + \frac{s}{k})^{-1}$$

$$= \frac{5}{6}\bar{n}_2^{-1}, \tag{8}$$

well below the n_2 trigger point of $\bar{n}_2^{-1}$.

Let H_n be the history of the process (where every ball goes) through time n, so that H_n characterizes the random variable Γ_{n+1}. We may view the actual value of Γ_{n_2} as

$$\Gamma_{n_2} = \Gamma_{n_1} + \sum_{n=n_1}^{n_2-1}(\Gamma_{n+1} - \Gamma_n)$$

$$= \Gamma_{n_1} + \sum_{n=n_1}^{n_2-1}(\Gamma_{n+1} - \mathbb{E}(\Gamma_{n+1} \mid H_n)) + \sum_{n=n_1}^{n_2-1}(\mathbb{E}(\Gamma_{n+1} \mid H_n) - \Gamma_n). \tag{9}$$

Everything but the first summation in (9) was already treated in (8). Writing Y_n for the random variable $\Gamma_{n+1} - \mathbb{E}(\Gamma_{n+1} \mid H_n)$, the values Y_n form a martingale difference sequence. Our assumption (used in (7) and (9)) that, throughout, $\frac{1}{2\bar{n}} \leq \Gamma_n \leq \frac{2}{\bar{n}}$, can fail only if for some time $n \in \{n_1, n_2 - 1\}$, Γ_n deviates from its expectation by at least $\frac{1}{4}n_1^{-1}$, which is to say only if the martingale $\sum_{i=n_1}^{n} Y_i$ deviates from 0 by this amount. Similarly, Γ_{n_2} is below the triggering level unless there is a martingale deviation of at least $\frac{1}{7}\bar{n}_1^{-1}$. From time n_1 to n_2-1, the $s = O(n_1^{1/2})$ steps of the martingale each have bounded deviation $c = O(\frac{\varepsilon}{\bar{n}}) = O(\bar{n}_1^{-3/2})$, and by standard martingale inequalities (see for example [McD98]), the probability of a deviation of $t = \frac{1}{7}n_1^{-1}$ or more has exponentially small probability:

$$\Pr\left(\left|\max_{n=n_1}^{n_2-1} \sum_{i=n_0}^{n} Y_i\right| \geq t\right) \leq 2e^{-2t^2/(sc^2)}$$

$$= e^{-\Omega\left(n_1^{-2}/(n_1^{1/2}n_1^{-3})\right)}$$

$$= e^{-\Omega(n_1^{1/2})}. \tag{10}$$

Thus, with high probability, for every time n between n_1 and n_2, $\frac{1}{2\bar{n}} \leq \Gamma_n \leq \frac{2}{\bar{n}}$, and at time n_2, $\Gamma_{n_2} \leq \frac{1}{\bar{n}_2}$.

Writing $n_1^{(1)} := n_1$ and $n_2^{(1)} = n_2$, we may now repeat the same argument for the first trigger point $n_1^{(2)}$ (if any) after n_2 and its completion time $n_2^{(2)}$, the next trigger time $n_1^{(3)}$ after $n_2^{(2)}$, and so forth. By the union bound, the probability of a failure in any stage is at most the sum of the failure probabilities in (10), which by an appropriate choice of n_0 can be made arbitrarily small (in particular, strictly less than 1):

$$\sum_i e^{-\Omega\left((n_1^{(i)})^{1/2}\right)} \leq \sum_{j=n_0}^{\infty} e^{-\Omega(j^{1/2})}$$

$$< 1. \tag{11}$$

This completes the proof. □

The following corollary of Theorems 2 and 3 follows by the same simple logic as Lemma 2.3 in [CHJ03].

Theorem 4. *Let $c \leq m \leq k$. In the $PC(\gamma, k, c)$ Pólya-choice urn process with $\gamma > 1$, with positive probability exactly m urns collect all balls, and their occupancies are asymptotically equal.*

Proof. It follows from Theorem 2 that with positive probability the first c (and thus the first $m \geq c$) urns collect all the balls. (It may be more intuitive to consider that with positive probability the last $k - c$ urns get no balls.) Conditioned on the event A that the first m urns collect all the balls, at each step of the Pólya-choice process, the probability that a fixed set $\{i_1, \ldots, i_c\} \subseteq \{1, \ldots, m\}$ of urns is selected is proportional to $\prod_{j=1}^{c} n_{i_j}^{\gamma}$, the same as the probability if there were just m urns altogether. Thus, we can use Theorem 3 to infer that, conditioned on A, with positive probability the m urns obtain asymptotically equal numbers of balls. □

5 Open Questions

We have shown that in the Pólya-choice model with $\gamma > 1$, any subset of m urns, $c \leq m \leq k$, may acquire *all* the balls beyond the first k, with asymptotic balance among these m urns. An easy extension is that, for any specified occupancies of up to $m - c$ urns, those urns have precisely the specified (constant) occupancies for all sufficiently large times, with the remaining urns taking all further balls and staying in asymptotic balance amongst themselves.

What is not clear is whether all urn occupancies that tend to infinity must do so with asymptotic equality. Probably it is enough to answer the question for $c = 2$, $k = 3$. Per the remarks after Theorem 2, we can think of a leading pseudo-urn with occupancy x growing at rate $\frac{1}{2}x^\gamma$ while a trailing urn with occupancy y grows at rate $2y^\gamma$. This has a stable equilibrium when x and y are about equal, so the trailing urn grows faster, and the urns keep "swapping roles" to stay in synchrony. It also has an equilibrium where x is a constant times y (the constant is a function of γ) where the relative growths $\frac{1}{2}x^\gamma$ and $2y^\gamma$ are in the same proportion as x to y. Presumably the latter equilibrium is not stable: sooner or later random fluctuations will either lead x to race to infinity leaving y stuck at a constant, or y to catch up to x and restore asymptotic equality. This leads us to conjecture that an urn either goes to infinity in asymptotic equality with the greatest-occupancy urn, or stops growing at some finite value.

Our results so far are not quantitative, and it would be interesting to establish the likelihood of the various possible behaviors as functions of the parameters c, k, and γ. For example, if the conjecture above is true, what is the likelihood that m asymptotically balanced urns take all but a finite number of balls?

Acknowledgments

Richard Karp introduced me to the $\gamma = 1$ case of this model, and provided me with an early version of [LAS$^+$08]. Stefanie Gerke read a version of the manuscript carefully and was instrumental in improving it. The referees noted some points of confusion in the submission, and made a few very gratifying remarks.

References

[ABKU99]   Azar, Y., Broder, A.Z., Karlin, A.R., Upfal, E.: Balanced allocations. SIAM J. Comput. 29(1), 180–200 (1999) (electronic)

[CHJ03]   Chung, F., Handjani, S., Jungreis, D.: Generalizations of Polya's urn problem. Ann. Comb. 7(2), 141–153 (2003)

[DEM01]   Drinea, E., Enachescu, M., Mitzenmacher, M.: Variations on random graph models for the web, Tech. Report TR-06-01, Harvard University (2001)

[KLMadH96] Karp, R.M., Luby, M., Meyer, F., Meyer auf der Heide, F.: Efficient PRAM simulation on a distributed memory machine. Algorithmica 16(4-5), 517–542 (1996)

[LAS+08] Lin, H., Amanatidis, C., Sideri, M., Karp, R.M., Papadimitriou, C.H.: Linked decompositions of networks and the power of choice in Polya urns. In: SODA 2008: Proceedings of the Nineteenth Annual ACM–SIAM Symposium on Discrete Algorithms, Philadelphia, PA, USA. Society for Industrial and Applied Mathematics, pp. 993–1002 (2008) See also, http://www.eecs.berkeley.edu/~henrylin/covers_soda_full.pdf

[McD98] McDiarmid, C.: Concentration. Probabilistic methods for algorithmic discrete mathematics, Algorithms Combin. 16, 195–248 (1998)

[McD07] Personal communication (2007)

Corruption and Recovery-Efficient Locally Decodable Codes

David Woodruff

IBM Almaden
dpwoodru@us.ibm.com

Abstract. A (q, δ, ϵ)-*locally decodable code (LDC)* $C : \{0,1\}^n \to \{0,1\}^m$ is an encoding from n-bit strings to m-bit strings such that each bit x_k can be recovered with probability at least $\frac{1}{2} + \epsilon$ from $C(x)$ by a randomized algorithm that queries only q positions of $C(x)$, even if up to δm positions of $C(x)$ are corrupted. If C is a linear map, then the LDC is linear. We give improved constructions of LDCs in terms of the corruption parameter δ and recovery parameter ϵ. The key property of our LDCs is that they are *non-linear*, whereas all previous LDCs were linear.

1. For any $\delta, \epsilon \in [\Omega(n^{-1/2}), O(1)]$, we give a family of $(2, \delta, \epsilon)$-LDCs with length $m = \text{poly}(\delta^{-1}, \epsilon^{-1}) \exp\left(\max(\delta, \epsilon)\delta n\right)$. For linear $(2, \delta, \epsilon)$-LDCs, Obata has shown that $m \geq \exp(\delta n)$. Thus, for small enough constants δ, ϵ, two-query non-linear LDCs are shorter than two-query linear LDCs.
2. We improve the dependence on δ and ϵ of all constant-query LDCs by providing general transformations to non-linear LDCs. Taking Yekhanin's linear $(3, \delta, 1/2 - 6\delta)$-LDCs with $m = \exp\left(n^{1/t}\right)$ for any prime of the form $2^t - 1$, we obtain non-linear $(3, \delta, \epsilon)$-LDCs with $m = \text{poly}(\delta^{-1}, \epsilon^{-1}) \exp\left((\max(\delta, \epsilon)\delta n)^{1/t}\right)$.

Now consider a (q, δ, ϵ)-LDC C with a decoder that has n matchings $M_1, \ldots, M_n$ on the complete q-uniform hypergraph, whose vertices are identified with the positions of $C(x)$. On input $k \in [n]$ and received word y, the decoder chooses $e = \{a_1, \ldots, a_q\} \in M_k$ uniformly at random and outputs $\bigoplus_{j=1}^{q} y_{a_j}$. All known LDCs and ours have such a decoder, which we call a matching sum decoder. We show that if C is a two-query LDC with such a decoder, then $m \geq \exp\left(\max(\delta, \epsilon)\delta n\right)$. Interestingly, our techniques used here can further improve the dependence on δ of Yekhanin's three-query LDCs. Namely, if $\delta \geq 1/12$ then Yekhanin's three-query LDCs become trivial (have recovery probability less than half), whereas we obtain three-query LDCs of length $\exp\left(n^{1/t}\right)$ for any prime of the form $2^t - 1$ with non-trivial recovery probability for any $\delta < 1/6$.

1 Introduction

Classical error-correcting codes allow one to encode an n-bit message x into a codeword $C(x)$ such that even if a constant fraction of the bits in $C(x)$ are

A. Goel et al. (Eds.): APPROX and RANDOM 2008, LNCS 5171, pp. 584–595, 2008.
© Springer-Verlag Berlin Heidelberg 2008

corrupted, x can still be recovered. It is well-known how to construct codes C of length $O(n)$ that can tolerate a constant fraction of errors, even in such a way that allows decoding in linear time [1]. However, if one is only interested in recovering a few bits of the message, then these codes have the disadvantage that they require reading all (or most) of the codeword. This motivates the following definition.

Definition 1. *([2]) Let $\delta, \epsilon \in [0,1]$, q an integer. We say $C : \{0,1\}^n \to \{0,1\}^m$ is a (q,δ,ϵ)-locally decodable code (LDC for short) if there is a probabilistic oracle machine A such that:*

- *In every invocation, A makes at most q queries.*
- *For every $x \in \{0,1\}^n$, every $y \in \{0,1\}^m$ with $\Delta(y, C(x)) \leq \delta m$, and every $k \in [n]$, $\Pr[A^y(k) = x_k] \geq \frac{1}{2} + \epsilon$, where the probability is taken over the internal coin tosses of A. An algorithm A satisfying the above is called a (q,δ,ϵ)-local decoding algorithm for C (a decoder for short).*

In the definition above, $\Delta(y, C(x))$ denote the Hamming distance between y and $C(x)$, that is, the number of coordinates for which the strings differ. For a (q,δ,ϵ)-LDC, we shall refer to q as *the number of queries*, δ as *the corruption parameter*, ϵ as *the recovery parameter*, and m as *the length*. An LDC is *linear* if C is a linear transformation over $GF(2)$. Note that recovery probability $1/2$ (corresponding to $\epsilon = 0$) is trivial since the decoder can just flip a random coin.

There is a large body of work on locally decodable codes. Katz and Trevisan [2] formally defined LDCs, proved that 1-query LDCs do not exist, and proved super-linear lower bounds on the length of constant-query LDCs. We refer the reader to the survey [3] and the references therein.

All known constructions of LDCs with a constant number of queries are super-polynomial in length, and not even known to be of subexponential length. Thus, understanding the asymptotics in the exponent of the length of such codes is important, and could be useful in practice for small values of n. A lot of work has been done to understand this exponent for two-query linear LDCs [4,5,6,7]. Important practical applications of LDCs include private information retrieval and load-balancing in the context of distributed storage. Depending on the parameters of the particular application, δ and ϵ may be flexible, and our constructions will be able to exploit this flexibility.

We state the known bounds relevant to this paper. The first two concern LDCs for which $q = 2$, while the remaining pertain to $q > 2$.

Notation. $\exp(f(n))$ denotes a function $g(n)$ that is $2^{O(f(n))}$.

Theorem 1. *([8])[1] Any $(2,\delta,\epsilon)$-LDC satisfies $m \geq \exp(\epsilon^2 \delta n)$.*

For linear LDCs, a tight lower bound is known.

---

[1] This bound can be strengthened to $m \geq \exp\left(\epsilon^2 \delta n/(1-2\epsilon)\right)$ using the techniques of [6] in a relatively straightforward way. We do not explain the proof here, as our focus is when ϵ is bounded away from $1/2$, in which case the bounds are asymptotically the same.

Theorem 2. *([6,7]) Any linear $(2, \delta, \epsilon)$-LDC has $m \geq \exp(\delta n/(1 - 2\epsilon))$. Moreover, there exists a linear $(2, \delta, \epsilon)$-LDC with $m \leq \exp(\delta n/(1 - 2\epsilon))$.*

The shortest LDCs for small values of $q > 2$ are due to Yekhanin [9], while for large values one can obtain the shortest LDCs by using the LDCs of Yekhanin together with a recursion technique of Beimel, Ishai, Kushilevitz, and Raymond [10]. The following is what is known for $q = 3$.

Theorem 3. *([9]) For any $\delta \leq 1/12$ and any prime of the form $2^t - 1$, there is a linear $(3, \delta, 1/2 - 6\delta)$-LDC with $m = \exp\left(n^{1/t}\right)$. Using the largest known such prime, this is $m = \exp\left(n^{1/32582657}\right)$.*

Notice that this theorem does not allow one to obtain shorter LDCs for small δ and $\epsilon < 1/2 - 6\delta$, as intuitively should be possible.

Results. We give improved constructions of constant-query LDCs in terms of the corruption parameter δ and recovery parameter ϵ. A key property of our LDCs is that they are the first non-linear LDCs. Our main theorem is the following transformation.

Theorem 4. *Given a family of $(q, \delta, 1/2 - \beta\delta)$-LDCs of length $m(n)$, where $\beta > 0$ is any constant, and $\delta < 1/(2\beta)$ is arbitrary (i.e., for a given n, the same encoding function C is a $(q, \delta, 1/2 - \beta\delta)$-LDC for any $\delta < 1/(2\beta)$), there is a family of non-linear $(q, \Theta(\delta), \epsilon)$-LDCs of length $O(dr^2)m(n'/r)$ for any $\delta, \epsilon \in [\Omega(n^{-1/2}), O(1)]$, where $d = \max(1, O(\epsilon/\delta))$, $r = O((\epsilon + \delta)^{-2})$, and $n' = n/d$.*

As a corollary, for any $\delta, \epsilon \in [\Omega(n^{-1/2}), O(1)]$, we give a $(2, \delta, \epsilon)$-LDC with length $m = \text{poly}(\delta^{-1}, \epsilon^{-1}) \exp(\max(\delta, \epsilon)\delta n)$. Thus, by Theorem 2, as soon as δ and ϵ are small enough constants, this shows that 2-query non-linear LDCs are shorter than 2-query linear LDCs. This is the first progress on the question of Kerenidis and de Wolf [8] as to whether the dependence on δ and ϵ could be improved. Another corollary is that for any prime of the form $2^t - 1$ and any $\delta, \epsilon \in [\Omega(n^{-1/2}), O(1)]$, there is a family of non-linear $(3, \delta, \epsilon)$-LDCs with $m = \text{poly}(\delta^{-1}, \epsilon^{-1}) \exp\left((\max(\delta, \epsilon)\delta n)^{1/t}\right)$.

Next, we show that our bound for 2-query LDCs is tight, up to a constant factor in the exponent, for a large family of LDCs including all known ones as well as ours. Let C be a (q, δ, ϵ)-LDC with a decoder that has n matchings $M_1, \ldots, M_n$ on the complete q-uniform hypergraph whose vertices are identified with the positions of $C(x)$. On input $k \in [n]$ and received word y, the decoder chooses a hyperedge $e = \{a_1, \ldots, a_q\} \in M_k$ uniformly at random and outputs $\bigoplus_{j=1}^{q} y_{a_j}$. We call such a decoder a *matching sum decoder*, and show that if a 2-query LDC C has such a decoder then $m \geq \exp(\max(\delta, \epsilon)\delta n)$. Thus, our upper bound is tight for such LDCs. To prove that for *any* $(2, \delta, \epsilon)$-LDC, $m \geq \exp(\max(\delta, \epsilon)\delta n)$, our result implies that it suffices to transform any LDC into one which has a matching sum decoder, while preserving δ, ϵ, and m up to small factors.

Finally, as an independent application of our techniques, we transform the $(3, \delta, 1/2 - 6\delta)$-LDCs with $m = \exp(n^{1/t})$ of Theorem 3, into $(3, \delta, 1/2 - 3\delta - \eta)$-LDCs with $m = \exp(n^{1/t})$, where $\eta > 0$ is an arbitrarily small constant. In particular, we extend the range of δ for which the LDCs in Theorem 3 become non-trivial from $\delta \leq 1/12$ to $\delta < 1/6$. Moreover, there is no 3-query LDC with a matching sum decoder with $\delta \geq 1/6$. Indeed, if the adversary corrupts exactly $m/6$ hyperedges of M_i, the recovery probability can be at most $1/2$.

Techniques. Our main idea for introducing non-linearity is the following. Suppose we take the message $x = x_1, \ldots, x_n$ and partition it into n/r blocks $B_1, \ldots, B_{n/r}$, each containing $r = \Theta(\epsilon^{-2})$ different x_i. We then compute $z_j = \text{majority}(x_i \mid i \in B_j)$, and encode the bits $z_1, \ldots, z_{n/r}$ using a (q, δ, ϵ)-LDC C. To obtain x_k if $x_k \in B_j$, we use the decoder for C to recover z_j with probability at least $1/2 + \epsilon$. We should expect that knowing z_j is useful, since, using the properties of the majority function, $\text{Pr}_x[x_k = z_j] \geq \frac{1}{2} + \epsilon$.

This suggests an approach: choose $s_1, \ldots, s_\tau \in \{0,1\}^n$ for a certain $\tau = O(r^2)$, apply the above procedure to each of $x \oplus s_1, \ldots, x \oplus s_\tau$, then take the concatenation. The $s_1, \ldots, s_\tau$ are chosen randomly so that for any $x \in \{0,1\}^n$ and any index k in any block B_j, a $\frac{1}{2} + \epsilon$ fraction of the different $x \oplus s_i$ have the property that their k-th coordinate agrees with the majority of the coordinates in B_j. The length of the encoding is now τm, where m is the length required to encode n/r bits.

To illustrate how recovery works, suppose that C were the Hadamard code. The decoder would choose a random $i \in [\tau]$ and decode the portion of the encoding corresponding to the (corrupted) encoding of $x \oplus s_i$. One could try to argue that with probability at least $1 - 2\delta$, the chosen positions by the Hadamard decoder are correct, and given that these are correct, $(x \oplus s_i)_k$ agrees with the majority of the coordinates in the associated block with probability at least $\frac{1}{2} + \epsilon$. If these events were independent, the success probability would be $\geq (1 - 2\delta)(1/2 + \epsilon) + 2\delta(1/2 - \epsilon) = 1/2 + \Omega(\epsilon)$.

However, these events are very far from being independent! Indeed, the adversary may first recover x from the encoding, and then for any given k, determine exactly which $(x \oplus s_i)_k$ agree with the majority of the coordinates in the associated block, and corrupt only these positions. This problem is unavoidable. However, we observe that we can instead consider $r = \Theta(\delta^{-2})$. Then, if $\delta = \Omega(\epsilon)$, we can show the decoder's success probability is at least $1/2 + \Omega(\epsilon)$. If, on the other hand, $\epsilon = \Omega(\delta)$, we can first allow δ to grow to $\Theta(\epsilon)$ via a technique similar to the upper bound given in [6], reducing n to $n' = \delta n/\epsilon$. Then we can effectively perform the above procedure with $r = \Theta(\epsilon^{-2})$ and $n'/r = \Theta(\epsilon^2 n') = \Theta(\epsilon \delta n)$.

To show that this technique is optimal for LDCs C with matching sum decoders, we need to significantly generalize the quantum arguments of [8]. A general matching sum decoder may have matchings M_i with very different sizes and contain edges that are correct for a very different number of $x \in \{0,1\}^n$. If we recklessly apply the techniques of [8], we cannot hope to obtain an optimal dependence on δ and ϵ.

Given such a C, we first apply a transformation to obtain a slightly longer LDC C' in which all matchings have the same size, and within a matching, the average fraction of x for which an edge is correct, averaged over edges, is the same for all matchings. We then apply another transformation to obtain an LDC

C'' which increases the length of the code even further, but makes the matching sizes very large. Finally, we use quantum information theory to lower bound the length of C'', generalizing the arguments of [8] to handle the case when the average fraction of x for which an edge is correct, averaged over edges in a matching of C'', is sufficiently large.

Finally, we use an idea underlying the transformation from C' to C'' in our lower bound argument to transform the LDCs of Theorem 3 into LDCs with a better dependence on δ and ϵ, thereby obtaining a better upper bound. The idea is to blow up the LDC by a constant factor in the exponent, while increasing the sizes of the underlying matchings. Constructing the large matchings in the blown-up LDC is more complicated than it was in our lower bound argument, due to the fact that we run into issues of consistently grouping vertices of hypergraphs together which did not arise when we were working with graphs.

Other Related Work. Other examples where non-linear codes were shown to have superior parameters to linear codes include the construction of t-resilient functions [11,12], where it is shown [13] that non-linear Kerdock codes outperform linear codes in the construction of such functions. See [14] for a study of non-linearity in the context of secret sharing.

2   Preliminaries

The following theorem is easy to prove using elementary Fourier analysis. We defer the proof to the full version. Throughout, we shall let c be the constant $(2/\pi)^{3/4}/4$.

Theorem 5. *Let r be an odd integer, and let $f : \{0,1\}^r \to \{0,1\}$ be the majority function, where $f(x) = 1$ iff there are more $1s$ than $0s$ in x. Then for any $k \in [r]$, $\Pr_{x \in \{0,1\}^r}[f(x) = x_k] > \frac{1}{2} + \frac{2c}{r^{1/2}}$.*

We also need an approximate version of this theorem, which follows from a simple application of the probabilistic method.

Lemma 1. *Let r and f be as in Theorem 5. Then there are $\tau = O(r^2)$ strings $\mu_1, \mu_2, \ldots, \mu_\tau \in \{0,1\}^r$ so that for all $x \in \{0,1\}^r$ and all $k \in [r]$, $\Pr_{i \in [\tau]}[f(x \oplus \mu_i) = (x \oplus \mu_i)_k] \geq \frac{1}{2} + \frac{c}{r^{1/2}}$.*

In our construction we will use the Hadamard code $C : \{0,1\}^n \to \{0,1\}^{2^n}$, defined as follows. Identify the 2^n positions of the codeword with distinct vectors $v \in \{0,1\}^n$, and set the vth position of $C(x)$ to $\langle v, x \rangle$ mod 2. To obtain x_k from a vector y which differs from $C(x)$ in at most a δ fraction of positions, choose a random vector v, query positions y_v and $y_{v \oplus e_k}$, and output $y_v \oplus y_{v \oplus e_k}$. With probability at least $1 - 2\delta$, we have $y_v = \langle v, x \rangle$ and $y_{v \oplus e_k} = \langle v \oplus e_k, x \rangle$, and so $y_v \oplus y_{v \oplus e_k} = x_k$. It follows that for any $\delta > 0$, the Hadamard code is a $(2, \delta, 1/2 - 2\delta)$-LDC with $m = \exp(n)$.

Finally, in our lower bound, we will need some concepts from quantum information theory. We borrow notation from [8]. For more background on quantum information theory, see [15].

A *density matrix* is a positive semi-definite (PSD) complex-valued matrix with trace 1. A *quantum measurement* on a density matrix ρ is a collection of PSD matrices $\{P_j\}$ satisfying $\sum_j P_j^\dagger P_j = I$, where I is the identity matrix ($A^\dagger$ is the conjugate-transpose of A). The set $\{P_j\}$ defines a probability distribution X on indices j given by $\Pr[X = j] = \text{tr}(P_j^\dagger P_j \rho)$.

We use the notation AB to denote a bipartite quantum system, given by some density matrix ρ^{AB}, and A and B to denote its subsystems. More formally, the density matrix of ρ^A is $\text{tr}_B(\rho^{AB})$, where tr_B is a map known as the *partial trace* over system B. For given vectors $|a_1\rangle$ and $|a_2\rangle$ in the vector space of A, and $|b_1\rangle$ and $|b_2\rangle$ in the vector space of B, $\text{tr}_B(|a_1\rangle\langle a_2| \otimes |b_1\rangle\langle b_2|) \stackrel{\text{def}}{=} |a_1\rangle\langle a_2|\text{tr}(|b_1\rangle\langle b_2|)$, and $\text{tr}_B(\rho^{AB})$ is then well-defined by requiring tr_B to be a linear map.

$S(A)$ is the *von Neumann entropy* of A, defined as $\sum_{i=1}^d \lambda_i \log_2 \frac{1}{\lambda_i}$, where the λ_i are the eigenvalues of A. $S(A \mid B) = S(AB) - S(B)$ is the *conditional entropy* of A given B, and $S(A; B) = S(A) + S(B) - S(AB) = S(A) - S(A \mid B)$ is the *mutual information* between A and B.

3 The Construction

Let $C : \{0,1\}^n \to \{0,1\}^{m(n)}$ come from a family of $(q, \delta, 1/2 - \beta\delta)$-LDCs, where $\beta > 0$ is any constant, and $\delta < 1/(2\beta)$ is arbitrary (i.e., for a given n, the same function C is a $(q, \delta, 1/2 - \beta\delta)$-LDC for any $\delta < 1/(2\beta)$). For example, for any $\delta < 1/4$, the Hadamard code is a $(2, \delta, 1/2 - 2\delta)$-LDC, while Yekhanin [9] constructed a $(3, \delta, 1/2 - 6\delta)$-LDC for any $\delta < 1/12$.

Setup. Assume that $\delta, \epsilon \in [\Omega(n^{-1/2}), O(1)]$. W.l.o.g., assume n is a sufficiently large power of 3. Recall from Section 2 that we will use c to denote the constant $(2/\pi)^{3/4}/4$. Define the parameter $r = (\epsilon(1 + 2\beta c)/c + 2\beta\delta/c)^{-2} = \Theta((\epsilon + \delta)^{-2})$. Let $\tau = O(r^2)$ be as in Lemma 1. We define $d = \max(1, c\epsilon/\delta)$. Let $n' = n/d$. We defer the proof of the following lemma to the full version. The lemma establishes certain integrality and divisibility properties of the parameters that we are considering.

Lemma 2. *Under the assumption that $\delta, \epsilon \in [\Omega(n^{-1/2}), O(1)]$ and $\beta = \Theta(1)$, by multiplying δ and ϵ by positive constant factors, we may assume that the following two conditions hold simultaneously: (1) r and d are integers, and (2) $(rd) \mid n$.*

In the sequel we shall assume that for the given δ and ϵ, the two conditions of Lemma 2 hold simultaneously. If in this case we can construct a (q, δ, ϵ)-LDC with some length m', it will follow that for any δ and ϵ we can construct a $(q, \Theta(\delta), \Theta(\epsilon))$-LDC with length $\Theta(m')$.

Proof strategy. We first construct an auxiliary function $f : \{0,1\}^{n'} \to \{0,1\}^\ell$, where $\ell = \tau m(n'/r)$. The auxiliary function coincides with our encoding function $C' : \{0,1\}^n \to \{0,1\}^{m'(n)}$ when $d = 1$. When $d > 1$, then C' will consist of d applications of the auxiliary function, each on a separate group of n' coordinates of the message x. Recall that $d > 1$ iff $c\epsilon \geq \delta$, and in this case we effectively allow δ to grow while reducing n (see Section 1 for discussion). We will thus have $m'(n) = d\tau m(n'/r)$. We then describe algorithms $\mathsf{Encode}(x)$ and $\mathsf{Decode}^y(k)$ associated with C'. Finally, we show that C' is a (q, δ, ϵ)-LDC with length $m'(n)$. Note that we have ensured r, d, and $n'/r = n/(dr)$ are all integers.

An auxiliary function. Let $\mu_1, \ldots, \mu_\tau$ be the set of strings in $\{0,1\}^r$ guaranteed by Lemma 1. For each $i \in [\tau]$, let s_i be the concatenation of n'/r copies of μ_i. For each $j \in [n'/r]$, let B_j be the set $B_j = \{(j-1)r+1, (j-1)r+2, \ldots, jr\}$. The B_j partition the interval $[1, n']$ into n'/r contiguous blocks each of size r. We now explain how to compute the auxiliary function $f(u)$ for $u \in \{0,1\}^{n'}$. Compute $w_1 = u \oplus s_1, w_2 = u \oplus s_2, \ldots, w_\tau = u \oplus s_\tau$. For each $i \in [\tau]$, compute $z_i \in \{0,1\}^{n'/r}$ as follows: $\forall j \in [n'/r]$, $z_{i,j} = \mathrm{majority}(w_{i,k} \mid k \in B_j)$. Then $f(u)$ is defined to be, $f(u) = C(z_1) \circ C(z_2) \cdots \circ C(z_\tau)$, where $\circ$ denotes string concatenation. Observe that $|f(u)| = \tau m(n'/r)$.

The LDC. We describe the algorithm $\mathsf{Encode}(x)$ associated with our encoding $C' : \{0,1\}^n \to \{0,1\}^{m'(n)}$. We first partition x into d contiguous substrings $u_1, \ldots, u_d$, each of length n'. Then, $\mathsf{Encode}(x) = C'(x) = f(u_1) \circ f(u_2) \cdots \circ f(u_d)$. Observe that $|C'(x)| = m'(n) = d\tau m(n'/r)$. Next we describe the algorithm $\mathsf{Decode}^y(k)$. We think of y as being decomposed into $y = y_1 \circ y_2 \cdots \circ y_d$, where each y_h, $h \in [d]$, is a block of $m'(n)/d = \tau m(n'/r)$ consecutive bits of y. Let h be such that x_k occurs in u_h. Further, we think of y_h as being decomposed into $y_h = v_1 \circ v_2 \cdots \circ v_\tau$, where each v_i, $i \in [\tau]$, is a block of $m(n'/r)$ consecutive bits of y_h.

To decode, first choose a random integer $i \in [\tau]$. Next, let $j \in [n'/r]$ be such that $(k \bmod d)+1 \in B_j$. Simulate the decoding algorithm $A^{v_i}(j)$ associated with C. Suppose the output of $A^{v_i}(j)$ is the bit b. If the kth bit of s_i is 0, output b, else output $1 - b$. The following is our main theorem.

Theorem 6. *Given a family of $(q, \delta, 1/2 - \beta\delta)$-LDCs of length $m(n)$, where $\beta > 0$ is any constant, and $\delta < 1/(2\beta)$ is arbitrary (i.e., for a given n, the same encoding function C is a $(q, \delta, 1/2 - \beta\delta)$-LDC for any $\delta < 1/(2\beta)$), there is a family of non-linear $(q, \Theta(\delta), \epsilon)$-LDCs of length $O(dr^2)m(n'/r)$ for any $\delta, \epsilon \in [\Omega(n^{-1/2}), O(1)]$, where $d = \max(1, O(\epsilon/\delta))$, $r = O((\epsilon + \delta)^{-2})$, and $n' = n/d$.*

Proof. We show that C' is a (q, δ, ϵ)-LDC with length $m'(n) = d\tau m(n'/r)$.

First, observe that $\mathsf{Decode}^y(k)$ always makes at most q queries since the decoder A of C always makes at most q queries. Also, we have already observed that $|C'(x)| = m'(n) = d\tau m(n'/r)$. Now, let $x \in \{0,1\}^n$ and $k \in [n]$ be arbitrary. Let h be such that x_k occurs in u_h.

First, consider the case that $c\epsilon < \delta$, so that $h = d = 1$. Suppose k occurs in the set B_j. By Theorem 5 and the definition of r, for at least a $\frac{1}{2} + \frac{c}{r^{1/2}} = \frac{1}{2} + (1 + 2\beta c)\epsilon + 2\beta\delta$ fraction of the τ different z_i, we have $z_{i,j} = y_{i,k} = x_k \oplus s_{i,k}$. Since i is chosen at random by $\mathsf{Decode}^y(k)$, we have $\Pr_i[z_{i,j} = x_k \oplus s_{i,k}] > \frac{1}{2} + (1 + 2\beta c)\epsilon + 2\beta\delta$. In case that $z_{i,j} = x_k \oplus s_{i,k}$, we say i is *good*. Let $\mathcal{E}$ be the event that the i chosen by the decoder is good, and let G be the number of good i. We think of the received word $y = y_1$ (recall that $d = 1$) as being decomposed into $y = v_1 \circ v_2 \cdots \circ v_\tau$. The adversary can corrupt a set of at most $\delta m'(n)$ positions in $C'(x)$. Suppose the adversary corrupts $\delta_i m'(n)$ positions in $C(z_i)$, that is, $\Delta(C(z_i), v_i) \leq \delta_i m'(n)$. So we have the constraint $0 \leq \frac{1}{\tau} \sum_i \delta_i \leq \delta$.

Conditioned on $\mathcal{E}$, the decoder recovers $z_{i,j}$ with probability at least $\frac{1}{G} \sum_{\text{good } i} (1 - \beta\delta_i) = 1 - \frac{\beta}{G} \sum_{\text{good } i} \delta_i \geq 1 - \frac{\tau\beta\delta}{G} \geq 1 - 2\beta\delta$, where we have used that $G \geq \tau/2$. In this case the decoder recovers x_k by adding $s_{i,k}$ to $z_{i,j}$ modulo 2. Thus, the decoding probability is at least $\Pr[\mathcal{E}] - 2\beta\delta \geq \frac{1}{2} + (1 + 2\beta c)\epsilon + 2\beta\delta - 2\beta\delta > \frac{1}{2} + \epsilon$. Now consider the case that $c\epsilon \geq \delta$, so that d may be greater than 1. The number of errors in the substring $f(u_h)$ of $C'(x)$ is at most $\delta m'(n) = \delta d\tau m(n'/r) = \delta(c\epsilon/\delta)\tau m(n'/r) = c\epsilon|f(u_h)|$, so there is at most a $c\epsilon$ fraction of errors in the substring $f(u_h)$. Again supposing that $(k \bmod d) + 1 \in B_j$, by Theorem 5 we deduce that $\Pr_i[z_{i,j} = x_k \oplus s_{i,k}] > \frac{1}{2} + (1 + 2\beta c)\epsilon + 2\beta\delta$. We define a good i and the event $\mathcal{E}$ as before. We also decompose y_h into $y_h = v_1 \circ v_2 \cdots \circ v_\tau$. By an argument analogous to the case $d = 1$, the decoding probability is at least $\Pr[\mathcal{E}] - 2\beta c\epsilon > \frac{1}{2} + (1 + 2\beta c)\epsilon + 2\beta\delta - 2\beta c\epsilon > \frac{1}{2} + \epsilon$, as needed.

We defer the proofs of the next two corollaries to the full version, which follow by plugging in Hadamard's and Yekhanin's codes into Theorem 6.

Corollary 1. *For any $\delta, \epsilon \in [\Omega(n^{-1/2}), O(1)]$, there is a $(2, \delta, \epsilon)$-LDC of length $m = \text{poly}(\delta^{-1}, \epsilon^{-1}) \exp(\max(\delta, \epsilon)\delta n)$.*

Corollary 2. *For any $\delta, \epsilon \in [\Omega(n^{-1/2}), O(1)]$ and any prime of the form $2^t - 1$, there is a $(3, \delta, \epsilon)$-LDC with $m = \text{poly}(\delta^{-1}, \epsilon^{-1}) \exp((\max(\delta, \epsilon)\delta n)^{1/t})$.*

4   The Lower Bound

Consider a (q, δ, ϵ)-LDC C with length m which has a decoder that has n matchings $M_1, \ldots, M_n$ of edges on the complete q-uniform hypergraph, whose vertices are identified with positions of the codeword. On input $i \in [n]$ and received word y, the decoder chooses $e = \{a_1, \ldots, a_q\} \in M_i$ uniformly at random and outputs $\bigoplus_{j=1}^q y_{a_j}$. All known LDCs, including our non-linear LDCs, satisfy this property. In this case we say that C has a *matching sum decoder*.

Any linear $(2, \delta, \epsilon)$-LDC C can be transformed into an LDC with slightly worse parameters, but with the same encoding function and a matching sum decoder. Indeed, identify the m positions of the encoding of C with linear forms v, where $C(x)_v = \langle x, v \rangle$. Obata [6] has shown that such LDCs have matchings M_i of edges $\{u, v\}$ with $u \oplus v = e_i$, where $|M_i| \geq \beta\delta m$ for a constant $\beta > 0$. By replacing δ with $\delta' = \beta\delta/3$, the decoder can query a uniformly random edge in M_i and output the correct answer with probability at least $(\beta\delta m - \beta\delta m/3)/(\beta\delta m) \geq 2/3$. One can extend this to linear LDCs with $q > 2$ by generalizing Obata's argument.

Theorem 7. *Any $(2, \delta, \epsilon)$-LDC C with a matching sum decoder satisfies $m \geq \exp(\max(\delta, \epsilon)\delta n)$.*

Proof. For each $i \in [n]$, let the matching M_i of the matching sum decoder satisfy $|M_i| = c_i m$. We may assume, by relabeling indices, that $c_1 \leq c_2 \leq \cdots \leq c_n$. Let $\bar{c} = \sum_i c_i / n$ be the average of the c_i. For each edge $e = \{a, b\} \in M_i$, let $p_{i,e}$ be the probability that $C(x)_a \oplus C(x)_b$ equals x_i for a uniformly chosen $x \in \{0, 1\}^n$. The probability, over a random $x \in \{0, 1\}^n$, that the decoder outputs x_i if there are no errors is $\psi_i = \sum_{e \in M_i} p_{i,e} / |M_i|$, which is at least $1/2 + \epsilon$. But ψ_i is also at least $1/2 + \delta/c_i$. Indeed, otherwise there is a fixed x for which it is less than $1/2 + \delta/c_i$. For this x, say $e = \{a, b\}$ is *good* if $C(x)_a \oplus C(x)_b = x_i$. Then $\sum_{\text{good } e \in M_i} 1/|M_i| < 1/2 + \delta/c_i$. By flipping the value of exactly one endpoint of δm good $e \in M_i$, this probability drops to $1/2$, a contradiction.

We first transform the LDC C to another code C'. Identify the coordinates of x with indices $0, 1, \ldots, n-1$. For $j = 0, \ldots, n-1$, let π_j be the j-th cyclic shift of $0, \ldots, n-1$, so for $x = (x_0, \ldots, x_{n-1}) \in \{0, 1\}^n$, we have that $\pi_j(x) = (x_j, x_{j+1}, \ldots, x_{j-1})$. We define $C'(x) = C(\pi_0(x)) \circ C(\pi_1(x)) \cdots \circ C(\pi_{n-1}(x))$. Then $m' = |C'(x)| = n|C(x)|$. For $j, k \in \{0, 1, \ldots, n-1\}$, let $M_{j,k}$ be the matching M_k in the code $C(\pi_j(x))$. Define the n matchings $M'_0, \ldots, M'_{n-1}$ with $M'_i = \cup_{j=0}^{n-1} M_{j,i-j}$.

We need another transformation from C' to a code C''. For each $i \in \{0, \ldots, n-1\}$, impose a total ordering on the edges in M'_i by ordering the edges $e_1, \ldots, e_{|M'_i|}$ so that $p_{i,e_1} \geq p_{i,e_2} \cdots \geq p_{i,e_{|M'_i|}}$. Put $t = \lfloor 1/(2\bar{c}) \rfloor$, and let C'' be the code with entries indexed by ordered multisets S of $[m']$ of size t, where $C''_S(x) = \bigoplus_{v \in S} C'(x)_v$. Thus, $m'' = |C''(x)| = (m')^t$. Consider a random entry $S = \{v_1, \ldots, v_t\}$ of C''. Fix an $i \in \{0, 1, \ldots n-1\}$. Say S *hits* i if $S \cap (\cup_{e \in M'_i} e) \neq \emptyset$. Now, $|\cup_{e \in M'_i} e| = 2|M'_i| = 2\bar{c}m'$, so, $\Pr[S \text{ hits } i] \geq 1 - (1 - 2\bar{c})^t \geq 1 - e^{-2\bar{c}t} \geq 1 - e^{-1} > 1/2$. Thus, at least a $1/2$ fraction of entries of C'' hit i. We can group these entries into a matching M''_i of edges of $[m'']$ with $|M''_i| \geq m''/4$ as follows. Consider an S that hits i and let $e = \{a, b\}$ be the *smallest* edge of M'_i for which $S \cap \{a, b\} \neq \emptyset$, under the total ordering of edges in M'_i introduced above. Since S is ordered, we may look at the *smallest* position j containing an entry of e. Suppose, w.l.o.g., that $S_j = a$. Consider the ordered multiset T formed by replacing the j-th entry of S with b. Then, $C''_S(x) \oplus C''_T(x) = \bigoplus_{v \in S} C'(x)_v \oplus \bigoplus_{v \in T} C'(x)_v = 2 \bigoplus_{v \notin e} C'(x)_v \oplus (C'(x)_a \oplus C'(x)_b) = C'(x)_a \oplus C'(x)_b$. Given T, the smallest edge hit by T is e, and this also occurs in position j. So the matching M''_i is well-defined and of size at least $m''/4$.

We will also need a more refined statement about the edges in M''_i. For a random entry S of C'', say S *hits* i *by time* j if $S \cap \left(\cup_{\ell=1}^{j} \cup_{e \in M_{\ell, i-\ell}} e \right) \neq \emptyset$. Let $\sigma_j = \sum_{\ell=1}^{j} c_\ell$. Now, $|\cup_{\ell=1}^{j} \cup_{e \in M_{\ell, i-\ell}} e| = 2\sigma_j m = 2\sigma_j m'/n$. Thus,

$$\Pr[S \text{ hits } i \text{ by time } j] \geq 1 - \left(1 - \frac{2\sigma_j}{n}\right)^t \geq 1 - e^{-\frac{2\sigma_j t}{n}} \geq 1 - e^{-\frac{\sigma_j}{n\bar{c}}} \geq \frac{\frac{\sigma_j}{n\bar{c}}}{1 + \frac{\sigma_j}{n\bar{c}}},$$

where the last inequality is $1 - e^{-x} > x/(x+1)$, which holds for $x > -1$. Now, $\sigma_j/(n\bar{c}) = \sigma_j / \sum_{\ell=1}^{n} c_\ell \leq 1$, so $\Pr[S \text{ hits } i \text{ by time } j] \geq \sigma_j/(2n\bar{c})$.

For $\{S, T\} \in M''_i$, let $p''_{i, \{S, T\}}$ be the probability over a random x that $C''(x)_S \oplus C''(x)_T = x_i$. Then $p''_{i, \{S, T\}} = p_{i,e}$, where e is the smallest edge of M'_i hit by S

and T. We define $\psi_i'' = \frac{1}{|M_i''|} \sum_{\{S,T\} \in M_i''} p_{i,\{S,T\}}''$, which is the probability that the matching sum decoder associated with C'' with matchings M_i'' outputs x_i correctly for a random x, given that there are no errors in the received word. Let $\phi_{i,j}$ be the probability that the smallest edge $e \in M_i'$ hit by a randomly chosen edge in M_i'' is in $M_{j,i-j}$. Due to our choice of total ordering (namely, within a given $M_{j,i-j}$, edges with larger $p_{j,e}$ value are at least as likely to occur as those with smaller $p_{j,e}$ for a randomly chosen edge in M_i'', conditioned on the edge being in $M_{j,i-j}$), $\psi_i'' \geq \sum_j \phi_{i,j} \psi_j \geq \sum_j \phi_{i,j} \left(\frac{1}{2} + \max(\epsilon, \delta/c_j) \right) = \frac{1}{2} + \sum_j \phi_{i,j} \max(\epsilon, \delta/c_j)$. Observe that $\sum_{\ell=1}^{j} \phi_{i,\ell} \geq \sigma_j/(2n\bar{c})$, and since the expression $\max(\epsilon, \delta/c_j)$ is non-increasing with j, the above lower bound on ψ_i'' can be further lower bounded by setting $\sum_{\ell=1}^{j} \phi_{i,\ell} = \sigma_j/(2n\bar{c})$ for all j. Then $\phi_{i,j}$ is set to $c_j/(2n\bar{c})$ for all j, and we have $\psi_i'' \geq 1/2 + \max(\epsilon, \delta/\bar{c})/2$.

Let $\bar{r} = \max(\epsilon, \delta/\bar{c})/2$. We use quantum information theory to lower bound m''. For each $j \in [m'']$, replace the j-th entry of $C''(x)$ with $(-1)^{C''(x)_j}$. We can represent $C''(x)$ as a vector in a state space of $\log m''$ qubits $|j\rangle$. The vector space it lies in has dimension m'', and its standard basis consists of all vectors $|b\rangle$, where $b \in \{0,1\}^{\log m''}$ (we can assume m'' is a power of 2). Define $\rho_x = \frac{1}{m''} C(x)^\dagger C(x)$. It is easy to verify that ρ_x is a density matrix. Consider the $n + \log m''$ qubit quantum system XW: $\frac{1}{2^n} \sum_x |x\rangle\langle x| \otimes \rho_x$. We use X to denote the first system, X_i for its qubits, and W for the second subsystem. By Theorem 11.8.4 of [15], $S(XW) = S(X) + \frac{1}{2^n} \sum_x S(\rho_x) \geq S(X) = n$. Since W has $\log m''$ qubits, $S(W) \leq \log m''$, hence $S(X : W) = S(X) + S(W) - S(XW) \leq S(W) \leq \log m''$. Using a chain rule for relative entropy and a highly non-trivial inequality known as the strong subadditivity of the von Neumann entropy, we get $S(X \mid W) = \sum_{i=1}^{n} S(X_i \mid X_1, \dots, X_{i-1}, W) \leq \sum_{i=1}^{n} S(X_i \mid W)$. In the full version, we show that $S(X_i \mid W) \leq H(\frac{1}{2} + \frac{\bar{r}}{2})$. That theorem is a generalization of the analogous theorem of [8], as here we just have matchings M_i'' for which the average probability that the sum of endpoints of an edge in M_i'' is at least $\frac{1}{2} + \bar{r}$, whereas in [8] this was a worst case probability. Putting everything together, $n - \sum_{i=1}^{n} H\left(\frac{1}{2} + \frac{\bar{r}}{2}\right) \leq S(X) - \sum_{i=1}^{n} S(X_i \mid W) \leq S(X) - S(X \mid W) = S(X : W) \leq \log m''$. Now, $H(\frac{1}{2} + \frac{\bar{r}}{2}) = 1 - \Omega(\bar{r}^2)$, and so $\log m'' = \Omega(n\bar{r}^2)$. But $\log m'' = O(t) \log m' = O(t) \log nm = O(t \log m) = O\left(\frac{1}{\bar{c}} \log m\right)$. Thus, $m \geq \exp\left(n\bar{c}\bar{r}^2\right)$. If $\delta \geq \epsilon$, then $\delta/\bar{c} \geq \epsilon$, and so $\bar{r} \geq \delta/\bar{c}$. Thus, $\bar{c}\bar{r}^2 \geq \delta^2/\bar{c} \geq \delta^2$. Otherwise, $\epsilon > \delta$, and so $\bar{c}\bar{r}^2 \geq \max(\bar{c}\epsilon^2, \delta^2/\bar{c})$, which is minimized if $\bar{c} = \delta/\epsilon$ and equals $\epsilon\delta$. Thus, $m \geq \exp\left(\max(\delta, \epsilon)\delta n\right)$.

5   A Better Upper Bound for Large δ

We improve the dependence on δ of 3-query LDCs, while only increasing m by a constant factor in the exponent. The proof uses a similar technique to that used for constructing the auxiliary code C'' in the previous section.

Theorem 8. *For any $\delta > 0$ and any constant $\eta > 0$, there is a linear $(3, \delta, 1/2 - 3\delta - \eta)$-LDC with $m = \exp\left(n^{1/t}\right)$ for any prime $2^t - 1$.*

Proof. Let $\gamma > 0$ be a constant to be determined, which will depend on η. Let C be the linear $(3, \delta, 1/2 - 6\delta)$-LDC with $m = \exp\left(n^{1/t}\right)$ constructed in [9]. The LDC C has a matching sum decoder by definition [9]. We identify the positions of C with linear forms $v_1, \ldots, v_m$. We first increase the length of C - for each $j \in [m]$, we append to C both a duplicate copy of v_j, denoted a_j, and a copy of the zero function, denoted b_j. Thus, a_j computes $\langle v_j, x \rangle$ and b_j computes $\langle 0, x \rangle = 0$. Notice that the resulting code C' is a $(3, \delta/3, 1/2 - 6\delta)$-LDC with length $m' = 3m$, and that C' has a matching Z of m triples $\{v_j, a_j, b_j\}$ with $v_j \oplus a_j \oplus b_j = 0$. For each triple $\{v_j, a_j, b_j\}$, we think of it as a *directed cycle* with edges $(v_j, a_j), (a_j, b_j), (b_j, v_j)$. For any $\delta > 0$, the LDC C also has n matchings $M_1, \ldots, M_n$ of triples of $v_1, \ldots, v_m$ so that for all $i \in [n]$ and all $e = \{v_a, v_b, v_c\} \in M_i$, we have $v_a \oplus v_b \oplus v_c = e_i$, where e_i is the i-th unit vector. We prove the following property of C in the full version.

Lemma 3. *For all $i \in [n]$, $|M_i| \geq m/18$.*

Now, for each $i \in [n]$ and for each triple $\{a, b, c\} \in M_i$, we think of the triple as a directed cycle with edges $(a, b), (b, c), (c, a)$ for some arbitrary ordering of $a, b,$ and c. Define the parameter $p = \lceil 18 \ln 1/(3\gamma) \rceil$. We form a new linear code C'' indexed by all ordered multisets $S \subset [m']$ of size p. Let $m'' = |C''(x)| = (m')^p$. We set the entry $C''_S(x)$ equal to $\bigoplus_{v \in S} C'_v(x)$. For $i \in [n]$, arbitrarily impose a total order $\succeq$ on the triples in M_i. For a particular ordered multiset S_1, we say that S_1 *hits* M_i if there is a triple $e \in M_i$ for which $e \cap S_1 \neq \emptyset$. Then, $\Pr[S_1 \text{ hits } M_i] \geq 1 - \left(1 - \frac{3|M_i|}{m'}\right)^p \geq 1 - \left(1 - \frac{1}{18}\right)^p \geq 1 - e^{-\frac{p}{18}} \geq 1 - 3\gamma$. For any S_1 that hits M_i, let $\{a, b, c\}$ be the smallest triple hit, under the total ordering $\succeq$. Since S_1 is ordered, we may choose the smallest of the p positions in S_1 which is in $\{a, b, c\}$. Let j be this position. Suppose the j-th position contains the linear form a, and that $(a, b), (b, c),$ and (c, a) are the edges of the directed cycle associated with $\{a, b, c\}$. Consider the triple $\{S_1, S_2, S_3\}$ formed as follows.

Triple-Generation(S_1):

1. Set the j-th position of S_2 to b, and the j-th position of S_3 to c.
2. For all positions $k \neq j$, do the following,
   (a) If v_ℓ is in the k-th position of S_1, then put a_ℓ in the k-th position of S_2 and b_ℓ in the k-th position of S_3.
   (b) If a_ℓ is in the k-th position of S_1, then put b_ℓ in the k-th position of S_2 and v_ℓ in the k-th position of S_3.
   (c) If b_ℓ is in the k-th position of S_1, then put v_ℓ in the k-th position of S_2 and a_ℓ in the k-th position of S_3.
3. Output $\{S_1, S_2, S_3\}$.

Since $v_j \oplus a_j \oplus b_j = 0$ for all j, we have, $\left(\bigoplus_{v \in S_1} v\right) \oplus \left(\bigoplus_{v \in S_2} v\right) \oplus \left(\bigoplus_{v \in S_3} v\right) = a \oplus b \oplus c = e_i$. The elaborate way of generating S_2 and S_3 was done to ensure that, had we computed Triple-Generation(S_2) or Triple-Generation(S_3), we would also have obtained $\{S_1, S_2, S_3\}$ as the output. This is true since, independently for each coordinate, we walk along a directed cycle of length 3. Thus, we may partition the ordered sets that hit M_i into a matching M_i'' of $m''/3 - \gamma m''$ triples $\{S_1, S_2, S_3\}$ containing linear forms that sum to e_i.

Consider the following decoder for C'': on input $i \in [n]$ with oracle access to y, choose a triple $\{S_1, S_2, S_3\} \in M_i''$ uniformly at random and output $y_{S_1} \oplus y_{S_2} \oplus y_{S_3}$. If the adversary corrupts at most $\delta m''$ positions of C'', then at most $\delta m''$ triples in M_i'' have been corrupted, and so the recovery probability of the decoder is at least $\frac{|M_i''| - \delta m''}{|M_i''|} = \frac{\frac{m''}{3} - \gamma m'' - \delta m''}{\frac{m''}{3} - \gamma m''} = 1 - \frac{3\delta}{1 - 3\gamma} \geq 1 - 3\delta - \eta$, where the final inequality follows for a sufficiently small constant $\gamma > 0$. So C'' is a $(3, \delta, 1/2 - 3\delta - \eta)$-LDC. The length of C'' is $m'' = (3m)^p = m^{O(1)} = \exp\left(n^{1/t}\right)$. This completes the proof.

Acknowledgment. The author thanks T.S. Jayram and the anonymous referees for many helpful comments.

References

1. Sipser, M., Spielman, D.A.: Expander codes. IEEE Trans. Inform. Theory 42, 1710–1722 (1996)
2. Katz, J., Trevisan, L.: On the efficiency of local decoding procedures for error-correcting codes. In: STOC (2000)
3. Trevisan, L.: Some applications of coding theory in computational complexity. Quaderni di Matematica 13, 347–424 (2004)
4. Dvir, Z., Shpilka, A.: Locally decodable codes with two queries and polynomial identity testing for depth 3 circuits. SIAM J. Comput. 36(5), 1404–1434 (2007)
5. Goldreich, O., Karloff, H.J., Schulman, L.J., Trevisan, L.: Lower bounds for linear locally decodable codes and private information retrieval. Computational Complexity 15(3), 263–296 (2006)
6. Obata, K.: Optimal lower bounds for 2-query locally decodable linear codes. In: Rolim, J., Vadhan, S.P. (eds.) RANDOM 2002. LNCS, vol. 2483, pp. 39–50. Springer, Heidelberg (2002)
7. Shiowattana, D., Lokam, S.V.: An optimal lower bound for 2-query locally decodable linear codes. Inf. Process. Lett. 97(6), 244–250 (2006)
8. Kerenidis, I., de Wolf, R.: Exponential lower bound for 2-query locally decodable codes via a quantum argument. J. Comput. Syst. Sci. 69(3), 395–420 (2004)
9. Yekhanin, S.: Towards 3-query locally decodable codes of subexponential length. J. ACM 55(5) (2008)
10. Beimel, A., Ishai, Y., Kushilevitz, E., Raymond, J.F.: Breaking the $O(n^{\frac{1}{2k-1}})$ barrier for information-theoretic private information retrieval. In: FOCS (2002)
11. Chor, B., Goldreich, O., Håstad, J., Friedman, J., Rudich, S., Smolensky, R.: The bit extraction problem of t-resilient functions. In: FOCS, pp. 396–407 (1985)
12. Bennett, C.H., Brassard, G., Robert, J.M.: Privacy amplification by public discussion. SIAM J. Comput. 17(2), 210–229 (1988)
13. Stinson, D.R., Massey, J.L.: An infinite class of counterexamples to a conjecture concerning nonlinear resilient functions. J. Cryptology 8(3), 167–173 (1995)
14. Beimel, A., Ishai, Y.: On the power of nonlinear secret-sharing. In: IEEE Conference on Computational Complexity, pp. 188–202 (2001)
15. Nielsen, M.A., Chuang, I.: Quantum computation and quantum information. Cambridge University Press, Cambridge (2000)

Quasi-randomness Is Determined by the Distribution of Copies of a Fixed Graph in Equicardinal Large Sets

Raphael Yuster

Department of Mathematics, University of Haifa, Haifa, Israel
raphy@math.haifa.ac.il

Abstract. For every fixed graph H and every fixed $0 < \alpha < 1$, we show that if a graph G has the property that all subsets of size αn contain the "correct" number of copies of H one would expect to find in the random graph $G(n, p)$ then G behaves like the random graph $G(n, p)$; that is, it is p-quasi-random in the sense of Chung, Graham, and Wilson [4]. This solves a conjecture raised by Shapira [8] and solves in a strong sense an open problem of Simonovits and Sós [9].

1 Introduction

The theory of quasi-random graphs asks the following fundamental question: which properties of graphs are such that any graph that satisfies them, resembles an appropriate random graph (namely, the graph satisfies the properties that a random graph would satisfy, with high probability). Such properties are called *quasi-random*.

The theory of quasi-random graphs was initiated by Thomason [10,11] and then followed by Chung, Graham and Wilson who proved the fundamental theorem of quasi-random graphs [4]. Since then there have been many papers on this subject (see, e.g. the excellent survey [6]). Quasi-random properties were also studied for other combinatorial structures such as set systems [1], tournaments [2], and hypergraphs [3]. There are also some very recent results on quasi-random groups [5] and generalized quasi-random graphs [7].

In order to formally define p-quasi-randomness we need to state the fundamental theorem of quasi-random graphs. As usual, a *labeled copy* of an undirected graph H in a graph G is an injective mapping ϕ from $V(H)$ to $V(G)$ that maps edges to edges. That is $xy \in E(H)$ implies $\phi(x)\phi(y) \in E(G)$. For a set of vertices $U \subset V(G)$ we denote by $H[U]$ the number of labeled copies of H in the subgraph of G induced by U and by $e(U)$ the number of edges of G with both endpoints in U. A graph sequence (G_n) is an infinite sequence of graphs $\{G_1, G_2, \dots\}$ where G_n has n vertices. The following result of Chung, Graham, and Wilson [4] shows that many properties of different nature are equivalent to the notion of quasi-randomness, defined using edge distribution. The original theorem lists seven such equivalent properties, but we only state four of them here.

A. Goel et al. (Eds.): APPROX and RANDOM 2008, LNCS 5171, pp. 596–601, 2008.
© Springer-Verlag Berlin Heidelberg 2008

Theorem 1 (Chung, Graham, and Wilson [4]). *Fix any $1 < p < 1$. For any graph sequence (G_n) the following properties are equivalent:*

$\mathcal{P}_1$: *For an even integer $t \geq 4$, let C_t denote the cycle of length t. Then $e(G_n) = \frac{1}{2}pn^2 + o(n^2)$ and $C_t[V(G_n)] = p^t n^t + o(n^t)$.*

$\mathcal{P}_2$: *For any subset of vertices $U \subseteq V(G_n)$ we have $e(U) = \frac{1}{2}p|U|^2 + o(n^2)$.*

$\mathcal{P}_3$: *For any subset of vertices $U \subseteq V(G_n)$ of size $n/2$ we have $e(U) = \frac{1}{2}p|U|^2 + o(n^2)$.*

$\mathcal{P}_4$: *Fix an $\alpha \in (0, \frac{1}{2})$. For any $U \subseteq V(G_n)$ of size αn we have $e(U, V \setminus U) = p\alpha(1 - \alpha)n^2 + o(n^2)$.*

The *formal* meaning of the properties being equivalent is expressed, as usual, using ϵ, δ notation. For example the meaning that $\mathcal{P}_3$ implies $\mathcal{P}_2$ is that for any $\epsilon > 0$ there exist $\delta = \delta(\epsilon)$ and $N = N(\epsilon)$ so that for all $n > N$, if G is a graph with n vertices having the property that any subset of vertices U of size $n/2$ satisfies $|e(U) - \frac{1}{2}p|U|^2| < \delta n^2$ then also for any subset of vertices W we have $|e(W) - \frac{1}{2}p|W|^2| < \epsilon n^2$.

Given Theorem 1 we say that a graph property is p-quasi-random if it is equivalent to any (and therefore all) of the four properties defined in that theorem. (We will usually just say *quasi-random* instead of *p-quasi-random* since p is fixed throughout the proofs). Note, that each of the four properties in Theorem 1 is a property we would expect $G(n,p)$ to satisfy with high probability.

It is far from true, however, that any property that almost surely holds for $G(n,p)$ is quasi-random. For example, it is easy to see that having vertex degrees $np(1 + o(1))$ is not a quasi-random property (just take vertex-disjoint cliques of size roughly np each). An important family of *non* quasi-random properties are those requiring the graphs in the sequence to have the correct number of copies of a fixed graph H. Note that $\mathcal{P}_1(t)$ guarantees that for any *even* t, if a graph sequence has the correct number of edges as well as the correct number of copies of $H = C_t$ then the sequence is quasi-random. As observed in [4] this is not true for all graphs H. In fact, already for $H = K_3$ there are simple constructions showing that this is not true.

Simonovits and Sós observed that the standard counter-examples showing that for some graphs H, having the correct number of copies of H is not enough to guarantee quasi-randomness, have the property that the number of copies of H in some of the induced subgraphs of these counter-examples deviates significantly from what it should be. As quasi-randomness is a hereditary property, in the sense that we expect a sub-structure of a random-like object to be random-like as well, they introduced the following variant of property $\mathcal{P}_1$ of Theorem 1, where now we require all subsets of vertices to contains the "correct" number of copies of H.

Definition 1 ($\mathcal{P}_H$). *For a fixed graph H with h vertices and r edges, we say that a graph sequence (G_n) satisfies $\mathcal{P}_H$ if all subsets of vertices $U \subset V(G_n)$ satisfy $H[U] = p^r|U|^h + o(n^h)$.*

As opposed to $\mathcal{P}_1$, which is quasi-random only for even cycles, Simonovits and Sós [9] showed that $\mathcal{P}_H$ is quasi-random for any nonempty graph H.

Theorem 2. *For any fixed H that has edges, property $\mathcal{P}_H$ is quasi-random.*

We can view property $\mathcal{P}_H$ as a generalization of property $\mathcal{P}_2$ in Theorem 1, since $\mathcal{P}_2$ is just the special case $\mathcal{P}_{K_2}$. Now, property $\mathcal{P}_3$ in Theorem 1 guarantees that in order to infer that a sequence is quasi-random, and thus satisfies $\mathcal{P}_2$, it is enough to require only the sets of vertices of size $n/2$ to contain the correct number of edges. An open problem raised by Simonovits and Sós [9], and in a stronger form by Shapira [8], is that the analogous condition also holds for any H. Namely, in order to infer that a sequence is quasi-random, and thus satisfies $\mathcal{P}_H$, it is enough, say, to require only the sets of vertices of size $n/2$ to contain the correct number of copies of H. Shapira [8] proved that is it enough to consider sets of vertices of size $n/(h+1)$. Hence, in his result, the cardinality of the sets *depends* on h. Thus, if H has 1000 vertices, Shapira's result shows that it suffices to check vertex subsets having a fraction smaller than $1/1000$ of the total number of vertices. His proof method cannot be extended to obtain the same result for fractions larger than $1/(h + \epsilon)$.

In this paper we settle the above mentioned open problem completely. In fact, we show that for any H, not only is it enough to check only subsets of size $n/2$, but, more generally, we show that it is enough to check subsets of size αn for any fixed $\alpha \in (0, 1)$. More formally, we define:

Definition 2 ($\mathcal{P}_{H,\alpha}$). *For a fixed graph H with h vertices and r edges and fixed $0 < \alpha < 1$ we say that a graph sequence (G_n) satisfies $\mathcal{P}_{H,\alpha}$ if all subsets of vertices $U \subset V(G_n)$ with $|U| = \lfloor \alpha n \rfloor$ satisfy $H[U] = p^r |U|^h + o(n^h)$.*

Our main result is, therefore:

Theorem 3. *For any fixed graph H and any fixed $0 < \alpha < 1$, property $\mathcal{P}_{H,\alpha}$ is quasi-random.*

2   Proof of the Main Result

For the remainder of this section let H be a fixed graph with $h > 1$ vertices and $r > 0$ edges, and let $\alpha \in (0, 1)$ be fixed. Throughout this section we ignore rounding issues and, in particular, assume that αn is an integer, as this has no effect on the asymptotic nature of the results.

Suppose that the graph sequence (G_n) satisfies $\mathcal{P}_{H,\alpha}$. We will prove that it is quasi-random by showing that it also satisfies $\mathcal{P}_H$. In other words, we need to prove the following lemma which, together with Theorem 2, yields Theorem 3.

Lemma 1. *For any $\epsilon > 0$ there exists $N = N(\epsilon, h, \alpha)$ and $\delta = \delta(\epsilon, h, \alpha)$ so that for all $n > N$, if G is a graph with n vertices satisfying that for all $U \subset V(G)$ with $|U| = \alpha n$ we have $|H[U] - p^r |U|^h| < \delta n^h$ then G also satisfies that for all $W \subset V(G)$ we have $|H[W] - p^r |W|^h| < \epsilon n^h$.*

Proof: Suppose therefore that $\epsilon > 0$ is given. Let $N = N(\epsilon, h, \alpha)$, $\epsilon' = \epsilon'(\epsilon, h, \alpha)$ and $\delta = \delta(\epsilon, h, \alpha)$ be parameters to be chosen so that N is sufficiently large and $\delta \ll \epsilon'$ are both sufficiently small to satisfy the inequalities that will follow, and it will be clear that they are indeed only functions of ϵ, h, and α.

Now, let G be a graph with $n > N$ vertices satisfying that for all $U \subset V(G)$ with $|U| = \alpha n$ we have $|H[U] - p^r |U|^h| < \delta n^h$. Consider any subset $W \subset V(G)$. We need to prove that $|H[W] - p^r |W|^h| < \epsilon n^h$.

For convenience, set $k = \alpha n$. Let us first prove this for the case where $|W| = m > k$. This case can rather easily be proved via a simple counting argument. Denote by $\mathcal{U}$ the set of $\binom{m}{k}$ k-subsets of W. Hence, by the given condition on k-subsets,

$$\binom{m}{k}(p^r k^h - \delta n^h) < \sum_{U \in \mathcal{U}} H[U] < \binom{m}{k}(p^r k^h + \delta n^h). \tag{1}$$

Every copy of H in W appears in precisely $\binom{m-h}{k-h}$ distinct $U \in \mathcal{U}$. It follows from (1) that

$$H[W] = \frac{1}{\binom{m-h}{k-h}} \sum_{U \in \mathcal{U}} H[U] < \frac{\binom{m}{k}}{\binom{m-h}{k-h}}(p^r k^h + \delta n^h) < p^r m^h + \frac{\epsilon'}{2} n^h, \tag{2}$$

and similarly from (1)

$$H[W] = \frac{1}{\binom{m-h}{k-h}} \sum_{U \in \mathcal{U}} H[U] > \frac{\binom{m}{k}}{\binom{m-h}{k-h}}(p^r k^h - \delta n^h) > p^r m^h - \frac{\epsilon'}{2} n^h. \tag{3}$$

We now consider the case where $|W| = m = \beta n < \alpha n = k$. Notice that we can assume that $\beta \geq \epsilon$ since otherwise the result is trivially true. The set $\mathcal{H}$ of H-subgraphs of G can be partitioned into $h + 1$ types, according to the number of vertices they have in W. Hence, for $j = 0, \ldots, h$ let $\mathcal{H}_j$ be the set of H-subgraphs of G that contain precisely j vertices in $V \setminus W$. Notice that, by definition, $|\mathcal{H}_0| = H[W]$. For convenience, denote $w_j = |\mathcal{H}_j|/n^h$. We therefore have, together with (2) and (3) applied to V,

$$w_0 + w_1 + \cdots + w_h = \frac{|\mathcal{H}|}{n^h} = \frac{H[V]}{n^h} = p^r + \mu \tag{4}$$

where $|\mu| < \epsilon'/2$.

Define $\lambda = \frac{(1-\alpha)}{h+1}$ and set $k_i = k + i\lambda n$ for $i = 1, \ldots, h$. Let $Y_i \subset V \setminus W$ be a random set of $k_i - m$ vertices, chosen uniformly at random from all $\binom{n-m}{k_i-m}$ subsets of size $k_i - m$ of $V \setminus W$. Denote $K_i = Y_i \cup W$ and notice that $|K_i| = k_i > \alpha n$. We will now estimate the number of elements of $\mathcal{H}_j$ that "survive" in K_i. Formally, let $\mathcal{H}_{j,i}$ be the set of elements of $\mathcal{H}_j$ that have all of their vertices in K_i, and let $m_{j,i} = |\mathcal{H}_{j,i}|$. Clearly, $m_{0,i} = H[W]$ since $W \subset K_i$. Furthermore, by (2) and (3),

$$m_{0,i} + m_{1,i} + \cdots + m_{h,i} = H[K_i] = p^r k_i^h + \rho_i n^h \tag{5}$$

where ρ_i is a random variable with $|\rho_i| < \epsilon'/2$.

For an H-copy $T \in \mathcal{H}_j$ we compute the probability $p_{j,i}$ that $T \in H[K_i]$. Since $T \in H[K_i]$ if and only if all the j vertices of T in $V \setminus W$ appear in Y_i we have

$$p_{j,i} = \frac{\binom{n-m-j}{k_i-m-j}}{\binom{n-m}{k_i-m}} = \frac{(k_i - m) \cdots (k_i - m - j + 1)}{(n - m) \cdots (n - m - j + 1)}.$$

Defining $x_i = (k_i - m)/(n - m)$ and noticing that

$$x_i = \frac{k_i - m}{n - m} = \frac{\alpha - \beta}{1 - \beta} + \frac{\lambda}{1 - \beta}i$$

it follows that (for large enough graphs)

$$\left| p_{j,i} - x_i^j \right| < \frac{\epsilon'}{2} . \tag{6}$$

Clearly, the expectation of $m_{j,i}$ is $\mathrm{E}[m_{j,i}] = p_{j,i} |\mathcal{H}_j|$. By linearity of expectation we have from (5) that

$$\mathrm{E}[m_{0,i}] + \mathrm{E}[m_{1,i}] + \cdots + \mathrm{E}[m_{h,i}] = \mathrm{E}[H[K_i]] = p^r k_i^h + \mathrm{E}[\rho_i] n^h.$$

Dividing the last equality by n^h we obtain

$$p_{0,i} w_0 + \cdots + p_{h,i} w_h = p^r (\alpha + \lambda i)^h + \mathrm{E}[\rho_i] . \tag{7}$$

By (6) and (7) we therefore have

$$\sum_{j=0}^{h} x_i^j w_j = p^r (\alpha + \lambda i)^h + \mu_i \tag{8}$$

where $\mu_i = \mathrm{E}[\rho_i] + \zeta_i$ and $|\zeta_i| < \epsilon'/2$. Since also $|\rho_i| < \epsilon'/2$ we have that $|\mu_i| < \epsilon'$.

Now, (4) and (8) form together a system of $h + 1$ linear equations with the $h + 1$ variables $w_0, \ldots, w_h$. The coefficient matrix of this system is just the Vandermonde matrix $A = A(x_1, \ldots, x_h, 1)$. Since $x_1, \ldots, x_h, 1$ are all distinct, and, in fact, the gap between any two of them is at least $\lambda/(1-\beta) = (1-\alpha)/((h+1)(1 - \beta)) \geq (1 - \alpha)/(h + 1)$, we have that the system has a unique solution which is $A^{-1}b$ where $b \in R^{h+1}$ is the column vector whose i'th coordinate is $p^r (\alpha + \lambda i)^h + \mu_i$ for $i = 1, \ldots, h$ and whose last coordinate is $p^r + \mu$. Consider now the vector b^* which is the same as b, just without the μ_i's. Namely $b^* \in R^{h+1}$ is the column vector whose i'th coordinate is $p^r (\alpha + \lambda i)^h$ for $i = 1, \ldots, h$ and whose last coordinate is p^r. Then the system $A^{-1}b^*$ also has a unique solution and, in fact, we *know* explicitly what this solution is. It is the vector $w^* = (w_0^*, \ldots, w_h^*)$ where

$$w_j^* = p^r \binom{h}{j} \beta^{h-j} (1 - \beta)^j .$$

Indeed, it is straightforward to verify the equality

$$\sum_{j=0}^{h} p^r \binom{h}{j} \beta^{h-j} (1 - \beta)^j = p^r$$

and, for all $i = 1, \ldots, h$ the equalities

$$\sum_{j=0}^{h} \left(\frac{\alpha - \beta}{1 - \beta} + \frac{\lambda}{1 - \beta}i \right)^j p^r \binom{h}{j} \beta^{h-j} (1 - \beta)^j = p^r (\alpha + \lambda i)^h .$$

Now, since the mapping $F : R^{h+1} \to R^{h+1}$ mapping a vector c to $A^{-1}c$ is continuous, we know that for ϵ' sufficiently small, if each coordinate of c has absolute value less than ϵ', then each coordinate of $A^{-1}c$ has absolute value at most ϵ. Now, define $c = b - b^* = (\mu_1, \ldots, \mu_h, \mu)$. Then we have that each coordinate w_i of $A^{-1}b$ differs from the corresponding coordinate w_i^* of $A^{-1}b^*$ by at most ϵ. In particular,

$$|w_0 - w_0^*| = |w_0 - p^r \beta^h| < \epsilon.$$

Hence,

$$|H[W] - n^h p^r \beta^h| = |H[W] - p^r |W|^h| < \epsilon n^h$$

as required. ∎

References

1. Chung, F.R.K., Graham, R.L.: Quasi-random set systems. Journal of the AMS 4, 151–196 (1991)
2. Chung, F.R.K., Graham, R.L.: Quasi-random tournaments. Journal of Graph Theory 15, 173–198 (1991)
3. Chung, F.R.K., Graham, R.L.: Quasi-random hypergraphs. Random Structures and Algorithms 1, 105–124 (1990)
4. Chung, F.R.K., Graham, R.L., Wilson, R.M.: Quasi-random graphs. Combinatorica 9, 345–362 (1989)
5. Gowers, T.: Quasirandom groups (manuscript, 2006)
6. Krivelevich, M., Sudakov, B.: Pseudo-random graphs. In: Györi, E., Katona, G.O.H., Lovász, L. (eds.) More sets, graphs and numbers. Bolyai Society Mathematical Studies, vol. 15, pp. 199–262.
7. Lovász, L., Sós, V.T.: Generalized quasirandom graphs. Journal of Combinatorial Theory Series B 98, 146–163 (2008)
8. Shapira, A.: Quasi-randomness and the distribution of copies of a fixed graph. Combinatorica (to appear)
9. Simonovits, M., Sós, V.T.: Hereditarily extended properties, quasi-random graphs and not necessarily induced subgraphs. Combinatorica 17, 577–596 (1997)
10. Thomason, A.: Pseudo-random graphs. In: Proc. of Random Graphs, Poznań (1985); Karoński, M. (ed.): Annals of Discrete Math., 33, 307–331 (1987)
11. Thomason, A.: Random graphs, strongly regular graphs and pseudo-random graphs. In: Whitehead, C. (ed.) Surveys in Combinatorics. LMS Lecture Note Series, vol. 123, pp. 173–195 (1987)

Author Index

Abrahamson, Jeff   254
Adler, Micah   1
Ailon, Nir   512
Alon, Noga   266
Arvind, V.   276
Asadpour, Arash   10

Bădoiu, Mihai   21
Ben-Eliezer, Ido   266
Ben-Sasson, Eli   290
Bernasconi, Nicla   303
Blais, Eric   317
Bogdanov, Andrej   331
Bresler, Guy   343

Cardinal, Jean   35
Chlamtac, Eden   49
Chung, Kai-Min   357
Csaba, Béla   254

David, Matei   371
Demaine, Erik D.   21
Dinh, Hang   385
Dumitrescu, Adrian   63

Feige, Uriel   10
Fischer, Eldar   402
Fürer, Martin   416

Gabizon, Ariel   430
Guruswami, Venkatesan   77, 444
Gutfreund, Dan   455, 469

Hajiaghayi, MohammadTaghi   21
Halman, Nir   91
Heeringa, Brent   1

Jackson, Jeffrey C.   483
Jiang, Minghui   63

Karger, David R.   104
Kaufman, Tali   498
Khuller, Samir   165
Kortsarz, Guy   118
Kowalik, Łukasz   132
Krivelevich, Michael   266

Lachish, Oded   402
Lando, Yuval   146
Langberg, Michael   118
Lee, Homin K.   483
Lee, James R.   444
Levy, Eythan   35
Li, Chung-Lun   91
Liberty, Edo   512
Litsyn, Simon   498

Magen, Avner   523
Mastrolilli, Monaldo   153
Matsliah, Arie   402
Matthew McCutchen, Richard   165
Mittal, Shashi   179
Mossel, Elchanan   331, 343
Mucha, Marcin   132
Mukhopadhyay, Partha   276
Mutsanas, Nikolaus   153

Nagarajan, Viswanath   193
Narayanan, Hariharan   535
Newman, Ilan   402
Nguyen, Thành   207
Niyogi, Partha   535
Nutov, Zeev   118, 146, 219

Panagiotou, Konstantinos   303
Pitassi, Toniann   371
Prasad Kasiviswanathan, Shiva   416

Raghavendra, Prasad   77
Rao, Anup   549, 557
Ravi, R.   193
Rothblum, Guy N.   455
Russell, Alexander   385

Saberi, Amin   10
Safari, MohammadAli   233
Salavatipour, Mohammad R.   233
Schulz, Andreas S.   179
Scott, Jacob   104
Servedio, Rocco A.   483
Shaltiel, Ronen   430
Shokoufandeh, Ali   254

Sidiropoulos, Anastasios   21
Simchi-Levi, David   91
Singer, Amit   512
Singh, Gyanit   49
Sly, Allan   343
Sorkin, Gregory B.   571
Srinivasan, Aravind   247
Steger, Angelika   303
Svensson, Ola   153

Vadhan, Salil   331, 357, 469
Viderman, Michael   290
Viola, Emanuele   371

Wan, Andrew   483
Wigderson, Avi   444
Woodruff, David   584

Xie, Ning   498

Yahalom, Orly   402
Yuster, Raphael   596

Zadimoghaddam, Morteza   21
Zouzias, Anastasios   523
Zuckerman, David   557

Lecture Notes in Computer Science

Sublibrary 1: Theoretical Computer Science and General Issues

For information about Vols. 1– 4917
please contact your bookseller or Springer

Vol. 5234: V. Adve, M.J. Garzarán, P. Petersen (Eds.), Languages and Compilers for Parallel Computing. XV, 354 pages. 2008.

Vol. 5204: C.S. Calude, J.F. Costa, R. Freund, M. Oswald, G. Rozenberg (Eds.), Unconventional Computing. X, 259 pages. 2008.

Vol. 5201: F. van Breugel, M. Chechik (Eds.), CONCUR 2008 - Concurrency Theory. XIII, 524 pages. 2008.

Vol. 5191: H. Umeo, S. Morishita, K. Nishinari, T. Komatsuzaki, S. Bandini (Eds.), Cellular Automata. XVI, 577 pages. 2008.

Vol. 5171: A. Goel, K. Jansen, J.D.P. Rolim, R. Rubinfeld (Eds.), Approximation, Randomization and Combinatorial Optimization. XII, 604 pages. 2008.

Vol. 5170: O.A. Mohamed, C. Muñoz, S. Tahar (Eds.), Theorem Proving in Higher Order Logics. X, 321 pages. 2008.

Vol. 5169: J. Fong, R. Kwan, F.L. Wang (Eds.), Hybrid Learning and Education. XII, 474 pages. 2008.

Vol. 5165: B. Yang, D.-Z. Du, C.A. Wang (Eds.), Combinatorial Optimization and Applications. XII, 480 pages. 2008.

Vol. 5162: E. Ochmański, J. Tyszkiewicz (Eds.), Mathematical Foundations of Computer Science 2008. XIV, 626 pages. 2008.

Vol. 5156: K. Havelund, R. Majumdar, J. Palsberg (Eds.), Model Checking Software. X, 343 pages. 2008.

Vol. 5148: O. Ibarra, B. Ravikumar (Eds.), Implementation and Applications of Automata. XII, 289 pages. 2008.

Vol. 5147: K. Horimoto, G. Regensburger, M. Rosenkranz, H. Yoshida (Eds.), Algebraic Biology. XII, 245 pages. 2008.

Vol. 5133: P. Audebaud, C. Paulin-Mohring (Eds.), Mathematics of Program Construction. X, 423 pages. 2008.

Vol. 5132: P.J. Bentley, D. Lee, S. Jung (Eds.), Artificial Immune Systems. XIV, 436 pages. 2008.

Vol. 5130: J. von zur Gathen, J.L. Imaña, Ç.K. Koç (Eds.), Arithmetic of Finite Fields. X, 205 pages. 2008.

Vol. 5126: L. Aceto, I. Damgård, L.A. Goldberg, M.M. Halldórsson, A. Ingólfsdóttir, I. Walukiewicz (Eds.), Automata, Languages and Programming, Part II. XXII, 730 pages. 2008.

Vol. 5125: L. Aceto, I. Damgård, L.A. Goldberg, M.M. Halldórsson, A. Ingólfsdóttir, I. Walukiewicz (Eds.), Automata, Languages and Programming, Part I. XXIII, 896 pages. 2008.

Vol. 5124: J. Gudmundsson (Ed.), Algorithm Theory – SWAT 2008. XIII, 438 pages. 2008.

Vol. 5123: A. Gupta, S. Malik (Eds.), Computer Aided Verification. XVII, 558 pages. 2008.

Vol. 5117: A. Voronkov (Ed.), Rewriting Techniques and Applications. XIII, 457 pages. 2008.

Vol. 5114: M. Bereković, N. Dimopoulos, S. Wong (Eds.), Embedded Computer Systems: Architectures, Modeling, and Simulation. XVI, 300 pages. 2008.

Vol. 5104: F. Bello, E. Edwards (Eds.), Biomedical Simulation. XI, 228 pages. 2008.

Vol. 5103: M. Bubak, G.D. van Albada, J. Dongarra, P.M.A. Sloot (Eds.), Computational Science – ICCS 2008, Part III. XXVIII, 758 pages. 2008.

Vol. 5102: M. Bubak, G.D. van Albada, J. Dongarra, P.M.A. Sloot (Eds.), Computational Science – ICCS 2008, Part II. XXVIII, 752 pages. 2008.

Vol. 5101: M. Bubak, G.D. van Albada, J. Dongarra, P.M.A. Sloot (Eds.), Computational Science – ICCS 2008, Part I. XLVI, 1058 pages. 2008.

Vol. 5092: X. Hu, J. Wang (Eds.), Computing and Combinatorics. XIV, 680 pages. 2008.

Vol. 5090: R.T. Mittermeir, M.M. Sysło (Eds.), Informatics Education - Supporting Computational Thinking. XV, 357 pages. 2008.

Vol. 5084: J.F. Peters, A. Skowron (Eds.), Transactions on Rough Sets VIII. X, 521 pages. 2008.

Vol. 5083: O. Chitil, Z. Horváth, V. Zsók (Eds.), Implementation and Application of Functional Languages. XI, 272 pages. 2008.

Vol. 5073: O. Gervasi, B. Murgante, A. Laganà, D. Taniar, Y. Mun, M.L. Gavrilova (Eds.), Computational Science and Its Applications – ICCSA 2008, Part II. XXIX, 1280 pages. 2008.

Vol. 5072: O. Gervasi, B. Murgante, A. Laganà, D. Taniar, Y. Mun, M.L. Gavrilova (Eds.), Computational Science and Its Applications – ICCSA 2008, Part I. XXIX, 1266 pages. 2008.

Vol. 5065: P. Degano, R. De Nicola, J. Meseguer (Eds.), Concurrency, Graphs and Models. XV, 810 pages. 2008.

Vol. 5062: K.M. van Hee, R. Valk (Eds.), Applications and Theory of Petri Nets. XIII, 429 pages. 2008.

Vol. 5059: F.P. Preparata, X. Wu, J. Yin (Eds.), Frontiers in Algorithmics. XI, 350 pages. 2008.

Vol. 5058: A.A. Shvartsman, P. Felber (Eds.), Structural Information and Communication Complexity. X, 307 pages. 2008.

Vol. 5050: J.M. Zurada, G.G. Yen, J. Wang (Eds.), Computational Intelligence: Research Frontiers. XVI, 389 pages. 2008.

Vol. 5045: P. Hertling, C.M. Hoffmann, W. Luther, N. Revol (Eds.), Reliable Implementation of Real Number Algorithms: Theory and Practice. XI, 239 pages. 2008.

Vol. 5038: C.C. McGeoch (Ed.), Experimental Algorithms. X, 363 pages. 2008.

Vol. 5036: S. Wu, L.T. Yang, T.L. Xu (Eds.), Advances in Grid and Pervasive Computing. XV, 518 pages. 2008.

Vol. 5035: A. Lodi, A. Panconesi, G. Rinaldi (Eds.), Integer Programming and Combinatorial Optimization. XI, 477 pages. 2008.

Vol. 5029: P. Ferragina, G.M. Landau (Eds.), Combinatorial Pattern Matching. XIII, 317 pages. 2008.

Vol. 5028: A. Beckmann, C. Dimitracopoulos, B. Löwe (Eds.), Logic and Theory of Algorithms. XIX, 596 pages. 2008.

Vol. 5022: A.G. Bourgeois, S.Q. Zheng (Eds.), Algorithms and Architectures for Parallel Processing. XIII, 336 pages. 2008.

Vol. 5018: M. Grohe, R. Niedermeier (Eds.), Parameterized and Exact Computation. X, 227 pages. 2008.

Vol. 5015: L. Perron, M.A. Trick (Eds.), Integration of AI and OR Techniques in Constraint Programming for Combinatorial Optimization Problems. XII, 394 pages. 2008.

Vol. 5011: A.J. van der Poorten, A. Stein (Eds.), Algorithmic Number Theory. IX, 455 pages. 2008.

Vol. 5010: E.A. Hirsch, A.A. Razborov, A. Semenov, A. Slissenko (Eds.), Computer Science – Theory and Applications. XIII, 411 pages. 2008.

Vol. 5008: A. Gasteratos, M. Vincze, J.K. Tsotsos (Eds.), Computer Vision Systems. XV, 560 pages. 2008.

Vol. 5004: R. Eigenmann, B.R. de Supinski (Eds.), OpenMP in a New Era of Parallelism. X, 191 pages. 2008.

Vol. 5000: O. Grumberg, H. Veith (Eds.), 25 Years of Model Checking. VII, 231 pages. 2008.

Vol. 4996: H. Kleine Büning, X. Zhao (Eds.), Theory and Applications of Satisfiability Testing – SAT 2008. X, 305 pages. 2008.

Vol. 4988: R. Berghammer, B. Möller, G. Struth (Eds.), Relations and Kleene Algebra in Computer Science. X, 397 pages. 2008.

Vol. 4985: M. Ishikawa, K. Doya, H. Miyamoto, T. Yamakawa (Eds.), Neural Information Processing, Part II. XXX, 1091 pages. 2008.

Vol. 4984: M. Ishikawa, K. Doya, H. Miyamoto, T. Yamakawa (Eds.), Neural Information Processing, Part I. XXX, 1147 pages. 2008.

Vol. 4981: M. Egerstedt, B. Mishra (Eds.), Hybrid Systems: Computation and Control. XV, 680 pages. 2008.

Vol. 4978: M. Agrawal, D.-Z. Du, Z. Duan, A. Li (Eds.), Theory and Applications of Models of Computation. XII, 598 pages. 2008.

Vol. 4975: F. Chen, B. Jüttler (Eds.), Advances in Geometric Modeling and Processing. XV, 606 pages. 2008.

Vol. 4974: M. Giacobini, A. Brabazon, S. Cagnoni, G.A. Di Caro, R. Drechsler, A. Ekárt, A.I. Esparcia-Alcázar, M. Farooq, A. Fink, J. McCormack, M. O'Neill, J. Romero, F. Rothlauf, G. Squillero, A.Ş. Uyar, S. Yang (Eds.), Applications of Evolutionary Computing. XXV, 701 pages. 2008.

Vol. 4973: E. Marchiori, J.H. Moore (Eds.), Evolutionary Computation, Machine Learning and Data Mining in Bioinformatics. X, 213 pages. 2008.

Vol. 4972: J. van Hemert, C. Cotta (Eds.), Evolutionary Computation in Combinatorial Optimization. XII, 289 pages. 2008.

Vol. 4971: M. O'Neill, L. Vanneschi, S. Gustafson, A.I. Esparcia Alcázar, I. De Falco, A. Della Cioppa, E. Tarantino (Eds.), Genetic Programming. XI, 375 pages. 2008.

Vol. 4967: R. Wyrzykowski, J. Dongarra, K. Karczewski, J. Wasniewski (Eds.), Parallel Processing and Applied Mathematics. XXIII, 1414 pages. 2008.

Vol. 4963: C.R. Ramakrishnan, J. Rehof (Eds.), Tools and Algorithms for the Construction and Analysis of Systems. XVI, 518 pages. 2008.

Vol. 4962: R. Amadio (Ed.), Foundations of Software Science and Computational Structures. XV, 505 pages. 2008.

Vol. 4961: J.L. Fiadeiro, P. Inverardi (Eds.), Fundamental Approaches to Software Engineering. XIII, 430 pages. 2008.

Vol. 4960: S. Drossopoulou (Ed.), Programming Languages and Systems. XIII, 399 pages. 2008.

Vol. 4959: L. Hendren (Ed.), Compiler Construction. XII, 307 pages. 2008.

Vol. 4957: E.S. Laber, C. Bornstein, L.T. Nogueira, L. Faria (Eds.), LATIN 2008: Theoretical Informatics. XVII, 794 pages. 2008.

Vol. 4943: R. Woods, K. Compton, C. Bouganis, P.C. Diniz (Eds.), Reconfigurable Computing: Architectures, Tools and Applications. XIV, 344 pages. 2008.

Vol. 4942: E. Frachtenberg, U. Schwiegelshohn (Eds.), Job Scheduling Strategies for Parallel Processing. VII, 189 pages. 2008.

Vol. 4941: M. Miculan, I. Scagnetto, F. Honsell (Eds.), Types for Proofs and Programs. VII, 203 pages. 2008.

Vol. 4935: B. Chapman, W. Zheng, G.R. Gao, M. Sato, E. Ayguadé, D. Wang (Eds.), A Practical Programming Model for the Multi-Core Era. VI, 208 pages. 2008.

Vol. 4934: U. Brinkschulte, T. Ungerer, C. Hochberger, R.G. Spallek (Eds.), Architecture of Computing Systems – ARCS 2008. XI, 287 pages. 2008.

Vol. 4927: C. Kaklamanis, M. Skutella (Eds.), Approximation and Online Algorithms. X, 289 pages. 2008.

Vol. 4926: N. Monmarché, E.-G. Talbi, P. Collet, M. Schoenauer, E. Lutton (Eds.), Artificial Evolution. XIII, 327 pages. 2008.

Vol. 4921: S.-i. Nakano, M.. S. Rahman (Eds.), WALCOM: Algorithms and Computation. XII, 241 pages. 2008.

Vol. 4919: A. Gelbukh (Ed.), Computational Linguistics and Intelligent Text Processing. XVIII, 666 pages. 2008.